TREATISE ON INVERTEBRATE PALEONTOLOGY

Prepared under Sponsorship of
The Geological Society of America, Inc.

The Paleontological Society *The Society of Economic Paleontologists and Mineralogists*
The Palaeontographical Society *The Palaeontological Association*

RAYMOND C. MOORE
Founder

RAYMOND C. MOORE and CURT TEICHERT
Editors

JACK D. KEIM, LAVON MCCORMICK, ROGER B. WILLIAMS
Assistant Editors

Part T
ECHINODERMATA 2

VOLUME 1

By GEORGES UBAGHS, R. C. MOORE, H. WIENBERG RASMUSSEN, N. GARY LANE, ALBERT BREIMER, H. L. STRIMPLE, J. C. BROWER, RUSSELL M. JEFFORDS, JAMES SPRINKLE, R. E. PECK, D. B. MACURDA, JR., D. L. MEYER, MICHEL ROUX, HERTHA SIEVERTS-DORECK, R. O. FAY, and R. A. ROBISON

THE GEOLOGICAL SOCIETY OF AMERICA, INC.
and
THE UNIVERSITY OF KANSAS
BOULDER, COLORADO, and LAWRENCE, KANSAS
1978

Library of Congress Catalogue Card
Number: 53-12913
I.S.B.N. 0-8137-3021-X
TIVPA 1-1027 (1978)

Text Composed by
THE UNIVERSITY OF KANSAS PRINTING SERVICE
Lawrence, Kansas

Illustrations and Offset Lithography
THE MERIDEN GRAVURE COMPANY
Meriden, Connecticut

Binding
TAPLEY-RUTTER COMPANY
Moonachie, New Jersey

Published 1978

Distributed by the Geological Society of America, Inc., 3300 Penrose Place, Boulder, Colo., 80301, to which all communications should be addressed.

The *Treatise on Invertebrate Paleontology* has been made possible by (1) grants of funds from The Geological Society of America through the bequest of Richard Alexander Fullerton Penrose, Jr., for initial preparation of illustrations, and partial defrayment of organizational expenses in 1948-1957, and again since 1971, and from the United States National Science Foundation, awarded annually since 1959, for continuation of the *Treatise* project; (2) contribution of the knowledge and labor of specialists throughout the world, working in cooperation under sponsorship of The Geological Society of America, The Paleontological Society, The Society of Economic Paleontologists and Mineralogists, The Palaeontographical Society, and The Palaeontological Association; and (3) acceptance by The University of Kansas of publication without any financial gain to the University.

TREATISE ON INVERTEBRATE PALEONTOLOGY

PARTS

Parts of the *Treatise* are distinguished by assigned letters with a view to indicating their systematic sequence while allowing publication of units in whatever order each is made ready for the press. The volumes are cloth-bound with title in gold on the cover. Copies are available on orders sent to the Publication Sales Department, The Geological Society of America, 3300 Penrose Place, Boulder, Colorado 80301. Special discounts are available to members of sponsoring societies under arrangements made by appropriate officers of these societies, to whom inquiries should be addressed.

VOLUMES ALREADY PUBLISHED

(Previous to 1978)

Part C. PROTISTA 2 (Sarcodina, chiefly "Thecamoebians" and Foraminiferida), xxxi+900 p., 5311 fig., 1964.

Part D. PROTISTA 3 (chiefly Radiolaria, Tintinnina), xii+195 p., 1050 fig., 1954.

Part E. ARCHAEOCYATHA, PORIFERA, xviii+122 p., 728 fig., 1955.

Part E, Volume 1. ARCHAEOCYATHA, second edition (revised and enlarged), xxx+158 p., 871 fig., 1972.

Part F. COELENTERATA, xvii+498 p., 2700 fig., 1956.

Part G. BRYOZOA, xii+253 p., 2000 fig., 1953.

Part H. BRACHIOPODA, xxxii +927 p., 5198 fig., 1965.

Part I. MOLLUSCA 1 (Mollusca General Features, Scaphopoda, Amphineura, Monoplacophora, Gastropoda General Features, Archaeogastropoda, mainly Paleozoic Caenogastropoda and Opisthobranchia), xxiii+351 p., 1732 fig., 1960.

Part K. MOLLUSCA 3 (Cephalopoda General Features, Endoceratoidea, Actinoceratoidea, Nautiloidea, Bactritoidea), xxviii+519 p., 2382 fig., 1964.

Part L. MOLLUSCA 4 (Ammonoidea), xxii+490 p., 3800 fig., 1957.

Part N. MOLLUSCA 6 (Bivalvia), Volumes 1 and 2 (of 3), xxxviii+952 p., 6198 fig., 1969; Volume 3, iv+272 p., 742 fig., 1971.

Part O. ARTHROPODA 1 (Arthropoda General Features, Protarthropoda, Euarthropoda General Features, Trilobitomorpha), xix+560 p., 2880 fig., 1959.

Part P. ARTHROPODA 2 (Chelicerata, Pycnogonida, Palaeoisopus), xvii+181 p., 565 fig., 1955.

Part Q. ARTHROPODA 3 (Crustacea, Ostracoda), xxiii+442 p., 3476 fig., 1961.

Part R. ARTHROPODA 4 (Crustacea exclusive of Ostracoda, Myriapoda, Hexapoda), Volumes 1 and 2 (of 3), xxxvi+651 p., 1762 fig., 1969.

Part S. ECHINODERMATA 1 (Echinodermata General Features, Homalozoa, Crinozoa, exclusive of Crinoidea), xxx+650 p., 2868 fig., 1967 [1968].

Part U. ECHINODERMATA 3 (Asterozoans, Echinozoans), xxx+695 p., 3485 fig., 1966.

Part V. GRAPTOLITHINA, xvii+101 p., 358 fig., 1955.

Part V. GRAPTOLITHINA, second edition (revised and enlarged), xxxii+163 p., 507 fig., 1970.

Part W. Miscellanea (Conodonts, Conoidal Shells of Uncertain Affinities, Worms, Trace Fossils, Problematica), xxv+259 p., 1058 fig., 1962.

Part W. Miscellanea (Supplement 1). Trace Fossils, second edition (revised and enlarged), xxi+269 p., 912 fig., 1975.

THIS VOLUME

Part T. Echinodermata 2 (Crinoidea), Volumes 1-3, xxxviii + 1027 p., 4833 fig., 1978.

VOLUMES IN PREPARATION (1978)

Part A. Introduction.

Part B. Protista 1 (Chrysomonadida, Coccolithophorida, Charophyta, Diatomacea, etc.).

Part J. Mollusca 2 (Gastropoda, Streptoneura exclusive of Archaeogastropoda, Euthyneura).

Part M. Mollusca 5 (Coleoidea).

Part R. Arthropoda 4, Volume 3 (Hexapoda).

Part X. Addenda, Index.

Part E, Volume 2. Porifera (revised edition).

Part F. Coelenterata (supplement) (Anthozoa, Rugosa and Tabulata, revised edition).

Part G. Bryozoa (revised edition).

Part L. Ammonoidea (revised edition).

Part W. Miscellanea (Supplement 2) (Conodonts, revised edition).

CONTRIBUTING AUTHORS

[Arranged by countries and institutions; an alphabetical list follows. An asterisk preceding name indicates author working on revision of or supplement to a published *Treatise* volume.]

AFRICA

Shell Gabon (Port Gentil)
*R. A. Pohowsky

AUSTRALIA

South Australia Geological Survey (Adelaide)
N. H. Ludbrook

University of Adelaide
M. F. Glaessner

University of Queensland (Brisbane)
Dorothy Hill

AUSTRIA

Universität Wien (Paläontologisches Institut)
Adolph Papp

BELGIUM

Unattached
*Charles Grégoire (Bruxelles)

Université de Liège
Georges Ubaghs

CANADA

Geological Survey of Canada (Dartmouth, Nova Scotia)
G. L. Williams

Geological Survey of Canada (Ottawa)
J. A. Jeletzky, D. J. McLaren, G. W. Sinclair

Institute of Sedimentary & Petroleum Geology (Geological Survey of Canada, Calgary)
A. W. Norris

University of British Columbia (Vancouver)
V. J. Okulitch

University of Saskatchewan
W. A. S. Sarjeant

University of Toronto
D. H. Collins

DENMARK

Universitet København
*Eckart Håkansson, H. Wienberg Rasmussen

FRANCE

Université de Paris
Colette Dechaseaux (Laboratoire de Paléontologie des Vertébrés); Geneviève Lutaud (Laboratoire de Cytologie)

Université Paris-Sud (Orsay)
Michel Roux

GERMAN DEMOCRATIC REPUBLIC

Bergakademie Freiberg (Fachbereich Geowissenschaften)
A. H. Müller

GERMANY, FEDERAL REPUBLIC OF

Friedrich Wilhelms Universität (Bonn)
H. K. Erben, *K. J. Müller
Natur-Museum und Forschungs-Institut Senckenberg (Frankfurt)
Herta Schmidt, Wolfgang Struve
Philipps Universität (Marburg)
Gerhard Hahn, *Maurits Lindström, *Willi Ziegler
Unattached
Hertha Sieverts-Doreck (Stuttgart-Möhringen)
Universität Münster
Helmut Hölder
Universität Tübingen
*Jürgen Kullmann, *Adolf Seilacher
Universität Würzburg
Klaus Sdzuy

ITALY

Universitá Modena
Eugenia Montanaro Gallitelli
Universitá di Roma
Franco Rasetti

JAPAN

Saito Ho-on Kai Museum of Natural History (Sendai)
Kotora Hatai
University of Tokyo
Tetsuro Hanai

NETHERLANDS

Rijksmuseum van Natuurlijke Historie (Leiden)
L. B. Holthuis
Vrije Universiteit Amsterdam
A. Breimer, M. J. S. Rudwick

NEW ZEALAND

Auckland Institute and Museum
A. W. B. Powell
Dominion Museum (Wellington)
R. K. Dell
New Zealand Geological Survey (Lower Hutt)
C. A. Fleming
Unattached
J. Marwick

NORWAY

Unattached
Tron Soot-Ryen (Hosle)
Universitet Oslo
Gunnar Henningsmoen, Leif Størmer (Institutt for Geologi)

POLAND

Pánstwowe Wydawnictwo Naukowe (Warszawa)
Gertruda Biernat, Adolf Riedel

SAUDI ARABIA

Arabian American Oil Company (Dhahran)
A. L. Bowsher

SWEDEN

Naturhistoriska Museet Göteborg
Bengt Hubendick
Naturhistoriska Riksmuseet Stockholm
Valdar Jaanusson
Universitet Lund
Gerhard Regnéll
Universitet Stockholm
Ivar Hessland
Universitet Uppsala
R. A. Reyment

SWITZERLAND

Universität Basel
Manfred Reichel

UNITED KINGDOM

British Museum (Natural History) (London)
*P. L. Cook, Isabella Gordon, *M. K. Howarth, S. M. Manton, N. J. Morris, C. P. Nuttall
British Petroleum Company (Middlesex)
F. E. Eames
Institute of Geological Sciences (London)
Raymond Casey, R. V. Melville
Iraq Petroleum Company (London)
G. F. Elliott
Queen's University of Belfast
Margaret Jope, *R. E. H. Reid, A. D. Wright
Unattached
Dennis Curry (Middlesex); Sir James Stubblefield (London); R. P. Tripp (Sevenoaks, Kent); C. W. Wright (Dorset); Sir Maurice Yonge (Edinburgh)
University of Birmingham
*Anthony Hallam, L. J. Wills
University of Cambridge
H. B. Whittington
University College London
*J. H. Callomon, *D. T. Donovan
University College of Swansea (Swansea, Wales)
D. V. Ager, *J. S. Ryland

University of Durham
*G. P. Larwood
University of Glasgow
W. D. I. Rolfe, John Weir, Alwyn Williams
University of Hull
*M. R. House
University of Leicester
P. C. Sylvester-Bradley
University of Manchester
E. R. Trueman
University of Southampton
*R. L. Austin

UNITED STATES OF AMERICA

Academy of Natural Sciences of Philadelphia (Pennsylvania)
A. A. Olsson, Robert Robertson
American Museum of Natural History (New York)
R. L. Batten, W. K. Emerson, N. D. Newell
Appalachian State University (Boone, North Carolina)
*F. K. McKinney
Brown University (Providence, Rhode Island)
R. D. Staton
California Academy of Sciences (San Francisco)
Eugene Coan, Barry Roth, A. G. Smith
California Institute of Technology (Pasadena)
H. A. Lowenstam
Carnegie Museum (Pittsburgh, Pennsylvania)
Juan Parodiz
Cornell University (Ithaca, New York)
W. S. Cole, J. W. Wells
Exxon Production Research Company (Houston, Texas)
H. H. Beaver, R. M. Jeffords, S. A. Levinson, L. A. Smith, Joan Stough
Field Museum of Natural History (Chicago)
Fritz Haas, G. A. Solem
Florida State University (Tallahassee)
W. H. Heard
Getty Oil Company (Houston, Texas)
Lavon McCormick
Harvard University (Cambridge, Massachusetts)
Kenneth Boss, F. M. Carpenter, W. J. Clench, H. B. Fell, Bernhard Kummel, Ruth Turner
Illinois State Geological Survey (Urbana)
M. L. Thompson
Indiana Geological Survey (Bloomington)
R. H. Shaver
Kansas Geological Survey (Lawrence)
D. E. Nodine Zeller
Kent State University (Kent, Ohio)
A. H. Coogan
Louisiana State University (Baton Rouge)
W. A. van den Bold, H. B. Stenzel
New Mexico Institute Mining & Technology (Socorro)
Christina Lochman-Balk
New York State Museum (Albany)
D. W. Fisher
Ohio State University (Columbus)
*S. M. Bergström, Aurèle La Rocque, W. C. Sweet
Oklahoma Geological Survey (Norman)
T. W. Amsden, R. O. Fay
Oregon State University (Corvallis)
A. J. Boucot, J. G. Johnson
Paleontological Research Institution (Ithaca, New York)
K. V. W. Palmer
Pennsylvania State University (University Park)
*R. J. Cuffey
Princeton University (Princeton, New Jersey)
A. G. Fischer
Professional Geophysics, Inc. (Oklahoma City, Oklahoma)
J. A. Eyer
Queens College of the City of New York (Flushing)
*R. M. Finks
Radford College (Radford, Virginia)
R. L. Hoffman
St. Mary's College of California (St. Mary's)
A. S. Campbell
San Diego Natural History Museum (San Diego, California)
George Radwin
San Francisco State University (San Francisco, California)
Y. T. Mandra
Smithsonian Institution (Washington, D.C.)
F. M. Bayer, R. H. Benson, *R. S. Boardman, *A. H. Cheetham, A. H. Clarke, Jr., G. A. Cooper, Anita Epstein, T. G. Gibson, R. E. Grant, E. G. Kauffman, P. M. Kier, R.

B. Manning, David Pawson, H. A. Rehder

Southern Illinois University (Carbondale)
*John Utgaard

Southern Methodist University (Dallas, Texas)
A. L. McAlester

Southwest Missouri State University (Springfield)
*J. F. Miller

Stanford University (Stanford, California)
A. Myra Keen

State University of New York (Stony Brook)
A. R. Palmer

Syracuse University (Syracuse, N.Y.)
J. C. Brower, *O. B. Nye

Tulane University (New Orleans, Louisiana)
Emily Vokes, H. E. Vokes

Unattached
R. Wright Barker (Bellaire, Texas), J. W. Hedgpeth (Santa Rosa, Calif.), H. S. Puri (Tallahassee, Florida)

United States Geological Survey (Washington, D.C.)
J. M. Berdan, R. C. Douglass, Mackenzie Gordon, Jr., *O. L. Karklins, K. E. Lohman, N. F. Sohl, I. G. Sohn, E. L. Yochelson, Dwight Taylor (Menlo Park, Calif.), John Van Couvering (Denver)

University of Alaska (Fairbanks)
C. D. Wagner

University of California (Berkeley)
J. W. Durham

University of California (Los Angeles)
A. R. Loeblich, Jr., W. P. Popenoe, Helen Tappan

University of California (San Diego, La Jolla)
R. R. Hessler, W. A. Newman

University of Cincinnati (Ohio)
K. E. Caster, D. L. Meyer

University of Florida (Gainesville)
H. K. Brooks, F. G. Thompson

University of Illinois (Urbana)
*D. B. Blake, *Philip Sandberg, H. W. Scott

University of Indiana (Bloomington)
N. Gary Lane

University of Iowa (Iowa City)
W. M. Furnish, B. F. Glenister, *Gilbert Klapper, H. L. Strimple

University of Kansas (Lawrence)
A B. Leonard, R. A. Robison, A. J. Rowell, R. H. Thompson

University of Massachusetts (Amherst)
C. W. Pitrat

University of Miami (Florida)
William W. Hay, Donald Moore

University of Michigan (Ann Arbor)
J. B. Burch, R. V. Kesling, D. B. Macurda, C. P. Morgan, F. H. T. Rhodes

University of Minnesota (Minneapolis)
F. M. Swain

University of Missouri (Columbia)
R. E. Peck

University of North Carolina (Wilmington)
V. A. Zullo

University of Rochester (Rochester)
Curt Teichert

University of Texas (Arlington)
B. F. Perkins

University of Texas (Austin)
J. T. Sprinkle

University of Wisconsin (Madison)
*D. L. Clark

University of Wyoming (Laramie)
D. W. Boyd

Western Reserve University (Cleveland, Ohio)
F. G. Stehli

Western Washington University (Bellingham)
C. A. Ross, J. R. P. Ross

Wichita State University (Kansas)
Paul Tasch

Woods Hole Oceanographic Institute (Massachusetts)
W. A. Berggren

Wright State University (Dayton, Ohio)
*T. S. Wood

DECEASED

W. J. Arkell, R. S. Bassler, H. Boschma, M. N. Bramlette, C. C. Branson, O. M. Bulman, André Chavan, L. R. Cox, L. M. Davies, Harriet Exline, D. L. Frizzell, Julia Gardner, Walter Häntzschel, G. D. Hanna, H. J. Harrington, W. H. Hass, H. L. Hawkins, L. G. Hertlein, H. V. Howe, B. F. Howell, L. H. Hyman, J. B. Knight, M. W. deLaubenfels, Marius Lecompte, A. K. Miller, R. C. Moore, H. M. Muir-Wood, Alexander Petrunkevitch,

Chr. Poulsen, Emma Richter, Rudolf Richter, O. H. Schindewolf, W. K. Spencer, M. A. Stainbrook, L. W. Stephenson, E. C. Stumm, O. W. Tiegs, Johannes Wanner, J. M. Weller, T. H. Withers, Arthur Wrigley

Alphabetical List

Ager, D. V., London (Univ. College of Swansea)
Amsden, T. W., Norman, Okla. (Oklahoma Geol. Survey)
Arkell, W. J. (deceased)
*Austin, R. L., Southampton, Eng. (Univ. Southampton)
Barker, R. W., Bellaire, Texas (unattached)
Bassler, R. S. (deceased)
Batten, R. L., New York (American Museum Nat. History)
Bayer, F. M., Washington, D.C. (Smithsonian Inst.)
Beaver, H. H., Houston, Texas (Exxon Production Research Company)
Benson, R. H., Washington, D.C. (Smithsonian Inst.)
Berdan, J. M., Washington, D.C. (U.S. Geol. Survey)
Berggren, W. A., Woods Hole, Mass. (Woods Hole Oceanographic Inst.)
*Bergström, S. M., Ohio (Ohio State Univ.)
Biernat, Gertruda, Warszawa, Poland (Pánstwowe Wydawnictwo Naukowe)
*Blake, D. B., Urbana, Ill. (Univ. Illinois)
*Boardman, R. S., Washington, D.C. (Smithsonian Inst.)
Bold, W. A. van den, Baton Rouge, La. (Louisiana State Univ.)
Boschma, H. (deceased)
Boss, Kenneth, Cambridge, Mass. (Harvard Univ.)
Boucot, A. J., Corvallis, Ore. (Oregon State Univ.)
Bowsher, A. L., Dhahran, Saudi Arabia (Arabian American Oil Co.)
Boyd, D. W., Laramie, Wyo. (Univ. Wyoming)
Bramlette, M. N. (deceased)
Branson, C. C. (deceased)
Breimer, A., Amsterdam, Netherlands (Inst. Aardwetensch. Vrije Univ.)
Brooks, H. K., Gainesville, Fla. (Univ. Florida)
Brower, J. C., Syracuse, N.Y. (Syracuse Univ.)
Bulman, O. M. B. (deceased)
Burch, J. B., Ann Arbor, Mich. (Univ. Michigan)
*Callomon, J. H., London (Univ. College)
Campbell, A. S., St. Mary's, Calif. (St. Mary's College)
Carpenter, F. M., Cambridge, Mass. (Harvard Univ.)
Casey, Raymond, London (Inst. Geol. Sciences)
Caster, K. E., Cincinnati, Ohio (Univ. Cincinnati)
Chavan, André (deceased)
*Cheetham, A. H., Washington, D.C. (Smithsonian Inst.)
*Clark, D. L., Madison, Wis. (Univ. Wisconsin)
Clarke, A. H., Jr., Washington, D.C. (Smithsonian Inst.)
Clench, W. J., Cambridge, Mass. (Harvard Univ.)
Coan, Eugene, San Francisco, Calif. (California Acad. Sci.)
Cole, W. S., Ithaca, N.Y. (Cornell Univ.)
Collins, D. H., Toronto, Canada (Univ. Toronto)
Coogan, A. H., Kent, Ohio (Kent State Univ.)
*Cook, P. L., London (British Museum Nat. History)
Cooper, G. A., Washington, D.C. (Smithsonian Inst.)
Cox, L. R. (deceased)
*Cuffey, R. J., University Park, Pa. (Pennsylvania State Univ.)
Curry, Dennis, Middlesex, Eng. (unattached)
Davies, L. M. (deceased)
Dechaseaux, Colette, Paris (Laboratoire de Paléontologie des Vertébrés)
Dell, R. K., Wellington, N.Z. (Dominion Museum)
*Donovan, D. T., London, Eng. (Univ. College)
Douglass, R. C., Washington, D.C. (U.S. Geol. Survey)
Durham, J. W., Berkeley, Calif. (Univ. California)
Eames, F. E., Middlesex, Eng. (British Petroleum Co.)
Elliott, G. F., London (Iraq Petroleum Co.)
Emerson, W. K., New York (American Museum Nat. History)
Epstein, Anita, Washington, D.C. (Smithsonian Inst.)
Erben, H. K., Bonn, West Germany (Friedrich Wilhelms Univ.)
Exline, Harriet (deceased)
Eyer, J. A., Oklahoma City, Okla. (Professional Geophysics, Inc.)
Fay, R. O., Norman, Okla. (Oklahoma Geol. Survey)
Fell, H. B., Cambridge, Mass. (Harvard Univ.)
*Finks, R. M., Flushing, N.Y. (Queens College)
Fischer, A. G., Princeton, N.J. (Princeton Univ.)
Fisher, D. W., Albany, N.Y. (New York State Museum)
Fleming, C. A., Lower Hutt, N.Z. (New Zealand Geol. Survey)
Frizzell, D. L. (deceased)
Furnish, W. M., Iowa City, Iowa (Univ. Iowa)
Gardner, Julia (deceased)
Gibson, T. G., Washington, D.C. (Smithsonian Inst.)
Glaessner, M. F., Adelaide, S. Australia (Univ. Adelaide)
Glenister, B. F., Iowa City, Iowa (Univ. Iowa)
Gordon, Isabella, London (British Museum Nat. History)
Gordon, Mackenzie, Jr., Washington, D.C. (U.S. Geol. Survey)

Grant, R. E., Washington, D.C. (Smithsonian Institution)
*Grégoire, Charles, Bruxelles, Belgium (unattached)
*Håkansson, Eckart, København (Univ. København)
Haas, Fritz, Chicago, Ill. (Field Museum Nat. History)
Häntzschel, Walter (deceased)
Hahn, Gerhard, Marburg (Philipps Univ.)
*Hallam, Anthony, Oxford, Eng. (Univ. Birmingham)
Hanai, Tetsuro, Tokyo (Univ. Tokyo)
Hanna, G. D. (deceased)
Harrington, H. J. (deceased)
Hass, W. H. (deceased)
Hatai, Kotora, Sendai, Japan (Saito Ho-on Kai Museum Nat. History)
Hawkins, H. L., Reading, Eng. (deceased)
Hay, W. W., Miami, Fla. (Univ. Miami)
Heard, W. H., Tallahassee, Fla. (Florida State Univ.)
Hedgpeth, J. W., Santa Rosa, Calif. (unattached)
Henningsmoen, Gunnar, Oslo (Univ. Oslo)
Hertlein, L. G., San Francisco, Calif. (deceased)
Hessland, Ivar, Stockholm, Sweden (Univ. Stockholm)
Hessler, R. R., La Jolla, Calif. (Scripps Inst. Oceanography)
Hill, Dorothy, Brisbane, Australia (Univ. Queensland)
Hölder, Helmut, Münster, Germany (Univ. Münster)
Hoffman, R. L., Radford, Va. (Radford College)
Holthuis, L. B., Leiden, Netherlands (Rijksmuseum van Natuurlijke Historie)
*House, M. R., Kingston upon Hull, Eng. (Univ. Hull)
*Howarth, M. K., London (British Museum Nat. History)
Howe, H. V. (deceased)
Howell, B. F. (deceased)
Hubendick, Bengt, Göteborg, Sweden (Naturhistoriska Museet)
Hyman, L. H. (deceased)
Jaanusson, Valdar, Stockholm (Naturhistoriska Riksmuseet)
Jeffords, R. M., Houston, Texas (Exxon Production Research Company)
Jeletsky, J. A., Ottawa, Ontario, Canada (Geol. Survey Canada)
Johnson, J. G., Corvallis, Ore. (Oregon State University)
Jope, Margaret, Belfast, N. Ireland (Queen's Univ. of Belfast)
*Karklins, O. L., Washington, D.C. (U.S. Geol. Survey)
Kauffman, E. G., Washington, D.C. (Smithsonian Inst.)
Keen, A. Myra, Stanford, Calif. (Stanford Univ.)
Kesling, R. V., Ann Arbor, Mich. (Univ. Michigan)
Kier, P. M., Washington, D.C. (Smithsonian Inst.)
*Klapper, Gilbert, Iowa City, Iowa (Univ. Iowa)
Knight, J. B. (deceased)
*Kullmann, Jürgen, Tübingen, W. Germany (Univ. Tübingen)
Kummel, Bernhard, Cambridge, Mass. (Harvard Univ.)
Lane, N. Gary, Bloomington, Ind. (Univ. Indiana)
La Rocque, Aurèle, Columbus, Ohio (Ohio State Univ.)
*Larwood, G. P., Durham, Eng. (Univ. Durham)
Laubenfels, M. W. de (deceased)
Lecompte, Marius (deceased)
Leonard, A. B., Lawrence, Kans. (Univ. Kansas)
Levinson, S. A., Houston, Texas (Exxon Production Research Company)
*Lindström, Maurits, Marburg, Germany (Philipps Univ.)
Lochman-Balk, Christina, Socorro, N. Mex. (New Mexico Inst. Mining & Technology)
Loeblich, A. R., Jr., Los Angeles, Calif. (Univ. California)
Lohman, K. E., Washington, D.C. (U.S. Geol. Survey)
Lowenstam, H. A., Pasadena, Calif. (California Inst. Technology)
Ludbrook, N. H., Adelaide, S. Australia (South Australia Geol. Survey)
Lutaud, Geneviève, Paris (Laboratoire Cytologie, Univ. Paris)
McAlester, A. L., Dallas, Texas (Southern Methodist Univ.)
McCormick, Lavon, Houston, Texas (Getty Oil Company)
*McKinney, F. K., Boone, N.Car. (Appalachian State Univ.)
McLaren, D. J., Ottawa, Ontario, Canada (Geol. Survey Canada)
Macurda, D. B., Ann Arbor, Mich. (Univ. Michigan)
Mandra, Y. T., San Francisco, Calif. (San Francisco State University)
Manning, R. B., Washington, D.C. (Smithsonian Inst.)
Manton, S. M., London (British Museum Nat. History)
Marwick, J., Havelock North, N.Z. (Unattached)
Melville, R. V., London (Inst. Geol. Sciences)
Meyer, D. L., Cincinnati, O. (Univ. Cincinnati)
Miller, A. K. (deceased)
*Miller, J. F., Springfield, Mo. (Southwest Missouri State Univ.)
Montanaro Gallitelli, Eugenia, Modena, Italy (Univ. Modena)
Moore, Donald, Miami, Fla. (Univ. Miami, Inst. Marine Sci.)
Moore, R. C. (deceased)
Morgan, C. P., Ann Arbor, Mich. (Univ. Michigan)
Morris, N. J., London (British Museum Nat. History)

Müller, A. H., Freiberg, German Democratic Republic (Geol. Inst. Bergakad.)
*Müller, K. J., Bonn, West Germany (Friedrich Wilhelms Univ.)
Muir-Wood, H. M. (deceased)
Newell, N. D., New York (American Museum Nat. History)
Newman, W. A., La Jolla, Calif. (Scripps Inst. Oceanography)
Norris, A. W., Calgary, Alberta, Canada (Geol. Survey Canada)
Nuttall, C. P., London (British Museum Nat. History)
*Nye, O. B., Syracuse, N.Y. (Syracuse Univ.)
Okulitch, V. J., Vancouver, Canada (Univ. British Columbia)
Olsson, A. A., Coral Gables, Fla. (Acad. Nat. Sci. Philadelphia)
Palmer, A. R., Stony Brook, Long Island, N.Y. (State Univ. New York)
Palmer, K. V. W., Ithaca, N.Y. (Paleont. Research Inst.)
Papp, Adolph, Wien, Austria (Univ. Wien)
Parodiz, Juan, Pittsburgh, Pa. (Carnegie Museum)
Pawson, D., Washington, D.C. (Smithsonian Inst.)
Peck, R. E., Columbia, Mo. (Univ. Missouri)
Perkins, B. F., Arlington, Texas (Univ. Texas at Arlington)
Petrunkevitch, Alexander (deceased)
Pitrat, C. W., Amherst, Mass. (Univ. Massachusetts)
*Pohowsky, R. A., Port Gentil, Gabon, Afr. (Shell Gabon)
Popenoe, W. P., Los Angeles, Calif. (Univ. Calif.)
Poulsen, Chr. (deceased)
Powell, A. W. B., Auckland, N.Z. (Auckland Inst. & Museum)
Puri, H. S., Tallahassee, Fla. (unattached)
Radwin, George, San Diego, Calif. (San Diego Nat. History Museum)
Rasetti, Franco, Rome, Italy (Univ. Roma)
Rasmussen, H. Wienberg, København, Denmark (Univ. København)
Regnéll, Gerhard, Lund, Sweden (Univ. Lund)
Rehder, H. A., Washington, D.C. (Smithsonian Inst.)
Reichel, Manfred, Basel, Switz. (Univ. Basel)
*Reid, R. E. H., Belfast, N. Ireland (Queen's Univ. Belfast)
Reyment, R. A., Uppsala, Sweden (Univ. Uppsala)
Rhodes, F. H. T., Ann Arbor, Mich. (Univ. Michigan)
Richter, Emma (deceased)
Richter, Rudolf (deceased)
Riedel, Adolf, Warszawa, Poland (Pánstwowe Wydawnictwo Naukowe)
Robertson, R., Philadelphia, Pa. (Acad. Nat. Sci.)
Robison, R. A., Lawrence, Kans. (Univ. Kansas)
Rolfe, W. D. I., Glasgow, Scot. (Univ. Glasgow)
Ross, C. A., Bellingham, Wash. (Western Washington University)
Ross, J. R. P., Bellingham, Wash. (Western Washington University)
Roth, Barry, San Francisco, Calif. (California Acad. Sci.)
Roux, Michel, Orsay, France (Univ. Paris-Sud)
Rowell, A. J., Lawrence, Kans. (Univ. Kansas)
Rudwick, M. J. S., Amsterdam (Vrije Univ.)
*Ryland, J. S., Swansea, Wales (Univ. College)
*Sandberg, Philip, Urbana, Ill. (Univ. Illinois)
Sarjeant, W. A. S., Saskatoon, Canada (Univ. Saskatchewan)
Schindewolf, O. H. (deceased)
Schmidt, Herta, Frankfurt, Germany (Natur Museum u. Forsch.-Inst. Senckenberg)
Scott, H. W., Urbana, Ill. (Univ. Illinois)
Sdzuy, Klaus, Würzburg, Germany (Univ. Würzburg)
*Seilacher, Adolf, Tübingen, West Germany (Univ. Tübingen)
Shaver, R. H., Bloomington, Ind. (Indiana Geol. Survey & Univ. Indiana)
Sieverts-Doreck, Hertha, Stuttgart-Möhringen, Germany (unattached)
Sinclair, G. W., Ottawa, Ontario, Canada (Geol. Survey Canada)
Smith, A. G., San Francisco, Calif. (California Acad. Sci.)
Smith, L. A., Houston, Texas (Exxon Production Research Company)
Sohl, N. F., Washington, D.C. (U.S. Geol. Survey)
Sohn, I. G., Washington, D.C. (U.S. Geol. Survey)
Solem, G. A., Chicago, Ill. (Field Museum Nat. History)
Soot-Ryen, Tron, Hosle, Nor. (unattached)
Spencer, W. K. (deceased)
Sprinkle, J. T., Austin, Texas (Univ. Texas)
Stainbrook, M. A. (deceased)
Staton, R. D., Providence, R.I. (Brown Univ.)
Stehli, F. G., Cleveland, Ohio (Western Reserve Univ.)
Stenzel, H. B., Baton Rouge, La. (Louisiana State Univ.)
Stephenson, L. W. (deceased)
Størmer, Leif, Oslo (Univ. Oslo)
Stough, Joan, Houston, Texas (Exxon Production Research Company)
Strimple, H. L., Iowa City, Iowa (Univ. Iowa)
Struve, Wolfgang, Frankfurt, Germany (Natur-Museum u. Forsch.-Inst. Senckenberg)
Stubblefield, Sir James, London (unattached)
Stumm, E. C. (deceased)
Swain, F. M., Minneapolis, Minn. (Univ. Minnesota)
Sweet, W. C., Columbus, Ohio (Ohio State Univ.)
Sylvester-Bradley, P. C., Leicester, Eng. (Univ. Leicester)
Tappan, Helen, Los Angeles, Calif. (Univ. California)
Tasch, Paul, Wichita, Kans. (Wichita State Univ.)
Taylor, Dwight, Menlo Park, Calif. (U.S. Geol. Survey)

Teichert, Curt, Rochester, N.Y. (Univ. Rochester)
Thompson, F. G., Gainesville, Fla. (Univ. Florida)
Thompson, M. L., Urbana, Ill. (Illinois State Geol. Survey)
Thompson, R. H., Lawrence, Ks. (Univ. Kansas)
Tiegs, O. W. (deceased)
Tripp, R. P., Sevenoaks, Kent, Eng. (unattached)
Trueman, E. R., Manchester, Eng. (Univ. Manchester)
Turner, Ruth, Cambridge, Mass. (Harvard Univ.)
Ubaghs, Georges, Liège, Belgium (Univ. Liège)
*Utgaard, John, Carbondale, Ill. (Southern Illinois Univ.)
Van Couvering, John, Denver, Colo. (U.S. Geol. Survey)
Vokes, Emily, New Orleans, La. (Tulane Univ.)
Vokes, H. E., New Orleans, La. (Tulane Univ.)
Wagner, C. D., Fairbanks, Alaska (Univ. Alaska)
Wanner, Johannes (deceased)
Weir, J., Tayport, Fife, Scotland (Univ. Glasgow)
Weller, J. M. (deceased)
Wells, J. W., Ithaca, N.Y. (Cornell Univ.)
Whittington, H. B., Cambridge, Eng. (Univ. Cambridge)
Williams, Alwyn, Glasgow, Scot. (Univ. Glasgow)
Williams, G. L., Dartmouth, Nova Scotia (Geol. Survey Canada)
Wills, L. J., Birmingham, Eng. (Univ. Birmingham)
Withers, T. H. (deceased)
*Wood, T. S., Dayton, Ohio (Wright State Univ.)
Wright, A. D., Belfast, N. Ireland (Queen's Univ. Belfast)
Wright, C. W., Dorset, Eng. (unattached)
Wrigley, Arthur (deceased)
Yochelson, E. L., Washington, D.C. (U.S. Geol. Survey)
Yonge, Sir Maurice, Edinburgh, Scotland (unattached)
Zeller, D. E. Nodine, Lawrence, Kans. (Kansas Geol. Survey)
*Ziegler, Willi, Marburg, Germany (Philipps Univ.)
Zullo, V. A., Wilmington, N.C. (Univ. North Carolina)

EDITORIAL PREFACE

INTRODUCTION

The history of the volumes constituting Part T of the *Treatise on Invertebrate Paleontology* goes back as far as the beginnings of the *Treatise* project itself, that is, to the years 1948 and 1949. It seems that the earliest section of systematic text to be attacked was treatment of the Articulata, for in the academic year 1950-51 Dr. Hertha Sieverts-Doreck spent several months in Lawrence working with R. C. Moore on the first draft for the systematic descriptions of that group. Little or no progress seems to have been made on this or any other section of Part T, however, during the one-and-a-half decades following Dr. Sieverts-Doreck's visit. No doubt, during this period Dr. Moore must have worked sporadically on his own assignments, which included the Inadunata and Flexibilia.

In 1965, at Moore's request, Teichert conferred with Dr. Sieverts-Doreck in Stuttgart-Möhringen and it was decided to invite Dr. H. Wienberg Rasmussen of Copenhagen to assist in the completion of the description of the Articulata. These arrangements, however, did not come to fruition until about 10 years later when Wienberg Rasmussen finally undertook to complete the text on the Articulata himself.

The systematic chapter on the Flexibilia was completed by R. C. Moore in 1973. Dr. Georges Ubaghs completed the chapter on the Camerata in the same year. Moore's assignments also included the description of the Inadunata, a task in which he was ably assisted by N. Gary Lane and H. L. Strimple. In spite of grave illness Moore continued work on text for the Inadunata throughout 1973 and completed the task only a few months before his death in April, 1974.

Until 1972, little thought had been given to the organization of materials for the introductory part, although in 1968, at the invitation of Dr. Moore, Albert Breimer had delivered a manuscript describing the anatomy, physiology, and ecology of living crinoids.

In the spring of 1972, Teichert paid a visit to Ubaghs in Liège, Belgium, and the two of them drew up plans for preparation of materials required for the introductory sections of Part T and an outline of contents was prepared that was very closely adhered to in the cooperative efforts that followed. The editors enlisted the help of Georges Ubaghs to shape the chapter on skeletal morphology, of D. B. Macurda, Jr., D. L. Meyer, and Michel Roux the chapter on the crinoid stereom, and of N. Gary Lane, H. Wienberg Rasmussen, Albert Breimer, J. C. Brower, and H. L. Strimple for a variety of assignments as is apparent from the following text. Most of these manuscripts were completed only after

R. C. MOORE's death in April, 1974, and they were reviewed and edited by CURT TEICHERT and the editorial staff. The cutoff date for addition of information to the systematic descriptions was toward the end of 1976.

It is the purpose of the Introduction of the Editorial Preface to give a brief history of a particular Part of the *Treatise,* to explain the aims of the *Treatise* in general terms, and to make appropriate acknowledgments for support received. The bulk of the Editorial Preface then is devoted to an explanation of nomenclatural practices adopted in the *Treatise.* Although in the present volume much of this text has been copied unchanged from the prefaces of earlier editions, numerous sections have been thoroughly revised by CURT TEICHERT.

The aim of the *Treatise on Invertebrate Paleontology,* as originally conceived and consistently pursued, is to present the most comprehensive and authoritative, yet compact statement of knowledge concerning invertebrate fossil groups that can be formulated by collaboration of competent specialists in seeking to organize what has been learned of this subject up to the year of publication of each individual part. Such work has value in providing a most useful summary of the collective results of multitudinous investigations and thus constitutes an indispensable text and reference book for all persons who wish to know about remains of invertebrate organisms preserved in rocks of the earth's crust. This applies to neozoologists as well as paleozoologists and to beginners in study of fossils as well as to thoroughly trained, long-experienced professional workers, including teachers, stratigraphical geologists, and individuals engaged in research on fossil invertebrates. The making of a reasonably complete inventory of present knowledge of invertebrate paleontology is yielding needed foundation for future research.

The *Treatise* is divided into parts which bear index letters, each except the initial and concluding ones being defined to include designated groups of invertebrates. The chief purpose of this arrangement is to provide for independence of the several parts as regards date of publication, because it was judged desirable to print and distribute each segment as soon as possible after it is ready for press. Pages in each part bear the assigned index letter joined with numbers beginning with 1 and running consecutively to the end of the part. In numerous cases materials for individual parts were so voluminous that these parts had to be published in two or even three volumes. In such cases, pagination is continuous through successive volumes.

The outline of subjects to be treated in connection with each large group of invertebrates includes (1) description of morphological features, with special reference to hard parts, (2) ontogeny, (3) classification, (4) geological distribution, (5) evolutionary trends and phylogeny, (6) paleoecology, and (7) systematic description of genera, subgenera, and higher taxonomic units. Selected lists of references only were furnished in earlier parts of the *Treatise,* but since the mid-1960's the tendency has been to make these lists as comprehensive as possible.

Features of style in the taxonomic portions of this work have been fixed by the editors with aid furnished by advice from representatives of the societies which have undertaken to sponsor the *Treatise.* It is the editors' responsibility to consult with authors and coordinate their work, seeing that manuscript properly incorporates features of adopted style. Especially they are called on to formulate policies in respect to many questions of nomenclature and procedure. The subject of genus-group as well as family and subfamily names is reviewed briefly in a following section of this preface, and features of *Treatise* style in generic descriptions are explained.

A generous grant of $35,000 was made in 1948 by the Geological Society of America for initial work in preparing *Treatise* illustrations. Additional grants were made by The Geological Society of America in 1971 ($6,200), 1972 ($6,000), $7,000 each year for 1973 and 1974, and $20,000 each for 1975, 1976, and 1977. Administration of expenditures has been in charge of the editors and most of the work by photographers and artists has been done under their direction at the University of Kansas, but sizable parts of this program have also been carried forward in Washington, London, Ottawa, and many other places.

In December, 1959, the National Science Foundation of the United States, through its Division of Biological and Medical Sciences and the Program Director for Systematic Biology, made a grant in the amount of $210,000 for the purpose of aiding the completion of yet-unpublished volumes of the *Treatise*. Payment of this sum was provided to be made in installments distributed over a five-year period, with administration of disbursements handled by the University of Kansas. An additional grant (No. GB 4544) of $102,800 was made by the National Science Foundation in January, 1966, for the two-year period 1966-67, and this was extended for the calendar year 1968 by payment of $25,700 in October, 1967. This grant was extended further by payments of $57,800 in 1968 for calendar year 1969, and $66,000 each for calendar years 1970-72. For the years 1973-77 grants totaled $197,400. These funds are used primarily to maintain editorial operations at the University of Kansas and to provide assistance to authors needed in preparation of manuscripts and illustrations. Grateful acknowledgment to the Foundation is expressed on behalf of the societies sponsoring the *Treatise,* the University of Kansas, and innumerable individuals benefited by the *Treatise* project.

ZOOLOGICAL NAMES

Many questions arise in connection with zoological names, especially including those that relate to their acceptability and to alterations of some which may be allowed or demanded. Procedure in obtaining answers to these questions is guided and to a large extent governed by regulations published (1961) in the *International Code of Zoological Nomenclature*[1] (hereinafter cited simply as the *Code*). The prime object of the *Code* is to promote stability and universality in the scientific names of animals, ensuring also that each name is distinct and unique while avoiding restrictions on freedom of taxonomic thought or action. Priority is a basic principle, but under specified conditions its application can be modified. This is all well and good, yet nomenclatural tasks confronting the zoological taxonomist are formidable. They warrant the complaint of some that zoology, including paleozoology, is the study of animals rather than of names applied to them.

Several ensuing pages are devoted to aspects of zoological nomenclature that are judged to have chief importance in relation to procedures adopted in the *Treatise*. Terminology is explained, and examples of style employed in the nomenclatural parts of systematic descriptions are given.

TAXA GROUPS

Each taxonomic unit (taxon, pl., taxa) belongs to a rank in the adopted hierarchy of classificatory divisions. In part, this hierarchy is defined by the *Code* to include a species-group of taxa, a genus-group, and a family-group. Units of lower rank than subspecies are excluded from zoological nomenclature and those higher than superfamily of the family-group are not regulated by the *Code*. It is natural and convenient to discuss nomenclatural matters in general terms first and then to consider each of the taxa groups separately. Especially important is the provision that within each taxa group classificatory units are coordinate (equal in rank), whereas units of different taxa groups are not coordinate.

FORMS OF NAMES

All zoological names are divisible into groups based on their form (spelling). The first-published form (or forms) of a name is defined as original spelling (*Code,* Art. 32) and any later-published form (or forms) of the same name is designated as subsequent spelling (Art. 33). Obviously, original and subsequent spellings of a given name may or may not be identical and this affects consideration of their correctness. Further, examination of original spellings of names shows that by no means all can be distinguished as correct. Some are incorrect, and the same is true of subsequent spellings.

Original Spellings

If the first-published form of a name is consistent and unambiguous, the original spelling is defined as correct unless it contravenes some stipulation of the *Code* (Arts.

[1] N. R. Stoll et al. (ed. comm.), *International Code of Zoological Nomenclature, adopted by the XV International Congress of Zoology,* xvii + 176 p. (International Trust for Zoological Nomenclature, London, 1961).

26-31), or the original publication contains clear evidence of an inadvertent error, in the sense of the *Code,* or, among names belonging to the family-group, unless correction of the termination or the stem of the type-genus is required. An original spelling that fails to meet these requirements is defined as incorrect.

If a name is spelled in more than one way in the original publication, the form adopted by the first reviser is accepted as the correct original spelling, provided that it complies with mandatory stipulations of the *Code* (Arts. 26-31).

Incorrect original spellings are any that fail to satisfy requirements of the *Code,* or that represent an inadvertent error, or that are one of multiple original spellings not adopted by a first reviser. These have no separate status in zoological nomenclature and therefore cannot enter into homonymy or be used as replacement names and they call for correction. For example, a name originally published with a diacritic mark, apostrophe, diaeresis, or hyphen requires correction by deleting such features and uniting parts of the name originally separated by them, except that deletion of an umlaut from a vowel is accompanied by inserting "e" after the vowel.

Subsequent Spellings

If a name classed as a subsequent spelling is identical with an original spelling, it is distinguishable as correct or incorrect on the same criteria that apply to the original spelling. This means that a subsequent spelling identical with a correct original spelling is also correct, and one identical with an incorrect original spelling is also incorrect. In the latter case, both original and subsequent spellings require correction (authorship and date of the original incorrect spelling being retained).

If a subsequent spelling differs from an original spelling in any way, even by the omission, addition, or alteration of a single letter, the subsequent spelling must be defined as a different name (except that such changes as altered terminations of adjectival specific names to obtain agreement in gender with associated generic names, of family-group names to denote assigned taxonomic rank, and corrections for originally used diacritic marks, hyphens, and the like are excluded from spelling changes conceived to produce a different name). In certain cases species-group names having variable spellings are regarded as homonyms as specified in Art. 58 of the *Code.*

Altered subsequent spellings other than the exceptions noted may be either intentional or unintentional. If demonstrably intentional, the change is designated as an emendation. Emendations are divisible into justifiable and unjustifiable ones. Justifiable emendations are corrections of incorrect original spellings, and these take the authorship and date of the original spellings. Unjustifiable emendations are names having their own status in nomenclature, with author and date of their publication; they are junior objective synonyms of the name in its original form.

Subsequent spellings that differ in any way from original spellings, other than previously noted exceptions, and that are not classifiable as emendations are defined as incorrect subsequent spellings. They have no status in nomenclature, do not enter into homonymy, and cannot be used as replacement names. It is the purpose of the following chapters to explain in some detail the implications of various kinds of subsequent spellings and how these are dealt with in the *Treatise.*

AVAILABLE AND UNAVAILABLE NAMES

Available Names

An available zoological name is any that conforms to all mandatory provisions of the *Code.* Such names are classifiable in groups which are recognized in the *Treatise,* though not explicitly differentiated in the *Code.* They are as follows:

1) So-called *"inviolate names"* include all available names that are not subject to alteration from their originally published form. They comprise correct original spellings and commonly include correct subsequent spellings, but include no names classed as emendations. Here belong most genus-group names (including those for collective groups), some of which differ in spelling from others by only a single letter.

2) Names may be termed *"perfect names"* if, as originally published they

meet all mandatory requirements, needing no correction of any kind, but nevertheless are legally alterable in such ways as changing the termination (e.g., many species-group names, family-group names). This group does not include emended incorrect original spellings (e.g., *Oepikina,* replacement of *Öpikina*).

3) *"Imperfect names"* are available names that as originally published contain mandatorily emendable defects. Incorrect original spellings are imperfect names. Examples of emended imperfect names are: among species-group names, *guerini* (not *Guérini*), *obrienae* (not *O'Brienae*), *terranovae* (not *terra-novae*), *nunezi* (not *Nuñezi*), *Spironema rectum* (not *Spironema recta,* because generic name is neuter, not feminine); among genus-group names, *Broeggeria* (not *Bröggeria*), *Obrienia* (not *O'Brienia*), *Maccookites* (not *McCookites*); among family-group names, Oepikidae (not Öpikidae), Spironematidae (not Spironemidae, incorrect stem), Athyrididae (not Athyridae, incorrect stem). The use of "variety" for named divisions of fossil species, according to common practice of some paleontologists, gives rise to imperfect names, which generally are emendable (*Code,* Art. 45e) by omitting this term so as to indicate the status of this taxon as a subspecies.

4) *"Vain names"* are available names consisting of unjustified intentional emendations of previously published names. The emendations are unjustified because they are not demonstrable as corrections of incorrect original spellings as defined by the *Code* (Art. 32c). Vain names have status in nomenclature under their own authorship and date. They constitute junior objective synonyms of names in their original form. Examples are: among species-group names, *geneae* (published as replacement of original unexplained masculine, *geni,* which now is not alterable), *ohioae* (invalid change from original *ohioensis*); among genus-group names, *Graphiodactylus* (invalid change from original *Graphiadactyllis*); among family-group names, Graphiodactylidae (based on junior objective synonym having invalid vain name).

5) An important group of available zoological names can be distinguished as *"transferred names."* These comprise authorized sorts of altered names in which the change depends on transfer from one taxonomic rank to another, or possibly on transfers in taxonomic assignment of subgenera, species, or subspecies. Most commonly the transfer calls for a change in termination of the name so as to comply with stipulations of the *Code* on endings of family-group taxa and agreement in gender of specific names with associated generic names. Transferred names may be derived from any of the preceding groups except the first. Examples are: among species-group names, *Spirifer ambiguus* (masc.) to *Composita ambigua* (fem.), *Neochonetes transversalis* to *N. granulifer transversalis* or vice versa; among genus-group names, *Schizoculina* to *Oculina (Schizoculina)* or vice versa; among family-group names, Orthidae to Orthinae or vice versa, or superfamily Orthacea derived from Orthidae or Orthinae; among suprafamilial taxa (not governed by the *Code*), order Orthida to suborder Orthina or vice versa. The authorship and date of transferred names are not affected by the transfers, but the author responsible for the transfer and the date of his action is generally recorded in the *Treatise.*

6) Improved or *"corrected names"* include both mandatory and allowable emendations of imperfect names and of suprafamilial names, which are not subject to regulation as to name form. Examples of corrected imperfect names are given with the discussion of group 3. Change from the originally published ordinal name Endoceroidea (Teichert, 1933) to the presently recognized Endocerida illustrates a "corrected" suprafamilial name. Group 6 names differ from those in group 5 in not being dependent on transfers in taxonomic rank or assignment, but some names are classifiable in both groups.

7) *"Substitute names"* are available names expressly proposed as replacements for invalid zoological names, such as junior homonyms. These may be classifiable also as belonging in groups 1, 2, or 3. The glossary appended to the *Code* refers to these as "new names" *(nomina nova)* but they are better designated as substitute names, since their newness is temporary and relative. The first-published substitute name

that complies with the definition here given takes precedence over any other. An example is *Marieita* LOEBLICH & TAPPAN, 1964, as substitute for *Reichelina* MARIE, 1955 (*non* ERK, 1942).

8) *"Conserved names"* include a relatively small number of species-group, genus-group, and family-group names which have come to be classed as available and valid by action of the International Commission on Zoological Nomenclature exercising its plenary powers to this end or ruling to conserve a junior synonym in place of a rejected "forgotten" name *(nomen oblitum)* (Art. 23b). Currently, such names are entered on appropriate "Official Lists," which are published from time to time.

It is useful for convenience and brevity of distinction in recording these groups of available zoological names to employ Latin designations in the pattern of *nomen nudum* (abbr., *nom. nud.*) and others. Thus we recognize the preceding numbered groups as follows: 1) *nomina inviolata* (sing., *nomen inviolatum*, abbr., *nom. inviol.*), 2) *nomina perfecta (nomen perfectum, nom. perf.)*, 3) *nomina imperfecta (nomen imperfectum, nom. imperf.)*, 4) *nomina vana (nomen vanum, nom. van.)*, 5) *nomina translata (nomen translatum, nom. transl.)*, 6) *nomina correcta (nomen correctum, nom. correct.)*, 7) *nomina substituta (nomen substitutum, nom. subst.)*, 8) *nomina conservata (nomen conservatum, nom. conserv.)*. It should be noted that the *Code* does not differentiate between different kinds of subsequent intentional changes of spelling, all of which are grouped as "emendations" (see below).

Additional to the groups differentiated above, the *Code* (Art. 17) specifies that a zoological name is not prevented from availability a) by becoming a junior synonym, for under various conditions this may be re-employed, b) for a species-group name by finding that original description of the taxon relates to more than a single taxonomic entity or to parts of animals belonging to two or more such entities, c) for species-group names by determining that it first was combined with an invalid or unavailable genus-group name, d) by being based only on part of an animal, sex of a species, ontogenetic stage, or one form of a polymorphic species, e) by being originally proposed for an organism not considered to be an animal but now so regarded, f) by incorrect original spelling which is correctable under the *Code*, g) by anonymous publication before 1951, h) by conditional proposal before 1961, i) by designation as a variety or form before 1961, j) by concluding that a name is inappropriate (Art. 18), or k) for a specific name by observing that it is tautonymous (Art. 18).

Unavailable Names

All zoological names which fail to comply with mandatory provisions of the *Code* are unavailable names and have no status in zoological nomenclature. None can be used under authorship and date of their original publication as a replacement name *(nom. subst.)* and none preoccupies for purposes of the Law of Homonymy. Names identical in spelling with some, but not all, unavailable names can be classed as available if and when they are published in conformance to stipulations of the *Code* and they are then assigned authorship and take date of the accepted publication. Different groups of unavailable names can be discriminated as follows.

9) *"Naked names"* include all those that fail to satisfy provisions stipulated in Article 11 of the *Code,* which states general requirements of availability. In addition they include names, if published before 1931, that were unaccompanied by a description, definition, or indication (Arts. 12, 16), as well as names published after 1930 that lacked accompanying statement of characters purporting to serve for differentiation of the taxon, or definite bibliographic reference to such a statement, or that were not proposed expressly as replacement (nom. subst.) of a pre-existing available name (Art. 13a) or that were unaccompanied by definite fixation of a type species by original designation or indication (Art. 13b). Examples of "naked names" are: among species-group taxa, *Valvulina mixta* PARKER & JONES, 1865 (=*Cribrobulimina mixta* CUSHMAN, 1927, available and valid); among genus-group taxa, *Orbitolinopsis* SILVESTRI, 1932 (=*Orbitolinopsis* HENSON, 1948, available but classed as invalid junior synonym of *Orbitolina* D'ORBIGNY, 1850); among family-group taxa, Aequilateralidae D'ORBIGNY,

1846 (lacking type-genus), Hélicostègues d'ORBIGNY, 1826 (vernacular not latinized by later authors, Art. 11e(iii)), Poteriocrinidae AUSTIN & AUSTIN, 1843 (=fam. Poteriocrinoidea AUSTIN & AUSTIN, 1842) (neither 1843 or 1842 names complying with Art. 11e, which states that "a family-group name must, when first published, be based on the name then valid for a contained genus," such valid name in the case of this family being *Poteriocrinites* MILLER, 1821).

10) *"Denied names"* include all those that are defined by the *Code* (Art. 32c) as incorrect original spellings. Examples are: Specific names, *nova-zelandica, mülleri, 10-brachiatus;* generic names, *M'Coyia, Størmerella, Römerina, Westgårdia;* family name, Růžičkinidae. Uncorrected "imperfect names" are "denied names" and unavailable, whereas corrected "imperfect names" are available.

11) *"Impermissible names"* include all those employed for alleged genus-group taxa other than genus and subgenus (Art. 42a) (e.g., supraspecific divisions of subgenera), and all those published after 1930 that are unaccompanied by definite fixation of a type species (Art. 13b). Examples of impermissible names are: *Martellispirifer* GATINAUD, 1949, and *Mirtellispirifer* GATINAUD, 1949, indicated respectively as a section and subsection of the subgenus *Cyrtospirifer; Fusarchaias* REICHEL, 1949, without definitely fixed type species (=*Fusarchaias* REICHEL, 1952, with *F. bermudezi* designated as type species).

12) *"Null names"* include all those that are defined by the *Code* (Art. 33b) as incorrect subsequent spellings, which are any changes of original spelling not demonstrably intentional. Such names are found in all ranks of taxa.

13) *"Forgotten names"* are defined (Art. 23b) as senior synonyms that have remained unused in primary zoological literature for more than 50 years. Such names are not to be used unless so directed by ICZN.

Latin designations for the discussed groups of unavailable zoological names are as follows: 9) *nomina nuda* (sing., *nomen nudum,* abbr., *nom. nud.*), 10) *nomina negata (nomen negatum, nom. neg.),* 11) *nomina vetita (nomen vetitum, nom. vet.),* 12) *nomina nulla (nomen nullum, nom. null.),* 13) *nomina oblita (nomen oblitum, nom. oblit.).*

VALID AND INVALID NAMES

Important distinctions relate to valid and available names, on one hand, and to invalid and unavailable names, on the other. Whereas determination of availability is based entirely on objective considerations guided by Articles of the *Code,* conclusions as to validity of zoological names partly may be subjective. A valid name is the correct one for a given taxon, which may have two or more available names but only a single correct name, generally the oldest. Obviously, no valid name can also be an unavailable name, but invalid names may include both available and unavailable names. Any name for a given taxon other than the valid name is an invalid name.

A sort of nomenclatorial no-man's-land is encountered in considering the status of some zoological names, such as *"doubtful names," "names under inquiry,"* and *"forgotten names."* Latin designations of these are *nomina dubia, nomina inquirenda,* and *nomina oblita,* respectively. Each of these groups may include both available and unavailable names, but the latter can well be ignored. Names considered to possess availability conduce to uncertainty and instability, which ordinarily can be removed only by appealed action of ICZN. Because few zoologists care to bother in seeking such remedy, the "wastebasket" names persist.

SUMMARY OF NAME GROUPS

Partly because only in such publications as the *Treatise* is special attention to groups of zoological names called for and partly because new designations are here introduced as means of recording distinctions explicitly as well as compactly, a summary may be useful. In the following tabulation valid groups of names are indicated in boldface type, whereas invalid ones are printed in italics.

DEFINITIONS OF NAME GROUPS

nomen conservatum (nom. conserv.). Name unacceptable under regulations of the *Code* which is made valid, either with original or altered spelling, through procedures specified by the *Code* or by action of ICZN exercising its plenary powers.

nomen correctum (nom. correct.). Name with intentionally altered spelling of sort required or allowable by the *Code* but not dependent on transfer from one taxonomic rank to another ("improved name"). (*See Code,* Arts. 26b, 27, 29, 30a(i)(3), 31, 32c(i), 33a; in addition change of endings for suprafamilial taxa not regulated by the *Code.*)

nomen imperfectum (nom. imperf.). Name that as originally published meets all mandatory requirements of the *Code* but contains defect needing correction ("imperfect name"). (*See Code,* Arts. 26b, 27, 29, 32c, 33a.)

nomen inviolatum (nom. inviol.). Name that as originally published meets all mandatory requirements of the *Code* and also is not correctable or alterable in any way ("inviolate name").

nomen negatum (nom. neg.). Name that as originally published constitutes invalid original spelling, and although possibly meeting all other mandatory requirements of the *Code,* cannot be used and has no separate status in nomenclature ("denied name"). It is to be corrected wherever found.

nomen nudum (nom. nud.). Name that as originally published fails to meet mandatory requirements of the *Code* and having no status in nomenclature, is not correctable to establish original authorship and date ("naked name").

nomen nullum (nom. null.). Name consisting of an unintentional alteration in form (spelling) of a previously published name (either available name, as *nom. inviol., nom. perf., nom imperf., nom. transl.*; or unavailable name, as *nom. neg., nom. nud., nom. van.,* or another *nom. null.*) ("null name").

nomen oblitum (nom. oblit.). Name of senior synonym unused in primary zoological literature in more than 50 years, not to be used unless so directed by ICZN ("forgotten name").

nomen perfectum (nom. perf.). Name that as originally published meets all mandatory requirements of the *Code* and needs no correction of any kind but which nevertheless is validly alterable by change of ending ("perfect name").

nomen substitutum (nom. subst.). Replacement name published as substitute for an invalid name, such as a junior homonym (equivalent to "new name").

nomen translatum (nom. transl.). Name that is derived by valid emendation of a previously published name as result of transfer from one taxonomic rank to another within the group to which it belongs ("transferred name").

nomen vanum (nom. van.)[1]. Name consisting of an invalid intentional change in form (spelling) from a previously published name, such invalid emendation having status in nomenclature as a junior objective synonym ("vain name").

nomen vetitum (nom. vet.). Name of genus-group taxon not authorized by the *Code* or, if first published after 1930, without definitely fixed type species ("impermissible name").

Except as specified otherwise, zoological names accepted in the *Treatise* may be understood to be classifiable either as *nomina inviolata* or *nomina perfecta* (omitting from notice *nomina correcta* among specific names) and these are not discriminated. Names which are not accepted for one reason or another include junior homonyms, senior synonyms classifiable as *nomina negata* or *nomina nuda,* and numerous junior synonyms which include both objective *(nomina vana)* and subjective types; rejected names are classified as completely as possible.

NAME CHANGES IN RELATION TO TAXA GROUPS

SPECIES-GROUP NAMES

Detailed consideration of valid emendation of specific and subspecific names is unnecessary here because it is well understood and relatively inconsequential. When the form of adjectival specific names is changed to obtain agreement with the gender of a generic name in transferring a species from one genus to another, it is never needful to label the changed name as a *nom. transl.* Likewise, transliteration of a letter accompanied by a diacritical mark in manner now called for by the *Code* (as in changing originally published *bröggeri* to *broeggeri*) or elimination of a hyphen (as in changing originally published *cornu-oryx* to *cornuoryx*) does not require *"nom. correct."* with it.

GENUS-GROUP NAMES

So rare are conditions warranting change of the originally published valid form of generic and subgeneric names that lengthy discussion may be omitted. Only elimination of diacritical marks of some names

[1] CHORN & WHETSTONE (*Jour. Paleontology,* in press, 1978) have called attention to the fact that the term *nomen vanum* was first proposed by SIMPSON (*Bull. Am. Museum Nat. History,* v. 85, Oct., 1945, p. 27, 30) for taxa properly proposed but "which cannot be properly defined either on the basis of the original diagnosis or from the available specimens" (SIMPSON, *ibid.,* v. 91, Art. 1, July, 1948, p. 31). It appears, however, that at the time of SIMPSON's writings the term *nomen dubium* was already in use for the kind of name SIMPSON had in mind (R. RICHTER, *Einführung in die zoologische Nomenklatur durch Erläuterung der Internationalen Regeln,* Senck. Naturf. Gesellsch., Frankfurt/Main, 1943, p. 102; H. M. SMITH, *Science,* v. 102, no. 2648, Aug., 1945, p. 186). As early as 1913, the International Commission on Zoological Nomenclature had referred to a genus based on unrecognizable specimens as *genus dubium* (ICZN, Op. 54). The term *nomen dubium* was later used by authors from SCHENK & MCMASTERS (*Procedure in Taxonomy,* 2nd edit., Stanford Univ. Press, Stanford, 1948, p. 9) to LEHMANN (*Geologisches Wörterbuch,* Ferd. Enke, Stuttgart, 1964, p. 196). We, therefore, regard *nomen vanum sensu* SIMPSON (1945) as a synonym of *nomen dubium* of authors. The term *nomen vanum* for intentional, unjustified emendations of names was first defined by MOORE (Editorial Preface, *Treatise,* Part E, June, 1955, p. xi) and this use was continued in all later *Treatise* volumes. It is here maintained.—CURT TEICHERT.

in this category seems to furnish basis for valid emendation. It is true that many changes of generic and subgeneric names have been published, but virtually all of these are either *nomina vana* or *nomina nulla.* Various names which formerly were classed as homonyms are not now, for two names that differ only by a single letter (or in original publication by presence or absence of a diacritical mark) are construed to be entirely distinct.

Examples in use of classificatory designations for genus-group names as previously given are the following, which also illustrate designation of type species as explained later.

Paleomeandron PERUZZI, 1881, p. 8 [**P. elegans*; SD HÄNTZSCHEL, 1975, p. *W*91] [=*Palaeomeandron* FUCHS, 1885, p. 395 *(nom. van.)*].

Stichophyma POMEL, 1872 [**Manon turbinatum* RÖMER, 1841; SD RAUFF, 1893] [=*Stychophyma* VOSMAER, 1885 *(nom. null.)*; *Sticophyma* MORET, 1924 *(nom. null.)*].

Vacuocyathus OKULITCH, 1950, p. 392 [**Coelocyathus kidrjassovensis* VOLOGDIN, 1937, p. 478 *(nom. nud.)*; 1939, p. 237; OD] [=*Coelocyathus* VOLOGDIN, 1934, p. 502 *(nom. nud.)*; 1937, p. 472 *(nom. nud.)*].

Cyrtograptus CARRUTHERS, 1867, p. 540 [*nom. correct.* LAPWORTH, 1873 *(pro Crytograpsus* CARRUTHERS, 1867), ICZN Op. 650, 1963] [**Cyrtograpsus murchisoni*; OD].

It is in many cases difficult to decide whether a change in spelling of a name by a subsequent author was intentional or unintentional, that is, whether it should be classified as *nomen vanum* or *nomen nullum,* and the decision will often have to be arbitrary.

FAMILY-GROUP NAMES; USE OF *"NOM. TRANSL."*

The *Code* specifies the endings only for subfamily (-inae) and family (-idae) but all family-group taxa are defined as coordinate, signifying that for purposes of priority a name published for a taxon in any category and based on a particular type genus shall date from its original publication for a taxon in any category, retaining this priority (and authorship) when the taxon is treated as belonging to a lower or higher category. By exclusion of -inae and -idae, respectively reserved for subfamily and family, the endings of names used for tribes and superfamilies must be unspecified different letter combinations. These, if introduced subsequent to designation of a subfamily or family based on the same nominate genus, are *nomina translata,* as is also a subfamily that is elevated to family rank or a family reduced to subfamily rank. In the *Treatise* it is desirable to distinguish the valid alteration comprised in the changed ending of each transferred family-group name by the abbreviation *"nom. transl."* and record of the author and date belonging to this alteration. This is particularly important in the case of superfamilies, for it is the author who introduced this taxon that one wishes to know about rather than the author of the superfamily as defined by the *Code,* for the latter is merely the individual who first defined some lower-rank family-group taxon that contains the nominate genus of the superfamily. The publication of the author containing introduction of the superfamily *nomen translatum* is likely to furnish the information on taxonomic considerations that support definition of the unit.

Examples of the use of *"nom. transl."* are the following.

Subfamily STYLININAE d'Orbigny, 1851

[*nom. transl.* VERRILL, 1864 (*ex* Stylinidae D'ORBIGNY, 1851)]

Superfamily ARCHAEOCTONOIDEA Petrunkevitch, 1949

[*nom. transl.* PETRUNKEVITCH, 1955 (*ex* Archaeoctonidae PETRUNKEVITCH, 1949)]

Superfamily ANCYLOCERATACEAE Meek, 1876

[*nom. transl.* WRIGHT, 1957 (*ex* Ancyloceratidae MEEK, 1876)]

FAMILY-GROUP NAMES; USE OF *"NOM. CORRECT."*

Valid name changes classed as *nomina correcta* do not depend on transfer from one category of family-group units to another but most commonly involve correction of the stem of the nominate genus; in addition, they include somewhat arbitrarily chosen modification of ending for names of tribe or superfamily. Examples of the use of *"nom. correct."* are the following.

Family STREPTELASMATIDAE Nicholson, 1889

[*nom. correct.* WEDEKIND, 1927 (*pro* Streptelasmidae NICHOLSON, 1889)]

Family PALAEOSCORPIIDAE Lehmann, 1944

[*nom. correct.* PETRUNKEVITCH, 1955 (*pro* Palaeoscorpionidae LEHMANN, 1944)]

Family AGLASPIDIDAE Miller, 1877

[*nom. correct.* STØRMER, 1959 (*pro* Aglaspidae MILLER, 1877)]

Superfamily AGARICIICAE Gray, 1847

[*nom. correct.* WELLS, 1956 (*pro* Agaricioidae VAUGHAN & WELLS, 1943, *nom. transl.* WELLS, 1956, *ex* Agaricidae GRAY, 1847)]

FAMILY-GROUP NAMES; USE OF "*NOM. CONSERV.*"

It may happen that long-used family-group names are invalid under strict application of the *Code*. In order to retain the otherwise invalid name, appeal to ICZN is needful. An example of use of *nom. conserv.* in this connection, as cited in *Treatise* style, is the following.

Subfamily OMPHALOTROPIDINAE Thiele, 1927

[*nom. conserv.*, ICZN (pending)] [=Realiinae PFEIFFER, 1858, *nom. correct.*, KOBELT, 1906 (*ex Realiea* PFEIFFER, 1858)]

FAMILY-GROUP NAMES; REPLACEMENTS

Family-group names are formed by adding letter combinations (prescribed for family and subfamily) to the stem of the name belonging to genus (nominate genus) first chosen as type of the assemblage. The type genus need not be the oldest in terms of receiving its name and definition, but it must be the first-published as name-giver to a family-group taxon among all those included. Once fixed, the family-group name remains tied to the nominate genus even if its name is changed by reason of status as a junior homonym or junior synonym, either objective or subjective. Seemingly, the *Code* (Art. 39) requires replacement of a family-group name only in the event that the nominate genus is found to be a junior homonym, and then a substitute family-group name is accepted if it is formed from the oldest available substitute name for the nominate genus. Authorship and date attributed to the replacement family-group name are determined by first publication of the changed family group-name, but for purposes of the Law of Priority, they take the date of the replaced name. Numerous long-used family-group names are incorrect in being *nomina nuda,* since they fail to satisfy criteria of availability (Art. 11e). These also demand replacement by valid names.

The aim of family-group nomenclature is greatest possible stability and uniformity, just as in case of other zoological names. Experience indicates the wisdom of sustaining family-group names based on junior subjective synonyms if they have priority of publication, for opinions of different workers as to the synonymy of generic names founded on different type species may not agree and opinions of the same worker may alter from time to time. The retention similarly of first-published family-group names which are found to be based on junior objective synonyms is less clearly desirable, especially if a replacement name derived from the senior objective synonym has been recognized very long and widely. To displace a much-used family-group name based on the senior objective synonym by disinterring a forgotten and virtually unused family-group name based on a junior objective synonym because the latter happens to have priority of publication is unsettling.

Replacement of a family-group name may be needed if the former nominate genus is transferred to another family-group. Then the first-published name-giver of a family-group assemblage in the remnant taxon is to be recognized in forming a replacement name.

FAMILY-GROUP NAMES; AUTHORSHIP AND DATE

All family-group taxa having names based on the same type genus are attributed to the author who first published the name for any of these assemblages, whether tribe, subfamily, or family (superfamily being almost inevitably a later-conceived taxon). Accordingly, if a family is divided into subfamilies or a subfamily into tribes, the name of no such subfamily or tribe can antedate the family name. Also, every family containing differentiated subfamilies must have a nominate *(sensu stricto)* subfamily, which is based on the same type genus as that for the family, and the author and date set down for the nominate subfamily invariably are identical with those of the family, without reference to whether the author of the family or some subsequent author introduced subdivisions.

Changes in the form of family-group names of the sort constituting *nomina correcta,* as previously discussed, do not affect authorship and date of the taxon concerned, but in the *Treatise* it is desirable to record the authorship and date of the correction.

SUPRAFAMILIAL TAXA

International rules of zoological nomenclature as given in the *Code* (1961) are limited to stipulations affecting lower-rank categories (infrasubspecies to superfamily). Suprafamilial categories (suborder to phylum) are either unmentioned or explicitly placed outside of the application of zoological rules. The *Copenhagen Decisions on Zoological Nomenclature*[1] (1953, Arts. 59-69) proposed to adopt rules for naming suborders and higher taxonomic divisions up to and including phylum, with provision for designating a type genus for each, hopefully in such manner as not to interfere with the taxonomic freedom of workers. Procedures for applying the Law of Priority and Law of Homonymy to suprafamilial taxa were outlined and for dealing with the names for such units and their authorship, with assigned dates, when they should be transferred on taxonomic grounds from one rank to another. The adoption of terminations of names, different for each category but uniform within each, was recommended.

The Colloquium on zoological nomenclature which met in London during the week just before the XVth International Congress of Zoology convened in 1958 thoroughly discussed the proposals for regulating suprafamilial nomenclature, as well as many others advocated for inclusion in the new *Code* or recommended for exclusion from it. A decision which was supported by a wide majority of the participants in the Colloquium was against the establishment of rules for naming taxa above family-group rank, mainly because it was judged that such regulation would unwisely tie the hands of taxonomists. For example, if a class or order was defined by some author at a given date, using chosen morphologic characters (e.g., gills of bivalves), this should not be allowed to freeze nomenclature, taking precedence over another later-proposed class or order distinguished by different characters (e.g., hinge-teeth of bivalves). Even the fixing of type genera for suprafamilial taxa might have small value, if any, hindering taxonomic work rather than aiding it. At all events, no legal basis for establishing such types and for naming these taxa has yet been provided.

[1] Francis Hemming, ed., *Copenhagen Decisions on Zoological Nomenclature. Additions to, and modifications of, the Règles Internationales de la Nomenclature Zoologique*, xxix + 135 p. (International Trust for Zoological Nomenclature, London, 1953).

The considerations just stated do not prevent the editors of the *Treatise* from making "rules" for dealing with suprafamilial groups of animals described and illustrated in this publication. At least a degree of uniform policy is thought to be needed, especially for the guidance of *Treatise*-contributing authors. This policy should accord with recognized general practice among zoologists, but where general practice is indeterminate or nonexistent our own procedure in suprafamilial nomenclature needs to be specified as clearly as possible. This pertains especially to decisions about names themselves, about citation of authors and dates, and about treatment of suprafamilial taxa which on taxonomic grounds are changed from their originally assigned rank. Accordingly, a few "rules" expressing *Treatise* policy are given here, some with examples of their application.

1) The name of any suprafamilial taxon must be a Latin or latinized uninominal noun of plural form, or treated as such, a) with a capital initial letter, b) without diacritical mark, apostrophe, diaeresis, or hyphen, and c) if a component consisting of a numeral, numerical adjective, or adverb is used, this must be written in full (e.g., Stethostomata, Trionychi, Septemchitonina, Scorpiones, Subselliflorae). No uniformity in choice of ending for taxa of a given rank is demanded (e.g., orders named Gorgon*acea,* Millepor*ina,* Rugos*a,* Scleractin*ia,* Stromatopor*oidea,* Phalang*ida*).

2) Names of suprafamilial taxa may be constructed in almost any way, a) intended to indicate morphological attributes (e.g., Lamellibranchiata, Cyclostomata, Toxoglossa), b) based on the stem of an included genus (e.g., Bellerophontina, Nautilida, Fungiina), or c) arbitrary combinations of letters, (e.g., Yuania), but none of these can be allowed to end in -idae or -inae, reserved for family-group taxa. No suprafamilial name identical in form to that of a genus or to another published suprafamilial name should be employed (e.g., order Decapoda Latreille, 1803, crustaceans, and order Decapoda Leach, 1818, cephalopods; suborder Chonetoidea Muir-Wood, 1955, and genus *Chonetoidea* Jones, 1928). Worthy of notice is the classificatory and nomenclatural distinction between suprafamilial and family-group taxa which respectively are

named from the same type genus, since one is not considered to be transferable to the other (e.g., suborder Bellerophontina Ulrich & Scofield, 1897; superfamily Bellerophontacea M'Coy, 1851; family Bellerophontidae M'Coy, 1851). Family-group names and suprafamilial names are not coordinate.

3) The Laws of Priority and Homonymy lack any force of international agreement as applied to suprafamilial names, yet in the interest of nomenclatural stability and the avoidance of confusion these laws are widely applied by zoologists to taxa above the family-group level wherever they do not infringe on taxonomic freedom and long-established usage.

4) Authors who accept priority as a determinant in nomenclature of a suprafamilial taxon may change its assigned rank at will, with or without modifying the terminal letters of the name, but such change(s) cannot rationally be judged to alter the authorship and date of the taxon as published originally. a) A name revised from its previously published rank is a "transferred name" *(nom. transl.)*, as illustrated in the following.

Order CORYNEXOCHIDA Kobayashi, 1935

[*nom. transl.* MOORE, 1959 (*ex* suborder Corynexochida KOBAYASHI, 1935)]

b) A name revised from its previously published form merely by adoption of a different termination, without changing taxonomic rank, is an "altered name" *(nom. correct.)*. Examples follow.

Order DISPARIDA Moore & Laudon, 1943

[*nom. correct.* MOORE, in MOORE, LALICKER, & FISCHER, 1952 (*pro* order Disparata MOORE & LAUDON, 1943)]

Suborder AGNOSTINA Salter, 1864

[*nom. correct.* HARRINGTON & LEANZA, 1957 (*pro* suborder Agnostini SALTER, 1864)]

c) A suprafamilial name revised from its previously published rank with accompanying change of termination (which may or may not be intended to signalize the change of rank) is recorded as *nom. transl. et correct.*

Order ORTHIDA Schuchert & Cooper, 1932

[*nom. transl. et correct.* MOORE in MOORE, LALICKER, & FISCHER, 1952, p. 220 (*ex* suborder Orthoidea SCHUCHERT & COOPER, 1932, p. 43)] [*emend.* WILLIAMS & WRIGHT, 1965]

5) The authorship and date of nominate subordinate and superordinate taxa among suprafamilial taxa are considered in the *Treatise* to be identical since each actually or potentially has the same type. Examples are given below.

Subclass ENDOCERATOIDEA Teichert, 1933

[*nom. transl.* TEICHERT in TEICHERT, *et al.*, 1964, p. *K*128 (*ex* superorder Endoceratoidea SHIMANSKIY & ZHURAVLEVA, 1961, *nom. transl.* TEICHERT in TEICHERT *et al.*, 1964, p. *K*128, *ex* order Endoceroidea TEICHERT, 1933)]

Order ENDOCERIDA Teichert, 1933

[*nom. correct.* TEICHERT in TEICHERT *et al.*, 1964, p. *K*165 (*pro* order Endoceroidea TEICHERT, 1933)]

Suborder ENDOCERINA Teichert, 1933

[*nom. correct.*, herein, *ex* Endoceratina SWEET, 1958 (suborder)]

TAXONOMIC EMENDATION

Emendation has two distinct meanings as regards zoological nomenclature. These are 1) alteration of a name itself in various ways for various reasons, as has been reviewed, and 2) alteration of taxonomic scope or concept in application of a given zoological name. The *Code* (Art. 33a and Glossary p. 148) concerns itself with only the first type of emendation, applying the term to either justified or unjustified changes, both intentional, of the original spelling of a name. These categories are identified in the *Treatise* as *nomina correcta* and *nomina vana,* respectively. The second type of emendation primarily concerns classification and inherently is not associated with change of name. Little attention generally has been paid to this distinction in spite of its significance.

Most zoologists, including paleozoologists, who have signified emendation of zoological names refer to what they consider a material change in application of the name such as may be expressed by an importantly altered diagnosis of the assemblage covered by the name. The abbreviation *"emend."* then may accompany the name, with statement of the author and date of the emendation. On the other hand, many workers concerned with systematic zoology think that publication of *"emend."* with a zoological name is valueless, because more or less alteration of taxonomic sort is introduced whenever a subspecies, species, genus, or other assemblage of animals is incorporated under or removed from the coverage of a given zoological name. Inevitably associated with such classificatory expansions and restrictions is some degree of emendation affecting diagnosis. Granting this, still it is true that now and then somewhat radical revisions are put forward, generally with published statement of reasons for changing the application of a name. To erect a signpost at such points of most significant change is worthwhile, both as aid to subsequent workers in taking account of the altered nomenclatural usage and as indication that not to-be-overlooked discussion may be found at a particular place in the literature. Authors of contributions to the *Treatise* are encouraged to include records of all specially noteworthy emendations of this nature, using the abbreviation *"emend."* with the name to which it refers and citing the author and date of the emendation.

Examples from *Treatise* volumes are:

Order ORTHIDA Schuchert & Cooper, 1932

[*nom. transl. et correct.* MOORE in MOORE, LALICKER, & FISCHER, 1952, p. 220 (*ex* suborder Orthoidea SCHUCHERT & COOPER, 1932, p. 43)] [*emend.* WILLIAMS & WRIGHT, 1965]

Subfamily ROVEACRININAE Peck, 1943

[Roveacrininae PECK, 1943, p. 465; *emend.* PECK in MOORE & TEICHERT, eds., 1978, p. *T*921]

STYLE IN GENERIC DESCRIPTIONS

CITATION OF TYPE SPECIES

The name of the type species of each genus and subgenus is given next following the generic name with its accompanying author, date, and page reference or after entries needed for definition of the name if it is involved in homonymy. The originally published combination of generic and trivial names for this species is cited, accompanied by an asterisk (*), with notation of the author and date of original publication. An exception in this procedure is made, however, if the species was first published in the same paper and by the same author as that containing definition of the genus which it serves as type; in such case, the initial letter of the generic name followed by the trivial name is given without repeating the name of the author and date. Examples of these two sorts of citations are as follows:

Diplotrypa NICHOLSON, 1879 [**Favosites petropolitanus* PANDER, 1830].

Chainodictyon FOERSTE, 1887 [**C. laxum*].

If the cited type species is a junior synonym of some other species, the name of this latter also is given, as follows:

Acervularia SCHWEIGGER, 1819 [**A. baltica* (=**Madrepora ananas* LINNÉ, 1758); M].

It is desirable to record the manner of establishing the type species, whether by original designation or by subsequent designation.

Fixation of type species originally. The type species of a genus or subgenus, according to provisions of the *Code,* may be fixed in various ways in the original publication or it may be fixed in specified ways subsequent to the original publication. Fixation of the type species of a genus or subgenus in an original publication is stipulated by the *Code* (Art. 68) in order of precedence as 1) *original designation* (in the *Treatise* indicated as OD) when the type species is explicitly stated or (before 1931) indicated by "n. gen., n. sp." (or its equivalent) applied to a single species included in a new genus, 2) defined by use of *typus* or *typicus* for one of the species included in a new genus (adequately indicated in the *Treatise* by the specific name), 3) established by *monotypy* if a new genus or subgenus includes only one originally included species (in the *Treatise* indicated as M), and 4) fixed by *tautonymy* if the genus-group name is identical to an included species name not indicated as type belonging to one of the three preceding categories.

Fixation of type species subsequently. The type species of many genera are not determinable from the publication in which the generic name was introduced and therefore such genera can acquire a type species only by some manner of subsequent designation. Most commonly this is established by publishing a statement naming as type species one of the species originally included in the genus, and in the *Treatise* fixation of the type species in this manner is indicated by the letters "SD" accompanied by the name of the subsequent author (who may be the same person as the original author) and the date of publishing the subsequent designation. Some genera, as first described and named, included no mentioned species and these necessarily lack a type species until a date subsequent to that of the original publication when one or more species are assigned to such a genus. If only a single species is thus assigned, it automatically becomes the type species and in the *Treatise* this subsequent monotypy is indicated by the letters "SM." Of course, the first publication containing assignment of species to the genus which originally lacked any included species is the one concerned in fixation of the type species, and if this named two or more species as belonging to the genus but did not designate a type species, then a later "SD" designation is necessary. Examples of the use of "SD" and "SM" as employed in the *Treatise* follow.

Hexagonaria GÜRICH, 1896 [**Cyathophyllum hexagonum* GOLDFUSS, 1826; SD LANG, SMITH & THOMAS, 1940].

Muriceides STUDER, 1887 [**M. fragilis* WRIGHT & STUDER, 1889; SM WRIGHT & STUDER, 1889].

Another mode of fixing the type species of a genus is action of the International Commission on Zoological Nomenclature using its plenary powers. Definition in this way may set aside application of the *Code* so as to arrive at a decision considered to be in the best interest of continuity and stability of zoological nomenclature. When made, it is binding and commonly is cited in the *Treatise* by the letters "ICZN," accompanied by the date of announced decision and reference to the appropriate numbered Opinion.

It should be noted that *subsequent designation* of a type species is admissible only for genera established prior to 1931. A new genus-group name established after 1930, and not accompanied by fixation of a type species through original designation or original indication, is invalid (*Code,* Art. 13b). Effort of a subsequent author to "validate" such a name by subsequent designation of a type species constitutes an original publication making the name available under authorship and date of the subsequent author. This provision of the *Code* has not been consistently applied in all earlier *Treatise* volumes, but is rigidly adhered to in the present volume.

Type species of synonyms. In about 1969 a decision was made by the editors to include the names of type species of genera that were placed in subjective synonymy. Such species are simply identified as "type." An example is:

Trachycardium MÖRCH, 1853 [**Cardium isocardia* LINNÉ, 1758; SD VON MARTENS, 1870] [=*Kathocardia* TUCKER & WILSON, 1932 (type, *Cardium (K.) aclinense*; OD)].

HOMONYMS

Most generic names are distinct from all others and are indicated without ambiguity by citing their originally published spelling accompanied by name of the author and date of first publication. If the same generic name has been applied to two or more distinct taxonomic units, however, it is necessary to differentiate such homonyms, and this calls for distinction between junior homonyms and senior homonyms. Because a junior homonym is invalid, it must be replaced by some other name. For example, *Callopora* HALL, 1851, introduced for Paleozoic trepostome bryozoans, is invalid because GRAY in 1848 published the same name for Cretaceous-to-Holocene cheilostome bryozoans, and BASSLER in 1911 introduced the new name *Hallopora* to replace HALL's homonym. The *Treatise* style of entry is:

Hallopora BASSLER, 1911 [*nom. subst. pro Callopora* HALL, 1851 (*non* GRAY, 1848)].

In like manner, a needed replacement generic name may be introduced in the *Treatise* (even though first publication of generic names otherwise in this work is generally avoided). The requirement that an exact bibliographic reference must be given for the replaced name commonly can be met in the *Treatise* by citing a publication recorded in the list of references, as shown in the following example.

Mysterium DE LAUBENFELS, herein [*nom. subst. pro Mystrium* SCHRAMMEN, 1936 (ref., p. 60) (*non* ROGER, 1862)] [**Mystrium porosum* SCHRAMMEN, 1936].

Otherwise, no mention of the existence of a junior homonym generally is made.

Synonymous homonyms. An author sometimes publishes a generic name in two or more papers of different date, each of which indicates that the name is new. This is a bothersome source of errors for later workers who are unaware that a supposed first publication which they have in hand is not actually the original one. Although the names were separately published, they are identical and therefore definable as homonyms; at the same time they are absolute synonyms. For the guidance of all concerned, it seems desirable to record such names as synonymic homonyms and in the *Treatise* the junior one of these is indicated by the abbreviation "jr. syn. hom."

Identical family-group names not infrequently are published as new names by different authors, the author of the later-introduced name being ignorant of previous publication(s) by one or more other workers. In spite of differences in taxonomic concepts as indicated by diagnoses and grouping of genera and possibly in assigned rank, these family-group taxa are nomenclatural homonyms, based on the same type genus, and they are also synonyms. Wherever encountered, such synonymic homonyms are distinguished in the *Treatise* as in dealing with generic names.

SYNONYMS

Citation of synonyms is given next following record of the type species and if two or more synonyms of differing date are recognized, these are arranged in chronological order. Objective synonyms are indicated by accompanying designation "(obj.)," others being understood to constitute subjective synonyms, of which the types are also indicated. Examples showing *Treatise* style in listing synonyms follow.

Calapoecia BILLINGS, 1865 [**C. anticostiensis*; SD LINDSTRÖM, 1883] [=*Columnopora* NICHOLSON, 1874; *Houghtonia* ROMINGER, 1876].

Staurocyclia HAECKEL, 1882 [**S. cruciata* HAECKEL, 1887] [=*Coccostaurus* HAECKEL, 1882 (obj.); *Phacostaurus* HAECKEL, 1887 (obj.)].

Graphiocrinus DE KONINCK & LE HON, 1854, p. 115 [**G. encrinoides*; M] [=*Scaphiocrinus* HALL, 1858b, p. 550 (type, *S. simplex*; OD)].

A synonym which also constitutes a homonym is recorded as follows:

Lyopora NICHOLSON & ETHERIDGE, 1878 [**Palaeopora? favosa* M'COY, 1850] [=*Liopora* LANG, SMITH & THOMAS, 1940 (*non* GIRTY, 1915)].

Some junior synonyms of either objective or subjective sort may take precedence desirably over senior synonyms wherever uniformity and continuity of nomenclature are served by retaining a widely used but technically rejectable name for a generic assemblage. This requires action of ICZN using its plenary powers to set aside the unwanted name and validate the wanted one, with placement of the concerned names on appropriate official lists.

STRATIGRAPHIC DIVISIONS

Classification of rocks forming the geologic column as commonly cited in the *Treatise* in terms of units defined by concepts of time is reasonably uniform and firm throughout most of the world as regards major divisions (e.g., series, systems, and rocks representing eras) but it is variable and unfirm as regards smaller divisions (e.g., substages, stages, and subseries), which are provincial in application. Users of the *Treatise* have suggested the desirability of publishing reference lists showing the stratigraphic arrangement of at least the most commonly cited divisions. Accordingly, a tabulation of European and North American units, which broadly is applicable also to some other continents, is given here. No stage subdivisions of the Tertiary series are given here because they are not used in these volumes.

Generally Recognized Divisions of Geologic Column

EUROPE
CAINOZOIC ERATHEM
QUATERNARY SYSTEM
Holocene Series
Pleistocene Series
TERTIARY SYSTEM[1]
Pliocene Series
Miocene Series
Oligocene Series
Eocene Series
Paleocene Series

MESOZOIC ERATHEM
CRETACEOUS SYSTEM
Upper Cretaceous Series
Maastrichtian Stage[3]
Campanian Stage[3]
Santonian Stage[3]

NORTH AMERICA
CENOZOIC ERATHEM
QUATERNARY SYSTEM
Holocene Series
Pleistocene Series
TERTIARY SYSTEM[1]
Pliocene Series
Miocene Series
Oligocene Series
Eocene Series
Paleocene Series

MESOZOIC ERATHEM
CRETACEOUS SYSTEM[2]
Gulfian Series (Upper Cretaceous)
Navarroan Stage
Tayloran Stage
Austinian Stage

Coniacian Stage[3]
Turonian Stage
Cenomanian Stage

Lower Cretaceous Series

Albian Stage (Gault)
Aptian Stage
Barremian Stage[4]
Hauterivian Stage[4]
Valanginian Stage[4]
Berriasian Stage[4]

JURASSIC SYSTEM

Upper Jurassic Series

Tithonian Stage
Kimmeridgian Stage
Oxfordian Stage

Middle Jurassic Series

Callovian Stage[5]
Bathonian Stage
Bajocian Stage

Lower Jurassic Series (Liassic)

Toarcian Stage
Pliensbachian Stage
Sinemurian Stage
Hettangian Stage

TRIASSIC SYSTEM

Upper Triassic Series

Rhaetian Stage
Norian Stage
Carnian Stage

Middle Triassic Series

Ladinian Stage
Anisian Stage

Lower Triassic Series

Scythian Stage

PALEOZOIC ERATHEM

PERMIAN SYSTEM

Upper Permian Series

Tatarian Stage[6]
Kazanian Stage[7]
Kungurian Stage

Lower Permian Series

Artinskian Stage[8]
Sakmarian Stage
Asselian Stage

CARBONIFEROUS SYSTEM

Upper Carboniferous Series

Stephanian Stage

Westphalian Stage

Namurian Stage

Lower Carboniferous Series

Visean Stage

Eaglefordian Stage
Woodbinian (Tuscaloosan) Stage

Comanchean Series (Lower Cretaceous)

Washitan Stage
Fredericksburgian Stage
Trinitian Stage

Coahuilan Series (Lower Cretaceous)

Nuevoleonian Stage
Durangoan Stage

JURASSIC SYSTEM

Upper Jurassic Series

Portlandian Stage
Kimmeridgian Stage
Oxfordian Stage

Middle Jurassic Series

Callovian Stage[5]
Bathonian Stage
Bajocian Stage

Lower Jurassic Series (Liassic)

Toarcian Stage
Pliensbachian Stage
Sinemurian Stage
Hettangian Stage

TRIASSIC SYSTEM

Upper Triassic Series

Rhaetian Stage
Norian Stage
Carnian Stage

Middle Triassic Series

Ladinian Stage
Anisian Stage

Lower Triassic Series

Scythian Stage

PALEOZOIC ERATHEM

PERMIAN SYSTEM

Upper Permian Series

Ochoan Stage
Guadalupian Stage

Lower Permian Series

Leonardian Stage
Wolfcampian Stage

PENNSYLVANIAN SYSTEM

Upper Pennsylvanian Series

Virgilian Stage
Missourian Stage

Middle Pennsylvanian Series

Desmoinesian Stage
Atokan Stage

Lower Pennsylvanian Series

Morrowan Stage

MISSISSIPPIAN SYSTEM

Upper Mississippian Series

Chesterian Stage
Meramecian Stage

Tournaisian Stage	**Lower Mississippian Series**
	Osagian Stage
	Kinderhookian Stage
DEVONIAN SYSTEM	**DEVONIAN SYSTEM**
Upper Devonian Series	**Chautauquan Series (Upper Devonian)**
Famennian Stage	Bradfordian Stage[10]
	Cassadagan Stage[10]
Frasnian Stage	**Senecan Series (Upper Devonian)**
	Chemungian Stage[10]
	Fingerlakesian Stage[10]
Middle Devonian Series	**Erian Series (Middle Devonian)**
Givetian Stage	Taghanican Stage[10]
	Tioughniogan Stage[10]
Couvinian Stage[9]	Cazenovian Stage[10]
Lower Devonian Series	**Ulsterian Series (Lower Devonian)**
Emsian Stage	Onesquethawan Stage[10]
Siegenian Stage	Deerparkian Stage[10]
Gedinnian Stage	Helderbergian Stage[10]
SILURIAN SYSTEM	**SILURIAN SYSTEM**
Pridolian Series	**Cayugan Series[12] (Upper Silurian)**
Ludlovian Series	**Niagaran Series[12] (Middle Silurian)**
Wenlockian Series	**Alexandrian Series[12] (Lower Silurian)**
Llandoverian Series	
ORDOVICIAN SYSTEM	**ORDOVICIAN SYSTEM**
	Cincinnatian Series (Upper Ordovician)
Ashgillian Series	Richmondian Stage
	Maysvillian Stage
	Edenian Stage
	Champlainian Series (Middle Ordovician)
	Mohawkian Stage
	Trentonian Substage
Caradocian Series	Blackriveran Substage
Llandeilian Series	Chazyan Stage
Llanvirnian Series	Whiterockian Stage
Arenigian Series	**Canadian Series (Lower Ordovician)**
Tremadocian Series[11]	
CAMBRIAN SYSTEM	**CAMBRIAN SYSTEM**
Upper Cambrian Series (Merioneth)	**Croixian Series (Upper Cambrian)**
	Trempealeauan Stage
	Franconian Stage
	Dresbachian Stage
Middle Cambrian Series (St. David)	**Albertan Series (Middle Cambrian)**
Lower Cambrian Series (Comley)	**Waucoban Series (Lower Cambrian)**
ROCKS OF PRECAMBRIAN ERAS	**ROCKS OF PRECAMBRIAN ERAS**
PROTEROZOIC ERATHEM	**PROTEROZOIC ERATHEM**
Dalradian, Eocambrian, Vendian, Riphean, and equivalents	Algonkian, Beltian, Hadrynian, Helikian, Aphebian, and equivalents

R. C. Moore and Curt Teichert

1 For convenience Miocene and Pliocene are often grouped as Neogene, Paleocene, Eocene, and Oligocene as Paleogene subsystems.

2 Follows essentially Gulf Coast usage.

3 Classed as division of Senonian Subseries.

4 Classed as division of Neocomian Subseries.

5 Included in Upper Jurassic by some authors.

6 Equivalent to upper Thuringian (Zechstein) deposits.

7 Equivalent to lower Thuringian (Zechstein) deposits.

8 Equivalent to upper Autunian and part of Rotliegend deposits.

9 Also known as Eifelian.

10 Applies essentially to eastern United States; in western North America European stage terminology is used.

11 Tremadocian placed in Cambrian by some authors.

12 Applies essentially to eastern North America only. Berry and Boucot have advocated use of the English standard scale everywhere in North America (Geol. Soc. America, Spec. Paper 102, 1970).

ABBREVIATIONS

Abbreviations used in this division of the *Treatise* are explained in the following alphabetically arranged list.

Abhandl., *Abhandlung(en)*
abstr., abstract
Abt., *Abteilung*
aff., *affinis* (related to)
Afr., Africa,-an
Ala., Alabama
Alb., Albian
Alg., Algeria
Alta., Alberta
A.M., *Artium Magister* (Master of Arts)
Am., America,-n
Anis., Anisian
ant., anterior
append., appendix
Apt., Aptian
Arenig., Arenigian
Arg., Argentina
Ariz., Arizona
Ark., Arkansas
art., article
Artinsk., Artinskian
Atl., Atlantic
auctt., *auctorum* (of authors)
Aug., August
Aus., Austria
Austral., Australian
Avd., *Avdelingen*

Bajoc., Bajocian
Barrem., Barremian
Bathon., Bathonian
B.C., British Columbia
Bd., *Band*
Beil., *Beilage*
Belg., *Belgique*, Belgium
Bend., Bendian
Blackriver., Blackriveran
Boh., Bohemia
Bol., *Boletim, Boletín*, Bolivia
Brit., Britain, British
Bulg., Bulgaria
Bull., Bulletin

C., Centigrade, Central
ca., *circa*
Calif., California
Callov., Callovian
Cam., Cambrian
Campan., Campanian
Can., Canada
Caradoc., Caradocian
Carb., Carboniferous
Carib., Caribbean
Carn., Carnian
cat., catalogue
Cauc., Caucasus
cc., cubic centimeter(s)
Cenoman., Cenomanian
cf., *confer* (compare)
Chazy., Chazyan
Chemung., Chemungian
Chester., Chesterian
Cincinnat., **Cincinnati.**, Cincinnatian
cm., centimeter(s)
Co., Company, County
Coll., Collection(s)
Colo., Colorado
Colom., Colombia
commun., communication
Coniac., Coniacian
correct., *correctum*
cosmop., cosmopolitan
Couvin., Couvinian
Cret., Cretaceous
Czech., Czechoslovakia

Dan., Danian
Dec., decade, December
Denm., Denmark
Desmoines., Desmoinesian
Dev., Devonian
diag., diagram
diagram., diagrammatic, diagrammatical
Doc., Document
Dol., Dolomite

E., East
ed., edited, editor
edit., edition
eds., editors
e.g., *exempli gratia* (for example)
emend., *emendatus(-a)*, emended
Ems., Emsian
Eng., England, English
enl., enlarged
Eoc., Eocene
equiv., equivalent
Erforsch., *Erforschung*
Est., Estonia
et al., *et alii* (and others, persons)
etc., *et cetera* (and others, objects)
Eu., Europe
Ex., Executive
ext., exterior

F., Formation
fam., family
Feb., February
fig., figure(s)
Fla., Florida
Frasn., Frasnian
Ft., Fort

Ga., Georgia
G.Brit., Great Britain
gen., genus
Geol., Geological, *Geologicheskikh, Geologische, Geologiya*, Geology
Géol., *Géologique*
Ger., German, Germany
Givet., Givetian
Gotl., Gotland
Gr., Great, Group
Greenl., Greenland
Guadalup., Guadalupian

Hamilton., Hamiltonian
Handl., *Handling(ar)*
Hauteriv., Hauterivian
Heers., Heersian
Helderberg., Helderbergian
Hettang., Hettangian
Hist., History
Holo., Holocene
Hung., *Hungarica*, Hungary

I., Island, Isles
Ia., Iowa
ICZN, International Commission on Zoological Nomenclature
i.e., *id est* (that is)
Ill., Illinois
illus., illustrated,-ions
incl., inclined, including
incompl., incomplete
Ind., Indiana
indet., *indeterminata*, indeterminate
Ind. O., Indian Ocean
Indon., Indonesia
Inst., *Institut*, Institute, Institution
int., interior
Internatl., International
Ire., Ireland
Is., Island(s)

Jahr., *Jahrgang*
jr., junior
Jur., Jurassic

K., **Kongl.**, *Königlich, Kongelige, Kongliga, Koninklijk*
K.K., *Kaiserlich Königlich*
Kans., Kansas

Kimmeridg., Kimmeridgian
Kinderhook., Kinderhookian
Ky., Kentucky

L., Low., Lower
La., Louisiana
Ladin., Ladinian
Lancash., Lancashire
lat., lateral
Lias., Liassic
Llandeil., Llandeilian
Llandov., Llandoverian
Llanvirn., Llanvirnian
long., longitudinal, longitude
Ls., Limestone
Ludlov., Ludlovian

m., meter(s)
M, monotypy
M., Middle
Maastricht., Maastrichtian
Madag., Madagascar
Mbr., Member
Md., Maryland
Medit., Mediterranean
Meramec., Meramecian
Mesoz., Mesozoic
Mex., Mexico
Mich., Michigan
Minn., Minnesota
Mio., Miocene
Miss., Mississippi, Mississippian
Missour., Missouri., Missourian
mm., millimeter(s)
Mo., Missouri
mod., modified
Mohawk., Mohawkian
Mont., Montana
Morrow., Morrowan
Moscov., Moscovian
MS., manuscript
Mtg., Meeting
Mts., Mtns., Mountains
Mus., Museum

n., n, new
N., North
N. Am., North America(n)
Namur., Namurian
Nat., Natural
Natl., National
NC., north central
N.Car., North Carolina
NE., Northeast
Neb., Nebraska
Neocom., Neocomian
Neth., Netherlands
Nev., Nevada
Newf., Newfoundland
Niag., Niagaran
N.J., New Jersey
no., number
Nom., Nomenclator
nom. conserv., *nomen conservatum* (conserved name)
nom. correct., *nomen correctum* (corrected or intentionally altered name)
nom. imperf., *nomen imperfectum* (imperfect name)
nom. nud., *nomen nudum* (naked name)
nom. null., *nomen nullum* (null, void name)
nom. subst., *nomen substitum* (substitute name)
nom. transl., *nomen translatum* (transferred name)
nom. van., *nomen vanum* (vain, void name)
nom. vet., *nomen vetitum* (impermissible name)
Nomencl., Nomenclature
nov., *novum* (new)
NW., Northwest
N.Y., New York

O., Ocean
obj., objective
Oct., October
OD, original designation
Okla., Oklahoma
Oligo., Oligocene
Ont., Ontario
Op., Opinion
opp., opposite
Ord., Ordovician
Ore., Oregon
orig., original
Osag., Osagian
Oxford., Oxfordian

p., page(s)
Pa., Pennsylvania
Pac., Pacific
Palaeoz., Palaeozoic
Paläont., *Paläontologie*
Paleoc., Paleocene
Paleont., Paleontological, *Paleontologicheskiy*
Paleoz., Paleozoic
part., partial
Penn., Pennsylvanian
Perm., Permian
pers., personal
Ph.D., *Philosophiae Doctor* (Doctor of Philosophy)
Philip., Philippines
pl., plate(s), plural
platf., platform
Pleist., Pleistocene
Pliensbach., Pliensbachian
Plio., Pliocene
Pol., Poland
Port., Portugal
post., posterior
Prag., Pragian
prov., province(s)
prox., proximal
pt., part(s)
publ., *publicacion,* publication(s), published

Quat., Quaternary
Que., Quebec
Queensl., Queensland

Rec., recent, record(s)
reconstr., reconstructed, -ion
ref., reference(s)
reg., region
Rept., Report
Rev., Review, *Revista, Revue*
Richmond., Richmondian
Riksmus., Riksmuseum

S., Sea, South
S.Am., South America
Santon., Santonian
S.Car., South Carolina
Sc.D., *Scientiae Doctor* (Doctor of Science)
Sci., Science, Scientific
Scot., Scotland
SD, subsequent designation
SE., Southeast
sec., *seccion(es),* section(s)
Sept., September
ser., serial, series, *seriya*
sér., *séries*
sess., session, *sessiya*
Sh., Shale
Sib., Siberia
Siegen., Siegenian
Sil., Silurian
Sinemur., Sinemurian
SM, subsequent monotypy
sp., species (spp., plural)
spec., special, specification, specimen
Ss., Sandstone
Ste., *Sainte*
subj., subjective
subtrop., subtropical
suppl., supplement(s), supplementary
SW., Southwest
Swed., Sweden
Switz., Switzerland
syn., synonym, synonymous

Tasm., Tasmania
tech., technical
temp., temperate
Tenn., Tennessee

Terr., Territory,-ies
Tert., Tertiary
Tithon., Tithonian
Toarc., Toarcian
Tournais., Tournaisian
Trans., Transactions
transl., translation
transv., transverse
Tremadoc., Tremadocian
Trenton., Trentonian
Trias., Triassic
trop., tropical
Turon., Turonian

U., **Up.**, Upper
Univ., *Universidad, Università, Universität, Université, Universitet,* University
unpubl., unpublished
Urgon., Urgonian
U.S., United States
U.S.A., United States of America
USSR, Union of Soviet Socialist Republics

v., **vol.**, volume(s)
Va., Virginia
Valangin., Valanginian
var., variety
vert., vertical
Virgil., Virgilian
vyp., *vypusk*

W., West
Wenlock., Wenlockian
Westphal., Westphalian
Wis., Wisconsin
Wolfcamp., Wolfcampian
Word., Wordian
Wyo., Wyoming

Yorks., Yorkshire
Yugo., Yugoslavia

Z., Zone
Zeitschr., *Zeitschrift*
Zool., Zoological, *Zoologicus, Zoologie, Zoologisch,* Zoology

REFERENCES TO LITERATURE

Each part of the *Treatise* is accompanied by a list, or lists, of references to paleontological literature. In *Treatise* parts published in the 1950's and early 1960's these lists were highly selective, consisting primarily of recent and comprehensive monographs, but also including some older works recognized as outstanding in importance. In time, however, *Treatise* authors and readers pressed for more exhaustive documentation, and for volumes published from about 1964 to 1965, this has been as comprehensive as possible. Since that time the aim has been to provide documentation, complete with author, publication year, and page number, for all taxa described anywhere in the text, as well as for all illustrations copied or adapted from preexisting publications. In other words, the lists of references contain the full titles and places of publication of all books, monographs, and serial articles to which reference is made in the text.

The following is a statement of the full names of serial publications which are cited in abbreviated form in the lists of references in the present volume. The information thus provided should be useful in library research work. The list is alphabetized according to the serial titles which were employed at the time of original publication. Those following in brackets are those under which the publication may be found currently in the *Union List of Serials,* the United States Library of Congress listing, and most library card catalogues. The names of serials published in Cyrillic are transliterated; in the reference lists these titles, which may be abbreviated, are accompanied by transliterated authors' names and titles, with English translation of the title. The place of publication is added (if not included in the serial title).

The method of transliterating Cyrillic letters that is adopted as "official" in the *Treatise* is that suggested by the Geographical Society of London and the U.S. Board on Geographic Names. It follows that names of some Russian authors in transliterated form derived in this way differ from other forms, possibly including one used by the author himself. In *Treatise* reference lists the alternative (unaccepted) form is given enclosed by square brackets (e.g., Chernyshev [Tschernyschew], T.N.).

List of Serial Publications

Academia Naturae Curiosorum. Halle. (*See* Deutsche Akademie der Naturforscher.)

Academia Republicii Popülare Române, Buletin Stiintific. Bucuresti.

Académie Impériale des Sciences, St. Pétersbourg, Mémoires; Recueil des actes de la seance publique (Akademiya Nauk SSSR Leningrad).

Académie Malgache, Mémoires. Tananarive, Malagasy Republic.

Académie Royale de Belgique, Class des Sciences, Bulletin; Mémoires. Bruxelles.

Académie des Sciences de Paris, Comptes Rendus; Mémoires. Paris.

Académie des Sciences de l'URSS, Comptes Rendus; Institut Paléontologique, Travaux; Institut Paléozoologique, Travaux [Akademiya Nauk SSSR, Doklady]. Leningrad.

Academy of Natural Sciences of Philadelphia, Journal; Proceedings.
Academy of Science of St. Louis, Bulletin; Memoirs; Transactions.
Accademia Gioenia delle Scienze Naturali di Catantia, Atti; Bollettino.
Acta et Commentationes Universitatis Tartuensis (Dorpatensis). Tartu Riiliku Ülikooli Toimetused, Uchenye Zapiski; Tartu Ülikooli Geologia-Instituudi Toimetused.
Acta Geologica (Academiae Scientiarum Hungaricae). Budapest.
Acta Geologica Polonica. Warszawa.
Acta Helvetica. Basel.
Acta Palaeontologica Polonica [Polska Akademia Nauk, Komitet Geologiczny]. Warszawa.
Acta Palaeontologica Sinica. Peking. (*See* Ku Sheng Wu Hsüeh Pao.)
Acta Zoologica. Stockholm.
[K.] Akademie van Wetenschappen te Amsterdam, Jaarboek; Proceedings; Verslagen en Mededeelingen.
Akademie der Wissenschaften, physikalisch-mathematische Klasse, Abhandlungen; Monatsberichte. Berlin.
Akademie der Wissenschaften zu München, mathematisch-physikalische Klasse, Denkschriften; Sitzungsberichte.
[K.] Akademie der Wissenschaften zu Wien, mathematisch-naturwissenschaftliche Klasse, Denkschriften; Sitzungsberichte.
Akademiya Nauk Kazakhskoy SSR, Institut Geologicheskikh Nauk, Trudy. Alma Ata.
Akademiya Nauk SSSR, Doklady; Izvestiya; Trudy. Moskva, Leningrad.
Akademiya Nauk SSSR. Geologicheskiy Muzei, Trudy. Leningrad.
Akademiya Nauk SSSR, Izvestiya, Seriya Biologicheskaya; Byulletin, Seriya Geologicheskaya. Moskva.
Akademiya Nauk SSSR Leningrad. Paleontologicheskiy Zhurnal. Moskva, Leningrad.
Akademiya Nauk SSSR. Sibirskoe Otdelenie, Institut Geologii i Geofiziki, Trudy. Novosibirsk.
Akademiya Nauk Tadzhikskoy SSR, Doklady. Dyushanbe.
Albany Institute, Proceedings; Transactions. Albany, N.Y.
Allgemeine Schweizerische Gesellschaft für die gesamten Naturwissenschaften, Neue Denkschriften. Zürich.
American Academy of Arts and Sciences, Memoirs; Proceedings. Boston.
American Association for the Advancement of Science, Proceedings; Publications. Washington, D.C.
American Association of Petroleum Geologists, Bulletin. Tulsa, Okla.
American Geologist. Minneapolis, Minn.
American Journal of Science. New Haven, Conn. (American Journal of Science and Arts, 1819-79).
American Midland Naturalist. Notre Dame, Ind.
American Museum of Natural History, Novitates; Micropaleontology; Memoirs; Bulletins. New York.
American Naturalist. Lancaster, Pa.
American Philosophical Society, Proceedings; Memoirs; Transactions. Philadelphia, Pa.
Anales de la Sociedad Cientifica Argentina. Buenos Aires.
Annales de Géologie et de Paléontologie. Palermo.
Annales de Paléontologie. Paris.
Annales Universitatis Saraviensis. Saarbrücken.
Annals and Magazine of Natural History. London.
Archiv für Anatomie, Physiologie und Wissenschaftliche Medizin. Leipzig.
Archiv für Naturgeschichte. Berlin, Leipzig.
Archiv für die Naturkunde Liv-, Ehst- und Kurlands. Dorpat.
Archives de Musée Teyler. Haarlem.
Arkansas, Geological Survey of, Bulletin. Little Rock.
Arkiv för Kemi, Mineralogi och Geologi. Uppsala.
Arkiv för Zoologi. Uppsala.
Association Française pour l'Avancement des Sciences, Compte Rendu. Reims, Paris.
Aus der Heimat. Stuttgart.
[The] Atlantic [Atlantic Monthly]. Boston.
Australian Museum, Memoirs; Records. Sydney.
Bayerische Staatssammlung für Paläontologie und Historische Geologie, Mitteilungen. München.
Beiträge zur Naturkundlichen Urwelt.
Beiträge zur Naturkundlichen Forschung in Südwestdeutschland. Karlsruhe.
Beiträge zur Paläontologie und Geologie Österreich-Ungarns und des Orients. Wien.
Beiträge zur Petrefacten-Kunde. Bayreuth.
Bijdragen tot de Dierkunde. Leiden.
Biological Reviews. Cambridge, Eng. (*See* Cambridge Philosophical Society.)
Biological Society of Washington, Proceedings. Washington, D.C.
Boston Society of Natural History, Memoirs; Proceedings. Boston, Mass.
Breviora (Museum of Comparative Zoology, Harvard University). Cambridge, Mass.
Brigham Young University, Geology Studies. Provo, Utah.
Bristol Naturalists' Society, Proceedings. Bristol.
British Association for the Advancement of Science, Reports. London.
British Museum (Natural History), Geology; Zoology; Bulletin. London.
Buffalo Society of Natural Science, Bulletin. Buffalo, N.Y.
Bulletin of American Paleontology, Ithaca, N.Y.
Bulletin of Zoological Nomenclature. London.
Bureau of Geology. Norman, Okla.
Cahiers Paléontologiques. Paris.

California, University of, Geological Sciences, Bulletins; Publications. Berkeley, Los Angeles.

Cambridge Philosophical Society, Biological Reviews and Biological Proceedings. Cambridge, Eng.

Canada, Geological Survey of, Department of Mines and Resources, Mines and Geology Branch, Bulletin; Contributions to Canadian Palaeontology; Memoir; Museum Bulletin; Victoria Memorial Museum Bulletin. Ottawa.

Canadian Field Naturalist, Transactions. Ottawa. (Formerly Ottawa Field Naturalists' Club and Ottawa Naturalist 1887-1919.)

Canadian Journal of Industry, Science and Art. Toronto.

Canadian Naturalist and Geologist. Montreal, Canada.

Canadian Petroleum Geology, Bulletin. Alberta Society of Petroleum Geologists. Calgary.

Canadian Record of Science. Montreal.

Carnegie Institution of Washington, Papers; Publications. Washington, D.C.

Carnegie Museum, Annals. Pittsburgh, Pa.

Carte Géologique Detailée de France, Mémoires. Paris.

Časopis Moravského Zemského Musea. Brno.

Centralblatt für Mineralogie, Geologie, Paläontologie. Stuttgart. (*See* also Zentralblatt.)

Česká Akademie Ved a Uměni v Praze, Třída II, Matematicko-přírodnicka, Rozpravy. (*See* Palaeontographica Bohemiae.)

[K.] České Společnost Nauk Prague, Třída Mathematicko-Přírodovědecká; Rozpravy; Věstník.

Challenger. Report on the Scientific Results of the Exploring Voyage of HMS Challenger, Zoology. Edinburgh.

Chicago Academy of Sciences, Bulletin; Natural History Survey Bulletin; Transactions.

Chicago, University of, Abstracts of Theses, Science Series. Chicago, Ill.

China, Geological Survey of, Palaeontologia Sinica, Bulletin; Memoirs. Peking.

Cincinnati, Quarterly Journal of Science. Cincinnati, Ohio.

Cincinnati Society of Natural History, Journal. Cincinnati, Ohio.

Cincinnati, University of, University Museum. Cincinnati, Ohio.

Colorado School of Mines, Professional Contribution; Quarterly. Golden.

Comité Géologique, Mémoires. St. Petersbourg. (*See* Geologicheskiy Komitet, Trudy.)

Commission Géologique du Portugal, Travaux. Lisbon.

Connecticut Academy of Arts and Sciences, Memoirs; Transactions. New Haven.

Consejo Nacional de Investigaciones Cientificas y Tecnicas. Madrid.

Contributions from the Walker Museum. Chicago.

Current Science. Indian Institute of Science. Bangalore, India.

Danmarks Geologiske Undersøgelse, Skrifter. København.

Dansk Geologisk Forening, Meddelelser, Bulletin. København.

[K.] Danske Videnskabernes Selskabs, Matematisk-Fysiske, Forhandlinger; Skrifter. København.

Davenport Academy of Sciences, Proceedings. Davenport, Iowa.

Decheniana. Verhandlungen des Naturhistorischen Vereins der Preussischen Rheinlande und Westfalens. Bonn.

Denison University, Scientific Laboratories, Bulletins; Journals. Granville, Ohio.

Deutsche Akademie der Naturforscher zu Halle. (Nova Acta Leopoldina.)

Deutsche Geologische Gesellschaft, Zeitschrift. Berlin, Hannover.

Earth Science. Chicago. (Earth Science Digest.)

Earth and Planetary Science, Annual Reviews, Letters. Amsterdam.

Eclogae Geologicae Helvetiae (*see* Schweizerische Geologische Gesellschaft). Basel.

Edinburgh Geological Society, Transactions.

Embryologia. Nagoya University, Biological Institute, Faculty of Science. Nagoya, Japan.

Endeavour. London.

Erdöl und Kohle. Hamburg, Ger.

Estudios Geológicos y Paleontólogicos sobre la Cordillera Oriental de Colombia. Bogota.

Evolution. Lancaster, Pa.

Ezhegodnik Russkago Paleontologicheskago Obshchestva. Petrograd.

Ezhegodnik Vsesoyuznogo Paleontologicheskogo Obshchestvo. Moskva, Leningrad.

Fauna d'Italia. Bologna.

Field Columbian Museum, Publications. Chicago. (Field Museum of Natural History, 1905-43.)

Field Museum of Natural History, Bulletin; Publication; Geological Series; Zoological Series. Chicago.

Fieldiana, Geology. Chicago.

Fortschritte in der Geologie von Rheinland und Westfalens. Krefeld.

Fortschritte der Paläontologie. Berlin.

[K.] Fysiografiska Sällskapet i Lund, Förhandlingar; Handlingar.

Géobios. Lyon.

Geochimica et Cosmochimica Acta. Oxford, New York, Braunschweig.

Geologica Hungarica; Series Geologica; Series Palaeontologica. Budapest.

Geological Magazine. London, Hertford.

Geological Society of America, Bulletin; Memoir; Special Paper. Boulder, Colo.

Geological Society of Glasgow, Transactions.

Geological Society of London; Memoir; Proceedings; Quarterly Journal; Geological Journal; Transactions. (The Geological Society.)

Geologicheskiy Komitet, Trudy. Leningrad. (*See* Comité Géologique, Mémoires. St. Petersbourg.)

Geologie. Berlin. (*See* Zeitschrift für das Gesamt-

gebiet der Geologie und Mineralogie sowie der angewandten Geophysik.)
[K.K.] Geologische Bundesanstalt Wien, Abhandlungen; Jahrbuch; Verhandlungen.
[K.K.] Geologische Reichsanstalt Wien, Abhandlungen; Jahrbuch; Verhandlungen. (*See* Geologische Bundesanstalt Wien.)
Geologische Rundschau, Internationale Zeitschrift für Allgemeine Geologie. Geologische Vereinigung. Leipzig, Stuttgart.
Geologiska Föreningens i Stockholm, Förhandlingar.
Geologist. London.
Geologists' Association, Proceedings. London.
Gesellschaft der Geologie- und Bergbaustudenten, Mitteilungen. Wien.
Gesellschaft Naturforschender Freunde Berlin, Magazin, Sitzungsberichte.
Gesellschaft der Wissenschaften zu Göttingen, Mathematisch-physikalische Klasse, Abhandlungen; Nachrichten.
Giornale di Geologia, Annali del Museo Geologia di Bologna.
Glavnogo Geologo-Razvedochnogo Upravleniya, Izvestiya; Trudy. Moskva.
Great Britain, Geological Survey of, Palaeontology, Bulletin; Memoirs. London.
Hamburg, Geologisches Staatsinstitut, Mitteilungen.
Handbuch der Mikroskopie in der Technik. Frankfurt.
Harvard University, Museum of Comparative Zoology, Breviora; Bulletin; Memoirs; Special Publications. Cambridge, Mass.
Helgoländer Wissenschaftliche Meeresunterschungen. Biologischen Anstalt Helgoland. List, Sylt.
Hessisches Landesamt für Bodenforschung; Abhandlungen; Notizblatt. Wiesbaden.
Hokkaido University, Journals of the Faculty of Science. Sapporo, Japan.
Hydro-Lab Journal. Freeport, Grand Bahama Island.
Illinois State Academy of Science, Transactions. Springfield.
Illinois State Geological Survey, Reports of Investigations; Bulletin; Memoirs; Monographs. Urbana.
Illinois State Museum of Natural History, Bulletins. Springfield.
India, Geological Survey of, Bulletin; Memoirs (Palaeontologia Indica); Publications; Records. Calcutta.
Indian Museum, Memoirs; Records. Calcutta.
Indiana Academy of Science, Proceedings. Brookville.
Indiana Department of Geology and Natural History, Annual Report. Bloomington.
Indiana Department of Geology and Natural Resources, Annual Report. Bloomington.
Institut Océanographique de Monaco, Annales; Bulletin. Monte Carlo, Paris.
Instituta Prikladnoy Mineralogii i Tsvetnoy Metallurgii, Trudy. Moskva.
Instituto Geológico y Minero de España; Boletín; Memorias; Notas y Comunicaciones. Madrid.
[The] Intellectual Observer. London.
International Geological Congress, 24th Session, Abstracts. Montreal.
Iowa Academy of Science, Proceedings. Des Moines.
Iowa Geological Survey, Annual Report; Report of Investigations. Iowa City.
Iowa, University of, Laboratory of Natural History Bulletin. Iowa City.
Iowa University Studies (in Natural History). Iowa City.
International Geology Review. American Geological Institute. Washington, D.C.
Ireland, Department of Agriculture and Technical Instruction for Ireland, Fisheries Branch, Scientific Investigations. Dublin.
Israel Journal of Earth Sciences. Jerusalem.
Istituto Veneto di Scienze Lettere ed Arti, Classe di Scienze Mathematiche e Naturali, Atti. Venezia (Venice).
Japanese Journal of Geology and Geography. Tokyo.
Japanese Journal of Zoology. Tokyo.
Jenaische Zeitschrift für Naturwissenschaften. Jena.
Journal of General Physiology. Baltimore.
Journal of Geology. Chicago.
Journal of Morphology. Philadelphia.
Journal of Paleontology. Tulsa, Okla.
Journal of Parasitology. Colorado Springs, Colo.
Journal de Physique, de Chimie, d'Histoire Naturelle et des Arts. Paris.
Journal of Sedimentary Petrology. Tulsa, Okla.
Kansas Academy of Science, Transactions. Topeka, Kans.
Kansas City Scientist. Kansas City, Mo.
Kansas State Geological Survey, Bulletin; Publications; Volumes. Lawrence, Kans.
Kansas, The University of, Paleontological Contributions, Article; Paper. Lawrence, Kans.
Kansas University Quarterly. Lawrence, Kans.
Kansas University Science Bulletin. Lawrence, Kans.
Kazan Universitet. Obshchestvo Estestvoispytatelei, Trudy.
Kentucky Geological Survey, Reports. Lexington.
Ku Sheng Wu Hsüeh Pao. Acta Palaeontologica Sinica. Peking.
Leidse Geologische Mededeelingen. Leiden.
Leningrad Universitet, Vestnik.
[K.] Leopoldischen Carolinischen Deutschen Akademie der Naturforscher (Nova Acta Leopoldina), Abhandlungen. Halle.
Lethaia. Oslo.
Leyden Museum, Notes.
Linnean Society of London (Botany), Journals; Transactions.
Linnean Society of London (Zoology), Journals; Proceedings; Transactions.

Linnean Society of New South Wales, Journal and Proceedings. Sydney.

Lucrarile Institutului de Speologie "Emil Racovita." Bucharest.

Lunds Geologisk-Mineralogiska Institution, Meddelanden.

Lunds Universitet, Årsskrift.

Lvovskoe Geologicheskoe Obshchestvo pri Gosudarstvennyi Universitet Ivana Franko, Geologicheskiy Sbornik, Mineralogicheskiy Sbornik, Paleontologicheskiy Sbornik, Trudy. Lvov.

Madagascar. Service Géologique des Mines, Annales. Tananarive.

Maden Tetkik ve Arama Enstitüsü, Yayinlari. Ankara.

Magazine of Zoology and Botany. Edinburgh, London.

Marine Biology. Berlin, New York.

Marine Science of the Gulf and Caribbean, Bulletin. Miami, Fla.

Maryland Geological Survey. Baltimore.

Materialy po Geologii i Poleznym Iskopaemym Altaya i Kazakhstana. Moskva.

Materialy po Istorii Fauny i Flory Kazakhstana (Akademiya Nauk Kazakhskoy SSR). Alma-Ata.

Meyniana. Kiel Universität, Geologisches Institut.

Miami University, Institute of Marine Science, Studies in Tropical Oceanography. Coral Gables, Fla.

Michigan, University of, Museum of Paleontology, Contributions. Ann Arbor.

Micropaleontology. American Museum of Natural History. New York.

Mijnwezen in Nederlandsch-Oost-Indië, Jaarboek; Verhandlungen. s'Gravenhage.

Ministerstvo Geologii RSFSR. Severo-Vostochnoe Ordena Trudovogo Krachnogo Znameni Geologicheskoe Upravlenie.

Minnesota, Geological and Natural History Survey, Annual Reports; Geology of Minnesota. Minneapolis.

Missouri Bureau of Geology and Mines. Jefferson City.

Missouri Geological Survey and Water Resources, Bulletins; Reports of Investigations. Rolla, Mo.

Missouri, University of, School of Mines and Metallurgy, Bulletins; Technical Series. Rolla.

Missouri, University of, Studies. Columbia.

Moskovskoe Obshchestvo Ispytatelei Prirody, Byulletin (formerly Société Impériale des Naturalistes de Moscou). Moskva.

Musée Royal d'Histoire Naturelle de Belgique, Annales; Bulletins; Mémoires (continued as Institut Royal des Sciences Naturelles de Belgique). Bruxelles.

Museo de Historia Natural de Mendoza, Revista.

Museo Libico Storia Naturale, Annali. Tripoli.

Muséum (National) d'Histoire Naturelle, Annales; Bulletin; Mémoires. Paris.

Museum Senckenbergianum. Frankfurt. (*See* Senckenbergische Naturforschende Gesellschaft.)

Natur und Museum. Senckenbergische Naturforschende Gesellschaft. Frankfurt.

Natur und Volk. Senckenbergische Naturforschende Gesellschaft. Frankfurt.

Natural History. American Museum of Natural History. New York.

Natural History Review. London.

Natural History Society of Glasgow, Transactions.

Natural Science. Edinburgh.

Nature. London.

Naturforschende Gesellschaft zu Freiburg im Breisgau, Berichte.

Naturforschende Gesellschaft zu Görlitz, Abhandlungen.

Naturforschende Gesellschaft zu Halle, Abhandlungen.

Naturhistorische Gesellschaft zu Hannover, Berichte.

Naturhistorischen Gesellschaft zu Nürnberg, Abhandlungen.

Naturhistorischer Verein der Preussischen Rheinlande und Westfalens, Sitzungsberichte; Verhandlungen. Bonn.

[K.K.] Naturhistorisches Hofmuseum, Annalen, Wien.

Naturwissenschaftliche Monatshefte für den Biologischen, Chemischen, Geographischen und Geologischen Unterricht. Leipzig.

Naturwissenschaftlicher Verein, Elberfeld, Jahresberichte; Verhandlungen.

Nauchno-issledovatelskii Institut Geologii Arktiki (NIIGA), Trudy. Leningrad.

Nebraska Geological Survey, Bulletin; University Studies. Lincoln.

Nebraska State Museum, Bulletin. Lincoln.

[K.] Nederlandsch Geologisch-Mijnbouwkundig Genootschap, Verhandelingen, Geologische Serie, Deel. s'Gravenhage.

Nederlandsch-Oost-Indië Dienst van den Mijnbouw, Wetenschappelijke Mededeelingen. Den Haag.

[K.] Nederlandse Akademie van Wetenschappen, Afdeeling Natuurkunde, Verhandelingen; Verslagen en Mededeelingen. Amsterdam.

Neues Jahrbuch für Geologie und Paläontologie, Abhandlungen; Beilage-Bände, Monatshefte. Stuttgart. (Before 1950, Neues Jahrbuch für Mineralogie, Geologie, und Paläontologie.)

Neues Jahrbuch für Mineralogie, Abhandlungen; Beilage-Bände, Monatshefte. Stuttgart. (Before 1950, Neues Jahrbuch für Mineralogie, Geologie, und Paläontologie.)

New Jersey, Geological Survey of, Paleontology. Trenton.

New South Wales, Geological Survey of, Ethnology, Memoirs; Geology, Memoirs; Paleontology, Memoirs; Records. Sydney.

New York Academy of Science, Annals.

New York State Cabinet of Natural History, Annual Report. Albany.

New York State Geological Survey, Annual Report; Natural History of New York; Palaeontology of New York. Albany.
New York State Museum (of Natural History), Annual Report; Bulletin; Circular; Memoirs. Albany.
North Carolina Geological Survey, Bulletin; Report. Raleigh.
Nova Acta Regiae Societatis Scientiarium Upsaliensis. Upsaliae (Upsala).
Nyt Magazin for Naturvidenskaberne. Kristiania (Oslo).
Oberrheinischer Geologischer Verein, Jahresberichte und Mitteilungen. Stuttgart.
Ohio, Geological Survey of, Bulletin; Report. Columbus.
Ohio Journal of Science. Columbus.
Ohio State Academy of Sciences, Proceedings. Columbus.
Oklahoma, Geological Survey, Bulletin; Circular; Geological Notes. Norman.
Oklahoma Geology Notes. Norman.
Ontario Department of Mines, Annual Report. Toronto.
Ottawa Field Naturalists' Club, Transactions. Ottawa.
Ottawa Naturalist. (*See* Canadian Field Naturalist.)
Palaeobiologica. Wien.
Palaeogeography, Palaeoclimatology, Palaeoecology. Amsterdam.
Palaeontographia Italica. Pisa.
Palaeontographica. Stuttgart, Kassel.
Palaeontographica Americana. Ithaca, N.Y.
Palaeontographica Bohemiae. Praze. (Česká Akademie Ved a Uměni v Praze.)
Palaeontographical Society, Monograph. London.
Palaeontologia Indica (Memoirs of the Geological Survey of India). Calcutta.
Palaeontological Society of Japan, Transactions and Proceedings. Tokyo.
Paläontologie von Timor. Stuttgart.
Paläontologische Abhandlungen. Jena.
Paläontologische Zeitschrift. Berlin, Stuttgart.
Palaeontology. Palaeontological Association, London.
Paleobiology. Menlo Park, Calif.
Paleontologicheskiy Zhurnal. Moskva. (*See* Akademiya Nauk SSSR.)
Pan-American Geologist. Des Moines, Iowa.
Panjab University, Geological Bulletin. Lahore.
Peabody Museum of Natural History, Bulletin. New Haven, Conn.
Pennsylvania Academy of Science, Proceedings. Harrisburg.
Philosophical Magazine. London.
Portland Society of Natural History, Proceedings. Portland, Maine.
Prace Museum Ziemi, Instytut Geologiczny Museum Ziemi. Warszawa.
[K.] Preussische Akademie der Wissenschaften, Abhandlungen; Monatsberichte. Berlin.
[K.] Preussische Geologische Landesanstalt, Abhandlungen; Jahrbuch. Berlin.
Priroda. Akademiya Nauk SSSR. Moskva.
Publicaciones Alemanas (Extranjeras) sobre Geologia de España. Madrid.
Quarterly Journal of Microscopical Science, Memoirs. London.
Queensland Museum, Memoirs. Brisbane.
Recueil Zoologique Suisse, Annales. Geneva.
Reichsstelle (Reichsamt) für Bodenforschung, Abhandlungen; Bericht. Wien.
República de Colombia, Ministerio de Industrias y Trabajo, Estudios Geológicos y Paleontológicos sobre la Cordillera Oriental de Colombia. Bogotá.
Revue Magasin de Zoologie. Paris.
(La) Revue Scientifique, Année. Paris.
Rheinische Heimatpflege. Düsseldorf.
Rivista Italiana di Paleontologia e Stratigraphia. Milano.
Royal Dublin Society, Scientific Proceedings.
Royal Geological Society of Cornwall, Transactions. Penzance.
Royal Irish Academy, Proceedings. Dublin.
Royal Ontario Museum Paleontology, Contributions. Toronto.
Royal Society of Edinburgh, Memoirs; Transactions.
Royal Society of London, Philosophical Transactions; Proceedings.
Royal Society of New South Wales, Journal and Proceedings. Sydney.
Royal Society of New Zealand, Journal; Proceedings; Transactions and Proceedings. Wellington.
Royal Society of Western Australia, Journal. Perth.
Sborník Geologických Věd, Paleontologie. Praha.
Schul-Corsus 1848 an dem Gymnasium zu Trier, Jahresbericht.
Schweizerische Geologische Gesellschaft, Verhandlungen. Basel.
Schweizerische Naturforschende Gesellschaft, Denkschriften. Zürich.
Schweizerische Paläontologische Gesellschaft, Abhandlungen. Zürich.
Science. American Association for the Advancement of Science. Washington, D.C.
Scientific Reports, John Murray Expedition, 1933-34. London.
Scottish Journal of Geology. Edinburgh.
Senckenbergiana Lethaea (Senckenbergische Naturforschende Gesellschaft Wissenschaftliche Mitteilungen). Frankfurt. ("Lethaea" added to title, 1954.)
Senckenbergische Naturforschende Gesellschaft, Abhandlungen. Frankfurt.
Service de la Carte Géologique de l'Algérie, Bulletin. Algiers.
Service de la Carte Géologique de la Tunisie, Mémoires. Tunis.
Service Géologique de l'Indochine, Bulletin; Mémoires. Hanoi-Haiphong.

Service des Mines et de la Carte Géologique du Maroc, Notes et Mémoires. Rabat.
Sibirskogo Nauchno-Issledovatelskogo Instituta Geologii, Geofiziki i Mineralnogo Syrya, Trudy. Moskva.
Siboga Expedition, Monographs. Leiden.
Skandinaviske Naturforskeres Møte, Forhandlinger. Kristiania.
Smithsonian Miscellaneous Collections. Washington, D.C.
[R.] Sociedad Española de Historia Natural, Boletín. Madrid.
Sociedade Brasileira de Geologia, Boletim. Sao Paulo.
Societá Geológica Italiana, Bolletino; Memorie. Roma.
Société d'Agriculture, d'Histoire Naturelle et Arts Utiles de Lyon, Annales.
Société Belfortaine d'Émulation, Bulletin. Belfort.
Société d'Émulation du Doubs, Mémoires. Besançon.
Société Géologique de Belgique, Annales; Bulletin; Mémoires. Liège.
Société Géologique de France, Compte Rendu des Séances; Bulletin; Mémoires. Paris.
Société Géologique et Minéralogique de Bretagne, Mémoires. Rennes.
Société Géologique du Nord, Annales; Mémoires. Lille.
Société d'Histoire Naturelle de Toulouse, Bulletin.
Société Hollandaise des Sciences, Verhandlingen. Haarlem.
Société Impériale des Naturalistes de Moscou, Bulletin. [Moskovskoe Obshchestvo Ispytatelei Prirody.]
Société Linnéenne de Normandie, Bulletin; Mémoires. Caen.
Société Neuchâteloise Sciences Naturelles, Mémoires.
Société Paléontologique de la Russie, Annuaire. Leningrad.
Société Paléontologique de la Suisse, Mémoires. Zürich. (*See* Schweizerische Paläontologische Gesellschaft.)
Société Philomathique de Paris, Bulletin.
Société de Physique et d'Histoire Naturelle de Genève, Mémoires.
Société Royale des Sciences de Liège, Mémoires. Bruxelles.
Société des Sciences Historiques et Naturelles de l'Yonne, Bulletins. Auxerre, France.
Société des Sciences Naturelles de Tunisie, Bulletin. Tunis.
Société Vaudoise des Sciences Naturelles, Bulletin; Mémoires.
Société Zoologique de France, Bulletin; Mémoires. Paris.
Southeastern Geology. Duke University. Durham, N.C.
Stazione Zoologica di Napoli, Monografía.
[K.] Svenska Vetenskapsakademien, Arkiv för Mineralogi och Geologi; Arkiv för Zoologi; Handlingar. Stockholm.
Systematic Zoology. Lawrence Kans.
Temminckia. Leyden.
Texas, University of, Bulletin; Publications. Austin.
Tissue and Cell. Edinburgh.
Tohoku University, Science Reports. Sendai, Japan.
Tschechische Akademie der Wissenschaften, Mitteilungen. Prag.
Tuatara. Victoria University College, Biological Society, Journal. Wellington, N.Z.
United States Geological and Geographical Survey of the Territories, Annual Report. Washington, D.C.
United States Geological Survey, Bulletin; Monograph; Professional Paper. Washington, D.C.
United States National Museum, Bulletin; Proceedings. Washington, D.C.
Université de Paris, Laboratoire de Paléontologie, Travaux. Paris.
University of Missouri Studies. Columbia.
Uppsala, University of, Geological Institution, Bulletins.
Upravleniya Geologii Soveta Ministrov Tadzhikskoy SSR, Trudy. Moskva.
Ústředního Ústavu Geologickeho, Rozpravy; Sborník; Věstník. Praha.
Věda Přírodní. Praha.
Verein für Vaterländische Naturkunde in Württemberg, Jahreshefte. Stuttgart.
Vetenskapsakademiens Förhandlingar, Öfversigt. Stockholm.
Victoria, Geological Survey, Bulletin; Memoirs; Records. Melbourne.
Videnskabelige Meddelelser fra Dansk Naturhistorisk Forening. København.
Videnskabs-Selskabet i Kristiania, Forhandlinger; Skrifter. Oslo.
Virginia Academy of Science, Proceedings. Richmond.
Voprosy Paleobiologii i Biostratigrafii, Vsesoiuznoe Paleontologicheskoe Obshchestvo, Trudy Sessii. Moskva.
Voprosy Paleontologii. Leningrad.
Vsesoyuznyy Nauchno Issledovatelskiy Geologicheskiy Institut, Informatsii Sbornik, Trudy (VSEGEI). Moskva.
Vsesoyuznyy Nauchno-Issledovatelskiy Institut Geofizicheskikh Metodov Razvedki, Trudy. Moskva.
Wagner Free Institute of Science of Philadelphia, Bulletin.
Washington Academy of Sciences, Journal; Proceedings. Washington, D.C.
West Riding Geological and Polytechnic Society, Proceedings.
Western Australia, Geological Survey of, Bulletin. Perth.
Wisconsin, Geological and Natural History Survey, Bulletin. Madison.
Wisconsin, Geological Survey of, Annual Report; Bulletin. Madison.

Wissenschaftliche Arbeiten aus dem Burgenland. Eisenstadt. (Burgenlaendisches Landesmuseum.)
Wissenschaftliche Erforschung des Balatonsees, Resultate. Budapest.
Wissenschaftliche Ergebnisse der Deutschen Tiefsee-Expedition auf dem Dampfer "Valdivia," 1898-1899. Jena.
Wissenschaftliche Ergebnisse der Schwedischen Südpolarexpedition, 1901-1903. Stockholm.
Yokohama Kokuritsu Daigaku Science Reports, Section II, Biological and Geological Sciences. Kamakura, Japan.
Zabaikalskii Filial, Geograficheskoe Obshchestva SSR, Zapiski. Chita, Sib.
Zeitschrift für das Gesamtgebiet der Geologie und Mineralogie sowie der Angewandten Geophysik. Berlin.
Zeitschrift für die Gesamte Naturwissenschaft. Braunschweig.
Zeitschrift für Geschiebeforschung und Flachlandsgeologie. Berlin. (Formerly Zeitschrift für Geschiebeforschung.)
Zeitschrift für Induktive Abstammungs- und Vererbungslehre. Berlin.
Zeitschrift für Wissenschaftliche Zoologie. Leipzig.
Zentralblatt für Geologie und Paläontologie. Stuttgart. (Before 1950, Zentralblatt für Mineralogie, Geologie, und Paläontologie.)
Zentralblatt für Mineralogie. Stuttgart. (Before 1950, Zentralblatt für Mineralogie, Geologie, und Paläontologie.)
Zoological Society of London, Proceedings; Transactions.
Zoologische Jahrbücher, Abteilung 2, Anatomie und Ontogenie. Jena.
Zoologischer Anzeiger. Leipzig.
Zoologischer Jahresbericht. Berlin.
Zoologiska Bidrag från Uppsala. Uppsala, Stockholm.
Zoology. New York.

SOURCES OF ILLUSTRATIONS

At the end of figure captions names and dates are given to supply record of the authors of illustrations used in the *Treatise,* reference being made to publications cited in reference lists. Previously unpublished illustrations are marked by the letter "n" (signifying "new") with the name of the author and, where appropriate, the museum catalogue number of the specimen figured.

PART T
ECHINODERMATA 2
CRINOIDEA

By Georges Ubaghs, R. C. Moore, H. Wienberg Rasmussen, N. Gary Lane, Albert Breimer, H. L. Strimple, J. C. Brower, Russell M. Jeffords, James Sprinkle, R. E. Peck, D. B. Macurda, Jr., D. L. Meyer, Michel Roux, Hertha Sieverts-Doreck, R. O. Fay, and R. A. Robison

VOLUME 1

CONTENTS

[Volume 1, p. i-xxxviii, *T*1-*T*402; Volume 2, p. *T*403-*T*812; Volume 3, p. *T*813-*T*1027]

DEDICATION

By Raymond C. Moore

The *Treatise* volumes on Crinoidea here published are dedicated appropriately to four paleontologists who have advanced knowledge of fossil representatives of these echinoderms most greatly. These outstanding contributors include one American, Frank Springer (1848-1927); one Britisher, F. A. Bather (1863-1934); and two Germans, Otto Jaekel (1863-1929), and Johannes Wanner (1878-1956). (Portraits by Raymond C. Moore.)

FRANK SPRINGER

Frank Springer was born at Wapello, Iowa, on June 17, 1848. He was graduated from the University of Iowa in 1867 and then, following his father, who was a distinguished judge prominent in gaining statehood for Iowa and in writing its constitution, studied law, followed by admission to the bar in 1869. At Burlington, some 30 miles south of his birthplace, young Springer was chosen as county prosecuting attorney and served until 1873, when interest in opportunities offered by the rapidly expanding Southwest led him to move to Cimarron, New Mexico; he transferred residence to Las Vegas, New Mexico, and considered this city as home throughout the remainder of his life. Soon after his arrival in Cimarron he was chosen as attorney for a company that had acquired title to a Mexican land grant of more than 1,700,000 acres. Validation of the title required extensive litigation, however, leading ultimately to the U.S. Supreme Court where Springer won a favorable decision. Also, he had to appear before committees of the U.S. House and Senate in order to obtain passage of needed legislation. In 1891, Springer was elected president of the company, which by then had been proved to control huge reserves of valuable coking coal. Meanwhile, in 1878, he became counsel for the Santa Fe railway when westward construction of the line reached New Mexico; his chief duty was the arranging of needed rights-of-way.

As a member of the New Mexico Legislative Council (1880, 1901) and Constitutional Convention (1889), SPRINGER promoted excellence in both lower and higher education within the State.

The scientific side of FRANK SPRINGER's career covers a span of approximately 63 years, running concurrently with his interest in the law and business activities, for it began early in his university student days and constantly was augmented up to the time of his death at age 79. During successive summers before his graduation from the University, the Iowa State Geologist, Dr. C. A. WHITE, and his assistant, ORESTES ST. JOHN, both paleontologists, left him in charge of the Survey's fossil collections and library while they were engaged in field work. Young FRANK used this opportunity to study the collections ardently and to search fossiliferous Devonian and Mississippian outcrops near Iowa City for specimens discovered by himself. When he moved to Burlington for the beginning of law practice, he continued paleontological work by making firsthand acquaintance with the crinoid-rich Burlington Limestone. Also, he met CHARLES WACHSMUTH, who had already made large collections of crinoids from the area. Together they undertook to compile information concerning all described fossil crinoids, especially from Paleozoic deposits of North America and Europe. This led to their first major written contribution entitled *Revision of the Palaeocrinoidea,* published in Proceedings of the Philadelphia Academy of Natural History (1880-86), which contained a comprehensive new classification of then-known species, genera, and families of these fossils.

In vacation periods after SPRINGER took up residence in New Mexico, he returned to Iowa for the purpose of continuing crinoid studies with WACHSMUTH. A very major result of this collaboration was publication in 1897 of the outstanding three-volume royal quarto monograph on *North American Crinoidea Camerata* (Harvard University). This is a landmark addition to paleontology which WACHSMUTH did not live to see, for he died in 1896. Its very many handsome illustrations consist of drawings made by CHARLES R. KEYES, a fellow student of SPRINGER at Iowa, who became State Geologist of Missouri in 1894-96.

Besides several large and important articles by SPRINGER on fossil crinoids, such as *Uintacrinus, its Structure and Relations* (1901), *The Crinoid Fauna of the Knobstone Formation* [Lower Mississippian of Kentucky] (1911), *Some New American Fossil Crinoids* (1911), *On a Trenton* [Middle Ordovician] *Echinoderm Fauna at Kirkfield, Ontario* (1911), *The Fossil Crinoid Dolatocrinus and its Allies* (1921), and *Unusual Forms of Fossil Crinoids* (1926), as well as the section on Crinoidea in ZITTEL-EASTMAN's *Textbook of Paleontology* (1913), he produced two large monographs published by the Smithsonian Institution. These are *The Crinoidea Flexibilia* (2 volumes, 1920), and *American Silurian Crinoids* (1926), both of them comprehensive and authoritative contributions of great value.

The crinoids gathered by WACHSMUTH prior to 1874, when a program of joint studies with SPRINGER was arranged, were sold in that year to Professor LOUIS AGASSIZ for the Museum of Comparative Zoology at Harvard University. Collections made by SPRINGER then were transferred to the fireproof brick-built laboratory located in the back yard of WACHSMUTH's home in Burlington and libraries of the two men were consolidated there. In 1911, SPRINGER gave both the collection and library to the Smithsonian Institution in Washington for display and further research in the United States National Museum. The collection of approximately 100,000 specimens, which included very many types and other illustrated, specially studied specimens, weighed more than 12 tons and may be valued conservatively at $250,000. SPRINGER continued to augment the collections by purchases of fossil crinoids, especially from European localities, and by intensive collecting carried on by assistants working in many North American regions. The latter effort included the opening of quarries for the sole purpose of finding well-preserved crinoids, many consisting of complete crowns with attached stems, some with their distal holdfasts.

SPRINGER was named as Associate in Paleontology at Harvard University in 1901 and given the same title at the U.S. National Museum in 1914. He was chosen Foreign Correspondent of the Geological Society of London in 1906. Also, he was given an honorary Sc.D. degree by George Washington University in 1921 and a Ph.D. degree by the Universität Bonn, Germany, in 1924.

It is reasonable to judge that no one in the world excels FRANK SPRINGER in adding to knowledge of fossil crinoids. He died in 1927 at the home of one of his daughters in Overbrook, Pennsylvania, a suburb of Philadelphia.

FRANCIS ARTHUR BATHER

A world leader in the study of fossil echinoderms, especially crinoids, was FRANCIS ARTHUR BATHER, whose entire career was based on his association with the British Museum (Natural History) in London. Born at Winchester, England, in 1863, he was educated in the Public School there and at New College, Oxford University, where he earned a bachelor's degree with first class honors in natural history (1886) and an A.M. degree (1890). His first scientific paper (on Jurassic rocks of Oxfordshire) was published (1886) in the *Journal of the Geological Society of London* and in 1887 he was chosen for curatorial work in geology and paleontology at the British Museum. The prolific nature of BATHER's pen was evidenced by the appearance of 16 publications by him in the four-year period 1887-90. Indeed, this pace was increased since the 48 years (1886-1934) of his entire career are represented by a total of 213 large and small contributions. Most important of these are his monographs on *The Crinoidea of Gotland* (Kongl. Svenska Vetenskapsakad., Handl., v. 25, 180 p., 10 pl., 1893), *Triassic Echinoderms of Bakony* [Hungary] (Resultate d. Wissenschaftl. Erforsch. d. Balatonsees, v. 1, pt. 1, Paläont. Anhang, 280 p., 18 pl., 1909), and *Caradocian Cystidea from Girvan* [Scotland] (Roy. Soc. Edinburgh, Trans., v. 49, pt. 2, p. 359-529, pl. 1-6, 1913). Also of special interest are BATHER's *Phylogenetic Classification of the Pelmatozoa* (Rept. British Assoc. 1898, p. 916-923, 1899) and *The Echinoderma* (E. R. Lankaster's Treatise on Zoology, viii + 344 p., 309 fig., Black, London, 1900).

BATHER received the Sc.D. degree *(honoris causa)* from Oxford University in 1909. He became a Fellow of the Geological Society of London and of the Royal Society. He had been president of the Geological Society of London, the Museum Associates, and the British Association for the Advancement of Science, and a correspondent of the Geological Society of America. In addition, he was awarded the Rolleston Prize of the Universities of Cambridge and Oxford, for researches in biology and the Lyell Medal of the Geological Society of London.

BATHER died at his home in London on March 20, 1934, aged 71.

OTTO JAEKEL

A brilliant paleontologist who contributed innovatively to syntheses of knowledge concerning fossil crinoids was OTTO JAEKEL. Born on February 21, 1863, at Neusalz an der Oder, now known as Nowa Sól, in western Poland, he received training at an academy in nearby Liegnitz where he graduated in 1883 with intent to become a geologist. Studies then were undertaken at the Universität Berlin under FERDINAND ROEMER (1883-85) and Universität München under KARL VON ZITTEL, the most eminent paleontologist of Germany, who supervised completion of JAEKEL's work for the doctorate which he received in 1886. At once he was appointed as an assistant of E. W. BENECKE at the University of Strasburg where he continued to work (except for a lengthy visit to London) until in 1890 he was made a Professor of Geology at the Universität Berlin. He remained there until 1903 when he became Professor of Paleontology at the Universität Wien, but after only one year accepted a similar position at the Universität Greifswald in the northeastern part of present East Germany, not far from his birthplace. On his retirement from the faculty at age 65, he was invited to continue teaching by transference of his work to Sun-Yat Sen University in Peking, China. This he did in 1928. He failed to complete a single academic year, however, for after a sudden illness he died on March 6, 1929.

Although JAEKEL's primary interest in paleontology was focused on fossil vertebrates, 27 of his publications dealt with echinoderms, chiefly crinoids. Among these, the most important are *Beiträge zur Kenntnis der paläozoischen Crinoiden Deutschlands* (Paläont. Abhandl., no. 7, p. 1-116, 29 fig., 10 pl., 1895) and *Phylogenie und System der Pelmatozoen* (Paläont. Zeitschr., v. 3, p. 1-128, fig. 1-114, 1918). The latter contains a new suprageneric classification of crinoids accompanied by diagnoses and illustrations of many new species and genera. A complete list of JAEKEL's long and short papers has a total of 211 titles (OTHENIO ABEL, *Veröffentlichungen von Otto Jaekel,* Palaeobiologica, v. 2, p. 156-186, 1929).

JOHANNES WANNER

JOHANNES WANNER was born in the village of Scheidegg im Allgäu, southern Bavaria, on April 21, 1878, son of a high school teacher belonging to a well-known Allgäu family. His academic training included study as a pupil of the world-renowned paleontologist Professor KARL VON ZITTEL at the Universität München where he received his Ph.D. degree in 1901, based partly on a thesis describing Cretaceous fossils from Libya.

WANNER's work in subsequent years is divisible into two quite separate but partly complementary careers, one in petroleum geology and the other in paleontology. In 1902-05 he made his first visit to the East Indies for geological exploration wanted by the Dutch firm of Bataafsche Petroleum Maatschapij. He not only attained objectives for guiding search for oil deposits but made observations on several islands which advanced fundamentally knowledge of the geologic structure of the archipelago and its development. On his return to Germany, he was appointed in 1906 as chief assistant of Professor GUSTAV STEINMANN at the Universität Bonn and published (1907) an important paper entitled *Hebungen und Senkungen der Erdkruste*. He then embarked on a second trip to the East Indies for reconnaissance of petroleum-development possibilities. On Timor he discovered exceptionally rich Permian and Triassic fossil localities. In 1911 a Bonn University expedition to Timor for the purpose of collecting fossils included WANNER with two other geologists and a zoologist. Their efforts were rewarded very richly.

During World War I, WANNER thought that because of his deafness he would not be called for military duty and thus he could devote himself to work on Timor crinoids. Instead, he was designated as a military geologist and assigned to mapping work in Alsace.

After the war, WANNER was elected in 1920 as Haniel Distinguished Professor of Applied Geology at Bonn Universität with permanent tenure, but in practice he supervised studies in paleontology and regional geology.

A monument signalizing WANNER's scholarly work in the study of crinoids and other fossils is the series of monographs and long articles which he initiated in 1916 under the general title of *Paläontologie von Timor*. A host of new crinoid species, genera, and families is described and illustrated in this series (1916-29, 495 pages, 34 plates, 158 figures), in Netherlands publications (1920-40, 554 pages, 39 plates, 117 figures), and in *Palaeontographica* in the 1940's. In addition, several papers are devoted to Devonian crinoids of Germany. The entire list of WANNER's publications, chiefly on fossil crinoids, contains 83 titles.

In 1952 when I had the privilege of spending some days with him in Scheidegg, Professor Emeritus WANNER was in good health, but four years later on July 31, 1956, death overtook him.

INTRODUCTION

By RAYMOND C. MOORE and CURT TEICHERT
[University of Kansas]

Echinoderms belonging to the class Crinoidea (from Gr. *krinós*, lily) are among the most complexly organized, highly varied forms of all marine invertebrates. They are assigned to a subphylum variously named Crinozoa or Pelmatozoa. Both names are here regarded as synonyms, but preference of usage is given to the name Crinozoa to preserve continuity with previously established *Treatise* usage (see Part S).

Beyond an initial free-swimming larval stage, most fossil crinoids throughout life were attached by a stem to the sea bottom or, rarely, floating objects; however, most living species of crinoids are stemless. Only the Antedonidae are active swimmers, whereas the comatulids are sluggish crawlers. These are the feather stars, whereas stem-bearing types, because of fancied resemblance to flowering plants, are known as sea lilies. Actually, the stemless forms are also fixed by a stem during part of their very early ontogeny, but by breaking away from this anchorage, they become able to crawl or swim about freely.

The crinoids are distinguished especially by their general form and by the structure of their skeleton. A relatively small disk-shaped or globular body enclosed by an armor of symmetrically arranged calcareous plates, forming the cup or calyx, bears radially outspread food-gathering appendages, which generally are branched. These appendages, called arms, and the stem are composed of many calcareous segments, joined together in a manner generally permitting differential movement and providing a degree of flexibility.

Adult crinoids range in size from a few millimeters, including length of stem and arms, to 18 m. or more; one Cretaceous species had arms at least 120 cm. long. Modern stemless crinoids have arms as much as 25 cm. in length.

Crinoids have worldwide distribution in present-day oceans, not only in tropical and temperate belts but in frigid waters of the Arctic and Antarctic. They are found at depths ranging from a few meters below sea level to a depth of about 9,000 m., though most stalked crinoids live between 180 and 5,000 meters. The occurrence of fossil crinoids generally suggests moderately shallow water as the habitat preferred by most species. None are found in freshwater deposits.

The number of living species of crinoids is approximately 650, only 80 of which are stalked. To date, about 5,500 fossil species of crinoids have been described, representing more than 1,000 genera (N. Gary Lane, pers. commun., 1976).

Remains of crinoids are widely distributed in many rock formations ranging in age from Middle Cambrian to Cenozoic, although commonly the hard parts are more or less dissociated. At many places, sedimentary deposits ranging in thickness to 30 m. or more are composed largely of crinoidal debris, with or without fossils consisting of articulated crinoid hard parts. Study of these remains has paleontological importance because the variety of fossils is extraordinarily great and the stratigraphic range of most individual species is very short.

Mass occurrences of crinoids are well known in present seas where Clark (1915a) described the dredging of tens of thousands of individuals from single localities. Crinoid colonies, patches, and stands, sometimes also referred to as "gardens" or "meadows" have been described by Lane (1973) and earlier workers. Rocks that consist predominantly of crinoid remains have been called "criquinite" (Bissell & Chilingar, 1967).

The distribution of crinoid limestones through time has been described for central and western Europe by Deecke (1915). Crinoids occur as rock builders in many places from Ordovician to Jurassic time. Best known perhaps is the *Trochitenkalk* of the German Muschelkalk (Martin Schmidt, 1928). An excellent and detailed summary and annotated bibliography of crinoidal limestones, mainly from the Mississippian of the Midcontinent, was given by Laudon (1957). Useful general discussions of formation and distribution of crinoidal limestones have been published by Cayeux (1931) and Johnson (1951). In places, crinoidal fragments have accumulated in biohermal masses, especially in Mississippian time (Harbaugh, 1957; Carozzi & Soderman, 1962).

Crinoids have been persistent reef dwellers since at least Late Ordovician time, the most notable associations of this kind being known from the Silurian of central North America and the Baltic region of northern Europe (Gotland, Saaremaa), and the Devonian of central Europe (Hadding, 1933, 1941; Lowenstam, 1948; Ingles, 1963; Lane, 1971; Ruhrmann, 1971). Crinoids lived in vast numbers not only on these reefs but also in the offreef and interreef environments (Manten, 1970).

Crinoids are important indicators of depositional environments (Cain, 1968) and their generally long, slender stems are excellently suited to serve as paleocurrent indicators (Klähn, 1929; Wiman, 1933; Schwarzacher, 1963; Anderson, 1968).

The biostratonomy of recent and fossil crinoid associations has been studied by Teichert (1949), Carozzi & Soderman (1962), Ruhrmann (1971), Lane (1973), Liddell (1975), and others. Preservation of entire cups of crinoids, other than camerates, indicates either very rapid burial or sedimentation in very quiet water.

Progressive modifications of crinoid structures during geologic time provide rich material for research on the nature of evolutionary trends. A point deserving stress is that completeness of fossil crinoid specimens is not a measure of their worth to paleontology, because a surprisingly large volume of precise and useful knowledge can be gained from the study of crinoid fragments (MOORE, JEFFORDS, & MILLER, 1968; MOORE & JEFFORDS, 1968).

Because of their characteristic microstructure and optical properties, crinoid fragments are generally easy to recognize in thin sections under the microscope (CAYEUX, 1931; HOROWITZ & POTTER, 1971).

Finally, crinoidal limestones have been found to be of commercial interest as reservoir rocks for oil and gas (IMBT & MCCOLLUM, 1950).

GENERAL FEATURES OF CRINOIDEA

By GEORGES UBAGHS, ALBERT BREIMER, N. GARY LANE, H. WIENBERG RASMUSSEN, R. C. MOORE, J. C. BROWER, HERTHA SIEVERTS-DORECK, H. L. STRIMPLE, D. B. MACURDA, JR., D. L. MEYER, and MICHEL ROUX

GENERAL MORPHOLOGY

RECENT CRINOIDS[1]

By ALBERT BREIMER
[Instituut voor Aardwetenschappen der Vrije Universiteit, Amsterdam, The Netherlands]

CONTENTS

INTRODUCTION

Much of our knowledge of recent Crinoidea is based upon specialized forms generally known as the feather stars and systematically grouped as the Comatulida. This group of crinoids, which is still flourishing, is regarded by paleontologists as not very typical, because they are vagile, stalkless crinoids, lacking a complete armoring

[1] The genus *Neocrinus* as used in this chapter is elsewhere placed in synonymy of *Chladocrinus*, except for the species *N. blakei* which is placed in *Isocrinus* (see p. T851, T857).—Eds.

of tegmen and ambulacra and in some (e.g., Comasteridae) having an excentric mouth and only partly developed ambulacra.

As a matter of fact, in zoological literature the comatulids are considered to be typical crinoids. This is not surprising, taking account of their occurrence in areas in which scientific zoology for almost 150 years could deal with them. They form the bulk of the extant crinoids, which partly live at moderate depths and are regularly caught at marine biological stations, biological expeditions, and even on fishermens' lines. Consequently they are readily available in zoological collections. As a result, the species *Antedon bifida* has developed in literature (Cuénot, 1948) as a model crinoid for zoologists.

In contrast, paleontological literature is almost exclusively devoted to sessile, stalked crinoids, generally known as sea lilies, which to the mind of paleontologists are better regarded as typical crinoids. They almost invariably have a complete armoring of the tegmen and ambulacra, a centrally located mouth, and fully developed ambulacra. However, extant stalked crinoids have the handicap of being rare organisms living mostly at appreciable depth, probably in a state of approaching extinction, and, almost exclusively, available specimens have been collected by zoological expeditions. Only a very few species inhabit shallow water. One of these is *Neocrinus decorus* in Caribbean waters. This species is fairly well represented in zoological collections and may occasionally be caught at marine biological stations. It has long been recorded in zoological literature and even has been the subject of detailed anatomical and systematic studies. For use of paleontologists, *N. decorus* could very well serve as a model crinoid.

The aim of this chapter is to present information on the anatomy and morphology of recent crinoids with special emphasis on stalked forms belonging to the Isocrinida, Millericrinida, Bourgueticrinida, and Cyrtocrinida, which best may serve for comparison with similar fossil forms.

The section on anatomy is largely based on literature data, taken from older works on the subject. More recent and full accounts on anatomy have been published by Cuénot (1948) and Hyman (1955). Morphological descriptions are largely based on my own observations. In preparing the text I have been able to study important collections of recent crinoids (Clark collection, U.S. National Museum, Washington; Agassiz collection, Museum of Comparative Zoology, Harvard University, Cambridge, Mass.; and the Challenger collection, British Museum of Natural History, London). Financial support for this study was received from the Netherlands Organization for the Advancement of Pure Research (Z.W.O.) in The Hague, The Netherlands.

Acknowledgments are due to N. Gary Lane (University of Indiana) and H. Wienberg Rasmussen (Universitet København, Denmark), who critically read the manuscript and made many valuable comments.

New drawings in this chapter are *camera lucida* drawings made by the author *ad natura delicta.* A. Heine (Amsterdam) prepared the drawings for final publication.

ANATOMY

BODY WALL

The body cavity in crinoids is enclosed within a firm, more or less leathery body wall (Fig. 1), which is usually heavily calcified in many parts. Only few openings perforate this wall: mouth and anus as openings for the digestive tract and hydropores or ciliated funnels in the interambulacral parts of the tegmen, as openings to serve the water-vascular system.

In parts of the body wall that are not heavily calcified, this wall consists only of a thin outer **epithelium** of one cell layer in thickness, a layer of **connective tissue,** which gives the wall most of its substance and support, and a thin, inner **coelomic endothelium,** which lines the body cavity. This condition is found in naked tegmens, in lateral parts of the theca between the arms, and at the oral surfaces of arms and pinnules outside the ambulacra.

Most parts of the body wall are heavily calcified, however. When skeletal elements

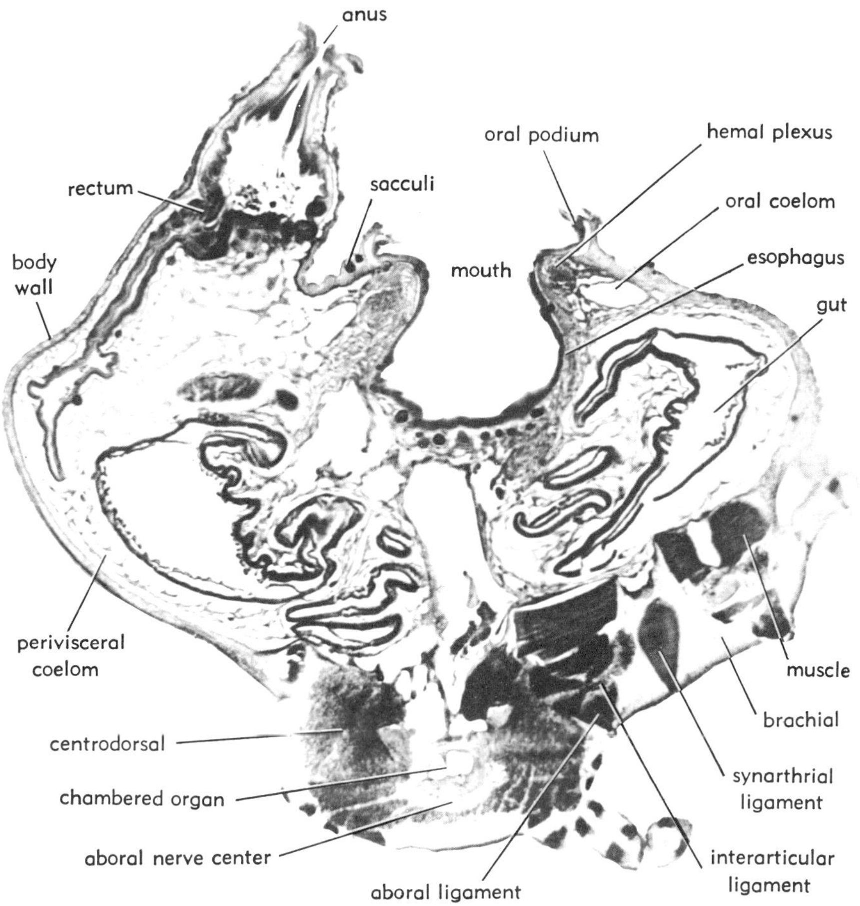

FIG. 1. Vertical sagittal section through theca of *Antedon bifida* [from preparations in the British Museum (Natural History), London] (Breimer, n).

are present, they have partially or entirely lost their epithelial covering over their aboral surfaces. If at all preserved, the epidermis is only a delicate cuticle covering these elements and difficulty separable from the underlying mesenchyme.

Highly differentiated and ciliated epithelia occur in the ambulacra and in the ciliated funnels or hydropores. In the ambulacra a firm and thick **ambulacral epidermis** is formed of very tall, slender cells which serve for support and highly attenuated sensory cells. The ambulacral epidermis has the nuclei of its component cells arranged at several different levels. The hydropores or ciliated funnels in the tegmen possess an epithelium composed of slender ciliated cells.

By far the most important part of the body wall is formed by mesenchyme cells, which may be differentiated into connective tissue, muscles, and elastic fibers (ligaments) or skeleton-secreting cells. Formation of the echinoderm skeleton has recently been studied by OKAZAKI (1960) from sea urchin larvae and by HEATFIELD (1971) from echinoid spines. It is thought that skeletal growth occurs in crinoids in a similar way.

First a rod or spicule of calcite is formed within a skeleton-secreting mesenchyme cell and later this calcitic particle comes to occupy nearly the whole cell. The cell then divides and starts to form a syncytial cell mass, which allows the contained skeletal element to grow. It does not grow as a solid block, for growth increments are added preferentially in many different directions, so that a crystal meshwork is formed. The syncytial cell mass containing the nuclei occupies the spaces within and around the calcareous meshwork. This mode of growth allows the skeletal element ultimately to assume any shape needed. Each separate element forms a single crystal.

BODY CAVITY

During individual development of the crinoid the body cavity arises from the larval left and right somatocoels, which—after rotation upon metamorphosis—take an adoral and aboral position, respectively. In the juvenile crinoid the adoral and aboral somatocoels meet along an equatorial plane and form a horizontal septum. Both somatocoels are crescentic in shape, with blind ends meeting at the posterior side, where they form the adoral and aboral vertical mesenteries, respectively. The aboral somatocoel is by far the largest of the two and surrounds the larval enteric sac. During postmetamorphic growth stages it retains this position and grows out as a perivisceral coelom. The smaller adoral somatocoel is situated around the adoral surface of the enteric sac and directly underlies the hydrocoel, to which it closely corresponds in form and position. During the postmetamorphic growth stages it follows growth patterns essentially similar to those of the hydrocoel and ambulacra. It grows out until it definitely takes a subambulacral position in the mature crinoid.

In the mature crinoid two major parts of the body cavity can be distinguished, namely an **adoral or subambulacral** coelomic compartment and an **aboral or perivisceral** coelomic compartment. The total body cavity in the theca is occupied for the most part by the extremely voluminous digestive tract. The aboral coelom is in contact with the aboral body wall and touches some interambulacral spaces of the oral or tegminal body wall. The hydropores or ciliated funnels in the interambulacral parts of the tegmen and lateral body wall open into the aboral coelom. In general, the aboral coelom surrounds the intestine (not the esophagus) and rectum. It is lined with coelomic endothelium and is filled in with numerous

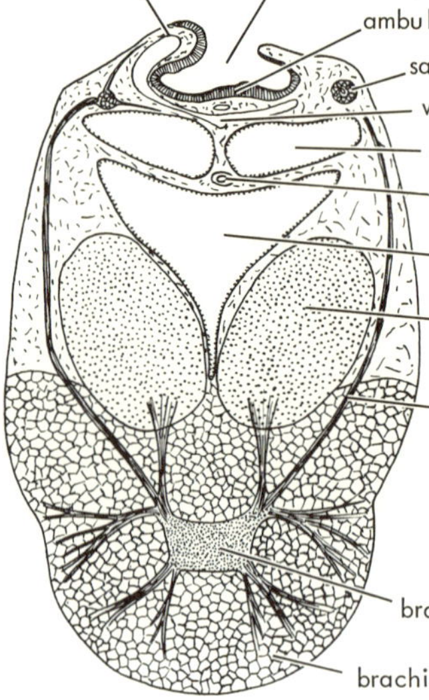

FIG. 2. Transverse section through arm of crinoid (after Hyman, 1955, and Hamann, 1889; from *The Invertebrates* by Libbie H. Hyman, copyright 1955, McGraw-Hill Book Company. Used with permission of McGraw-Hill Book Company).

strands and webs of connective tissue, thus giving it a spongy appearance. It is easily recognizable in anatomical sections. The aboral coelom continues into the arms (Fig. 2) and pinnules as **aboral coelomic canals.** In the pinnules the aboral walls of these canals are provided with typical **ciliated pits,** formed by differentiated endothelial cells, apparently for provisions of some movement in the coelomic fluid or for passage of coelomocytes.

The adoral coelom is distinguishable from the aboral one by its lack of spongy appearance. Some filaments of connective tissue do occur in it, but it is essentially an uninterrupted and unobstructed open space. It is separated from the aboral coelom by membranes of endothelium and connective tissue in which different organ systems (axial, genital, hemal) are developed. These membranes are apparently developed from the larval horizontal membrane between adoral and aboral somatocoels. They do not form a perfect separation of the two major coelomic compartments, for many pores and funnels interconnect the compartments. The adoral coelom has a large central or axial space at the left and anterior sides of the esophagus, with an extension to the central axis of the body cavity, just behind the axial organ. This aboral extension of the adoral coelom has erroneously been described by some authors as a separate coelomic compartment known as the axial sinus. From the main central space of the adoral coelom five large **adoral coelomic** canals are formed, everywhere underlying the water-vascular and ambulacral systems. The coelomic compartment is here called subambulacral (subtentacular in older literature) for this reason. The adoral coelomic canals continue in arms and pinnules. In the arm portions the canals are paired by a vertical septum. This septum is not present in the tegminal and pinnular parts of the adoral coelomic canals.

A minor coelomic canal is present between the radial water vessels and the ambulacra, directly beneath the epithelial nerve tract of the ambulacra. It is known as the **hyponeural sinus.**

Another compartment, also of coelomic nature, known as the **chambered organ,** is found in the central axis of the calyx and in the stalk. It is lodged in a rosette-like structure formed by the basal or infrabasal circlets and is composed of five chambers, which are radially disposed in dicyclic crinoids, but interradially in monocyclic forms. The chambers are formed by mesenteries and adorally end blindly. There is a central strand from which the axial gland arises. Aborally the chambered organ continues into the axial canal of the stalk as a central tube with five compartments around the central cord. The compartments are continuous with the divisions of the chambered organ. At each nodal columnal cirrus canals branch off from the axial canal of the column and extend to tips of the cirri. The chambered organ ontogenetically develops as an evagination of the aboral coelom, from which it becomes entirely separated.

AMBULACRAL AND DIGESTIVE SYSTEMS

The functions of food gathering and food conveying are carried out by the ambulacra and podia distributed along them. In extant stalked crinoids ambulacra are present on all of the pinnules, on the arms, and on the tegmen. In certain comatulid species, however, the pinnules are specialized into groups distinguished as oral pinnules (with tactile function), genital pinnules, and distal pinnules. In such forms ambulacral grooves are absent on the oral pinnules and rudimentary on the genital pinnules; only the distal pinnules have well-developed ambulacra and podia. Several comasterid species entirely lack ambulacra in some of the arms, especially those belonging to the *C* and *D* rays.

The ambulacra are typically formed by a differentiated epithelium (Fig. 2), here spoken of as **ambulacral epidermis,** which is the only well-developed epithelium in the crinoid body wall. It consists of two sorts of cells: 1) very tall and slender ciliated cells, which serve for support, and 2) highly attenuated sensory cells, almost hairlike in form with a bulging nucleus. The ambulacral epidermis is fairly thick and has the nuclei of its component cells arranged at several different levels. The epidermis contains numerous mucus-gland cells, which secrete mucus as a conveying medium for the food particles. The ambulacral epidermis is everywhere underlain by a thick nervous layer to which the sensory cells of the epidermis connect.

The ambulacra are everywhere accompanied by the podia (Fig. 2), placed in triads along both sides of the ambulacral grooves. The typical ambulacral epidermis and underlying nerve tract pass laterally onto the adambulacral sides of the podia, which in this way have a thicker epithelium on the inner side than on the outer. Groups of four or five tall attenuated cells project beyond the podial wall to form papillae with presumably a combined glandular and sensory function. Some of its cells are gland cells, secreting mucus as an adhesive for food particles which happen to come to contact with the podia. Nerve and muscle fibrils are also present in the papillae.

The podia or tube feet are arranged in triads, each individual podium of a group differing in length from the others. The longest tube feet are the main food-gathering organs. They possess muscle cells on their adoral sides and are thus able to bend in toward the ambulacrum and pass food particles to the smaller tube feet. The tube

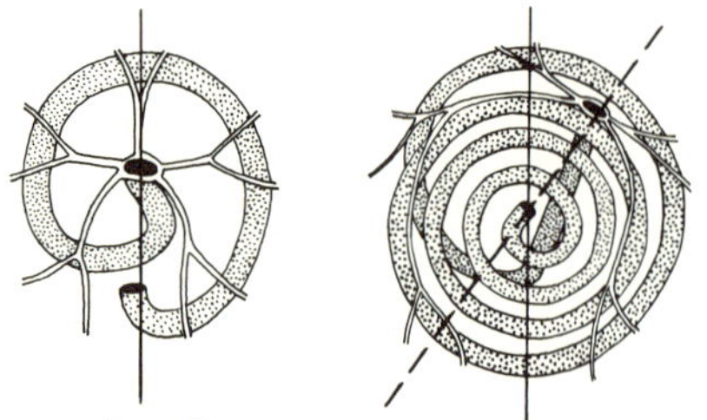

FIG. 3. Digestive systems in crinoids; examples of endocyclic and exocyclic conditions (after Ubaghs, 1953, and Carpenter, 1884a). [Vertical line indicates normal crinoid plane of symmetry; broken line indicates secondary axis.]

feet of intermediate length have muscle cells on their adoral and aboral sides and thus are able to move toward the ambulacra or away from it, receiving food particles from the longer podia and sticking them into the ambulacral groove. The smallest podia have muscles all around their walls and are capable of lateral movement in any direction. They probably serve for conveying food along the ambulacra. The muscle cells in the tube feet are innervated by branches from the hyponeural lateral nerves in the arms and pinnules. The podia embody the terminal branches of the water-vascular system. It is logical to suppose that erection of the podia (which occurs as feeding starts) is brought about by regulation of the water pressure in the water vessels. NICHOLS (1960, 1966) has described the possible mechanism that could be responsible for water-pressure regulation in the water-vascular system (see section on water-vascular system).

The mouth of crinoids usually is placed at the center of the tegmen, this position being defined as **endocyclic** (Fig. 3). In some comasterids, however, the mouth is displaced to a position in the *AB* interray and location of the mouth then is called **exocyclic.** In this condition the anus is placed at the center of the tegmen.

The mouth leads to a long or short **esophagus**, which runs obliquely in an aboral direction under the *C* ambulacrum. The ambulacral epithelium and the underlying nervous layer of the ectoneural system continue down the esophagus (Fig. 4). In this way the esophageal and intestinal epithelia consist of tall attenuated cells, which are ciliated everywhere in the intestine, except in the rectum. The esophagus is sheathed by a nervous layer and by well-developed circular muscle fibers forming a sphincter around the esophagus. The same is true for the rectum.

In endocyclic forms the esophagus is followed by a voluminous **intestine** of larger diameter than the esophagus, which in part is lobed and plicated. It describes one complete spiral volution clockwise if viewed from the adoral side. The terminal part of the intestine is vertical and forms the **rectum,** which opens through the anus on the tegmen. The intestine has several branched **diverticula.**

In exocyclic forms the digestive tract describes up to four spiral windings or

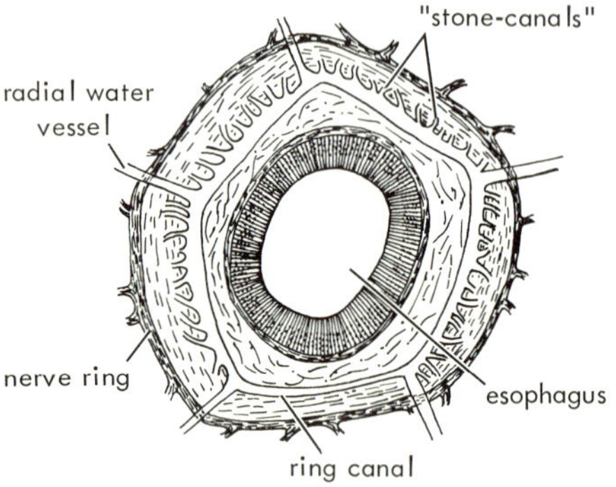

FIG. 4. Ring canal of water-vascular system with "stone-canals" along four sides (after Hamann, 1889).

volutions. Such a digestive tract has the same diameter as the esophagus, is not lobed or plicated and has no diverticula.

The anus may be situated on a special elevation of the tegmen, known as the **anal tube.** This is (frequently reported to be) engaged in rhythmic pulsations by which water is intermittently brought into the rectum and again ejected. The interpretation of this is probably twofold: by water movements the elimination of fecal pellets probably is aided and at the same time a kind of rectal respiration is carried out.

WATER-VASCULAR SYSTEM

In adult extant crinoids the water-vascular system is internally discontinuous because of lack of a direct connection between its two component parts: a system of hydropores, mainly located in the tegmen and a system of definite canals everywhere accompanying the ambulacral grooves and podia.

The distribution of **hydropores** in various genera may be different. In primitive forms such as *Rhizocrinus lofotensis* the juvenile condition of five hydropores persists during maturity, one in each interray, that located in the *CD* interray apparently being the original hydropore formed by the axocoel. In this genus the five hydropores are connected with the ring canal by five stone canals.

In stalked as well as nonstalked extant crinoids large numbers of hydropores occur in the four normal interrays and five interbrachial areas (see Fig. 33). Hydropores are positively absent in the *CD* interray in *Neocrinus, Endoxocrinus* and *Calamocrinus*. They may be present in narrow zones bordering the posterior side of the *C* and *D* ambulacra in stalked crinoids (e.g., *Saracrinus*), and are reported to occur in the anal tube and arm bases of comatulid genera. The hydropores penetrate the tegminal wall and are lined with a heavily ciliated and very vibratile epithelium (Fig. 5). The hydropores open into the aboral peri-intestinal coelomic compartment.

The central element of the water vascular system is the **ring canal** (Fig. 4). It is circumesophageal in position and directly underlies the tegmen. In perradial position five main **radial canals** originate from it. Also, many small canals or tubes are given off from the ring canal and on four of the interradial sides many (up to 30) so-called **stone canals** emerge. These open into the axial space of the adoral coelomic compartment. In the accompanying illustration (Fig. 4) attention may be called to the lack of stone canals along one interradial side, which presumably is the *CD* interray. Absence of stone canals in one interray coincides with absence of hydropores in the *CD* interray. No correlation exists between the number of stone canals and the number of hydropores in each interray, the latter being far more numerous. The number of hydropores is variable in adult crinoids, ranging from approximately 500 to 1,500. Apparently in many extant stalked crinoids, after reduction of the original posterior

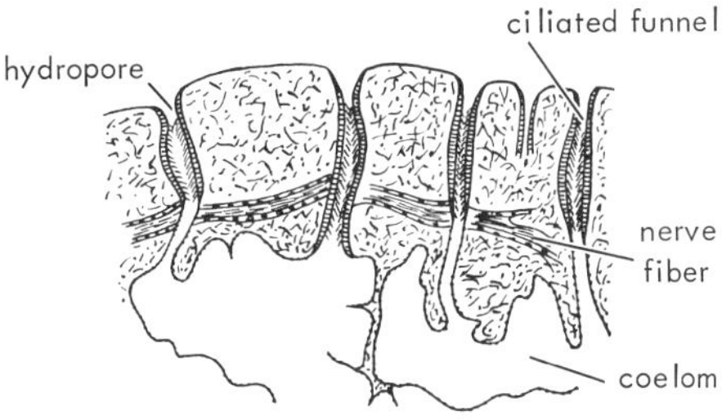

FIG. 5. Section through adoral body wall with ciliated water canals (after Hamann, 1889).

stone canal no additional stone canals and hydropores are evaginated from the posterior side of the ring canal, nor are additional hydropores formed in the posterior tegminal wall.

Adorally, the ring canal gives off many delicate tubes leading into the labial podia surrounding the ambulacral depression around the mouth. Five main **radial** or **brachial water canals** diverge from the ring canal. These canals penetrate into the arms and pinnules, closely accompanying the ambulacra which they immediately underlie, only separated from them by the epithelial ambulacral nerve tract. Where ambulacra are tegminal, the water canals occur immediately beneath the tegmen, following the ambulacral tract. In each arm the brachial water canal has a similar subambulacral position and is situated between the ambulacral groove and the two coelomic canals. The brachial water canal has branches to each of the podia bordering the ambulacral groove, and to each of the pinnules in alternating order. From the pinnular water canal terminal branches lead to the groups of three podia, which are placed in alternating order along the pinnular ambulacral groove. As the number of podial groups is generally the same as the number of side plates or lappets on each pinnular and as side plate numbers per pinnular may vary, individual pinnulars may bear different numbers of podia. Consequently, on each pinnular various terminal water canals are given off to the podial groups in alternating order. Each group of three podia receives one terminal water canal, which branches to become the lumen of each podium.

The wall of the water-vascular canals is made up of coelomic endothelium and some connective tissue. Circular muscles do not occur in their walls but some longitudinal muscle cells are present. Muscle cells also are found to traverse the lumen of the water canal obliquely. The wall of the water canals may be described as weakly musculated. Apparently it is not capable of much contraction of the canals.

The function of the water-vascular system in crinoids is still subject to some uncertainty. Recently, NICHOLS (1960, 1966) has made new studies on this subject. As the podia or tube feet are capable of contraction by muscular activity, another mechanism is needed as an antagonist to protract them. In echinoderms this mechanism is generally present in the form of a regulation system that controls the water pressure in the tube feet. The radial water canals of crinoids are compartmented for this purpose, and each compartment may be contracted by the muscles traversing the lumen of the compartment. Contraction of the compartment sends water into the tube feet, and makes them protract.

HEMAL SYSTEM

The hemal system in crinoids is not formed by definite canals, but by many intercommunicating spaces in the filaments and mesenteries of connective tissue which are present in the body cavity. The system is therefore best described as a **blood-lacunar system.** It is only known from anatomical studies. The blood substance is a colorless liquid rich in proteins, which in preparation coagulates and hence becomes visible. It is reported not to possess any typical cells other than coelomocytes.

Surrounding the esophagus, not far from the ring canal, is a **periesophageal plexus** in the form of an irregular network of anastomosing lacunae. From this plexus many branches pass aborally to connect with the intestinal wall. These are absorbing lacunae and are especially frequent along the inner wall of the first half of the digestive tract. The blood lacunae have an intimate relation to the intestinal wall. The strands and webs of connective tissue in which the blood substance circulates is continuous with the connective tissue of the intestinal wall. The same is true for the endothelial covering. The periesophageal plexus sends branches to another circular plexus of blood lacunae, which is situated immediately beneath the tegmen. This plexus is called the **subtegminal plexus.** Five radially disposed hemal canals are given off from the subtegminal plexus which continue into the arms. The hemal canal is situated in the center of the arm between the two adoral coelomic canals and the single aboral coelomic canal. This canal lodges the genital cord and is so called the genital tube. It is the only part of the hemal system in the arms.

In the posterior (*CD*) interray the periesophageal plexus is connected with the **spongy organ,** which consists of lacunae with much thickened walls, while the lacunar cavities are filled in with many rounded cells, leaving hardly any open space. This organ extends aborally and has an intimate relation with the axial organ, to which it is connected by a network of blood lacunae. In *Neocrinus decorus* it extends aborally all along the posterior side of the axial organ, but in comatulids it covers only the oral part of the gland. In older literature the spongy organ is interpreted as a labial plexus, but its status as an ordinary hemal plexus is doubtful. CUÉNOT interpreted it as a lymphoid organ. The periesophageal plexus also has a direct connection with the axial gland from which it receives a mesh of efferent blood lacunae.

HOLLAND (1970) described the fine structure of the axial organ and the extracellular hemal fluid from *Nemaster*. The gland cells of the glandular tubules of the axial organ have the characteristic fine structure of protein exporting cells and may produce granular and filamentous components of the hemal fluid.

In the literature (CUÉNOT, 1948; NICHOLS, 1962), part of the coelomic canal system is interpreted as a perihemal system. The first to draw attention to the presence of a perihemal system in crinoids was HAMANN (1889). In his view the small coelomic canal between the epithelial nerve band of the ambulacra and the radial water canal is a schizocoelous canal, which he compared to the perihemal canals of asteroids. This canal, termed hyponeural sinus by HYMAN (1955), in rare cases has been observed to continue into a subtegminal ring canal. Later, CUÉNOT (1948) drew further at-

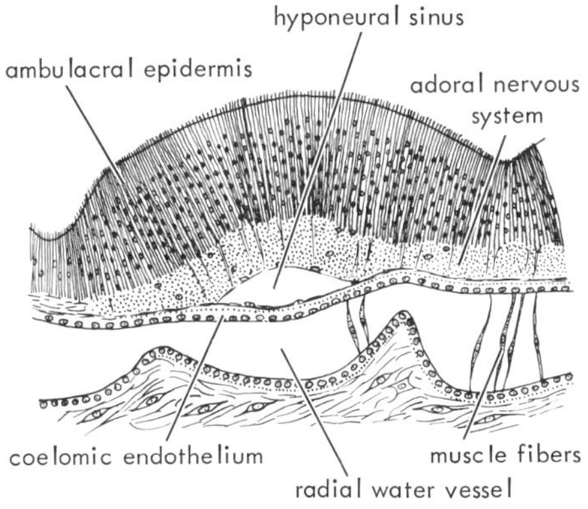

FIG. 6. Transverse section through ambulacrum (after Hamann, 1889).

tention to possible perihemal structures in crinoids. He interpreted the sheath surrounding the genital canal in the arms as a perihemal sheath. He also stated that all the organs traversing the body cavity as the digestive tract, the water-vascular system and the blood-lacunar system are ensheathed. NICHOLS (1962) interpreted these sheaths as perihemal lacunae.

NERVOUS SYSTEM

The nervous system in crinoids is complex and divisible according to its ontogenetic development, to its position in the adult animal, and to its function.

According to ontogenetic development, two different nervous systems can be distinguished. The first is of ectodermal origin, developed in an early larval stage as a thickening of the ectoderm along the course of the hydrocoel and podia. In juvenile and adult crinoids this nervous system is adorally disposed. It is known as the **adoral** or **ectoneural system.** This superficial nervous system is present as a band of nerve cells and longitudinal fibers just beneath the strongly differentiated epidermis of the ambulacral grooves (Fig. 6). It continues to the internal sides of the podia. It is in contact with the sensory cells in the ambulacral epidermis and the sensory papillae on the podia. The five ectoneural bands coming from the primary arms proceed on the aboral side of the tegmen and meet around the mouth. They continue downward as an epithelial nerve sheath around the esophagus. In contrast to other echinoderm groups, the ectoneural system is weakly developed in crinoids.

The second nervous system is of mesodermal origin and probably appears very late in ontogeny, as larval stages possess only an ectoneural nervous system. The ontogenetic development of the mesodermal nervous system is poorly known. According to its position in adult crinoids the mesodermal nervous system is divisible into two parts: 1) an **adoral** or **hyponeural system** and 2) an **aboral** or **entoneural system,** which are strongly intercommunicating.

The hyponeural system is located in the connective tissues of the tegmen and arms. Its center is a pentagonal circumesophageal nerve ring in connective tissues of the tegmen. The nerve ring is situated lateral to the circumesophageal rings of other organ systems. From the nerve ring some smaller nerves run to the labial podia on the tegmen. Larger nerves are given off directly from the ring to serve the internal organs, among which are two prominent nerves leading to the anal tube. Ten main nerve branches depart from the nerve ring to the arm bases. They continue in each arm as two longitudinal nerves, located laterally in the connective tissues of the adoral surface of arms and pinnules. These longitudinal nerves innervate the musculature of the wall of the radial water canal and the external sides of the podia.

The entoneural system (Fig. 7) has its center in a compact, cup-shaped, ganglionated nerve mass in the central axis of the body cavity, embedding the base of the chambered organ. A sheath of nervous tissue descends from this mass into the stem surrounding each of the compartments of the axial canal and the coelomic cirral canals. From the aboral cuplike nerve mass ten small trunks are given off, which soon unite in five radial trunks. The five radial trunks are interconnected by a pentagonal commissure lodged in the calcareous bodies of the radials. Five main brachial nerve trunks depart from the edges of the pentagonal commissure, leading into the arms. In each axillary plate this aboral nerve trunk bifurcates. The two trunks produced by the bifurcation are interconnected by one

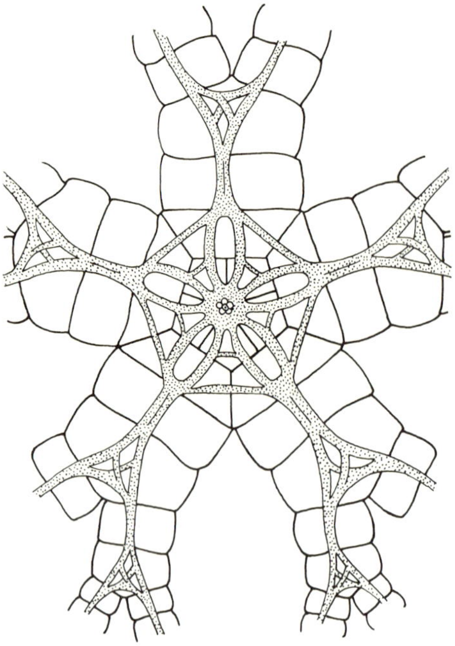

FIG. 7. Aboral nervous system in *Neocrinus decorus* (after Reichensperger, 1905).

transverse commissure and two crossing commissures. The aboral brachial nerve trunk traverses the skeletal elements (brachials) through a nerve canal. This canal has been formed during growth by lateral ingrowth of skeletal material of a deep groove which originally housed part of the coelomic canals of the arm. The aboral brachial nerve trunk has a thickening in the center of each brachial. From this node nerve branches are given off to innervate the articular muscles. Other branches connect the aboral brachial nerve trunk with the orally disposed longitudinal nerves of the hyponeural system. In this way a firm interconnection joins the hyponeural and ectoneural systems.

The entoneural system in crinoids is by far the most important of the three. Experiments have shown that this system is to be regarded as the motor nervous system. The cup-shaped nerve mass in the central axis apparently controls and coordinates the movements of arms and pinnules, stem and cirri, necessary for swimming and feeding. If the aboral nerve center is damaged, movement of arms and cirri stops. On the contrary, if parts of the hyponeural or ectoneural systems are damaged, no such an effect is produced. A purely sensory function is attributed to the ectoneural system.

HOLLAND (1970) described the fine structure of all constitutent cell types in the axial organ of *Nemaster*. The neurons (perikarya and axons) of the axial organ may possibly be neurosecretory, since they are filled with electron-dense granules.

REPRODUCTIVE SYSTEM

The genital system in crinoids consists of a system of genital cords, located in genital tubes, which are sheathed by genital canals. This system is mainly located in the arms and pinnules. The **genital canal** is situated in the arms between the two adoral coelomic canals and the single aboral coelomic canal at the junction of the horizontal and vertical mesenteries. The genital canal is part of the perihemal system and has many interconnections with the coelomic canals. The genital canal contains a **genital tube** suspended in it by filaments of connective tissue. The genital tube is part of the hemal system and the only blood lacuna in the arms. In the axis of the genital tube is a **genital cord** composed of cells with large nuclei, identical with young genital cells. At the level of each pinnule the genital cord gives off a lateral branch, which has an enormous volume and constitutes a **gonad** or **genital cavity,** occupying almost all of the aboral coelomic space in the pinnule. It is probable that the cells of the genital cords multiply and displace themselves along the genital cords and only arrive at maturity in the gonads, which are rather storage compartments for mature sex cells. In a few crinoids the gonads are lodged in the arms (e.g., *Holopus*) and are exceptionally found in the tegmen, evidently along the tract of the genital cord. Toward the tegmen the course of the genital cords can be traced only with much difficulty. The genital canal meets the subtegminal plexus but the genital cords are ultimately lost in the meshes of this plexus.

In some comatulid species the gonads are situated in special **genital pinnules,** much inflated structures in breeding time, which more or less completely lose their ambulacra and podia, but in connection with this the

sterile distal pinnules have strongly developed ambulacra and podia. The gonads usually lack an opening to the external medium; only in *Notocrinus* is such an opening present in the testes at the base of the pinnules. Usually the sex cells escape by rupture of the pinnular wall and the eggs are stuck to the pinnules by the secretion of cement glands present as a longitudinal band of mucus-secreting cells on the external side of the pinnular wall. In some species rupture occurs through preformed spots on elevated areas of the pinnules. It is recorded that rupture of the ovarian wall is stimulated by the spawning products of male individuals, thus providing for fertilization immediately after rupture of the ovarian wall. Several Antarctic species are known to brood their eggs. In these species a **brood pouch** or **marsupium** is present on the base of the pinnules, and in them fertilized eggs develop to the pentacrinoid larval stage.

The ontogenetic development of the gonads in crinoids is insufficiently known. Russo (1902) has discovered in the larva of a comatulid the primary gonad, a cellular network situated in the *CD* interray in the horizontal mesentery close to the primary stone canal. Dawydoff (1948) has described a similar primary gonad in the aboral vertical mesentery. From these primary gonads sex cells migrate into the arms and later atrophy. Much speculation is found in the literature concerning the original primary gonad. As the axial gland also appears to arise in the aboral vertical mesentery, this gland is sometimes (Perrier, 1886-90) interpreted as a sterile genital stolon. Reichensperger (1905) has interpreted the spongy organ of *Neocrinus decorus* as a primary gonad. Speculation on the phylogenetic significance of a supposed single primary gonad in crinoids has spread among textbooks. This primary gonad is believed to be significant in interpretation of a single pore of the *CD* interray of some cystoids and primary crinoids (e.g., *Hybocystites, Hybocrinus, Porocrinus*). This pore is interpreted as a gonopore on the weak grounds of ontogenetic development of primary gonads in recent crinoids. However, a gonopore is not shown to exist in crinoid ontogeny.

MORPHOLOGY

GENERAL STATEMENT

Morphologically the crinoid organism is divisible into several major parts: the stalk or **column,** the **theca** which may be divided into an aboral cup or **calyx,** and an oral membrane or **tegmen,** and pinnule-bearing arms or **brachia.** The crinoid organism consists of a very voluminous body wall and only a relatively small body cavity. By far the largest part of the body cavity is lodged in the theca and only extensions of it consisting of small canals penetrate the stalk, arms, and pinnules. The bulk of the volume of the several internal organ systems is found in the central body cavity of the theca. The crinoid body wall is supported by a voluminous mesodermal endoskeleton, the form of which determines the total shape of the organism. Describing the crinoid form is practically the same as describing the form of its skeleton.

Traditionally the crinoid skeleton has been described as consisting of a primary skeleton, a secondary or perisomic skeleton, and a visceral skeleton. In this chapter, however, the crinoid skeleton is divided into an **aboral skeleton,** an **adoral** or **perisomic skeleton,** and a **visceral skeleton.**

1) The aboral skeleton comprises elements distinguished as **columnals, centrodorsal, infrabasals, basals, radials, brachials,** and **pinnulars.** Definition given in this way differs from that of the primary skeleton in that the orals are not included. Also, because adult recent crinoids lack an anal plate or plates, such elements are unmentioned. The new distinction is purely topologic, and not ontogenetic, as the traditional one. Topologic criteria derived from adult crinoid specimens are thought to be best suited for describing crinoid morphology. The aboral skeleton is by far the largest part of the total skeletal mass; as a rule it is always well developed. It is the main supporting skeleton and determines the overall body form.

2) The adoral or perisomic skeleton com-

prises the **orals, tegmen plates** or **scales, ambulacral** and **adambulacral plates,** and **interradial** and **interbrachial plates** or **scales.** Thus defined, the perisome is the adoral body wall, penetrating between the arm bases to complete the central thecal cavity in conjunction with the aboral skeleton. The adoral skeleton is not always well developed. In the Isocrinida the perisome is generally highly calcified and almost completely occupied by skeletal plates or scales, but in the Comatulida the perisome is reinforced only with incoherent calcareous spicules and rods and appears as a leathery body wall.

3) The visceral skeleton is formed by spicules and rods of calcite together with smaller calcareous meshworks found in the connective tissues around the visceral mass. Complete calcification of internal membranes to form structures such as the so-called convoluted organ in some fossil crinoids (e.g., *Teleiocrinus,* Miss., N. Am.) are unknown in extant crinoids.

For descriptive purposes we have distinguished between adoral and aboral sides of the crinoid organism. It is agreed with HYMAN (1955) that use of the terms dorsal and ventral for the description of crinoids is conjectural. The position of the mouth in crinoids is always very well observable. In a vast majority of the forms the mouth is located in the center of the tegmen. Hence the mouth is the starting point for considerations on orientation and symmetry.

In describing orientation and symmetry of the crinoid body, it is convenient to distinguish a **central axis,** which is the line connecting the centers of the adoral and aboral skeletons. Through this axis pass five reference planes, each of which is called a **radius.** A radius is expressed in the skeleton by the structures forming a **ray** (see Glossary for definition). Ideally the radius passes through the central axis and the median line of the ray-structures.

Orientation of modern crinoids is usually only possible from the adoral surface, on which mouth and anus are invariably located. The anus, which usually is placed interradially, generally is recognized to indicate the **posterior side.** The ray or radius opposite to it is termed **anterior. Left and right sides** of the crinoid then correspond to the left and right sides of the observer if the tegmen is directed upward. In the system of designations introduced by CARPENTER (1884a), the five rays are termed *A* (anterior), *B, C, D,* and *E,* respectively, in a clockwise direction when the crinoid is viewed from the adoral side. The anus thus is in the *CD* interradius.

The obvious symmetry of the adult crinoid is **pentaradial,** as is evident from both the aboral and the adoral skeletons. In the aboral skeleton the five radii are expressed by the five prominent series of proximal arm segments, and in the adoral skeleton by the five ambulacra that radiate from the mouth. In aboral view nothing disturbs the pentaradial symmetry because the interradial areas normally are not differentiated.

The posterior interradius can be recognized in aboral view of the larval stages of some recent crinoids (e.g., *Promachocrinus*), where anal plates may exist for some time in postmetamorphic stages of development. It is also known to exist in postmetamorphic growth stages (e.g., *Thaumatocrinus*), where a sixth or posterior "arm" exists in the posterior interradius. In adoral view, too, it is noted that the crinoid symmetry is not perfectly pentaradial, for the anus is situated in one of the interradial areas of the tegmen. This disturbs the pentaradial symmetry and in fact gives the organism a **bilateral symmetry.** This is the original type of symmetry in crinoids, since embryonic growth stages lack any trace of pentaradial symmetry. During the larval growth stages a strong pentaradial symmetry is steadily imposed on the original bilateral symmetry. In the adult crinoid bilateral symmetry is expressed by a sagittal plane through the *A* radius, mouth, anus, and aboral pole of the calyx. This plane is known as the **crinoid plane of symmetry.** The symmetry pattern described is normally present in representatives of the orders of the stalked crinoids and also in many Comatulida.

Two marked deviations from the normal crinoid plane of symmetry are known. The first is represented by the comatulid family Comasteridae. The mouth is excentric in these crinoids and displaced some distance in the direction of the *AB* interradius. Consequently the radius opposite to it is modified. In this way a symmetry plane is created which is at 36 degrees to the normal crinoid plane of symmetry. The second aberration in symmetry is exemplified by the genus *Holopus*. In this form the five arms are unequally developed. Two small arms form the so-called **bivium** and the three larger arms are clustered to make the **trivium.**

In paleontology, morphological descriptions of crinoids conventionally are organized in the order of the major morphological parts of the organism: stem, calyx, tegmen, arms, and pinnules. In zoology, it is hardly desirable to follow this arrangement, since major morphological parts—very well defined in fossil crinoids—may not be differentiated or may even be nonexistent in recent crinoids. The stem is not present in most adults of the latter group, since all comatulids lack it. Also, the calyx is frequently very difficult to identify in recent crinoids, for such well-differentiated calices as characterize Camerata and Flexibilia are absent. The only recent crinoid having a well-marked calyx as well as a tegmen, together forming a complete skeletal case, is *Calamocrinus diomedae*. The radials in this form are specially differentiated to form the largest part of the calyx, and the orals are preserved to form the major part of the tegmen. The spacious theca thus formed has the same rigid nature as is found in camerates. Commonly the radials—and in some crinoids the orals—are the essential components of a theca. Good development of the radials—and to a lesser degree the basals—does not exist in the Isocrinida and the Comatulida, but may be found in the Millericrinida, Bourgueticrinida, and the Cyrtocrinida. Hence calices are identifiable in the three last-mentioned groups, but a tegmen is hardly developed and a theca still poorly defined. In the comatulids a theca is found, composed of a more or less inflated adoral membrane or tegmen, which is usually not calcified, sitting on the central part of the aboral skeleton, thus providing the necessary space for the voluminous digestive tract. Comatulids are crinoids which almost exclusively are composed of their arms.

In this chapter morphological descriptions are arranged according to location of the elements in the aboral and adoral skeletons.

ABORAL SKELETON

COLUMN

In most crinoids other than extant forms, the stem or stalk, designated as the **column,** is a very characteristic part of the body. It serves the purpose of fixing the animal to the sea bottom or to any available or suitable object that might serve as a substrate. For example, telegraph cables are sometimes found to have crinoid specimens attached to them.

The column is composed of many different skeletal elements called **columnals.** Some of these may bear short and unbranched appendages named **cirri.** Such cirrus-bearing columnals are termed **nodals,** in contrast to the non-cirrus-bearing **internodals.** The cirri, like the stem, are supported by a row of skeletal pieces, called **cirrals.** Both the stem and the cirri are pierced by an **axial canal,** containing extensions of the coelom **(chambered organ)** and the aboral nervous system. Columnals and cirrals are bound together by bundles of elastic fibrils or **ligaments,** which provide the column and cirri with a degree of flexibility. Muscles are entirely absent in crinoid stems and cirri.

The distal extremity of the stalk has special modifications for fixation to the bottom or a substrate. Primarily—and even ontogenetically—attachment is obtained by a disclike columnal, which may be retained during further life as an **attachment disc** (e.g., *Phrynocrinus*). In some genera (e.g., *Democrinus, Rhizocrinus*) the distal columnals bear special rootlike, unbranched appendages for fixation (Fig. 8,*1*). These are named **radicular cirri.** In the Isocrinida (e.g., *Neocrinus decorus*) the normal cirri may also serve for attachment. The British Museum (Natural History) in London has a specimen attached to a telegraph cable with its cirri. CARPENTER (1884a) figured the terminal part of a stem of *Hypalocrinus naresianus* with something like an attachment disc (Fig. 8,*2*). A very special way of attachment occurs in *Holopus,* for it is cemented to rocky substrates by means of its highly modified basal and radial circlets, which form a sort of foot. Permanent attachment to the bottom is abandoned in the Comatulida. These crinoids may freely

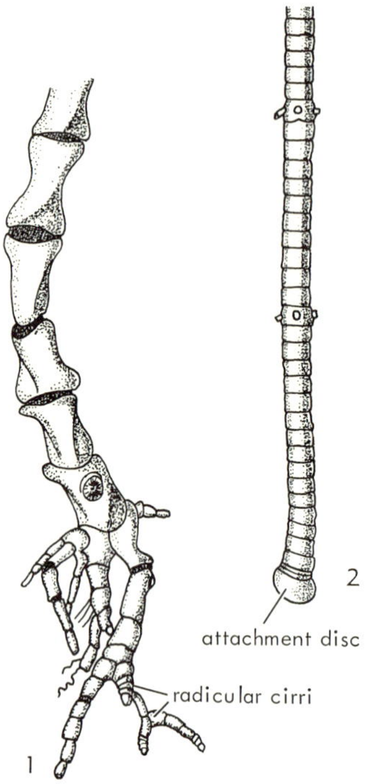

FIG. 8. Mode of attachment of crinoid columns.——*1.* Radicular cirri of *Rhizocrinus lofotensis* (Breimer, n).——*2.* Attachment disc of *Hypalocrinus naresianus* (Carpenter, 1884a).

swim, crawl, or creep with aid of the arms, but throughout most of their lives they are sedentary, fixing themselves to the bottom or a substrate by their cirri or even by some of their arms (e.g., in some comasterids).

One may confidently expect that, at least ontogenetically, a stalk occurs in all crinoids. Development of a stalk is known to occur in the ontogeny of the nonstalked feather stars (e.g., *Antedon, Promachocrinus*). In these crinoids the juvenile stem is cut off by autotomy (except in *Thiolliericrinus*), but is preserved in all other recent crinoids.

Insofar as known, the column is a truly primary organ which develops very early in crinoid embryology. The first stem joint (columnal) is formed directly after the left somatocoel has assumed its aboral position by rotation of the internal organs. This first columnal is located aborally to the basals. As a rule, every new columnal is introduced by becoming inserted between the basals (or infrabasals) and the next preceding stem joint. Insertion of new columnals occurs regularly one after another, so that each individual columnal assumes a more distal position in the stalk as it grows.

Mature development of the stalk may produce quite different products. Among recent stalked crinoids two types of stalks are distinguished and discussed below: 1) a very simple type, present in the Bourgueticrinida, and 2) a much more complex one in the Isocrinida.

The simple type of stalk observed in the Bourgueticrinida, an order containing primitive, delicate, and simple-stalked crinoids, consists of simple elements, all more or less alike. They become introduced in the stalk as described above. The newly formed columnals in the proximal part of the stalk, adjacent to the calyx, have undifferentiated

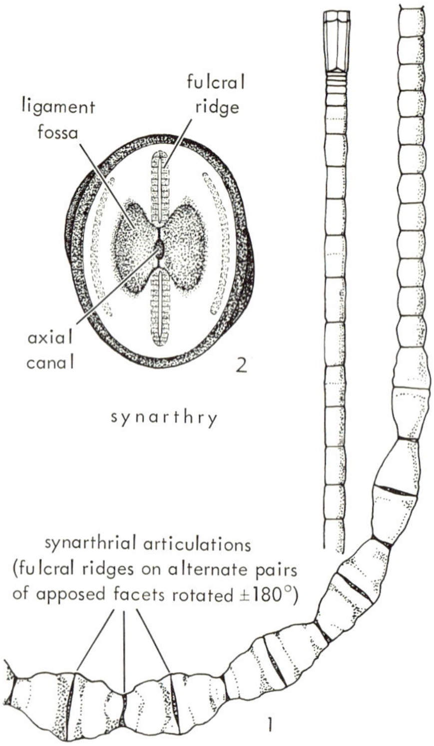

FIG. 9. Growth and scheme of columnar synarthries.——*1.* Column of *Naumachocrinus hawaiiensis,* with columnals united by synarthries (after Clark, 1915).——*2.* Synarthrial joint face (schematic) (after Carpenter, 1884a).

cylindrical forms. The full-grown columnals in the distal part of the stalk have undergone a certain differentiation in form. During its individual growth each columnal adapts its form to the relatively enormous development of the elliptical articular surfaces at its proximal and distal extremities (Fig. 9). The stem growth is confined to insertion of more columnals if needed. The stem does not expand very much in diameter during growth. Its distal part is produced in radicular cirri or it terminates in a calcareous plate. The bourgueticrinid type of stalk normally lacks differentiation into nodals and internodals and is generally devoid of cirri.

Columnal articulation in the Bourgueticrinida is entirely ligamentary and nonmuscular. Some proximal columnals may have closely fitting interfaces, connected with very short and few ligament fibers. Such elements are said to be united by **synostosis.** The regularly occurring type of articulation in the Bourgueticrinida is named **synarthry.** Columnals united by synarthrial ligamentary articulation have typically elliptical articular surfaces, the longer axes of which are placed at different angles to one another on the proximal and distal faces of each individual columnal. The articular surface itself is **bifascial,** for a median fulcral ridge separates two fossae that lodge the ligament bundles. Synarthrial articulation allows differential movement of the columnals in two directions. The stem of the Bourgueticrinida is able to accommodate to almost every passive and horizontal movement to which it is exposed, especially in such cases where the synarthrial articulations are placed at various different angles along the length of the stalk.

Exceptionally, nodals occur in the stems of some Bourgueticrinina (e.g., *Monachocrinus*). Where present, they display no regular distribution along the stalk and the cirri which they bear may be deciduous.

A complex type of stem is developed in the Isocrinida (e.g., *Neocrinus*). The complication consists in differentiation of the columnals into nodals and internodals and in their different mode of growth. The nodal columnals are placed at regular intervals along the stalk, closely spaced near the crown but more distant from one another away from the crown. The distances between successive pairs of nodals in mature parts of the column may be subequal or they may increase distally. The stem may attain great length, with nodals normally bearing a circlet of five cirri, but some with cirri reduced to two, or three, as in *Endoxocrinus alternicirrus.* The cirri fit movably into sockets on sides of the nodals. They are composed of 20 to 50 cirrals, all more or less alike, cylindrical or elliptical in section and connected by simple ligamentary articulations. Cirri of the Isocrinida are known to assist in attachment of the crinoid. Unlike some comatulid cirri, they are smooth sided. Fixation to the bottom in isocrinids takes place in juvenile specimens through a terminal attachment disc (Fig. 8).

The growth of the isocrinid stem is more complex than in stalks of bourgueticrinid type. Nodals are introduced in the stem just below the lowermost plate circlet of the theca. Series of internodals are intercalated between the nodals. This occurs as follows. Between two cirrus-bearing columnals, which thus are recognizable as nodals, one internodal is introduced as a primary internodal (Fig. 10). Soon, secondary internodals are introduced between the nodals and the primary internodal. Tertiary internodals then are inserted next above and below the secondary internodals, quaternary internodals above and below each of the tertiary internodals, and so on until a characteristic number of internodals is formed. The nodal and contiguous suc-

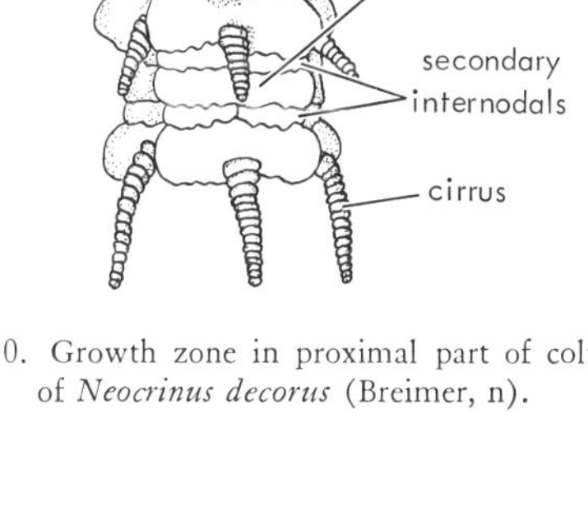

FIG. 10. Growth zone in proximal part of column of *Neocrinus decorus* (Breimer, n).

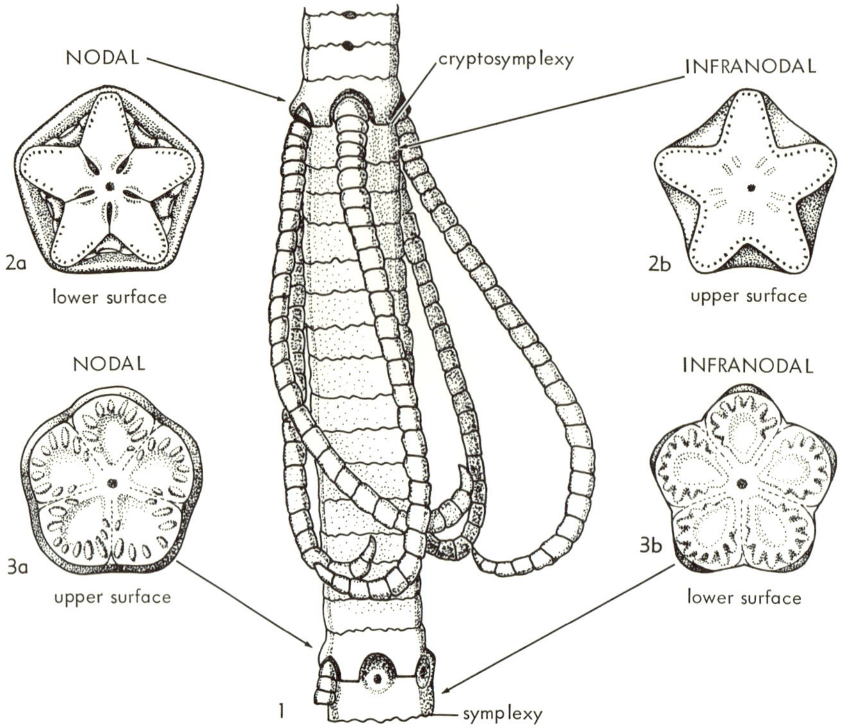

FIG. 11. Portion of adult column of *Neocrinus decorus* with schematic enlargements of columnal articular facets (Breimer, n).

1. Complete noditaxis consisting of nodal with attached cirri (3rd columnal from top) and 15 subjacent internodals, uppermost of which is classed as infranodal.

2. Cryptosymplexy between nodal *(2a)* and infranodal *(2b)*.

3. Symplexy at upper surface of nodal *(3a)* and lower surface of infranodal *(3b)*.

cession of internodals on the proximal side of the nodal are defined as a **noditaxis.** The number of columnals in successive noditaxes is a constant character in most isocrinid species, among which all internodals in a full-grown noditaxis are similarly pentagonal in transverse outline and subequal in size. In still incomplete noditaxes in the proximal part of the stem the internodals show readily seen differences in height, and to some extent in diameter, indicating their order of appearance between the nodals.

The articular facets of isocrinid-type columnals are all nonmuscular and ligamentary. Two types exist (Fig. 11). 1) One type of ligamentary articulation occurs between a nodal and the internodal next above it, and between internodals; they are of a type called **symplexy.** On one joint face culmina (ridges) interlock with crenellar grooves on the opposite joint face, marked externally by crenulate sutures. This type of articulation allows very little movement between the columnals so united. Consequently the stem in the isocrinids is rather stiff and only capable of gentle bending. 2) The other ligamentary joint faces between a nodal and the internodal next below it, called **infranodal,** articulate as a synostosis, frequently, but incorrectly, indicated as a syzygy. Externally the synostosis is marked by a straight suture between nodal and infranodal. The synostosial connection allows nodal and infranodal to have moderately flexible differential movement in all directions.

A very special type of ligamentary articulation between columnals is found in *Ptilocrinus brucei.* Its cylindrical columnals are united by ten ligament bundles, located in niches in the intercolumnar sutures (Fig. 12). These bundles are not situated between the joint faces, but are located near the outer edge of the columnals and, hence, fully exposed to the exterior. They provide the stem a considerable degree of flexibility. The columnals of *Ptilocrinus* are formed by five to ten different segments (according to the species) and in further growth these merge into a single plate as they assume positions in the distal part of the stalk.

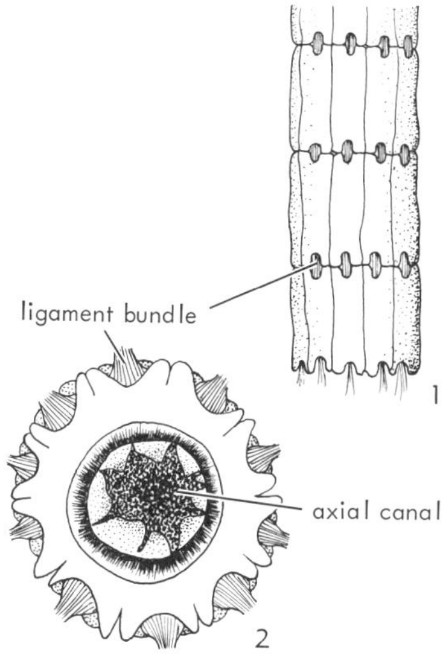

FIG. 12. Column of *Ptilocrinus brucei* (Breimer, n).——*1.* External aspect of few columnals, showing ligament bundles.——*2.* Interarticular face of columnal with axial canal and ten peripheral ligament bundles, enl.

PROXIMALE AND CENTRODORSAL

It was a hypothesis of CLARK (1915a) that the stem of all recent and Mesozoic crinoids possesses a skeletal element known as the **proximale,** or a structure strictly homologous to it. The proximale normally is the last columnal to be formed and, as no further columnal formation occurs, it becomes intimately attached to the calyx by a union exactly similar to that between basals and radials (synostosis). The proximale primarily denotes maturity of the column and the completion of stem growth. It takes the shape of the dorsal part of the calyx and assumes the function of a centrodorsal plate. The enlargement of the proximale affects also the columnal just below it, the proximal face of which, becomes closely approximated to the distal face of the proximale, uniting with it in exactly the same way as the proximale unites with calyx plates. This union is known in zoological literature as the "stem syzygy," but the union between the plates in reality is a synostosis. The nodal plates of the Isocrinida are interpreted by CLARK as reduplicated proximales, having the same synostosial connections with the infranodals as the proximales have with the underlying columnals.

Though interesting, CLARK's hypothesis is conjectural. The proximale may indeed denote the end of column formation, itself being the last columnal formed (as is said of *Phrynocrinus*). In the Isocrinida no end of growth in the stem seems ever to occur. No single specimen of an isocrinid, to my knowledge, has ever been found with a fully developed noditaxis immediately beneath the lowermost thecal circlet. If CLARK's hypothesis were right, a single isocrinid noditaxis would be homologous to a complete stem in the Millericrinida, and Bourgueticrinida, which remains to be seen. Theoretically, it is also possible that the proximale is ontogenetically not the last but the first columnal to be formed, retaining its original contact with the basals and after it all other columnals are formed successively.

Adult comatulids, after shedding their juvenile stalks, are characterized by the presence of a skeletal element called **centrodorsal:** a huge hemispherical or conical plate at the aboral pole of the skeleton commonly provided with a high number of cirri. Ontogenetic evidence from *Antedon* seems to suggest that it originates as a single element and is the modified topmost columnal of the juvenile pentacrinoid stalk, no other elements from the column entering its construction. In this interpretation the centrodorsal is supposed to be homologous with the proximale of other crinoids. However, phylogenetic evidence from fossil comatulids (see RASMUSSEN's section on Evolution of Comatulida, this *Treatise* volume) seems to suggest that the conical centrodorsal may be composite and formed by several nodals, which are articulated and not fused. The continued formation of new parts with cirrus sockets and cirri at the upper edge of the centro-

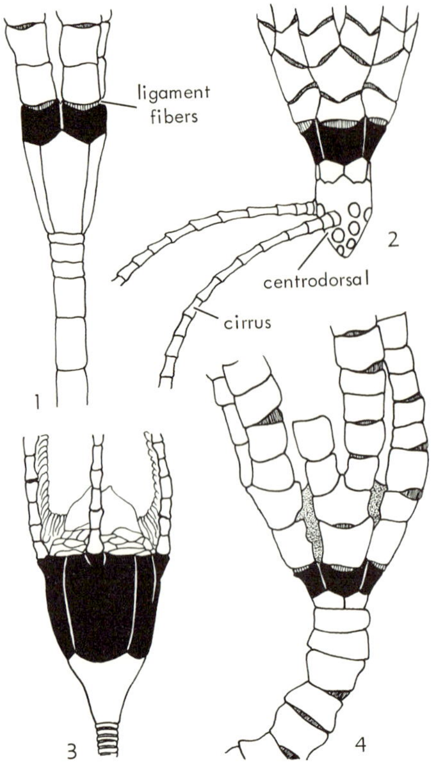

FIG. 13. Radial and basal circlets in various recent crinoids (radials solid black).——*1. Democrinus conifer* (Breimer, n).——*2. Atelecrinus anomalus* (after Clark, 1915a).——*3. Hyocrinus bethellianus* (after Carpenter, 1884a).——*4.* Presumed *Zeuctocrinus* from Antarctic (Breimer, n).

dorsal seems to correspond to the normal proximal insertion of new nodals at the top of the column, although these parts in most recent comatulids are fused with the centrodorsal.

LUDWIG (1877) showed that on the upper or adoral surface of intact centrodorsals of *Antedon* there are five radial depressions lodging the ends of five blind, radial, coelomic sacs extending from the coelom. A. M. CLARK (1973a) found for some fossil and recent comatulids that these coelomic extensions may completely penetrate the centrodorsal.

INFRABASALS

In the fossil crinoids the proximal cycles of plates in the aboral skeleton—infrabasals, basals and radials—are involved in the formation of a well-defined aboral cup or calyx. Contrary to most fossil crinoids, recent forms only exceptionally develop an aboral cup. It is well defined in some primitive genera (e.g., *Calamocrinus; Holopus; Hyocrinus,* Fig. 13,*3; Ptilocrinus*) and of sufficient extent to enclose the entire visceral mass. Among isocrinids the aboral cup (Fig. 10) is nothing but a platform supporting only the central portion of the visceral mass. In the comatulids an aboral cup is altogether absent. The visceral mass is supported by the centrodorsal and by the proximal parts of the arms, rather than by the radial and basal circlets.

The **infrabasals** of crinoids normally form a closed circlet of five small plates at the aboral pole of the theca. They are disposed in radial position and rest upon the topmost columnal. In later fossil and recent crinoids the infrabasals are greatly reduced and functionless, or altogether absent. By application of the Law of WACHSMUTH and SPRINGER the recent crinoids are shown to be **dicyclic** and it is assumed that infrabasals are normally present in juvenile growth stages but become resorbed during later ontogeny. For this reason the recent crinoids lacking infrabasals are termed **cryptodicyclic** or **pseudomonocyclic.**

Infrabasals have been conclusively demonstrated to occur in some Comatulida (e.g., *Antedon, Promachocrinus*) but only as transitory skeletal elements in the juvenile growth stages. *Promachocrinus kerguelensis* has five infrabasals, which for some time in the ontogeny are large and well developed. In postmetamorphic growth stages the greatly reduced infrabasals become concealed by the proximale and finally either fuse with it or atrophy. This condition occurs in all recent comatulids in which infrabasals have been observed. The large plate formed by the coalition of infrabasals with the proximale therefore in reality might be a double structure.

Among the Isocrinida infrabasals are present in some Mesozoic forms, but as a rule, they are absent in recent adult isocrinids. They are stated (CLARK, 1908a) to be rudimentary in *Metacrinus.* In the extant Millericrinida, Bourgueticrinida, and Cyrtocrinida no evidence whatever of the presence of infrabasals is found and also no evidence that they ever existed in any of their ancestors. Since these crinoids have stems with circular outlines, without nodals and nodal cirri, application to them of the Law of WACHSMUTH and SPRINGER is impossible. Strictly speaking, these crinoids can only be called **monocyclic,** rather than pseudomonocyclic or cryptodicyclic.

BASALS

The **basals** are primarily five in number, forming a circlet near the aboral pole of the theca between the circlet of infrabasals and the circlet of radials, with both of which they alternate, being interradial in position. The basals have synostosial sutural contacts with both infrabasals and radials. In nearly all recent crinoids the basals are modified in some way. In some genera they are well developed, in lateral contact with one another all around the basal circlet, forming a fair portion of the calyx (Fig. 13,*1,3*). In *Hyocrinus* and *Ptilocrinus* their number is reduced to three. In *Democrinus* (Fig. 13, *1*) the basals may be greatly elongated, forming the largest part of the calyx; they may be more or less fused together showing irregular or weakly defined interbasal sutures. The recent isocrinids generally have an interrupted basal circlet, the basals are externally visible only as five swollen triangular plates between the aboral parts of the radials (Fig. 10). Only a few isocrinid genera possess a laterally uninterrupted basal circlet (e.g., *Annacrinus, Hypalocrinus*).

Among recent comatulids the Atelecrinidae are the only ones in which the basals persist as normal basals instead of becoming modified. They are visible externally (Fig. 13,*2*). In all other recent comatulids the basals in the adult become metamorphosed into a peculiar plate, termed the **rosette** (Fig. 14,*3*). This element is an internal centrally perforated calcareous disc or diaphragm covering the cavity of the centrodorsal. The structure is decagonal in outline and has five triangular processes directed to the sutures between the five radials, interradial in position; also, it bears five spoutlike processes in radial position. The interradial processes of the rosette may connect with five rodlike structures, known as **basal rays** (Fig. 14,*1*), lying on the oral

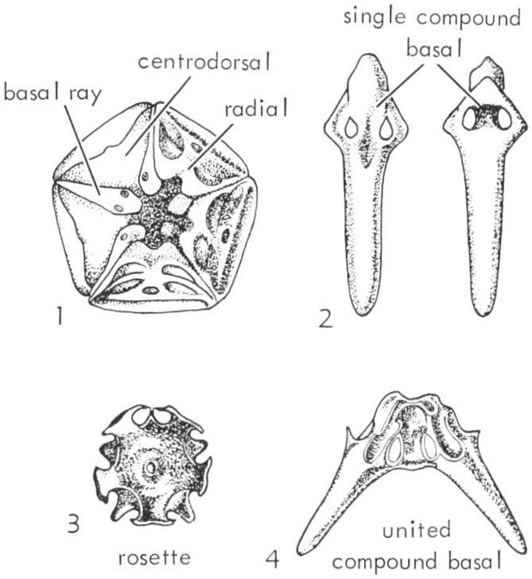

FIG. 14. Basal structures of comatulids (after Carpenter, 1884a).——*1*. Ventral view of centrodorsal with basal rays and three radials in place. ——*2*. Two single compound basals.——*3*. Ventral view of rosette.——*4*. Side view of united compound basal.

surface of the centrodorsal in interradial position. In some genera the rosette can be seen at the external surface between the radials and the centrodorsal. If a basal ray unites with an interradial process of the rosette a single compound basal may be formed (Fig. 14,*2*). Two basal rays may fuse with two interradial processes from the rosette to form a united compound basal (Fig. 14,*4*).

RADIALS

Regularly crinoids have five **radials** arranged in a circle in the cup. The radial circlet may be interrupted by primary interradials and by an anal plate, as in *Thaumatocrinus.* The radials alternate with the basals and are connected to them with sutural contacts of synostosial type. Each radial is followed by a series of brachial plates, with the first of which it has a muscular articulation. Traditionally the radials have been considered as first brachial plates, but in modern work they are more frequently regarded as calyx plates. Probably, they are both. In stalked crinoids they may indeed be very large (e.g., *Hyocrinus,* Fig. 13,*3*) and substantially contribute to the formation of an aboral cup. In the comatulids the radials compose the

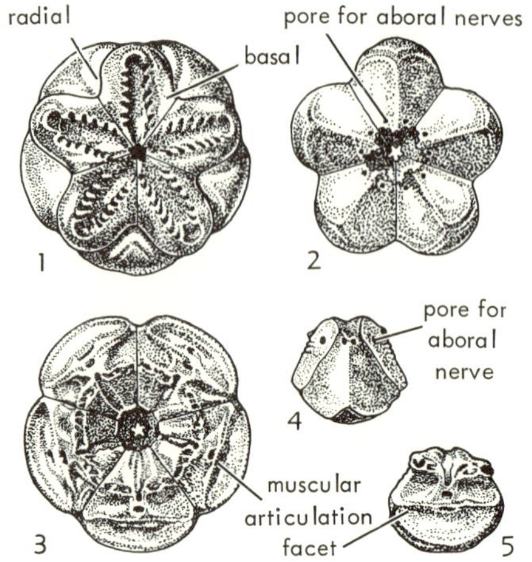

FIG. 15. Structure of the radials in *Neocrinus decorus* (after Carpenter, 1884a).——*1.* Lower surface of radial and basal circlets showing stem impression.——*2.* Upper surface of basal circlet showing basal-radial facets.——*3.* Upper surface of radial circlet showing straight muscular articulation on distal facet of radials.——*4.* Lower facets of radial.——*5.* Upper face of radial.

only circlet of plates forming an aboral cup in the adult stage. As the basals are modified into a rosette, the radials are in contact with the centrodorsal; this contact is also a synostosis.

In two genera of the comatulids, *Promachocrinus* and *Thaumatocrinus,* multiplication of radials occurs; besides the five regular radials an extra set of five radials is formed, giving the adult crinoid ten radials, all of which are alike and each bearing one arm. In the early larval stages of these crinoids only five regular radials occur in the usual radial position. At a later stage in ontogeny narrow, linear interradial plates are introduced between the "radial radials." These elements develop into plates morphologically indistinguishable from the five regular radials. They are "interradial radials" or **pararadials.**

The internal surfaces of each radial typically shows five small, rounded openings leading to the interior and serving for passage of the chief cords of the aboral nervous system (Fig. 15,*2,4*). A pair of openings is present also on the lower sides of the radials for reception of the interradial nerve trunks from the basals; the lateral sides of the radials have one opening for passage of the radial nerve commissure to the adjoining radials and the upper articular surface of the radial has a large central canal, just oral to the transverse ridge, for passage of the radial nerve trunk to the brachials.

The upper facets of the radials bear complicated articular facets to which the lower ends of the first brachials are joined (Fig. 15,*3,5*). These facets may have different positions on the radial. They may be inclined at an angle of 45° to the polar axis of the animal, being trapezoidal or even triangular in shape. In comatulids they may even be parallel to the polar axis and become practically oblong. The articular facets of the radials represent what is known as **straight muscular articulation.** This may serve as a basis for consideration of all articulations between the brachials.

The articular facets are divided by a more or less prominent ridge, which extends transversely across the joint face. It serves as the fulcrum upon which the motion at the articulation is accommodated. The ridge is known as the **fulcral ridge.** Just adoral to it the **central canal** is found. Aboral to the fulcral ridge is an unpaired **aboral ligament fossa** lodging the extensor ligament bundle, which in articulation antagonizes the flexor muscles. Near the adoral margin of the articular facet a pair of **adoral muscle scars** is located, lodging the flexor muscles. Between the adoral muscle scars and aboral ligament fossa a pair of **interarticular ligament fossae** is present. The straight muscular articulation is chiefly characterized by its internal bilateral symmetry. Marked asymmetries may occur in the muscular articulations, mainly due to unevenness of the interarticular ligaments.

ANALS AND "INTERRADIALS"

Adult recent crinoids normally lack any plate or plates in the posterior interradius or other interradii. The pentagonal symmetry is not disturbed by differentiation of the posterior interradius. In exceptional cases interradially disposed skeletal elements are found separating the radials; these are not true perisomic plates, but probably belong to the aboral skeleton. In juvenile growth stages of various comatulids anal plates are known to occur. Gen-

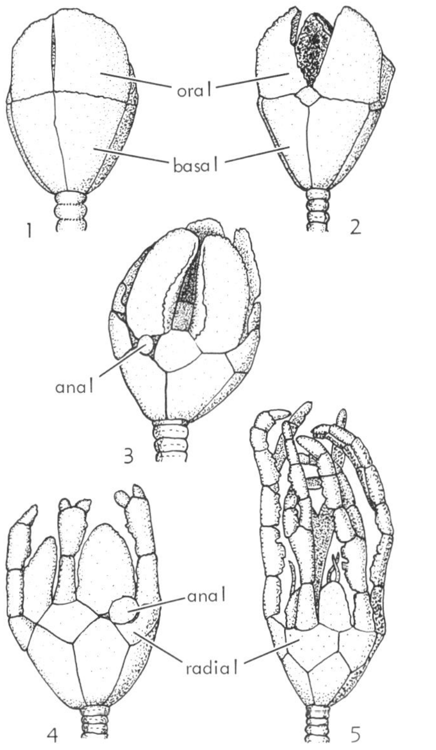

FIG. 16. Transitory development of anal plates in early ontogeny of *Hathrometra prolixa* (after Clark, 1915a).——*1-3*. Successive early to late cystidean stages.——*4-5*. Early and subsequent pentacrinoid juvenile stages.

erally they never persist, but are resorbed before maturity.

In the pentacrinoid stage of comatulids (e.g., *Antedon, Comactilia, Comactinia, Promachocrinus, Thaumatocrinus*) an anal plate occurs for some time in the posterior interradius between the *C* and *D* radials (Fig. 16,*1-5*). From the ontogenetic development of *Promachocrinus kerguelensis* it is known that this anal plate appears before any of the radials is introduced. The *C* radial appears soon after the anal plate at its right side. The radial grows much faster than the anal plate so as to extend partly beneath it. The anal plate then is pushed out of the radial circlet as the *C* and *D* radials establish lateral contacts. After extrusion the anal plate is resorbed.

CLARK (1915a) concluded that the first-formed anal plate is homologous to the radianal of fossil crinoids and not to anal *X*, as was the traditional interpretation of his days. BATHER (1915) and MORTENSEN (1920), however, interpreted the first-formed anal plate as the true *C* radial pushed leftward. Eventually this was replaced by a secondarily formed new plate which, in their view, took the form and place of a true right posterior (= *C*) radial and assumed its functions. Such a horizontal replacement of one radial by another is entirely unknown from crinoid paleontology. In a modern interpretation of fossil inadunates (PHILIP, 1965), however, the *C* radial and adjoining radianal are interpreted as superradial and inferradial, respectively. In a phylogenetic succession of inadunate genera the *C* inferradial may become entirely excluded from the aboral cup. It seems reasonable to explain the two plates successively formed in the posterior interradius of the comatulids as homologues of the infer- and superradials of fossil crinoids.

If the first-formed anal plate of some extant crinoids is to be interpreted as the morphological equivalent of an inferradial (radianal), an anal *X* should be expected to occur also. CLARK (1915a) found the anal *X* and the radianal side by side between the *C* and *D* radials in juvenile pentacrinoid growth stages of *Promachocrinus kerguelensis*. A large, modified, posterior interradial (anal *X*) does not exist in adult recent crinoids. Interradial plates are known, however. These may occur in all five interrays with exactly the same degree of development, or all five are absent; a plate in the posterior (*CD*) interray alone never occurs.

In the juvenile pentacrinoid growth stage of *Antedon bifida* five interradials develop between the orals and the basals, separating the radials from each other all around the aboral cup. The one in the posterior interradius may be the homologue of the anal *X* of fossil crinoids. Interradial plates ordinarily do not develop, but if they appear, are soon resorbed. In *Promachocrinus* and *Thaumatocrinus*, however, they increase in size as the radials move farther and farther apart and ultimately they assume the characters of radials adjoining them. These interradials develop into pararadials and

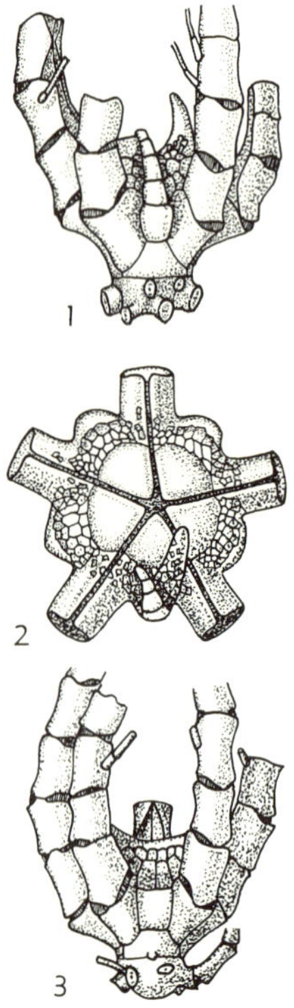

FIG. 17. Interrupted radial circlet in six-ray juvenile stage of *Promachocrinus abyssorum* (after Carpenter, 1884a).——*1*. Side view of *CD* interray with posterior armlet (center).——*2*. Oral aspect of calyx.——*3*. Interradial aspect showing primary interradials in contact with basals.

give rise to a series of plates which form arms in no way distinguishable from those arising from the true radials. In these ten-rayed genera, however, armless interradials may occur and such plates appear as simple interradials separating the radials. Commonly six-rayed forms are found among representatives of these otherwise ten-armed genera. Moderately numerous individuals of *P. kerguelensis* are six-rayed, the extra ray being located between *C* and *D* rays, resting on an anal *X*. The other interradials either have been resorbed or suppressed, as in five-rayed genera. The interesting crinoid described by CARPENTER (1884a) as *Thaumatocrinus renovatus* (Fig. 17,*1-3*) was reinterpreted by CLARK (1915a) as such a six-rayed form of the ten-armed species *P. abyssorum,* with which it was found associated. The posterior "arm" is the first to be formed among the pararadial arms, but apparently development sometimes does not proceed beyond this stage.

ARMS

NUMBER AND COMPONENTS

Normally, the arms of recent crinoids are densely pinnulated complex structures that primarily serve the function of feeding. For this purpose a structure of ambulacral grooves is developed on the adoral sides of the arms and pinnules. The microorganisms on which crinoids feed are caught by podia bordering the ambulacra. Only secondarily have the arms assumed other functions than feeding and then special means for carrying out the other functions have been adopted.

An important secondary function of the arms in recent crinoids, found exclusively in the nonstalked feather stars, is locomotion, either by swimming or creeping. Antedontid comatulids have reasonable swimming capacities, in which the arms play a main role. In the comasterid comatulids cirri are absent and the rather clumsy arm structure is specially adapted for their creeping mode of locomotion. In temporary fixation, the posterior arms are used; these are devoid of an ambulacral groove, but do have genital pinnules.

The number of arms in extant crinoids is extremely variable. Some forms have only five simple arms, whereas multibrachiate forms may possess no less than 200 free arm ends. No matter what may be the number of free arm ends, the arm bases are practically always laterally in contact because of their incorporation in the theca. The tegmen is situated at a fairly high level above the radials. Since the tegmen is flexible and the arm bases themselves have flexible connections, the arm bases are not included into a rigid calyx, as in the camerates.

Arms of crinoids are supported by skeletal elements located in longitudinal series all along the arm, such elements being termed **brachials.** Inasmuch as the arms commonly divide in several places, the brachials constituting various division series (called **brachitaxes**) are given ranking designations (e.g., **primibrachs, secundibrachs, tertibrachs,** etc.). The number of

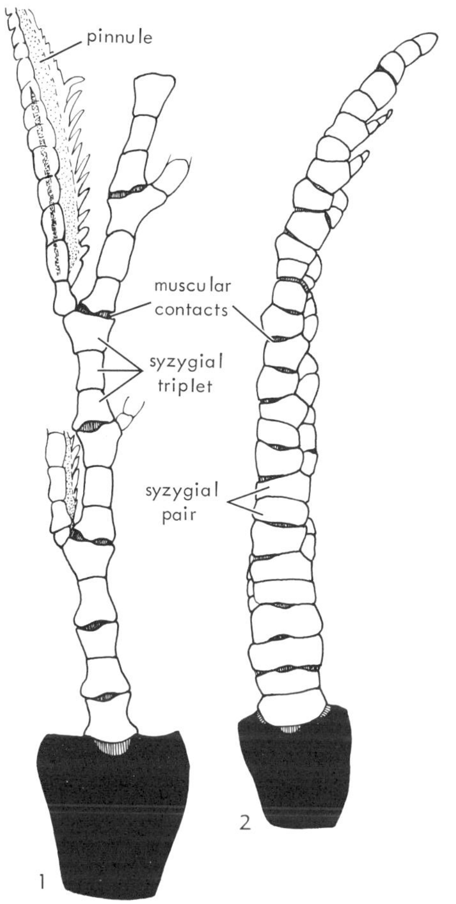

FIG. 18. Arm structures in primitive five-armed crinoids (Breimer, n). *1. Hyocrinus bethellianus* showing muscular and syzygial articulations of brachials.——*2. Ptilocrinus pinnatus.*

plates in a brachitaxis is variable. The most proximal brachitaxis is composed of a primibrach or primibrachs and the radial, which for this purpose may be regarded as an arm plate (although most paleontologists exclude the radial because undoubtedly, it is a cup plate). The plates terminating a brachitaxis distally are either the last-formed brachials of the free arm ends or axillary brachials upon which the arms divide. They are ranked like the brachials proximal to them as **primaxils, secundaxils, tertaxils,** and so on.

Primitively, crinoids have only five unbranched arms. Among recent crinoids this five-armed condition is preserved in such stalked forms as *Rhizocrinus, Ptilocrinus,* and *Hyocrinus* (Fig. 18). Normally, the five arms in five-armed forms are all alike, and have the same development on the five radial sides of the body. *Holopus* (Fig. 19), which is adapted to life in reef environments where it is cemented at its base to hard rocky bottoms, has the most peculiar

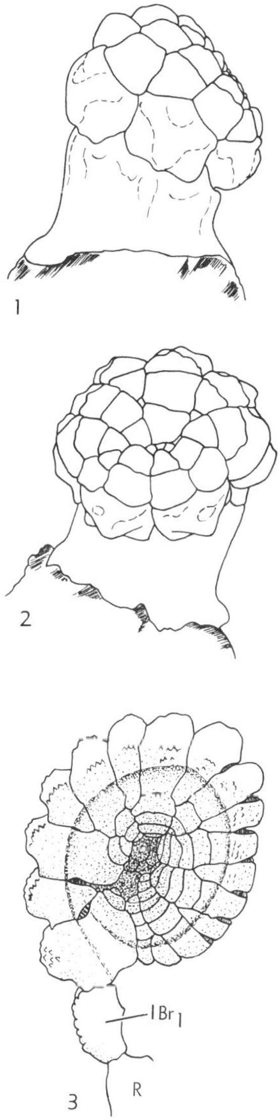

FIG. 19. Organization and arm structure in *Holopus rangi* (Breimer, n).——*1,2.* Side views showing undivided peduncle beneath radials and incurved short arms.——*3.* Side view of single inwardly coiled arm, enl.

arms of any recent crinoid. The ten arms are divisible into smaller arms called a **bivium** and larger ones called a **trivium.**

Ten-armed crinoids very strongly tend to have only two primibrachs, the second one axillary and called a **primaxil.** This condition is exclusive in the comatulids. Only in isocrinid genera, such as *Metacrinus,* are more than two primibrachs seen in each ray. In most multibrachiate comatulids the brachitaxes between the first and the last have either two or four brachials. Multibrachiate isocrinids (e.g., *Cenocrinus*) may have more than four elements in brachitaxes between the first and the last.

ARM GROWTH

Growth of the arms in extant crinoids has been studied and described especially by CLARK (1915a, 1921) and GISLÉN (1924). Early in crinoid ontogeny the arms are formed as radial extrusions of the body on five different sides of the embryological theca, shortly after the radial plates have been introduced. The brachial elements are formed in line with the radials; they are the product of terminal growth and skeletal secretion at the tip of the growing arm. In five-armed crinoids such terminal growth is the only mode of arm formation observed. It is known in literature as the **direct** or **continuous mode of arm formation.** The ten-armed stage in crinoids, whether transitory or ultimate, is brought about in the same direct way. Two primibrachs in each ray are produced by terminal secretion, the second being axillary and called the primaxil. In direct arm growth the primaxil is thus a primarily formed plate. After secretion of a primaxil, two growth sectors appear distally from it. These start to produce a pair of separate arms equal to each other in form and size. The resultant ten-armed pentacrinoid larval stage is so widespread among recent crinoids as to be almost universal. It occurs in all pentacrinoid growth stages of the recent crinoids as far as known. To be remembered here is discussion of such ten-armed comatulids as *Promachocrinus* and *Thaumatocrinus,* which developed their ten arms without the intervention of primaxils, but with intercalation of five pararadials. No ten-armed crinoids are known to possess primaxils formed in other than the direct mode of growth.

Crinoids with more than ten free arm ends are termed **multibrachiate.** Several different ways of bringing about a multibrachiate condition are observed in recent crinoids. One is by direct or continuous growth, just described. It is supposed to occur in the isocrinid *Neocrinus.* Studies by CARPENTER (1884a) have led me to conclude that the secundaxils and tertaxils of this genus are formed by terminal growth. If so, *Neocrinus* is the only genus with any indication of reaching a multibrachiate arm structure in this way.

A second way of producing a multibrachiate arm pattern is also direct and continuous but not by terminal growth of arms. Instead, it involves **differentiation of pinnules.** This mode of growth is found in some isocrinids (e.g., *Metacrinus,* Fig. 20,*1*) and millericrinids (e.g., *Calamocrinus*). The best observations have been made by GISLÉN (1924), who observed that the formation of secundaxils to quartaxils in *Metacrinus* is accomplished as follows. At the free arm ends of the ten-armed pentacrinoid growth stage, certain pinnules begin to gain strength and become larger. New small pinnules then arise on the sides of the enlarged pinnules, which gradually attain the same length and girth as the main arm. The place for thickening of pinnules in this way is found in the region between immature and mature pinnules of the arm. Since this region is rather near the distal extremity of the arm, it is clear that the arm ramification at a secundaxil will occur fairly near the tip of the growing arm. Therefore, a moderately large number of brachials is always included in successive, different brachitaxes. The secundaxils, tertaxils, and so on, hence are formed by modification of a preexisting normal brachial plate. It is unknown whether the primaxil in *Metacrinus* is formed in the same way. The large number of primibrachs (7), which in part are pinnulated, makes one suspect that it is. GISLÉN counted the number of plates in the secundibrach-to-quartibrach series in several different species of *Metacrinus* and found that an overwhelming majority of the brachitaxes contained even numbers of brachials (taking syzygial pairs as units). He ex-

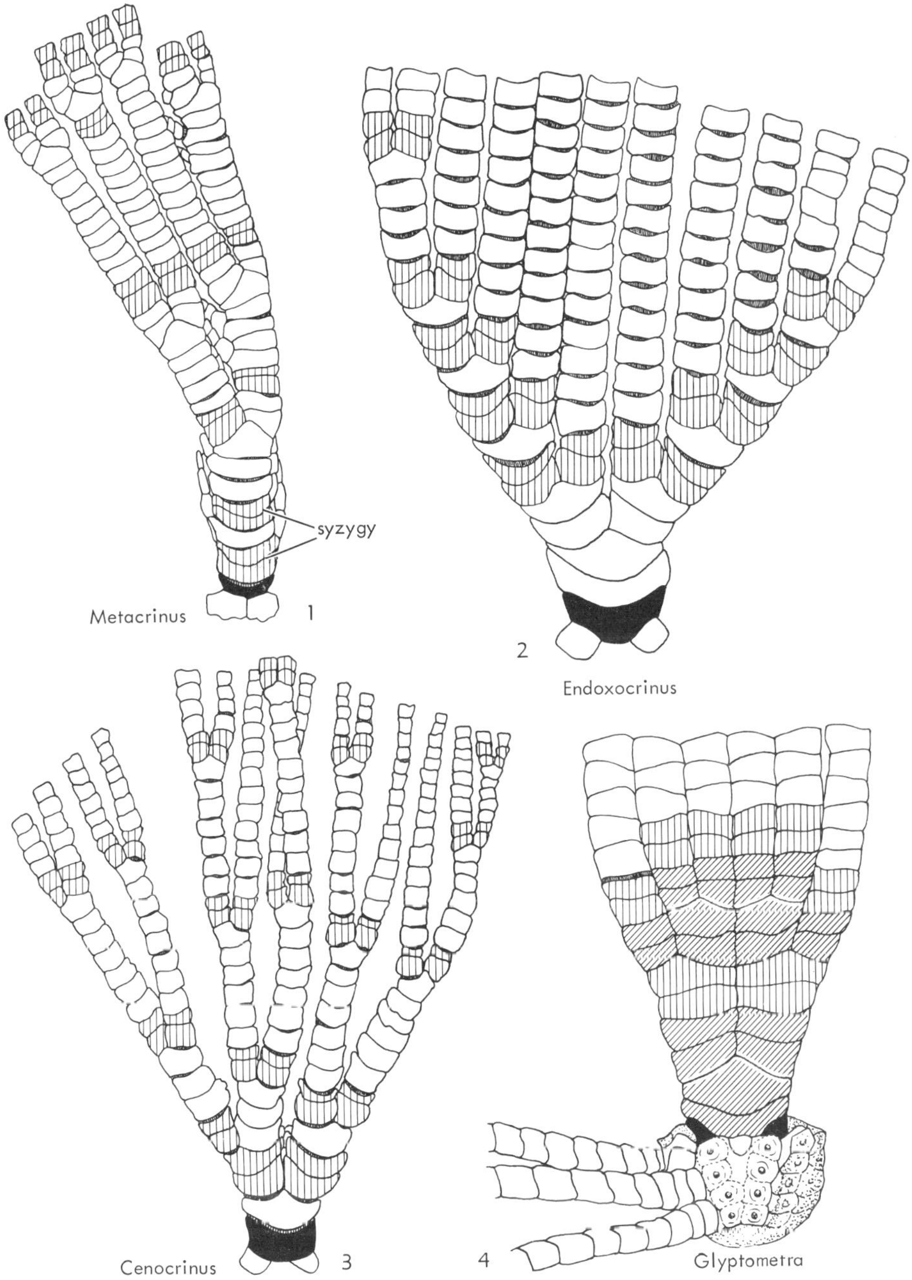

FIG. 20. Types of arm ramification in recent crinoids with distribution of synarthrial (oblique ruled) and syzygial (vertical ruled) brachial pairs.

1. Dichotomy (or isotomy) in *Metacrinus wyvillei* (Breimer, n).

2. Endotomy in *Endoxocrinus parrae* (Breimer, n).

3. Heterotomy in *Cenocrinus asteria* (Breimer, n).

4. Exotomy in *Glyptometra investigatoris* (after Clark, 1921).

plained this satisfactorily by noting that the first pinnule is always borne by the first brachial unit on the abmedial side of the arm, so that the new arm is formed from a pinnule on the admedial side of the arm, where it is more protected during its further growth than on the outer side.

Among comatulids still another method of developing the multibrachiate condition exists. This was discovered by Minckert (1905a) and is now known as the **indirect or discontinuous mode** of arm ramification. During ontogeny of the comatulids the ten-armed stage is reached by continuous terminal growth. One or more of the ten arms may be shed by **autotomy** at a proximal syzygial interface, probably by degeneration of the elastic fibers connecting the brachials, for syzygies are articulations of low resistance where arms may break easily. At this level the autotomy is exclusively produced. The hypozygal segment is prepared for **regeneration** of the shed arm. It may happen that the hypozygal regenerates an axillary plate and two arms, instead of a new epizygal and single arm. When an axillary brachial is generated, it provides a new arm ramification and augments the number of free arms; hence it is called **augmentative regeneration.** When a non-axillary brachial is generated, it introduces no arm-pattern change, and this is termed **reproductive regeneration.** Discontinuous arm duplication is also reported to exist in the isocrinid genera *Endoxocrinus* and *Teliocrinus.*

In all types of arm ramification described so far, the arm divisions borne by a given axillary plate are equal as compared to one another. In *Calamocrinus,* however, the arms borne by the axillaries are unequal, smaller lateral arms or **ramules** being formed alternately toward left and right on the asymmetrical axillary plates. Both the main arm and the ramules are pinnulated. In this genus the arms are probably formed in the manner described for *Metacrinus,* except that as a rule the first pinnule and first ramule come off toward the left in *Calamocrinus.*

MATURE PATTERNS OF RAY STRUCTURES

The different ways of forming arm ramifications which have been described produce various patterns of ray structures in mature crinoids. The number of free arm ends and the branching pattern may be characteristic for species or for genera. In some this pattern is highly stable but in others extremely variable.

In *Metacrinus* (Fig. 20,*1*) a regular dichotomous branching is achieved on relatively prominent axillaries, each brachitaxis containing relatively numerous elements. This is an isotomy, producing two equal arms on each axillary plate, with no arms undivided.

Quite another pattern of branching is displayed by *Endoxocrinus* (Fig. 20,*2*) in which regularly the admedial brachitaxis remains undivided and the abmedial one forks again, leaving very short brachitaxes behind. This produces an endotomous pattern of branching, which is highly typical for the genus. Without exception the axillaries following the primaxil are located in the abmedial brachitaxes.

An irregular pattern of branching, which could be indicated as bilateral heterotomous is found in *Cenocrinus* (Fig. 20,*3*), in which some brachitaxes may remain unbranched, while their fellows on the same axillary plate give rise to another ramification. The distribution of the undivided brachitaxes is irregular, for either the admedial or abmedial brachitaxis may remain undivided. In parts of the total ray structure a bilateral symmetry may be produced, though by no means do the two half-rays necessarily display a bilateral symmetry.

The opposite of the endotomous pattern of *Endoxocrinus* is theoretically conceivable and would give an exotomous pattern of ray structure. Actually, however, it scarcely exists among recent crinoids. Only in the comatulid family Charitometridae are several genera (e.g., *Glyptometra, Pachylometra, Zygometra*) that have six-armed rays, with secundaxils on the admedial brachitaxis, whereas the abmedial one remains undivided (Fig. 20,*4*). Unlike *Endoxocrinus,* the axillary brachials above the primaxil have an admedial position in the half ray.

The arm structure of *Calamocrinus* as described above could be called a holotomous pattern, because a pinnule grows out alternately to left and right into a ramule, which never gains the same strength and size as the original arm.

BRACHIAL ARTICULATIONS

All recent crinoids are grouped in the subclass named Articulata because of the elaborate articular faces of their brachials. The articulations between brachial plates are of two types, designated respectively as muscular and ligamentary. The first of these is most common and the second less so. In recent crinoids musculation is confined to the arms and pinnules, for the stalks and the cirri lack any musculature. Slight indications that some ligament fibers in stalks and cirri may have contractile power are found, however. The collagenous nature of problematical ligaments is described by MEYER (1971). The emphasis on musculation and its development in the recent crinoids is quite variable. Sessile stalked forms have weakly developed musculature, as have creeping forms among the comatulids. Antedontacean comatulids, especially the swimming antedontids, have very powerful muscles. Strength and effect of musculation is also determined by the number of ligamentary connections between the brachials; stiff arms may be produced where many ligamentary connections are present in such arms. This is particularly strong in some stalked crinoids. In *Hyocrinus,* for instance, the number of ligamentary articulations may exceed the muscular ones. No examples are found among the recent crinoids of arms fully devoid of muscular articulations, such crinoids being represented only by fossil forms. The opposite, however, occurs in the recent crinoids. The undivided arms of the specialized reef crinoid *Holopus* have exclusively muscular articulations, ligamentary ones being altogether absent.

Muscular Articulations. Most articulations between the brachials of recent crinoids are of muscular type, for in addition to the presence of ligaments, muscles importantly aid arm movements. As stated previously, musculation is confined to the arms, which function primarily for feeding. Since the arms extend outward while feeding and flex inward while resting, a mechanism for effecting these movements is required. Therefore, the brachials are provided with paired flexor muscles located on the inner (adoral) parts of their articular facets and an extensor ligament on the outer (aboral) parts of the facets. The muscles

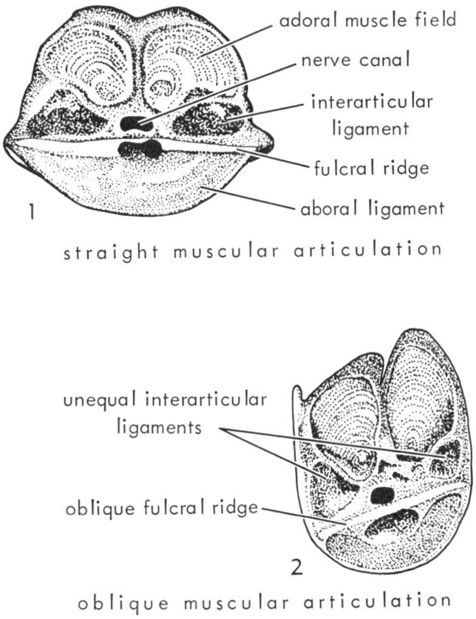

FIG. 21. Muscular brachial articulations.——*1.* Straight muscular articulation in radial of *Endoxocrinus alternicirrus* (after Carpenter, 1884a).——*2.* Oblique muscular articulation on brachial of *Neocrinus decorus* (Breimer, n).

and ligament are on opposite sides of a transverse fulcral ridge that serves as a fulcrum when the antagonizing muscles and ligament induce differential movement of contiguous pairs of brachials. Between the fulcral ridge and the adorally placed muscle areas is a pair of interarticular ligaments attached to the facets on opposite sides of a crest or groove running from the transverse ridge to the inner (adoral) margin of each facet.

When the described features of the articular facets are subequal on left and right sides, they display bilateral symmetry of the halves divided by the crest or groove running normal to the transverse ridge. The plane of such facets is disposed at a right angle to the longitudinal axis of the arm and its edge, seen either from the outer (aboral) or inner (adoral) side of the brachials, runs transversely straight across the arm. Articulation of this type is known as **straight muscular articulation** (Fig. 21,*1*).

In many recent crinoids, features of the brachial articular facets lack bilateral symmetry and their planes are more or less oblique to the arm's longitudinal axis. The

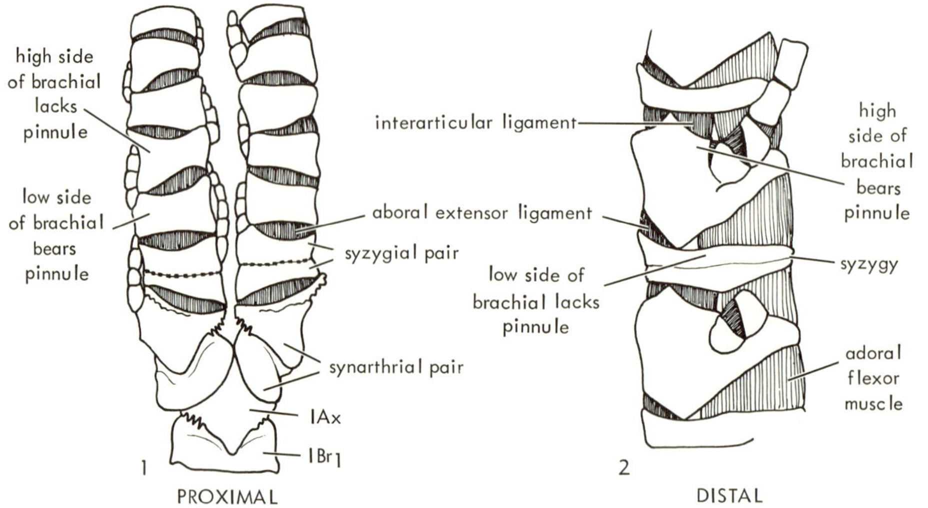

FIG. 22. External obliqueness of brachials and the reversion phenomenon in *Florometra serratissima* (Breimer, n).

1. Proximal parts of free arms with pinnules given off from lower side of brachial. *2.* Distal parts of free arms with pinnules given off from higher side of brachial.

articulation then is designated as **oblique muscular articulation** (Fig. 21,*2*) and from different aspects can be characterized as possessing **external obliquity** and **internal obliquity.** The former refers to the oblique disposition of the facetal planes as viewed from either the outer (aboral) or inner (adoral) sides of the arm. The latter takes account of inequality of size and asymmetry in arrangement of paired muscle and ligament areas of the facets, features which cannot be observed without disarticulating successive brachials from their neighbors. The nature of both external and internal obliquity needs to be studied in relation to pinnule sockets located alternately on left and right upper edges of successive brachials. If one observes the adoral side of the arms from which all soft tissues except the ligaments have been removed, he finds the inner terminations of the facetal grooves or crests between the two muscle fields perfectly aligned with the longitudinal axis of the arm. Therefore, it is proper to account for features of internal obliquity in oblique muscular articulations by stating that the enlarged muscle and interarticular ligament fields on the left or right sides of the facets have differentially displaced the transverse ridge toward the outer (aboral) margin of the facet more than the opposite smaller muscle and interarticular ligament areas adjacent to them beyond the median groove or crest. The ligament on the aboral side of the transverse ridge and the enlarged one on the adoral side may operate as a sort of bifascial or synarthrial articulation (explained in discussing ligamentary articulation, below) disposed more or less obliquely.

External obliquity seen in oblique muscular articulations is marked by inclined attitudes of the outer (aboral) and inner (adoral) margins of the facetal planes with respect to the longitudinal axis of the arm. The distal articular facet of each brachial slopes gently to somewhat strongly upward to left or right sides of the arm, with greatest height of each brachial on the side which bears a pinnule socket and least height opposite to this. Thus, along either the left or right sides of an arm, viewed from its aboral or adoral side, successive brachials alternately are taller and lower. Usually the sockets for attachment of pinnules are located on the higher side of the brachials, at least in middle and distal portions of the arms. In some crinoids (e.g., *Florometra serratissima,* Fig. 22) the opposite may be true of proximal brachials, pinnules being given off from the lower side of each brachial. External obliquity of this sort has been termed **reversion** by GISLÉN (1924). It is known only in comatulids, among which such swimmers as the antedontids display reversion most clearly and some creeping comasterids only moderately or very little. The phenomenon has not yet been explained satisfactorily but probably it has to do with the "synarthrial capacities" of the oblique muscular articulations. By reversion of the articulations at least two different longer axes of oblique synarthrial articulation are superposed in the arm. This tends to increase the flexibility of the arm when it is exposed to lateral pressure.

Internal obliquity of the oblique muscular articulations is mainly caused by pinnulation. Forms with defective pinnulation (e.g., *Atelecrinus*), which lack pinnules on proximal brachials of the free arms, have straight articular musculations on such pinnuleless brachials. Proximal brachials of the free arms by invasion of the pinnule socket, which lies either between the muscle and the interarticular ligament or adorally from the muscle fields. Distal brachials of the free arms may have the pinnule sockets situated either on their lateral or on their adoral surfaces, completely separated from the articular facet, but still causing obliquity of that face.

It should be kept in mind that no sharp distinction may be found between straight and oblique muscular articulations, for the two types intergrade with one another. Perhaps the only perfect straight muscular articulation is to be found in the interface between radials and the first primibrachs. As a rule, oblique muscular articulation is most strongly developed immediately after the axillaries; in distal direction they first tend to become straight but farther on they become somewhat more oblique again.

Ligamentary Articulations. Ligamentary articulations known to occur in the arms of articulate crinoids are chiefly of three types: synostosial, syzygial, and synarthrial.

Ligamentary contacts in the arms of *Calamocrinus* and *Ptilocrinus* are of such a simple type that they are best indicated as **synostosial.** The brachials so united have practically flat interfaces held together by short ligament fibers rather evenly distributed over the joint face.

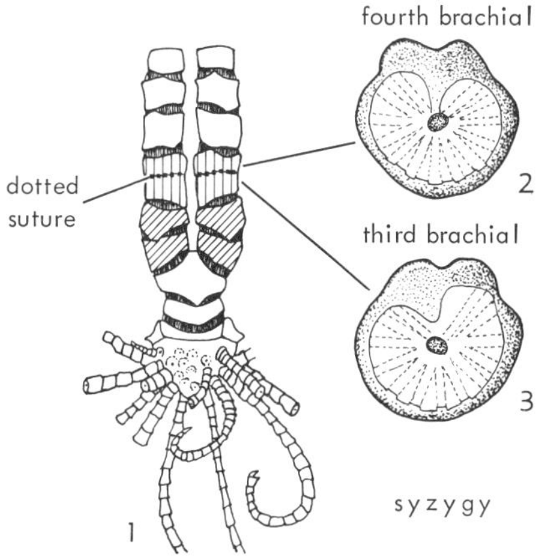

FIG. 23. Proximal arm portions *(1)* of *Eumorphometra hirsuta,* showing synarthrial articulations (oblique ruled) between 1st and 2nd secundibrachs and syzygial ones (vertical ruled) between 3rd and 4th secundibrachs; schematic views of distal facet of 3rd secundibrach *(3)* and proximal facet of 4th secundibrach *(2)*, enl. (after Carpenter, 1888).

Syzygial articulations are confined to the arms and are even specially developed to serve the mobility of arms in vagile crinoids. What is known in the literature as stem syzygy is in fact a synostosis or cryptosymplexy. On the other hand, synarthries are not confined to crinoid arms for they occur in the stalks of Bourgueticrinida.

The sort of ligamentary articulation most frequently found in the arms of recent crinoids, but typically developed only in the comatulids, is syzygial. It may be defined as a ligamentary articulation in which the opposed joint faces bear a number of culmina that radiate from the axial canal, the culmina of apposed faces meeting one another instead of being interlocked by fitting into crenellae. When the syzygial suture is observed from the outer (aboral) side of the arm, it appears as a characteristically dotted line. The ligament fibers at the syzygial contact are very short and located mainly in the depressions (crenellae) in between the radiating culmina and therefore appear externally as dots along the suture (Fig. 23). The syzygial contact allows the plates so connected slight mobility in all directions. The syzygy may be regarded as an extremely helpful articulation since it may prevent extended arms

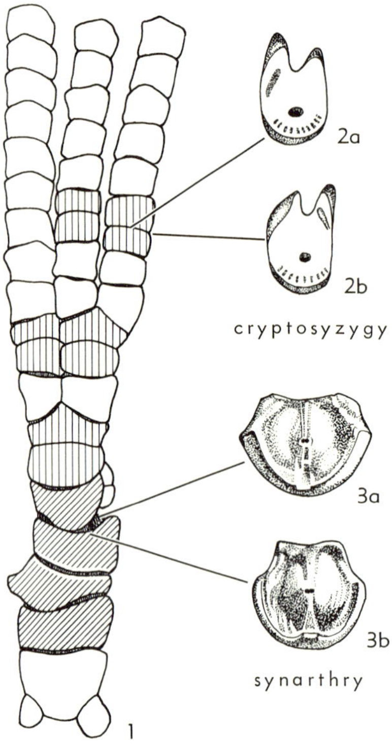

FIG. 24. Ligamentary articulations in arms of *Neocrinus decorus* (Breimer, n).——*1.* Outer (aboral) face of arm showing distribution of brachials united by synarthrial (oblique ruled) and syzygial (vertical ruled) articulations.——*2.* Facing articular facets of tertibrach syzygial brachial pair (cryptosyzygy); *2a,* proximal facet of 4th tertibrach; *2b,* distal facet of 3rd tertibrach.——*3.* Facing articular facets of synarthrial brachial pair; *3a,* proximal facet of 4th primibrach; *3b,* distal facet of 3rd primibrach.

from breaking when exposed to torsion. The plates united by syzygy are sometimes indicated as a syzygial pair. It is stressed however, that the term syzygy is introduced to designate the type of ligamentary articulation, in order to discriminate it from other such articulations. The plate below the syzygial contact is called **hypozygal** and the one above it **epizygal.** Very commonly axillary brachials are joined to plates next preceding them by syzygy.

In stalked crinoids syzygies are far less developed than in the comatulids. The radiating culmina of their articular facets are very short and may even be interrupted to form rows of tubercles; they fail to

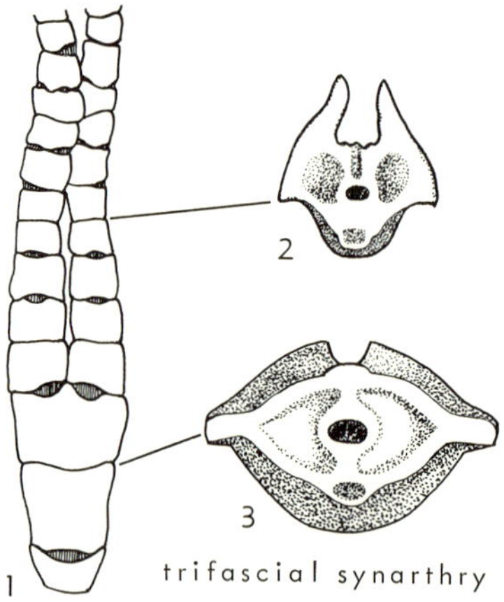

FIG. 25. Trifascial synarthrial brachial articulation in *Bathycrinus aldrichianus.*——*1.* Outer (aboral) side of arm showing locations of facets illustrated (Breimer, n).——*2.* Distal facet of 4th secundibrach (after Carpenter, 1884a).——*3.* Distal facet of 1st primibrach (after Carpenter, 1884a).

reach the aboral surface of the brachials and do not show a dotted suture. Such weakly developed syzygies, known as **cryptosyzygies,** occur in the isocrinids (Fig. 24). The mobility allowed by such contacts is probably very low.

A third type of ligamentary articulation, known as **synarthrial,** evidently developed to allow the brachials so united a differential flexible movement in two directions. Brachial synarthries are fundamentally of the same kind as are found in the stalks of some crinoids (compare Fig. 24 with Fig. 9), inasmuch as two larger ligament bundles are present on either side of an articulatory bar or fulcral ridge running in adoral-aboral direction. The central nerve canal is located on this ridge. Traditionally —and rightly—this type of synarthy has been called a **bifascial articulation.** Another type of synarthry possessing three ligament bundles has been termed **trifascial articulation.** It occurs in certain ten-armed Bourgueticrinida (e.g., *Bathycrinus;* see Fig. 25). In *Bathycrinus* such synarthrial pairs fully replace syzygial pairs.

Synarthries, like syzygies, may be weakly developed and these are indicated as **cryptosynarthries.** They occur in creeping types of comatulid crinoids (Comasteridae). The power of flexibility is practically reduced to nil.

A rather special type of ligamentary articulation is found in *Neocrinus blakei* and *Hypalocrinus naresianus* (Fig. 26). This is known as **symmorphial,** a ligamentary contact in which a prominent transverse culmination of the epizygal brachial fits into a corresponding depression of the hypozygal; peripheral culmina and crenulae may occur near the aboral margins of the symmorphic joint faces, just as in cryptosyzygies. For this reason the symmorphy may be considered as a modified sort of syzygy. Symmorphial contacts allow only very slight possibility of differential movement of the brachials in two directions, if any at all.

Among all ligamentary contacts the synostosis is morphologically the most simple and phylogenetically the oldest type. Cryptosynarthries, cryptosyzygies, and cryptosymplexies are all very close to synostosis and sometimes in literature are indicated as such. It seems logical to think that synarthries, syzygies, and symplexies are derived from synostoses. Indeed, as soon as a concentration of ligament fibers into bundles occurs and the remaining spots are filled in with calcareous ridges, striae, or culmina, we get tendencies toward synarthries, syzygies, and symplexies, which might be called cryptosynarthry, cryptosyzygy and cryptosymplexy, respectively.

The distribution of synarthrial and syzygial contacts in recent crinoid arms has been an object of study by CLARK (1908e). His conclusions have been criticized by GISLÉN (1924). From observations of these workers some regularities in distribution of the ligamentary articulation types are known. The most regular occurrence of synarthrial and syzygial pairs throughout the arms is found in the comatulids. These crinoids develop only two primibrachs in each ray, with articulation between them of synarthrial type (except in the Zygometridae where it is a syzygial). All brachitaxes between the primaxil and most distal axillary include either two or four brachials. In case only two brachials are present in a brachitaxis, they are united by synarthry. If, on the contrary, four brachials constitute the brachitaxis, numbers *1* and *2* are united by synarthry, whereas *3* and *4* are joined by syzygy. The free-ending arms (or terminal brachitaxes) invariably start with a synarthrial pair followed by a syzygial pair. This so-called distal synarthrial pair in comatulids very generally is more elaborate, larger than the following plates, and not uncommonly ornamented (see Fig. 22, *Florometra serratissima*). Syzygies may be repeated at regular intervals in the free arms. The extent of such intervals is characteristic of genera or species. Synarthries are not repeated in the free arms.

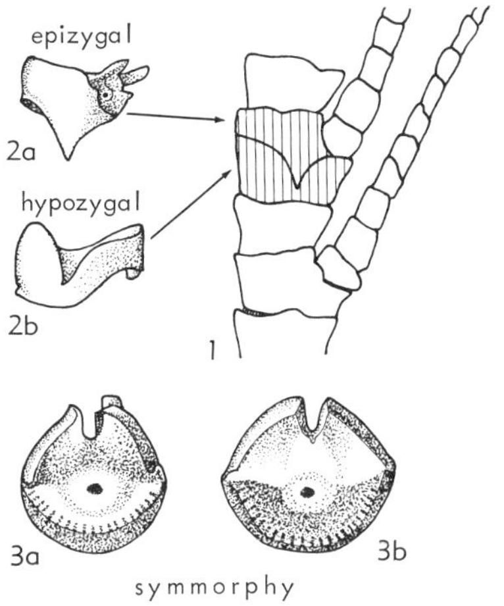

FIG. 26. Symmorphic brachial articulations in *Neocrinus blakei* (after Carpenter, 1884a).——*1*. Side view of part of arm with pinnules at right, showing location of symmorphially articulated brachials (vertical ruled).——*2*. Side views of distal *(2a)* and proximal *(2b)* units of brachial pair.——*3*. Facetal views of symmorphial brachial pair illustrated in *2; 3a,* distal facet of hypozygal; *3b,* proximal facet of epizygal, enl.

In Isocrinida the distribution patterns or synarthries and syzygies are less regular than in comatulids. Isocrinids with two primibrachs show them usually united by syzygy, but in *Neocrinus* and *Hypalocrinus* by synarthry. Isocrinids may have more than two primibrachs. For example, *Metacrinus* has either four primibrachs: a synarthrial pair followed by a syzygial pair, just as in comatulids; or seven primibrachs, with first a synarthrial pair and then a syzygial pair midway in the brachitaxis. The composition of the free-ending arms is different from the comatulids (except for *Hypalocrinus*). As a rule, synarthries do not occur in the free-ending arms, and even

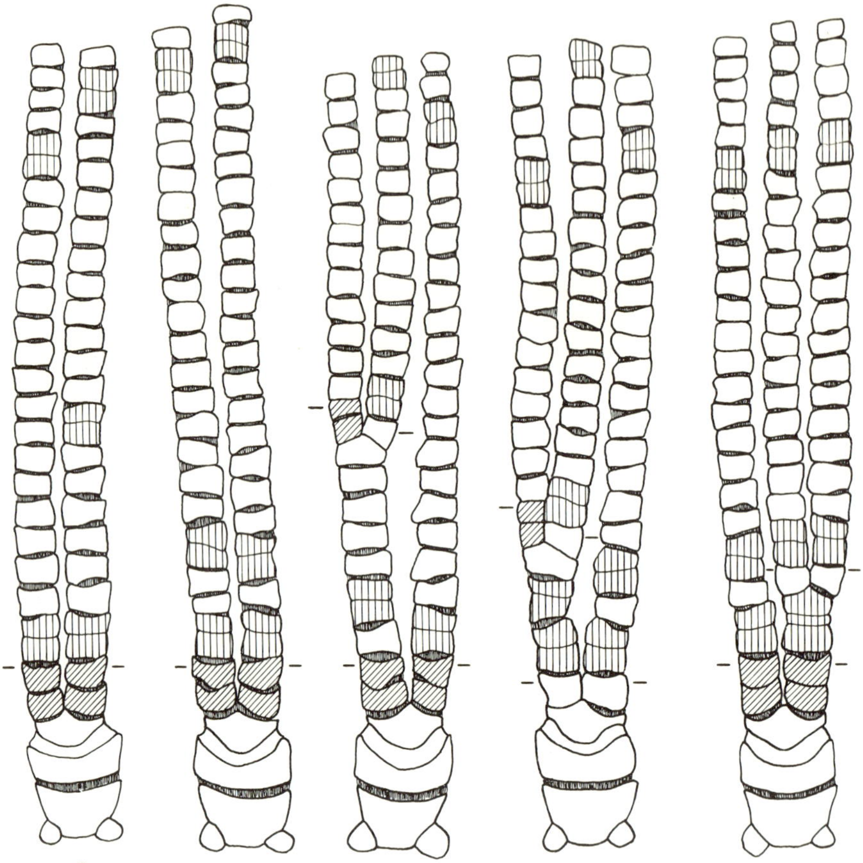

FIG. 27. Semidiagrammatic representation of arm ramification and distribution of synarthrial (oblique ruled) and syzygial (vertical ruled) pairs of brachials in a specimen of *Neocrinus decorus* (Breimer, n).

syzygies may be altogether absent there (e.g., *Endoxocrinus*). The distribution of ligamentary articulations in *Neocrinus* is shown in Figure 27. Brachitaxes between the primaxil and last axillaries have four elements as a rule: a synarthrial pair and a syzygial pair. Exceptionally, the number of such brachitaxes may be three or five. Free-ending arms invariably begin with a synarthrial pair followed by a syzygial pair if they arise from the primaxil. If they arise from a higher-rank axil, the synarthrial pair is commonly suppressed to a single plate, which is followed by syzygy, or it is present but not followed by a syzygy. In many cases it is also observed that the first two brachials do not form a synarthrial pair, but brachials *3* and *4* form a syzygial pair. Compared to the comatulids, this means a strong reduction of synarthries in the free-arm ends, and, consequently a lower degree of lateral flexibility in the arms.

In five-armed Bourgueticrinida (e.g., *Rhizocrinus, Democrinus*) synarthries are absent. The arms are composed of regularly alternating syzygial pairs. In ten-armed bourgueticrinids (e.g., *Bathycrinus,* Fig. 25) syzygies are absent, being replaced by synarthries of trifascial type; only secundibrachs *3, 6,* and *9* have muscular articulations at both ends, which means that the arms are formed mostly by synarthrial pairs.

In the Cyrtocrinida, synarthries are altogether absent, but syzygies are widespread.

In *Hyocrinus,* for instance, syzygial triplets (Fig. 18,*1*) regularly follow one another in the distal parts of the arms. A very peculiar flexible arm occurs in *Holopus,* for all of its brachial articulations are of muscular type.

PINNULATION

Modern crinoid arms are densely pinnulated structures. A **pinnule** is a lateral appendage of the arm supported by serially arranged skeletal segments called **pinnulars.** The pinnulars are interconnected by ligament and muscle fibers, as are the brachials. Parts of the coelomic, water-vascular and nervous systems penetrate into the pinnules. On their adoral sides pinnules bear ambulacra. The morphology and function of the pinnules is largely the same as that of the main arms.

As a rule, pinnules are distributed alternately along left and right sides of the arm, with a single pinnule to each brachial. More precisely, each pinnule is given off by a **pinnular arm unit,** for this may be composed of more than one brachial. Pinnulation is subject to several strict rules. Normally, every brachial with muscular articulations on its proximal and distal facets comprises a pinnular arm unit, but this may consist of brachial pairs or even triplets united together by syzygy or synarthry. Such multiple-brachial pinnular arm units invariably give rise to a pinnule from their distal component, and never from their proximal or intermediate brachial. Thus epizygals are pinnule-bearing, whereas hypozygals lack pinnules. No axillary arm plate of a recent crinoid is known to give rise to a pinnule, even when muscular articulations characterize both its proximal and distal facets.

Absence of pinnules other than as just mentioned constitutes **defective pinnulation.** This is known in many species of both stalked and stalkless recent crinoids. Defective pinnulation occurs when pinnules are absent at the base of a brachitaxis in places where they could be expected. In the hyocrinid genera *Calamocrinus* and *Ptilocrinus* the first pinnule is to be found on the left side of the fourth brachial viewed aborally, while pinnules on more proximal brachials are missing. In the five-armed bourgueticrinid genera *Democrinus* and *Rhizocrinus,* the first pinnule is found on the left side of the fourth pinnular arm unit, more proximal pinnular arm units lacking pinnules. In these genera up to nine brachials may constitute these four pinnular arm units (e.g., *Democrinus campbellianus*). Ten-armed bourgueticrinids, such as *Bathycrinus,* have even more defective pinnulations; the first pinnule is developed on the admedial sides of the sixth pinnular arm units (which may include up to nine brachials) in the secundibrachitaxis.

The only known case of **complete pinnulation** is found in the ten-armed cyrtocrinid *Holopus.* Each of its secundibrachs is a pinnular arm unit because ligamentary connections between the brachials are absent. The first pinnule occurs on the first secundibrach.

Isocrinida and Comatulida tend to have more complete pinnulation than most millericrinids, bourgueticrinids, and cyrtocrinids. Even forms with many primibrachs (e.g., *Metacrinus*) have the pinnulation as complete as possible. The first pinnule is located on the second primibrach on the right side of the basal synarthrial pair. The first three or four pinnules on the primibrachs are entirely or at least partly fixed to the thecal wall; such pinnules are known as **fixed pinnules.**

Multibrachiate isocrinids have brachitaxes (apart from the free arms) made up of two to four brachials, but in no case is more than one pinnule present. If only two plates form a brachitaxis, these are invariably interconnected with ligaments; hence the proximal brachial has no pinnule and the second has none either, because it is axillary. If four plates form a brachitaxis, it contains a synarthrial pair with one pinnule and a syzygial pair without a pinnule, for the fourth brachial again is axillary.

The free-arm ends of isocrinids generally have no basal synarthrial pair. Commonly, as in *Annacrinus, Cenocrinus,* and *Endoxocrinus,* the free arms begin with a syzygial pair of brachials. The first pinnule is on the abmedial side of this basal syzygy. Accordingly, pinnulation of the free arms borne by any one axil is **symmetrical.** In *Teliocrinus* the basal syzygy of the free arms may occur between either the second and third or third and fourth brachials. The first pinnule is always located on the first brachial and uniformly

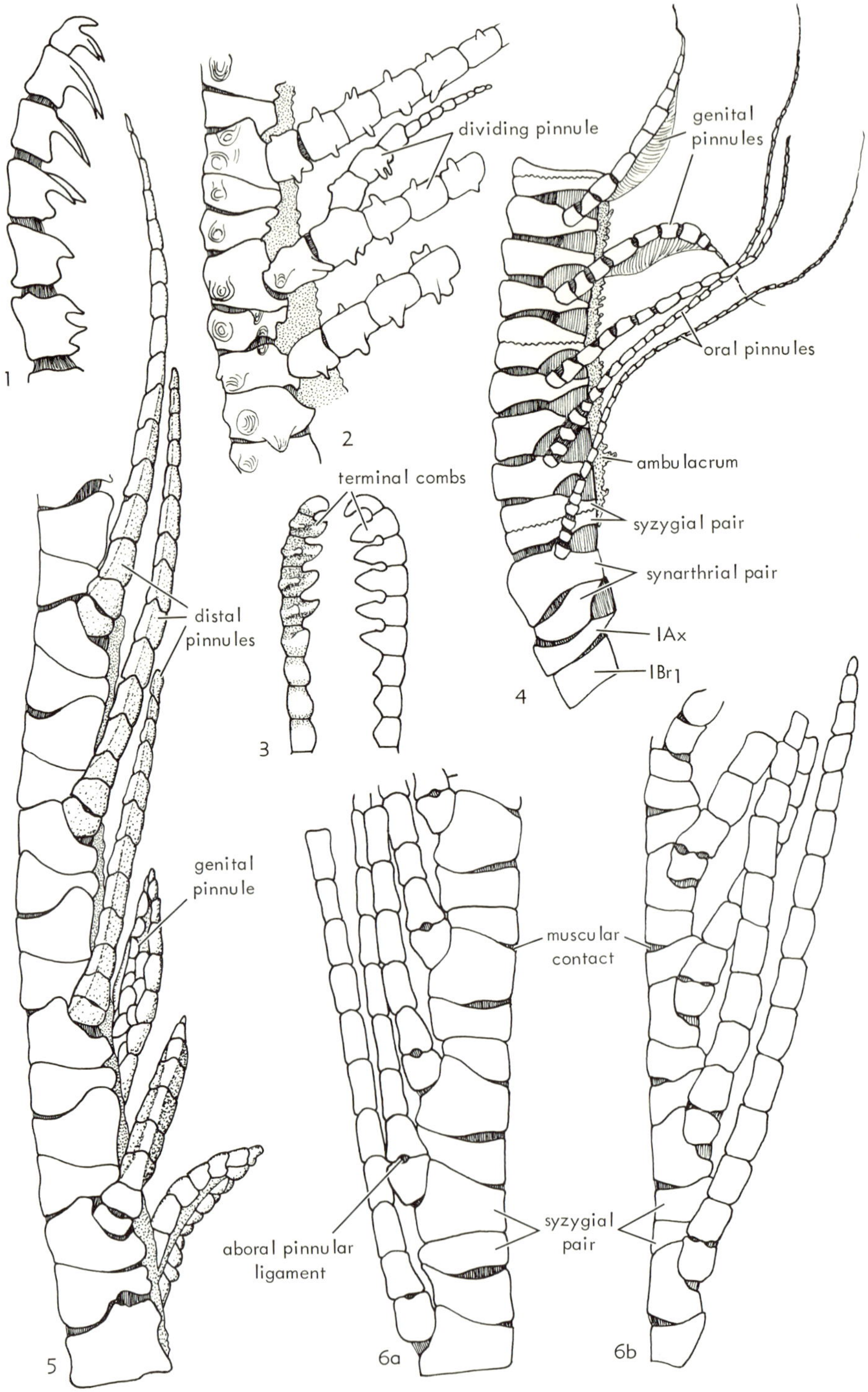

FIG. 28. Pinnulation and pinnular structures. *(Explanation on facing page.)*

either abmedial or admedial in direction. This makes pinnulation of the free arms arising from any given axil **identical.**

The free arms of the isocrinid *Hypalocrinus* and all comatulids begin with a synarthrial pair of brachials followed by a symmorphial pair. The first pinnule is found on the abmedial side of the synarthrial pair and the second pinnule on the admedial side of the syzygial pair. The pinnulation pattern is mostly symmetrical.

The form and organization of pinnules is variable according to specialized functions which they may assume. Very generally those of comatulids are differentiated into oral, distal, and genital pinnules.

The **oral pinnules** (Fig. 28,*4*) typically have become transformed into tactile organs or modified for defense of the naked, noncalcified tegmen of the comatulids. These pinnules are associated with the first synarthrial and syzygial pairs of brachials in the free arms on which they invariably occur. Several additional oral pinnules may follow, however. For effective performance of their function the oral pinnules have been developed as long, slender structures composed of many small pinnulars, which give them enormous flexibility. Ambulacra are absent on these pinnules.

Oral pinnules are restricted to the comatulids, those of comasterids being curiously specialized. Their distal pinnulars, which may be produced into arrowhead- or spadelike shapes, form the terminal comb (Fig. 28,*3*). Probably they serve to aid in fixation. Comasterids may use their posterior arms to curve around objects (see Fig. 41) for fixation; the terminal combs of the oral pinnules are helpful, since comasterids have few or no cirri.

Stalked crinoids with armored tegmens do not seem to need defensive organs to protect the vulnerable ambulacral ducts on the tegmen.

Following the oral pinnules are several **genital pinnules** (Fig. 28,*4,5*) which bear the gonads. In the spawning and breeding season these pinnules appear as swollen structures, for the gonads are packed with mature gametes. Genital pinnules are generally very much shorter than oral ones, being composed of only about ten pinnulars. An ambulacral groove that may be present on them is functional for food-collecting outside of the breeding season when the pinnule is not swollen.

The **distal pinnules** (Fig. 28,*5*), including all beyond the oral and genital pinnules, serve in gathering food particles, and hence their ambulacral grooves are particularly well developed. Throughout the recent crinoids they are very uniformly composed of about 20 segments. The total length of functional ambulacra of individuals is determined mainly by the number and length of their distal pinnules and by the number of their free arms. Clark (1921) has observed that ten-armed crinoids possess longer distal pinnules than multibrachiate species, and that relative length of the pinnules decreases in direct correlation with increase in the number of arms.

The growth and development of pinnules is known from the ontogeny of several different genera of comatulids. Pinnulation develops only very late in ontogeny. Juvenile comatulids of the ten-armed growth stage may have free-arm ends built of as many as 20 secundibrachs without a single pinnule present. Pinnules seem to become introduced in a distal-proximal sequence. The first oral pinnule is formed only after about half a dozen distal pinnules are well developed. This distal-proximal order of pinnule development may be understood by bearing in mind that the distal pinnules serve as food-collecting appendages, and as such apparently precede reproductive and tactile pinnules in order of appearance. The delay in development of oral pinnules may become appreciable enough to cause defective pinnulation.

In multibrachiate species that have surpassed the ten-armed growth stage, a terminal growth of pinnules is observed at

1. Side view of spiny pinnular tips in *Comatella nigra* (after Clark, 1921).

2. Dividing pinnules of *Ptilocrinus antarcticus* seen from side (Breimer, n).

3. Terminal combs on comasterid pinnules (after Clark, 1921).

4. Oral and genital pinnules of *Promachocrinus* in side view (Breimer, n).

5. Genital and distal pinnules of *Democrinus rawsoni* in side view (Breimer, n).

6. Pinnulation in two proximal fragments *(6a,b)* of *Ptilocrinus brucei* (Breimer, n).

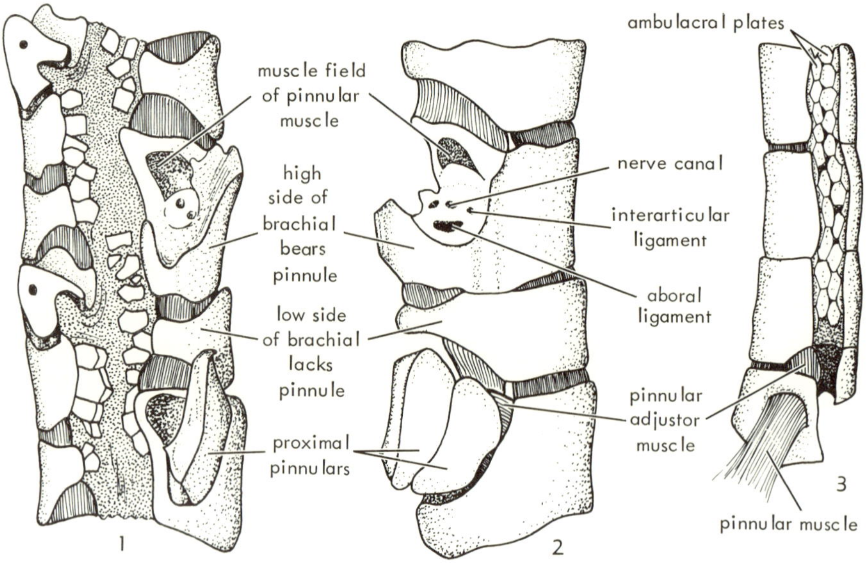

FIG. 29. Pinnular articular structures in *Neocrinus decorus* (Breimer, n).

1. Adoral side of arm fragment showing proximal pinnulars joined to alternating brachials on opposite sides of arm.

2. Side view of same fragment corresponding to right side of *1.*

3. Four proximal pinnulars with ambulacral covering plates.

the growing arm tips. The order of introducing new elements at the growth tip is: 1) a brachial, 2) a first pinnular, 3) another brachial, and 4) a first pinnular on the opposite side as the latter. This means that in multibrachiate crinoids pinnules are introduced in alternating order directly after appearance of the brachial to which they belong.

As mentioned previously, pinnules may contain a variable number of pinnulars, which generally are all alike, apart from the proximal two pinnulars. In the genital and distal pinnules these two pinnulars are always shorter than succeeding ones. The first pinnular is usually two or three times wider than high and more or less crescentic in outline; second pinnular is trapezoidal in shape, with width about equal to height. The two pinnulars are specialized in form because they serve special functions in movement of the pinnules and are provided with well-developed muscular-articulation facets.

The contacts between brachial and first pinnular, and also between first and second pinnulars are of a straight muscular type, with a large aboral ligament, and on the opposite side of the transverse ridge two equal interarticular ligaments and a pair of equal adoral muscles on either side of a crest or groove running normal to the transverse ridge.

The articulation between the first pinnular and the brachial to which it is attached is a straight muscular one, with the transverse ridge of the facet placed at an angle of about 45° to the longitudinal axis of the arm. The muscles connecting the first pinnular to its brachial may be displaced to the lateral side of the first pinnular (see Fig. 29,*3*). These muscles serve for adjusting the position of the pinnule with respect to the arm, moving the pinnule toward or away from the arm. The muscles may be regarded as **pinnular adjustor muscles.**

The articulation between the first and second pinnulars is a strongly developed muscular articulation, with the transverse ridge at right angles to the preceding one. The transverse ridge separates two strongly beveled sides of the articular surface. Contraction of the adoral muscles between the two pinnulars causes flexion of the pinnule.

The articulation between all pinnulars beyond the first pair is of a specialized type, entirely confined to the pinnular interfaces; it is termed **pinnular articulation.** It is very uniformly constructed throughout recent crinoids, consisting of a pair of small adoral muscles and a pair of large aboral ligaments on either side of a fulcral ridge running in adoral-aboral direction. It is therefore no longer a true transverse ridge. The pinnular fulcral ridges are oriented at right angles to the transverse ridge between the first and second pinnulars and all have the same orientation throughout the pinnule. The pinnular articulation allows mobility in lateral direction only. The pinnular muscles are unable to contribute either to adjustment or to flexion of the pinnule. Musculation in the pinnules is mostly rather weak and consequently these appendages have a degree of rigidity. Musculation is better developed in the oral pinnules, which may even have extreme flexibility.

ADORAL SKELETON

COMPONENT PARTS

The **adoral or perisomic skeleton** has been topologically defined (p. *T*19) as including every skeletal element which may be present in the adoral body wall. This wall has a main central part around the mouth, formerly referred to as the disc, but presently indicated as the **tegmen.** In forms with an advanced ontogeny the adoral body wall extends between the arm bases, so as to assist in forming lateral parts of the theca. The wall is mainly devoted to the functions of food-gathering, food-conveying, and food-receiving. Accordingly, the organ systems involved in these functions largely determine its form and shape. The water-vascular system also is involved in food-gathering. Its entrances, the **hydropores,** are distributed over the central and interradial parts of the adoral body wall and its terminal parts, the **podia or tube feet,** accompany the ambulacra all along their course. The ambulacral system is involved in food-conveying. Its main elements are the **ambulacra,** descending from the pinnules and arms and leading over the tegmen toward the mouth. The entrance to the digestive system and the exit from it invariably are located in the adoral body wall.

DEVELOPMENT

The degree of calcification of the adoral body wall is far smaller than in the aboral body wall. Consequently, the adoral skeleton contains much less definite and characteristic elements than the aboral skeleton. Unlike the latter, it does not contribute substantially to overall body form. A function of skeleton-secretion in the adoral wall is to provide internal coherence. This may lead to quite different products. In some comatulids the adoral wall appears to be a leathery integument entirely devoid of skeletal plates. But even such apparent leathery walls actually possess microscopic carbonate deposits in the form of rods, spicules, and smaller meshworks. In some isocrinids (e.g., *Neocrinus decorus,* Fig. 30) the tegmen is wholly calcified and plated. The need for calcification of the adoral body wall, and especially of structures bordering the ambulacra, depends on the requirement for protection of the food-conveying system, which is vital to the crinoid organism. In comatulids with naked tegmens the proximal pinnules are differentiated either into long and slender, very flexible, tactile structures or rigid, spinelike structures that serve for defense of the vulnerable tegmen. In isocrinids, however, the proximal pinnules are not differentiated, necessary protection of the ambulacra on the tegmen being provided by strong calcification of structures bordering the ambulacra.

Ontogenetically the adoral body wall becomes individualized only after rotation of the coelomic compartments. Just before this rotation a part of the lateral body wall of the crinoid embryo invaginates so as to form the vestibule, which later is cut off as a separate cavity and assumes an oral position after rotation. The roof of the vestibular cavity is the embryologic adoral body wall. It is supported by five large skeletal elements called **orals,** and these are the first elements to be formed in the skeleton of the adoral body wall. The orals correspond in interradial position to the basals in the aboral skeleton and are formed at about the same time. The floor of the vestibular cavity is primitively situated at the upper level of the radial circlet as soon as the radials have been secreted. The roof

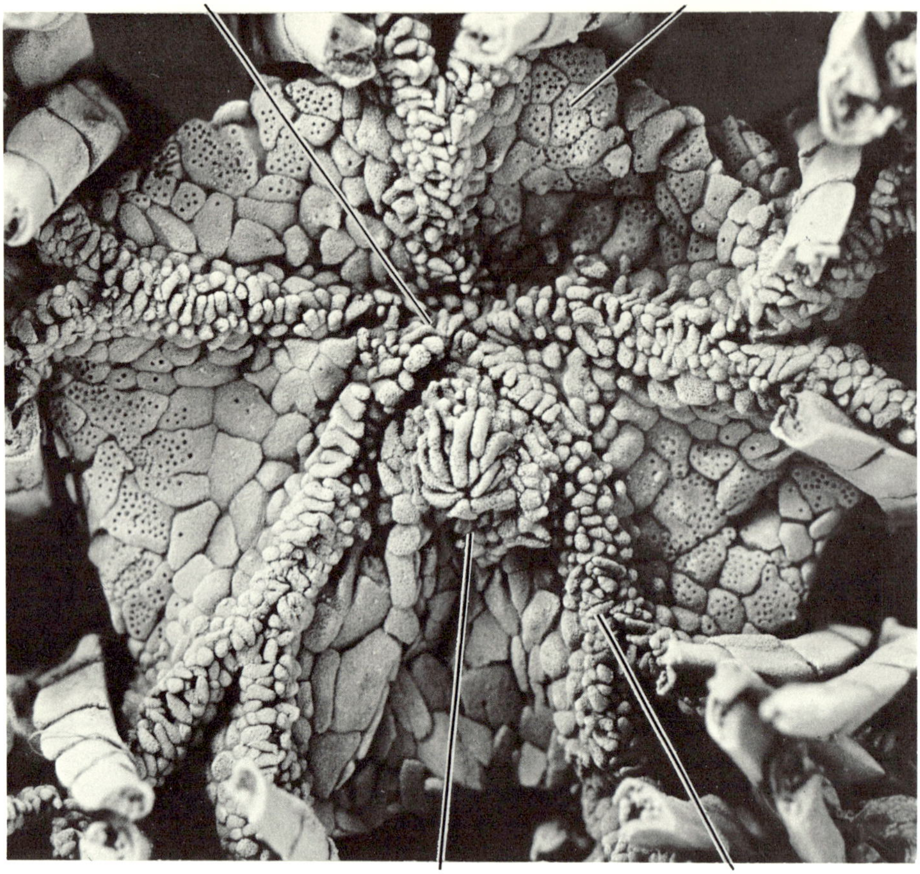

FIG. 30. Tegmen of *Neocrinus decorus*, showing hydropores on interambulacral plates except in *CD* interray (Breimer, n, through courtesy of Dr. Porter M. Kier, Washington).

of the cavity opens by formation of slits between the orals. The circlet of oral plates is split up into five lappets, which give them the appearance of five large valves. Shortly before opening of the roof of the vestibular cavity, the mouth has been formed in the bottom of this cavity. It is functional immediately after opening of the vestibular cavity and the orals are preserved for some time as protection for the juvenile mouth. In some forms with primitive ontogenetic development the orals persist during maturity and the mouth retains its position beneath them. The orals are fully preserved in *Hyocrinus* and *Holopus* (Fig. 31), in which the ambulacra run beneath edges of the orals. These five plates clearly protect the ambulacra. In the ontogeny of most comatulids, however, the orals are resorbed and the roof of the vestibular cavity disappears entirely. The bottom of the cavity forms the tegmen and during further life assumes the role of the adoral body wall. It is perhaps not justified to define the position of the mouth in crinoids with large valvate orals as subtegminal, since the tegmen may be recognized as an outgrowth of the bottom of the vestibular cavity.

TEGMEN

In mature crinoids the location of the

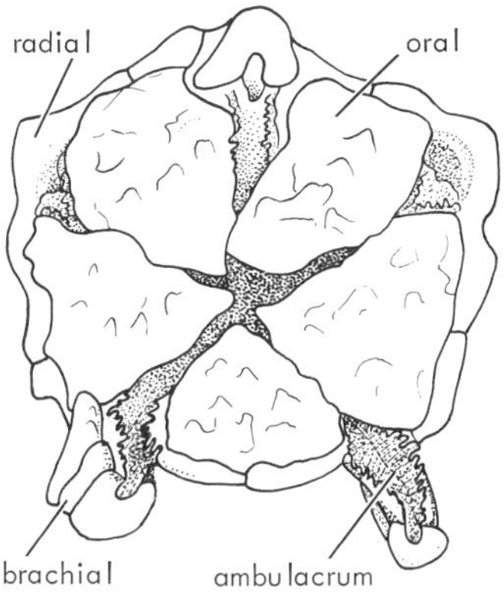

FIG. 31. Tegmen of *Hyocrinus* sp., roofed over by five large oral valves (Breimer, n).

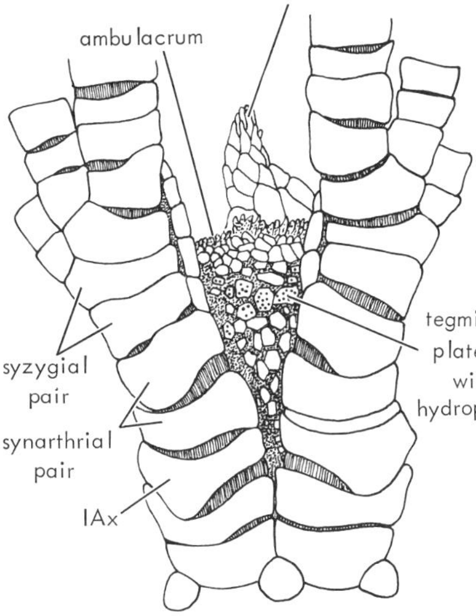

FIG. 32. *DE* interray of *Neocrinus decorus,* showing perisomic plates with hydropores (Breimer, n).

tegmen is variable. Primitively, it is developed as a flat-lying structure at the upper edge of the radial circlet. This position is retained in *Hyocrinus* and *Holopus,* whereas in *Calamocrinus* and *Ptilocrinus* a series of small scales becomes intercalated between edges of the radials and orals. Here the tegmen is a much inflated structure standing on the calyx. In all other recent crinoids the tegmen is elevated to a much higher level. In the bourgueticrinids it is located at the level of the fourth brachial element beyond the radial circlet; in five-armed forms above four primibrachs and in ten-armed crinoids above the two primibrachs and the proximal two secundibrachs. In the isocrinids, which generally are multibrachiate (except *Hypalocrinus*), the tegmen may be so elevated as to reach the bases of the free arm ends. Generally, the lower tertibrachs are in contact with the tegmen. In comatulids the tegmen is raised also. Commonly it is above the two primibrachs and may even be higher than the fourth secundibrach.

Elevation of the tegmen beyond the upper edge of the radial circlet provides extension of the central body cavity and hence enlargement of the theca. To maintain the theca as a closed space, the bases of the arms must contribute to its formation. This may be achieved by lateral abutment of the proximal brachials, as in bourgueticrinids, or by further expansion of the body wall in regions between the arms, as in isocrinids. The body wall in these places may be reinforced with perisomic plates or scales known as **interbrachial plates** (Fig. 32). These plates merge with others developed in the interradial parts of the tegmen.

Extension of the central body cavity is caused entirely by marked increase in volume of the gut as the animal grows. To provide the necessary space for it, the adoral body wall migrates upward between the arm bases. The ultimate location of the tegmen is determined by the volume of the gut, which in turn is determined by the length and efficiency of the ambulacral grooves. This length depends on such factors as richness of food and feeding mechanism. Valuable remarks on this subcct have been made by GISLÉN (1924).

The mature tegmen evidently and naturally is divided into **ambulacral and interambulacral areas.** The ambulacra converge on the mouth in a perradial position. The alternate intervening regions are indicated as interambulacral areas. Perhaps it would be better to indicate these regions simply as the **interradial areas of the tegmen,** since they may be occupied almost completely by the ambulacral grooves from fixed pinnules. This is particularly evident in *Metacrinus,* which has as many as eight primibrachs, four of which bear pinnules. These four pinnules are fixed to the theca and have their ambulacral tracts running independently over the tegmen toward the mouth (Carpenter, 1884a, pl. 39, fig. 2).

The interradial parts of the tegmen commonly are calcified in some degree. A strong development of interradial plates in the tegmen occurs in the isocrinids, where a complete pavement of small, irregularly arranged plates forms interrays of the tegmen. The presence of these scaly plates allows observation of **hydropores** in the tegmen, at least in dried specimens. The hydropores do not show up when submersed in alcohol. In naked tegmens they are only observable in anatomical sections of the body wall (Fig. 32). They are present in interradial areas of the tegmen, but may be absent from the posterior interray, as in *Neocrinus, Endoxocrinus,* and *Calamocrinus.* In *Metacrinus* posterior hydropores occur only in narrow zones bordering the *C* and *D* ambulacra. In crinoids with orals preserved in mature growth stages, the hydropores may perforate the orals. Up to 20 hydropores can be seen in each of the five orals of *Holopus,* whereas orals of *Hyocrinus,* on the contrary, lack any hydropores. In this genus the orals are valvate structures, the hydropores may pierce the tegminal wall below the orals, whereas in *Holopus* the orals closely abut each other and apparently are not capable of much movement. Cuénot (1948) stated that *Rhizocrinus* preserves five orals, each of which is pierced by a single hydropore. If this is true, the posterior one should be interpreted as the primary hydropore.

The radial parts of the tegmen are formed by the ambulacra. These extend from arm bases to the mouth and distally follow all divisions of the arms, running to the very tips of the pinnules. **Ambulacra** may be absent in the oral pinnules of comatulids and in the posterior arms of some comasterids. The structure of the ambulacra is rather simple. Their bottom is the ambulacral groove, formed out of the ambulacral epithelium. It never seems to be calcified and is perhaps the only part of the crinoid body wall entirely devoid of calcareous deposits. The sides of the ambulacra are produced into series of lappets, at the internal sides of which are groups of three podia or tube feet. These are terminal extensions of the water-vascular system. The podia may eventually contain one or more calcareous spicules, but calcite secretion in them is hardly worthy of mention. It would be contrary to their food-gathering function to stiffen them with much calcite secretion. The lappets bordering and in some crinoids covering the ambulacra commonly are calcified, however.

The plates deposited in the lappets are of two sorts: **ambulacral plates** and **adambulacral plates.** The ambulacral plates are present in the top of the lappets or occupy all the lappet in case adambulacral plates are absent. In most stalked crinoids four series of plates are developed along the ambulacra: two rows of adambulacral plates and two rows of ambulacral plates (Fig. 33,*1*,*2*). In case adambulacral plates are present, these are mostly notched for the reception of saccules, which may be seen to alternate with the adambulacral plates. The latter commonly are observed in terminal parts of the pinnules. Toward the bases of pinnules they may be absent, even if present at the pinnule tips. The lateral body wall of pinnules may be so much extended, especially in genital ones during the breeding season, that the distance between the ambulacral plates and pinnulars becomes large enough to break up the series of adambulacral plates, with replacement of them by a lateral pinnule wall reinforced with many irregularly distributed scales and plates of calcite. The genital pinnules of *Poecilometra acoela* are very short and swollen structures with a complete adoral armor of abutting adambulacral plates. Ambulacra are absent (Fig. 33,*4*).

In *Neocrinus decorus* (Fig. 33,*1*) the ambulacra are confined to the narrow strip on the adoral side of pinnules, with ambulacral plates, adambulacral plates, and pinnulars in contact along the entire length of the pinnule. In many comatulids (e.g., *Glyptometra,* Fig. 33,*3*) the ambulacral plates are very large, rounded valvate structures imbricating on one another. They act as a shield below which the podial groups hide. Down along the arms and on the tegmen the arrangement of adambulacral and ambulacral plates may be less regular than on the pinnules (Fig. 30). It is questionable whether the ambulacral plates on the tegmen of *Neocrinus decorus* ever open. These have become much elongated

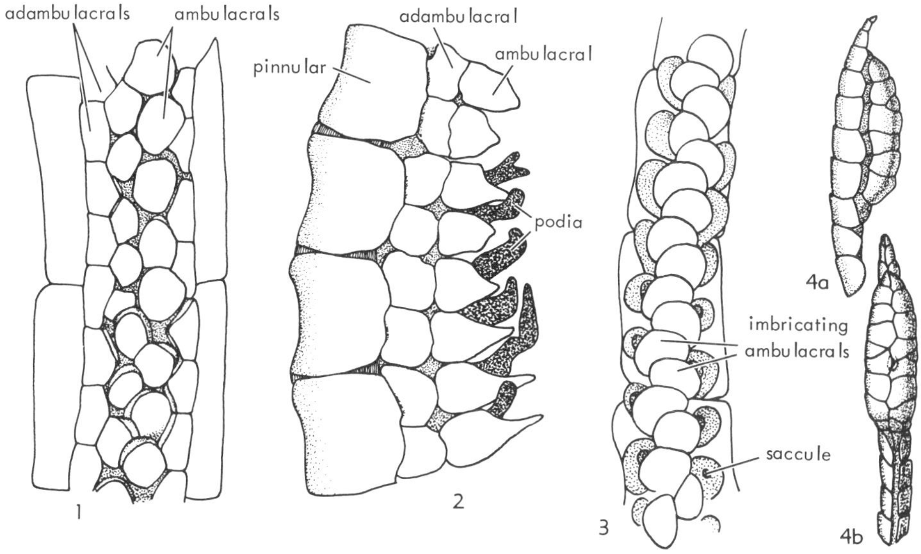

FIG. 33. Ambulacral and adambulacral plates.

1. Adoral aspect of two pinnulars of *Neocrinus decorus* with ambulacral plates closed (Breimer, n).
2. Lateral aspect of four pinnulars of *Ptilocrinus antarcticus* with ambulacral plates opened and podia exposed (Breimer, n).
3. Adoral aspect of three pinnulars of *Glyptometra selena* with imbricating ambulacrals (Breimer, n).
4. Genital pinnule of *Poecilometra acoela* in lateral *(4a)* and adoral *(4b)* aspects where elements of adoral skeleton contribute to form complete harness around the swollen gonads (after Carpenter, 1888).

and form a sort of palisade roof over the tegminal ambulacra for protection of the rich food stream. The palisade appears to be rather permanent.

The area immediately surrounding the **mouth** is never fully calcified. This part of the tegmen is kept flexible for opening and closing the mouth. The ambulacral lappets form a five-sided area around the mouth called the **peristome,** in which groups of labial podia are observed in naked tegmens of the comatulids; the ring canal of the water-vascular system is located directly underneath the peristome. The **anus** very commonly is situated on a protuberance of the posterior part of the adoral body wall. This structure, known as the **anal tube,** commonly is supported by elongated skeletal pieces, which form a sort of cone (Fig. 30).

VISCERAL SKELETON

The **visceral skeleton** is not formed of definite and coherent plates or scales, as in the adoral and aboral skeletons. It is mainly composed of numerous calcareous rods, spicules, and small meshwork platelets distributed more or less plentifully in strands and webs of connective tissue. It is to be remembered that especially the aboral or perivisceral coelom has a spongy structure characterized by the presence of numerous strands and webs of connective tissue (Fig. 1, anatomical section *Antedon*). These structures may be calcified to a greater or lesser extent. Calcification may also occur in the membrane separating the adoral and aboral coelomic canals in the arms. Complete calcification of such internal and intercoelomic membranes as known in some fossil crinoids (e.g., *Teleiocrinus,* Miss., N.Am.) is not known in recent crinoids. In some articulates (e.g., *Bathycrinus, Rhizocrinus*) the visceral mass may be so extended that it suppresses the perivisceral coelom and comes into contact with the body wall (CARPENTER, 1884a, pl. VIIb, fig. 6-8; pl. VIIIa, fig. 8). The suppressed coelom may be traced as a calcified film of connective tissue connecting the body wall with the wall of the visceral mass.

MORPHOGENESIS

The oldest information on morphogenesis of living crinoids, quite incomplete, has been derived from the studies of the North Atlantic species *Antedon bifida,* found along the coasts of France and England. More comprehensive and detailed work has been done by European authors (W. B. Carpenter, 1866; Perrier, 1886-90; Barrois, 1888; and Bury, 1888) during the latter half of the last century on the Mediterranean species *A. adriatica* and *A. moroccana.* An exhaustively precise description of the development of *A. adriatica* was published by Seeliger (1892), and his paper is still basic for our knowledge in the field. General echinoderm and crinoid literature (Clark, 1921; Cuénot, 1948; Hyman, 1955) has always been guided by his description.

Complete description of a crinoid ontogeny is available only for *Antedon.* Information on the ontogeny of other comatulid genera is very scattered and incomplete. During the first half of the present century, a series of papers by Mortensen included data on growth and development of many comatulid genera. John (1938) provided information on the development of Antarctic crinoids. Our knowledge on the morphogenesis of the stalked crinoids is so inadequate as to be practically nil. A few, but valuable, remarks have been made by Sars (1868) on juvenile growth stages of *Rhizocrinus lofotensis.* His description has attracted little attention, however.

Main stages in the life history of comatulid crinoids are 1) the **embryonic stage,** lasting about five days in the egg membrane, either lying on the sea bottom, attached to the mother crinoid, or kept in a special brood pouch. Rupture from the egg membrane then initiates the second life stage, which is 2) the free-swimming **larval stage.** It is very short, lasting only a few hours or days. The larva settles and now **metamorphosis** will lead to 3) a stalked juvenile **cystidean stage.** In this life stage the organism quickly learns how to feed and to perform other essential functions and it develops the necessary organ systems for them. The definite body form is only gradually assumed during 4) the juvenile **pentacrinoid stage,** during which the arms grow out. The pentacrinoid stage may last for several months, terminating when the young animal sheds its stalk and becomes free-moving again. The juvenile crinoid reaches sexual maturity shortly before the breeding season. When it first spawns it is about one year old. Upon spawning it attains 5) the **adult** life stage, which may last four or five years.

EMBRYO

The life span during which the crinoid is called an embryo starts with fertilization of the oogametes and ends with escape of the developing crinoid from the egg membrane. The embryonic period is variable in duration, in *Antedon* lasting five days.

Crinoids are normally oviparous animals. Egg production usually is very prolific. The eggs are yolky and thus the embryo is able to develop for a considerable period without feeding. Fell (1945, 1948) considered the production of a yolky egg to be a primitive condition, which has ever been retained by crinoids. Fertilization normally is external and takes place immediately after discharge of the oogametes by rupture of the ovarian wall, stimulated by the male spawning products. The zygote may fall to the sea bottom and become attached to some substrate, but usually pinnules of the mother crinoid serve for attachment. Partly this may be explained by ready availability of the pinnules for anchorage of the embryos and partly by the fact that the external pinnular walls may be provided with cement glands for fixation of the eggs. The fertilized egg is enclosed in a definite membrane, probably a fertilization membrane, the fine structure of which is described by Holland (1973). In most crinoids development takes place inside the membrane until the free-swimming larval stage is reached.

Some crinoids are known to diverge from the normal pattern. For example, various Antarctic species (e.g., *Isometra vivipara, Notocrinus* sp., *Phryxometra nutrix*) are viviparous. In such forms egg production is low, for usually only a few of them are yielded by individuals. Fertilization is internal in *I. vivipara;* sperm have been observed in the genital tube of the ovary, but it is unexplained how they work their way into the ovary. The eggs of viviparous crinoids develop in a **brood chamber** or **marsupium** located in the pinnules or in the arm at the pinnule bases. The marsupium is an extension of the adoral body wall adjacent to the gonad. The eggs get into the marsupium by rupture of the wall separating the ovary from the marsupium and the embryo escapes by an external opening in the marsupium.

Embryologic development begins with a short period of cleavage of the zygote. The blastomeres are subequal in size and placed in line with one another; denoting that the cleavage is holoblastic and radial. The result of cleavage is production of an 128-celled blastula of coeloblastula type. The *Antedon* egg reaches this stage in about seven hours after fertilization.

Gastrulation begins as an invagination at one pole of the blastula, but gradually it becomes embolic. The invagination forms the anlage of the entomesoderm. Cells at the far end of the invagination opposite the blastopore proliferate to give off numerous cells into the blastocoel. The cells wander about and assume the functions of mesenchyme. The invagination is cut off as a closed sac at one side of the blastocoel. The axis through the blastopore may be regarded as the anteroposterior axis. The archenteral sac lies in the posterior part of the embryo, which otherwise is filled with the entodermal mesenchyme. Gastrulation is completed when the blastopore has become closed off completely. In *Antedon* this takes place after 36 hours.

The stage in embryonic development during the next 48 hours is mainly characterized by repeated division of the archenteral sac, in which several different coelomic compartments are formed as anlages of the main internal organ systems to be developed later in ontogeny. The divisions of the archenteral sac are shown schematically in Fig. 34.

The archenteral sac first divides into two equal parts called **somatocoel** and **enterohydrocoel** (Fig. 34,*1a,b*). The somatocoel, which has an elongated form, is located at

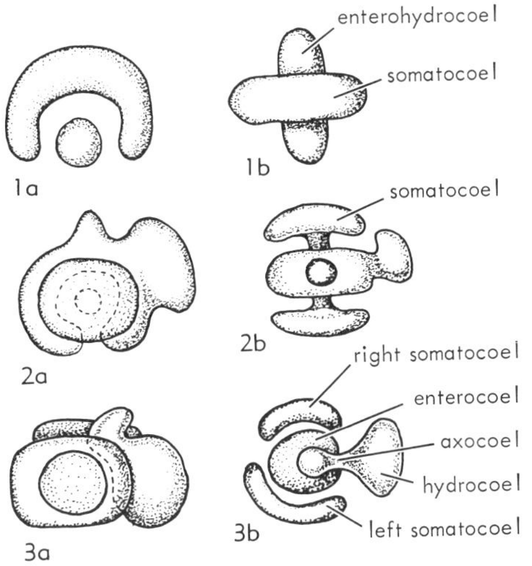

FIG. 34. Coelomic divisions in embryo of *Antedon* (after Heider, 1912).——*1-3.* Early, intermediate, and late larval stages (*1a, 2a, 3a,* side views; *1b, 2b, 3b,* dorsal views).

the posterior side of the embryo and is embraced by the crescentic enterohydrocoel, located in an anterior position. The somatocoel prepares for a further division, expanding its left and right extremities into the form of a dumbbell with constricted central portion (Fig. 34,*2a,b*). Soon the somatocoel divides into two parts: the left and right somatocoels. The enterohydrocoel also prepares for further division by evaginating its ventral and dorsal extremities. The ventral evagination which has considerable size and volume is separated as the **hydrocoel.** The much smaller dorsal evagination becomes the **axocoel.** The remainder of the enterohydrocoel is the **enterocoel** (Fig. 34,*3a,b*). It will develop as the future digestive system, but for quite awhile it remains a closed sac without entrance or exit to the external world. The axocoel and the hydrocoel both contribute to formation of the water-vascular system later in ontogeny. For this purpose the hydrocoel assumes a crescentic form and evaginates in five primary lobes. The crinoid axocoel and hydrocoel represent the left first and second coelomic compartments as compared to general echinoderm embryology. The right first and second compartments do not develop in crinoids.

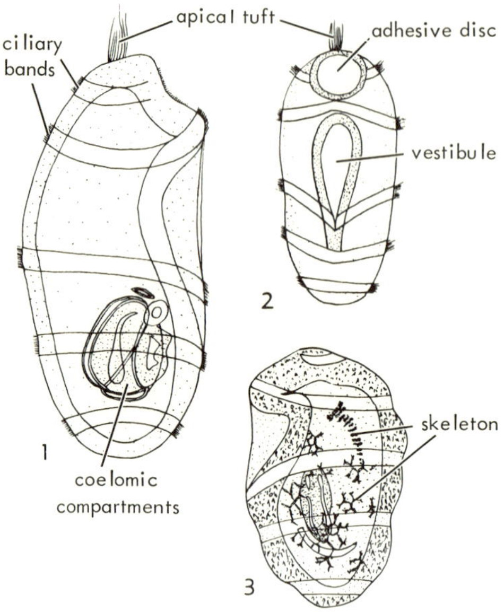

FIG. 35. Doliolaria of *Antedon* (after Seeliger, 1892).——*1*. Position of coelomic compartments in posterior part of larva.——*2*. Exterior ventral aspect of larva.——*3*. Skeleton of larva.

Through divisions of the archenteral sac the crinoid embryo assumes its primary bilateral symmetry and primary orientation. It now has well-defined anterior, posterior, ventral, and dorsal sides. The anteroposterior axis runs through the site of the former blastopore located posteriorly. The ventral and dorsal sides are indicated by the positions of the hydrocoel and axocoel, respectively. Left and right sides are indicated by the left and right somatocoels (Fig. 35).

Figure 35 shows that the coelomic compartments occupy only a rather small area in the posterior part of the embryo, exactly the same in position as the original archenteral sac. The larva now has assumed a much more elongated form, the external appearance of which is determined by further development of the ectoderm and mesenchyme. Part of the ectoderm is differentiated into nervous tissue, occurring in several different bands along the embryo. Accompanying them are four external bands of cilia and an apical tuft of cilia at the anterior extremity of the embryo (Fig. 35, *1,2*). In part, the ectoderm also differentiates into glandular tissue, mainly concentrated in a small spot near the anterior pole of the embryo, forming there the adhesive pit. Elongation of the embryo is also caused by the further development of wandering mesenchyme cells, formation of which continues in this stage of development. The ventral side of the embryo is determinable externally by the placement of an elongated depression of the body wall, known as the **vestibule** (Fig. 35,*2*). This important feature in further development of the crinoid is situated between the first and second ciliary bands.

At the end of its embryonic life the crinoid differentiates some of its mesenchyme into skeleton-secreting cells. Larval-skeleton secretion in echinoderms has been described by OKAZAKI (1960) and others from the developing sea urchin larva. This seems pertinent to the mode of secreting skeletal elements in crinoids, for development of crinoid hard parts is believed to proceed in essentially the same way as in echinoids. If this is true, formation of the crinoid skeleton is intracellular, deposition of calcium carbonate occurring in syncytial cell-masses and being initiated simultaneously in several different spots of the embryonic body wall. Two-dimensional meshworks of the carbonate are thus formed, arranged in two circles, protecting the coelomic compartments at the posterior extremity of the embryo (Fig. 35,*3*). These circles each contain five future skeletal plates, five to be developed as basals and five others as orals; they are longitudinally in line with one another. A columnar pile of circular carbonate meshes is formed in the anterior half of the embryo. These represent future stem segments or **columnals.** The embryo now has completed its development and is ready to escape from the egg membrane. It takes a developing *Antedon* larva five days to reach this moment.

LARVA

The embryonic part of crinoid development is completed in about five days. The embryo ruptures the egg membrane and begins a short life span as a free-swimming larva. The larva, commonly described as a **doliolaria,** has a length of about 0.25 mm.

when it emerges from the membrane. Its free-swimming period is variable in length, known to vary from five to 108 hours. Free-swimming larval stages are more or less suppressed in crinoids that breed their young. In *Phrixometra nutrix* the free-swimming larval stage is even completely suppressed. The embryos develop directly into the stalked pentacrinoid stage and attach themselves to the mother animal (Fig. 36).

Throughout its entire larval life the crinoid is unable to feed, for it still lacks a mouth. The larva is autotrophic. For its supplies of energy it must depend completely on its built-in food reserve, which contains many vitelline substances. The enteric sac is observed to be crowded with cells, interpreted (Mortensen, 1920; Hyman, 1955) as mesenchyme cells wandering into the enteric sac as amoeboid cells, which may have acquired food by phagocytizing degenerated tissues elsewhere in the larva.

During the free-swimming larval period no further development of internal organs occurs. The larva only grows in size and volume, with continuous increase in size and weight. Skeletal plates on the posterior side of the larva develop further. The larva swims with its anterior pole in front and is able to rotate along its longer axis.

At the end of its free-swimming larval life the crinoid prepares to become attached by selecting a suitable spot for fixation with the help of its anterior apical tuft. When such a spot is detected, the larva settles on its adhesive disc, which directly underlies a calcitic attachment disc at the end of the pile of columnal plates. Attachment sites for crinoids with calcite attachment discs generally are stony and firm substrates.

METAMORPHOSIS

Directly upon settlement of the larva a short period (about 5 days) of accelerated development sets in, introducing radical change in total organization of the larva and in its life habits. The fundamental phenomena dominating the metamorphosis are 1) loss of original bilateral symmetry of the larva by a temporary shift in the position of internal organs and by preponderant growth of the left side of the larva as com-

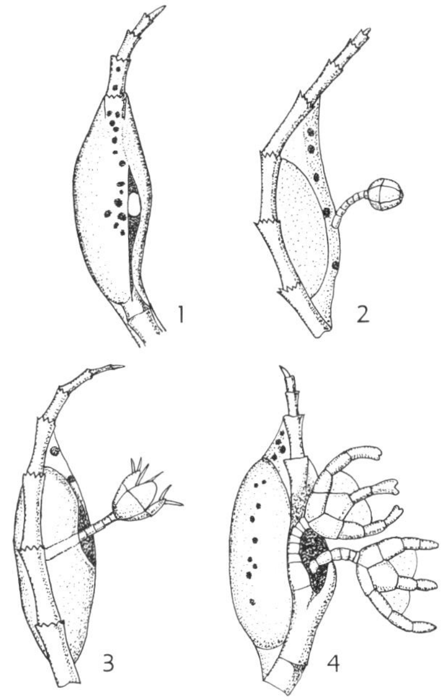

Fig. 36. Genital pinnules of viviparous comatulid *Phrixometra nutrix* (after Mortensen, 1920).——*1.* Egg in marsupium.——*2.* Early cystidean stage.——*3.* Late cystidean stage.——*4.* Pentacrinoid growth stage, attached to marsupium.

pared to its right side; 2) a rotation of the internal organs amounting to 90 degrees introduces a pentaradial symmetry with its own new orientation; and 3) physiological loss of autotrophic feeding and preparation for heterotrophic feeding, which implies further development of the organ systems that function in feeding.

When just settled, the larva is about 1 mm. long. After settlement, the vestibule up to now a simple depression of the ventral body wall—becomes invaginated and constricted into a large internal sac, which is destined to play an important role in metamorphosis. Some of the internal organs have temporarily shifted their position. The enteric sac still lies in the posterior part of the larva, and the hydrocoel in ventral position between the vestibule and enteric sac. The axocoel temporarily has an anterior location; it has produced an elongated evagination that opens to the exterior as the hydropore in the ventral part of the body wall. The right somatocoel, developed into a bowl-shaped vesicle partly

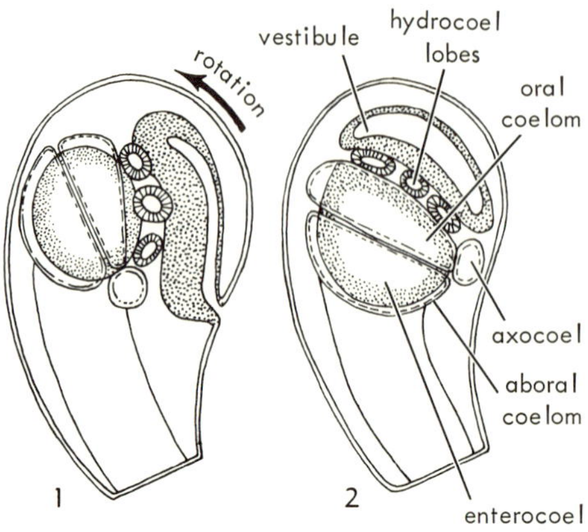

FIG. 37. Rotation of internal organs in *Antedon* larva (after Barrois, 1888).——*1.* Before rotation. ——*2.* After rotation.

surrounding the enteric sac, has temporarily assumed a position on the dorsal side of the enteric sac. The left somatocoel is located between the right somatocoel and hydrocoel. It is crescentic in shape, embracing part of the enteric sac with blind ends meeting on its anterior side.

Soon after constriction of the vestibule, a rotation of the internal organs completely changes orientation of the larva (Fig. 37, *1,2*). A 90-degree rotation brings the vestibule to the free end of the larva opposite the attachment disc. The other organs rotate with the vestibule, retaining their prerotation mutual positions. The former posterior side of the larva now becomes its top and develops as the free oral pole of the animal. The left somatocoel becomes the adoral coelom and the right somatocoel the aboral coelomic compartment. The position of the hydropore denotes the future posterior side. The elongated fixed part of the larva, formerly anterior, now develops as a stalk and the former posterior part of the larva develops as a theca. The theca is completely closed, with its body wall supported by closely abutting circles of five basals and five orals. The basals are aligned longitudinally with the orals, which cover the roof of the vestibule, the basals supporting the lower part of the theca, which includes the enteric sac.

The next step taken by the larva in metamorphosis is to prepare for heterotrophic feeding. Previously, the enteric sac of the larva has possessed no external openings, the mouth and anus still being absent. For feeding the larva needs to develop primitive food-catching organs and a mouth for receiving the food. In order to achieve this, the central part of the bottom of the vestibule begins to form an evagination that connects with the enteric sac, thus providing a passageway from the sac to the vestibule; these are the future esophagus and mouth. The esophagus is embraced by the five-lobed crescentic hydrocoel, which previously had formed a projection to meet the axocoel and unite with it as the primary stone canal. The central part of the hydrocoel now completely surrounds the esophagus as the future ring canal of the water-vascular system. The five lobes of the hydrocoel first form five primary podia and later on develop as the ring canal of the water-vascular system. The hydrocoel lobes then develop so as to push themselves upward through the bottom of the vestibule, thus forming five primary podia, which are lined with ectodermal epithelium. The larva now prepares to give the external world access to the just-formed mouth and podia. To this end the roof of the vestibule ruptures between the five oral plates, which now open as large valves, giving access to the just-formed mouth and podia. The bottom of the vestibule grows out as the tegmen with the mouth at its center as located originally. The posterior end of the enteric sac makes an evagination located near the primary hydropore on the posterior side. This opens to the exterior as the anus.

Opening of the vestibule occurs about five days after attachment of the larva. During the accelerated development of these days it has not grown very much, for its length is only 1.25 mm. as the vestibule opens.

CYSTIDEAN STAGE

The stalked crinoid possessing five oral valves and feeding with the help of primary podia is the oldest crinoid growth stage with definite pentaradial symmetry. This is termed the **cystidean growth stage** in zoological literature for some quite obscure reason (no cystoid being known to have the same organization as this crinoid de-

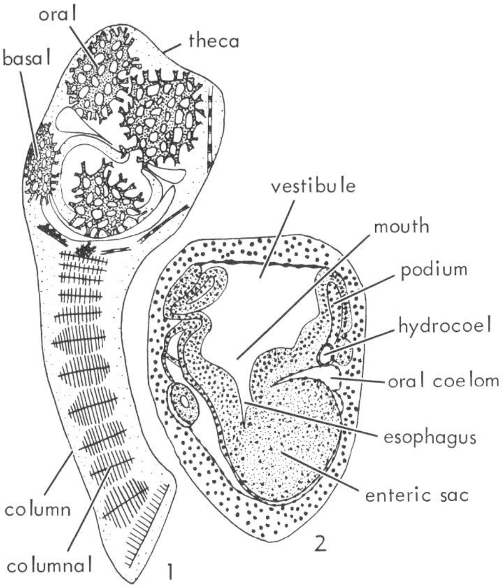

FIG. 38. Cystidean growth stage of *Antedon* (after Seeliger, 1892).——*1.* External form and skeleton. ——*2.* Anatomical section through anterior free end of larva.

velopmental stage). The juvenile crinoid, morphologically characterized by the presence of a stalk and a theca, and by the absence of arms, remains for about six weeks in the so-called cystidean growth stage (Fig. 38). Physiologically it is characterized by its primitive feeding mechanism of primary podia. During the cystidean stage the development of internal organs goes on but remains incomplete. The genital system and aboral nervous system develop only in later growth stages.

The skeleton of the cystidean crinoid still consists mainly of large basals and orals (Fig. 39,*1*). Eventually infrabasals and anal plates may become introduced but radial plates do not make appearance until shortly before the pentacrinoid stage.

The podia, once formed, elongate further and divide into groups of three podia situated in alternation with the oral valves (Fig. 39,*2*). In line with the orals the hydrocoel extends five paired evaginations that develop into five short interradial podia. This brings the total podia up to 25. Each of the longer ones in the triads of radial podia has conspicuous sensory papillae, which in part have a tactile function, serving for the detection and capture of food particles. They contain some very tiny nerve fibrils, a large muscle fiber, and some mucus-secreting gland cells. The nerve fibrils connect with the ectoneural nervous system, which is the only nervous system active in this growth stage of the crinoid. The ectoneural nervous system follows the course of the hydrocoel and the podia and is well developed there.

The enteric sac grows out into a curved stomach after it has been provided with mouth and anus. The main part of the aboral somatocoel still surrounds the enteric sac and is still crescentic in shape. A small part of the aboral somatocoel is cut off from the main somatocoel mass after its five extensions have been sent off into the stem, where they form the chambered organ and axial canal of the stalk. The remaining part of the aboral somatocoel develops further into the perivisceral coelom. The oral coelom is modified in essentially the

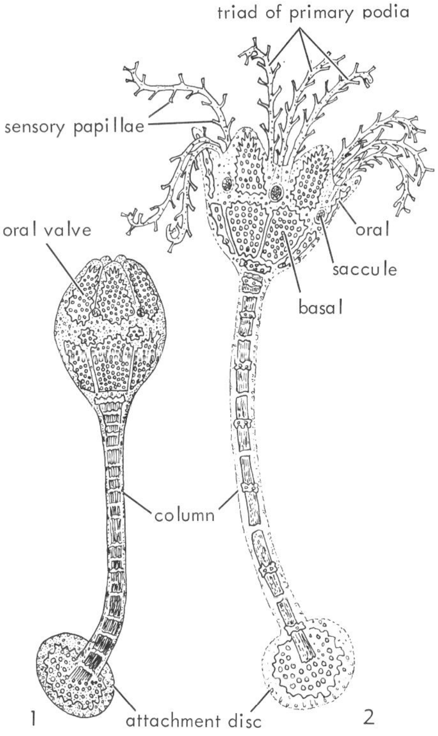

FIG. 39. Morphogenesis of cystidean stages in *Antedon*.——*1.* Early growth stage (after Thomson, 1865).——*2.* Late growth stage (after Clark, 1921).

same way as the hydrocoel and its future development everywhere follows the course of the hydrocoel, underlying it as the subambulacral coelom. Both of the coelomic compartments are now crescentic in form with their blind ends meeting on the posterior side to form a vertical membrane, which is thought to play an important role in the anlage of several different internal organ systems. The two coelomic compartments also meet in an equatorial plane so as to produce a horizontal mesentery at the place of their junction.

The axial organ is introduced shortly after separation of the chambered organ from the aboral somatocoel. This gland is reported to develop from the aboral part of the mesentery, which connects with the center of the chambered organ. The traditional opinion is that the primary gonad originates from the axial gland. This may be erroneous, however, since DAWYDOFF (1948) has stated that the definitive gonad originates as an elongated strand of compact cells located in the aboral part of the vertical mesentery. This strand is closely associated with the axial gland. The primary gonad seems to be transitory only, since it is reported to disappear after some time in the developing crinoid.

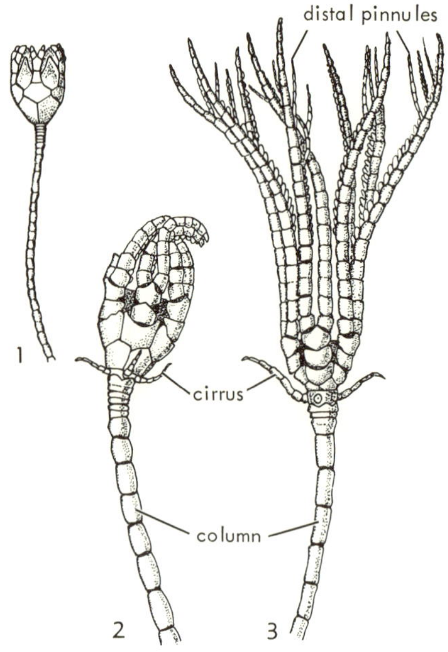

FIG. 40. Morphogenesis of pentacrinoid stages in *Antedon* (Breimer, n).——*1-3*. Early, middle, and late growth stages showing *(3)* absence of pinnules in proximal arm portions of late pentacrinoid stage, shortly before take-off from column.

PENTACRINOID STAGE

After six weeks in the cystidean growth stage, the juvenile crinoid reaches what is called the **pentacrinoid stage,** so named for its resemblance to stalked pentacrinoids. The pentacrinoid stage is characterized by its development of arms and pinnules (Fig. 40), with consequent adaptational outgrowth of the hydrocoel as the water-vascular system. During all of this stage the crinoid remains attached by its stalk. In comatulids the pentacrinoid stage ends when the crinoid breaks away from its stalk and starts a free-swimming life. Comatulids may remain for several months in the pentacrinoid stage. It is doubtful whether one should distinguish a pentacrinoid stage in the development of stalked crinoids, since they do not break free but remain attached by the stem throughout their life span.

The first sign of a crinoid's arrival at the pentacrinoid stage is appearance in the theca of five **radial plates** arranged in alternation with the basals (Fig. 40,*1*). Once the radials are introduced, the theca starts formation of the arms as five large radial extensions. Each radial extension soon becomes supported by skeletal pieces called **brachials.** The internal organization of the radial extensions or **arms** requires a rather drastic change in the further outgrowth of the internal organs. Inasmuch as the arms are radial extensions of the body, all internal organs of the body must prepare for extending themselves into the arms. This implies a partial reduction of morphologic organization realized in the cystidean growth stage.

Feeding of the pentacrinoid larva is effected by the arms, which bear ambulacra and podia for this purpose. Hence, the most drastic changes are made in the hydrocoel and its evaginating podia. Most of the original 25 podia disappear and only the five radially disposed primary podia persist. The arms are initiated as radial evaginations of the bottom of the former vestibule, which comprises the five primary podia (Fig. 40,*2*). These now grow out as radial canals of the water-vascular system. The adoral and aboral coeloms also form five radial extensions which penetrate into the growing arms as the adoral and aboral coelomic canals. Their contact is the horizontal membrane, which is continuous with the horizontal membrane in the central body cavity. In this mesentery parts of the blood-lacunar and genital systems eventually will be lodged. The vertical mesentery atrophies.

Definite podia are formed on the adoral surface of the arms in groups of three as terminal extensions of the water-vascular system. These triads of podia are placed in alternating order on both sides of a band of differentiated ectoderm, which now develops as the ambulacral epithelium. It differentiates some of its cells into mucus-secreting gland cells and others into attenuated supporting cells provided with cilia. The ectoneural nervous system everywhere underlies the ambulacrum.

Internally the water-vascular system changes its organization also. Up to now the hydrocoel ring has had only one opening to the exterior, obtained by its fusion with the elongated axocoel, which opened outward by means of the primary hydropore. Four other such stone canals are added to the original one, one in each interradius. They are formed by fusion of ingrowing canals from the bottom of the former vestibule, and by outgrowing canals from the hydrocoel ring. The five stone canals connect five hydropores in the tegmen with the five interradial sides of the ring canal. Normally these tubular connections are broken and the primary hydropores disappear, but in *Rhizocrinus* a stage with five hydropores seems to be definitive. The small tubes hanging from the ring canal persist. To these free hanging stone canals many others are added in the four regular interradial sides of the ring canal. In the interradial parts of the tegmen many more hydropores may be formed, however, but without direct tubular connections with the stone canals and without a direct correlation in number between pores and stone canals.

When arms grow out, the orals are resorbed, the oral valves atrophy, and the vestibule bottom becomes an adoral membrane or tegmen in which mouth and anus are located. The final stages in the development of the pentacrinoid are reached by further development of the arms and the first appearance of **pinnules.** The five- and ten-armed crinoids all develop these numbers of arms by direct terminal growth of the radial extrusions of the theca. In comatulids the arms have terminal growth sectors that divide after two primibrachs have been formed at the base of the growing arm. Ten arms are now developed, provided with ambulacra and groups of podia on the adoral side. The ambulacra become flanked by a fringed border giving rise to lappets under which the podial triads may hide. Saccules alternate with the lappets. The lappets may be reinforced with skeleton deposition. For a long time the arms remain devoid of pinnules. Up to 20 or 30 brachials may be formed in the free-arm ends before the first pinnule develops. The first pinnules to make appearance are distal ones which provide additional possibilities for food-gathering as the crinoid increases in volume (Fig. 40,*3*).

At end of the pentacrinoid stage the organism may be a few centimeters long, including the length of its stalk. The topmost columnal or **proximale** has been differentiated with accompanying development of one or more circles of cirri. The proximale unites with the infrabasals as the **centrodorsal.** When its time has come, the pentacrinoid breaks from the stalk and starts swimming around.

ADULT STAGE

When the juvenile crinoid has left its stalk it is not yet a mature and adult animal. Its juvenile status is easily recognized by the absence of genital pinnules. The **adult stage** is reached when genital pinnules have become differentiated and packed with ripe gametes. The animal is now prepared to take part actively in the reproduction of its species. As most crinoids have only short breeding seasons once every year, it is inferred that juvenile crinoids reach sexual maturity in their first breeding season after birth, hence at an age of about one year old.

The anlage and development of the genital system of crinoids is insufficiently known. As already stated, the primordial sex cells are probably formed in the vertical mesentery in what could be called a primary gonad. This vanishes, however, as does the entire vertical membrane later in ontogeny. Ripening sex cells appear in the horizontal membrane of mature crinoids. It is believed that the cells are formed in genital cords of the genital tubes and are transported toward the genital pinnules when ripening. The genital pinnules serve only for storage of the gametes. They are not true ovaries and testes.

Almost nothing is known about the first appearance and development of the aboral nerve system in crinoids. This is surprising, for the aboral nerve system is the main nervous system of crinoids and very conspicuous, present in such well-studied places as the base and arms. All one may speculate is to say that apparently the aboral nervous system develops very late in morphogenesis.

SKELETAL MORPHOLOGY OF FOSSIL CRINOIDS

By Georges Ubaghs
[Université de Liège]

CONTENTS

GENERAL FEATURES

The body of most crinoids consists of two main parts: the **crown** (corona) and the **pelma** (Fig. 41). The pelma comprises the **stem** (column or stalk), its appendages or **cirri** (if any), and the anchorage structures or **holdfast.** The crown is formed by a central mass that contains the main bulk of the viscera, and by five (simple or branched) **arms** (brachia), which are radial extensions of the central mass and are continuous skeletally with its supporting skeleton, or **theca.**

The theca consists of the **cup** or **calyx,** below the free arms, and the **tegmen,** which extends inward from bases of the free arms and forms a roof over the body cavity. The cup is joined to the topmost segment **(columnal)** of the stem and is primarily composed of lowest ray plates **(radials)** accompanied below by a circlet of interradially disposed plates designated as **basals.** Such cups are termed **monocyclic** because the radials overlie a single circlet of plates. All other crinoid cups possess a second plate circlet (named **infrabasals**) beneath the radials and consequently are classed as **dicyclic.**

All these plates, primary components of the cup, may combine with overlying plates, which are suturally joined more or less firmly together. These latter are identifiable as arm segments **(fixed brachials)** and in many crinoids between-arm plates **(interbrachials).** Supplementary plates in the posterior interray are classed as **anals.** The boundary between lower cup plates of the calyx and higher fixed plates is not prominent, since it is determinable only by the recognition of plates belonging to the radial circlet, the summits of which are not accentuated.

The tegmen may contain relatively large interradially placed plates named **orals** and associated with them **ambulacral plates** that cover the ambulacra and plates that lay between the ambulacra. In addition, part of the tegmen may be elevated as an **anal pyramid, anal sac,** or **anal tube.**

Each of these features, which taken together characterize the crinoids, can show extraordinary modifications or even be lacking. For instance, many fossil crinoids have

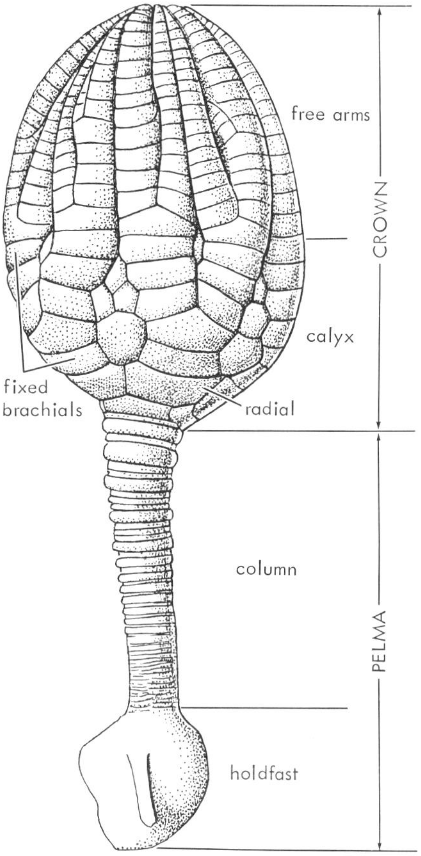

FIG. 41. *Calpiocrinus intermedius* SPRINGER, M. Sil., Eng., complete specimen of flexible crinoid, ×3 (Springer, 1920).

no holdfast, the stem itself assuming the function of a root. Some have no stem at all, their crown being entirely free or attached directly to the substratum. Others have lost infrabasals or radials and arms, or have retained only one or two brachia. Some have a body comprised of less than a dozen plates, whereas others have several hundred thousand and possibly even more than two million skeletal elements. In size, these echinoderms range from microcrinoids with a theca only one or two millimeters in diameter to relative giants with a theca ten centimeters or more in height and breadth. The arms of some crinoids are as much as one meter in length and

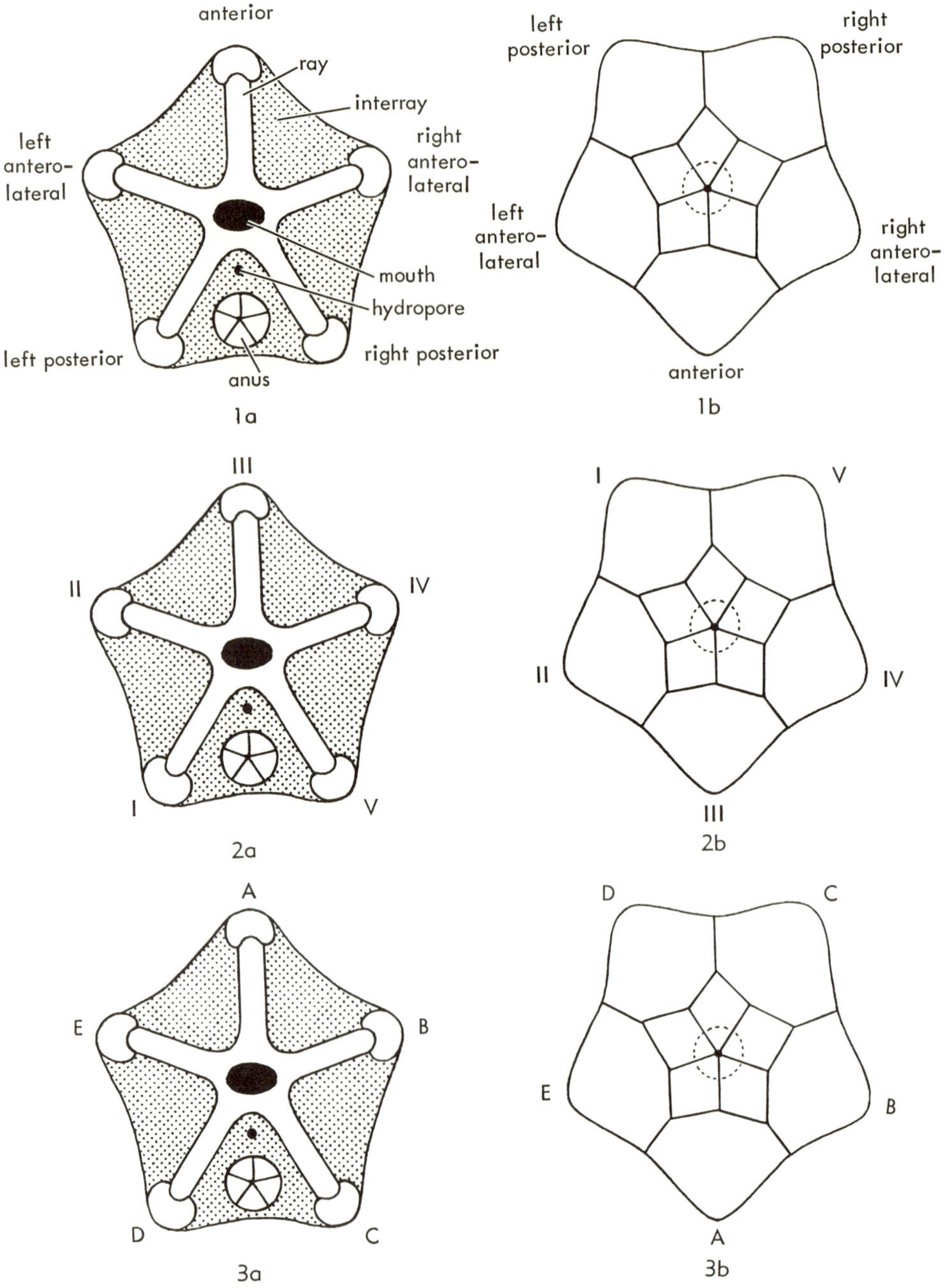

FIG. 42. Orientation of the theca in Crinoidea, in adoral and aboral views according to the systems of: *1a,b*, BATHER; *2a,b*, JAEKEL; *3a,b*, CARPENTER (Ubaghs, n).

crinoid columns 20 meters or more in length have been recorded. When account is taken of all these variations, one must judge that crinoids are an extraordinarily plastic group of invertebrates, and this diversity itself is certainly not the least of their most distinctive features.

In the course of time the crinoids evolved

in many directions and inhabited diverse marine environments. Because they include many distinctive short-lived forms and are widely distributed in marine deposits of the world, they may be useful, at least in some cases, for age determination and stratigraphic correlation.

ORIENTATION AND SYMMETRY

The position of the mouth in or near the center of the tegmen in typical crinoids allows recognition of an **adoral** (ventral or actinal) side and an **aboral** (dorsal, abactinal, or adapical) side of the organism. One may therefore distinguish an **adoral skeleton** essentially pertaining to the adoral side, and an **aboral skeleton** located on the aboral side. Such a distinction, although purely topological, is more convenient for descriptive purpose than a distinction between a **primary** skeleton, first developed in ontogeny, and a **secondary** or **supplementary** skeleton, the elements of which are added to or intercalated between primary plates.

For any structure located on the aboral side of the body, the terms proximal and distal refer to directions respectively toward the aboral pole of the cup or away from it. Thus plates of the first (lowest) circlet in the cup or calyx are the proximal elements of the crown, and the top ossicle of the stem is the proximal element of the stem. For adoral elements of the theca, however, it is the center of the tegmen that is considered to be the proximal point.

As in most radiate echinoderms, the skeletal individual elements, termed **ossicles,** of most crinoids have a quinqueradiate arrangement. Thus, the crown may be divided into five **rays** alternating with five **interrays.** The rays consist of the radials and all structures (mainly arm plates) supported by them. The midline of each ray is a **radius;** it corresponds to the trace of a plane passing through the oral-aboral axis of the body and dividing the ray into two equal parts. The body elements bisected by such a plane are **radial,** or more precisely **perradial,** in position. In similar manner, the interrays, that is to say the five sectors comprised between the five rays, are bisected by five **interradii,** and the structures located between the rays are said to be **interradial** in position.

For designating individual radii and interradii (or rays and interrays) most workers on fossil crinoids have used a method initiated by WACHSMUTH & SPRINGER (1889b), though generally attributed to BATHER (1890a, 1900a), or directly derived from it (Fig. 42,*1a,b*). The key for use of this method is furnished by position of the anus, which is usually situated on the tegmen in one of the interrays. Accordingly this interray is conventionally designated as **posterior** and the opposite ray as **anterior.** Now with the crinoid in its natural position, the mouth directed upward and the anal side toward the observer, it is possible to distinguish right and left sides, which respectively correspond to the right and left sides of the observer. To preserve this orientation when the crinoid is viewed from above, the anal side must be nearest the observer (downward in a figure); when viewed from below, the anal side must be away from the observer (upward in a figure). According to their position, the rays are respectively called **anterior, right anterolateral, right posterior, left posterior,** and **left anterolateral;** similarly, the interrays are designated as **right anterior, right posterolateral, posterior, left posterolateral,** and **left anterior** (Fig. 42, *1a,b*) or, as suggested by MOORE (1952a), anteroright, posteroright, posterior, posteroleft, and anteroleft. However, many authors, including BATHER in his first works, have used a more simple nomenclature and called both lateral rays and interrays respectively **right anterior, right posterior, left posterior,** and **left anterior.**

An alternative method of orientation, judged more natural and less confusing by its author, was proposed by JAEKEL (1918) (Fig. 42,*2a,b*). Starting from the interray which in cystoids contains the hydropore and gonopore and which corresponds generally to the anal interray of crinoids, he designated the ray at left of the pores by the Roman numeral I and the remaining rays counted in clockwise direction (in adoral view) by the four succeeding numerals II-V, so that the hydropore and gonopore (and in crinoids the anus) come to

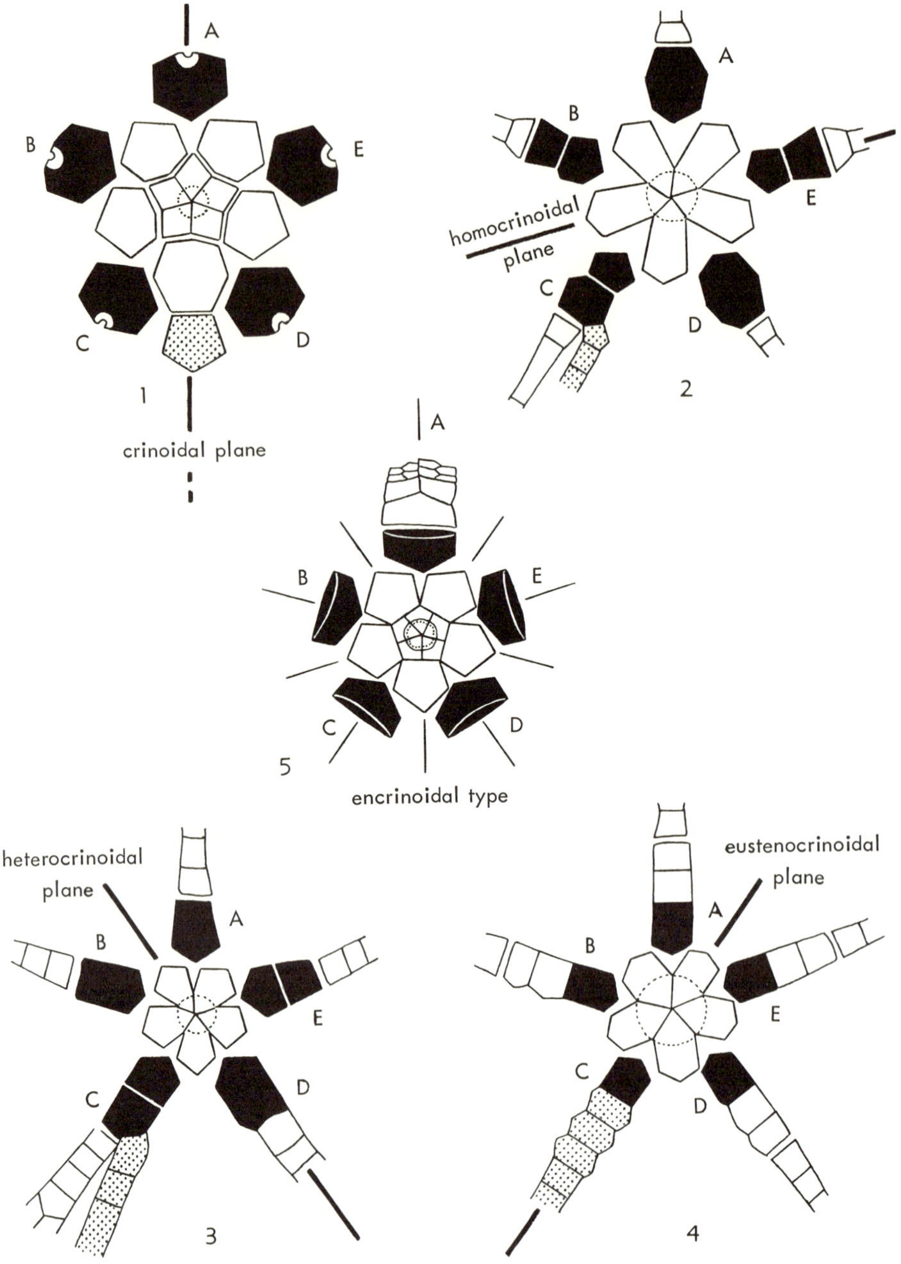

FIG. 43. Orientation of bilateral symmetry plane in some inadunate crinoids (mod. from Moore, 1962b).
1. Crinoidal plane, *Cyathocrinites.*
2. Homocrinoidal plane, *Homocrinus.*
3. Heterocrinoidal plane, *Heterocrinus.*
4. Eustenocrinoidal plane, *Eustenocrinus.*
5. Encrinoidal planes, *Sinocrinus.*

lie in V-I interray. This system, in spite of simplicity and suitability for the purpose of comparison among echinoderms, has practically never been applied to crinoids.

In this *Treatise* the nomenclature introduced by P. H. CARPENTER (1884a) has been accepted (Fig. 42,*3a,b*). The letter *A* is assigned to the ray opposite the interray containing the anus, and *B, C, D,* and *E* to other rays, in clockwise succession in adoral view, and counter-clockwise sequence in aboral view. The interrays are designated by the letters of rays that bound them (*AB, BC, CD, DE* and *EA*).

Quinqueradial symmetry is rarely perfect in crinoids, for usually it is disturbed by a bilateral symmetry, which is determined typically by position of the anus in one of the interrays of the tegmen and generally associated with this in most Paleozoic forms by the presence of additional plates and other distinctive features in one of the interrays of the cup. The plane of bilateral symmetry passing through the oral-aboral axis, the anus and hydropore (if present), is called the **crinoidal plane of symmetry** (Fig. 43,*1*). It is particularly obvious in Paleozoic crinoids, and may be considered as the normal crinoid plane of bilateral symmetry.

Other more or less dominant planes of bilateral symmetry may exist in crinoids. One is exemplified by the recent comatulid family Comasteridae (see above, p. T20). Others are illustrated by disparid inadunates, in which the structure of the body and particularly the presence and arrangement of undivided and compound radials allow recognition of three different planes of bilateral symmetry: 1) **homocrinoidal plane,** which passes through the *E* ray and *BC* interray (Fig. 43,*2*); 2) **heterocrinoidal plane,** which coincides with the *D* ray and *AB* interray (Fig. 43,*3*); 3) **eustenocrinoidal plane,** which bisects the *C* ray and *EA* interray (Fig. 43,*4*). According to LANE & WEBSTER (1967), the homocrinoidal plane corresponds to the larval dorsoventral symmetry plane.

These alterations from normal pentameral symmetry are not the only ones found among crinoids. Some others may be mentioned, such as produced by: 1)**reduplication of rays,** as in recent *Promachocrinus,* which has ten radials and ten arms at the adult stage; 2) **reduction of rays** from the normal five to four, three, two, or even a single ray; 3) **inequality of rays** in size; 4) **grouping of rays** into a **bivium** and **trivium,** as in the recent *Holopus;* and 5) **torsion of rays,** as in the flexible *Mespilocrinus.* Some of these modifications may depend on ecology, for instance, in rheophilic species, which may lose their radial symmetry and acquire a purely adaptive bilateral symmetry.

Finally, in what has been called the **encrinoidal type** of bilateral symmetry (MOORE, 1962b) the cup may show perfect pentameral symmetry, so that none of the rays can be distinguished from others and each one coincides with a plane of bilateral symmetry (Fig. 43,*5*).

ABORAL SKELETON

The **aboral skeleton** includes every skeletal element located in the teguments that form the surface of the body opposite to that containing the mouth and ambulacra. It comprises 1) all the ossicles of the column, with its appendages and anchorage structures, 2) the plates that compose the aboral cup or calyx (except interbrachials), and 3) the ossicles that support the arms and the pinnules. Morphologically the interbrachials belong to the adoral skeleton, but since they are an important component of the calyx of many Paleozoic crinoids, it has been judged convenient to treat them along with the other elements that form this part of the theca.

STEM

GENERAL FEATURES

Most adult fossil crinoids possessed a stem. Only in a few species the stem became more or less atrophied or lacking altogether in the adult growth stage. Generally, the stem was well developed, and displayed wide diversity in size and shape. Its characters are diagnostic for some Paleozoic species and genera, and for a large percentage of Mesozoic and Cenozoic species.

The diameter of the stem in fossil crinoids ranges from less than 1 mm. to more than 10 cm. According to BOUŠKA (1946), a stem fragment composed of several col-

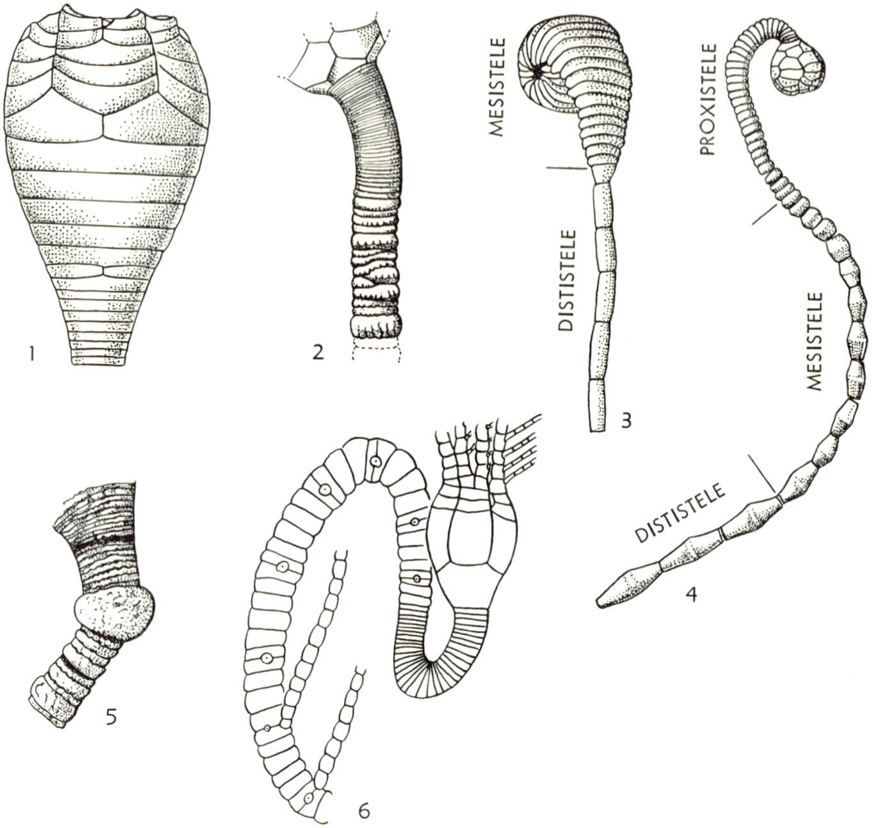

FIG. 44. Xenomorphic crinoid columns.

1. *Apiocrinites parkinsoni* (VON SCHLOTHEIM), M.Jur., France, with enlarged proxistele, ×0.7 (de Loriol, 1882-84).
2. *Onychocrinus diversus* MEEK & WORTHEN, Miss., USA(Iowa), proxistele and part of mesistele, ×1 (Springer, 1920).
3. *Ammonicrinus doliiformis* WOLBURG, M.Dev., Ger., with abrupt change between coiled mesistele and dististele, *ca.* ×1 (mod. from Wolburg, 1938b).
4. *Mespilocrinus konincki* HALL, L.Miss., USA (Iowa), with well-differentiated proxistele, mesistele, and dististele, ×1 (Springer, 1920).
5. *Nevadacrinus geniculatus* LANE & WEBSTER, Perm., Nevada, with bent column and greatly enlarged columnal between proxistele and mesistele, ×2 (Lane & Webster, 1966).
6. *Camptocrinus crawfordvillensis* SPRINGER, Miss., USA(Ind.), diagrammatic representation of part of crown, proxistele, and portion of mesistele, ×3 (Van Sant in Van Sant & Lane, 1964).

umnals (doubtfully attributed to *Pernerocrinus*) from the Lower Devonian of Bohemia has a maximum diameter of 11.5 cm. In many crinoids the stem tapers or expands distally or remains essentially constant in width throughout most of its length.

The total length of crinoid stems belonging to different species is rarely known. It must have been enormous (more than 20 m.) in some Mesozoic pentacrinids, but this is exceptional. Most species had colums probably not exceeding 1 m. and generally much shorter. For instance, in the famous Crawfordsville fauna (Mississippian) found in Indiana it appears that one group of crinoids lived with crowns raised 20 to 25 cm. above the sea floor and another group had stems 60 to 100 cm. long (LANE, 1963b).

In cross section, the majority of fossil crinoid stems are circular, many are pen-

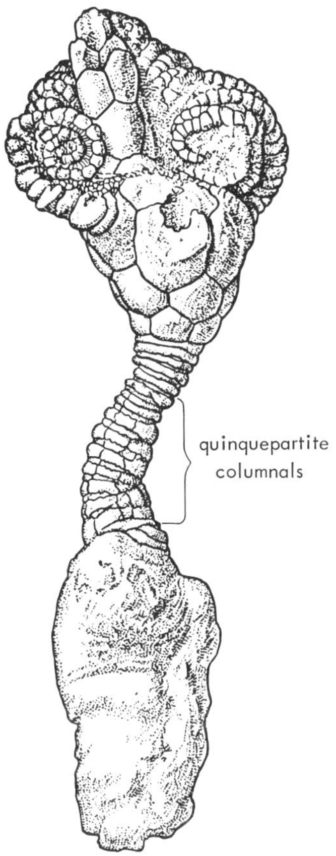

FIG. 45. *Euspirocrinus spiralis* ANGELIN, Sil., Gotl., with partially quinquepartite column, ×2 (Bather, 1893a).

tagonal, stelliform or elliptical, and a few are quadrangular or crescentic.

In longitudinal aspects, the stem may be uniform or almost uniform from end to end **(homeomorphic)**, but it may also be more or less differentiated into distinct regions (**xenomorphic**, Fig. 44), according to shape, size, proportions, arrangement and ornamentation of its component **ossicles (columnals)**. Frequently, a proximal part **(proxistele)** may be distinguished from an intermediate part **(mesistele)** and a distal part **(dististele)**; in some stems more than three regions are recognizable. Differences between these parts have an ontogenetic significance, since in any crinoid column the proxistele is last-formed and the dististele, earliest-formed.

Any part of a stem comprising two or more columnals in anatomical connection is termed a **pluricolumnal.**

Each columnal generally is composed of a single ossicle. In some Paleozoic crinoids, however, it consists of five separate pieces **(pentameres)**. Ordinarily the pentameres of successive columnals do not alternate, so that the stem is quinquepartite, that is, divided into five sectors by five longitudinal sutures (Fig. 45). All known species provided with quinquepartite stems belong to inadunate genera or to the camerate genus *Cleiocrinus*. Very different in origin is the secondary fragmentation of columnals into more or less irregular pieces, as illustrated, for example, by a highly specialized representative of the Crotalocrinitidae *(Pernerocrinus)*.

The stem is pierced by a longitudinal, usually central, **axial canal,** which in recent crinoids contains coelomic and nervous extensions from each of the five lobes of the chambered organ. It is possible that in some Paleozoic crinoids with an extremely wide canal, the axial canal lodged other organs or served other functions also, but what organs or functions is not known.

A relation occurs between orientation of the stem and that of the proximal cup plates (Fig. 46). This relation is expressed by the so-called "law of WACHSMUTH & SPRINGER" (1885, p. 229), which in fact is an empirical rule, liable to exceptions, and applicable only to quinquepartite or pentagonal stems or axial canals. According to this rule, in dicyclic crinoids (Fig. 46,*1*) the columnal pentameres and outer angles of the stem are directed interradially and therefore alternate with the infrabasals, whereas the longitudinal sutures, sides of the stem, angles of the axial canal, and attachments of cirri are located perradially. On the contrary, in monocyclic crinoids (Fig. 46,*2*) the columnal pentameres and outer angles of the stem are perradial, whereas the longitudinal sutures, sides of the stem, angles of the axial canal, and placement of cirri are interradial.

Exceptions are known in regard to all of these features. Some are only apparent and result from the fact that in some crinoids the infrabasals, which are present in

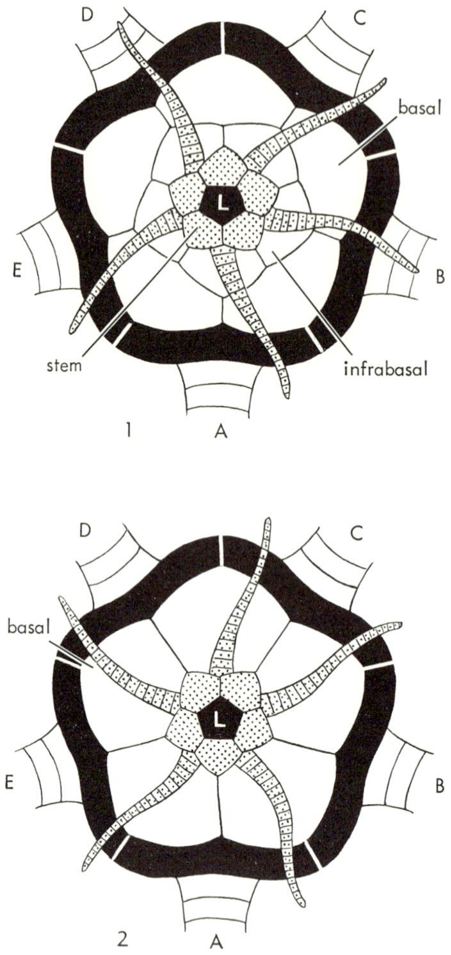

FIG. 46. Diagrammatic aboral views of dicyclic *(1)* and monocyclic *(2)* crinoids, showing orientation of various columnal features with reference to "law" of WACHSMUTH & SPRINGER (Ubaghs, n). [Explanation: Rays identified by letter symbols of CARPENTER system. Basals, infrabasals and arm-plates unshaded; radials and lumen *(L)* of axial canal of stem black; transverse section of column and cirri stippled.]

juveniles or near ancestors, are invisible or absent in adults. Such crinoids, monocyclic in appearance, are really dicyclic, and for them consideration of stem orientation allows recognition of concealed or former presence of infrabasals. But other exceptions cannot be explained in this way. For instance, positions of the pentameres generally alternate with those of the axial canal angles (Fig. 47,*1*). Some Ordovician species, however, including *Aethocrinus moorei* and *Grenprisia billingsi,* differ from all other species by having both pentameres and axial canal angles identically directed (Fig. 47,*2*). Moreover, in the first-mentioned species, the pentameres, instead of alternating with proximal thecal plates[1] in accordance with general rule, are identical with them in orientation (Fig. 48,*1,3*). The reasons of such differences may have a phylogenetic significance, but they are still obscure. Other exceptions concern the exterior angles of the stem, which usually correspond with pentameres but in some crinoids do not. The axial canal angles may be differently oriented as they ought to be according to "law." BATHER (1893a, p. 44) reported that a stem occurring in the Upper Silurian of Gotland has angles of both stem and axial canal and the longitudinal sutures all with the same orientation. In the recent comatulid *Antedon bifida,* the first-formed cirri are radial, but second-formed cirri are interradial. These diversities of plan suggest that actual changes of orientation can take place or secondary structures can obscure primitive arrangements.

It is generally agreed that the crinoid stalk originated from either an aboral evagination of a more or less globose theca (BATHER, 1900a) or a constriction of the posterior part of an elongated wormlike body (JAEKEL, 1918).[2] In either case, it is supposed that the primitive stem included a spacious cavity surrounded by an integument of numerous irregularly arranged skeletal pieces. Such a primitive stage has not been found among crinoids, but it occurs in several Cambrian Crinozoa, such as the eocrinoid *Gogia,* the lepidocystoid *Lepidocystis,* and the crinozoan *Echmatocrinus* discovered by SPRINKLE in 1967 in

---

[1] According to PHILIP & STRIMPLE (1971), these plates (herein called infrabasals) should be considered as homologous with columnal pentameres. Reasons for rejecting this alternative interpretation have been given by UBAGHS (1969, 1972).

[2] Another origin of the crinoid stem has been postulated by A. H. CLARK (1910c, p. 213), who supposed the columns of the recent and most fossil crinoids to be homologous to the central plate of the presumed common ancestor of crinoids and echinoids; this plate would have gradually become thickened and elongated and developed transverse fractures which later would have been modified into definite articulations.

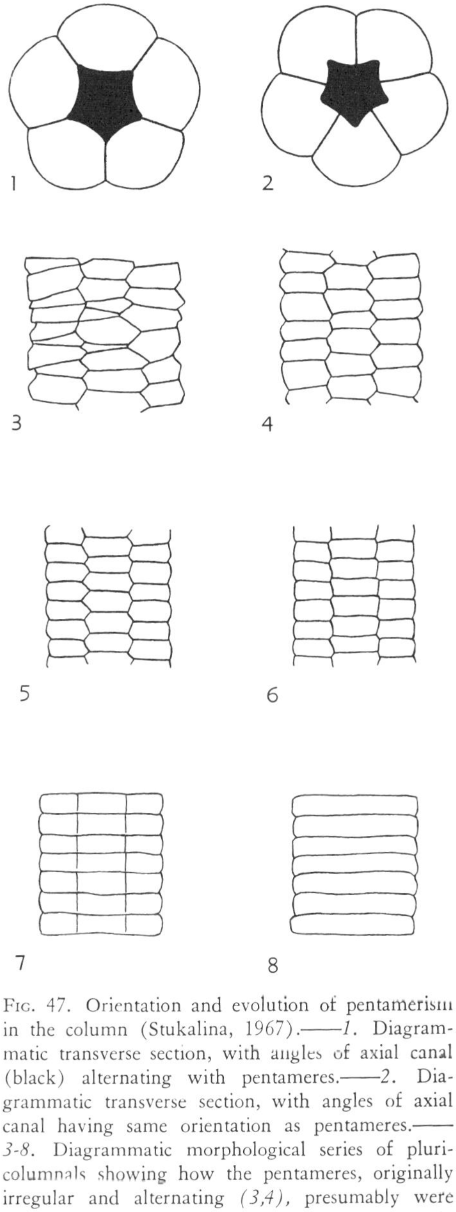

FIG. 47. Orientation and evolution of pentamerism in the column (Stukalina, 1967).——*1.* Diagrammatic transverse section, with angles of axial canal (black) alternating with pentameres.——*2.* Diagrammatic transverse section, with angles of axial canal having same orientation as pentameres.——*3-8.* Diagrammatic morphological series of pluricolumnals showing how the pentameres, originally irregular and alternating *(3,4)*, presumably were arranged in horizontal rows *(6,7)*, and eventually fused to form solid columnals *(8)*.

the Burgess Shale, Middle Cambrian of British Columbia (SPRINKLE, 1973a). Moreover, its former existence in crinoid ancestors is suggested by features of some primitive inadunates, particularly *Aethocrinus moorei* from the Lower Ordovician of France (Fig. 48). These features are: 1) irregular interlocking of plates forming the skeletal support of the stem, 2) similar and apparently correlative irregular arrangement of thecal plates and columnals, 3) merging in outer form of the cup into the stem, 4) merging of the thecal cavity into the very large axial canal of the stem, 5) similar ornamentation of the stem and cup plates, 6) presence of numerous irregular pieces in the distal region of the stem, that is to say, in the earliest formed part of the stem during development.

The next stage in evolution of the crinoid stem probably was characterized by progressive introduction of pentamerism (Fig. 47,*3-8*). At some time the individual hexagonal plates were supposedly arranged in alternating circlets of five around a very wide axial canal. Then the plates or pentameres no longer alternated, but were arranged in horizontal rows divided by five longitudinal sutures. Finally the pentameres of each row became fused to form monolithic columnals, at first still retaining wide axial canals or a quinqueradial arrangement of markings on the joint faces, or both. Such an evolution is not entirely hypothetical, for it may be traced in the limits of single stems in which the quinquepartite columnals of the distal region are replaced by monolithic columnals in a proximal direction (Fig. 45).

In most crinoid groups, the changes described above were accomplished before Ordovician time or during the first part of this period, for in Ordovician rocks the number of columnals with pentameres placed at different levels was already very small compared with that of monolithic columnals or with columnals composed of pentameres placed at same levels. During the following periods, the quinquepartite columnals disappeared progressively: seven genera with such columnals have been reported from the Silurian, three from the Devonian, two from the Mississippian, and only one from the Permian (SIEVERTS-DORECK, 1957b).

NODALS AND INTERNODALS

Columnals may be all alike throughout a crinoid stem or most of it in xenomorphic stems, or they may differ in shape

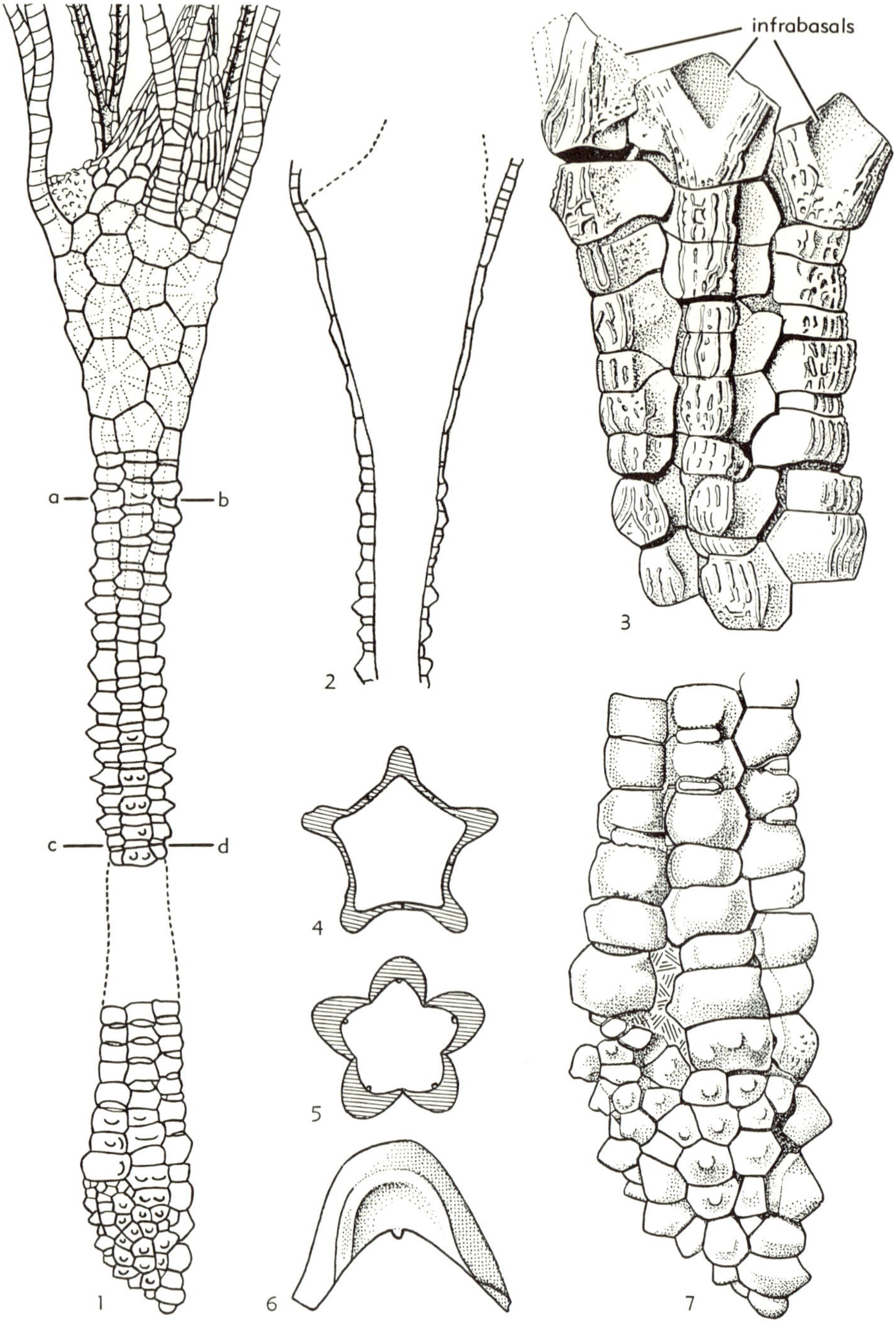

FIG. 48. *Aethocrinus moorei* UBAGHS, L.Ord., France (Ubaghs, 1969).

1. Reconstruction of an almost complete specimen, *CD*-interray view, ×1.5.

2. Diagrammatic sagittal section showing how the thecal cavity progressively merges into the large axial canal of the column, ×1.5 (Ubaghs, n).

3. Proximal portion of column and infrabasals, ×10.

4,5. Transverse sections of column at levels marked *a-b* and *c-d* in Fig. 1, ×3.

6. Distal face of a columnal pentamere, ×10.

7. Distal end of column, ×3.

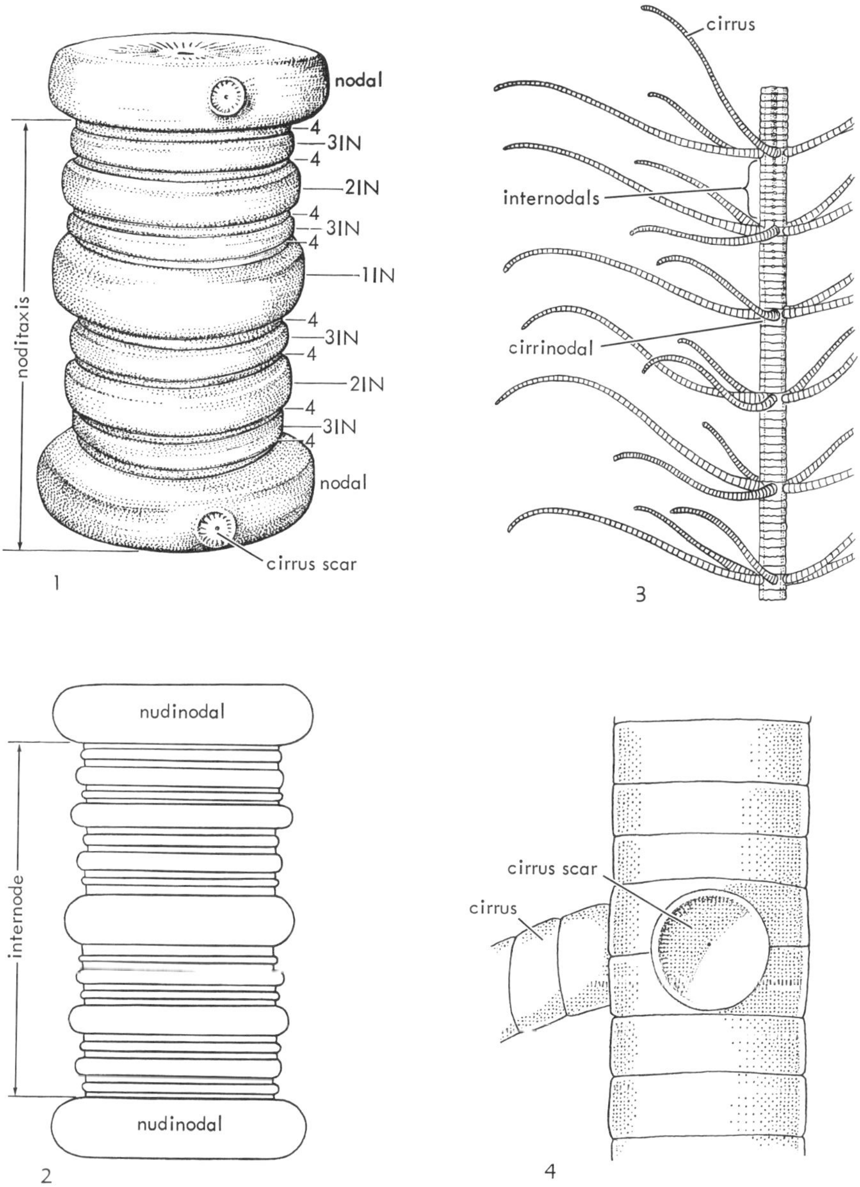

FIG. 49. Morphology of heteromorphic column.

1. Complete noditaxis and additional cirrinodal, 15 internodals *(IN)* divisible into four orders (Moore, Jeffords, & Miller, 1968).

2. Complete internode and two nudinodals, 31 internodals divisible into five orders (Ubaghs, n).

3. *Teliocrinus springeri* (A. H. CLARK), recent, India, part of column, with long cirri arranged in whorls of five, ×2 (Clark, 1915a).

4. *Austinocrinus mexicanus* (SPRINGER), U.Cret., Mexico, part of column and cirrus, with cirrus scar on two nodals (compound nodals), ×2.5 (Springer, 1922b).

and height at various levels. In the first case, the column (or parts of it in a xenomorphic stem) is said to be **homeomorphic,** in the second, **heteromorphic.** Notice must

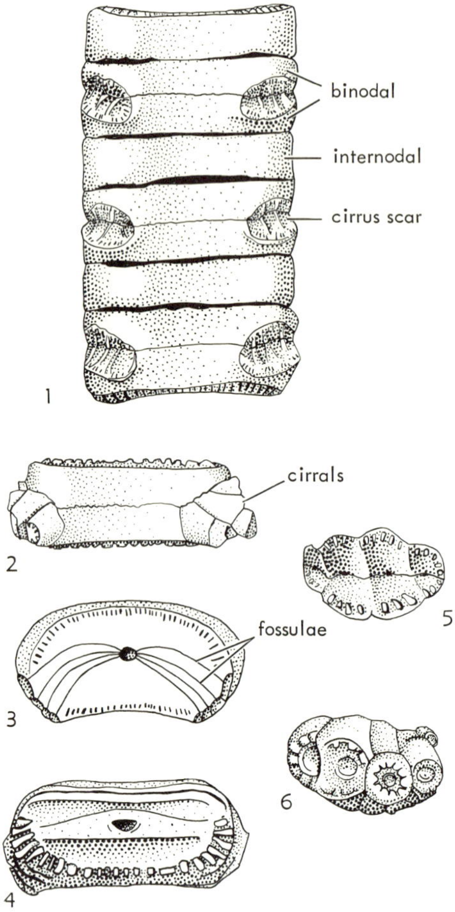

FIG. 50. *Camptocrinus compressus* WRIGHT, L. Carb., Scot. (Ubaghs, n).——*1.* Portion of column seen from cirrus-bearing side, ×6.5.——*2.* Binodal, with proximal cirrals still attached to them, ×6.5.——*3.* Synostosial articulation between binodals, ×6.5.——*4.* Synarthrial articulation between binodal and internodal, ×6.5.——*5.* Compound cirrus scar, ×13.——*6.* Distal aspect of proximal cirrals showing articula, ×13.

be made of apparently homeomorphic, but actually heteromorphic, stems in which longitudinal sections reveal the existence of thin and narrow columnals, which are not visible externally.

In heteromorphic stems the largest and most conspicuous columnals are called **nodals** and the smaller columnals between them are termed **internodals.** All internodals between a pair of sequent nodals form an **internode,** and a nodal plus the internode on its proximal side is a **noditaxis** (Fig. 49,*1-3*).

In recent crinoids, each nodal is characterized by bearing a whorl of generally five articulated appendages, or **cirri** (Fig. 49,*3*). Many fossil crinoids have the same arrangement, or their nodals are provided with a smaller number of cirri down to a single one. On the other hand, many fossil species have no cirri. One may therefore distinguish between **cirrinodals** (having cirri) and **nudinodals** (lacking cirri) (Fig. 49,*2*). In some species, two or more columnals may share in bearing a cirrus; such columnals may be defined as **compound nodals** (Fig. 49,*4*). In a few fossil crinoids, like the camerate genus *Camptocrinus,* the cirri are equally carried by closely united columnals with a more or less vanished intercolumnal suture; these closely paired nodals are called **binodals** (Fig. 50).

All internodals of an internode may be identical, but they may also differ in diameter and height, according to their order of appearance and stage of growth (Fig. 49,*1,2*). Those first formed after the limiting nodals are the largest: they are distinguished as first-order internodals or **priminternodals;** those formed next, and therefore next in size, are second-order internodals or **secundinternodals;** in the same way, third-order internodals **(tertinternodals),** fourth-order internodals, and so on, may be recognized. In a very few stems, differences in height and diameter among cirrinodals have been also reported; this suggests either existence of two orders of nodals or secondary development of cirri on some internodals. The distinction between nudinodals and internodals, as well as between internodals of different orders, is always more obvious in proximal than in distal parts of the stem, where it tends to disappear and commonly vanishes completely.

PROXIMALE AND CENTRODORSAL

In a number of Articulata, at the top of the column occurs a large, undivided ossicle called **proximale,** which is formed by one enlarged columnal or by two or several fused proximal columnals (Fig. 51). It is permanently attached to the theca by an immovable articulation.

The proximale in Thiollierierinidae and

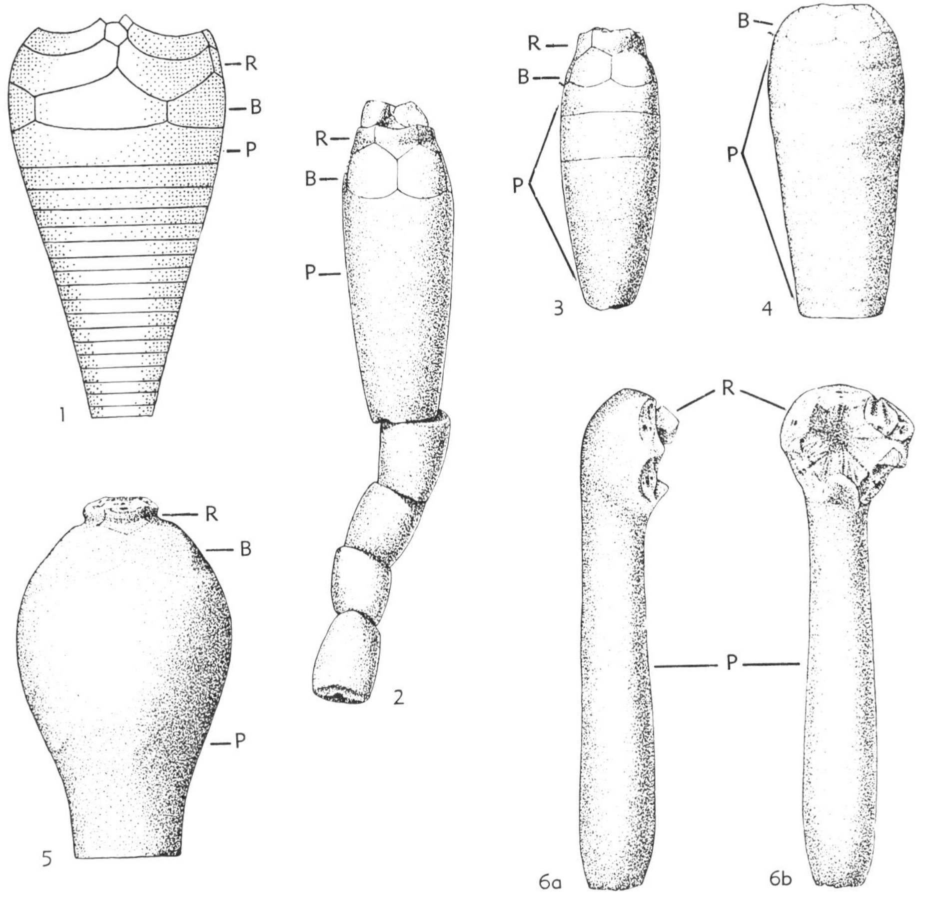

FIG. 51. Morphology of proximale. [Explanation: *B*, basal; *P*, proximale; *R*, radial.] (All Rasmussen, 1961, except *1*, from de Loriol, 1882-84.)

1. Apiocrinites meriani DESOR, ×0.5.
2. Bourgueticrinus danicus NIELSEN, U.Cret., Denmark, ×5.
3. Same, with distinct sutures in the proximale, ×5.
4. Bourgueticrinus cylindricus M'COY, U.Cret., Eng., ×2.5.
5. Bourgueticrinus papilliformis GRIFFITH & BRYDONE, U.Cret., Eng., ×5.
6a,b. Hemicrinus canon (SEELEY), L.Cret., Eng., lateral and adoral views of cup (fused with proximale), ×1.25.

Comatulida is known as the **centrodorsal,** an unusually large, discoidal, hemispheric, columnal, or a more or less conical ossicle, which forms the aboral portion of the theca and generally bears numerous cirri, except on its aboral central area (Fig. 52). In some Comasteridae, the centrodorsal is reduced to a thin noncirriferous stellate plate, and in certain specimens of the Thiolliericrinidae cirri are rudimentary or altogether lacking. The ontogeny of recent comatulids shows that only the top columnal is involved in formation of the centrodorsal, to which, however, are also incorporated the infrabasals in species having infrabasals introduced during development. The centrodorsal has a more or less well-developed central concavity **(centrodorsal cavity)** for reception of the chambered organ and associated structures (Fig. 52,5,6); its inner wall is studded with minute pores, inner openings, or passageways leading to axial canals of the cirri. Five depressions or perforations **(radial pits)** that harbor

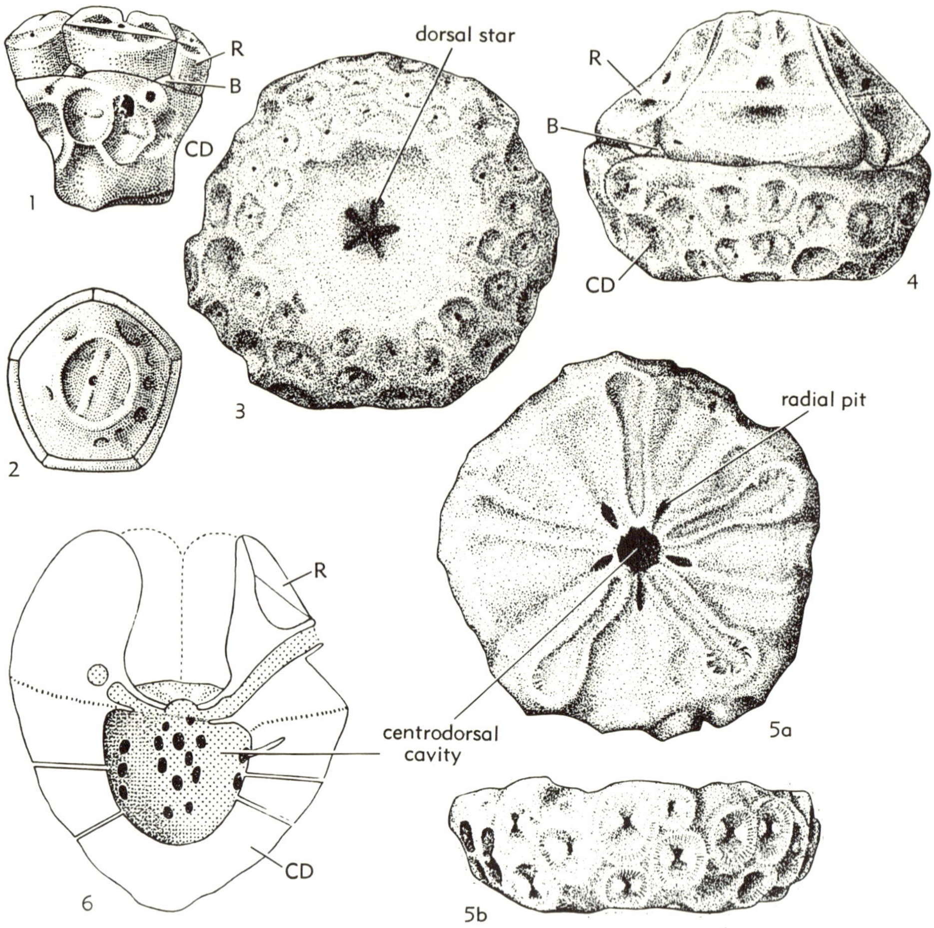

FIG. 52. Morphology of centrodorsal. [Explanation: *B*, basal; *CD*, centrodorsal; *R*, radial.]

1. Thiolliericrinus ribeiroi LORIOL, U.Jur., Portugal, side view, ×2 (Bather, 1900a).

2. Thiolliericrinus favieri (ÉTALLON), U.Jur., France, aboral view with articulum, ×2 (Bather, 1900a).

3,4. Glenotremites aequimarginatus (CARPENTER), L.Cret., England, aboral and side views, ×5 (Rasmussen, 1961).

5a,b. Glenotremites discoidalis GISLÉN, U.Cret., Czechoslovakia, adoral and side views of centrodorsal, ×5 (Rasmussen, 1961).

6. Florometra asperrima (A. H. CLARK), recent, Alaska, vertical section through centrodorsal and radial circlet, enl. (Clark, 1915a).

coelomic extensions may occur on the adoral surface outside the aperture of the central cavity unless that aperture is markedly lobate, in which case they emerge at the extremities of the lobes. They are lacking in the majority of recent comatulids, but are present in many fossil members of that group and in some extant species. In some forms they penetrate deeply into the skeleton of the centrodorsal and even may divide and anastomose. In no recent comatulids do these canals appear to open on the dorsal surface, the body wall being continuous over their aboral end, but in some fossil species they perforate the centrodorsal completely and have aboral radial openings (possibly due to erosion, at least in some cases) commonly situated in the lobes of a stellate hollow around the dorsal pole. This is the so-called **dorsal star** of authors

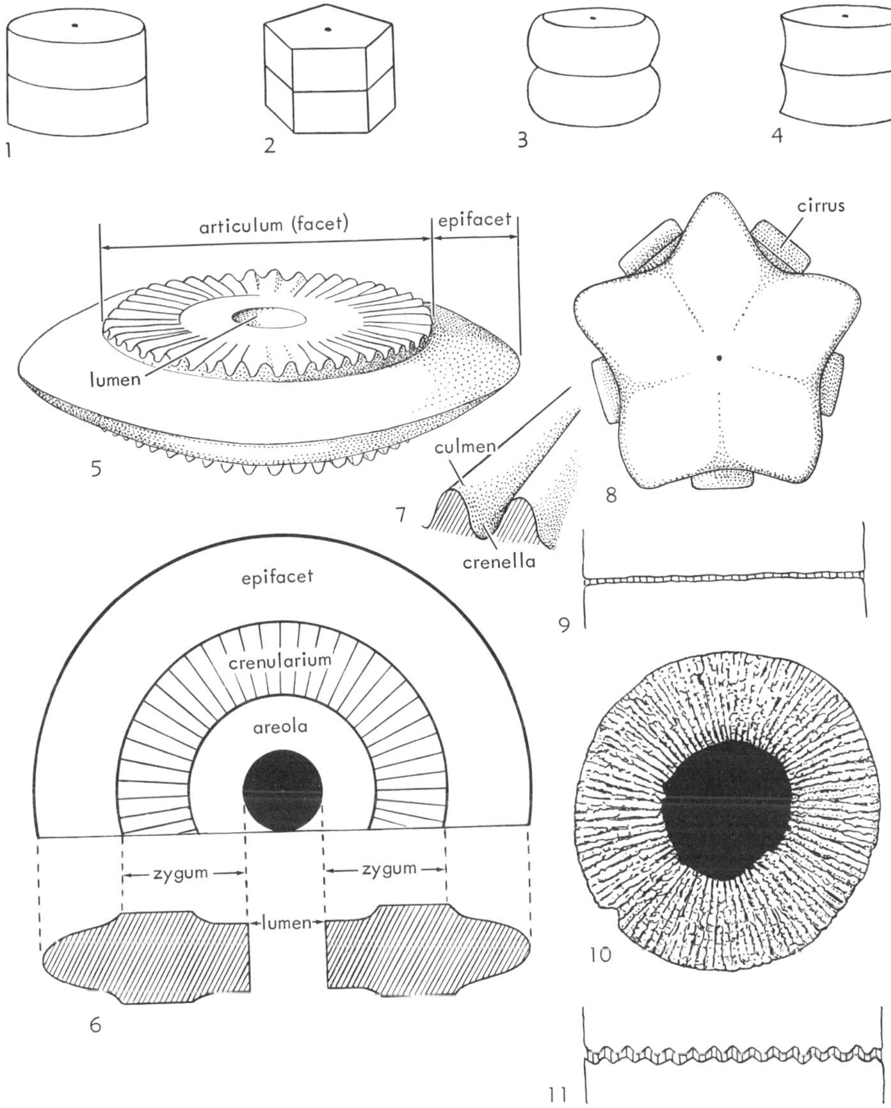

FIG. 53. Morphological features of crinoid columnals.

1-4. Various shapes of columnals (diagrammatic) (Bather, 1909a).

5,6. Downward and facetal views of a columnal with (below) median longitudinal section (Moore, Jeffords, & Miller, 1968).

7. Crenula formed of a culmen (ridge) and a crenella (groove) (Ubaghs, n).

8. Isocrinus nodal showing smooth surface of distal face as example of synostosial type of articulation (Moore, Jeffords, & Miller, 1968).

9. Diagrammatic transverse profile of juxtaposed synostosial facets (Moore, Jeffords, & Miller, 1968).

10. Lomalegnum hormidium MOORE & JEFFORDS, L.Miss., USA(Iowa), facetal view as example of symplectial type of articulation (after Moore & Jeffords, 1968).

11. Diagrammatic transverse profile of juxtaposed symplectial facets (crenulate suture) (Moore, Jeffords, & Miller, 1968).

(Fig. 52,*3*). The real functions of these coelomic extensions into the centrodorsal is unknown (A. M. Clark, 1972). In the Thiolliericrinidae the aboral side of the centrodorsal is occupied by a perforated articulation, which allows flexible union with the uppermost stem ossicle (Fig. 52,*2*). In the Comatulida, the articulation between the centrodorsal columnal next below it breaks at some time during ontogeny, and the animal becomes stemless and free. Exceptionally, remains of this articulation seem to have persisted during life in the shape of an indistinct petaloid impression on the aboral side of the centrodorsal. Not to be confounded with this figure is the so-called dorsal star, which has been just mentioned (Fig. 52,*3*).

LATERA

A median longitudinal section of a columnal shows that the outward facing sides or **latera** (sing., **latus**) may be straight, convex, or (rarely) concave (Fig. 53). If the latera are straight and the transverse section circular, the stem is cylindrical (Fig. 53,*1*). It is basaltiform, if the latera are straight and the transverse section pentagonal (Fig. 53,*2*), and moniliform (Fig. 53,*3*), if the latera are markedly convex and the transverse section circular.

The term **epifacet** is given to the projecting latus that surrounds the **facet** or **articulum** (pl., **articula**) of a columnal, that is to say, the surface of a columnal serving for articulation with a contiguous columnal (Fig. 53,*5*). The epifacet may be considerably extended outward.

The latera of crinoid columns are variously ornamented by spines, nodes, ridges, grooves, and scars serving for attachment of cirri or other outgrowths.

Pores may also be present on the latera of columnals. If they are connected with the axial canal of the stem by passageways through the columnals, they may correspond to abortive cirri, for it happens that some of them are still associated with small cirrus scars. [Such pores are not to be confounded with the so-called **interarticular radial pores** (or interarticular pores of P. H. Carpenter, 1884a) visible between the columnals in re-entrant angles of the quinquestellate stems of isocrinid crinoids; such pores are produced by the apposition of joint grooves radiating outward on the joint-faces but without communicating with the axial canal.]

ARTICULA

Any face of a columnal, or more generally of a pelma element that serves for articulation with a contiguous skeletal element is a joint-face or articular facet, technically called **facet** or **articulum** (pl., **articula**). It corresponds to proximal and distal facets of columnals or cirrals (component ossicles of cirri) as well as to cirrus scars on nodals. It is divisible into **lumen** or intercept of the axial canal in the plane of an articulum and **zygum,** which is the entire area of an articulum outside of the lumen (Fig. 53,*6*).

The lumen may be extremely minute, or, on the contrary, so large that the surrounding zygum is reduced to a narrow band. Generally, it consists of a single opening but it may be accompanied by generally five, more rarely three or four, accessory perforations partially or wholly separated from it and serving, like the central opening itself, as passageways for extensions of the coelom and nervous system. The shape of the lumen commonly is circular, but it may be elliptical, pentagonal, quadri- or quinquelobate, quadri- or quinquestellate (Fig. 54,*1-5*). The margin of the lumen is flush with the general surface of the joint-face, or raised to form a moderately broad elevated rim around it **(perilumen);** this may be smooth, granulose, tuberculate, or vermiculate (Fig. 55,*3,4*).

Surrounding the lumen (and perilumen if present) is generally a smooth or granular, circular or quinquestellate area **(areola),** which is even with the general floor of the joint-face, depressed below it, or slightly raised above it (Fig. 53,*6;* 55,*1-4*).

Between the areola and periphery of columnal articula a remaining space is generally differentiated as a tract called **crenularium,** characterized by radially disposed grooves **(crenellae)** and ridges **(culmina).** Combined crenellae and culmina are collectively designated as **crenulae** (Fig. 53,*7*). Some columnals have a narrow peripheral ridge termed **articular rim.** In some Mesozoic and Cenozoic articulate crinoids, the crenularium is replaced by a rosette **(petalodium)** composed of five lobed divisions **(petals)** each of which consists of a median

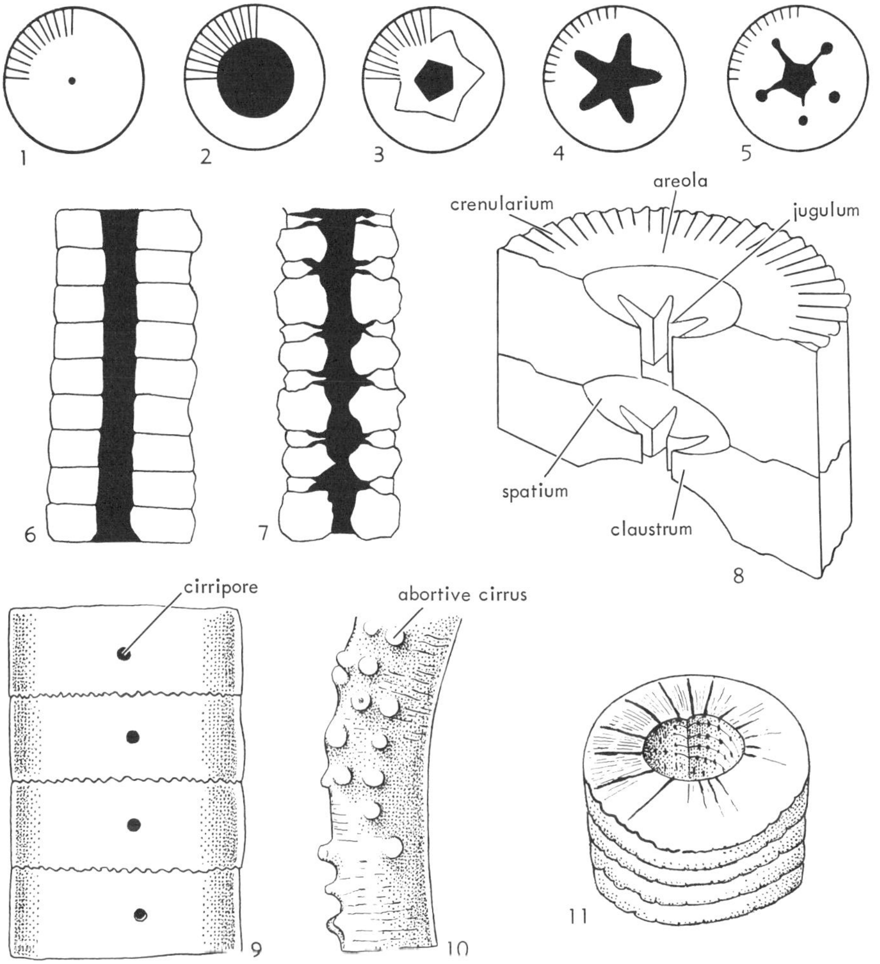

FIG. 51. Morphology of the axial canal in stems of crinoids (all from Moore, Jeffords, & Miller, 1968 (*1-5*, mod.), except *9, 10,* Moore & Jeffords, 1968, and *11,* Ubaghs, n).

1-5. Diagrammatic facetal views illustrating diversity in shape and relative size of lumen.

6. Floricyclus granulosus MOORE & JEFFORDS, U. Penn., USA(Texas), median longitudinal section of pluricolumnal with simple axial canal, ×2.

7. Heterostelechus texanus MOORE & JEFFORDS, L. Perm., USA(Texas), median longitudinal section of pluricolumnal with complex axial canal, ×2.

8. Diagrammatic oblique view of medially sectioned columnals illustrating features and terminology of axial canal.

9. Mooreanteris waylandensis MOORE & JEFFORDS, U.Penn., Texas, side view of pluricolumnal with cirripores, ×8.

10. Eurax eugenes MOORE & JEFFORDS, L.Dev., USA(Tenn.), side view of pluricolumnal with abortive cirri, ×2.35.

11. Trybliocrinus flatheanus GEINITZ, L.Dev., Spain, diagrammatic oblique view of pluricolumnal showing radiating grooves on articular facet and their openings into axial canal, ×1.3.

ovoid **floor** surrounded by crenulae; the floor is mostly flush with the articular surface of the columnal, but it may be depressed or (rarely) slightly elevated. The petals may be contiguous, and in that case the perradial crenulae **(radial ridge groups)** of adjacent petals alternate, or are gable-shaped or rectilinear; they may also be

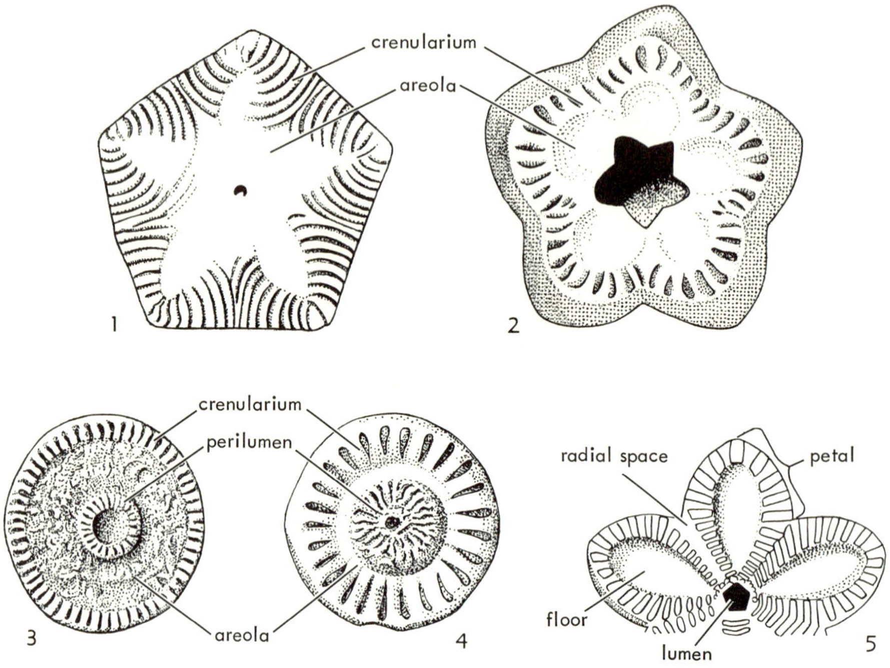

FIG. 55. Articula of crinoid columnals (all Moore & Jeffords, 1968, except *5,* from Bather, 1909a).

1. Pentagonostipes petaloides MOORE & JEFFORDS, M.Dev., USA(N.Y.), symplectial articulation with petaloid crenularium, areola, and small rounded lumen, ×8.5.

2. Floripila florealis MOORE & JEFFORDS, M.Dev., USA(N.Y.), symplectial articulation with narrow crenularium, petaloid areola, and large star-shaped lumen, ×8.5.

3. Dierocalipter doter MOORE & JEFFORDS, L.Miss., USA(Ky.), symplectial articulation with narrow crenularium, wide areola, and crenulate perilumen, ×3.5.

4. Flucticharax undatus MOORE & JEFFORDS, L. Miss., USA(Ky.), symplectial articulation with wide crenularium, narrow areola, and prominent perilumen, ×3.5.

5. Isocrinid internodal, diagram of three sectors of articulum.

separated from one another by smooth areas known as **radial spaces** (Fig. 55,*5*).

MAIN TYPES OF COLUMNAL ARTICULATIONS

Only ligamentary articulations are known in the pelma of crinoids, although distinct movements of the cirri and also of the stalk have been recorded in recent crinoids by many authors.

The simplest type of articulation, which is not very common, is characterized by essentially plane articular surfaces devoid of any sculptured features. Such juncture is termed **synostosis** and this type of articulation is called **synostosial** (Fig. 50,*3;* 53,*8*). The corresponding **sutures,** which are the externally visible edges of articula, are straight (Fig. 53,*9*). Short dermal fibers unite the apposed joint-faces, allowing small movement in all directions. Synostosis occurs between the nodal and infranodal joints of pentacrinids and between paired nodals **(binodals)** in the camerate genus *Camptocrinus* (Fig. 50,*1-3*). When addition of small amounts of calcareous deposits to articula reduce mobility practically to nothing, the synostosis becomes a **zygosynostosis** and the contact between apposed ossicles is a **close suture.** The zygosynostosis is rarely present in the stem itself, but exists commonly, along with complete fusion of ossicles **(ankylosis),** in the holdfast of crinoids. Opposite modification of synostosial articulation toward increased flexibility occurs in the stem of some primitive

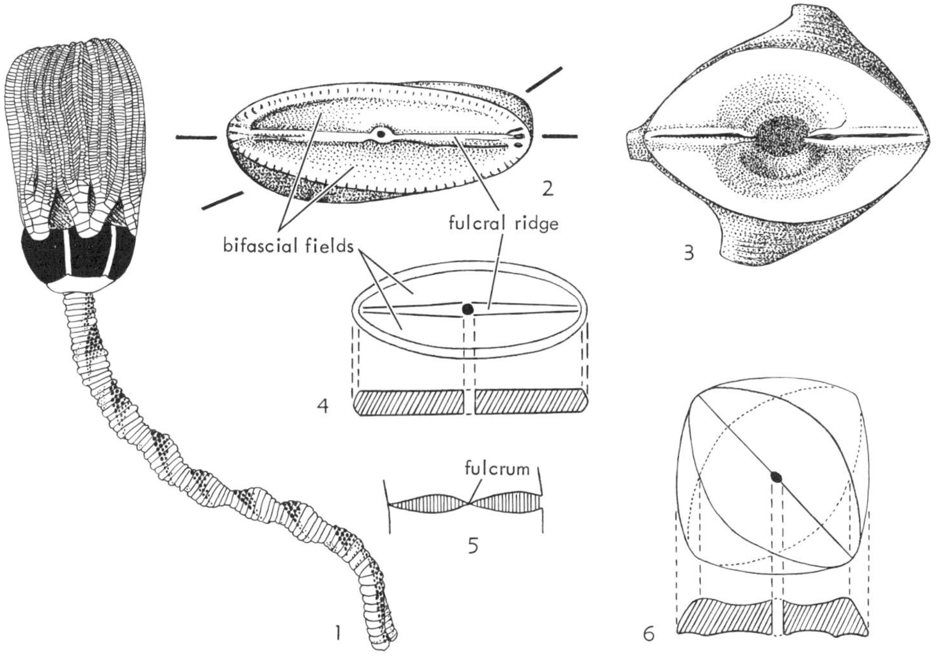

FIG. 56. Morphology of synarthrial type of articulation.

1. *Platycrinites symmetricus* (WACHSMUTH & SPRINGER), Miss., USA(Iowa), crown with twisted ribbonlike column, ×1 (Wachsmuth & Springer, 1897).
2. *Platycrinites* sp., L. Carb., Belg., with elliptical articular facets oriented differently on opposite sides (Ubaghs, n).
3. *Bourgueticrinus danicus* NIELSEN, U.Cret., Denmark, nodal, articular facet, with bifascial fields depressed into pits, ×8.3 (Nielsen, 1913).
4. Diagrammatic facetal view and median longitudinal section of elliptical columnal (Moore, Jeffords, & Miller, 1968).
5. Diagrammatic transverse profile of juxtaposed synarthrial articular facets (Moore, Jeffords, & Miller, 1968).
6. Quadrangular columnal with elliptical articular facets oriented differently on opposite sides (Moore, Jeffords, & Miller, 1968).

inadunates, in which the articula of columnal pentameres are distinctly concave (Fig. 48,6).

The most common type of articulation in cylindrical and basaltiform fossil crinoid stems is characterized by apposed faces that bear interlocking grooves **(crenellae)** and ridges **(culmina),** and their external sutures are **crenulate.** This type of union is called **symplexy** (adj., **symplectial**) (Fig. 53,*10,11;* 55,*1-4*). It allows very slight movement between columnals. [In the past, symplexy has been confused frequently with syzygy —an articulation common in the arms of some crinoids, but unknown in the pelma. In syzygy the culmina of apposed articula meet and the intervening crenellae are occupied by ligaments, so that the external suture has a finely beaded (not crenulate) appearance.] Weakly marked symplectial articulations are called **cryptosymplexies** (adj., **cryptosymplectial**).

Columnals with elliptical or diamond-shaped outlines in views normal to their articula and some with circular cross section are distinguished by **synarthrial** (also called **bifascial**) **articulations.** In a **synarthry,** each joint-face bears a median **fulcral ridge** along the major axis of elliptical and rhomboidal columnals, with equal ligamentary shallow depressions **(bifascial fields)** flanking sides of the ridge (Fig. 56,*4,6*). The latter are smooth and generally broad and shallow, but in some Articulata (e.g., *Bourgueticrinus, Rhizocrinus*) they are depressed into moderately

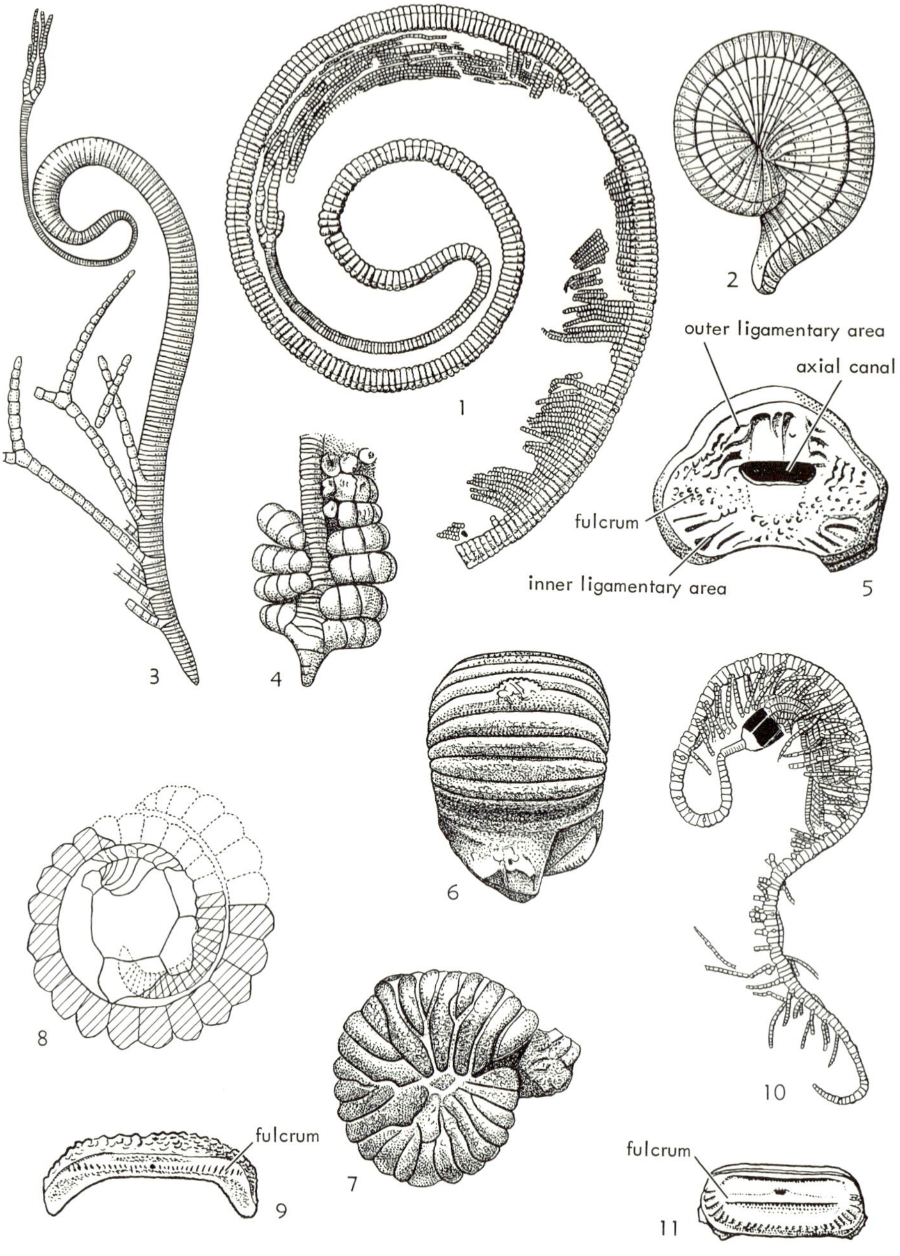

Fig. 57. Morphological features of coiled stems of crinoids.

1. Herpetocrinus fletcheri Salter (Inadunata), U. Sil.(Gotl.), Sweden, slender crown and recurved coiled stem with inward directed cirri, ×0.75 (Bather, 1900a).

2. Myelodactylus ammonis (Bather) (Inadunata), U.Sil., USA(Tenn.), closely enrolled specimen with cirri entirely concealing crown, ×1 (Springer, 1926b).

3. Crinobrachiatus brachiatus (Hall) (Inadunata), M.Sil., USA(N.Y.), specimen with very small crown, tenuous proxistele, thick mesi- and dististele, and strong branching cirri, ×1.5 (Springer, 1926b).

4. Brachiocrinus nodosarius Hall (Inadunata), L. Dev., USA(N.Y.), distal part of stem terminating in a point, with ponderous cirri, ×1 (Springer, 1926b).

deep pits (Fig. 56,*3*). There is no crenularium, or if any, it is rudimentary. A narrow marginal rim may be present. The fulcral ridges of apposed faces are aligned with one another and provide for movements in direction normal to the fulcral ridges. Each columnal has a more or less marked skew, so that the ridge of the proximal face lies at an angle to that of the distal face. The extreme is reached in some square columnals (as viewed from their articular sides) with the ridges at right angles to each other. This progressive or abrupt shifting in orientation of the fulcral ridges enables the stem to bend in different directions. When the columnals are elliptical and the skew of each columnal is slight, the stalk may look like a twisted ribbon (Fig. 56,*1*).

Modified synarthrial articulations occur in the coiled stem (Fig. 57) of the extraordinary homeomorphs, the inadunate *Myelodactylus,* the flexible *Ammonicrinus,* and the camerate *Camptocrinus.* In such crinoids, the stem is flattened or concave on the inner side of the coil throughout the greater part of its length and takes on a bilateral symmetry; in cross section, the columnals are elliptical or crescentic, and their articula bear fulcra separating two unequally developed areas, which, during life, were presumably occupied by ligamentary (and possibly contractile) tissues (Fig. 57,*5,9,11*).

AXIAL CANALS

All columnals and cirrals are pierced by an axial canal. Generally (but not invariably) it occupies the center of the columnal. Its transverse shape and relative size as shown by outline and size of the lumen on joint-faces have been described above (p. *T*65) (Fig. 54,*1-5*).

In median longitudinal section, the axial canal of columnals may be **simple** (straight sided) (Fig. 54,*6*) or **complex** (provided with expansions or constrictions) (Fig. 54, *7*). If present, the expansions **(spatia)** are located between the columnals, while the constriction **(claustra)** are placed at mid-height of the columnals (Fig. 54,*8*). As seen in longitudinal section, the spatia may be low, or tall and laterally rounded, truncate, or pointed; the claustra, clavate, truncate, or lanceolate. Adaxial surfaces of claustra may be convex, flat, or concave, and smooth or denticulate. The part of an axial canal limited by the adaxial faces of a claustrum is called **jugulum;** it corresponds to a distinct local narrowing of the axial canal, narrowing that may be longitudinally very short to moderately long and transversely circular or pentagonal to strongly quinquestellate (Fig. 54,*8*). The adaxial part of a claustrum bordering a jugulum may be thickened and form what is designated as a **jugular rampart.** The jugulum and the central part of spatia were probably occupied during life by coelomic and nervous extensions from the chambered organ, whereas the remaining part of the spatia was filled with dermal fibers.

Radially disposed simple or branched canals may be given off from the axial canal. If these side canals penetrate stereom and lead to cirrus scars on latera, they are termed **canaliculae.** If they are located between apposed joint-faces they are called **fossulae** (Fig. 50,*3*). In some stems these passageways emerge on the outer surface of columnals as small pores **(cirripores)** of uncertain function (Fig. 54,*9*) or they terminate in diminutive nodicirral sockets or pimplelike protuberances **(abortive cirri)** (Fig. 54,*10*). In the camerate *Trybliocrinus* the joint-faces of columnals show radiating

(Continued from facing page.)

5. Myelodactylus canaliculatus (Goldfuss) (Inadunata), M.Dev., Ger., columnal articulum, ×3.2 (Sieverts-Doreck, 1954).

6,7. Ammonicrinus wanneri Springer (Flexibilia), M.Dev., Ger., outer side and lateral views of tightly coiled specimen, ×1 (Krause, 1927).

8. Same, diagrammatic representation of a specimen with stem (longitudinally sectioned) completely enveloping crown, ×2 (Ubaghs, 1952).

9. Ammonicrinus sulcatus Kongiel, M.Dev., Poland, columnal articulum, ×2 (Kongiel, 1958).

10. Camptocrinus multicirrus Springer (Camerata), Miss., USA(Ill.), complete specimen, ×1 (Wachsmuth & Springer, 1897).

11. Camptocrinus compressus Wright, L.Carb., Scot., columnal articulum, ×5 (Ubaghs, 1953).

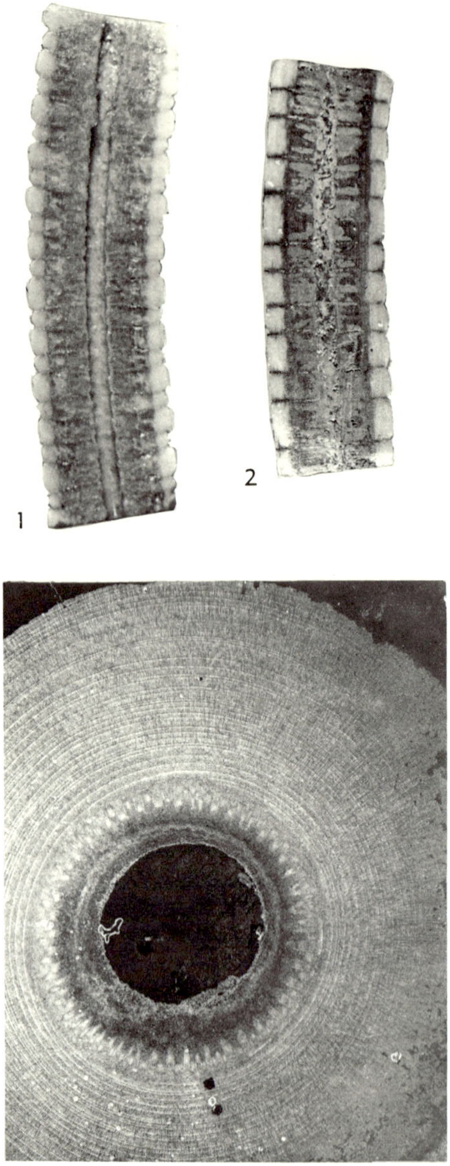

FIG. 58. Growth of crinoid stems indicated by microstructure.

1. Longitudinal median section of *Elytroclon elimatus* MOORE & JEFFORDS, L.Miss.(Osagian), New Providence F., USA(Ky.), showing dense light-colored cortical (outer) zone and spongy dark medulla surrounding axial canal, ×1.7 (Moore & Jeffords, 1968).

2. Longitudinal median section of *Ilematerisma enamma* MOORE & JEFFORDS, L.Miss.(Osagian), New Providence F., USA(Ky.), similarly showing well-marked cortical and medullary zones, ×3.5 (Moore & Jeffords, 1968).

3. Transverse section of columnal belonging to *Cyclocaudex plenus* MOORE & JEFFORDS, U.Penn.(Virgilian), Wayland Shale, USA(Texas), showing very wide cortex with concentric microstructure resembling fine tree rings and thin dark medulla next to large circular axial canal, ×7 (Moore, Jeffords, & Miller, 1968).

grooves which open into the axial canal through small vertical slits and terminate near the periphery without communicating with the exterior (Fig. 54,*11*). A similar canal system occurs in the grapnel-like root of the inadunate *Ancyrocrinus*. Apparently these passages between or through the columnals served to transmit the nutrient fluid to cells secreting the stereom and to aerate the same fluid by bringing it near the surface.

CORTEX AND MEDULLA

Sections of columnals may reveal the existence of more or less clear differentiation of the stereom into outer **(cortical)** and inner **(medullary)** portions, which may be very distinct or grade into each other (Fig. 58). The **cortex** shows generally a dense calcitic microstructure, whereas the **medulla** is more spongy or microlamellate to microreticulate. The medulla itself may be differentiated into a narrow adaxial zone of dense calcite (correlated with the perilumina of articular faces) and a relatively broad next abaxial zone of more spongy nature (corresponding with areolae of the joint-faces).

INDICES OF COLUMNALS

Description of columnals and pluricolumnals may require various measurements of morphological elements and means of expressing interrelationships between them. The latter can be expressed as **ratios,** which conveniently are multiplied by 100 in order to convert them to whole numbers. Such numbers are termed **indices** (MOORE & JEFFORDS, 1968).

Determination of columnal indices is simple in dealing with circular stem parts or features, but it needs computation of mean diameters consisting of the sum of measurements along two opposite radii when features of the columnals possess

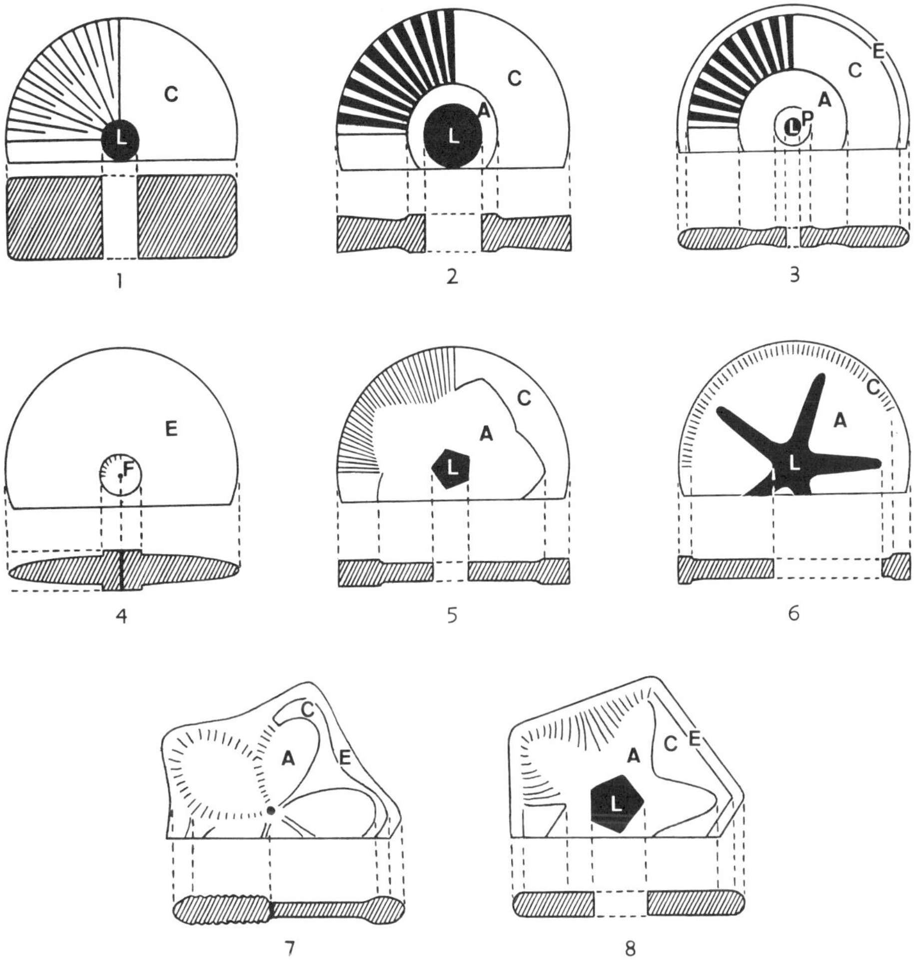

FIG. 59. Diagrammatic facetal views and median longitudinal sections of columnals designed to illustrate determination of various columnal indices (Moore, Jeffords, & Miller, 1968). [Explanation: lumen, black; columnals in sections, oblique ruled; *A*, areola; *C*, crenularium; *E*, epifacet; *F*, facet; *L*, lumen; *P*, perilumen.]

1. Circular columnal with wide crenularium lacking areola and perilumen.
2. Circular columnal with narrow areola, no perilumen.
3. Circular columnal with well-developed areola, perilumen and narrow epifacet.
4. Circular columnal with very wide epifacet and small articulum.
5,6. Circular columnals with differently shaped areolae and lumina.
7,8. Pentagonal columnals with petaloid and quinquestellate areolae.

pentagonal, quinquelobate or quinquestellate outlines (Fig. 59). In elliptical columnals, maximum and minimum radii (disposed at right angles to one another) are chosen for determining the mean diameter (Fig. 56,*4*,*6*).

The main indices distinguished by MOORE & JEFFORDS (1968) are ratios all multiplied by 100, as follows:

areolar index: ratio of total width of areola to that of articular facet.

articular facetal index: ratio of total width of articular facet to that of entire columnal.

crenularial index: ratio of total width of crenularium to that of columnal articular facet.

epifacetal index: ratio of total width of

epifacet to that of entire columnal.

height index: ratio of height to total width of columnal.

internodal index: ratio of total height of internode to height of noditaxis containing it.

jugular index: ratio of total width of jugulum to that of columnal articular facet.

luminal index: ratio of total width of lumen to that of columnal articular facet.

nodal index: ratio of height of nodal to that of noditaxis containing it.

periluminal index: ratio of total width of perilumen to that of columnal articular facet.

zygal index: ratio of total width of zygum to that of columnal articular facet.

STEM GROWTH

The growth of the stem in crinoids is a product of two factors: formation of new columnals and increase in size of individual columnals.

In homeomorphic stems new columnals are introduced at the proximal end of the stalk, so that proximal columnals are youngest and distal ones oldest. In heteromorphic stems it is the nodals that develop next to the cup, the internodals being intercalated between them subsequently at differing distances from the cup. Nodals are in contact with one another near the cup and just below are separated by one internodal. Next a new internodal is added on each side of the first one. Then two other internodals are inserted next below and above each of the just-formed internodals, and so on. Internodals present at successive growth stages would therefore number 0-1-3-7-15 . . . at least theoretically; in fact, the rate at which generations of internodals are introduced may be somewhat irregular (Jeffords & Miller, 1968).

After a few generations, a characteristic number of columnals in each internode commonly is reached and formation of new internodals ceases, but the accretionary growth of individual columnals continues. This allows recognition of three stages (called ages by Termier & Termier, 1949) in development of the columnals, each one being located in a more or less distinct region of the stem. The first stage characterizes a usually short region located immediately below the cup, in which the nodals and first internodals are formed, and all columnals have a juvenile aspect (Fig. 60,*1*). The second stage corresponds to part of the stem in which various cycles of internodals are introduced, size of the columnals increases, and cirri (if any) develop. The third stage includes the rest of the stem in which no new internodals are inserted, and distinction between different generations of columnals, so obvious near the cup, tends to be lost (Fig. 60,*2*).

This mode of growth has a direct effect on flexibility of the stem. Since new columnals are introduced near the cup as very thin discs and then are enlarged by accretionary growth in axial direction, the flexibility, high near the cup, diminishes progressively toward the root.

All crinoid stems do not develop according to this pattern, however. In some forms, such as the cyrtocrinid articulates *Eugeniacrinites* and *Cyrtocrinus,* the stem is short and entirely composed of long, cylindrical segments, so that the whole stem must certainly have acted as a rather rigid support of the crown. In Jurassic Pentacrinitidae, such as *Seirocrinus,* the number of columnals in each internode is not limited and continues to increase geometrically in a distal direction, becoming enormous in species provided with a very long stem. Moreover, as alternation of smaller and larger columnals is more pronounced and diameter of the stem becomes narrower away from the cup, flexibility of the stem increases toward the root, rather than decreases. This trend does not persist indefinitely, however, for complete specimens show a rapid diminution of internodal intercalation near the distal end, along with a concentration of nodals, so that in the terminal stem region a dense tuft of cirri is produced and acts as a holdfast (Seilacher *et al.,* 1968).

If nodals are introduced immediately below the base of the cup in most crinoids, the incorporation of the topmost columnal (or columnals) in the theca in some post-Paleozoic genera, producing a proximale or centrodorsal, stops formation of new columnals (at least above it). In some Paleozoic crinoids also, it seems possible that a

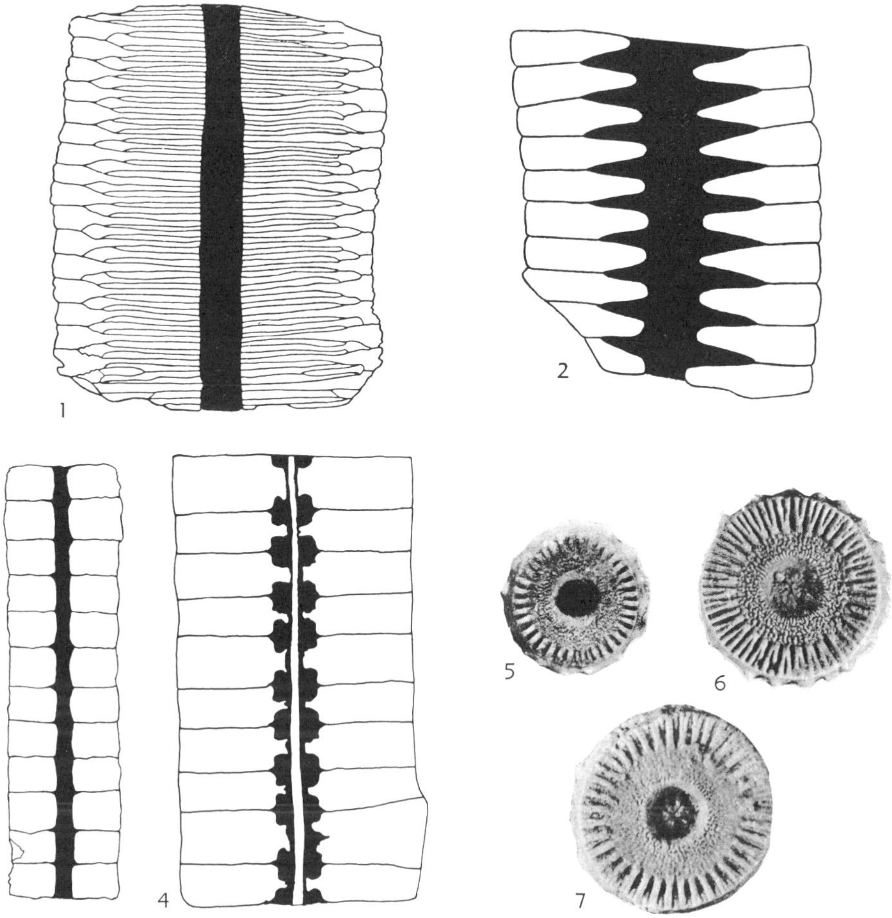

FIG. 60. Growth of crinoid column.

1,2. Barrandeocrinus sceptrum ANGELIN, U.Sil., Gotl., Sweden, median longitudinal sections showing shape of columnals and nature of axial canal in proximal (juvenile) and middle (mature) portions of stem, ×6.5 (Ubaghs, 1956).

3,4. Preptopremnum rugosum MOORE & JEFFORDS, U.Penn., USA(Texas), median longitudinal sections showing nature of axial canals in juvenile and mature portions of columns, ×2.65 (Jeffords & Miller, 1968).

5-7. Same, articula showing increase in numbers of crenulae with increase in diameter of columnal, ×2.65 (Jeffords & Miller, 1968).

few of the most proximal columnals may be related to the theca in such a way as to prevent intercalation of new columnals between them and the cup. Examples are offered by species having a basally concave theca so shaped as to render captive columnals occurring within it. The name **basilarids** has been proposed by STRIMPLE (1963a) for these "captured" columnals.

Growth of individual columnals takes place in both axial and transverse directions. This is clearly shown by growth lines (discernible in cross sections) (Fig. 58,*3*) and by observable changes in size, proportions, shape, nature of articular facets and other features of columnals along the stem from its youngest proximal part to its oldest distal region. Most of these changes can be described in terms of a few growth gradients, which seem to control a variety of adaptive modifications of the crinoid stem (SEILACHER *et al.*, 1968). This is well illustrated by transverse growth, diameter of the nodals being controlled primarily by size of the generating area at base of the cup, size of the nodals in turn determining diameter of the initial internodals. If no other factor than cup growth intervened, diameter of the stem would increase from root to cup, but accretionary growth of the columnals transversely brings about secondary modifications, which work in the opposite sense and tend to produce tapering of the stem from root toward cup. The resultant of these two gradients is shown by stems which taper in proximal direction or in distal direction, or thicken at both ends.

Turning to shape, one may observe that in many crinoids the longitudinal growth of columnals is greater initially at the rim than in the center; this produces a saucer-shaped depression in the joint-face, which is filled by developing internodals (Fig. 60,*1*). Then follows a gradual flattening of the articular facet as a consequence of accretionary growth, and a tendency of columnals of different generations to become similar (Fig. 60,*2*).

Considering articular surfaces, growth may be reflected by progressive enlargement of the lumen, expansion of the areola, and multiplication of the culmina (Fig. 60,*5-7*). Generally also, the axial canal is simpler in juvenile parts of the stem than in mature regions where it may show a great complexity (Fig. 60,*3,4*).

CIRRI

The cirri are jointed, generally undivided, appendages that arise from specialized columnals (nodals) along the stem of many crinoids, or are carried by the centrodorsal in most Thiolliericrinidae and Comatulida. They are formed by a row of skeletal ossicles called **cirrals,** which, like the columnals, are pierced by a central tubular canal that branches from the axial canal of the stem and contains coelomic and nervous extensions. Two or several cirrals in anatomical connection comprise a **pluricirral.** The most proximal segment of a cirrus that articulates with a columnal is distinguished as a **zygocirral.** In the distal region of the column, the cirri are replaced by the so-called **radicular cirri,** which form the root-like anchorage of many crinoids and will be described later.

Along the stem, the cirri may be arranged 1) without any definite order, 2) in vertical rows, but not in whorls, 3) in verticils of three, four, five, or six cirri, 4) in a helix (rarely), and 5) in two vertical rows along one side of the stem. They may be present throughout the length of the stem, or restricted to some part of it, generally the proximal or distal one. Their number may or may not vary from one region to another in a single stem. In Flexibilia, so far as known, cirri, except radicular cirri, are missing. This is equally true for most representatives of other Paleozoic crinoid subclasses. Among Camerata, though cirriferous species are known from Late Ordovician onward, they never become very common. In the Inadunata they first appeared in Silurian times, but were not common until the Carboniferous period. Among stalked post-Paleozoic forms, they are observed in some Encrinidae, well established in the Isocrinida, Thiolliericrinidae, and Comatulida, but lacking in Millericrinida and Cyrtocrinida.

On the centrodorsal of Thiolliericrinidae the cirri are arranged in one or several circles, or absent (in mature specimens) according to species. The centrodorsals of Comatulida bear closely placed cirri arranged in more or less distinct vertical rows or irregularly scattered. In the Comasterina, the cirri are rudimentary or lacking.

Regarding the formation of cirrals, it appears that new joints are added at the distal end of the cirri (and not at the base of the cirri, as sometimes erroneously stated). As no intercalation of cirrals takes place, the most distal ossicle is the last to be formed. The ontogeny of recent comatulids shows that the cirrals did not develop like the columnals. They appear at first as simple, round, fenestrated plates without a central perforation; this is formed later by absorption of the central part of the plate (Mortensen, 1920, p. 77).

The **cirrus sockets** (also called cirrus scars or cirrus facets), serving for attachment of cirri to a crinoid stem or centrodorsal are generally rounded, less commonly polygonal or elliptical in outline. They may be confined to sides of one columnal (nodal)

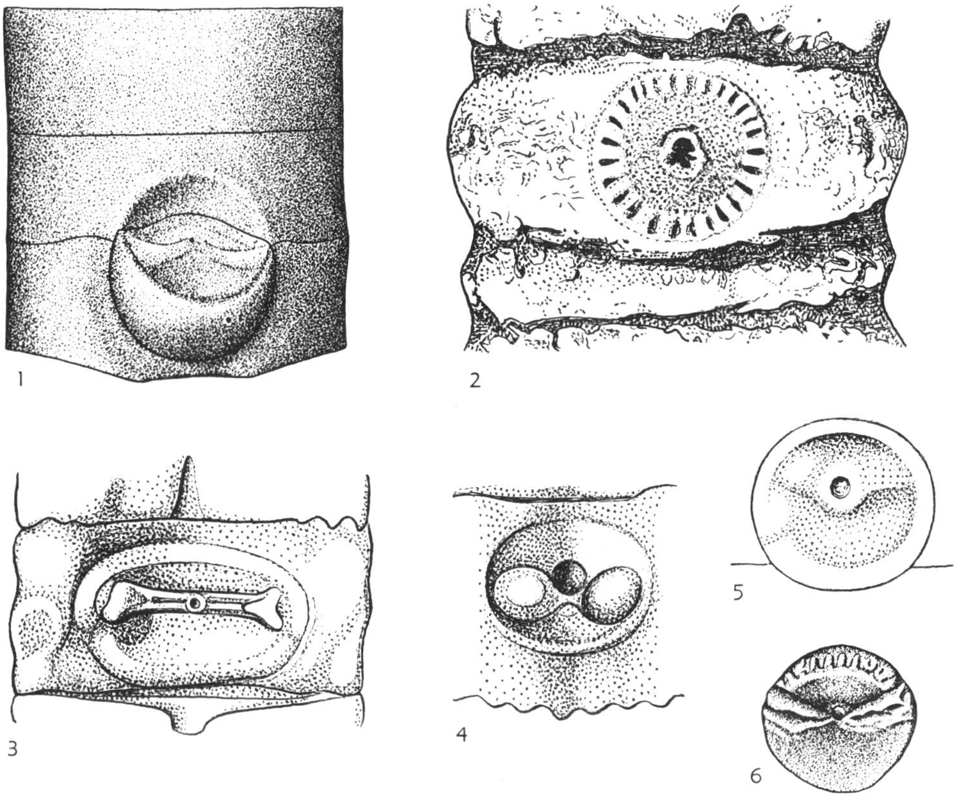

Fig. 61. Morphology of cirrus sockets and cirral articula.

1. *Austinocrinus rothpletzi* Stolley, U.Cret., Ger., pluricolumnal with cirrus socket and first cirral directed upward, ×5 (Rasmussen, 1961).
2. *Preptopremnum rugosum* Moore & Jeffords, U.Penn., USA(Texas), pluricolumnal with cirrus socket on nodal, ×7 (Moore & Jeffords, 1968).
3. *Isocrinus hercuniae* Bather, Trias., Hungary, cirrus socket, ×15 (Bather, 1909a).
4,5. *Isocrinus tyrolensis major* Bather, Trias., Hungary, cirrus socket, and distal articulum of cirral 2, ×30 (Bather, 1909a).
6. *Austinocrinus rothpletzi* Stolley, U.Cret., Ger., articulum of a cirral, ×10 (Rasmussen, 1961).

(Fig. 49,*1*), or be equally shared by two successive nodals (compound nodals and binodals) (Fig. 49,*4*; 50,*1*) or be impressed on several contiguous columnals (3 to 10 or more). When confined to single columnals, they are mostly placed at mid-height of the columnals, but in some stems they are much nearer to one joint-face of the columnal than the other, or may even project beyond the surface of the joint-face and occupy a notch in the next columnal (Fig. 61,*1*). They may be directed straight outward, upward, or downward, the corresponding cirri being more or less perpendicular to the column or growing along the column. In the camerate species *Camptocrinus compressus,* from the Lower Carboniferous of Scotland, the cirrus sockets on binodal columnals are compound, that is, divided into several facets, each of which carries a cirrus (Fig. 50,*5*).

The cirrus sockets are concave. Each one (or each facet of a compound socket) is provided with a more or less central axial pore (lumen) which communicates with the axial canal of the stem through intracolumnal passageways (canaliculae) or intercolumnal passageways (fossulae). The cirrus sockets are smooth or provided with radiate crenulae, short and confined to the margin (Fig. 61,*2*) or extending to or near the lumen. They may also have a more or

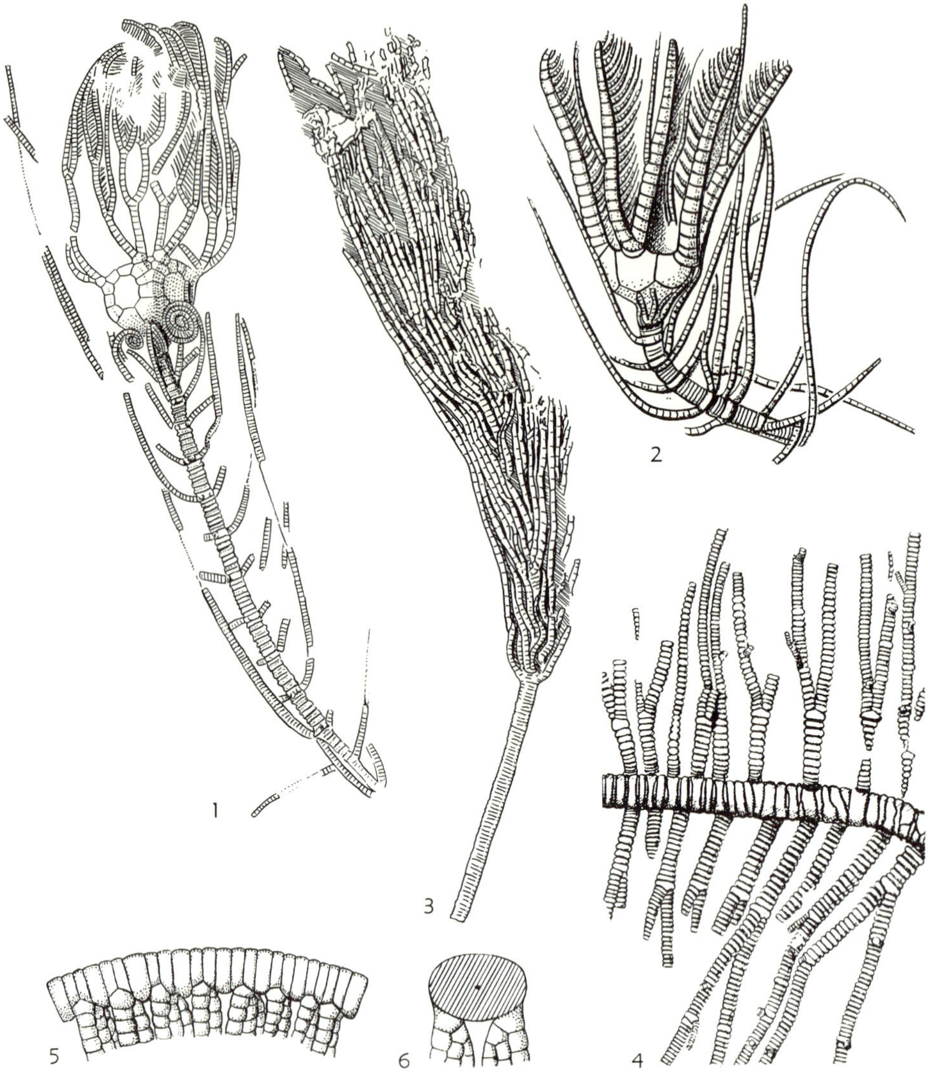

FIG. 62. Morphology of cirri in some Paleozoic crinoids.

1. *Clarkeocrinus troosti* GOLDRING, M.Dev., USA (N.Y.), crown and part of column, with circinate proximal cirri, ×0.7 (Goldring, 1923).
2. *Cordylocrinus plumosus* (HALL), L.Dev., USA (N.Y.), crown and part of column, with cirri longer than crown, ×2.6 (Goldring, 1923).
3. *Dichocrinus angustus* WHITE, L.Miss., USA (Iowa), extremely long cirri growing upward and enveloping small theca, ×1 (Springer, 1926b).
4. *Dicirrocrinus ramulosus* W. E. SCHMIDT, L.Dev., Ger., pluricolumnal with branching cirri, ×2 (Schmidt, 1942).
5,6. *Camptocrinus myelodactylus* WACHSMUTH & SPRINGER, L.Miss., Indiana, lateral view (×3) and transverse section (×6) of pluricolumnal, showing mode of branching of cirri (Springer, 1926b).

less distinct articular transverse ridge, commonly with a tubercle at each end (Fig. 61,*1,3*); or the ridge is reduced to these two lateral tubercles (Fig. 61,*4*).

A detailed knowledge of the cirri of Paleozoic crinoids is lacking. They seem generally to be composed of short, cylindrical, rarely barrel-shaped ossicles. In at least some species the lumen for passage of the axial canal was bean or dumbbell shaped, or even split into pairs. The way they are preserved and the aspect of their external sutures suggest that their articula did not allow a great amount of movement. In a few species, however, such as the Devonian camerate *Clarkeocrinus troosti* (Fig. 62,*1*), the cirri were able to coil in a plane as if they had properties similar to those of articulate crinoids. In the latter, articula of the cirrals (Fig. 61,*5,6*) are provided with transverse ridges which allow considerable motion in planes passing through the longitudinal axis of the body.

The length of cirri among fossil crinoids is quite variable. They may be extremely long, to the point of enveloping the theca wholly and part of the arms (Fig. 62,*2,3*). In the Silurian-Devonian Myelodactylidae (Fig. 57,*1-5*) and the Mississippian *Camptocrinus* (Fig. 57,*10,11*) with coiled stems, part of the column was provided with long cirri disposed in such manner as to conceal and protect the crown. Generally, however, the cirri are moderately developed, rarely exceeding a few centimeters in length. In recent comatulids, they show great diversity in form and size, usually correlated with the mode of life and habitat of the animals.

Typically, the cirri are simple undivided jointed appendages, but fossil species are known in which the cirri are branched. Examples are furnished by cladid inadunates such as the Silurian *Brachiocrinus,* the Devonian *Dicirrocrinus* (Fig. 62,*4*) and by some species of the Mississippian camerate *Camptocrinus* in which the first or first two cirrals are axillary (Fig. 62,*5,6*).

Abortive cirri, dwarfed in size and composed of a few small cirrals or even a single rounded ossicle without axial canal, occur in some Paleozoic crinoids. Stems of other crinoids have no actual sideward outgrowths but merely pores (cirripores) connected by passageways leading to the axial canal of the stalk (Fig. 54,*9*). Still other columnals possess irregular projections **(pseudocirri),** which resemble cirri in having an axial canal but differ from ordinary cirri in showing no division into component cirrals (Fig. 54,*10;* 63,*1*). Minute pimplelike ossicles of uncertain nature occur on some stems; when they fall, they leave on the columnal latera concave scars similar to cirrus scars except that they are not perforated (Fig. 63,*2*). In a genus based solely on stem parts, *Musivocrinus,* from the Permian of Tunisia, the stem apparently was covered with platelike contiguous unperforated ossicles, attached to double trans-

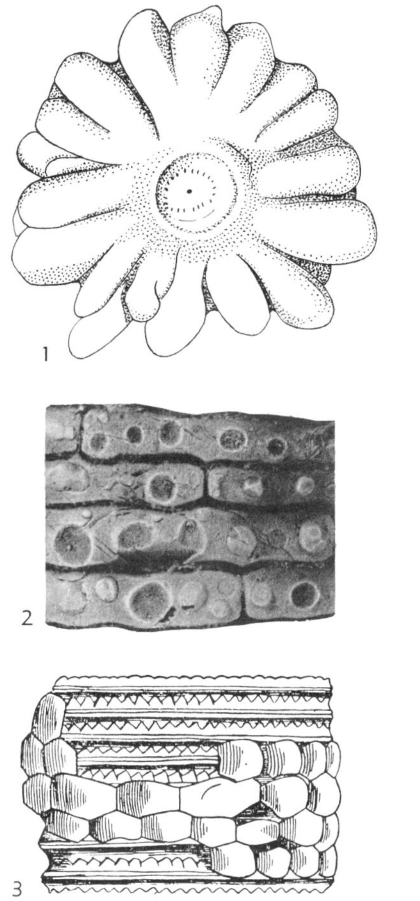

FIG. 63. Pseudocirri and structures of uncertain nature carried on latera of columnals.——*1. Hyperexochus immodiscus* MOORE & JEFFORDS, L.Dev., USA(Tenn.), facetal view of columnal provided with pseudocirri, ×2.8 (Moore & Jeffords, 1968).——*2. Polypeltes granulatus* ANGELIN, U.Sil., Gotl., lateral view of pluricolumnal showing pimplelike ossicles and concave nonperforated scars, ×6.6 (Ubaghs, 1956).——*3. Musivocrinus* sp. TERMIER & TERMIER, Perm., Tunisia, lateral view of pluricolumnal with partially preserved covering of small platelike ossicles (Termier & Termier, 1949).

verse ridges on latera of the columnals; the origin of such structures is presently unknown (Fig. 63,*3*).

HOLDFASTS

Any anchorage structure at and near the distal end of a crinoid column is called a **holdfast.** This term is applied chiefly in a functional sense, for the structures serving for attachment of crinoid stems are various and not necessarily homologous.

The attached end of the larval stalk of most recent comatulids expands into a soft disc inside which a fenestrated plate is located. This **terminal stem plate,** sometimes designated as the dorsocentral (a term not to be confused with centrodorsal) is always circular in very young larvae. During development it may acquire lobate, stellate, or digitiform outlines prior to detachment of the stem from the topmost columnal, which is incorporated in the theca. Whether such a terminal plate occurs in the ontogeny of stalked crinoids and what relation it could have with holdfasts of adults is not known.

From a purely morphological point of view, all crinoid holdfasts, in spite of their diversity, may be assigned to four fundamental types: 1) the so-called *"Hohlwurzel"* type, 2) the cirrus-bearing type, 3) the discoid type, and 4) the encrusting type.

1) As far as known, the *"Hohlwurzel"* type has been observed only in the archaic inadunate *Aethocrinus.* It consists of a hollow mass of small, irregular polygonal plates clustered around the distal extremity of the stem (Fig. 48,*1,7*). It may well represent the most primitive sort of stem termination among all crinoids (Ehrenberg, 1929), for a similar ending of the body occurs in primitive crinozoans such as the Lower Cambrian *Lepidocystis* and the Middle Cambrian *Gogia.* It is probable that it serves as anchor in soft muddy bottom sediment or at least helped the crinoid to maintain its balance by increasing weight of the distal region of the stem.

2) In the cirrus-bearing type, the distal end of the stem resembles what is usually called a "root" **(radix).** This root may consist of a single main trunk—the stem itself—which gives off branches **(radicular cirri)** at irregular intervals and attaches the crinoid by penetrating into sea-bottom sediment or by creeping on the substratum (Fig. 64,*1*). Or the main trunk bifurcates and breaks into branches, which may divide further to form a cluster of rootlets given off from the dististele at short intervals or at approximately the same level (Fig. 64,*2*). The radicular cirri may be very long. They are immovable and formed of variously shaped segments having different sizes, united by synostosis, symplexy, or ankylosis, and invariably provided with an axial canal which, like the axial canal of the stem, contains coelomic and nervous extensions. The scars or facets left on the stem trunk or its branches may be restricted to the latus of a single ossicle or extended to latera of two or more consecutive joints. Their distal end may carry fingerlike processes which become fastened to firm objects encountered during growth.

3) The discoid type of holdfast consists of a relatively wide disc-shaped expansion or an inverted bowl-shaped body (Fig. 64, *3-7*). It provides fixation on rocky bottoms or solid foreign objects such as shells of brachiopods and mollusks, as well as stems of other crinoids. Such holdfasts may spread symmetrically in all directions and assume subcircular outlines, or grow chiefly on one side, and acquire more irregular contours. The margin may be entire or cut into lobes or short digitations. On their upper convex side, the region about the stem-scar may be depressed or flush with the surrounding surface. Many holdfasts of this type, like those of the recent *Calamocrinus* and *Phrynocrinus,* or of the fossils *Encrinus* and *Aspidocrinus* (Fig. 64,*3-5*) seem to be entirely massive. Some show a more complex structure. For instance, the Ordovician holdfast called *Lichenocrinus* comprises a plated upper surface, supported by numerous radiating lamellae, which rise vertically from a distinct floor plate; in the center of the upper surface, a deep plated crater contains the distal extremity of the stem (Fig. 64,*6,7*).

4) Roots primarily belonging either to the cirrus-bearing or discoid type may be covered by an unjointed crust of secondary stereom that extends onto the substratum and tends to transform the whole structure into a solid calcareous mass. All such en-

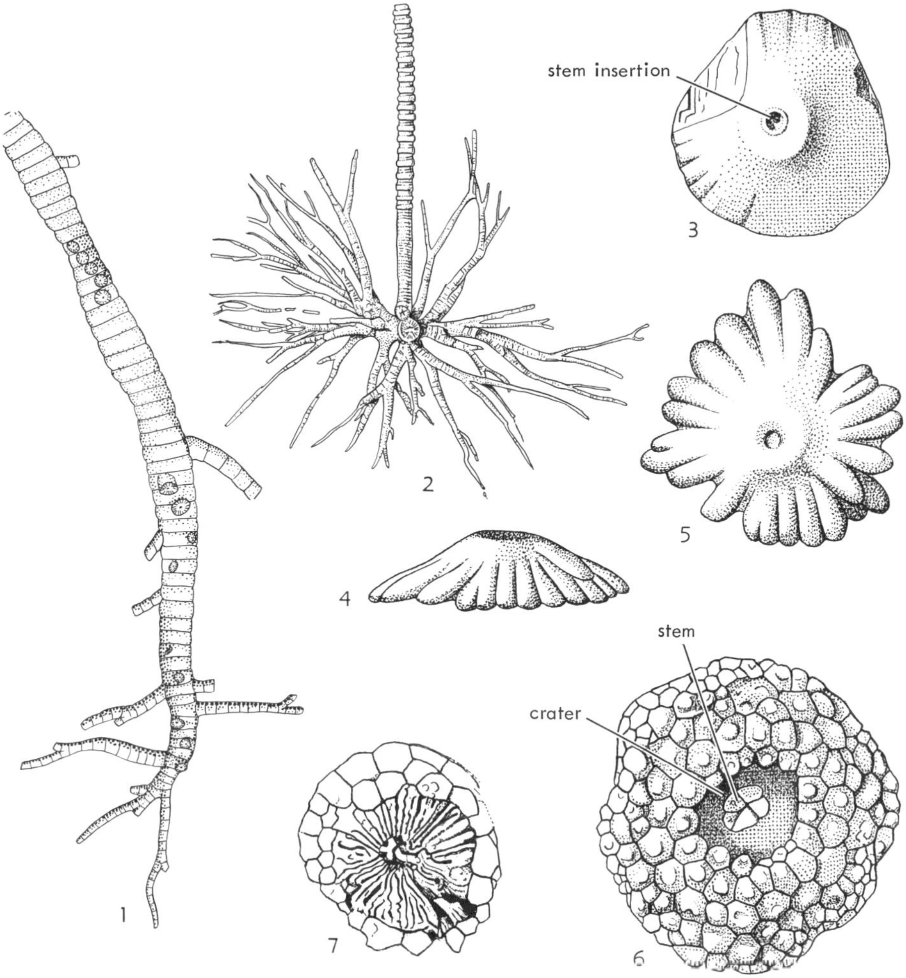

FIG. 64. Morphology of cirrus-bearing type *(1,2)* and discoidal type *(3-7)* of crinoid holdfasts.

1. Platycrinites regalis (HALL), Miss., USA(Iowa), distal portion of stem with radicular cirri given off at irregular intervals, ×0.7 (Wachsmuth & Springer, 1897).

2. Eucalyptocrinites ovalis (HALL), M.Sil.(Niag.), N.Am., clustered radicular cirri at distal end of column, ×0.7 (Wachsmuth & Springer, 1897).

3. Aspidocrinus scutelliformis HALL, L.Dev., USA (N.Y.), discoidal holdfast, upper side, ×0.7 (Goldring, 1923).

4,5. Aspidocrinus digitatus HALL, L.Dev., USA (N.Y.), discoidal holdfast, lateral and upper side views, ×0.7 (Goldring, 1923).

6. Lichenocrinus milleri FABER, U.Ord., USA (Ind.), discoidal holdfast, upper side showing deep crater at center for stem attachment, ×2.5 (Faber, 1929).

7. Lichenocrinus nodosus FABER, U.Ord., USA (Ind.), discoidal holdfast, upper side with part of outer plated covering removed to show inner radiating lamellae, ×2.5 (Faber, 1929).

crusting growths cause a more or less distinct deformation of primary root forms, which in extreme cases are entirely hidden. Encrusting roots occur chiefly in reef-dwelling crinoids and in forms living in moving waters. Typical examples are found in the Crotalocrinitidae, Cupressocrinitidae (Fig. 65,*1*), Apiocrinitidae (Fig. 65,*2*), Millericrinidae (Fig. 65,*3,4*), Sclerocrinidae, and Eugeniacrinitidae.

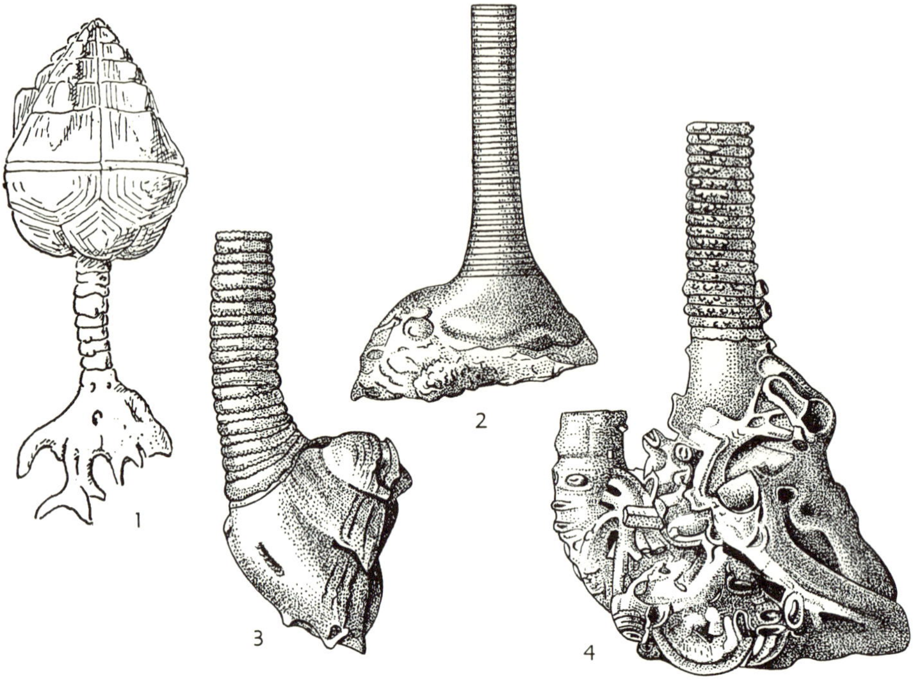

FIG. 65. Morphology of encrusting holdfasts.

1. *Cupressocrinites abbreviatus* GOLDFUSS, M.Dev., Ger., complete specimen, ×0.7 (Jaekel, 1918).
2. *Apiocrinites parkinsoni* (VON SCHLOTHEIM), Jur., W.Eu., dististele and root, ×0.7 (d'Orbigny, 1858).
3. *Millericrinus knorri* DE LORIOL, U.Jur., W.Eu., dististele and root, ×0.7 (de Loriol, 1882-84).
4. *Millericrinus horridus* D'ORBIGNY, U.Jur., France, two dististeles and roots encrusted with serpulid tubes, ×0.7 (de Loriol, 1882-84).

Holdfasts of some fossil crinoids are highly specialized. Already mentioned is *Lichenocrinus,* and still more curious are *Ancyrocrinus* and *Camarocrinus.*

The name *Ancyrocrinus* was given by HALL (1862) to grapnel-like bodies found in Devonian rocks of New York, and recently in the Givetian of Brittany (France) (MORZADEC, 1967) (Fig. 66,*1-5*). The discovery of a stem and crown attached to the grapnel has demonstrated that the latter served as the holdfast of a botryocrinid genus (GOLDRING, 1942). The grapnel consists of the distal portion of the stem and true cirri covered over by a crust of secondary stereom. In some specimens the primary root, which is composed of numerous small polygonal plates and very short stumplike encrusting radicular cirri, is incorporated into the grapnel (Fig. 66,*1,2*); in other specimens no trace of the original root is present (Fig. 66,*3-5*), which suggests that in such specimens the stem must have broken off between the original root and the undermost cirrinodal (or cirri nodals) included in the grapnel (MCINTOSH & SCHREIBER, 1971). A secondary canal system, branching at irregular intervals along the axial canal of the stem and cirri, developed within the grapnel and its stereomic crust up to the surface where they were sealed by a thin layer of calcite. The grapnel of *Ancyrocrinus* functioned probably less as a drag and ballast (KIRK, 1911) than as an anchor (EHRENBERG, 1929), giving the crinoid a secure and relatively stable base in a low- to moderate-energy environment (MCINTOSH & SCHREIBER, 1971).

HALL (1879b) also introduced the name *Camarocrinus* for large, bulbous, chambered structures found at several localities in Lower Devonian rocks of North America, but also known since the middle of the 19th century from the Upper Silurian of

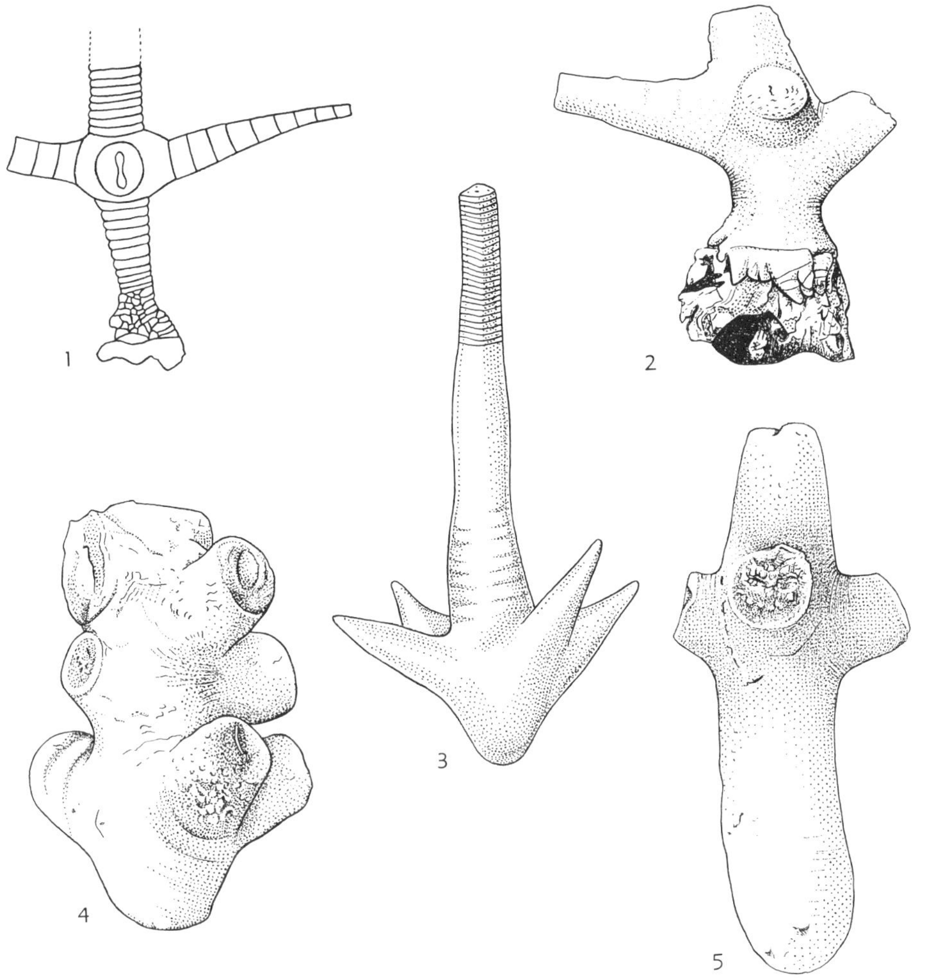

FIG. 66. Morphology of the grapnel-like holdfast of *Ancyrocrinus bulbosus* HALL, M.Dev., E.N.Am. (all McIntosh & Schreiber, 1971, except *3*).

1. Primary root system of an immature specimen before build-up of secondary stereom; root attached probably to a brachiopod shell, ×3.
2. Grapnel with radicular cirri of primary root attached to a coral, ×2.
3. Typical grapnel without trace of primary root, ×2 (Hall, 1862).
4. Grapnel showing three distinct cirri levels, ×2.
5. Grapnel with long segment of stem incorporated into it, but without any trace of primary root, ×2.

Bohemia under the vernacular French name of *lobolithes* (Fig. 67, 68). HALL recognized their real nature as holdfasts of a crinoid, which now is determined almost certainly to belong to the camerate *Scyphocrinites*. Some bulbs reach or even exceed 20 cm. in diameter. In the center of their proximal face is fixed the distal end of the stem, which at this level divides into primary roots. By repeated divisions these roots produce innumerable cirral elements that form the outer covering of the bulbs, as well as the partition-walls of inner chambers. *Camarocrinus,* therefore, belongs to the radicular cirri-bearing type of crinoid holdfasts. Two main, closely related sorts of bulbs have been distinguished (HAUDE, 1972). In the first one, known as the cirrus type (Fig. 67), the walls consist of a dense three-layered lattice work of numer-

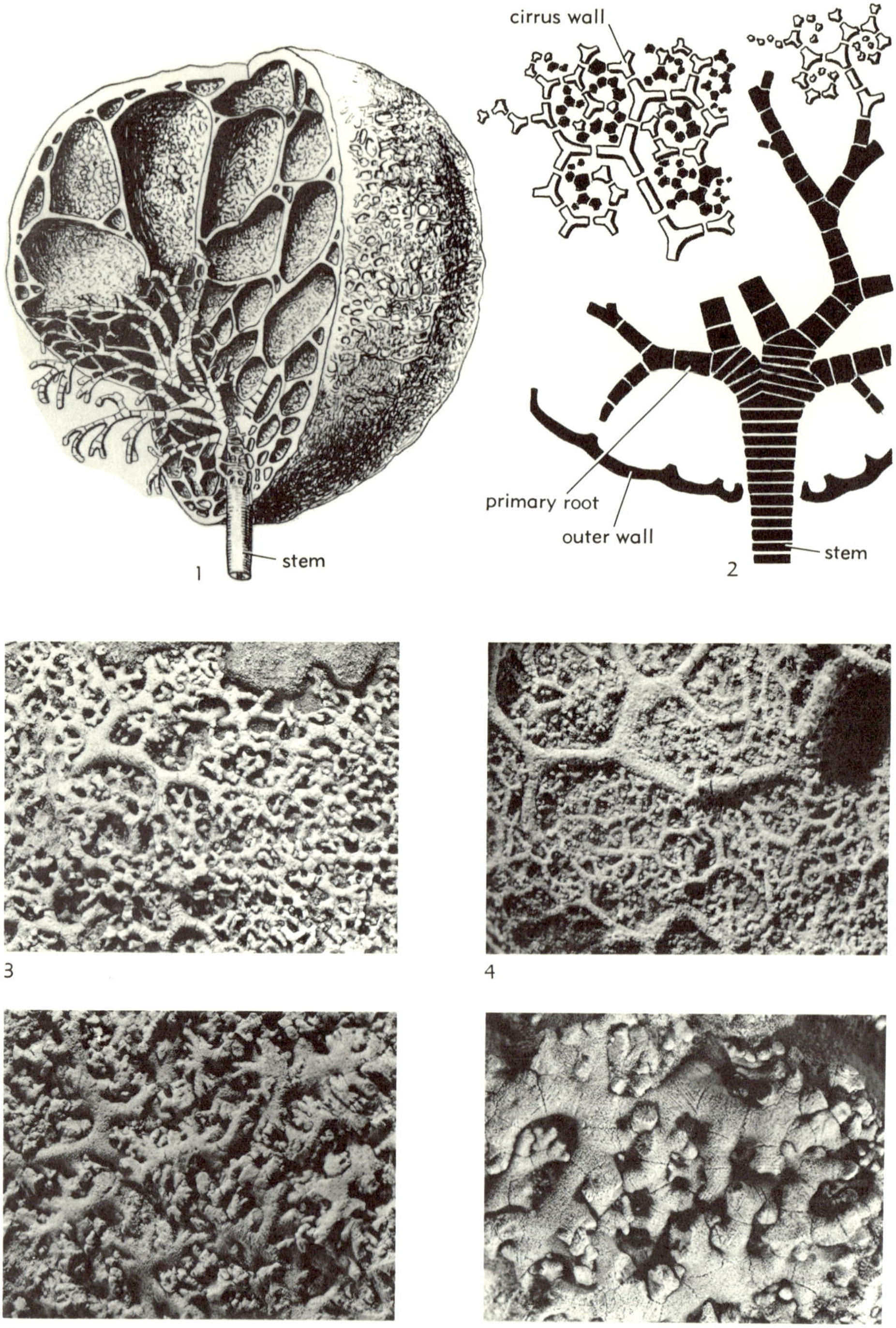

FIG. 67. Morphology of the cirrus type of the bulbous holdfast called *Camarocrinus* (all from Haude, 1972, modified in *1* and *2*).

ous spiculelike skeletal elements (cirrals), the largest of which lie in the middle layer (Fig. 67,*5,6*), whereas the smallest fill remaining spaces to form the cover layers (Fig. 67,*3,4*); the outer cover of the bulb closes up against the distal end of the stem; the chambers have no openings, are arranged without distinct order, and range in number from a few to very many.

The second kind of bulb (Fig. 68,*1*) differs from the first in its plated appearance (Fig. 68,*2*), the occurrence of a short projecting collar around the stem base and presence of primary roots, associated with fewness of chambers (usually 4 to 7, apparently 11 at most), and existence of an opening to each chamber in the axil of the primary roots. The collar consists of a single layer of plates, but the remainder of the bulb is composed of a three-layered wall with a middle network of minute irregular cirrals over- and underlain by layers of plates, which are nothing else than thickened portions of cirrals belonging to the middle network (Fig. 68,*3,5*).

The bulbous holdfasts of *Scyphocrinites* have been interpreted as bodies of organisms belonging to an independent class of echinoderms (Barrande, 1887), as cystoids (Sun & Szetu, 1947), as genital sacs (Jaekel, 1904), brood-pouches or receptacles (Haeckel, 1896a), pathologic cysts induced by myzostomids (Haeckel, 1896a; Ehrenberg, 1926), inflated roots serving for permanent or temporary settlement on muddy bottoms (Springer, 1917), and floating organs (Hall, 1879b; Schuchert, 1904; Yakovlev, 1953; Stukalina, 1967; Haude, 1972). The last hypothesis seems to be the most plausible by far, for it is consistent with morphological characteristics of the bulbs, their worldwide distribution, approximation to a mechanical model (paradigm) for their function, and calculation of their buoyancy as attachments to living crinoids which floated on the surface or at shallow depths in widespread ancient seas (Haude, 1972).

Not all stalked crinoids have special terminal stem organs serving for fixation. Some appear to have effected attachment by nonradicular cirri, the stem lying down partly on the sea floor, or having lost its normal holdfast, or as in comatulids, having been discarded except for its topmost columnal transformed into a centrodorsal. Some others were fastened in place by the stem itself, primary fixation having been accidentally or normally abandoned. In the last cited possibility generally but not universally, the stem tapers distally (Fig. 69,*3*). At the same time, there is a marked tendency toward looping or coiling of the distal stem region, enrollment taking place upon itself or around foreign bodies. The coil may comprise one, two or several whorls, which touch or are separated from each other (Fig. 69,*4*). The coil may be in a simple plane, as in *Diamenocrinus* (Fig. 69,*2*), or form a conical structure as in *Acanthocrinus* (Fig. 69,*1*) and *Thallocrinus,* which probably functioned as ballast for the crinoid, maintaining it in an approximately vertical position. A similar function may be ascribed to terminal knobs produced by secretion of stereom, supposedly induced by and following disruption of the column.

STEMLESS CRINOIDS

Sporadically in some crinoid groups and invariably in others the whole stem or most of it is lacking in adults. Five or six groups of such crinoids may be recognized, chiefly representing physiological developments, since they include widely divergent types. 1) One group contains forms which show a strong tendency to shorten the stem accompanied by more or less resorption of remaining columnals, as illustrated by the inadunate *Hoplocrinus estonus* Öpik (Fig. 70,*1*) and the articulate *Millericrinus prattii* (Gray) (Fig. 70,*2*), in which the stem is reduced to a small conical knob or even to a single flat ossicle. 2) In a second

(Continued from facing page.)

1. Model of bulb partially open to show inner structures; stem downward according to presumed position during life.

2. Diagram showing clockwise and counterclockwise spiral arrangement of branching cirri that form the wall.

3. Aspect of outer wall, ×4.

4. Aspect of chamber wall, ×4.

5,6. Middle layer with major branches and spaces filled by smaller branchlets, ×4.

group, the larval stem is discarded during ontogeny, except for the most proximal columnal, which fuses with infrabasals to produce the centrodorsal of the adult crinoid (see above, p. *T*71); this group corresponds to the articulate order Comatulida. 3) Somewhat similar in appearance, but very different in origin and structure is the Permian reef-dweller of Timor named *Timorocidaris* (Fig. 70,*3*); as interpreted by BATHER (1900a) and WANNER (1940, 1950), the skeleton of this strange crinoid is reduced to a single hemispherical calcite crystal which carries three (rarely two) arm facets on a small neck in the center of its flat surface. 4) Crinoids of a fourth group are characterized by the complete absence of a stem or of any anchoring structure, and by the presence of a central pentagonal apical plate in the cup; this plate is called **centrale** and, to the group belong the inadunate *Cryphiocrinus* (Fig. 70,*4*), members of the articulate order Uintacrinida (Fig. 70,*5,6*), and the doubtful Saccocomidae (*Pseudosaccocoma,* Fig. 70,7). 5) Crinoids of the next group have no stem, no centrodorsal, and no centrale; examples of this condition are furnished by the rhodocrinitid camerate *Monstrocrinus,* by the inadunate *Agassizocrinus* (Fig. 70,*9*), the proximal cup plates of which fuse partly or completely, with accompanying secondary deposition of stereom, and by most representatives of the articulate order Roveacrinida (*Poecilocrinus,* Fig. 70,*8*) (as interpreted by RASMUSSEN, 1961). 6) The members of a sixth group do not possess a jointed column, but as a rule are cemented to some solid object by a mass of stereom, the composition of which is conjectural; they comprise the inadunate *Pilidiocrinus* (Fig. 70,*10,11*), such flexibles as *Edriocrinus* (Fig. 70,*12*), *Permobrachypus, Calycocrinus* (Fig. 70,*13*), or *Palaeoholopus* (Fig. 70,*14*), and articulates like *Cotylederma, Eudesicrinus, Hemibrachiocrinus, Dibrachiocrinus, Cyathidium* (Fig. 70,*15,16*), and *Holopus.*

CUP (OR CALYX)

GENERAL FEATURES

The cup (equivalent to dorsal cup or aboral cup) or calyx[1] is part of the crinoid theca located between the stem and origins of the free arms. Most fossil crinoids, unlike the majority of recent ones, have a well-developed calyx that includes the main and central part of the skeleton surrounding soft structure of the visceral mass.

As we have seen (p. *T*59), the cup in its simplest form comprises two or three circlets of five plates, with those of one circlet alternating with those in adjacent circlets (Fig. 71). The uppermost cup plates, except one or two on the side defined as posterior, are directly in line with the arms and therefore are designated as **radials (radialia).** The interradially located plates next below the radials are termed **basals (basalia),** since in many crinoids they form the base of the cup or calyx and rest on the column. In some crinoids a circlet of perradially disposed plates called **infrabasals (infrabasalia)** occurs beneath the basals. A base composed of one circlet of plates between radials and stem is termed **monocyclic;** it is said to be **dicyclic** if it comprises two plate circlets. In a wider sense, a crinoid with a monocyclic base of the cup or calyx is designated as monocyclic, and is called dicyclic in a crinoid provided with a dicyclic base.

The radials, basals, and infrabasals are the primary skeletal elements of the cup. Together they form what is sometimes called the **patina,** a term proposed by JAEKEL (1891). The addition of various elements to this basic structure produces an extension of the cup. Thus, in many crinoids the proximal ossicles of the arms are firmly incorporated as **fixed brachials** into the calyx. Generally, the fixed portions of the arms are joined laterally together by plates located between the rays and between branches of the rays **(interbrachials).** Usually also, extra plates, called **anals,** occur in the posterior interray, contributing to the width. In this way, a complex structure is

[1] The term "calyx," herein taken as a synonym of "cup," has been used, and is still used, by some authors for designating the whole theca (cup plus tegmen). The proposal recently made by MOORE & STRIMPLE (1973), to call "cup" that part of the theca that comprises only the lowermost two or three circlets of plates (including anal plate if present) next above the stem of stalked crinoids or at base of the skeleton in stemless crinoids and to designate as "calyx" a crinoid cup joined firmly with fixed plates above the radial circlet, is judged by UBAGHS to be artificial and unnecessarily conflicting with common usage of these terms in both zoology and paleontology.

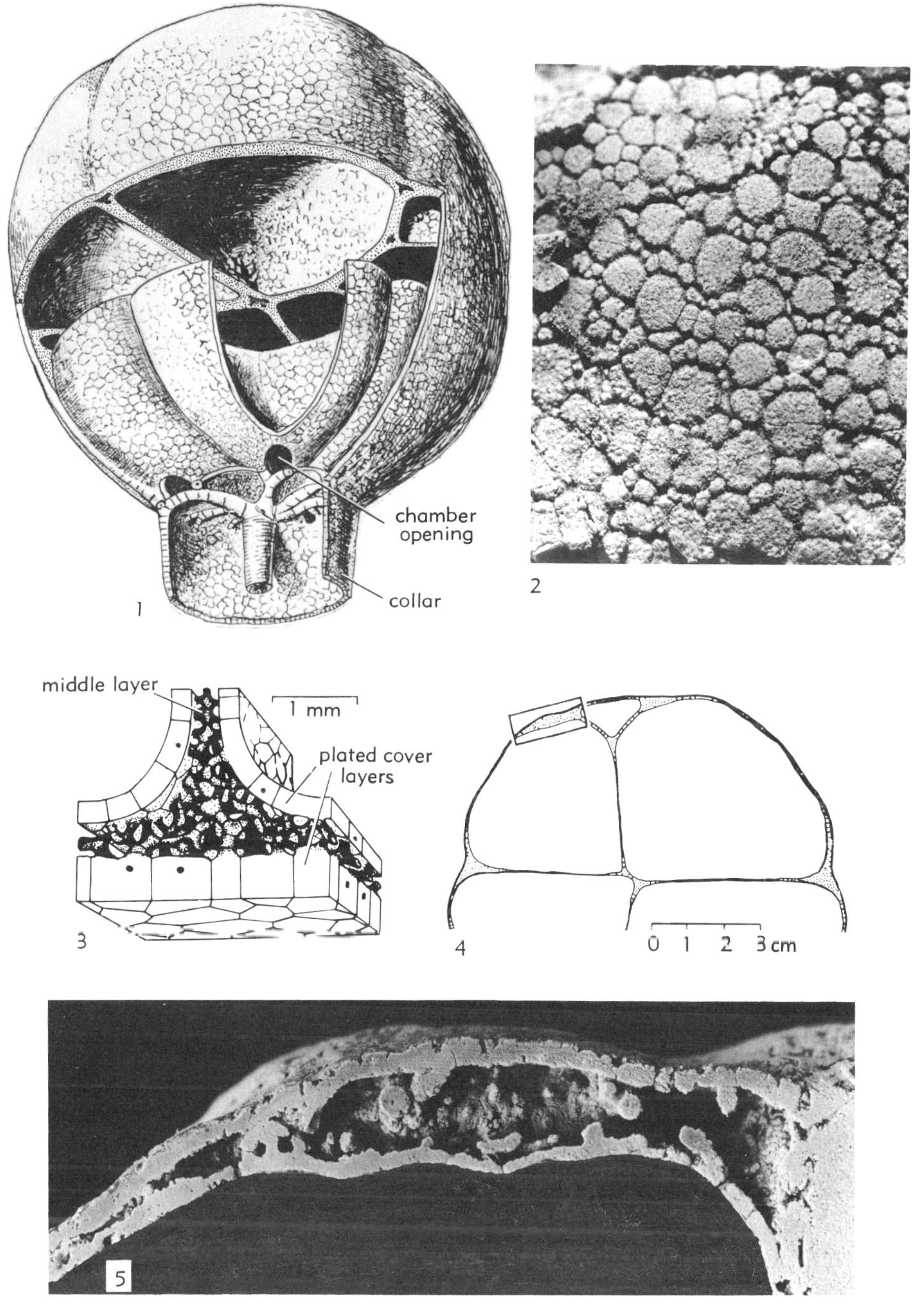

FIG. 68. Morphology of the plated type of the bulbous holdfast called *Camarocrinus* (all from Haude, 1972).

1. Model of bulb partially open to show inner structures, stem downward according to presumed position during life (modified).

2. Aspect of plated outer wall, ×4.

3. Diagrammatic model of wall structure.

4. Equatorial section of bulb.

5. Detail of region limited by a rectangle in Fig. 4.

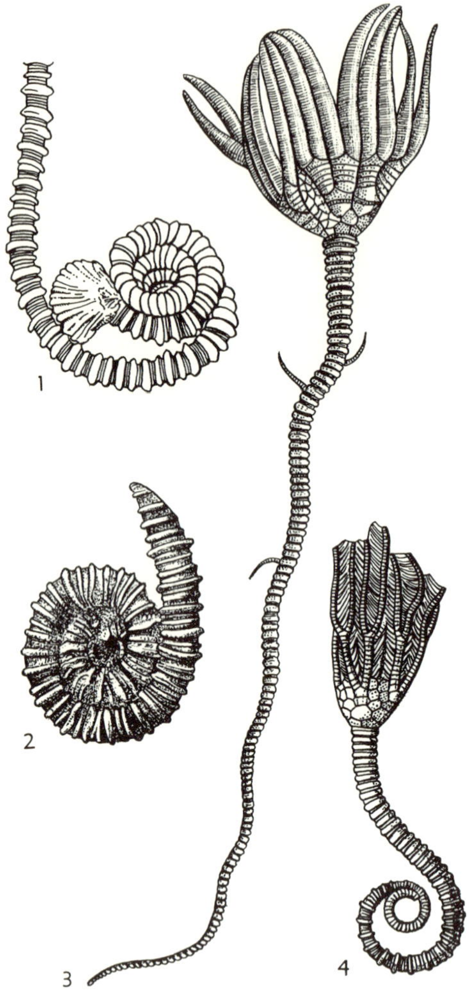

formed, involving the participation of a varyingly large number of skeletal pieces of different origins.

It has been thought generally that all types of crinoids pass through an early phylogenetic stage in which the calyx is reduced to the patina. This, however, does not seem to accord with progressive decrease in size of the visceral mass that crinoids as a whole are judged to have undergone during their history, nor does it agree with the fact that ordinarily the cup of oldest crinoids has a more complex structure than that of later members. This is best illustrated by the inadunates. Contrary to typical representatives of this subclass, characterized by a cup that usually includes only infrabasals (if any), basals, radials, and one or more anal plates, several disparid inadunates among those judged to be most primitive (e.g., *Eustenocrinus, Dystactocrinus, Peniculocrinus, Ohiocrinus*) have one or two circlets or arm plates firmly incorporated with cup plates as parts of a small calyx; the Ordovician perittocrinids have numerous accessory plates inserted between basals, radials, and anals, and the calyx of such primitive cladids as *Aethocrinus, Cupulocrinus,* and *Ottawacrinus* contains fixed brachials which may be accompanied by well-developed interbrachials. It is therefore suggested that the calyx of the crinoid, like the column, originally was composed

FIG. 69. Root-functions of the stem.

1. Acanthocrinus rex JAEKEL, L.Dev., Ger., distal portion of stem closely rolled into a conical coil, ×1 (Jaekel, 1895).

2. Diamenocrinus jouani OEHLERT, Dev., France, distal portion of column closely rolled into a planispiral coil, ×1 (Oehlert, 1891).

3. Woodocrinus macrodactylus DE KONINCK, U. Carb.(L.Namurian), Eng., complete specimen with tapering stem, ×0.3 (de Koninck, 1854).

4. Pycnocrinus dyeri (MEEK), U.Ord., USA(Ohio), stem terminating in a loose coil, ×1 (Wachsmuth & Springer, 1897).

FIG. 70. Reduction and complete disappearance of crinoid stem.

1. Hoplocrinus estonus ÖPIK, Ord., Estonia, oblique side view of theca with stem reduced to small conical knob, ×1.2 (Öpik, 1935).

2. Millericrinus prattii (GRAY), M.Jur., Eng., side view of theca with stem reduced to short stump of five columnals, ×3 (Carpenter, 1882a).

3. Timorocidaris sphaeracantha WANNER, Perm., Timor, oblique side view, ×1.65 (Lakeman, 1950).

4. Cryphiocrinus girtyi KIRK, U.Miss., USA(W. Va.), basal view of cup with small centrale concealing minute infrabasals, ×2 (Kirk, 1929c).

5. Uintacrinus socialis GRINNELL, U.Cret., N.Am., theca from side, ×0.6 (Rasmussen, 1961).

6. Marsupites testudinarius (VON SCHLOTHEIM), U.Cret., Eu., stemless cup from the side, ×0.7 (Bather, 1900a).

7. Pseudosaccocoma strombergense REMEŠ, U.Cret., Italy, theca from aboral side showing perforated centrale, ×2 (Rasmussen, 1961).

8. Poecilocrinus signatus (PECK), L.Cret.-U.Cret., USA(Texas), theca from lateral side, ×15 (Rasmussen, 1961).

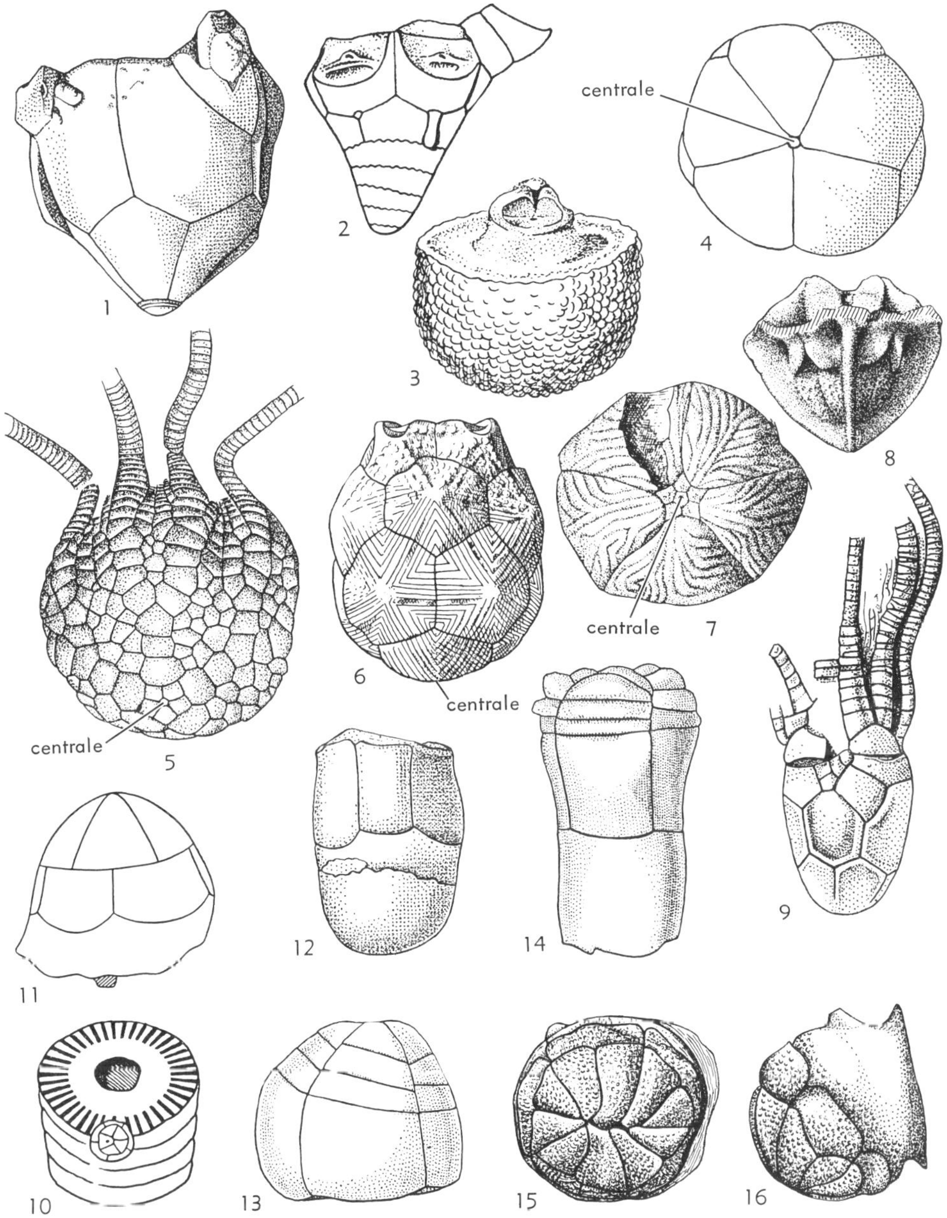

FIG. 70. *(Explanation continued from facing page.)*

9. Agassizocrinus laevis (ROEMER), U.Miss., N. Am., *CD* view of holotype, intrabasals not completely fused, stemless, ×1 (Springer, 1926b).

10. Pilidiocrinus hackeri ARENDT, L.Perm., USSR (Urals), specimen attached to crinoid pluricolumnal, ×3 (Arendt, 1970a).

11. Same, complete specimen from the side, ×8 (Arendt, 1970a).

12. Edriocrinus pocilliformis HALL, L.Dev., N. Am., *CD* interray view of stemless theca, ×2 (Goldring, 1923).

13. Calycocrinus perplexus WANNER, Perm., Timor, crown from side, ×2.4 (Wanner, 1930b).

14. Palaeoholopus pretiosus WANNER, Perm., Timor, crown from side, ×1 (Wanner, 1929a).

15,16. Cyathidium holopus STEENSTRUP, U.Cret., Eu., crown in adoral and lateral views, ×2 (Nielsen, 1913).

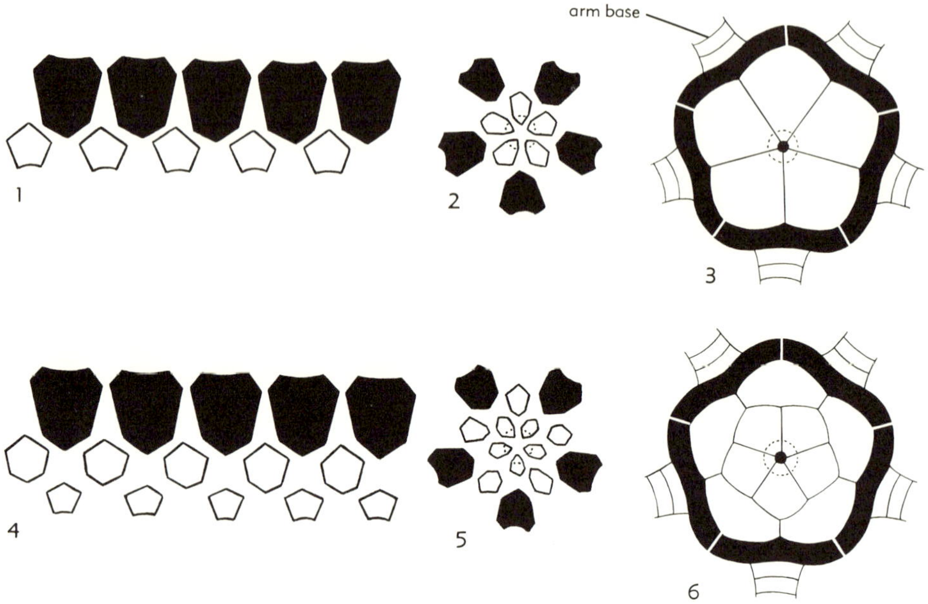

FIG. 71. Three different ways of representing basic calyx structures of crinoids with monocyclic *(1-3)* and dicyclic *(4-6)* bases (Ubaghs, n). [Explanation: radials black.]

of more or less numerous plates arranged without definite order; then, that it was progressively affected by a tendency toward pentamerous symmetry, as expressed by a gradual acquisition of definitive arrangement of plates, proximally in disposition of the successive plate circlets and distally in divergence of radial series extending beyond the theca as articulated outgrowths, which became the arm or brachia. Such a view seems to have recently received support from the discovery in the Middle Cambrian (Burgess Shale) of British Columbia of a crinoid-like form *(Echmatocrinus brachiatus)* with a conical calyx made up of numerous irregularly arranged plates and provided with an irregularly multiplated, stemlike outgrowth serving for attachment (SPRINKLE, 1973a).

SHAPE OF CUP OR CALYX

The theca of crinoids beneath the tegmen and free arm attachments varies considerably in shape. For the description of different forms it is convenient to use well-defined terms, such as those proposed by MOORE & PLUMMER (1940) (Fig. 72). To this end three different characters are considered: 1) main form of the cup or calyx, 2) ratio of its height to width, and 3) shape of its base.

1) *Main form.*—A cup or calyx may be cone, bowl, or globe shaped. It is said to be cone shaped when the sides flare upward with essentially uniform slope from the stem attachment, so that it reaches greatest width at the summit. It is defined as bowl shaped if the slope of its upward flaring sides becomes markedly steeper near the summit than adjacent to the stem attachment, greatest width remaining at the summit. It is described as globe shaped if the sides curve distinctly inward near the summit and if greatest width is below the summit.

2) *Ratio of height to width.*—The height of the cup or calyx is the distance measured along the aboral-adoral axis between the basal and summit planes. Width is the distance measured along a line perpendicular to the aboral-adoral axis and to the anteroposterior axis through the widest part of the cup or calyx. For each of the four main forms recognized one may distinguish

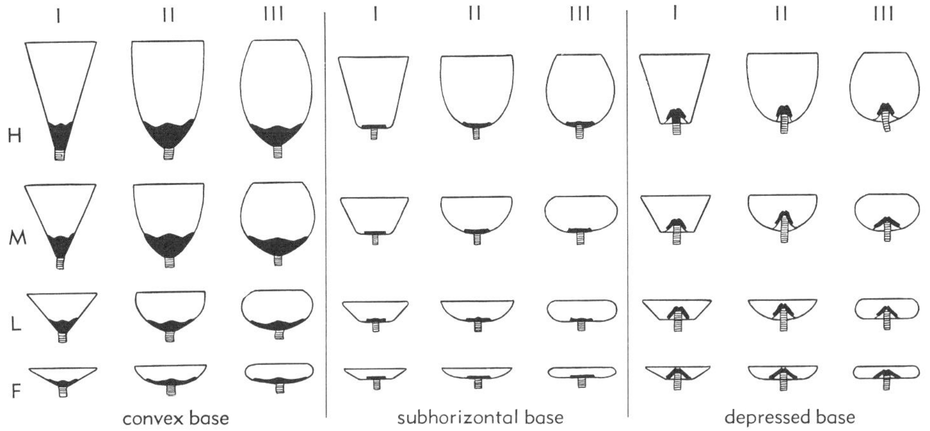

FIG. 72. Diagrams illustrating shapes of the calyx in crinoids (mod. from Moore & Plummer, 1940). [Explanation: *I*, cone shaped; *II*, bowl shaped; *III*, globe shaped; *H*, high; *M*, medium; *L*, low; *F*, flat; plates of base, black.]

1) a tall type with height greater than width, 2) a medium tall type with height approximately equal to width, 3) a low saucer-shaped type with height appreciably smaller than width, and 4) a flat or discoid type with height equal to or smaller than half of the width.

3) *Shape of base.*—The base may be convex and composed of plates flaring upward and visible in side view of the theca. It may be flat and composed of subhorizontal plates not visible in side view, with lowermost part of the cup or calyx appearing to be broadly truncated. The base may be depressed and the stem attachment concealed in a more or less deep concavity, the bottom of which is flat or concave.

ORNAMENTATION OF CUP OR CALYX

The outer surface of the cup or calyx plates may be smooth, finely granular, pustulose, covered with distinct lines or wrinkles, nodose, or spinose. Immovable spines may be present on some plates, and movable spines, attached to pitted tubercules, are known to occur on the calyx of some genera (e.g., *Arthroacantha*). According to WANNER (1942b), the paddle-shaped skeletal elements that envelop the proximal circlet of plates in the genus *Paradoxocrinus* are modified movable spines attached to basals. Many, mainly Lower Paleozoic crinoids have ridges radiating from the center of plates to the middle of the sutures (exceptionally to the angles), or following midlines of the rays and (in some camerates) the sagittal series of anal plates. All of these surface markings may distinguish large or small assemblages of crinoids. They have a varyingly important classificatory value.

INFRABASALS

The infrabasals are plates perradial in position that form the proximal circlet of the base in dicyclic types. Typically they are five in number and equal. This condition (probably primitive) is preserved in most dicyclic camerates and many inadunates. However, genera of these subclasses are known to have four, three or two infrabasals, or all of these plates fused into a solid skeletal piece. On the other hand, all but a few flexible crinoids have three unequal infrabasals—a small (azygous) plate almost invariably located in the *C* ray and two large (zygous) ones. Most certainly, this restriction in number results from ankylosis of sutures between adjoining infrabasals.

The infrabasals may be prominent and upflared, forming a significant part of the cup enclosing the visceral mass (Fig. 73,*1*; 74,*1*). In the Silurian flexibles *Calpiocrinus* and *Homalocrinus* their enlargement is

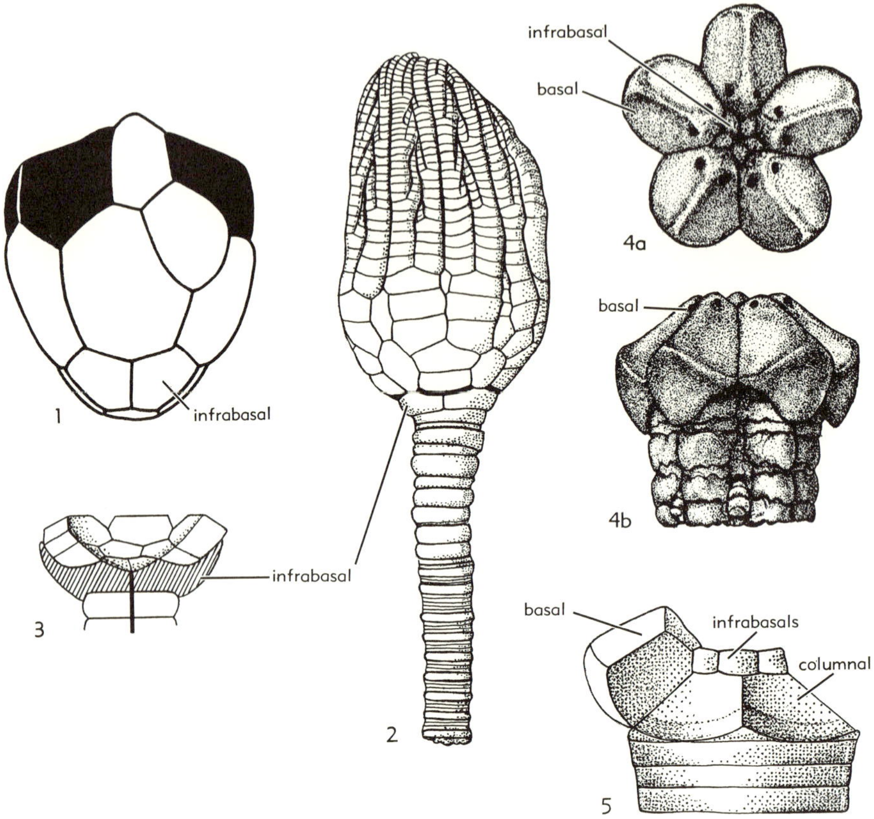

FIG. 73. Morphology of infrabasals. [Explanation: radials black.]

1. Ulocrinus fistulosus STRIMPLE & MOORE, Penn., Kans.; with upflaring infrabasals, ×1 (mod. from Strimple & Moore, 1971a).

2,3. Calpiocrinus rotundatus SPRINGER, U.Sil., Sweden(Gotland); *2,* crown and part of stem; *3,* diagrammatic vertical section of cup, basals and greater part of radials covered by hypertrophied infrabasals (redrawn from Springer, 1920).

4a,b. Cainocrinus tintinnabulum FORBES, Eoc., Eng.; reduced infrabasals, basals, and proximal columnals, in adoral and lateral views, ×10 (Rasmussen, 1972a).

5. Millericrinus polydactylus D'ORBIGNY, Jur., France; proximal columnals and reduced infrabasals after removal of all basals but one, lat. view, *ca.* ×6 (redrawn from de Loriol, 1882-84).

great enough to provide partial or entire concealment of basal and radial circlets by the infrabasals (Fig. 73,*2,3*). But this is an exceptional condition. Generally, they are small or medium-sized and take only a modest part in formation of the thecal wall. If they are recumbent or subhorizontal, the cup is truncated below and they are barely visible or invisible in side view, serving as support for overlying skeletal plates (Fig. 74,*2,3*). In cups or calyces with a basal concavity, they may be horizontal (Fig. 74,*4*) or downflaring (Fig. 74,*5*), according to their size and shape of the basal concavity; of course they are not visible in side view.

In some crinoids the infrabasals show a marked tendency to become atrophied, to be resorbed, or to lose their identity by fusing with other elements. This is particularly true of articulate crinoids, but may be observed also in other subclasses. In the flexible *Icthyocrinus,* for instance, they are reduced to a diminutive plug entirely covered by the column (Fig. 74,*3*), or infrabasals are completely resorbed. A nearly similar condition characterizes the infrabasal circlet of isocrinid genera (Fig. 73,

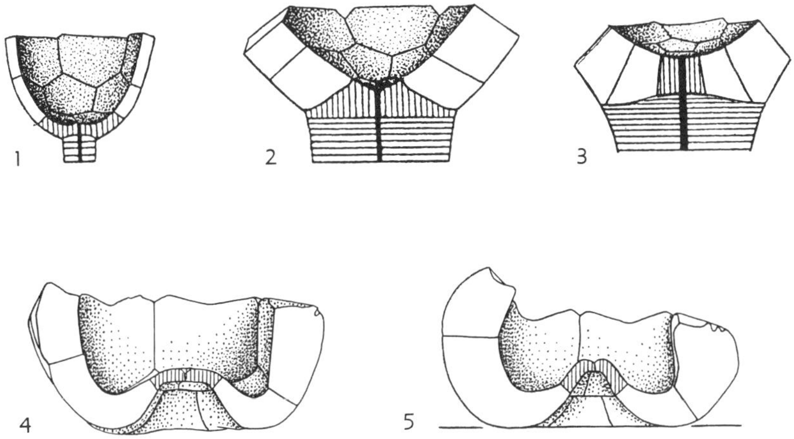

FIG. 74. Diagrammatic vertical sections of cups *(1,4,5)* and patinas of calyces *(2,3)* illustrating various conditions of infrabasals. [Explanation: infrabasals ruled vertically.]

1. Lecanocrinus, with upflaring infrabasals (Springer, 1920).
2. Forbesiocrinus, with subhorizontal infrabasals (Springer, 1920).
3. Icthyocrinus, with infrabasals reduced to a small plug, entirely covered by column (Springer, 1920).
4. Delocrinus major WELLER, L.Perm., USA (Texas); with horizontal infrabasals in basal concavity, ×1 (Moore & Plummer, 1940).
5. D. subhemisphericus MOORE & PLUMMER, Penn., Mo.; with downflaring infrabasals in basal concavity, ×2 (Moore & Plummer, 1940).

4,5). Among comatulids, these plates are present usually in the larvae as discrete elements but fuse later with the topmost columnal or columnals to form a proximale and take part in composing a centrodorsal. Such dicyclic crinoids, in which the infrabasals are not differentiated or absent in adults are said to be **cryptodicyclic** or **pseudomonocyclic.** It is very likely that among dicyclic fossil crinoids genera currently are classified as monocyclic because they offer no structural evidence to their dicyclic origin.

In various crinoids, however, the previous presence of infrabasals is suggested or definitely indicated by 1) orientation of the axial cords in the stem and of lobes in the chambered organ, invariably located perradially in pseudomonocyclic and dicyclic forms; 2) orientation of the columnal pentameres (liable to exceptions), outer angles of the stem, vertical sutures of the stem, angles of the lumen, and placement of cirri with reference to radial symmetry of the cup ("law" of WACHSMUTH & SPRINGER, see p. *T*61); 3) occurrence of infrabasals in immature growth stages; 4) paleontological demonstration of the former existence of these plates in ancestors; and 5) close relationship to genera that possess infrabasals.

CENTRALE

In a few dicyclic crinoids and pseudomonocyclic ones lacking stem and cirri, space enclosed by the proximal circlet of cup plates—either infrabasals or basals (if infrabasals are vestigial)—is occupied by a plate called **centrale.** Examples are the Upper Mississippian inadunate *Cryphiocrinus* (Fig. 70,*4*), the Jurassic and Lower Cretaceous articulate *Pseudosaccocoma* (Fig. 70,*7*), and the Upper Cretaceous articulates *Uintacrinus* (Fig. 70,*5*) and *Marsupites* (Fig. 70,*6*). The centrale, which is large in *Marsupites* but very small in other genera mentioned, is a nonperforated element (except in *Pseudosaccocoma*) lacking any sign of a stem attachment or partition into two or several pieces. It may conceal the infrabasals (as in *Cryphiocrinus*) or coalesce with them (as in some individuals of *Uintacrinus*).

The centrale has been the subject of much speculation and controversy (BATHER, 1896b, 1900a; CARPENTER, 1884a; CLARK, 1909, 1911a, 1915a; KIRK, 1911; SPRINGER, 1901). It cannot represent fused infrabasals

FIG. 75. *Ancyrocrinus bulbosus* HALL, M.Dev., USA(N.Y.); *CD* interray *(1)* and proximal *(2)* views of a calyx with lumen completely covered by polygonal plates after loss of column during life, ×4 (McIntosh & Schreiber, 1971).

(except perhaps in *Pseudosaccocoma*), or fused basals because it coexists with these plates in most of the cited genera. It is probably not the proximal stem ossicle, for it lies *within* the proximal circlet of cup plates, instead of abutting against them on their outer or aboral sides, as does the proximal columnal when it takes part (as among comatulids) in formation of the cup. Its homology with the supposed distal columnal of the stem (so-called "dorsocentral"), plus all columnals of the larval stem of comatulids, as suggested by CLARK, is purely speculative. It is certainly not a primary, primitive, constitutive element of the cup, since it occurs only in highly specialized genera, and never in early crinoids. Probably, therefore, it represents an entirely new element, introduced in genera characterized by a complete separation of crown and column, and serving to plug the opening through which the axial canal of the column communicates with the visceral cavity of the theca in stalked crinoids. A possible confirmation of this last interpretation seemingly is offered by some specimens of *Carabocrinus* and *Arachnocrinus* described by KIRK (1911), in which the introduction of a similar central plate within the infrabasal circlet seems to have been brought about by disruption of the crown and column.

One may compare the random apparition of such a plate with that of a pavement of small polygonal skeletal elements completely covering the lumen of the cups or calyx of crinoids that have lost their stem during life. An example of this condition has been reported by MCINTOSH & SCHREIBER (1971) for *Ancyrocrinus bulbosus* (Fig. 75).

BASALS

The basals are plates of a circlet typically located next proximal to radials. In the monocyclic type of base, such plates rest directly on the stem. In the dicyclic type, a circlet of infrabasals occurs on their aboral side. In both, the basals are interradial in position.

The basals of the camerate genera *Cleiocrinus* and *Spyridiocrinus,* instead of being below the radials, are inserted between them, so that the latter rest directly on infrabasals (Fig. 76,*1,2*). This arrangement is not judged to be fundamentally distinct from the normal dicyclic condition because species of dicyclic camerates (e.g., *Paulocrinus biturbinatus, Rhipidocrinus crenatus*) and inadunates (e.g., *Syndetocrinus bohemicus*) have some radials, either as a rule or as individual variations, which may enter into contact with infrabasals (Fig. 76,*3-7*).

Almost without exception in dicyclic crinoids the number of basals is five. In monocyclic crinoids, their primary number is also five, but a widespread tendency to fusion is observed, their number being reduced to four, three, two, or all coalesced into a single solid plate. The plates that fuse are not invariably the same. Consequently the position of interbasal sutures varies, allowing recognition of different kinds of bases (Fig. 77-81 illustrating most of these recognized as distinctive at generic and specific levels). For the sake of brevity and easiness, in these figures and in the following text, the individual basal plates

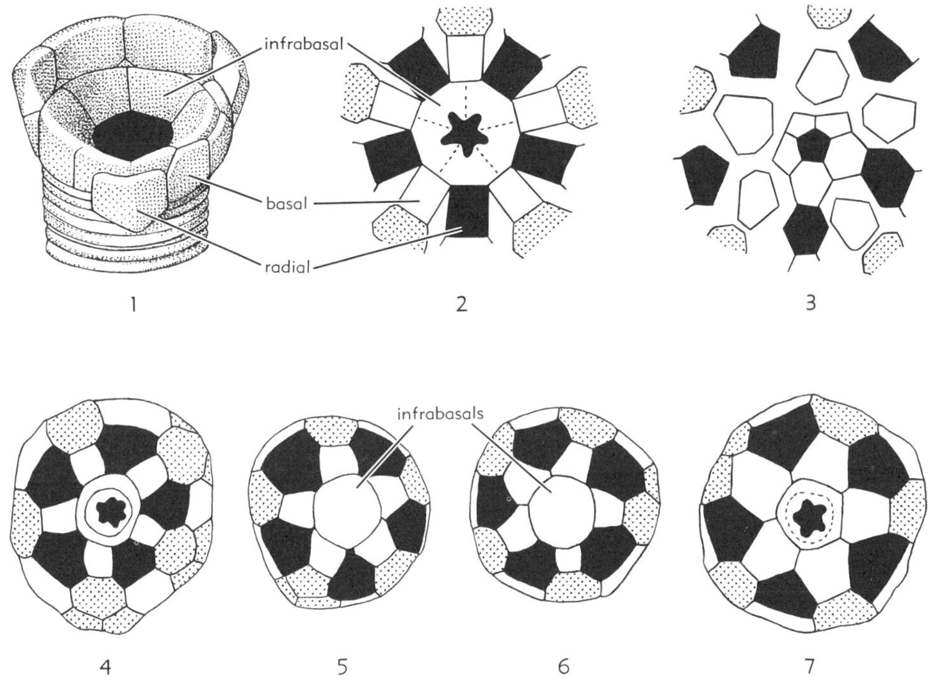

FIG. 76. Morphology of basals. [Explanation: radials and lumen of axial canal black; interbrachials stippled.]

1. Cleiocrinus regius BILLINGS, M.Ord., Can.; oblique lat. view of proximal parts of calyx and column, showing how basals and radials form a single circlet which surrounds the infrabasals, ×4 (Springer, 1905).

2. Spyridiocrinus cheuxi OEHLERT, L.Dev., France; basals and radials all in contact with infrabasals (Ubaghs, 1950).

3. Paulocrinus biturbinatus SPRINGER, M.Sil., USA (Ind.); two radials in contact with infrabasals (Springer, 1926a).

4-7. Rhipidocrinus crenatus (GOLDFUSS), M.Dev., Ger.; various arrangements of proximal cup plates showing how in some specimens the radials, or some of them, may enter into contact with infrabasals (Breimer, 1960).

will be designated arbitrarily by the small letters: *p* (for posterior) being the posterior or *CD* basal, and counterclockwise in dorsal view, *q* the *DE* basal, *r* the *AE* basal, *s* the *AB* basal, and *t* the *BC* basal (see orientation, p. *T*61).

The change in number of basals among monocyclic inadunates, as well as in monocyclic camerates provided with a five-sided, more or less pentagonal base (see below), seems to have proceeded through mere ankylosis of plates, with consequent production of unequally quadripartite, tripartite, bipartite, or undivided bases (Fig. 77).

In such crinoids, having a quadripartite pentagonal base, the *r* basal usually fuses with *s* (Fig. 77,*2*), more rarely with *q* (Fig. 77,*3*), but in the hybocrinid inadunate *Cornucrinus,* it is *p* and *t* basals that coalesce (Fig. 77,*4*).

A tripartite pentagonal base results from fusion of two pairs of original basal elements, giving two large (zygous) and one small (azygous) plate; the small plate is generally the *r* basal (Fig. 77,*5*), less commonly *s, p,* or *q* basals (Fig. 77,*6,7,8*), exceptionally (as in *Paradoxocrinus*) the *t* basal (Fig. 77,*9*). A bipartite pentagonal base has not been found among camerates, but it occurs in the disparid inadunate *Mycocrinus,* and is produced by the fusion of three $(p + q + r)$ and two $(r + s)$ basals (Fig. 77,*10*).

Much more complicated and difficult to explain is the position and relative size of basal plates in monocyclic camerates hav-

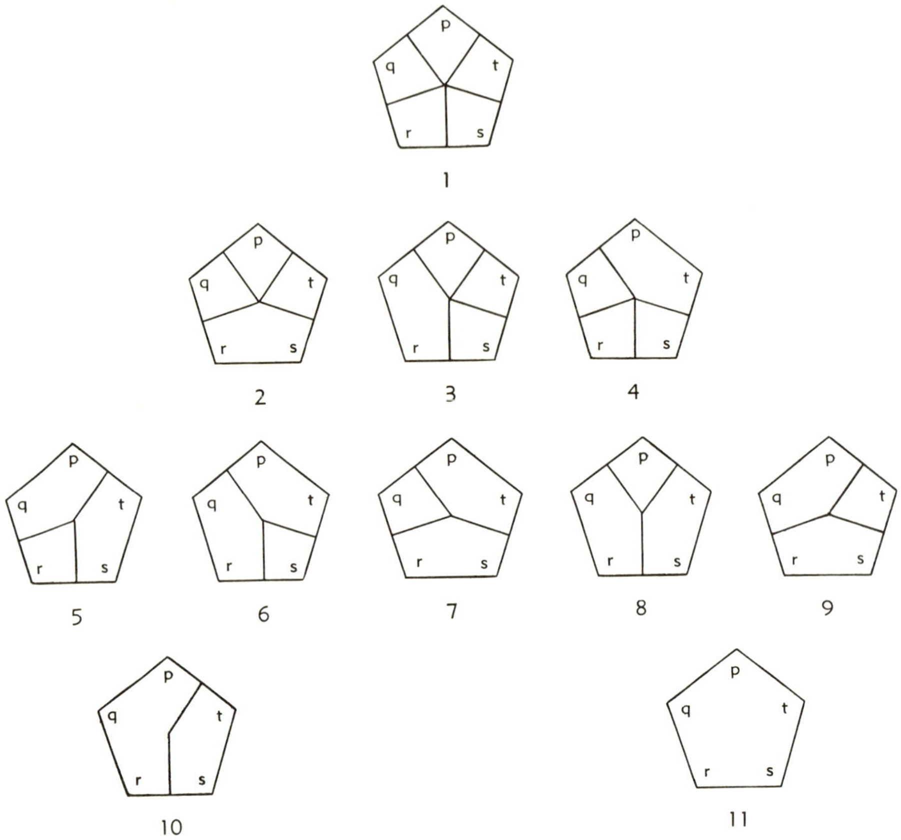

FIG. 77. Diagrams illustrating various arrangements of plates observed in monocyclic pentagonal bases (Ubaghs, n). [Explanation: *p-t,* individual basal elements as explained in text.]

ing an anal plate included in the radial circlet. In these the base is no longer five-sided or pentagonal, but becomes six-sided, that is to say more or less hexagonal in outline. The same condition is found in the curious disparid inadunate *Agostocrinus,* which has an equally tripartite hexagonal base supporting a circlet of six nearly equal plates composed of five radials and one anal plate.

According to WACHSMUTH & SPRINGER (1897) and WILSON (1916), the introduction of an anal plate into the radial circlet in camerates is secondary and therefore the pentagonal base of these crinoids is judged to be ancestral to the hexagonal base (Fig. 78,*1-3*). On the contrary, MOORE & LAUDON (1943a) and SPRENG & PARKS (1953) interpreted the pentagonal base of camerates as derivative from the hexagonal base through upward elimination of the anal plate from the radial circlet (Fig. 78, *4-6*). From these opposite views, the evolution of the hexagonal bases in camerates may be, and indeed has been, variously interpreted.

A quinquepartite hexagonal base (Fig. 78) is not known in these crinoids. Should it be discovered, it would probably show a large, posterior, truncated plate and four smaller subequal elements with pointed distal tips. The widening of the posterior plate *(p)* could have been achieved by enlargement of this plate either on one side (WACHSMUTH & SPRINGER) (Fig. 78,*2*) or on both sides (symmetrical enlargement) (WILSON) (Fig. 78,*3*); in either case, the interradial position of the basals would have been maintained. Quite different is the suggestion made by SPRENG & PARKS,

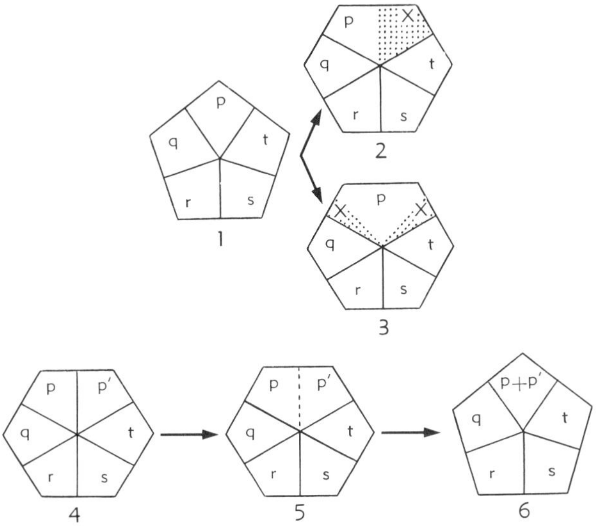

Fig. 78. Diagrams illustrating possible origin of the monocyclic hexagonal base in camerates, according to Wachsmuth & Springer *(1,2)*, Wilson *(1,3)* and Spreng & Parks *(4-6)* (Ubaghs, n). [Explanation: *p-p'*, individual basal elements as explained in text; stippled areas *(X)* inferred plate increments.]

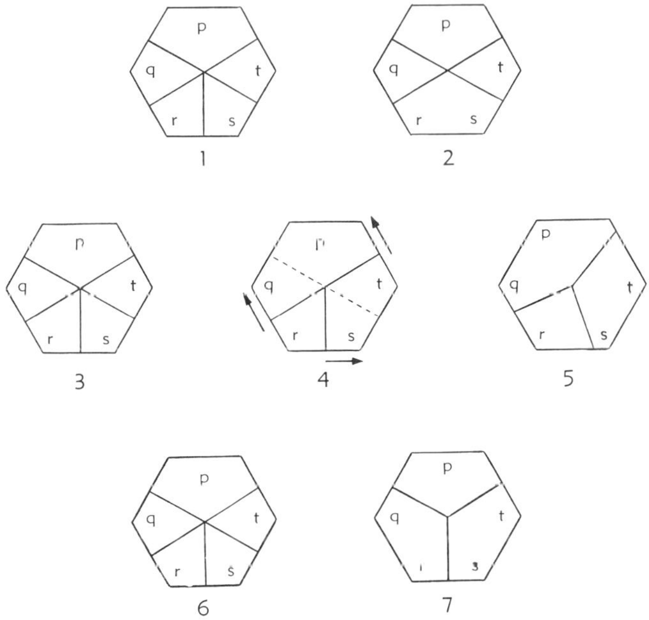

Fig. 79. Diagrams illustrating evolution of the quadripartite hexagonal base in camerates *(1,2)*, of the unequally tripartite hexagonal base in some camerates *(3-5)*, and of the equally tripartite hexagonal base of the inadunate *Agostocrinus* *(6,7)* (Ubaghs, n). [Explanation: *p-t*, individual basal elements as explained in text. Arrows running parallel to sides of bases indicate that the adjacent interbasal suture is being shifted in the direction of arrow.]

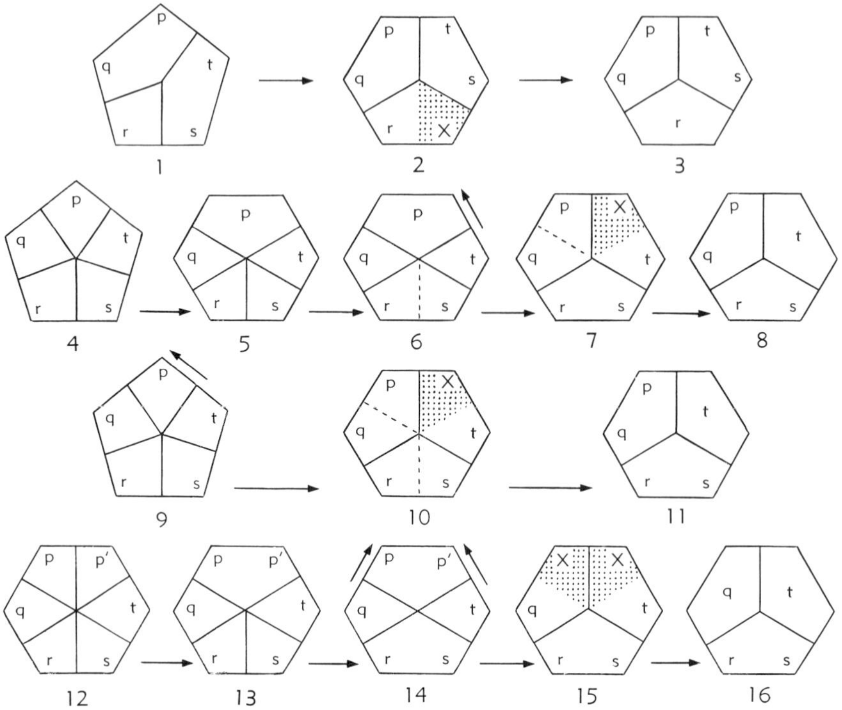

FIG. 80. Diagrams illustrating evolution of the unequally tripartite hexagonal base in camerates, according to WACHSMUTH & SPRINGER *(1-3)*, WILSON *(4-11)*, and SPRENG & PARKS *(12-16)* (Ubaghs, n). [Explanation: *p-p'*, individual basal elements as explained in text, stippled areas *(X)* inferred to represent plate increments. For interpretation of arrows running parallel to sides of bases, see Fig. 79.]

who supposed that the posterior basal in the quinquepartite hexagonal base is a compound plate originating through fusion of two small posterior basals *(p + p')* belonging to a hypothetical ancestor with a hexapartite hexagonal base (Fig. 78,*4-6*).

The quadripartite hexagonal base is easily derived from a quinquepartite hexagonal base, for it only requires fusion of the *r + s* basals (Fig. 79,*1,2*). The unequally tripartite hexagonal base found in *Desmidocrinus* and *Methabocrinus* could have been produced by fusion of two pairs of basals *(p + q, s + t)* in a quinquepartite hexagonal base (Fig. 79,*3-5*). Similarly the equally tripartite hexagonal base of the disparid *Agostocrinus* might result from fusion of the *q + r* basals and *s + t* basals in a quinquepartite hexagonal base (Fig. 79,*6,7*).

Much more puzzling is the problem of the origin of equally tripartite or bipartite hexagonal bases, because such bases in camerates have a posteriorly directed suture lacking in the other types of hexagonal bases. Let us consider first the equally tripartite hexagonal base. According to WACHSMUTH & SPRINGER (Fig. 80,*1-3*), this type of base derives from an unequally tripartite pentagonal base by 1) interpolation of an anal plate in the radial circlet and transformation of the pentagonal outline into an hexagonal outline; 2) enlargement of the right side of the *r* basal; and 3) shifting of the sutures separating the *r* and *s* basals and *p* and *t* basals. For WILSON (Fig. 80,*4-11*), who appeals to the process of truncation and widening of the posterior side as described above, the tripartite hexagonal base would derive from a quinquepartite pentagonal base either through intermediate quinquepartite and quadripartite hexagonal bases (Fig. 79,*5-6*), or directly from a pentagonal quinquepartite pentagonal base (Fig. 79,*9-11*); in both cases, the posteriorly directed suture would be homologous to the suture interposed be-

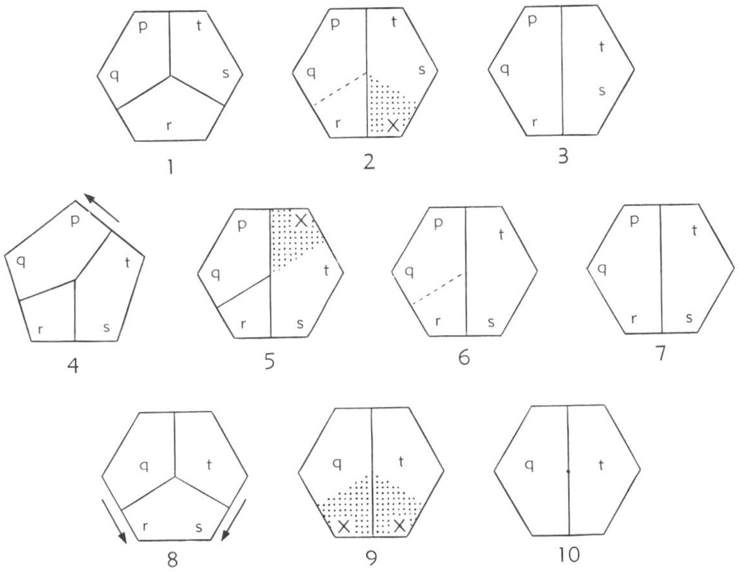

FIG. 81. Diagrams illustrating evolution of the bipartite hexagonal base in camerates, according to WACHSMUTH & SPRINGER *(1-3)*, WILSON *(4-7)*, and SPRENG & PARKS *(8-10)* (Ubaghs, n). [Explanation: *p-t,* individual basal elements as explained in text, stippled areas *(X)* inferred plate increments. For interpretation of arrows running parallel to sides of bases, see Fig. 79.]

tween *p* and *t* basals in the pentagonal base, which would have shifted its position through atrophy of the right half of the *p* basal and compensating hypertrophy of the left side of the *t* basal. The SPRENG & PARKS theory is different in supposing that the tripartite hexagonal base derives from a quadripartite hexagonal base through atrophy of the posterior *p* + *p′* basals and compensating overdevelopment of the adjacent *q* and *t* basals (Fig. 80,*12-16*).

The equally bipartite hexagonal base of camerates, according to WACHSMUTH & SPRINGER (Fig. 81,*1-3*) arises from a tripartite hexagonal base through ankylosis of the $p + q$ and *r* basals and growth of the compound $s + t$ basal on its left side, with consequent shifting of the suture between the $s + t$ and *r* basals. For WILSON (Fig. 81,*4-7*), the equally bipartite hexagonal base originates from a tripartite pentagonal base; this supposes the interpolation of an anal plate into the radial circlet, acquisition of an hexagonal outline of the base, shifting of the suture between the *p* and *t* basals to a posterior position, and fusion of the compound $p + q$ basal with the simple *r* basal on the same side. Finally, SPRENG & PARKS (Fig. 81,*3-10*), who believed that the bipartite hexagonal base derives from an equally tripartite hexagonal base, supposed nondevelopment of the compound $r + s$ basal and symmetrical enlargement of the *q* and *t* basals until they join along a newly directed anterior suture.

At present, the validity of any of these various interpretations is practically impossible to ascertain, except perhaps in some particular cases. The reason for such a situation is that knowledge of the phylogeny of camerate crinoids is still in its infancy, and therefore no means are given for checking relationships postulated by the diverse theories. For detailed discussion of this problem, see BEYRICH (1871) and BATHER (1898-99, 1917b), besides authors already mentioned.

If the basals present a great diversity in number and arrangement, they show as large a variety of shapes, sizes, and modes of growth, as will appear in the systematic part of the present *Treatise.* In some crinoids (e.g., the dicyclic camerate *Orthocrinus,* inadunate *Calceolispongia* [Fig. 82,*1*; see also Fig. 501] or the bourgueticrinid articulate *Democrinus* [Fig. 82,*2*]), they enlarge to the point of becoming the largest elements of the cup; in *Democrinus,* the

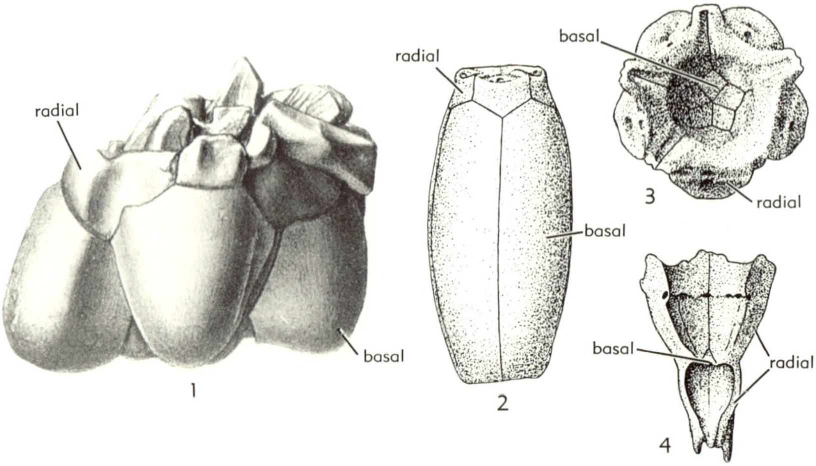

FIG. 82. Selected examples of crinoids with variously developed basals.

1. Calceolispongia mammeata (WANNER), Perm., Timor; *CD*-interray view, ×1 (Wanner, 1924).

2. Democrinus maximus (NIELSEN), U.Cret., Denmark; lat. view, ×5 (Rasmussen, 1961).

3,4. Roveacrinus communis DOUGLAS, U.Cret., Eng.; *3,* adoral view and *4,* longitudinal section of calyx, ×15 (Rasmussen, 1961).

height of the basal ring may be eight times its proximal diameter. Another noteworthy and rather frequent modification is the enlargement of the posterior basal generally in support of the anal plates—a feature particularly remarkable in taxocrinid flexible crinoids.

An opposite tendency is reduction in size of the basals. In some fossil and most recent isocrinids, instead of forming a continuous ring, they make an interrupted basal circlet, being externally visible as five small triangular plates between lower parts of the radials. In the camerate *Trybliocrinus,* as consequence of allometric growth of the proximal cup or calyx elements, they are transformed into an irregular structure entirely located inside the theca (BREIMER, 1962). According to RASMUSSEN (1961, 1971), the basals of the Roveacrinida, generally overgrown by the radials, form a thin wall separating a small proximal chamber from the main thecal cavity and may also take part in the formation of the lower part of the theca even in species where they are concealed by the radials (Fig. 82,*3,4*). In comatulids, the basals, which are still relatively well developed in earliest forms, are more and more reduced in the geologically younger ones, to the point of becoming in most recent representatives of this order a sort of calcareous diaphragm, called the **rosette,** that covers the centrodorsal cavity and represents the central part of the coalesced basals remaining at a larval stage of calcification (for details on this structure, see chapter on recent crinoids, p. *T27*). Finally, the basals are unknown in sessile flexibles (Fig. 70,*12-14*) and articulates (Fig. 70,*15,16*), as well as in short-stemmed cyrtocrinids, which have radials directly articulated with the column or fused with the proximale (Fig. 51,*6a,b*).

RADIALS

In most crinoids the **radials** (radialia) are easily recognized as the most proximal plates of the rays, and it is from their position with regard to the latter that they receive their name. They typically rest on the basals, with which they alternate. They are usually five in number and are followed directly by ray plates of free arms or by fixed brachials immediately above them. Very commonly, they are markedly larger than the succeeding arm plates, but in some crinoids they are approximately the same in size or distinctly smaller than the latter.

The radials are in lateral contact all around the cup or calyx so as to form an uninterrupted circlet or they may be sepa-

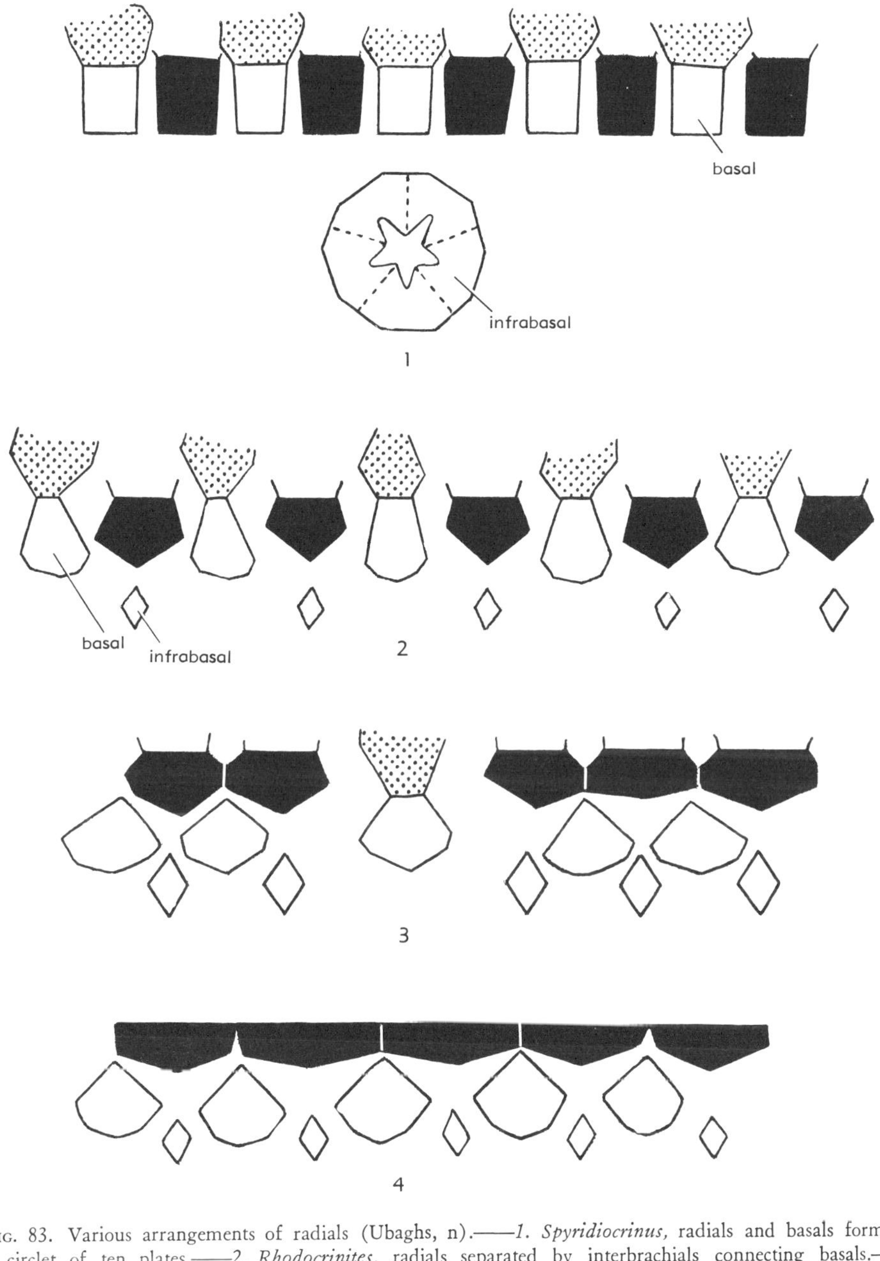

FIG. 83. Various arrangements of radials (Ubaghs, n).——*1. Spyridiocrinus,* radials and basals forming a circlet of ten plates.——*2. Rhodocrinites,* radials separated by interbrachials connecting basals.——*3. Dimerocrinites,* radials in lateral contact except on *CD* interray.——*4. Nyctocrinus,* all radials in lateral contact. [Explanation: interbrachials and anals stippled; radials black.]

rated from next neighbors at one or more places. Such separations may be accomplished in three different ways: 1) through intercalation of basals between radials, producing a circlet of ten plates directly overlying the infrabasals (e.g., dicyclic camerates *Cleiocrinus* and *Spyridiocrinus,* Fig. 83,*1*); 2) through complete separation of rays, including the radials, by interbrachial plates connecting with the basals—an arrangement that characterizes, for instance, members of the camerate superfamilies

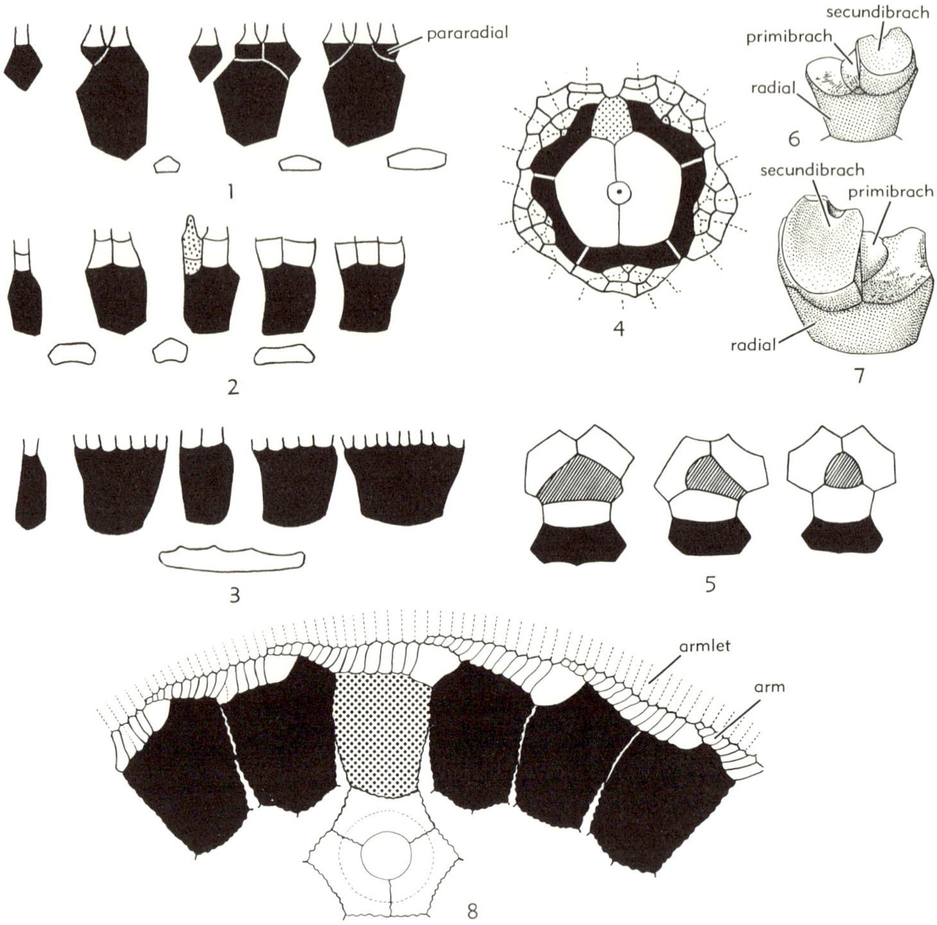

FIG. 84. Drawings suggesting how multiple-arms-bearing radials may have originated *(1-3)* by fusion of radials with pararadials, *(4-7)* by reduction of proximal brachitaxes so that several arms come to be borne by each radial, or *(8)* by fusion of radials with recumbent arms. [Explanation: radials black, primaxils oblique ruled, anal plate stippled.]

1. Calycanthocrinus decadactylus FOLLMANN, L.Dev., Ger. (Ubaghs, n).

2. Allagecrinus austinii CARPENTER & ETHERIDGE, L.Carb., Scot. (Ubaghs, n).

3. Neocatillocrinus incissus WANNER, Perm., Timor (Ubaghs, n).

4. Pterotocrinus pyramidalis LYON & CASSEDAY, Miss., USA(Ky.); basal view of cup, ×3 (Springer, 1926b).

5. Himerocrinus plenissimus (LYON), M.Dev., USA (Ky.); reduction of primaxil in three rays of the same specimen (Springer, 1921a).

6,7. Hexacrinites verrucosus FRAIPONT, U.Dev., Belg.; reduction of primibrachs, ×2 (Ubaghs, n).

8. Agostocrinus xenus KESLING & PAUL, M.Ord., USA(Va.); plate diagram of calyx and recumbent arms (Kesling & Paul, 1971).

Reteocrinidae and Rhodocrinitacea (Fig. 83,*2*), as well as some later genera of flexible crinoids; and 3) through separation of the two posterior radials *(C, D)* by one or several anal plates (Fig. 83,*3*). Such interrupted radial circlets are common among Paleozoic crinoids, but unknown in adults of post-Paleozoic forms (Articulata), which invariably possess uninterrupted radial circlets (Fig. 83,*4*).

In some crinoids, such as the disparid inadunate *Calycanthocrinus,* more than five arm-bearing plates occur in the radial circlet. These accessory "radials" have been

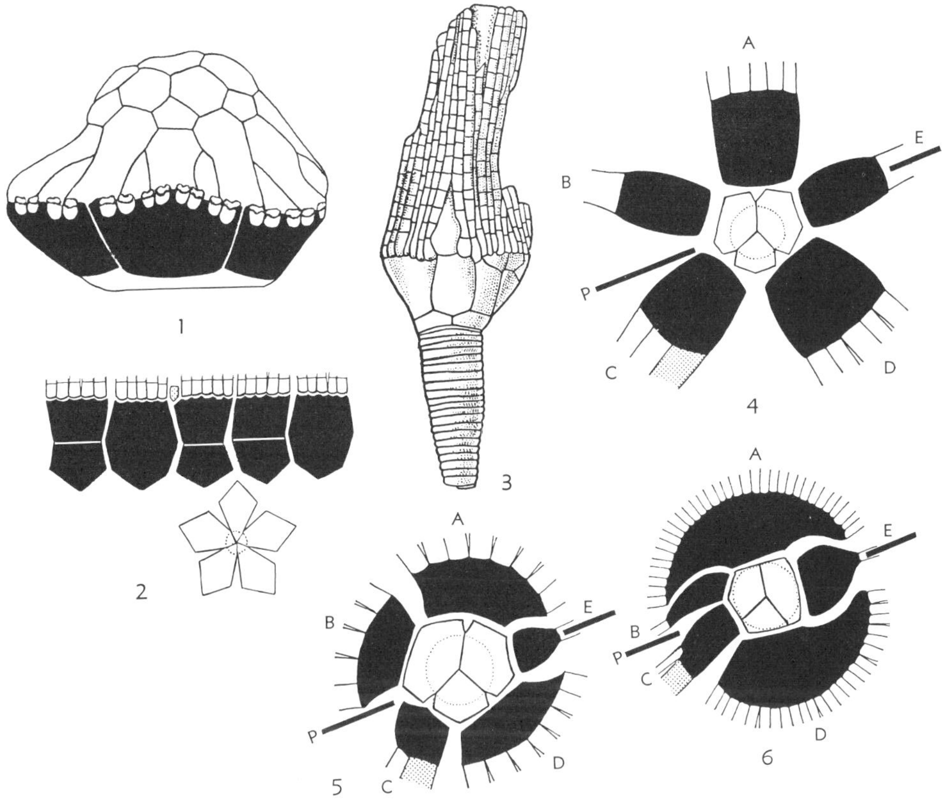

FIG. 85. Selected examples of crinoids provided with multiple-arm-bearing radials. [Explanation: *P*, plane of bilateral symmetry; rays identified by letter symbols of CARPENTER system; radials and biradials black, anal plate stippled.]

1. Parahexacrinus fungiformis SHEVCHENKO (Camerata, Parahexacrinidae), L.Dev., Siberia; lat. view of theca, ×1.5 (Shevchenko, 1967).

2. Anamesocrinus lutheri GOLDRING (Disparida, Anamesocrinidae), U.Dev., USA(N.Y.); plate diagram (mod. from Moore, 1962b).

3. Catillocrinus turbinatus SPRINGER (Disparida, Allagecrinidae), Miss., USA(Ky.-Tenn.); crown and proxistele, *CD*-interray view, ×2 (Springer, 1923).

4. Wrightocrinus MOORE (Disparida, Allagecrinidae), L.Carb., Scot., Perm., Timor; plate diagram (Moore, 1962b).

5. Allocatillocrinus WANNER (Disparida, Allagecrinidae), U.Miss.-L.Penn., USA, Scot.; plate diagram (Moore, 1962b).

6. Metacatillocrinus MOORE & STRIMPLE (Disparida, Allagecrinidae), L.Penn., USA(Okla.); plate diagram (Moore, 1962b).

named **pararadialia** by JAEKEL (1895) (Fig. 84,*1*). A somewhat similar condition, but possibly very different in origin, is shown by the recent comatulids *Promachocrinus* and *Thaumatocrinus,* in which five "interradial radials" are interpolated between the five primary radials at a relatively late stage of development, producing a circlet of ten identical arm-bearing plates in adults.

Whereas each radial of typical crinoids supports a branched or unbranched arm, some fossil representatives are characterized by the presence of two or more arms attached to individual radial plates. This peculiarity evolved independently in camerates (Parahexacrinidae) (Fig. 85,*1*) and some disparid inadunates (Calceocrinacea, Allagecrinacea) (Fig. 85,*2-6*). Very likely the multiple-arm-bearing radials are a product of fusion of radials either with pararadials (BATHER, 1900a; MOORE, 1962a,b) or with arm plates. Possible fusion with pararadials is illustrated in Figure 84,*1-3*. Fusion with arm plates may conceivably have

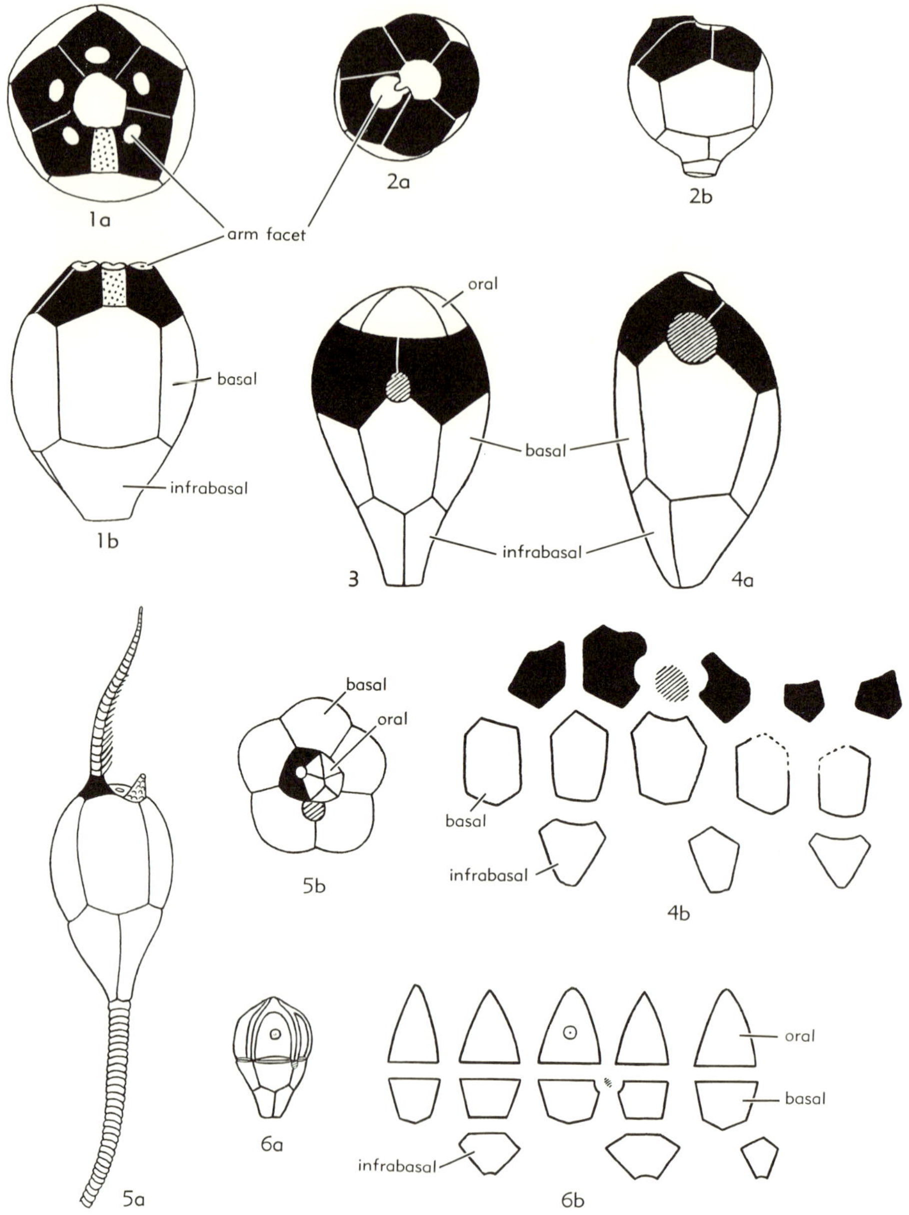

FIG. 86. Reduction of arms and radials in some Codiacrinacea. [Explanation: radials black, anal plate stippled; anus obliquely ruled. Circle with central dot in *6a,b* represents the hydropore, the gonopore, or a common hydropore-gonopore.]

1a,b. Hypocrinus schneideri BEYRICH, Perm., Timor; five radials and five arms, adoral and post. views, ×3.5 (Wanner, 1929a).

2a,b. Allosycocrinus pusillus WANNER, Perm., Timor; five radials but only one arm, adoral and post. lat. views, ×2 (Wanner, 1924).

3. Abrachiocrinus clausus (AUSTIN & AUSTIN), L. Carb., Eng.; five radials but no arm, ×7 (Wanner, 1920).

4a,b. Sycocrinites anapeptamenus AUSTIN & AUSTIN, L.Carb., Eng.; five unequal radials and no arm; *4a*, post. lat. side of cup, *ca.* ×5 (Wanner, 1920); *4b*, plate diagram (Arendt, 1970a).

been accomplished in two different manners: 1) by repeated branching of arms at very short intervals combined with extreme shortening of brachials, tending to and eventually reaching complete reduction of the series of arm plates resulting from each division, so that several arm plates are brought into contact with radials (Fig. 84,*4-7*); 2) by fusion of radials with uniramous series of arm plates totally recumbent on radials, each arm plate supporting a threadlike branchlet or armlet on its free side (Fig. 84,*8*).

An opposite trend to formation of multiple-arm-bearing radials consists of the simplification of ray structure and elimination of arms and even radials in some crinoids. This trend is well exemplified by the Codiacrinacea, a superfamily of dicyclic cyathocrinitid inadunates that contains a large number of generally small crinoids (microcrinoids) (Fig. 86). In this very specialized group, besides several genera normally provided with five arms and five radials (Fig. 86,*1*), are genera having five radials but less than five arms (Fig. 86,*2*), or five equal radials but no arms (Fig. 86,*3*), or five unequal armless radials (Fig. 86,*4*), or fewer than five radials (Fig. 86,*5*), or even no radials and no arms (Fig. 86,*6*). The disappearance of radials in such dicyclic crinoids results in the production of cups composed of two circlets of plates like a monocyclic cup. But it is important to notice that here the remaining circlets are the basals and infrabasals, rather than the radials and basals, as in the true monocyclic condition.

The five components of a radial circlet are rarely perfectly equal, but, except for slight differences in shape and size, they are generally very similar. To this rule, however, noticeable exceptions are produced in different ways: 1) loss of arms by some radials, as shown by several representatives of the Codiacrinacea (Fig. 86,*2,4*), a loss that brings with it disappearance of the brachial facet on these radials and not uncommonly their reduction in size or even complete atrophy (Fig. 86,*4,6*); 2) hypertrophy of certain radials, as illustrated by the Calceocrinacea and Allagecrinacea in relation with the number of arms borne by particular plates of the circlet (Fig. 85, *4-6*) and inequality of the arms (Fig. 87); 3) transverse or oblique bisection of certain radials by a suture, in contrast to other radials which are undivided (see below, biradials); 4) differentiation of the *C*-radial in relation with support of the anal structures (see below, anal plates). To a large extent, these inequalities in size and structure of the radials contribute to make particularly obvious the existence of planes of bilateral symmetry in crowns of the concerned crinoids (Fig. 85,*4-6*; Fig. 87,*3*). In some genera they also have a clear adaptive significance, as in the Calceocrinacea where they are related to bending of the crown on the stem (Fig. 87,*2,4*).

BIRADIALS

In many monocyclic disparid inadunates, some radials are transversely divided into two parts. Such bisected radials are diversely called **compound radials, multiple radials** (Moore, 1962b), or **biradials** (a term herein proposed by Moore, in a following glossary of morphological terms). The upper part of a biradial is named **superradial** and the lower part **inferradial,** two terms introduced by Bather (1892a). In the organization of a crinoid, a biradial occupies the position, and serves the function of an undivided radial. Consequently, its two components together, rather than the inferradial alone (as suggested by Moore, 1952a), are usually considered as equivalent to and thereby counted as a single radial plate. It is doubtful, however, that a real transverse bisection of certain radials ever occurred (Wilson, 1916). It seems more probable that the superradials and inferradials have always been distinct elements. If this is so, the simultaneous existence of biradials and undivided radials in some dis-

---

(Continued from facing page.)

5a,b. Monobrachiocrinus ficiformis granulatus Wanner, Perm., Timor; single radial with arm; *5a,* reconstruction, ×1; *5b,* adoral view of cup, ×1 (Wanner, 1924).

6a,b. Lageniocrinus seminulum de Koninck & Le-Hon, L.Carb., Belg.; no radials and no arms; *6a,* post. lat. view of cup, ×2; *6b,* plate diagram (Kirk, 1940e).

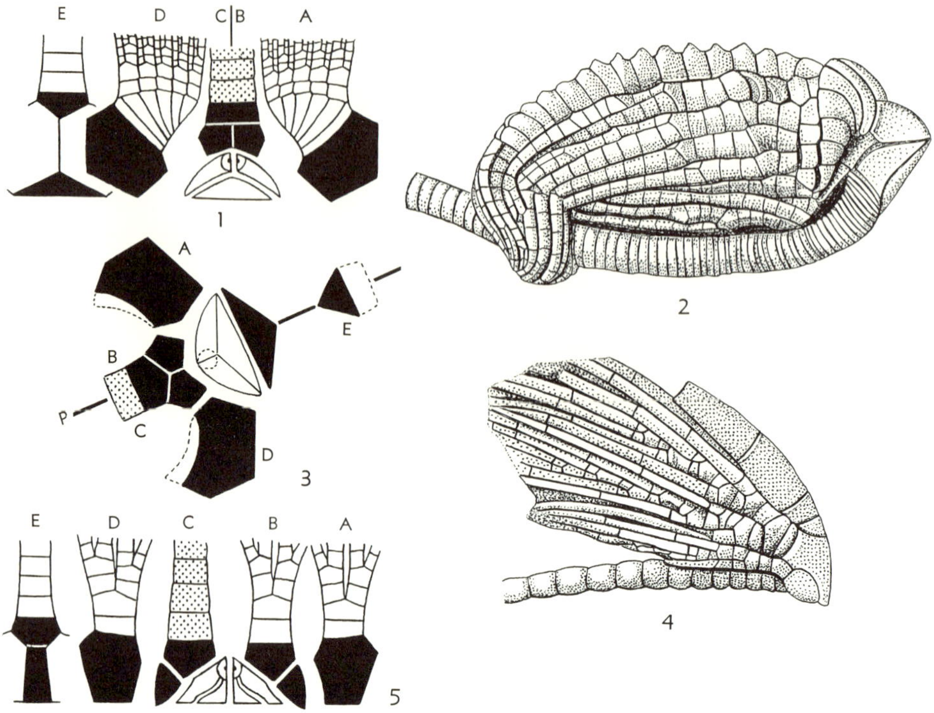

FIG. 87. Inequality of radials in relation to unequal arms in Calceocrinidae (Disparida). [Explanation: *P*, plane of bilateral symmetry; radials and biradials black; anal plates stippled; rays identified by letter symbols of CARPENTER system.]

1,3. Halysiocrinus sp. ULRICH, M.Dev.-L.Miss., USA; *1*, analysis of cup and proximal part of arms (Springer, 1926a); *3*, structure of cup (modified from Moore, 1962a).

2. Halysiocrinus nodosus (HALL), L.Miss., USA (Ind.); crown and proxistele, *A*-ray view, ×1 (Springer, 1926a).

4. Cremacrinus decatur SPRINGER, M.Sil., USA (Tenn.); crown and proxistele, *A*-ray view, ×1.5 (Springer, 1926a).

5. Cremacrinus sp. ULRICH, M.Ord.-U.Sil., N.Am.; analysis of cup and proximal part of arms (Springer, 1926a).

parid crinoids may conceivably be explained either by a marked inequality in size of the most proximal plate of each ray or by the fusion of a superradial and an inferradial resulting in the formation of a large undivided radial in some rays.

The biradials may occur in ?five rays, three rays *(B, C, E)*, two rays (*C* and *E* or *B* and *C*), or possibly one ray *(C)*. The first condition is difficult to prove, since it characterizes crinoids provided with five proximal ray plates that are perfectly similar. Nevertheless, the disparid genera *Eustenocrinus, Peniculocrinus* (Fig. 88,*1a,b*), *Ristnacrinus,* and the cladid genus *Ottawacrinus* have been interpreted in this way by MOORE (1962b)—an interpretation also proposed by KESLING & PAUL (1971) for the highly specialized monocyclic inadunate *Acolocrinus* (Fig. 88,*2*). According to MOORE (1962b, p. 12), the ray plates that immediately follow proximal ones in such genera "seem rather surely to correspond to the superradial elements of the so-called compound radials of homocrinids, heterocrinids, and some other disparid inadunate families, especially where the cup includes only ray-plate pairs."

The occurrence of biradials in the *B, C,* and *E* rays is a distinctive feature of at least 43 genera of disparid inadunates belonging to the superfamilies Homocrinacea (Fig. 88,*3*), Calceocrinacea (Fig. 87,*5*), Pisocrinacea, and Allagecrinacea (Fig. 85,

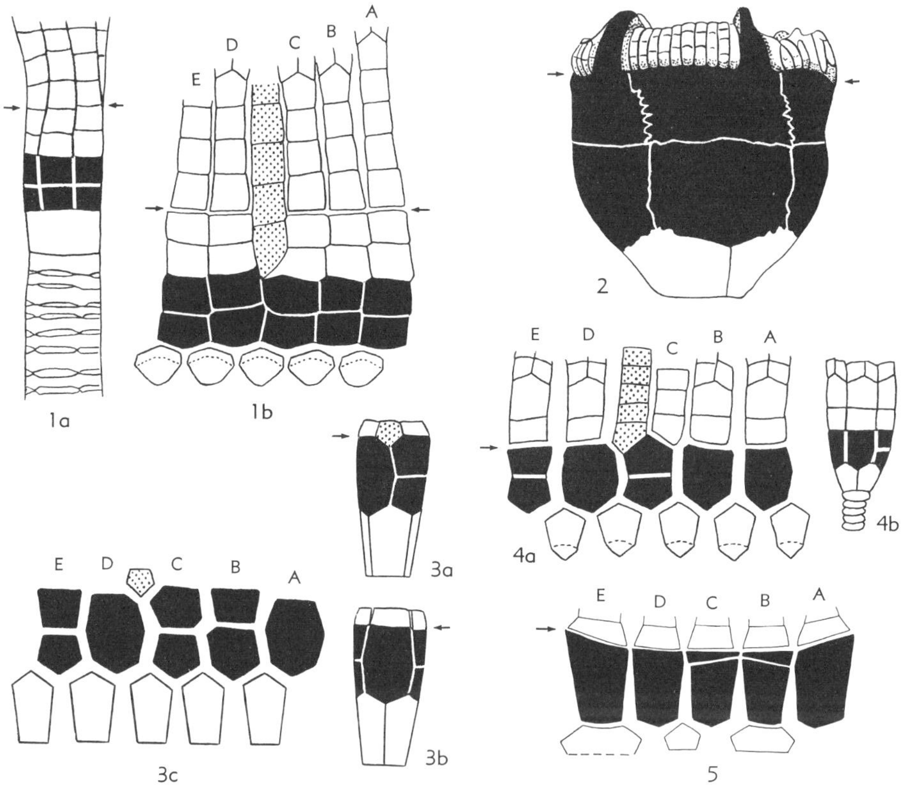

FIG. 88. Selected examples of disparid inadunates with biradials. [Explanation: radials and biradials black; anal plates stippled; top of calyx marked by small arrows; rays identified by letter symbols of CARPENTER system.]

1a,b. Peniculocrinus milleri (WETHERBY), M.Ord., USA(Ky.); *1a,* part of crown and stem, *A*-ray view, ×6; *1b,* analysis of cup and proximal part of arms and anal tube (Moore, 1962b).

2. *Acolocrinus hydraulicus* KESLING & PAUL, M. Ord., USA(Tenn.); crown, *E*-ray view, ×4 (Kesling & Paul, 1971).

3a-c. Homocrinus parvus HALL, Sil., USA(N.Y.); *3a,b, CD*-interray and *A*-ray views, ×8; *3c,* analysis of cup (Kirk, 1914).

4a,b. Columbicrinus crassus ULRICH, M.Ord., USA (Tenn.); plate diagram and *A*-ray view of proximal part of crown and stem (Moore, 1962b).

5. *Quiniocrinus* sp. SCHMIDT, M.Dev., Ger.; analysis of cup (Moore, 1962b).

2), whereas the presence of biradials in two rays distinguishes members of the disparid superfamilies Heterocrinacea (Fig. 88,*4*) and Anomalocrinacea, that have compound radials in the *C* and *E* rays, as well as the pisocrinid genus *Quiniocrinus* (Fig. 88,*5*), in which the compound radials are located in *C* and *B* rays.

The occurrence of a biradial in a single ray is not so evident, for it is a possibility that depends on interpretation given to proximal *C* ray plate in some crinoids. Problems that concern these plates in hybocrinid, perittocrinid, cladid inadunates, and flexible crinoids will be examined under anal plates (see below, p. *T*121). Here will be considered only the cases offered by the pisocrinid, iocrinid, and merocrinid crinoids.

In most pisocrinids, a family of disparid inadunates, two radials *(A, D)* are much larger than the others, and the *B* and *C* radials rest on a single plate, which with the two large radials forms the greater part of the cup (Fig. 89,*3,5*). This single plate has been interpreted by BATHER (1900a) as

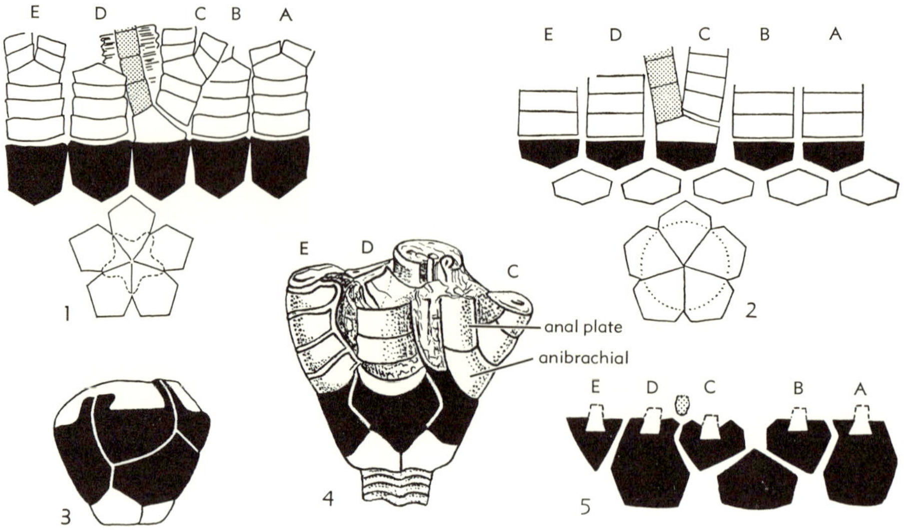

FIG. 89. Structure of pisocrinid, iocrinid, and merocrinid crinoids. [Explanation: radials and biradials black; anal plates stippled; rays identified by letter symbols of CARPENTER system.]

1. *Iocrinus* sp. HALL, ?L.Ord., Wales; M.Ord.-U.Ord., N.Am., Eu.; analysis of cup and proximal parts of arms and anal tube (Moore, 1962b).
2. *Merocrinus* sp. WALCOTT, M.Ord.-U.Ord., USA, Eng.; analysis of cup and proximal parts of arms and anal tube (Moore, 1962b).
3. *Pisocrinus pilula* DE KONINCK, U.Sil., Eu.; *C*-ray view of cup, ×4.3 (Bouška, 1956).
4. *Iocrinus crassus* (MEEK & WORTHEN), U.Ord., USA(Ohio); *D*-ray view of cup with attached parts of arms, anal tube, and stem, ×2 (Hall, 1872).
5. *Pisocrinus* sp. DE KONINCK, U.Sil., Eu., N.Am., Australia; analysis of cup (basals omitted) (Moore, 1962b).

the *C* inferradial displaced to the right into the *BC*-interray. If this is true, the pisocrinids are crinoids provided with a single biradial. If the plate below the *B* and *C* radials results from fusion of the *B* and *C* inferradials, however, as suggested by MOORE (1962b), they are classifiable with crinoids possessing biradials in two rays. The last interpretation is supported by structure of the pisocrinid genus *Quiniocrinus* (Fig. 88,*5*).

The Iocrinidae (Fig. 89,*1,4*) and Merocrinidae (Fig. 89,*2*) are inadunate crinoids, the former monocyclic and the latter dicyclic. Both have radial circlets composed of five equal undivided plates, one of which, located in the *C* ray, supports an axillary plate bearing an anal plate series on its left shoulder and an arm on its right shoulder. This axillary plate has been the subject of much controversy. In the view of some authors, because it has the same axillary function as the *C* superradial of such disparids as *Homocrinus* and *Heterocrinus,* it has to be considered to be a superradial or plate equivalent to a superradial, and accordingly the Iocrinidae and Merocrinidae are considered to be crinoids provided with a single biradial (BATHER, 1890a, 1900a; MOORE, 1950; RAMSBOTTOM, 1961; PHILIP, 1965). This axillary plate, however, 1) plainly has the appearance of an arm plate rather than radial; 2) is not included in the cup but located above the even summits of the five most-proximal ray plates (Fig. 89,*1,2,4*); 3) is supported by a

FIG. 90. Selected examples of crinoids with fixed ray plates. [Explanation: interbrachials and interpinnulars, stippled; radials, black.]

1. *Scyphocrinites elegans* ZENKER (Camerata), U. Sil., N.Am.; diagram of calyx structure (Springer, 1917, mod.).
2. *Icthyocrinus laevis* CONRAD (Flexibilia), Sil., USA(N.Y.); oblique basal view of crown; fixed brachials in lateral contact all around, ×1 (after Springer, 1920).
3. *Uintacrinus socialis* GRINNELL (Articulata), U.

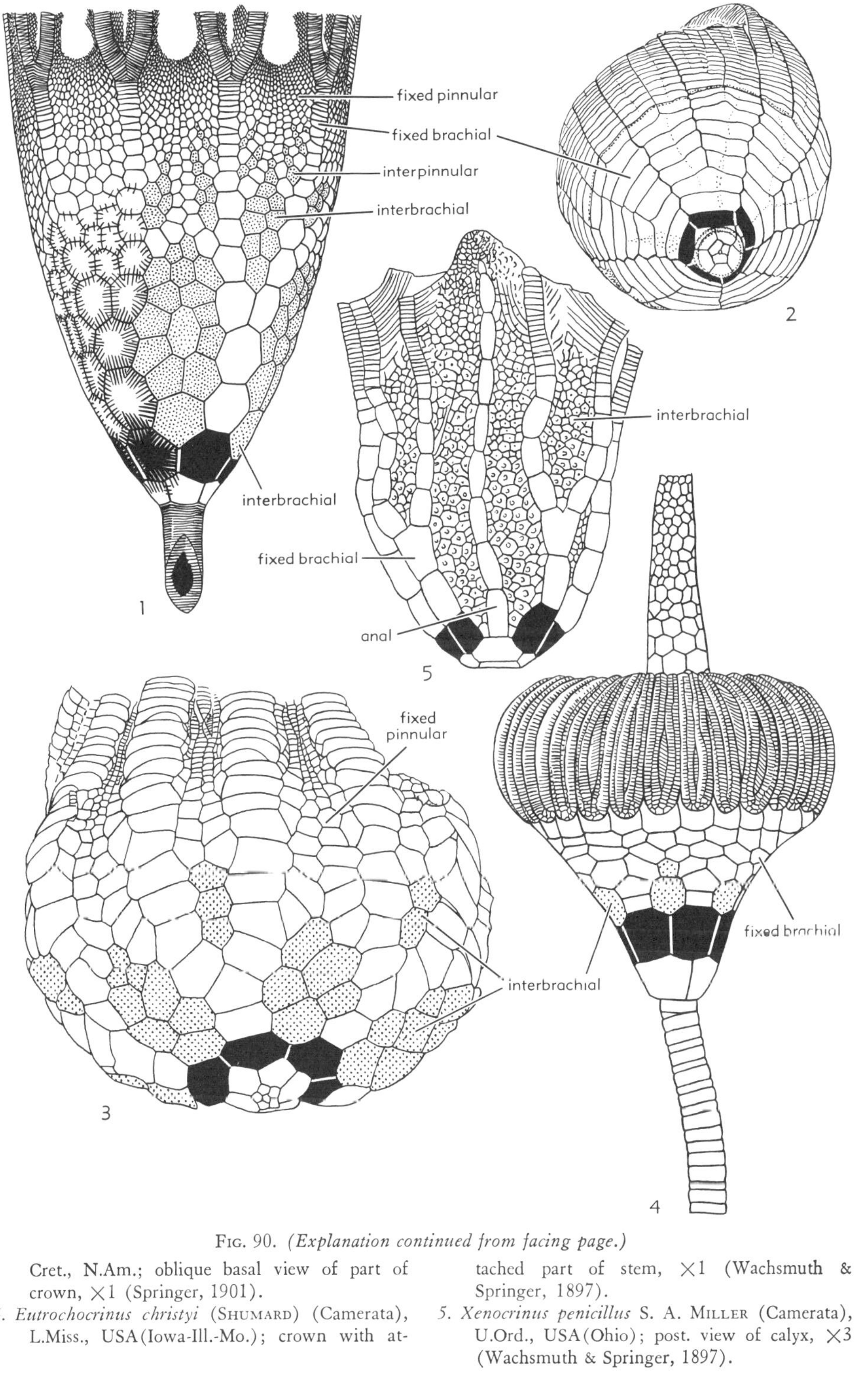

Fig. 90. *(Explanation continued from facing page.)*

Cret., N.Am.; oblique basal view of part of crown, ×1 (Springer, 1901).

4. *Eutrochocrinus christyi* (Shumard) (Camerata), L.Miss., USA(Iowa-Ill.-Mo.); crown with attached part of stem, ×1 (Wachsmuth & Springer, 1897).

5. *Xenocrinus penicillus* S. A. Miller (Camerata), U.Ord., USA(Ohio); post. view of calyx, ×3 (Wachsmuth & Springer, 1897).

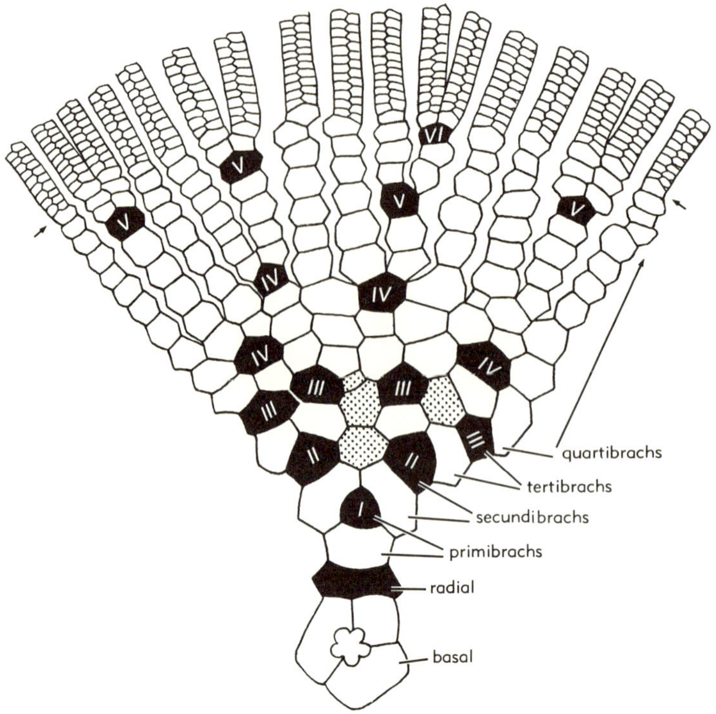

FIG. 91. Plate diagram of base and single ray of the camerate *Himerocrinus* SPRINGER, M.Dev., USA (Ind.) (after Springer, 1921a). [Explanation: radial black; axillary brachs black with Roman number indicating their respective order; interbrachials stippled; top of fixed part of ray marked by small arrows.]

cup plate with distal articular facet exactly like those of adjacent radials. For such reasons, it has been assumed by WACHSMUTH & SPRINGER (1879), UBAGHS (1953), and MOORE (1962b) that the Iocrinidae and Merocrinidae have no biradials but instead possess five undivided equal radials. Consequently, the plate supporting the anal plate series and an arm is a special brachial plate. The term brachianal, applied to it by MOORE (1962b, p. 28), is judged inappropriate, since no known crinoids other than those here discussed has a branched or unbranched arm borne by an anal plate. Instead, this axillary plate of iocrinids and merocrinids is designated here as an **anibrachial.** It is morphologically comparable to the aniradials of many inadunates (e.g., Homocrinacea, Allagecrinacea, Belemnocrinacea).

Just as contiguous inferradials may fuse (Pisocrinidae), coalescence of superradials appears to have occurred in the Calceocrinacea. During paleontological history of these specialized disparid crinoids, the superradials belonging to the *B* and *C* rays have fused to form a single plate that supports anal structures (Fig. 87,*1,2*).

FIXED BRACHIALS AND FIXED PINNULARS

In a large number of Paleozoic crinoids, mostly camerates (Fig. 90,*1,4,5*) and flexibles (Fig. 90,*2*), but also in some early disparid and cladid inadunates (Fig. 88, *1a,b*), as well as articulates like *Apiocrinites* and *Uintacrinus* (Fig. 90,*3*), the cup or calyx includes ray plates above the summits of radials or biradials. Morphologically, as shown by both comparative anatomy and ontogeny, such ray plates are either arm ossicles (**brachials,** more simply brachs) or pinnule ossicles **(pinnulars),** (the pinnules being slender, unbifurcated branchlets typically borne on alternate sides of successive brachials in some crinoids). Because ray

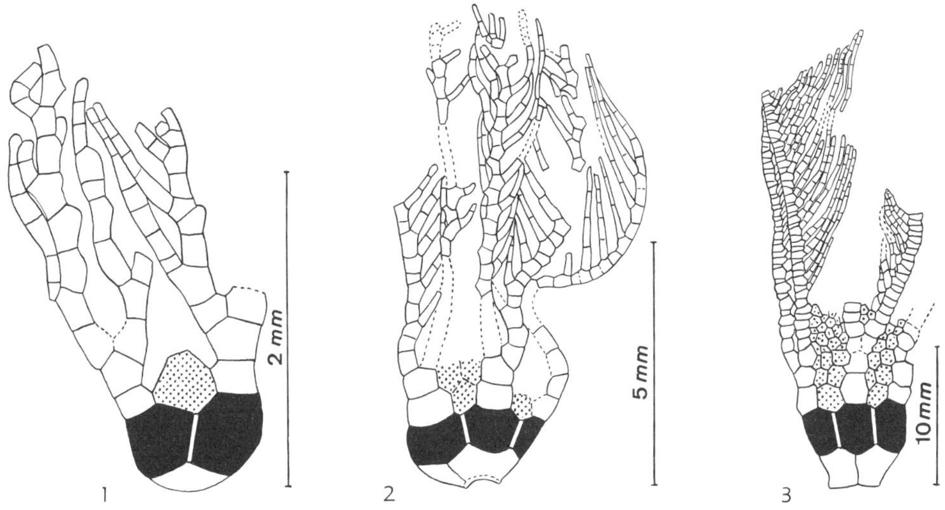

FIG. 92. *Eopatelliocrinus scythogracilis* BROWER, U.Ord., USA(Ill., Mo.). Progressive incorporation of free arms in calyx and multiplication of interbrachials during growth; *1,2* (Ill.), young stages; *3* (Mo.), adult stage (Brower, 1973). [Explanation: radials black; interbrachials stippled.]

plates that enter the composition of the cup or calyx are more or less firmly joined to neighboring thecal plates, they are identified as **fixed brachials and fixed pinnulars,**[1] whereas ray ossicles not incorporated in the theca are **free brachials** and **free pinnulars.** The transition between these two conditions may be very progressive (Fig. 90,*1-3*), or rather abrupt (Fig. 90,*4*). In early inadunates, in many camerates and flexibles, as well as in *Uintacrinus,* no clear distinction is found between fixed and free parts of the ray, and ray ossicles which are free in immature individuals may be fixed in full-grown individuals. In such early camerates as *Reteocrinus, Gaurocrinus, Canistrocrinus,* and *Xenocrinus,* the fixed ray plates stand in the most marked contrast with the minute, indefinitely arranged plates occupying the depressed areas between them, and their outer surfaces form strong, rounded ridges, whereas their inner side is deeply channelled (Fig. 90,*5*). Generally, however, the fixed ray plates share many characters with other plates of the calyx; they are united to each other and with adjacent interray plates by similar articulations and they are distinguished from other calyx plates mostly by their serial arrangement and, in many genera, by the presence of more or less strong median ridges leading to the free part of the arms or of the pinnules.

The supposed morphological distinction between fixed brachials and the proximal ray plates classed as inferradials, superradials, and undivided radials appear very artificial, especially in early crinoids, for all ray plates included in the cup or calyx of these crinoids are very much alike, and distinctions between them depend mainly on their respective positions (Fig. 88,*1b*). On the other hand, the occurrence of fixed brachials in earliest representatives of inadunates (Fig. 88,*1b*), flexibles, and camerates (Fig. 90,*5*) suggests that these plates, or at least the most proximal of them, could have been parts, like the radials, of the original plating of the theca and were under the control of similar morphogenetical factors.

The number of fixed brachials may be as few as one or two to each ray (e.g., early disparids and cladids, and some camerates), or as many as 140 in an individual ray (as in the camerate *Himerocrinus*) (Fig. 91). It grows during ontogeny until it reaches a mean value, characteristic for each species (Fig. 92). Among camerates, if in some

[1] "Cup-brachials" and "cup-pinnulars" are junior synonyms (MOORE, 1952a) that are not used here.

members (e.g., Hexacrinitacea, Platycrinitacea) the arms resemble those of inadunates in showing a strong tendency to become free immediately above the radials, an opposite trend is evident in other forms; for instance, in the actinocrinitid *Strotocrinus* the regularly branched arms may be incorporated in the calyx up to the twelfth bifurcation. Such important enlargement of the calyx by incorporation of very numerous fixed brachials unquestionably represents a specialization. On similar ground, the occurrence of fixed ray plates in the thecal walls of some articulates (e.g., *Apiocrinites, Uintacrinus*) is judged to be an advanced feature of these crinoids.

The fixed part of a ray may consist of a single series of plates or includes one or several bifurcations, the branches thus produced leading to free arms (Fig. 91). Each bifurcation is effected on a special brachial called an **axillary,** which differs from other brachs in having two distal sloping edges or shoulders, which serve to support the two branches resulting from the arm division. Starting from the radials (or biradials) brachials up to and including the first axillary plate **(primaxil)** are identified as **primibrachials** or **primibrachs;** the following ones in each branch are classed as **secundibrachs,** the last one (if further division occurs) being the **secundaxil;** then come **tertibrachs** (possibly including **tertaxils**), **quartibrachs** (possibly with **quartaxils**), and so on.

As stated previously, some pinnules, like the arms, may be incorporated in the calyx for a part or the totality of their length. This may be observed in some recent comatulids with proximal pinnules partly included in the thecal wall by a more or less plated integument. In such crinoids the fixed pinnulars are easily discriminated from neighboring ossicles. A similar condition occurs in many fossil crinoids (Fig. 90,*3*). In some genera, however, the fixed pinnulars, probably owing to loss of their normal function, growth of adjoining plates, and lack of space available for their accommodation, tend to lose their original shape, ornamentation, size, and linear arrangement. Eventually they become entirely similar to other thecal plates. The Silurian and Early Devonian genus *Scyphocrinites* is a good example of such an extreme modification (Fig. 90,*1*).

INTERBRACHIALS AND INTERPINNULARS

In disparid inadunates, some flexibles such as *Icthyocrinus* (Fig. 90,*2*), the camerate *Cleiocrinus,* and in some species of the articulate *Apiocrinites,* fixed parts of the rays are contiguous without plates lying between them, except generally in the *CD* interray. More commonly, however, the fixed parts of rays above radials are united by a more or less plated integument, which contributes with primary elements of the calyx and fixed brachials to enclose the thecal cavity. Thus, in many living crinoids, the proximal brachial plates are united by a flexible integument containing minute supplementary plates and a similar plated membrane may extend between the secundibrachs and tertibrachs of each single ray. A pliant integument studded with small, irregular plates or granules occurs between the rays and their divisions in some early camerates (Fig. 90,*5*) and in most representatives of the flexible order Taxocrinida. The small plates may increase in size and thickness sufficiently to form a well-defined pavement that binds together adjacent ray plates more or less firmly. Such a condition occurs in most camerates (Fig. 90,*1,4*), many members of the flexible order Sagenocrinida, early cladid inadunates such as *Aethocrinus* (see Fig. 94, *1a,b*) and *Cupulocrinus,* Mesozoic Apiocrinitidae, and the Cretaceous articulate *Uintacrinus* (Fig. 90,*3*). Calyx plates located between the rays (except anals, fixed pinnulars, and basals where such plates occur in the radial circlet) are classed as interbrachials, as well as those placed between brachs of a single ray (Fig. 90,*1,3-5*). More specifically, interbrachials located between fixed primibrachs of adjacent rays or above them are named **interprimibrachs,** interbrachials located between fixed secundibrachs or above them are called **intersecundibrachs,** higher interbrachials include **intertertibrachs, interquartibrachs,** and so on. Similarly the plates located between fixed pinnulars or between fixed pinnulars and fixed brachials of the same ray are called interpinnulars (Fig. 90,*1*). In spite of diversity of designation, all these secondary calyx elements are judged to be morphologically equivalent.

Generally, proximal interprimibrachs rest on radials (Fig. 90,*1,3,4*). In some camerates, however (e.g., Spyridiocrinidae, Rhodocrinitacea), they are supported by basals, thus contributing to separate the rays throughout their full length (Fig. 83,*1,2*). No other crinoids show the same arrangement, except sporadically or in some rays, but in some later species of the flexible genus *Taxocrinus,* a tendency of interprimibrachs to penetrate between radials and to connect by short faces with basals is observed. In the early camerate *Xenocrinus,* they sink deeply between the radials, but do not reach the basal circlet. On the contrary, in the patelliocrinid *Briarocrinus,* they rest on the contiguous first primibrachs or even primaxils.

In many crinoids provided with interbrachials, these plates merge with elements of the tegmen (Fig. 90,*5*). This probably represents the primitive condition. In more specialized genera, they are separated from the tegmen by fixed brachials and (or) fixed pinnulars (Fig. 90,*1,3,4*). This arching of interbrachials by ray plates may be produced in various ways, such as 1) increase in area occupied by the rays in distal parts of the calyx, 2) increase in number of arms, 3) reduction in size (with or without accompanying decrease in number) of interbrachials in distal direction, and so on.

The interbrachials may be very numerous and their arrangement ill defined. Examples of this condition are seen in such early camerates as *Reteocrinus* and *Xenocrinus* (Fig. 90,*5*). Generally, however, these plates are limited in number and regularity of placement (Fig. 90,*4*). Enlargement of individual elements and accompanying reduction in their number appear to be common evolutionary trends, and, in some genera, a single large interbrachial plate may occupy each interray of the calyx. An opposite tendency toward extension of the plated areas between rays and their branches may also have characterized the paleontological history of some lineages. Thus, the development of interbrachial plates shows a large range of variations, which constitutes an important taxonomic feature of many groups of crinoids.

During ontogeny, the number and size of interbrachials may change considerably. Species are known in which young individuals have few or no interbrachials, in contrast to adults which have many (Fig. 92). On the other hand, in the juvenile growth stages of some comatulids interbrachial plates, which are soon resorbed, develop between the radials. These observations, together with those just recorded about phylogeny, indicate that tendencies toward increase or reduction of interbrachials may occur among crinoids.

ANAL PLATES

The posterior or *CD* interray in most Paleozoic crinoids is easily distinguished from other interrays by the presence of special plates conveniently called **anal plates** (anals, analia) because they are directly or indirectly connected with the anus. In the cup or calyx these may be located between basals, radials, fixed brachials, or all three. They may be found also partly or entirely above the upper limit of the cup. The widening of the posterior interray resulting from their occurrence allows room for internal organs, particularly the rectum, and furnishes a support for overlying anal structures. In some crinoids their presence is accompanied by enlargement or an asymmetrical development of the *C* and *D* radials, or (and) by an enlargement of the posterior *(CD)* basal.

The anal plates may have different origin and it is far from proved that they are homologous throughout the class. We shall therefore consider them separately in 1) disparid inadunates; 2) hybocrinid and cladid inadunates and flexibles; 3) articulates; and 4) camerates.

In disparid inadunates, the anal plates have an intimate relationship with the adjacent *C* ray. They may rest 1) directly on the *C* radial or superradial, which, instead of supporting an arm, bears a long series of anal plates as sole appendage (e.g., *Eustenocrinus* (Fig. 93), *?Ramseyocrinus,* and the calceocrinid genera *Cremacrinus* and *Senariocrinus*) (Fig. 87,*5*); 2) directly on the *B* and *C* superradials, which have fused in a single plate and do not carry any arm (e.g., calceocrinid genera except those just

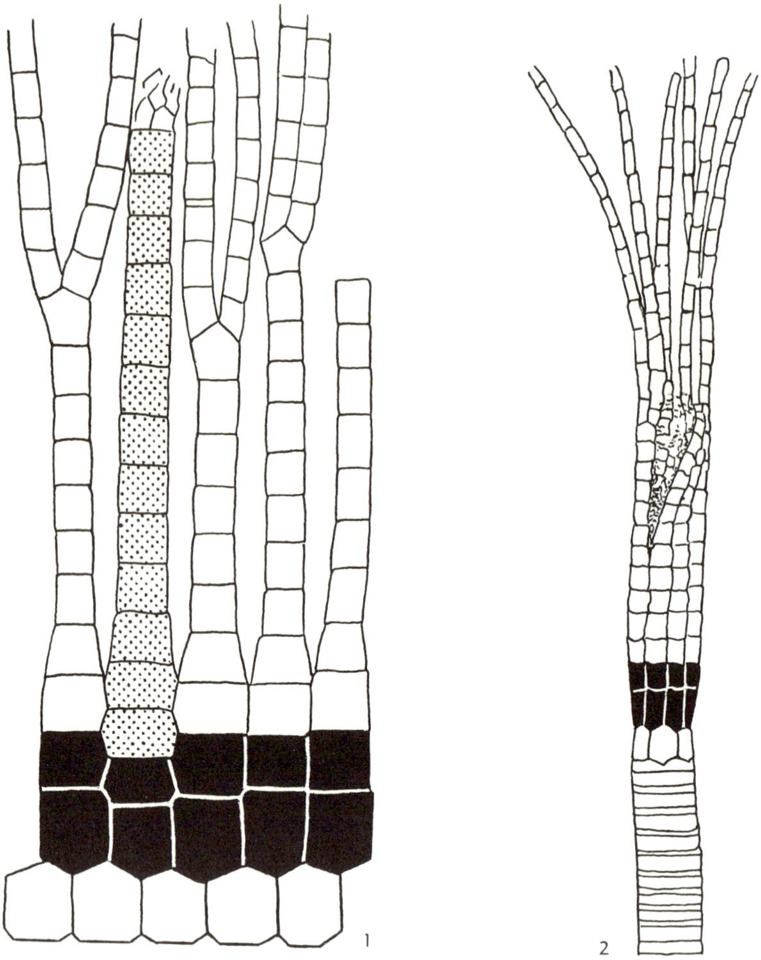

FIG. 93. *Eustenocrinus springeri* ULRICH (Disparida), M.Ord., Can.(Ont.); interpreted as having five compound radials (Ulrich, 1925).——*1.* Analysis of cup, arms, and anal tube.——*2.* *A*-ray view of crown with attached part of stem, *ca.* ×2. [Explanation: biradials black; anal plates stippled.]

mentioned) (Fig. 87,*1*); 3) on the left shoulder of an undivided *C* radial, which supports an arm on its right shoulder; as restricted by MOORE (1962b), the term **aniradial** may serve in a supplementary manner for designating a *C* radial or biradial that assumes such axillary functions (e.g., *Synbathocrinus,* most allagecrinid genera) (Fig. 84,*2*); 4) on the left shoulder of the second *C*-ray plate, which is either included in the calyx and then considered to be an axillary superradial (aniradial) (e.g., *Ectenocrinus, Heterocrinus, Homocrinus*) (Fig. 88,*3,4*), or is just above the upper rim of the cup having the appearance of an axillary primibrach (anibrachial) (see p. *T*116, *T*118) (e.g., *Iocrinus, Myelodactylus, Herpetocrinus*) (Fig. 89, *1,4*); 5) on the left shoulder of the third *C*-ray plate, which is axillary (anibrachial) (e.g., *Peniculocrinus*) (Fig. 88,*1b*). Alone among dicyclic inadunates, *Merocrinus* shows the same anal plate structure as *Iocrinus* (Fig. 89,*2*).

These privileged relations of the anal plates with *C*-ray plates in disparid crinoids and *Merocrinus* and the fact that in many of them the anal structures are supported by an elongate, segmented, armlike appendage have led to the supposition that

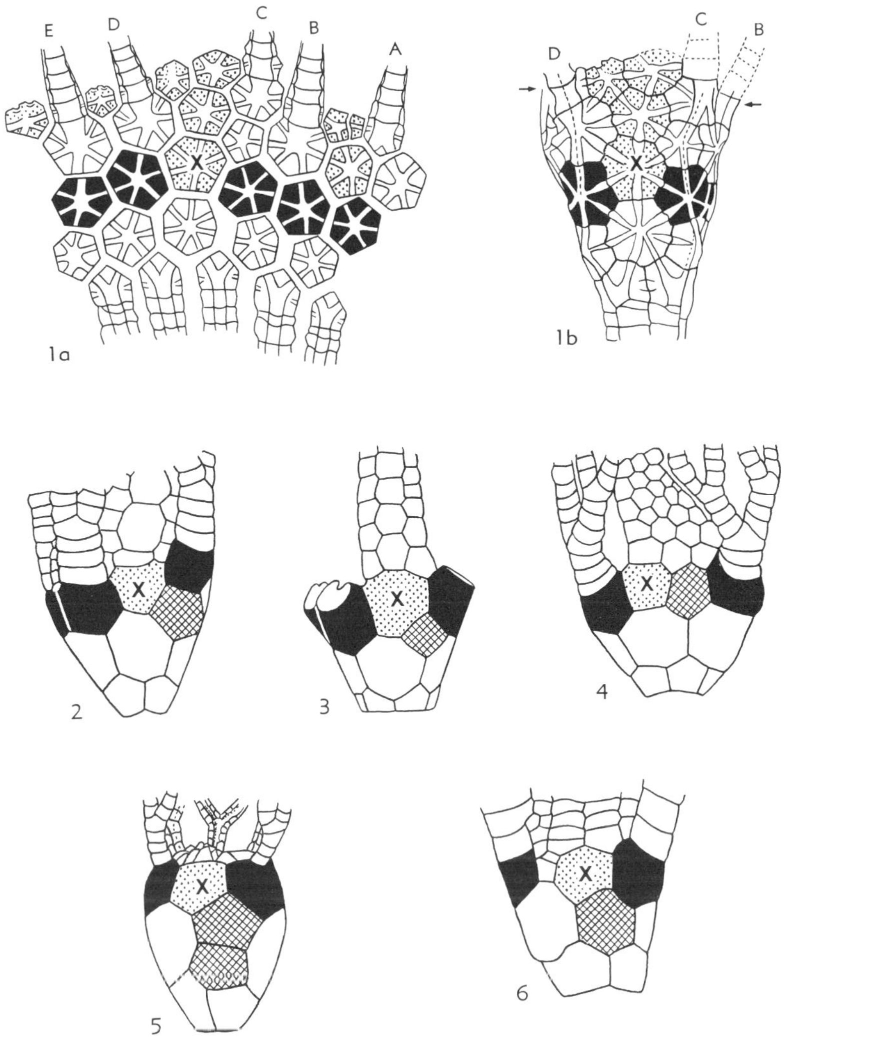

Fig. 94. Position and relationship of radianal in various cladid inadunate crinoids. [Explanation: radials black; radianal cross ruled; other anals and interbrachials stippled (*X* designates anal *X*); top of calyx marked by small arrows; rays identified by letter symbols of Carpenter system.]

1a,b. Aethocrinus moorei Ubaghs, L.Ord., France; most proximal *C*-ray plate (= radianal of other inadunates) not distinguishable from ordinary radials, ×2.7 (Ubaghs, 1969).

2. Cupulocrinus humilis (Billings), U.Ord., N. Am.; post. side, ×2.

3. Botryocrinus cucurbitaceus (Angelin), M.Sil., Sweden (Gotl.); post. side, ×3.

4. Parisocrinus crawfordsvillensis (Miller), Miss., USA (Ind.); post. side, ×1.5.

5. Carabocrinus Billings, M.Ord.-U.Ord., Can., USA, Estonia; post. side, *ca.* ×1.

6. Thenarocrinus gracilis Bather, M.Sil., Eng.; post. side, ×3.6 (*2-6*, Ubaghs, 1953, mod., from *Traité de Paléontologie*, v. 3, copyright 1953, courtesy Masson & Cie, Paris).

this anal appendage was produced by an evolutionary modification of the arm or left proximal branch of the arm belonging to the *C* ray. If this is so, then the first anal plate of these crinoids must have originated as a plate morphologically equivalent to an ordinary brachial.

In all monocyclic hybocrinid and peritto-

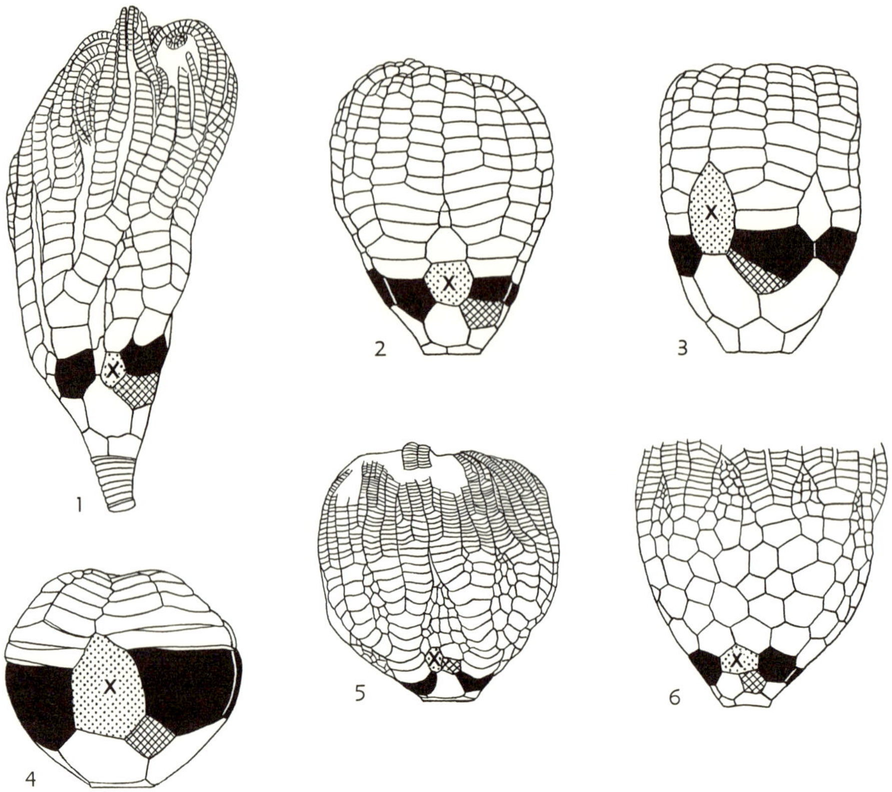

FIG. 95. Position and relationship of radianal in various flexible crinoids. [Explanation: radials black; radianal cross ruled; anal *X* plate stippled.]

1. *Protaxocrinus laevis* (BILLINGS), M.Ord., Can.; post. side of crown, ×2 (Springer, 1911b).
2. *Clidochirus pyrum* ANGELIN, U.Sil., Sweden (Gotl.); ×2.
3. *Anisocrinus greenei* (MILLER & GURLEY), M.Sil., USA(Ky.-Tenn.); ×2.2.
4. *Lecanocrinus pisiformis* (ROEMER), M.Sil., USA (Ky.); ×2.
5. *Forbesiocrinus wortheni* HALL, L.Miss., USA (Iowa); ×0.5.
6. *Sagenocrinites expansus* (PHILLIPS), M.Sil.-U.Sil., Eng., Sweden(Gotl.); ×1.1 (*2-6*, all post. side of crown, Ubaghs, 1953, mod., from *Traité de Paléontologie*, v. 3, copyright 1953, courtesy Masson & Cie, Paris).

crinid inadunates, in most dicyclic cladid inadunates (except *Merocrinus*) and flexibles, two elements especially characterize the anal series; the **radianal** and the so-called **anal X.**

The radianal is the most proximal *C*-ray plate, which migrated upward and leftward during phylogenesis, contributing to expand the posterior interray, and thus became an anal plate (Fig. 94, 95). Its name, proposed by BATHER (1890a), suggests its origin as a radial element and indicates its special relationship to anal plate structures.

The radianal may occupy various positions. In some early cladids (e.g., *Aethocrinus, Ottawacrinus*), the *C* radial is entirely similar to the other radials, except that, like the *D* radial, it is laterally in contact with proximal anal plate (Fig. 94,*1*); this condition is here judged to represent the state of the *C* ray before any differentiation of its most proximal plate as a radianal. In other early but perhaps less primitive cladids (e.g., *Dendrocrinus*) and in early flexibles (e.g., *Protaxocrinus*), the *C* radial has the appearance of a biradial; its upper

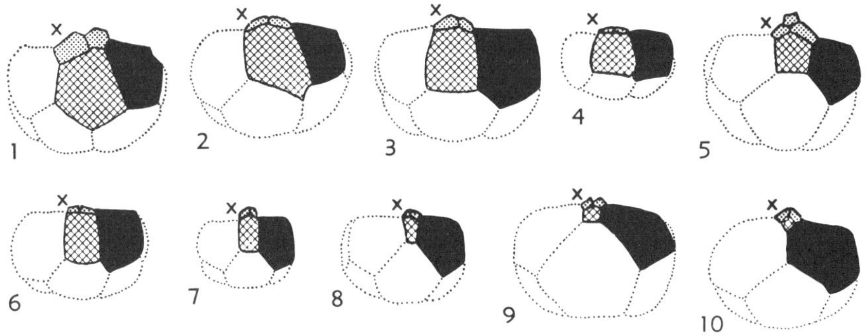

FIG. 96. Variations of the radianal in specimens of *Cadocrinus variabilis* (WANNER) (Cladida), Perm., Timor; *1-10,* selected specimens, post. view, ×0.87 (Moore, 1962b). [Explanation: *C* radial black; radianal cross ruled; anal *X* and other anal plates stippled.]

part is conveniently known as the *C* radial, its lower part as the radianal (Fig. 94,*2;* 95,*1,2*). From this position, in both flexibles and inadunates, as well as in hybocrinids and perittocrinids, the radianal appears to have migrated in the direction of the *CD* interray in such a manner that it came to support the *C* radial on its right margin and so-called anal *X* plate on its left side (Fig. 94,*3;* 95,*3,4*). It may even reach the same level as the *C* and *D* radials (Fig. 94,*4;* 95,*5*), but (except in small part) never lies above them or rises out of the cup. In some genera, it tends to be reduced and eventually disappears (Fig. 96). In a few forms (e.g., the inadunates *Thenarocrinus, Carabocrinus,* and the flexibles *Sagenocrinus, Homalocrinus*) the radianal is located between two basals and it enters into contact with the infrabasal circlet (Fig. 94,*5,6;* 95,*6*). Moreover, the radianal of *Carabocrinus* is transversely bisected and therefore said to be composed of an infer- and a superradianal (Fig. 94,*5*). A similar condition has been recorded in the flexible *Ammonicrinus* (UBAGHS, 1952) but, as only one cup of this genus is known, it is possible that it is an abnormal individual feature (WANNER, 1954).

Even in its most primitive condition (under *C* radial) the radianal is found invariably to adjoin the anal *X* plate. In its lateral migration, it plays a part in supporting other anal plates and functionally becomes the most proximal anal plate, but it never loses its connection with the *C* radial. With respect to neighboring plates, it may be relatively large, moderate in size or small. In some genera and even within some species, its dimensions and relationships to surrounding elements show a large amount of variation (Fig. 96).

The radianal, whatever its ultimate function and location, is a ray plate in origin; it constitutes the most proximal plate of the *C* ray, as shown by comparative morphology of the flexibles and cladid inadunates. It has been considered generally to be a special modified inferradial, though it differs from such plates in disparid inadunates by 1) its constant relation with the anal *X* plate, and 2) its marked tendency to migrate, losing its primary position and function as a ray plate—two features observed in no disparid inferradials. Another possibility is to regard it as the *C* radial itself (as suggested by the plate structure in *Aethocrinus* and *Ottawacrinus*), and to suppose that, after its move toward the posterior interray, it has been replaced and its original function assumed by the next-to-proximal *C*-ray plate. Similar changes seem to occur in the ontogeny of recent comatulids: their larval anal plate (herein judged to be a radianal) develops in the midline of the *C* ray like a true radial; then a new plate appears to its right side, which, during growth, acquires a radial position by pushing the anal plate to the left and becomes the definitive *C* radial. If such similarities are not mere analogies, they at least suggest that 1) the paleontological history of the *C* radial has been much alike in cladid, flexible, and articulate (and pos-

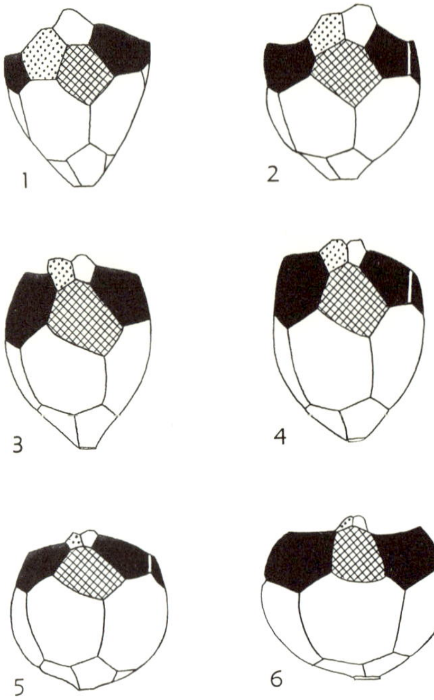

Fig. 97. Variations of anal plates in specimens of *Ureocrinus bockschii* (Geinitz) (Cladida), L.Carb., Scot.(Wight); *1-6*, selected specimens, post. views, ×1 (Wright, 1955-60). [Explanation: radials black; radianal cross ruled; anal *X* plate stippled.]

sibly hybocrinid and perittocrinid) crinoids; and 2) this history has been very different in disparid inadunates.

The anal *X* (or azygous plate of earlier authors) is located between *C* and *D* radials (Fig. 94, 95). It may rest on the radianal or on the posterior *(CD)* basal, or partly on both. Unlike the radianal, it may be above the upper limit of the radials and, if resorbed, it is not within the cup but above it (Fig. 97). Its origin is far less evident than that of the radianal, which is basically a ray plate, and indeed it has been variously considered by authors to be a secondary element suddenly introduced, a plate of the anal tube that gradually sank into the cup, a modified brachial plate homologous to the proximal anal plate of disparid crinoids, or an interbrachial. The fact that in the oldest known cladid inadunates (Fig. 94,*1a,b*), the anal *X* is already present as a large plate resting on the posterior basal, in lateral contact with the *C* radial, and unrelated in any way to arm plates supports none of these interpretations, except possibly its postulated status as a special interbrachial. This suggests at least that anal *X* originated as an interradially located element, which, along with other main components of the cup or calyx, may have become differentiated from original plating of the theca in correspondence with need for an enlargement of the posterior area.

Occurrence of anal plates in the cup or calyx of adult articulate crinoids has never been recorded, but in juvenile growth stages of various comatulids an anal plate develops and is resorbed before maturity. As explained in the chapter on recent crinoids (p. *T*29), the fact that this plate appears in the midline of the *C*-ray like a true radial, and then is pushed to the left by a new plate that becomes the definitive radial suggests strongly that it is homologous to the radianal (Clark, 1915a; Springer, 1921a) rather than to the anal *X* of fossil crinoids (Bather, 1918a; Mortensen, 1920), of which plate no certain equivalent has been found in the ontogeny of comatulids.

The anal plates of Camerata are very different from those of other crinoids. In their more typical aspect, they are arranged in a median series **(anitaxis)** which splits the posterior interray along the sagittal (crinoidal) plane of bilateral symmetry, whereas plates similar and surely homologous to ordinary interbrachials fill up the spaces on other sides (Fig. 98,*1,2*). In some early genera (e.g., *Reteocrinus, Tanaocrinus, Glyptocrinus*), the anitaxis, composed of many conspicuous plates in a linear series, bears a prominent ridge (anal ridge) analogous to the axial ridges of ray plates (Fig. 98,*3*). The connection of the anal ridge with ridges uniting the posterior basal with the *C* and *D* radials seems to indicate that a nerve cord passed along it (Bather, 1900a). Such structural features have led some authors (Moore & Laudon, 1943a; Spreng & Parks, 1953) to suggest that the anal series of camerates could have its origin in a sixth ray. Very early in some lineages (e.g., *Anthracocrinus, Deocrinus*), more belatedly in others (e.g., *Opsiocrinus*), the anal ridge disappeared. In some families, a marked tendency to-

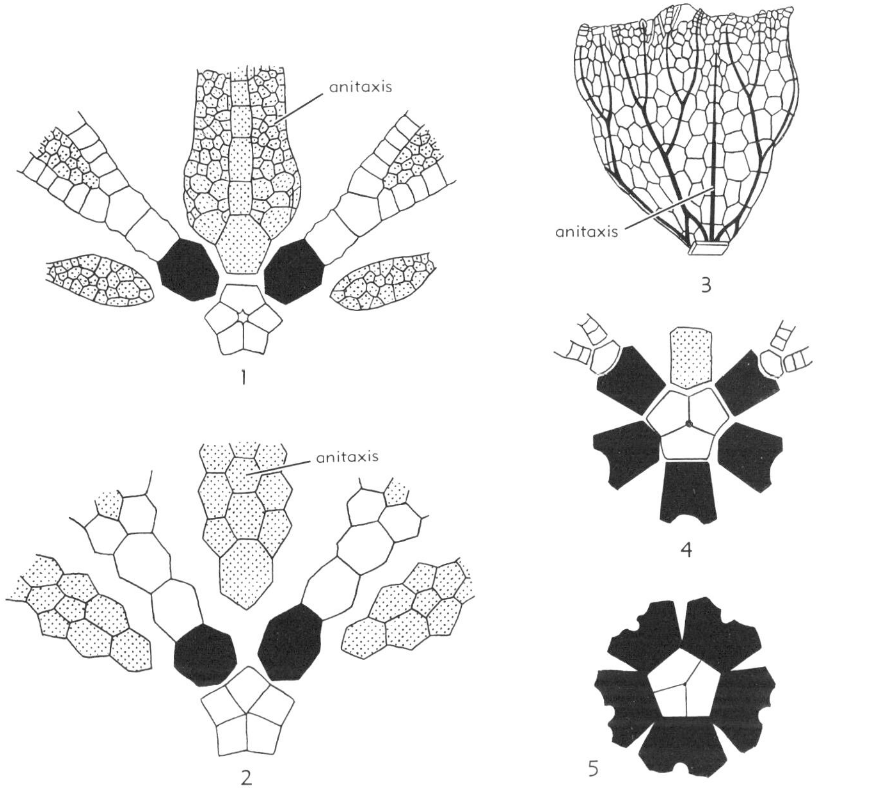

FIG. 98. Anal plate structure of camerates (Ubaghs, n). [Explanation: radials black; anals, interbrachials stippled.]

1. Canistrocrinus WACHSMUTH & SPRINGER, U.Ord., USA (Ohio); plate diagram.
2. Glyptocrinus HALL, M.Ord.-U.Sil., N.Am.; plate diagram.
3. Compsocrinus harrisi S. A. MILLER, U.Sil., N. Am.; calyx, with ray and anal ridges indicated by heavy lines, ×2.5.
4. Hexacrinites AUSTIN & AUSTIN, U.Sil.-U.Dev., Asia, Australia, Eu., N.Am.; plate diagram.
5. Platycrinites MILLER, ?U.Sil., L.Dev.-U.Perm., Eu., N.Am., E.Indies; plate diagram.

ward reduction of extra plates in the posterior interray led to forms in which a single anal plate was left (e.g., *Hexacrinites, Dichocrinus*) (Fig. 98,*4*) or no anal plate remained in the calyx (e.g., *Eucalyptocrinites, Platycrinites*) (Fig. 98,*5*). Although no hard and fast rule can be stated, it is observed that anal plates generally are poorly developed in species provided with a central anus or a central conspicuous anal tube or other tegminal structure (e.g., *Calliocrinus, Eucalyptocrinites*).

The most proximal anal plate is called **primanal,** a term proposed by JAEKEL (1918) and of which the term "tergal" (MOORE, 1952a), not used here, is a junior synonym. The primanal may be followed by a single secundanal or by more than one secundanal. Higher anal plates are designated as tertanals, quartanals, quintanals, etc. Among all these plates, the primanal is the most important, because its relations with other cup or calyx elements have a classificatory value. In dicyclic camerates, except *Cleiocrinus* and *Nyctocrinus*, the primanal is inserted between *C* and *D* radials resting on a basal or basals (Fig. 99,*1*,*2*). It may be similar to the first interbrachials, where these plates separate the radials, as in Rhodocrinitacea (Fig. 99,*1*), although ordinarily it supports three secundanals followed by other anals, instead

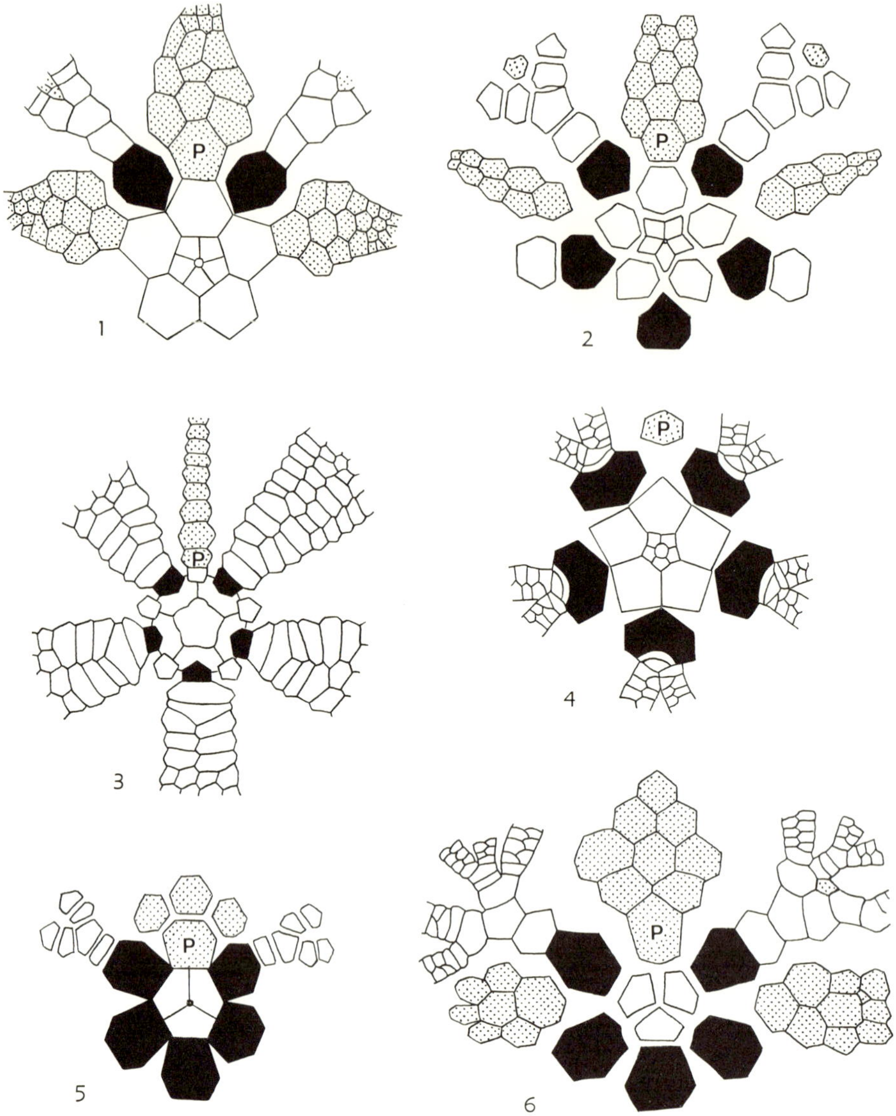

FIG. 99. Relations of primanal in camerates shown by plate diagrams (Ubaghs, n). [Explanation: radials black; anal plates, interbrachials stippled; *P*, primanal.]

1. Rhodocrinites MILLER, L.Carb., Eu., N.Am., Asia, ?Australia.
2. Dimerocrinites PHILLIPS, M.Sil.-M.Dev., Eu., N.Am.
3. Cleiocrinus BILLINGS, M.Ord., N.Am.
4. Nyctocrinus SPRINGER, M.Sil., N.Am., USA (Tenn.).
5. Batocrinus CASSEDAY, U.Miss., USA.
6. Actinocrinites Miller, L.Miss., ?Perm., Eu., N. Am., N.Afr., Australia, ?E.Indies.

of commonly two interbrachials above a most proximal interbrachial in the *AB, BC, DE,* and *EA* interrays (Fig. 98,*2*). In *Cleiocrinus,* the primanal is carried by the posterior basal, but because this plate occurs within the radial circlet, it stands on a level with the first primibrachs (Fig. 99,*3*). In *Nyctocrinus,* the primanal rests upon the *C* and *D* radials, and lacks contact with the basals (Fig. 99,*4*).

Turning now to the monocyclic camerates, it appears that in these crinoids two main types of anal-plate arrangements occur, providing basis for division into two suborders: Compsocrinina, with primanal interpolated in the radial circlet (Fig. 98,*1*); and Glyptocrinina, with primanal resting on two plates *(C, D)* of the radial circlet (Fig. 98,*2*). The primanal of compsocrinine camerates is succeeded by three secundanals in the next higher range (Fig. 99,*5*), except in the Amphoracrinidae, Actinocrinitidae, and advanced members of Paragaricocrinidae, where only two plates rest on the primanal (Fig. 99,*6*). Many camerates, belonging to various unrelated families (e.g., Spyridiocrinidae, Anthracocrinidae, Clonocrinidae, Eucalyptocrinitidae, Platycrinitidae), have the anal side scarcely recognizable in the calyx.

In many respects, the anal plates of camerates are unlike those of flexibles and inadunates, since they include no plate corresponding to the radianal, and lack any special relation with the *C*-ray or tendency to migrate upward-leftward in the direction of the *D*-ray. However, according to BROWER (1973, 1974b), comparison of the ontogeny of living comatulids and presumed phylogeny of camerates suggests a possible homology of the camerate primanal with the anal (radianal) plate of the young comatulids. But this homology is admittedly highly speculative.

Analogies which the primanal may show with the anal *X* plate of disparid and primitive cladid inadunates such as *Aethocrinus* and *Ottawacrinus* do not allow recognition of these elements as morphologically equivalent. Even within the camerates, homologies of the primanal still are uncertain, as likewise is its origin. This plate has been variously interpreted as 1) a plate originally belonging to a sixth ray (MOORE & LAUDON, 1943a; SPRENG & PARKS, 1953); 2) an ordinary interprimibrach; 3) as a plate homologous with the anal (radianal) plate of the larval comatulids (BROWER, 1973, 1975); 4) a posterior element remaining after loss of lateral plates, from a cycle of primary interbrachials interpolated between the radials (CARPENTER, 1886); 5) a new plate introduced in the radial circlet as progenitor of succeeding anals in linear series in manner similar to the "interradial radials" (pararadials) of recent *Thaumatocrinus* interpolated between the five primary radials; 6) one of "supplementary pieces developed as occasion arose in the position where they are found" (BATHER, 1900a); and 7) a new plate introduced between the latero-distal margins of the *C* and *D* radials, that is to say in such a position that a proximal growth and a distal inhibition of the plate would have pushed it into the radial circlet, whereas a distal growth with a proximal inhibition would have placed it above the radial circlet (WILSON, 1916). At present no evidence allows a satisfactory choice between these possibilities.

ACCESSORY PLATES

The calyx of Ordovician monocyclic perittocrinid inadunates includes about 14 accessory plates located between the basals, radials, or radials and anals (Fig. 100). The origin and homology of these elements is quite obscure. They have been considered as equivalent to the interbrachials of camerates (JAEKEL, 1902, 1918) or as plates possibly remaining from a multiplated primitive condition, in relation with the development of respiratory structures (UBAGHS, 1972).

Very different are the accessory plates inappropriately called "subradials" by WACHSMUTH & SPRINGER (1897) and "superbasals" by BATHER (1900a), but conveniently designated as **intercalaries** by MOORE & PLUMMER (1938) (Fig. 101). Known only in the camerate family Acrocrinidae, they form one or several (up to 25) circlets between the basals and radials. Their number ranges from six to about 700, according to the species. Those immediately above the basals are smallest, and apparently ones latest-formed during ontogeny; those located next below the radials are largest and seemingly first to be introduced during development. Typically, those in the posterior *(CD)* interray and, but not so generally, those in the anterior *(A)* ray form single series; the others are more or less irregularly arranged, some of them being radial in position and others interra-

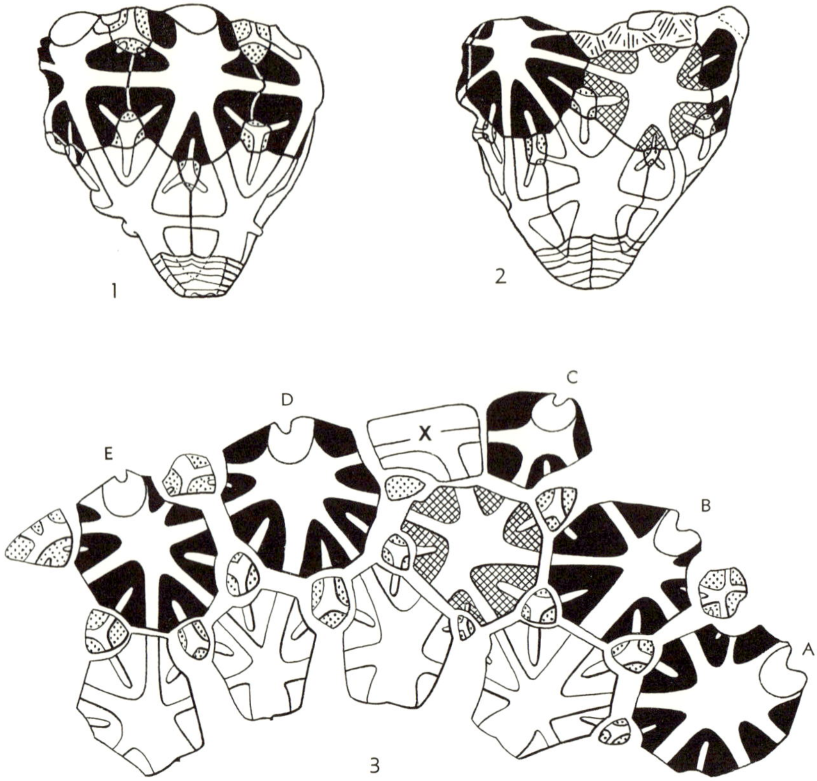

FIG. 100. Accessory plates of perittocrinid inadunates. [Explanation: radials black; radianal cross ruled; *X*, anal *X* plate, accessory plates stippled; rays identified by letter symbols of CARPENTER system.]——*1-3. Tetracionocrinus transitor* (JAEKEL), L.Ord., Baltic regions; *1,2,* calyx with attached portion of stem, in *A*-ray and *CD*-interray views, ×5; *3,* analysis of cup (Ubaghs, 1971a).

dially located. The structure of the Acrocrinidae has been interpreted (JAEKEL, 1918; SPRINGER, 1926b; MOORE & STRIMPLE, 1969) as indicating an evolutionary reversion from the simple theca of *Dichocrinus* to a cystoid-like form, in which the thecal plates lack any definite arrangement. The intercalaries are certainly new plates and not elements inherited from a multiplated ancestor. Their development results from a highly specialized trend, which was active in a dichocrinid stock during a long time period extending from the Early Mississippian to the Late Pennsylvanian.

ARTICULATIONS BETWEEN CALYX PLATES

The nature of articulations between plates of the calyx is one of the main features used in classification of crinoids. In articulates, inadunates, most camerates, and partly in flexibles, plates of the calyx typically have flat (rarely very slightly concave), smooth or weakly ornamented joint faces. Such articulations essentially belong to the **zygosynostosial** type. In some crinoids, radiating ridges (culmina) and furrows (crenellae) cover a part (ordinarily the outer part) of the joint faces and interlock with the ridges and furrows of the opposed joint faces, producing crenulate sutures. When the crenulae cover the whole surface of the facet, the articulation may be referred to as **symplexy.** Complete fusion of plates **(ankylosis),** with obliteration of sutures, occurs commonly in the proximal circlet of plates (basals or infrabasals), less generally in the radial circlet.

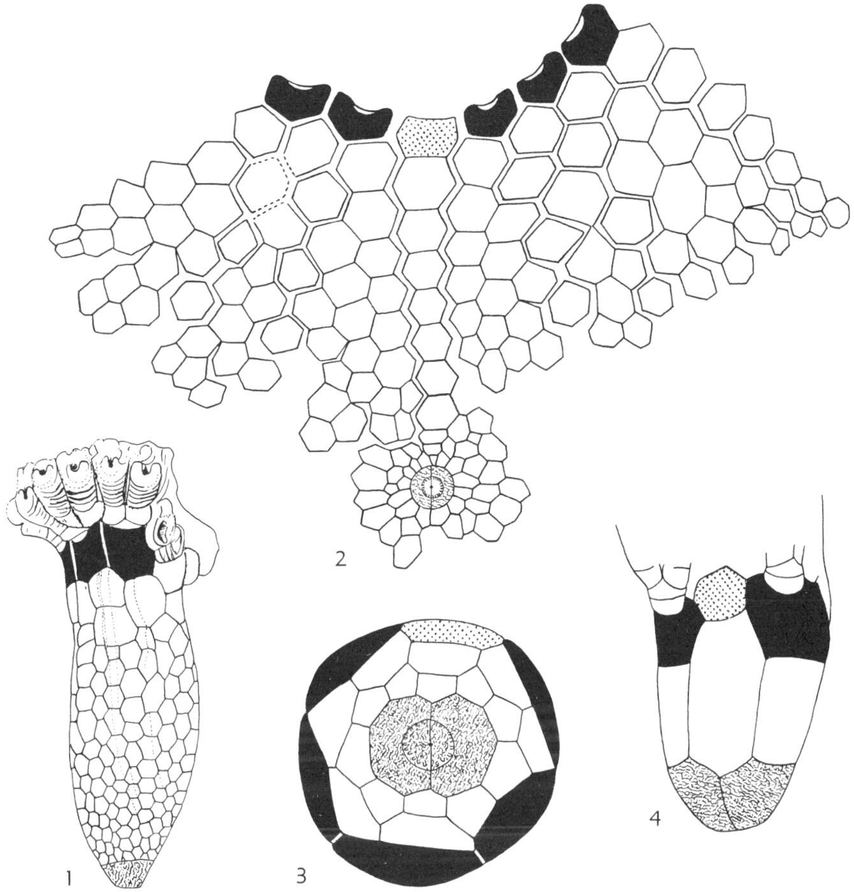

FIG. 101. Accessory plates (intercalaries) of acrocrinid camerates. [Explanation: basals shaded; primanal stippled; radials black.]

1. *Amphoracrocrinus amphora* (WACHSMUTH & SPRINGER), U.Miss., USA(Ala.); side view of calyx with arm bases, ×2 (Wachsmuth & Springer, 1897).
2. *Globacrocrinus pirum* (MOORE & PLUMMER), L.Perm., USA(Okla.); analysis of cup (Moore & Plummer, 1938).
3. *Planacrocrinus ambix* MOORE & STRIMPLE, L. Penn., USA(Okla.); aboral view of calyx, ×12 (Moore & Strimple, 1969).
4. *Protacrocrinus primitivus* (LAUDON & BEANE), L.Miss., USA(Iowa); *CD*-interray view of cup with arm bases, ×5 (Moore & Strimple, 1969).

All of these unions allow only slight movements, or none at all, and so are classified as immovable.

In most flexible crinoids, and some Pennsylvanian inadunates (STRIMPLE & MOORE, 1971), plates of the calyx are commonly connected by articula in which relatively deep and large depressions for the lodgement of connective tissue are surrounded by a generally crenulated rim that provides contact between the plates (Fig. 101). Within the fossae irregular elevated areas may rise nearly to the level of the marginal rim. To these basic elements may be added, in articulations between radials and first primibrachs and between successive fixed brachials, a weak transverse elevation separating an outer fossa from a single inner fossa or paired inner fossae (Fig. 102,*2,5*). Upon the proportion of size of fossae to

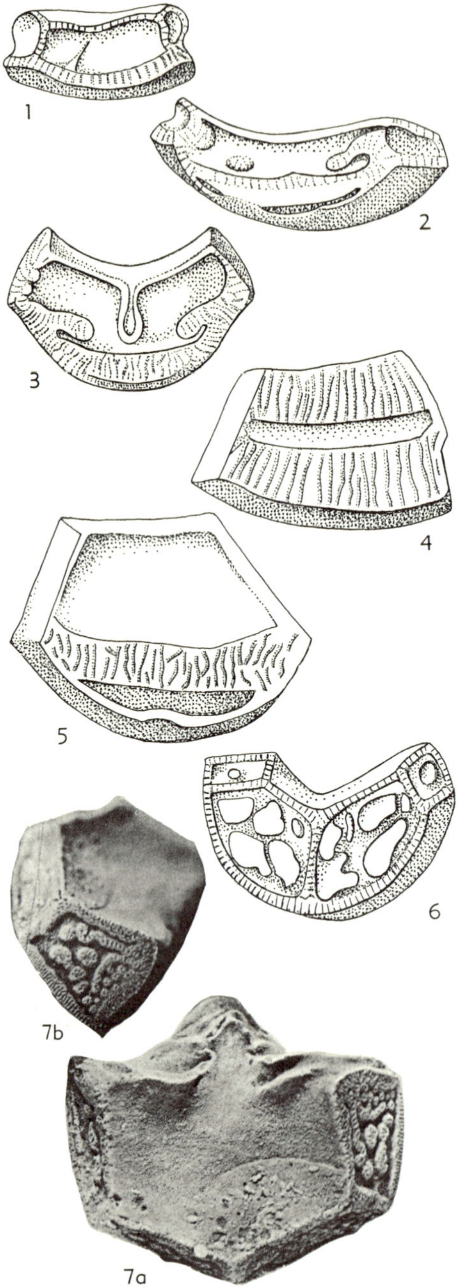

FIG. 102. Articula on selected calycinal plates of flexible *(1-6)* and inadunate *(7)* crinoids. [Explanation: aboral (outer) side of each ossicle directed downward.]——*1-3. Forbesiocrinus nobilis* DE KONINCK & LE HON, L.Carb., Belg.; *1,* proximal face of basal; *2,* distal face of radial; *3,* distal face of second primibrach, ×2.6 (Springer, 1920).——*4,5. Taxocrinus colletti* WHITE, U.Miss., USA (Ind.); *4,* proximal face of basal; *5,* proximal face of first primibrach, ×6.7 (Van Sant in Van Sant & Lane, 1964).——*6. Onychocrinus exculptus* LYON & CASSEDAY, Miss., USA(Iowa-Ill.); distal face of primaxil, ×6.7 (Wachsmuth & Springer, 1897).——*7. Delocrinus subhemisphericus* MOORE & PLUMMER, U.Penn., USA(Kans.); *7a,b,* dual plate consisting of both a radial and primibrach *1* viewed from interior and left lateral side, ×4.4 (Strimple & Moore, 1971d).

contact areas depended the amount of flexibility of the calyx wall. This type of articulation, though probably derived from a synostosial one, differs from typical synostosis in depth of fossae and rimmed peripheral edges. It has been inappropriately called "loose suture" (BATHER, 1900a; SPRINGER, 1920) or "pseudosysygy" (GISLÉN, 1924). The term **manosynostosis,** which literally means loose synostosis, is herein proposed to designate it.

A similar mode of union between calyx plates developed independently in the probably pelagic camerate genus *Scyphocrinites.* Each facet of its calyx plates shows one or several more or less large and deep depressions limited outward and inward by crenulated rims (Fig. 103). These rims, varyingly wide and sinuous, may be so reduced as to become almost nonexistent. Crenulae originate in the deepest part of the depressions, radiate outward, and cross the rim suggesting that this kind of articulation is a modified symplexy. In the distal region of the calyx, the joint faces of fixed brachials and adjacent fixed pinnulars have deeply inward-sloping fossae, which give the thecal wall the fallacious appearance of being composed of an inner pavement supporting thick pillars with constricted bases (Fig. 103,*4a-c*).

All of the articulations so far described are ligamentary. This means that plates of the calyx are or were held together by fibrils probably of primarily collagenous nature. Such tissue is reputed not to allow autonomous movements of the adjoining plates, but it may bestow a certain amount of flexibility to the thecal wall. Muscular articulations between plates of the calyx invariably are lacking except probably in the Calceocrinacea where a hingement, located between the basals and the radials,

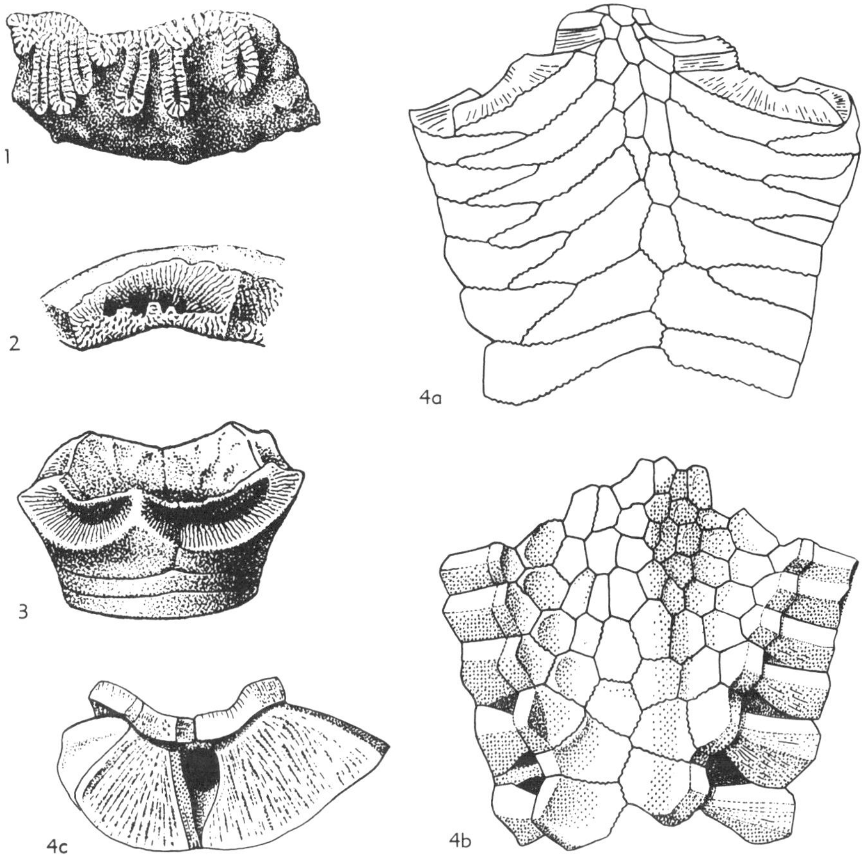

FIG. 103. Articula on selected cup plates of the camerate *Scyphocrinites* ZENKER, ?L.Sil., U.Sil.-L.Dev., Eu., N.Am., N.Afr., Asia, ×4.——*1.* Distal face of primibrach (Springer, 1917).——*2.* Proximal face of radial.——*3.* Distal face of basals.——*4a-c.* Aboral (outer), adoral (inner), and proximal views of distal fixed brachials and pinnulars (Ubaghs, n).

allowed the radial portion of the calyx to erect itself upon the basal portion (Fig. 104). This peculiar conformation—unique among crinoids—facilitated back-and-forth movements of the recumbent crown in a plane that coincides with the strongly marked bilateral symmetry of these disparid inadunates.

ARMS

GENERAL FEATURES

An **arm** or **brachium** is a jointed outgrowth of the central body mass (Fig. 105,*1*). Its aboral skeleton—a direct prolongation of the calycinal skeleton—carries extensions of the food grooves, water-vascular, ectoneural, hyponeural, entoneural, hemal, and genital systems, as well as adoral and aboral coelomic canals. These soft parts are directly connected with the organs and coelomic cavities of the theca. All or most of them penetrate into it through an orifice at the base of each arm **(arm opening)** (Fig. 105,*1*,*2*). A second but smaller orifice may be present; it serves for passage of the aboral motor nerve. Thus, a most intimate connection exists between

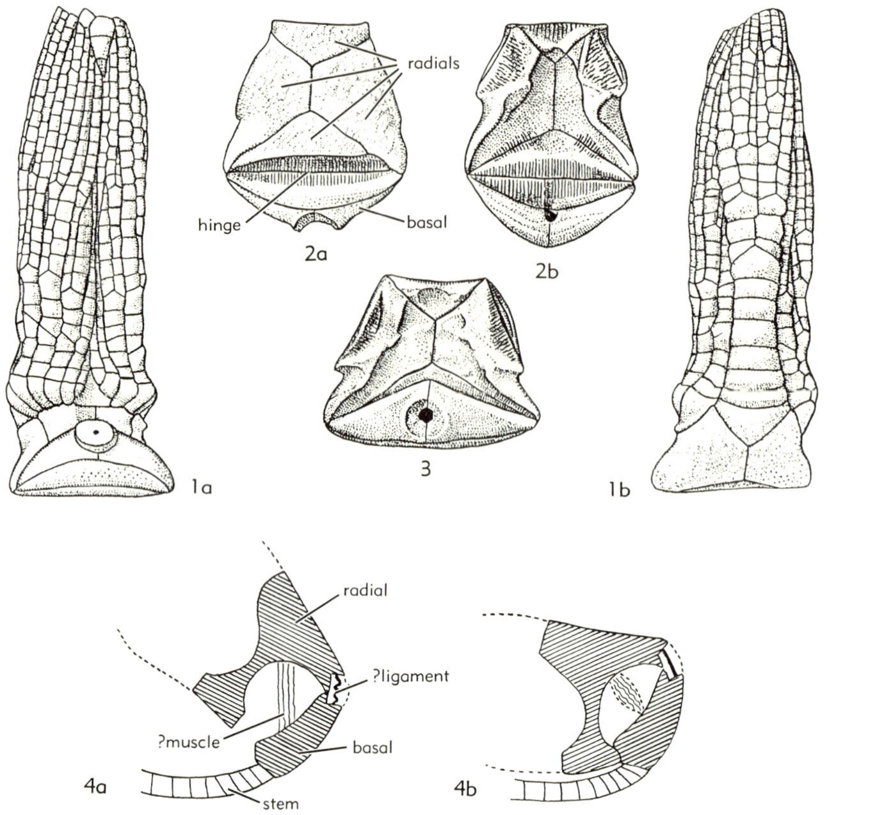

FIG. 104. Hingement in cup of calceocrinid inadunates.

1a,b. Halysiocrinus nodosus (HALL), Miss., USA; *BC* and *E*-ray views of crown, ×1 (Springer, 1926a).

2,3. Halysiocrinus perplexus (SHUMARD), Miss., USA; *2a,b,* outer and inner aspects of calyx with hinge between basals and radials fully extended; *3,* inner aspect of calyx with hinge partially closed (Springer, 1926a).

4a,b. Calceocrinus HALL, M.Ord.-U.Sil., Eu., N. Am.; sagittal section of calyx and proximal part of stem, illustrating articulation of base with radials and manner of erection of crown (Ringueberg, 1889, mod.).

brachia and theca—a feature judged to be highly distinctive of the Crinoidea.

In most other pelmatozoan echinoderms, the food-gathering appendages are exothecal processes, and not evaginations of the theca. Their skeletal support does not prolong the thecal series of plates, and there is no special opening at their base. They had no direct connection with the interior of the theca, and hence, could not have carried extensions from the main coelomic cavities, and from the hemal, genital, and entoneural systems (Fig. 105,*3*).

An arm of a crinoid is not necessarily equivalent to the free part of a ray. In allagecrinids, anamesocrinids, and parahexacrinids a varying number of independent arms may belong to a single ray (Fig. 84, *1-3,4,8*). In some batocrinid and coelocrinid camerates, two (rarely 3 or 4) arms (paired) commonly share a single arm opening (Fig. 106). In many camerates and flexibles, as well as in the articulate *Uintacrinus,* the divisions of the ray-plate series within the calyx increase the number of free arms per ray (Fig. 90, 91). Also, the major branches resulting from one or even several bifurcations of the ray just above the calyx are commonly considered arms. Interpreted in this way, some articulate and inadunate

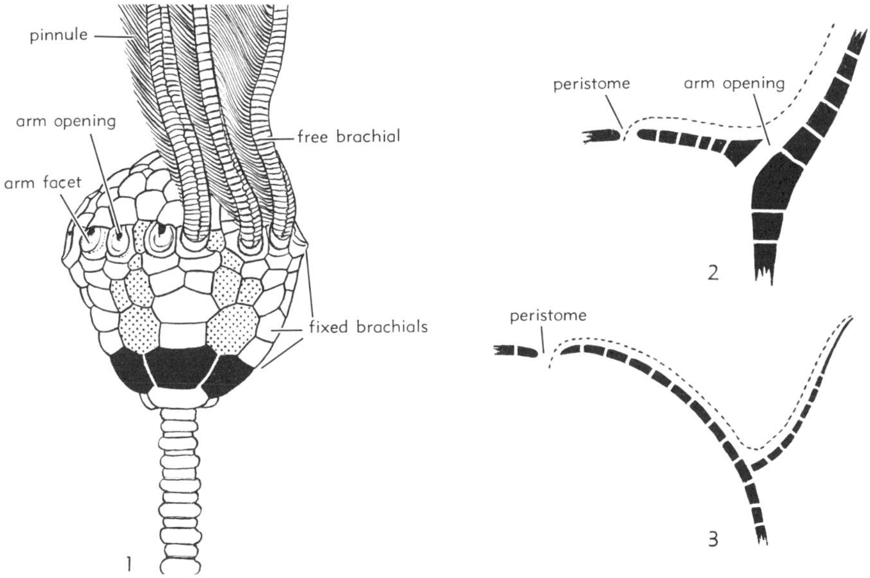

FIG. 105. Relationship of free arms with theca.——*1. Uperocrinus longirostris* (HALL), Miss., USA; theca with parts of free arms and column, ×1.3 (Ubaghs, n). [Explanation: radials black; interbrachials stippled.]——*2,3*. Diagrammatic vertical sections of parts of theca and arm or armlike appendage in a crinoid *(2)* and a noncrinoid pelmatozoan echinoderm *(3)* (mod. from Jaekel, 1898a).

genera are said to have 30 or more arms. Such practice is justified on the premise that an arm and its branches (including pinnules) have essentially the same constitution.

The aboral skeleton of the arms consists of a single or double series of ossicles, termed **brachials** or **brachs.** It is supported directly by the radials (e.g., most inadunates and articulates, but also some flexibles and camerates), or a variable number of arm ossicles are incorporated in the calyx (fixed brachials, see p. *T*119). The passage from the fixed to the free part of the ray may be gradually progressive or abrupt. If it is abrupt the attachment of each free arm to the theca is effected by means of an articulate surface called **arm facet** or, in case of attachment directly on a radial, **radial facet.** This facet will be described as **angustary, peneplenary,** or **plenary,** according to whether it is much narrower than, almost as wide, or as wide as the distal margin of the arm-bearing plate.

The arm ossicles of the free part of the arms **(free brachials)** are generally rounded on the outer or aboral (dorsal) face and grooved on the inner or adoral (ventral) face. This **ventral** or **adoral groove** contains all the extensions of the organic structures and coelomic canals that have been mentioned above, except that in some cases the entoneural cord lies in a special groove on the floor of the main groove or is included in a distinct, generally single, canal **(axial canal)** perforating the brachial ossicle longitudinally (Fig. 107). A brachial provided with an axial canal is said to be **perforate.** (The adoral groove of the arm skeleton is not to be confused with the ambulacral or food groove that is located on the adoral surface of the soft tissues of the arms. The ambulacral groove is but a minor structure housed by the adoral skeletal groove.)

The arms may remain undivided, but generally they bifurcate once or several times. There are various modes of arm branching (see below, p. *T*143). The most advanced condition occurs in crinoids that have a small undivided branch, called a **pinnule,** borne by each brachial. Such crinoids are said to be **pinnulate,** and the individual plates that form the aboral skeleton of the pinnules are termed **pinnulars.**

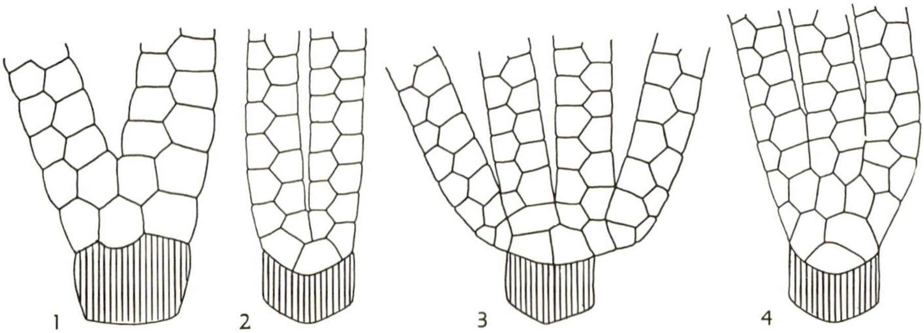

FIG. 106. Occurrence of two or several arms from a single arm-bearing fixed brachial (vertically ruled), to which a single arm opening corresponds (Van Sant in Van Sant & Lane, 1964).——*1. Abatocrinus grandis* (LYON & CASSEDAY), Miss., USA.——*2-4. Dizygocrinus indianensis* (LYON & CASSEDAY), Miss., USA.

The soft structures of the arms and of the pinnules are ordinarily protected by **covering plates,** also named **ambulacrals** (not homologous with ambulacral plates of echinozoan echinoderms). The covering plates belong to the adoral skeleton of the crinoid body and will be described subsequently.

For descriptive purposes, the two lateral sides of an arm may be distinguished as right and left, "right" and "left" corresponding to the right and left of the observer when the arm is viewed from its aboral surface with its distal end directed upward. In a branching arm the adradial branches are termed interior; those to the sides, exterior. The length of an arm (or a branch, or a pinnule) is defined as the distance from its proximal attachment to its distal tip; the width, as the distance from one lateral side to the other, perpendicular to length; the depth is equal to the diameter in an aboral-adoral direction and perpendicular to length and width.

INEQUALITY OF ARMS

The arms of a crinoid are generally alike in size, form, and mode of branching, albeit a certain amount of variability may be observed in almost every individual. Marked inequalities of arms appear, however, as a diagnostic feature of some families, genera, or species. Conspicuous examples are furnished by 1) the Holopodidae, with their arms arranged in a bivium and trivium; 2) the Calceocrinidae, in which one or commonly two arms have disappeared, whereas two of the remaining three are greatly developed (Fig. 87, p. *T*114); 3) the Silurian flexible *Cholocrinus,* which has two dwarfed rays (*B* and *E*) (Fig. 108,*1*); 4) the Mississippian inadunates *Anartiocrinus* and *Pentaramicrinus bimagnaramus,* in which half-rays are hypertrophied in *B* and *E* rays (Fig. 108,*2,3*); and 5) various reef-inhabiting forms, the morphology of which depend largely on ecology, and other factors. In many cases, such inequalities contribute to render planes of bilateral symmetry particularly obvious in the crown.

NUMBER OF ARMS

The number of free arms is primitively and normally five. This five-armed condition characterizes most disparids, some hybocrinids, cladids, and flexibles, and is still preserved in various articulates. On the other hand, in all camerates and many flexibles and articulates the number of free arms is equal to or greater than ten. If, however, some camerates seem to have only five arms, it is because each of their rays comprises two arms that either have fused together longitudinally (as in *Melocrinites*) or are so unequally developed that one of them is reduced to the size of a branchlet and does not look like an arm (as in *Cytidocrinus*).

The ten-armed or multiple-armed forms result generally from the bifurcation of the brachial-plate series within the calyx or at its upper limit or just above it. But it

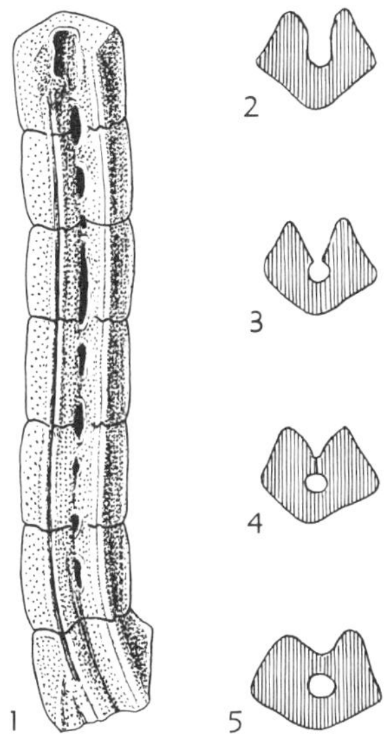

FIG. 107. Differentiation of axial canal from ventral groove.——*1. Gissocrinus* sp. cf. *G. goniodactylus* (PHILLIPS), Sil., Gotl., part of an arm in adoral view, showing axial canal still partly confluent with ventral groove, ×6 (Bather, 1893a).——*2-5. Gissocrinus goniodactylus* (PHILLIPS), Sil., Eng., Gotl.; stages in separation of an axial canal, ×8 (Bather, 1900a).

may also be produced by 1) occurrence of more than five arm-bearing plates in the radial circlet (as in *Calycanthocrinus*) (Fig. 84,*1*); 2) development of multiple-arm-bearing radials (as in Allagecrinidae, Anamesocrinidae, Parahexacrinidae) (Fig. 85); 3) fusion, with each radial, of a recumbent arm carrying a small arm on the free side of each of its component ossicles (as in *Agostocrinus*) (Fig. 84,*8*); 4) augmentative regeneration, as demonstrated by MINCKERT (1905a) for some comatulids; 5) sudden deviations from pentamerous symmetry affecting some individuals in normally five-rayed species.

The opposite condition, i.e., the tendency to reduction of the number of arms to four, three, two, one or even their complete disappearance, occurs in various families. This has already been illustrated in connection with the elimination of some radials in the Codiacrinacea (Fig. 86, p. *T*112). Another well-known example is furnished by the hybocrinid inadunates. Some of them have five radials but only three arms, or no arms at all *(Cornucrinus, Tripatocrinus)*; in these crinoids the atrophy and loss of arms seem to have been more or less compensated by the extension of the ambulacral grooves across the plates of the calyx or on the

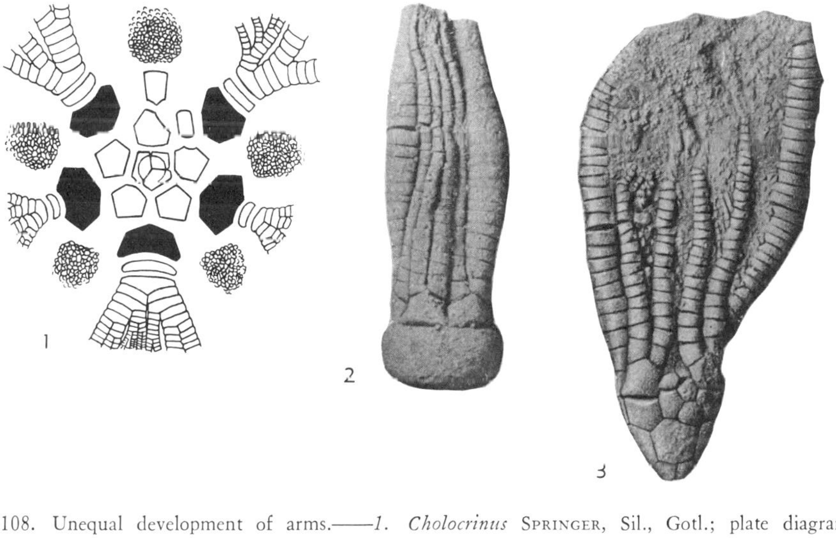

FIG. 108. Unequal development of arms.——*1. Cholocrinus* SPRINGER, Sil., Gotl.; plate diagram of calyx and proximal part of arms, showing enormously enlarged *A, C,* and *D* rays, and dwarfed *B* and *E* rays (Springer, 1920).——*2. Pentaramicrinus bimagnaramus* BURDICK & STRIMPLE, Miss., USA; crown in *A*-ray view, with greatly enlarged arms in *B* and *E* rays, ×1.3 (Burdick & Strimple, 1971).——*3. Anartiocrinus lyoni* KIRK, Miss., USA; crown in *CD* interray view, with hypertrophied arms in *B* and *E* rays, ×1.5 (Kirk, 1940).

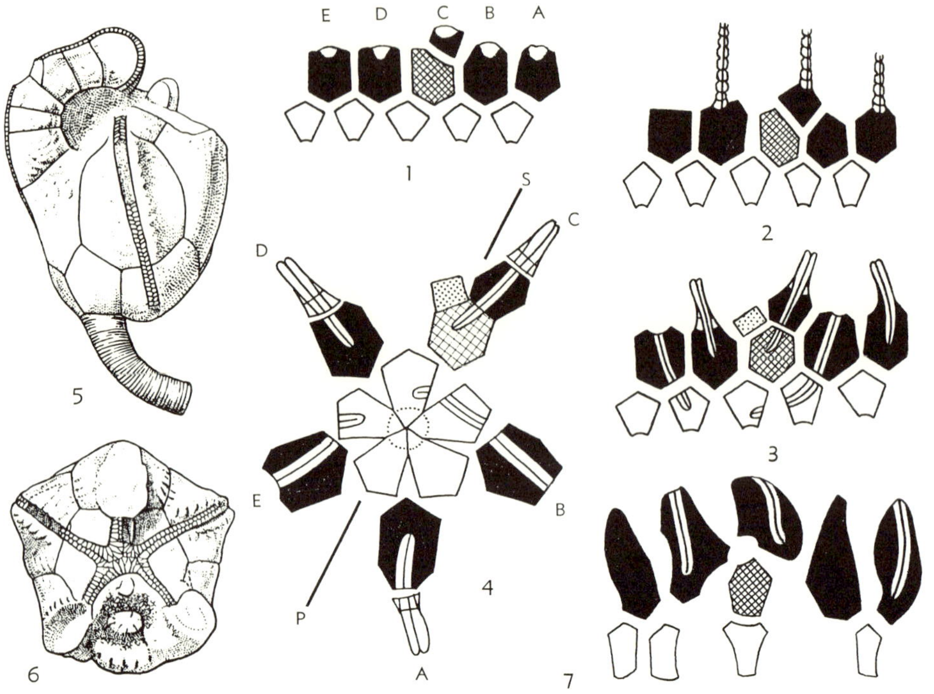

FIG. 109. Reduction of numbers of arms illustrated by hybocrinid inadunates. [Explanation: *P-S*, plane of bilateral symmetry; rays identified by letter symbols of the CARPENTER system; anal *X* stippled; radials black; radianal cross ruled.]

1. Hoplocrinus GREWINGK, M.Ord., Swed.; plate diagram of cup, five radials bearing arms (Regnéll, 1948a, mod.).

2. Baerocrinus VOLBORTH, M.Ord., USSR(Leningrad distr.); plate diagram of cup and parts of arms, only three arms (Regnéll, 1948a).

3,4. Hybocystites WETHERBY, Ord., N.Am.; *3*, plate diagram of cup and arms, showing three reduced arms, and five ventral grooves extending onto radials and basals (Regnéll, 1948a); *4*, plates radially arranged to show eustenocrinoidal plane of bilateral symmetry (Moore, 1962b).

5,6. Hybocystites eldonensis (PARKS), M.Ord., Ont.; *5*, crown and part of column, in *B*-ray view; *6*, cup in adoral view, arms not preserved, ×2 (Springer, 1911b).

7. Cornucrinus REGNÉLL, M.Ord., Swed.; plate diagram of cup, no arm, but ventral grooves on *A*, *C*, and *D* radials (Regnéll, 1948a).

aboral side of greatly reduced arms, giving them a superficial resemblance with some cystoids; to be noted is the relation of these modifications with the eustenocrinoidal plane of bilateral symmetry (Fig. 109). Equally noteworthy is the existence of only four arms in the primitive disparid inadunates *Eustenocrinus* and (?) *Ramseyocrinus*, the fifth arm of which is replaced by an anal tube (Fig. 93, p. *T*122). Many other examples could be cited, such as the four-rayed *Tiaracrinus*, *Nanocrinus*, and *Tetrapleurocrinus* (Fig. 110,*1a,b*), and the two-rayed *Scoliocrinus* (Fig. 110,*2a,b*), all from the Middle Devonian of Germany; the three-rayed *Holynocrinus* from the Middle Devonian of Bohemia (Fig. 110,*3a,b*); the four-armed *Parindocrinus* (Fig. 110,*4*) and three-armed *Sundacrinus* from the Permian of Timor; and the usually four-rayed *Tetracrinus* (which, however, may have 3 or 5 to 8 rays) from the Jurassic of Western Europe (Fig. 110,*5*). In these instances, the diminutive deviation from pentamerous symmetry appears mostly as a generic feature, but it may also occur as an individual variation in species normally provided with five rays.

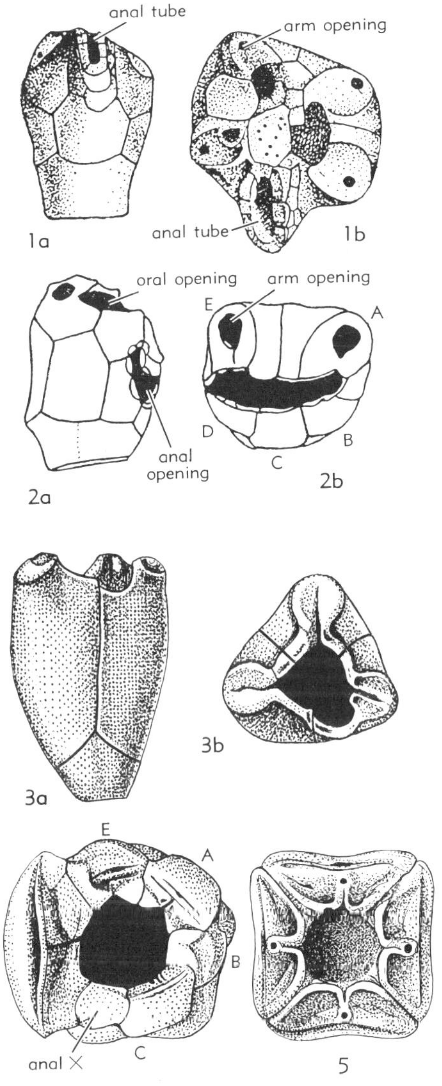

FIG. 110. Reduction of number of arms. [Explanation: rays identified by letter symbols of the CARPENTER system.]——*1. Tetrapleurocrinus eifelensis* WANNER, M.Dev., Ger.; four arm-bearing radials; *1a, CD*-interray view; *1b,* adoral view, ×2 (Wanner, 1943).——*2. Scoliocrinus eremita* JAEKEL, M.Dev., Ger.; five radials but only two arm-bearing; *2a,* lat. view; *2b,* adoral view, ×2 (Jaekel, 1895).——*3. Holynocrinus moorei* BOUŠKA, M.Dev., Boh.; three arm-bearing radials; *3a, CD*-interray view; *3b,* adoral view, ×20 (Bouška, 1948).——*4. Parindocrinus oyensi* WANNER, Perm., Timor; five radials, all of them arm-bearing except *B* radial; *D* radial hypertrophied, ×2.6 (Wanner, 1937).——*5. Tetracrinus moniliformis* (MÜNSTER), Jur., Eu.; only four radials, all of them arm-bearing, ×4 (de Loriol, 1882-84).

In many families, genera and species of crinoids, the number of free arms is fixed and therefore characteristic. But, in some cases, it varies, even within a species or a single specimen. This has been well shown by SPRINGER (1920), LANE (1963b), and others, who have demonstrated that the nonrecognition of this fact has led to the distinction of an illogical number of species in some genera. Figure 111 illustrates examples of such infraspecific variation, the coefficients of which may range from almost zero to seven or eight in different species. There is generally little consistent tendency for the number of arms to increase or decrease systematically through time within a genus. Within a species, the number of arms may be related to size of the specimen, or it may not; in the first instance, the increase in number of arms seems to be a growth phenomenon, in the second, it is either determined at an early ontogenetic stage or genetically controlled (LANE, 1963b). Within a specimen, the number of arms may vary from ray to ray; such variation may be at random, that is, without any definite order or system, or it may show a definite relation to the bilateral symmetry of the crinoid. For instance, in the camerate *Macrocrinus verneuilianus,* 12-armed specimens have two arms in each of the *A, B* and *E* rays, and three arms in each of the *C* and *D* rays; in 14-armed specimens, the arrangement is 2-3-3-P-3-3, and in 16-armed specimens it is 2-4-3-P-3-4 (P meaning posterior or *CD* interray).

ARM GROWTH

If the mode of growth of the arms in extant crinoids is rather well known (see p. *T*56, *T*134), it remains largely conjectural in fossil crinoids. It seems clear, however, that in most fossil crinoids the arms grew through addition of new brachials at their distal end and subsequent enlargement of the newly secreted brachials. Primary interpolation of new elements between those already formed seems to have occurred very rarely, if ever. WARN (1973) has recently described what looks like plate

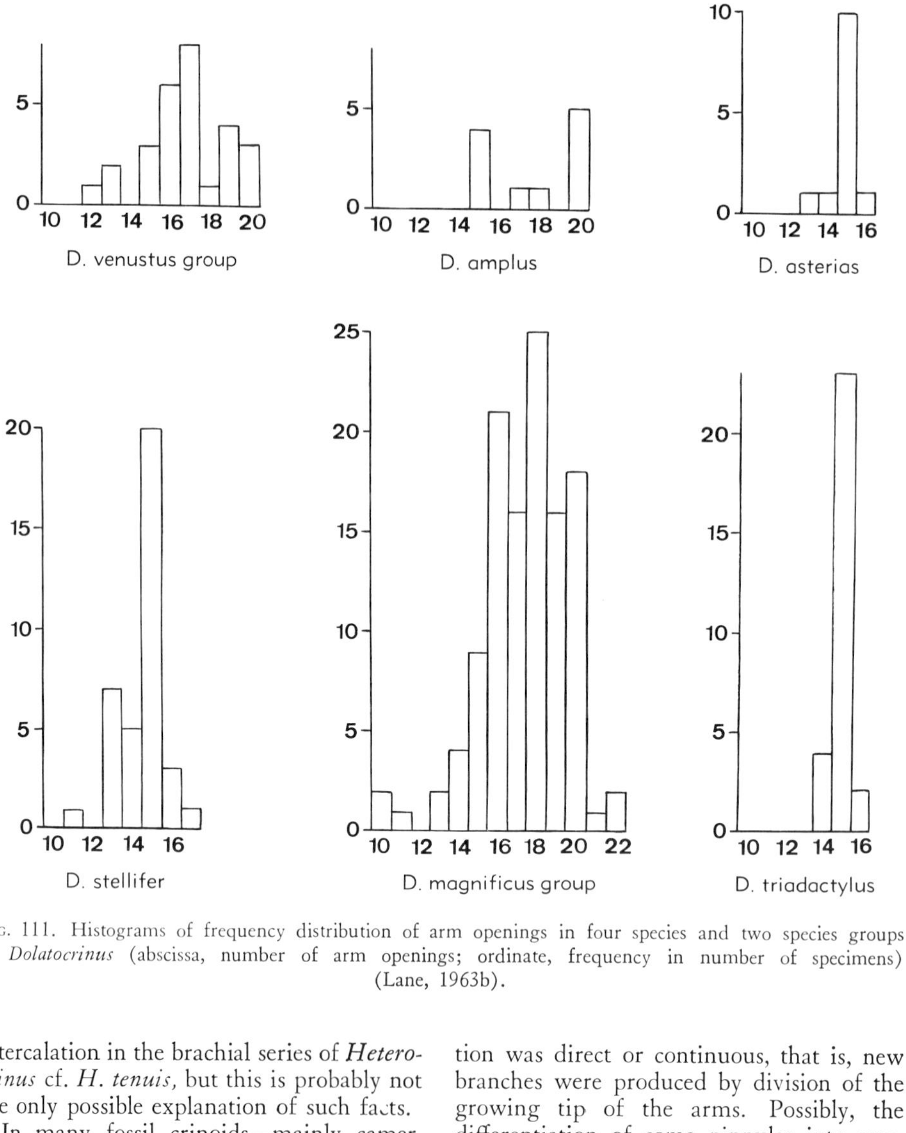

FIG. 111. Histograms of frequency distribution of arm openings in four species and two species groups of *Dolatocrinus* (abscissa, number of arm openings; ordinate, frequency in number of specimens) (Lane, 1963b).

intercalation in the brachial series of *Heterocrinus* cf. *H. tenuis*, but this is probably not the only possible explanation of such facts.

In many fossil crinoids—mainly camerates and flexibles—the growing tip of the arms is coiled upon itself (Fig. 112). Not infrequently its growth appears to have been slower than that of the adjoining branchlets or pinnules, so that it was more or less concealed and protected by these appendages.

Many fossil crinoids exhibit an increase in the number of arm branches per ray during growth. How these additional arm branches developed is not known. Presumably in most cases the mode of formation was direct or continuous, that is, new branches were produced by division of the growing tip of the arms. Possibly, the differentiation of some pinnules into armlets may also have occurred in some individuals; for instance, specimens of the camerate genera *Compsocrinus, Gaurocrinus*, and *Glyptocrinus* have been reported as having the first pinnule replaced by an arm. The existence of a discontinuous mode of formation as in living multibrachiate comatulids is another but yet unproved possibility; it would consist of the casting off of a series of distal arm plates by autotomy and their replacement by two branches.

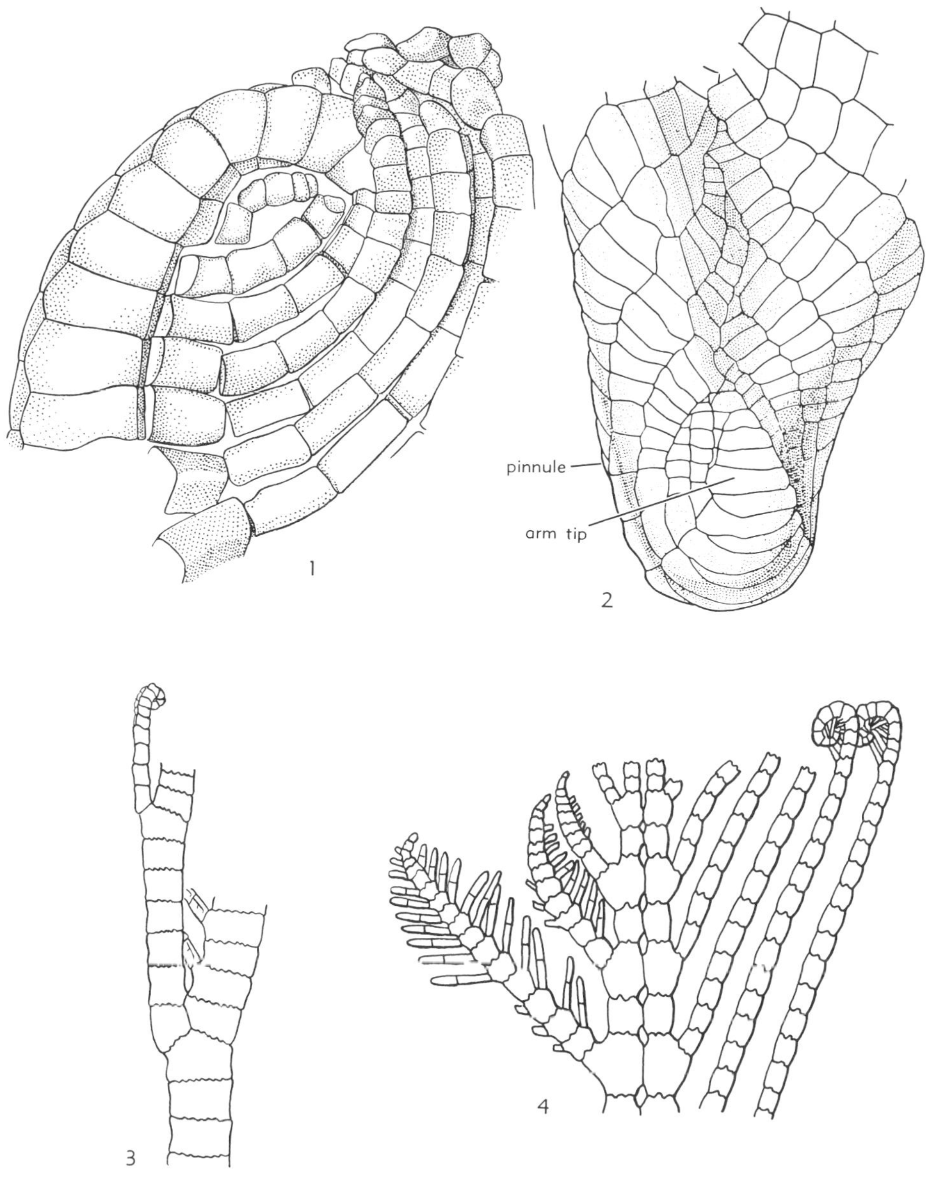

FIG. 112. Distal ends of arms in camerates——*1. Clonocrinus* sp., U.Sil., Gotl.; lat. view, coiled tip protected by pinnules, ×20 (Ubaghs, n).——*2. Barrandeocrinus sceptrum* ANGELIN, U.Sil., Gotl.; adoral view, coiled arm tip with interlocking pinnules forming pavement, ×15 (Ubaghs, 1956).——*3. Ctenocrinus gottlandicus* (PANDER), U.Sil., Gotl.; aboral view, coiled arm tip much smaller than adjacent ramules, ×10 (Ubaghs, 1958b).——*4. Melocrinites pulcher* (SPRIESTERSBACH), M.Dev., Ger., aboral view, main arm growing more slowly than adjacent ramules, enl. (Spriestersbach, 1919).

In most fossil crinoids, the arms, like those of recent forms, grew in an upward direction (away from the stem), and are commonly preserved close together (Fig. 113,*1*). This attitude is particularly well exemplified by the Flexibilia, which are usually found with the arms tightly coiled upon themselves over the tegmen (Fig.

FIG. 113. Various directions of growth in the arms. *(Explanation on facing page.)*

113,*2*). In some crinoids, however, the arms developed outward, spreading out more or less horizontally around the theca (Fig. 113,*3*), and in others they grew downward, assuming a permanent pendent or recumbent position during life (Fig. 113,*4-8*). This last arm type evolved independently in several camerate families (i.e., Acrocrinidae, Desmidocrinitidae, Patelliocrinidae, Platycrinitidae, and Rhodocrinitidae), but it is rare among inadunate crinoids (e.g., *Oklahomacrinus*) and unknown in flexible and articulate crinoids. This unusual arm habit is ordinarily associated with special structural features such as a projecting tegminal edge that gave a firm anchorage to the suspended arms, the downward orientation of the arm-facets, a particular shape and thickness of the calycinal plates supporting the arms, the fusion of proximal free brachials, and others.

Regeneration of arms or parts of arms occurred rather commonly among fossil crinoids. Many examples have been recorded, most of them of a reproductive type (Fig. 114,*1*). Augmentative regeneration seems to have occurred more rarely. In the example illustrated here (Fig. 114,*2*), the four inner arms of the *A* ray have been broken and the remaining portion of each of them has regenerated two arms, bringing to 10 the number of arms in this ray instead of six or seven as in the other rays.

ARM BRANCHING

The free arms of crinoids may be simple **(atomous)** or branched. Several types of arm-branching are known (Fig. 115). The simplest, and probably the most primitive, consists of the division of the arm into two branches **(dichotomy).** If the branches are equal or subequal, the division is called an **isotomy.** An arm affected by such bifurcation, constantly repeated in a regular manner, is said to be **regularly dichotomous** or **isotomous** (Fig. 115,*1*). The suppression of a bifurcation at definite points produces an **irregular dichotomy** or **heterotomy** (Fig. 115,*2*). In heterotomous arms the branches resulting from a bifurcation are commonly unequal in size. Such inequality leads to the formation of a main branch **(ramus),** which carries smaller branches (armlets or **ramuli**). The ramules may be borne on both sides of a main branch **(bilateral heterotomy)** (Fig. 115,*3*), or on one side only **(unilateral heterotomy),** and either on the inner (adradial) side **(endotomy)** (Fig. 115,*4*) or on the outer (abradial) side **(exotomy)** (Fig. 115,*5*). The ramules themselves may undergo a similar, but generally less elaborate, process of division. When they remain unbranched, and are regularly placed on alternate sides of successive brachials **(holotomy),** they are called **pinnules** (pl., **pinnulae**) and the arm is "pinnulate" (Fig. 115,*6*). This succession of stages, effectively traced in some fossil inadunate families, such as the Botryocrinidae and the Mastigocrinidae, suggests that the pinnules are the ultimate products of arm division. This interpretation is probably correct as far as it concerns most inadunates and articulates, but whether it is equally relevant for all pinnulate crinoids is still a matter of debate, as we shall see later.

A special type of arm structure, which seems to result from the hypertrophy and the very unequal heterotomous branching

1. *Abrotocrinus unicus* (Hall), Miss., USA(Ind.); complete crown in *A*-ray view, with arms erect, ×1 (Springer, 1926b).

2. *Pycnosaccus bucephalus* (Bather), M.Sil., Eng.; complete crown in *CD*-interray view with arms tightly coiled upon tegmen, ×1 (Springer, 1920).

3. *Petalocrinus mirabilis* Weller, M.Sil., Ia.; complete crown in aboral view with arms spreading out horizontally, ×2 (Springer, 1926a).

4. *Oklahomacrinus loeblichi* Moore, U.Penn., USA (Okla.); oblique view of crown, indicating pendent nature of arms, ×1.5 (Moore, 1939c).

5. ?*Macrostylocrinus recumbens* Springer, L.Dev., USA(Md.); crown with pendent arms partly removed, uncovering calyx and proxistele, *CD*-interray view, ×1 (Springer, 1926b).

6. *Barrandeocrinus sceptrum* Angelin, U.Sil., Gotl.; crown with recumbent arms completely enveloping the theca, ×1.5 (Ubaghs, 1956).

7. *Dichocrinus recurvibrachiatus* Van Sant, Miss., USA(Ind.); calyx with pendent arms and proxistele, lat. view, ×2 (Van Sant, in Van Sant & Lane, 1964).

8. *Amphoracrocrinus amphora* (Wachsmuth & Springer), Miss., USA(Ala.); calyx with downhanging arms and proxistele, ×1.5 (Wachsmuth & Springer, 1897).

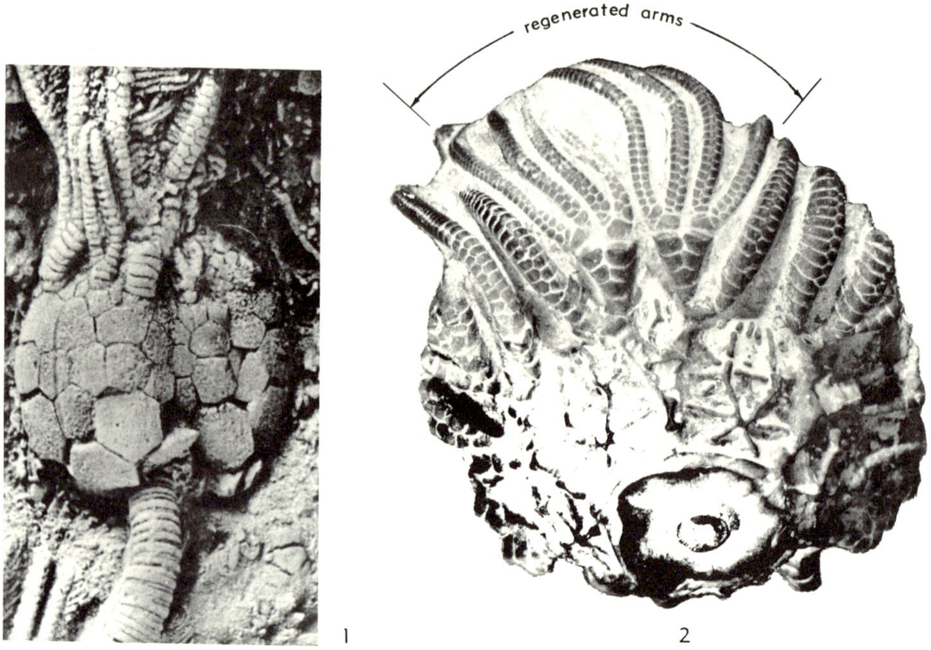

FIG. 114. Regeneration of arms.——*1. Rhodocrinites watersianus* (WACHSMUTH & SPRINGER), Miss., USA(Ia.); crown with two regenerated arms (reproductive regeneration), ×2.5 (Strimple & Beane, 1966).——*2. Gennaeocrinus mourantae* GOLDRING, M.Dev., Ont.; U.S. Natl. Museum S. 4533, oblique view of a crown that has regenerated 8 arms (instead of 4) in *A* ray (augmentative regeneration), ×2.4 (Ubaghs, n).

of the main arms, has been repeatedly produced among camerates. It consists of the occurrence of one or two powerful rami in each ray. These "arm trunks," as they are called, carry relatively slender ramuli, that are pinnulate, whereas the trunks usually (if not invariably) lack pinnules (Fig. 116). In advanced melocrinitids, the two arm trunks of each ray coalesce along their adradial margin—a phylogenetical process repeated in ontogeny that produces a compound ramule-bearing appendage (Fig. 117).

The pattern of the arm branching may be constant and diagnostic for a family or a genus, or it may vary. For instance, among dichocrinid camerates, the arms of *Dichocrinus* are isotomous, while those of *Paradichocrinus* are exotomous. In some forms different ray structures occur in a single individual. Thus, some calceocrinid inadunates have isotomous arms in the *E* ray and strongly heterotomous arms in the *A* and *D* rays.

AXILLARIES

In a branching arm, each brachial that supports two branches is an **axillary** (Fig. 118,*1*). An axillary, therefore, has two upper (distal) sloping shoulders or articular faces (facets), each notched by the adoral groove and pierced by the axial canal (if any), for the groove and the axial canal branch with the arm. The distal facets of an axillary may be equal or unequal, according to the relative sizes of the branches produced by a bifurcation. If the inequality is very important, as for instance when an axillary gives rise to an unbranched ramule or a pinnule on one facet and to a continuation of the arm, of which it is part, on the other, the brachial on which such unequal division takes place may be conveniently designated as a **subaxil** (Fig. 118,*2*). A pinnulate arm is made up of a succession of subaxils. It must be clearly understood, however, that a subaxil, as

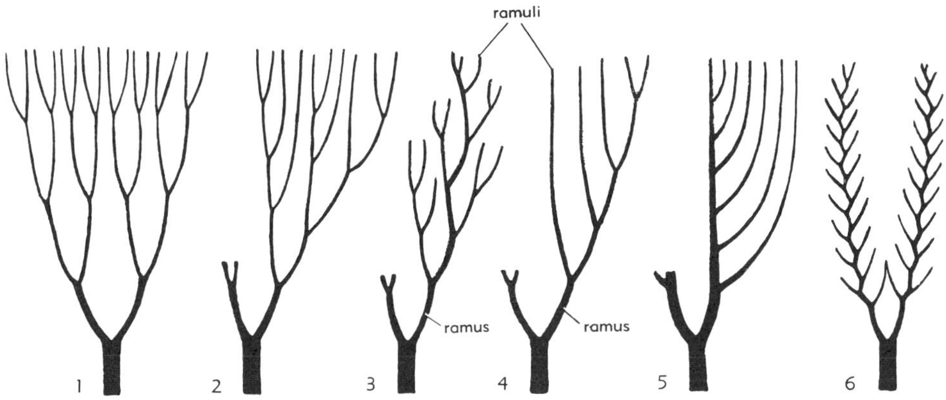

FIG. 115. Various types of arm branching.——*1,* isotomy; *2,* heterotomy; *3,* bilateral heterotomy; *4,* endotomy; *5,* exotomy; *6,* holotomy (mod. from Bather, 1900a).

here interpreted, is morphologically equivalent to an axillary. A pinnulate arm or a ramuli-bearing arm may bifurcate as a simple arm, though generally not to the same extent. The axillaries on which the rami or main branches of such pinnulate or ramule-bearing arms fork are similar to ordinary axillaries. In order to distinguish them from subaxils, they may be referred to as **main axils** (Fig. 118,*2,3*).

In a number of mainly Pennsylvanian and Permian cladid inadunates, the axillary plates are conspicuous; they are considerably larger than the other brachials and commonly bear nodes or spines (Fig. 119,*1*). Furthermore, in a number of species, the five axillaries corresponding to the first bifurcation of each arm are unequal; generally, the longest belongs to the *A* ray, and the shortest to the *B* ray (WANNER, 1949a) (Fig. 119,*2,3*).

As a rule, axillaries occur in all crinoids in which the arms bifurcate; however, in camerates with arms composed of two alternating series of brachs the axillaries are generally lacking, and the divisions result from the mere splitting of the rows of brachs immediately below each point of bifurcation (Fig. 120,*1*). On the contrary, in inadunates with similarly built arms, distinct axillaries are invariably present (Fig. 120,*2*). Another difference lies in the fact that as a rule in inadunates and articulates the axillary plates do not give off a pinnule, whereas in camerates axillary-like ossicles commonly bear such appendages (Fig. 120,*3,4*).

BRACHITAXIS

The terminology of the series of brachials that compose an arm has already been partially discussed under the treatment of fixed brachials in the calyx (p. *T*118). The general term **brachitaxis** is used for designating any undivided series of brachs that terminates in an axillary plate or remains undivided up to its distal extremity (Fig. 120,*2*). The plates of the proximal brachitaxis up to and including the first axillary or **primaxil** (if any) are termed **primibrachs** (also **primibrachials** or **primibrachialia**), those of the following brachitaxis with a **secundaxil** (if any) are **secundibrachs,** then succeed **tertibrachs, quartibrachs,** and so on. Numbering of brachials in any brachitaxis starts from the most proximal ossicle. Brachials of the distal rami that do not fork again are called **finials.**

Ordinarily such terminology is easily applied to nonpinnulate arms. In a pinnulate arm that branches, the primibrachs as a rule do not bear pinnules, and are therefore homologous with the primibrachs of a nonpinnulate arm. The pinnules generally begin with the next series. Of the brachs composing this series only the proximal one is strictly homologous with the secundibrach of a nonpinnulate arm with regard to the pinnule borne by it, and the next brachial

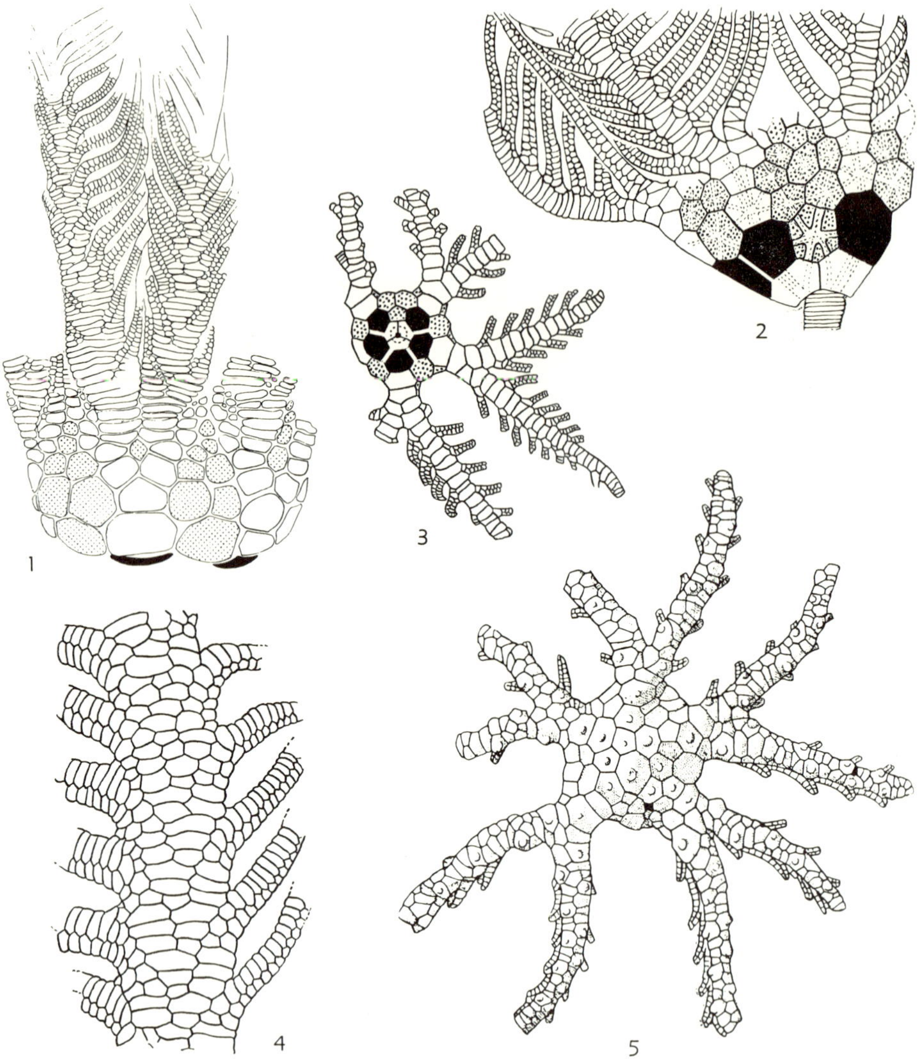

FIG. 116. Arm trunks in camerates.——*1. Rhipidocrinus* sp. cf. *R. perloricatus* W. E. SCHMIDT, M.Dev., Ger.; calyx and arm trunks of *E* ray, *ca.* ×*1* (Breimer, 1960).——*2. Cytidocrinus sculptus* (HALL), Miss., N.Am.; calyx and proximal portions of arm trunks, in *CD*-interray view, ×1 (Kirk, 1943a).——*3. Steganocrinus pentagonus* (HALL), Miss., N.Am.; calyx and parts of arm trunks, aboral view, ×1 (Kirk, 1943a).——*4. Trybliocrinus flatheanus* GEINITZ, L.Dev., Spain; part of an arm trunk, ×3.3 (Breimer, 1962).——*5. Eucladocrinus millebrachiatus immaturus* WACHSMUTH & SPRINGER, Miss., USA(Ia.); adoral side of theca and arm trunks, ×1 (Wachsmuth & Springer, 1897). [Explanation: interbrachials and anals stippled; radials black.]

represents the tertibrachs of a nonpinnulate arm. This makes extremely difficult the task of formulating a consistent terminology for pinnulate and nonpinnulate arms. An attempt made in that direction by BATHER (1892a) never gained acceptance. It is more practical, though unquestionably incorrect (if the pinnules are really the ultimate product of arm division), to consider the pinnule-bearing brachials of an arm not as markers of successive arm-divisions but as equivalent to nonpinnulate components.

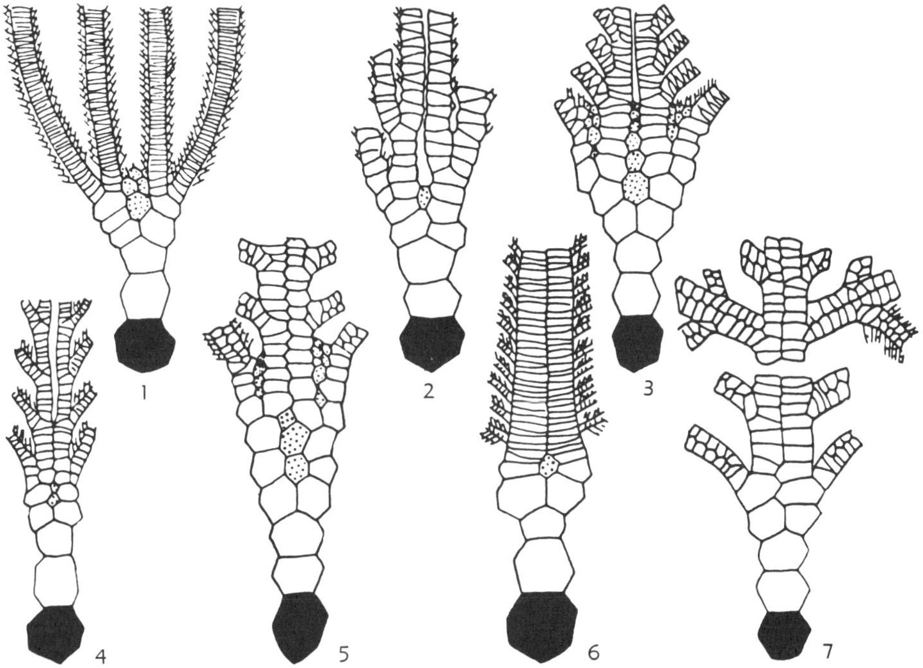

FIG. 117. Evolution of ray structure in Melocrinitidae, representation diagrammatic and not to scale (Ubaghs, 1953). [Explanation: interbrachials stippled; radials black.]——*1. Alisocrinus,* U.Ord.-U.Sil., N.Am.——*2. Promelocrinus fulminatus* (ANGELIN), U.Sil., Gotl.——*3. Promelocrinus anglicus* JAEKEL, M.Sil., Eng.——*4. Ctenocrinus gottlandicus* (PANDER), U.Sil., Gotl.——*5. Ctenocrinus nobilissimus* (HALL), L.Dev., USA(N.Y.).——*6. Melocrinites splendens* (GOLDRING), U.Dev., USA(N.Y.).——*7. Trichotocrinus harrisi* OLSSON, U.Dev., USA(N.Y.). (From *Traité de Paléontologie,* v. 3, copyright 1953, courtesy Masson & Cie, Paris.)

A similar procedure may be conveniently applied to arms that bear numerous unbranched ramules on every two or three brachs.

The complexity of arm branching in some crinoids is so great that a special nomenclature may be needed to designate any particular division in a clear manner. An example is furnished by the Calceocrinidae—a highly specialized disparid family characterized by the recumbent position of the crown along the stem and the consequent progressive acquisition of a bilateral symmetry (Fig. 87,*2,* p. *T*114). Each lateral (*A* and *D*) ray of most genera comprises a proximal ramus that is reduced to a series of axillaries (main axils), each one of which carries an arm, called an **axil-arm** (Fig. 121). The first (most proximal) axil-arm is the **primaxil-arm,** the next one, the **secundaxil-arm,** and so on. Each axil-arm forks at least once and generally several times. In any axil-arm the brachials of the proximal series are known as **alphabrachs,** those of the next series as **betabrachs,** of the third as **gammabrachs,** and so on. The axillary alphabrach of each axil-arm carries an unbranched ramule **(alpha-ramule)** on its abanal shoulder and a betabrach series on the other. In its turn the axillary betabrach bears a **beta-ramule** on its adanal facet and, on the other side, a gammabrach series that gives off a **gamma-ramule** on its abanal facet, and so on. This heterotomous pattern of branching extends to the distal end of the main ramus, the last axillary of which bears a branching axil-arm on its abanal side and an unbranched ramule (called **omega-ramule**) on its adanal side. This nomenclature has been devised by

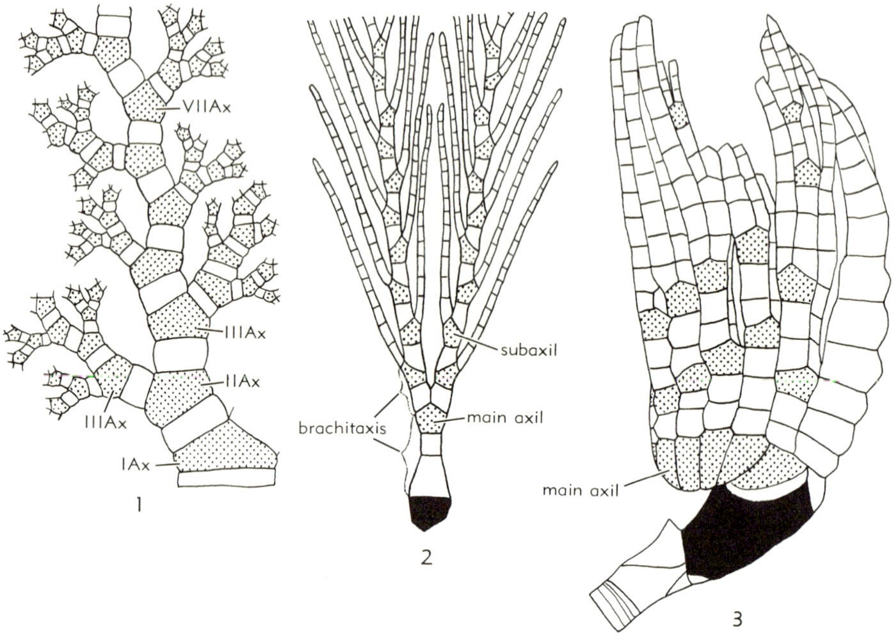

FIG. 118. Diagrammatic and not to scale representations of axillary brachials. [Explanation: *I Ax, II Ax, III Ax . . .* , axillaries of successive orders; radials black; axillaries stippled.]——*1. Barycrinus asteriscus* VAN SANT, Miss., USA(Ind.) (Van Sant in Van Sant & Lane, 1964).——*2. Logocrinus brandoni* SIGLER, WHITE, & KESLING, M.Dev., USA(Ohio); ×1 (Sigler, White, & Kesling, 1971).——*3. Halysiocrinus nodosus* (HALL), Miss., USA(Ind.) (drawn from Springer, 1926a).

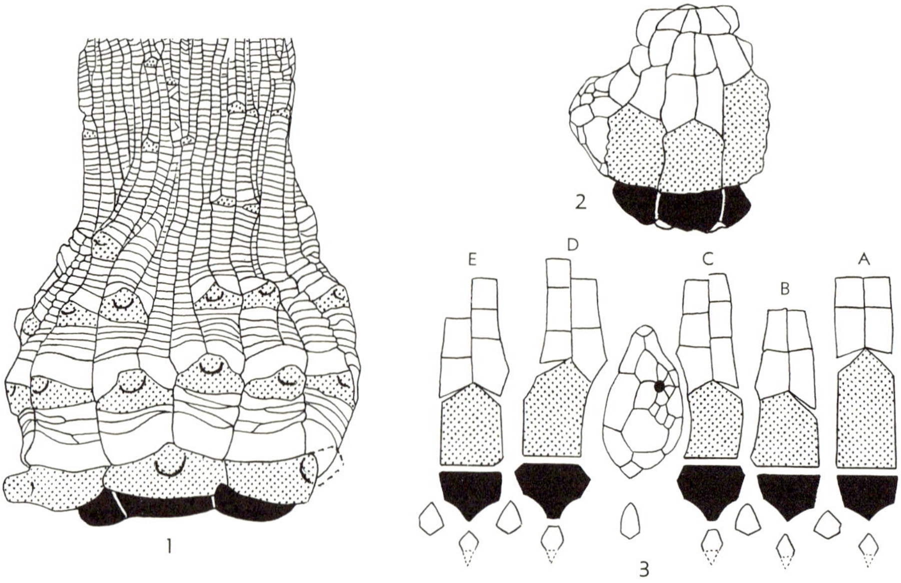

FIG. 119. Morphology of axillaries (axillaries stippled; radials black).——*1. Eirmocrinus grossus* STRIMPLE & WATKINS, Penn., USA(Texas); a species with conspicuous spiniferous axillaries, in *B*-ray

BATHER (1893a) and expanded by MOORE (1962a) (Fig. 121).

The number of primibrachs is an important taxonomic feature. Most camerate crinoids have two primibrachs per ray but some genera have three, four, or even five in each ray, whereas others have only a single one or even none. This reduction in the number of primibrachs in camerates has a quite different taxonomic value according to the group considered. In some families, it characterizes the whole assemblage; in others, some genera only. It may have a purely specific significance or be just a matter of individual variation; it may even be restricted to a ray or two in a single specimen. The reduction in the number of primibrachs in camerates does not seem to result, as sometimes suggested, from the ankylosis of the component ossicles, but mainly from the atrophy of the first one, more rarely of the second one (primaxil), and in some cases of both (Fig. 84,7, p. *T*110).

In flexible crinoids, all Ordovician and Silurian members (except *Meristocrinus*), most Devonian, and some Carboniferous genera have two primibrachs per ray, but from the Carboniferous onward three becomes the predominant number.

Among inadunate crinoids, the number of primibrachs per ray varies to a much larger extent than in other subclasses. In forms having branched arms, it may be as high as 12 or as low as one (possibly in some cases as a consequence of fusion of two elements). In Devonian and Mississippian genera, it is commonly two, but most Pennsylvanian and Permian inadunates have only one primibrach in each ray.

In articulate crinoids, the first brachitaxis commonly contains two primibrachs, but a much higher number occurs in several genera.

ARRANGEMENT OF BRACHIALS

Arms, arm branches, and pinnulae in which the component ossicles are arranged

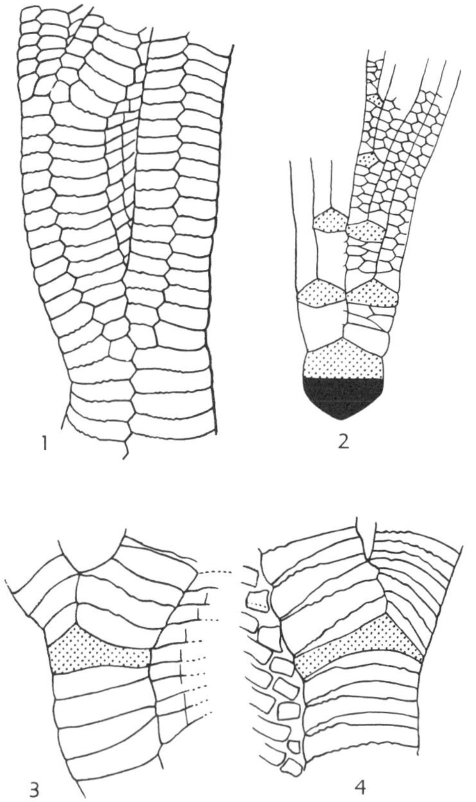

FIG. 120. Bifurcation of arms in camerates *(1, 3, 4)* and inadunates *(2)* (axillaries stippled; radial black).——*1. Abacocrinus* sp., U.Sil., Gotl.; no distinct axillaries, ×5 (Ubaghs, n).——*2. Hydreionocrinus goniodactylus* (DE KONINCK & WOOD), U.Carb., Eng.; conspicuous axillaries present, ×1 (Bather, 1911-12).——*3,4. Scyphocrinites* sp., U. Sil., Alg.; pinnule-bearing axillaries, ×7 (Ubaghs, n).

in a single series are said to be **uniserial.** Uniramous brachia and pinnulae are invariably uniserial. Their component elements have more or less parallel articular facets. In pinnulate arms the brachials tend to slope alternately to the right and left and thence to become wedge shaped (Fig. 122). Eventually, the brachials come to lie in two alternating rows with their pointed lateral sides meeting midway along a zig-

(Continued from facing page.)

view, ×1 (Strimple & Watkins, 1969).——*2,3. Prolobocrinus permicus* WANNER, Perm., Timor; a species with unequal primaxils; *2, B*-ray view, ×2; *3,* plate diagram of cup and proximal parts of arms (Wanner, 1943).

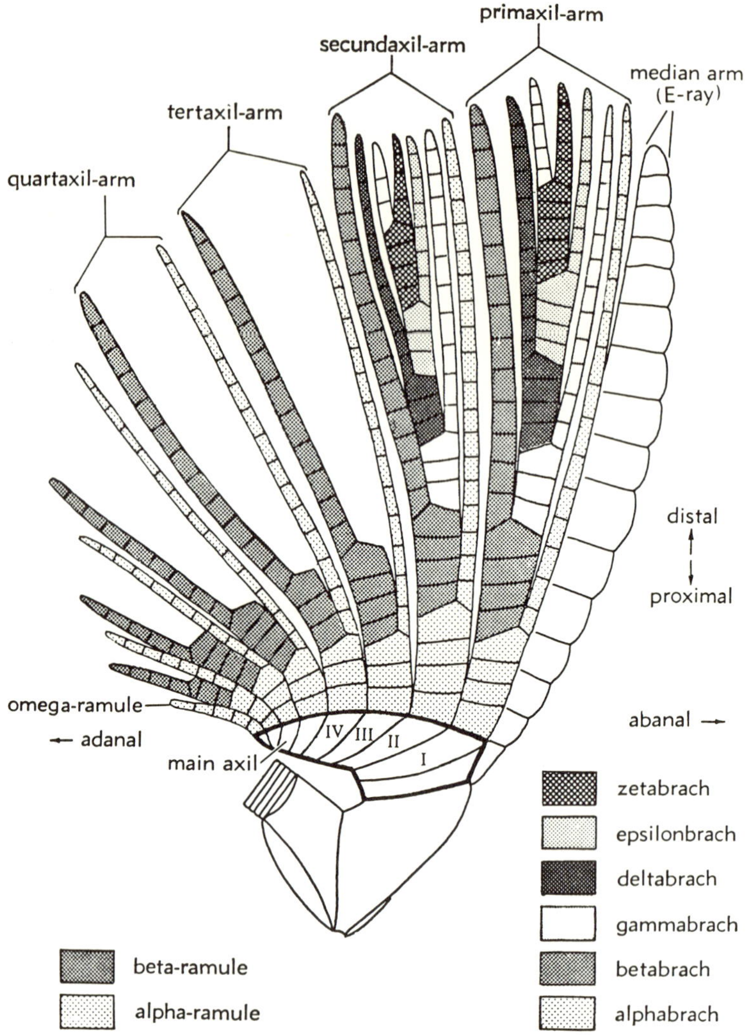

FIG. 121. Nomenclature applied to arm branching in lateral (*A* and *D*) rays of Calceocrinidae (Moore, 1962a).

zag suture line. Arms with such arrangement of the brachials are called **biserial.**

A ray is never entirely biserial. Its proximal portion is invariably uniserial, and the change, ordinarily progressive, into a biserial arrangement takes place at varying distances above the radials or the primaxils (Fig. 123,*3*). Also the few available data on the growing tips of biserial arms suggest that their short distal brachials preserve a uniserial arrangement (Fig. 123,*3*). The change from uniserial to biserial arrangement of the brachials may be advantageous in at least one respect; as it approximately doubles the number of pinnules in a given length, it serves to increase the food-gathering and other functions accomplished by the brachial system.

The above interpretation of the origin of biserial arms has gained almost general acceptance. It implies that the brachia of crinoids were primitively uniserial. This

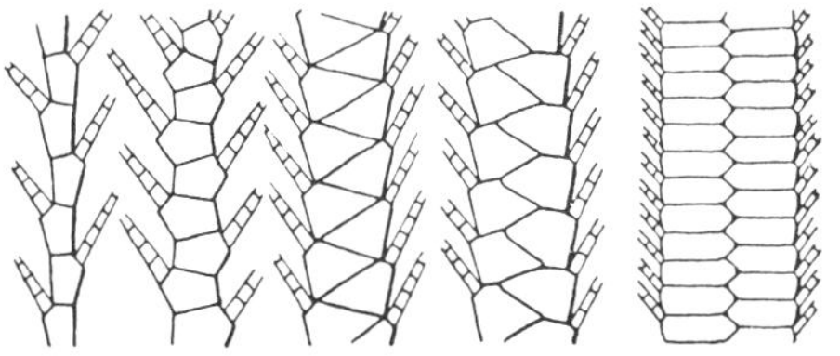

FIG. 122. Evolution from uniserial (left) to biserial arrangement of brachials, diagram. (Bather, 1900a).

conclusion is supported by the following evidence: 1) Most Ordovician crinoids have uniserial arms, and the number of species with biserial arms, both in camerates and in inadunates (the arms of flexibles are invariably uniserial), increased progressively during Paleozoic times. 2) Study of the phylogeny of some lineages, such as the melocrinitid camerates, shows that the uniserial condition precedes the biserial one. 3) Study of the ontogeny of

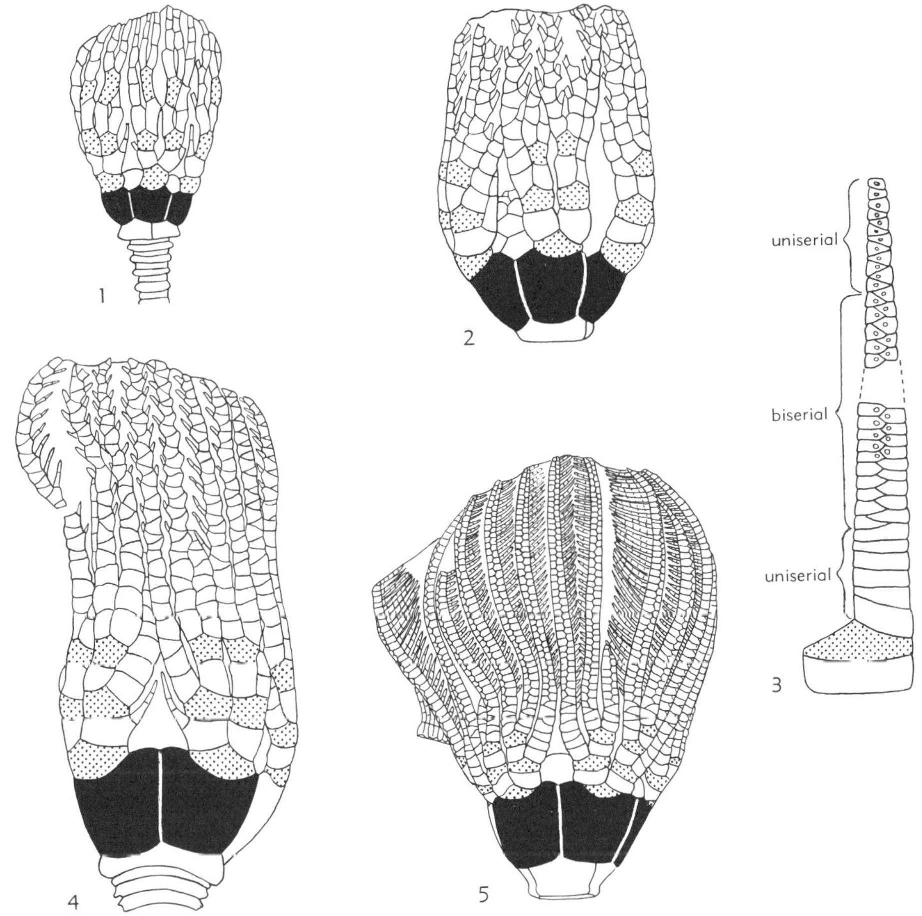

FIG. 123. Progressive change from uniserial to biserial arrangement of brachials in a single arm *(3)* and in ontogeny *(1,2,4,5)* (axillaries stippled; radials black).——*1,2,4,5.* Growth stages in *Platycrinites bozemanensis* (MILLER & GURLEY), Miss., N.Am.; *1,* very young specimen with elongate uniserial brachials, *ca.* ×8; *2,* young specimen with brachials becoming cuneiform distally, *ca.* ×7; *4,* somewhat older specimen with strongly cuneiform brachials in distal parts of arms, *ca.* ×7; *5,* adult specimen, with compactly biserial brachials, *ca.* ×3 (Laudon, 1967).——*3. Encrinus liliiformis* MILLER, Trias., Eu.; diagram. (Grabau, 1903).

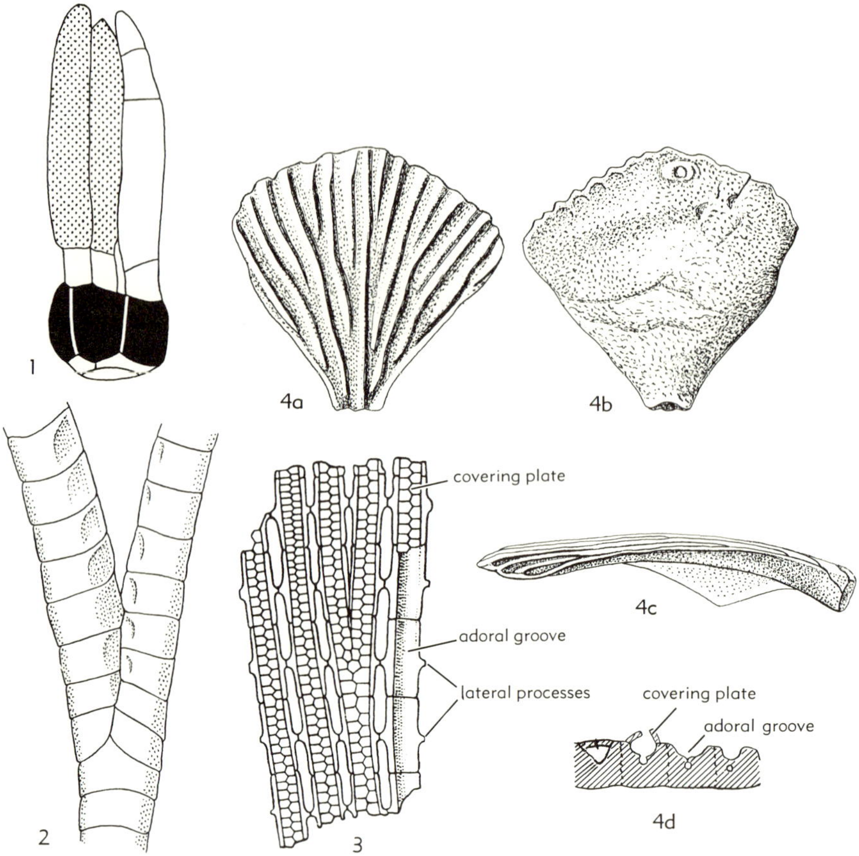

FIG. 124. Fusion of brachials.

1. *Isoallagecrinus strimplei* (KIRK), Penn., USA (Okla.); complete crown; with very long second brachial (stippled) in two arms, each such elongated ossicle possibly resulting from fusion of several elements, still distinct in other arms, ×4 (Strimple, 1972).
2. *Aethocrinus moorei* UBAGHS, L.Ord., France; part of a free arm showing partial lateral union of two branches, aboral side, ×6 (Ubaghs, 1969).
3. *Crotalocrinites rugosus* (MILLER), U.Sil., Eu.; adoral view of portion of arms united by lateral processes at repeated intervals, *ca.* ×4 (Wachsmuth & Springer, 1889a).
4. *Petalocrinus visbycensis* BATHER, U.Sil., Gotl.; *4a-d,* one arm-fan resulting from complete lateral fusion of arm branches, in adoral *(4a),* aboral *(4b),* and lat. *(4c)* views, ×3 (Bather, 1898); *4d,* sec. across four grooves of an arm-fan, ×5 (Bather, 1900a).

Paleozoic species with biserial arms reveals that an early uniserial stage develops before the biserial growth pattern is acquired (Fig. 123, *1,2,4,5*). 4) In full-grown individuals, the new brachials introduced at the growing tip of the arms appear to have generally, if not invariably, a uniserial arrangement (Fig. 123,*3*). All this suggests that the evolutionary sequence among crinoids is from a uniserial to a biserial condition. Or, in other words, that uniseriality is a primary—not a secondary—feature of the crinoid brachium.

It is a rule that in the pinnulate part of a ray the pinnules are borne on alternate sides of each brachial. However, in a number of camerate and inadunate species two or several pinnules may be inserted on a single brach. It has been generally accepted—though perhaps never clearly demonstrated

—that such **hyperpinnulation** results from the fusion of two or several adjacent right and left ossicles in a biserial arm, forming compound brachials. Thus, an arm comprising compound brachials, though uniserial in appearance, might be equivalent to a biserial one in reality. According to another explanation, the occurrence of two or more pinnules on a brach could have been produced through hypertrophy of this brach and development of one or several supplementary pinnules (W. E. Schmidt, 1942). This last interpretation implies, of course, a different origin for the pinnules than would result from the division of the arm on each brachial.

Fusion of brachials into large units certainly occur among crinoids. For instance, the brachials of the Devonian inadunate *Cupressocrinites* are reported to be composed each of several thin, discoid elements (Sieverts, 1934); those of the Pennsylvanian inadunate *Exocrinus* are commonly united by ankylosis; the allagecrinid *Isoallagecrinus strimplei* (Kirk) has in some rays a very long second primibrach, which may result from the fusion of two or three ossicles, still distinct in other rays (Fig. 124,*1*); the pendent arms of some Mississippian species of the camerate genus *Dichocrinus* are supported by coalesced proximal free brachials (Fig. 113,*7*, p. *T*142). Brachials belonging to adjacent arms may also fuse laterally. This happens commonly in Paleozoic crinoids when the two branches resulting from a bifurcation do not separate immediately (Fig. 124,*2*). In advanced Melocrinitidae, the main rami of each ray fuse laterally and give rise to powerful arm trunks (Fig. 117, p. *T*147). Still more remarkable appears to be the lateral union of the arms in the Crotalocrinitidae and Petalocrinidae. In these inadunates there is a marked tendency of the arm branches to be united by lateral processes from each brachial and to form a network (Fig. 124,*3*) or, in case of complete lateral union, compact blades (Fig. 124,*4a-d*).

PINNULES

Pinnules generally appear as slender, unbifurcated, appendages of the arms typically borne on alternate sides of successive brachials. Their aboral skeleton consists of uniserially arranged ossicles called **pinnulars.** So far as known, highly movable articulations generally connect brachials and first pinnulars, even in crinoids that possess only ligamentary articulations between brachials. A groove (ventral or adoral groove) is typically present on the adoral side of each pinnule; in extant crinoids it contains an ambulacral groove and extensions of water-vascular, nervous and coelomic systems, so that the pinnules have a constitution similar to that of the arms.

The basic rule of the arrangement of pinnules along the arms or **pinnulation** is that they are borne on alternate sides of each successive brachial. Many exceptions to this rule are known. 1) One or two newly added brachials at tip of arms lack pinnules. 2) In a number of camerates and inadunates more than one pinnule may be attached to a single brach (hyperpinnulation, see above). 3) In pinnulate branched arms, when there are no more than two elements in a brachitaxis, these elements generally (but not invariably) lack pinnules. 4) Axillary brachials have no pinnules, though in camerates brachs that have the appearance and function of axillary plates may carry pinnules (Fig. 120,*3,4*). 5) In articulate crinoids, pairs or triplets of brachs united by ligamentary (instead of muscular) articulations give rise to a single pinnule from their distal component, and never from their proximal or intermediate elements (see section on recent crinoids, p. *T*41). 6) At base of the arms (and commonly of the main branches) of many camerates, pairs or triplets of brachs occur that carry a pinnule on their distal ossicle only (Fig. 125); in spite of similarity with the preceding case, there is no evidence that such lack of pinnules in camerates is connected with a particular type of brachial articulation.

Pinnules may also be absent in places when their presence would be normally expected, a feature known as **defective pinnulation** and observed in many recent crinoids (see section on recent crinoids, p. *T*41). Defective pinnulation may also occur in fossil crinoids. For instance, in a number of poteriocrinitine genera, the proximal pinnules were suppressed, possibly as a consequence of the bulging of the tegminal pavement between the arms.

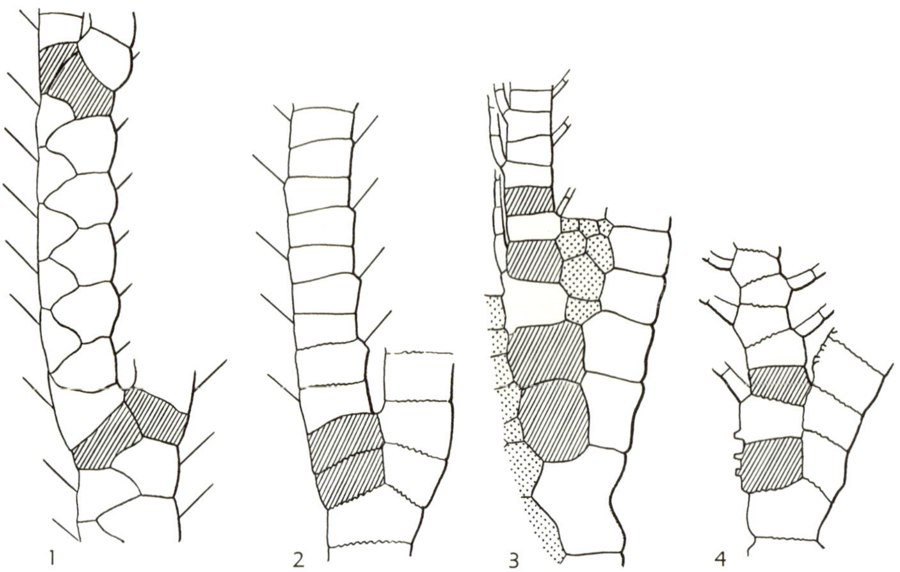

FIG. 125. Brachial units in camerates composed of two or three ossicles with the distal one alone provided with a pinnule (mod. from Ubaghs, 1958b). [Explanation: interbrachials stippled; single or double proximal ossicles without pinnule obliquely ruled. In fig. *1* and *2*, places of pinnules indicated by short oblique lines.]——*1. Paramelocrinus angelini* UBAGHS, U.Sil., Gotl., ×10.——*2,3. Promelocrinus radiatus* (ANGELIN), U.Sil., Gotl., ×10.——*4. Ctenocrinus gottlandicus* (PANDER), U.Sil., Gotl., ×10.

Among camerates, pinnules are generally (if not invariably) lacking on these hypertrophied rami called arm trunks; they fail also to develop on the inner (adradial) rami of the melocrinitid *Promelocrinus,* and on the widened, flattened, paddle-shaped distal portion of the arms of *Eretmocrinus* and allied genera (Fig. 126). These examples show that defective pinnulation may affect various parts of the arms and are probably caused by different inhibiting factors.

The pinnules may be subcylindrical, laterally flattened, or prismatic, with cross sections respectively subcircular, elliptical, or U-shaped (rarely subtriangular). Their ornamentation, extremely diversified, consists of spines, nodes, longitudinal median keel, comblike structures, and other features. Generally, the two lateral faces of a pinnule are similar, so that the transverse section is symmetrical. But it may also be asymmetrical, particularly in species provided with strongly imbricating pinnules. In such cases, it may be useful in descriptions to recognize a lateral **abthecal** side directed away from the theca and an opposite lateral **adthecal** side directed toward the theca.

The pinnules of some crinoids are set more or less wide apart, but usually, particularly in biserial arms, they are closely spaced. In some camerates, they are imbricate and provided with lateral and (or) aboral expansions or spines that overlap adjoining pinnules and served to compel these organs to move together (Fig. 127). Pinnules are generally inserted obliquely on the arms and directed upward and outward, but they may also be set at right angles to the arms and interfinger with pinnules of adjacent arms. In some fossil crinoids, as well as in some recent forms, they could be folded inward and laid back along the arms so that they formed a protecting covering over the soft brachial structures. The pinnules of each row of the Silurian camerate genus *Barrandeocrinus,* characterized by pendent arms, were attached together, their pinnulars interlocking and building an outer, probably rigid, pavement around the crown; water was introduced into this almost entirely

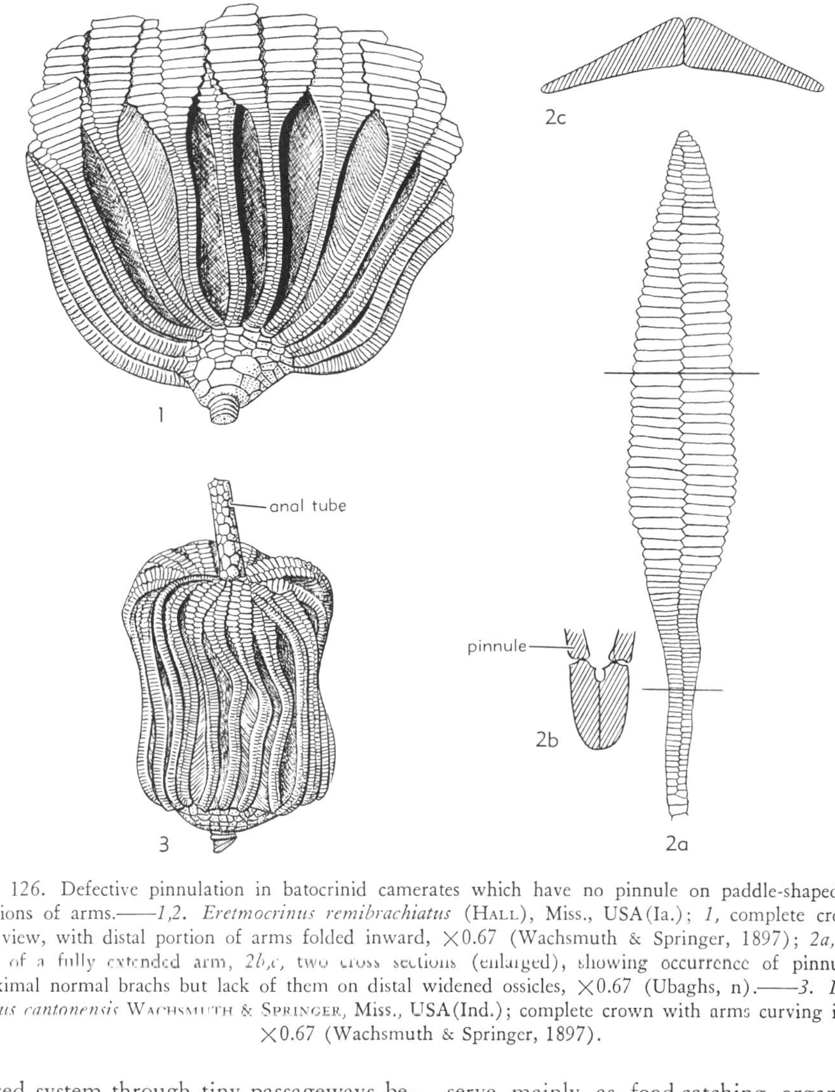

FIG. 126. Defective pinnulation in batocrinid camerates which have no pinnule on paddle-shaped distal portions of arms.——*1,2. Eretmocrinus remibrachiatus* (HALL), Miss., USA(Ia.); *1,* complete crown in lat. view, with distal portion of arms folded inward, ×0.67 (Wachsmuth & Springer, 1897); *2a,* aboral side of a fully extended arm, *2b,c,* two cross sections (enlarged), showing occurrence of pinnules on proximal normal brachs but lack of them on distal widened ossicles, ×0.67 (Ubaghs, n).——*3. Dizygocrinus cantonensis* WACHSMUTH & SPRINGER, Miss., USA(Ind.); complete crown with arms curving inward, ×0.67 (Wachsmuth & Springer, 1897).

closed system through tiny passageways between the distal portions (which remained free) of the pinnules, and expelled through a common aperture located at the apex of the crown (Fig. 128).

In recent crinoids, there are commonly very marked differences in the aspect of the pinnules of different parts of the arms. In comatulids especially, one or several proximal pinnules (oral pinnules) are modified and lack an ambulacral groove; they are followed by several pairs of genital pinnules that bear the gonads and have a rudimentary ambulacral groove, while the distal pinnules of the remaining part of each arm serve mainly as food-catching organs. In most fossil crinoids, such diversification of the pinnules along the arms does not occur, and all the pinnules are fundamentally alike. In some species, however, the proximal pinnules differ somewhat from the others. This may result from their total or partial incorporation in the calyx, the fixed pinnulars (see above, p. *T*153) tending to approach the adjacent cup plates in appearance and probably in function, or from specific modification of the first pinnule of each ray, as illustrated by some species of the camerate genera *Carpocrinus* (Fig. 129, *1,2*) and *Batocrinus* (Fig. 129,*3*). In at

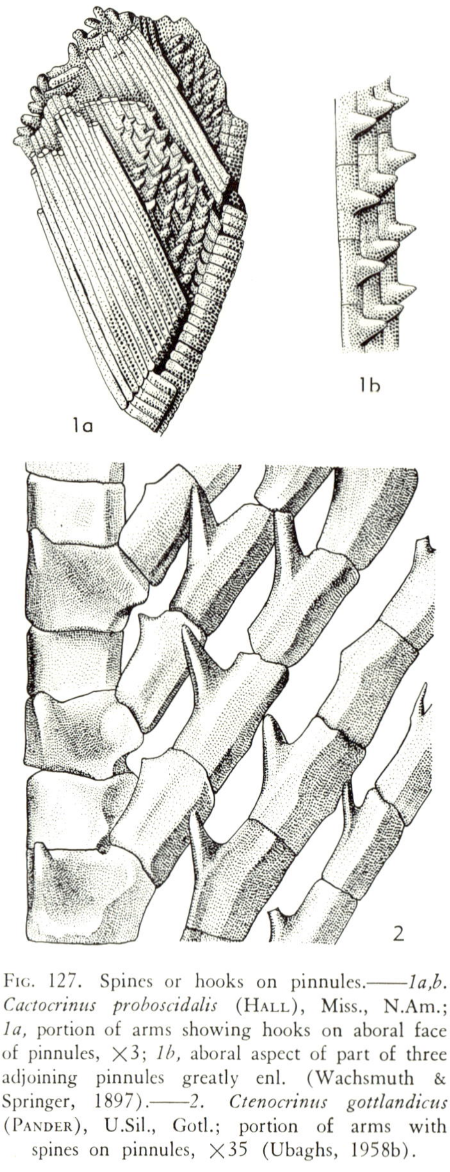

FIG. 127. Spines or hooks on pinnules.——*1a,b. Cactocrinus proboscidalis* (HALL), Miss., N.Am.; *1a,* portion of arms showing hooks on aboral face of pinnules, ×3; *1b,* aboral aspect of part of three adjoining pinnules greatly enl. (Wachsmuth & Springer, 1897).——*2. Ctenocrinus gottlandicus* (PANDER), U.Sil., Gotl.; portion of arms with spines on pinnules, ×35 (Ubaghs, 1958b).

least one species of the latter genus *(B. icosidactylus),* the two proximal ossicles of the first pinnule were considerably enlarged and served as a covering flap to the aperture through which the ambulacral groove and associated structures of the pinnule penetrated into the theca (Fig. 129,*3*). According to SPRINGER (1917), the huge tubular appendages of the rhodocrinitid camerate genus *Gilbertsocrinus* (Fig. 129,*4*) represented enormously enlarged and modified proximal pinnules—an opinion not accepted unreservedly (VAN SANT, in VAN SANT & LANE, 1964). As to the existence of genital pinnules in fossil crinoids, no unquestionable indication of it has been discovered as yet. JAEKEL (1895) thought that the enlarged distal portion of the pinnules of the Devonian camerate *Acanthocrinus rex* lodged the gonads, but neither the shape, structure, nor position of the ossicles composing this region made them appropriate to serve such a function. The proximal pinnules of another Devonian camerate *Hapalocrinus elegans* may show an undivided, variously wide, lateral extension that has been regarded by JAEKEL (1895) as a gonad receptacle, but this interpretation was rejected by W. E. SCHMIDT (1934).

In any pinnule, the pinnulars are never exactly similar. Ordinarily, the proximal pinnular is shorter and wider than the succeeding ones, and its proximal face differentiated into an articulation that serves to move the whole pinnule. In some crinoids, particularly the recent ones, it is the two first pinnulars that are specialized in this way and accomplish this function. The next pinnulars generally do not differ from one another except in size and proportions along the pinnule, which usually tapers gradually from the base to the tip. Distal pinnulars may be similar to the preceding ones (Fig. 130) or, relative to fulfillment of some special function, very different and specialized. Such is the case of the distal pinnulars of the oral pinnules of the comasterids. These are provided with toothlike projections that collectively form a terminal comb whose function is probably to aid fixation. Another example is furnished by various camerates (Fig. 131): their distal pinnulars are short, wide, and massive; they have a flattened or slightly rounded, sometimes spiniferous, aboral face and a strongly convex opposite side, which lacks a ventral groove; their transverse section is subtriangular, and they are perforated by an axial canal that ends blindly in the last segment. As a rule, the part of the pinnules that is composed of such differentiated pinnulars is flexed adorally in such a way that the two rows of pinnules of each arm coming into contact form a sort of

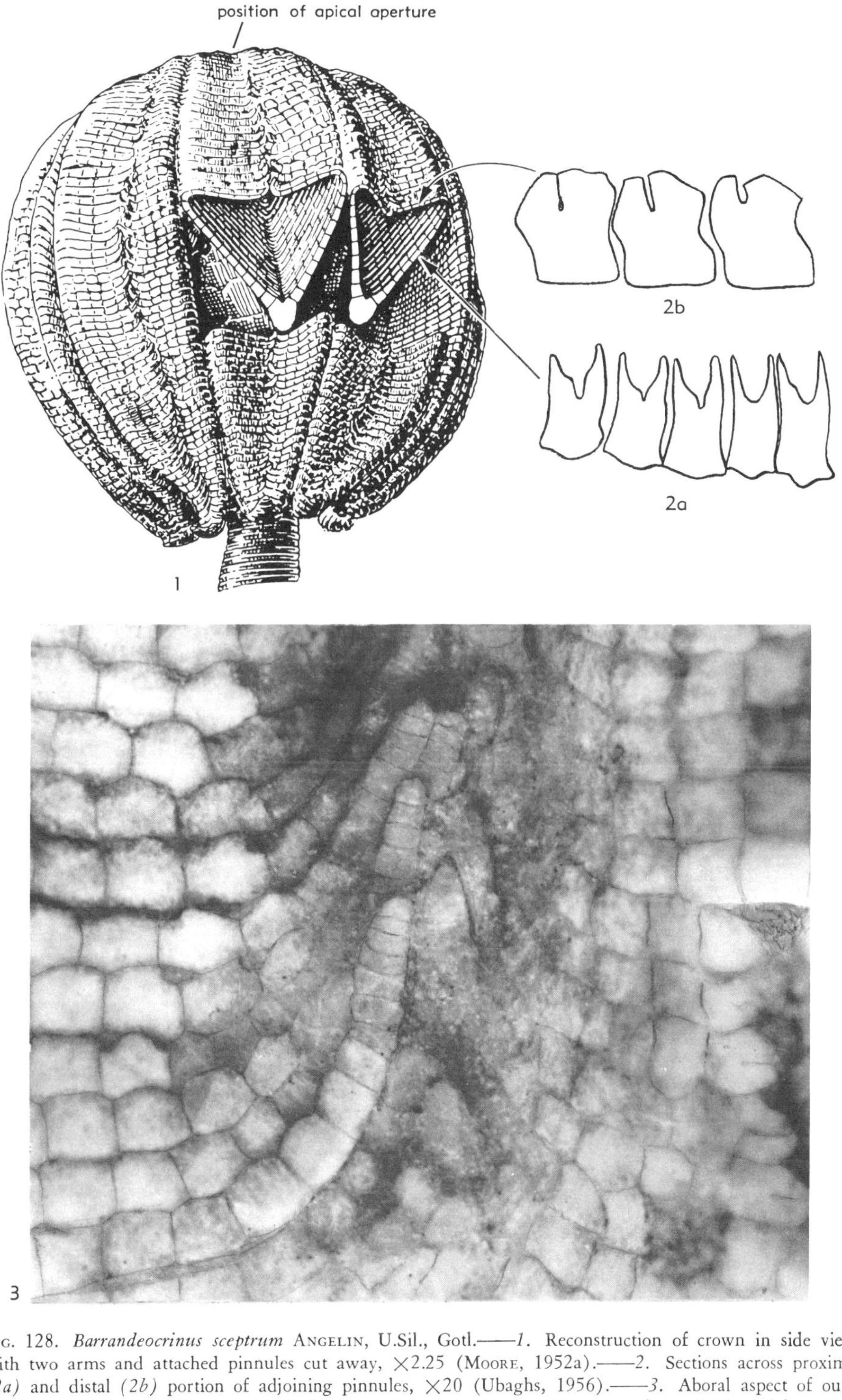

FIG. 128. *Barrandeocrinus sceptrum* ANGELIN, U.Sil., Gotl.——*1*. Reconstruction of crown in side view, with two arms and attached pinnules cut away, ×2.25 (MOORE, 1952a).——*2*. Sections across proximal *(2a)* and distal *(2b)* portion of adjoining pinnules, ×20 (Ubaghs, 1956).——*3*. Aboral aspect of outer pavement of crown showing interlocking, rigidly joined pinnulars and free distal ossicles, ×14 (Ubaghs, 1956).

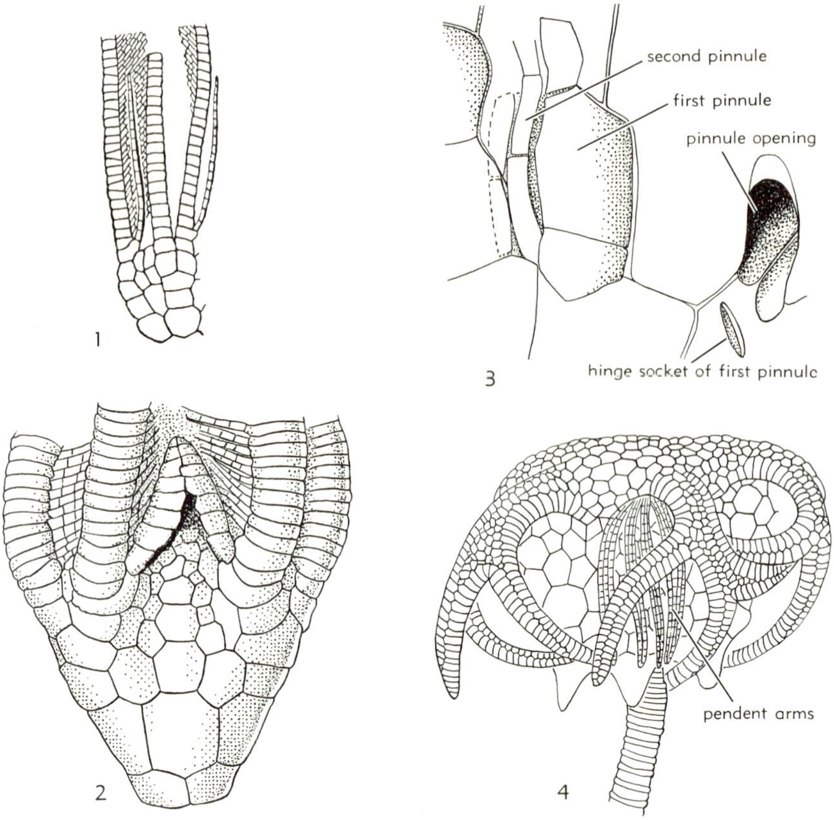

FIG. 129. Modified proximal pinnules in camerates.

1. *Carpocrinus simplex* (PHILLIPS), U.Sil., Eng.; large proximal pinnule (or atrophied ramus ?), ×1.3 (Bather, 1900a).
2. *Carpocrinus ornatus* (ANGELIN), U.Sil., Gotl.; Stockholm Mus. Ec. 11300, short and wide proximal pinnules on each side of *CD* interray, ×3 (Ubaghs, n).
3. *Batocrinus icosidactylus* CASSEDAY, Miss., USA (Ind.); two first pinnulars of first pinnule transformed into cover flap sealing pinnule aperture (shown at right side of figure), ×15 (Haugh, 1973).
4. *Gilbertsocrinus tuberosus* (LYON & CASSEDAY), Miss., N.Am.; diagram. sketch of crown and proxistele, showing large, tubular appendages (? proximal pinnules), ×0.8 (Van Sant, in Van Sant & Lane, 1964).

roof, which presumably protected the soft structures of the brachial system.

The pinnules of crinoids are typically undivided. A few cases of division into smaller branches have been reported, however, but they are not sufficiently numerous to invalidate the rule. Small, apparently uncalcified, threads occur along the adoral grooves of the pinnules of the Lower Devonian camerate *Macarocrinus semelfurcatus* (W. E. SCHMIDT, 1934), but they probably represent preserved tentacles or podia and not branchlets. MOORE & LAUDON (1943) have described, and called "pinnulets," long, articulated structures, seemingly attached on the pinnules (or ramules ?) of another camerate, the early Middle Ordovician rhodocrinitid *Trichotocrinus terranovicus;* they resemble cystoid brachioles, and apparently several of them arise from the adoral surface of each pinnular. Their exact nature and function are unknown.

In extant crinoids ambulacra are typically present on all of the pinnules, as well as on the arms. In some comatulids, however, they are absent on the oral pinnules and rudimentary on the genital pinnules; they may also be lacking in some of the arms of several comasterid species (see section on recent crinoids, p. *T*43). In fossil crinoids, their presence may be inferred from the existence of a groove on the pinnulars. Most pinnulars have such a groove, and therefore probably carried ambulacral extensions from the main ambulacra of the arms. The only significant exception to this rule concerns the massive pinnulars forming the distal differentiated portion of the pinnules of many camerates (Fig. 131). As described above, these ossicles have no ventral groove, and apparently served a protective rather than a food-collecting function.

The pinnules of the recent crinoids have a constitution very similar to that of the arms. This fact, the way these organs develop, and the occasional replacement of a pinnule by an arm indicate that the pinnules of the extant crinoids and more generally of all articulates are morphologically equivalent to the arms, which in many respects they duplicate on a small scale. The evolutionary development of the pinnulate condition from the closely spaced armlet condition observed in advanced dendrocrinine inadunates (particularly in the Bothryocrinidae and Mastigocrinidae) leads to the same conclusion concerning the pinnules of most cladid inadunates. It is not certain, however, that the pinnules of all fossil crinoids have arisen through a similar process of progressive specialization of arm branching (see above, p. *T*153). For instance, the arms of the Silurian dendrocrinine genus *Cyliocrinus* are provided with appendages that outwardly resemble pinnules and probably had the function of pinnules, but which are given off from the middle of brachials and occur in pairs instead of alternately, as if they were articulate outgrowths from the brachials rather than the products of an arm division (Fig. 132). Very surprisingly, also, the camerates from their earliest appearance have pinnulate arms, whereas this feature, which never occurred in flexibles or in monocyclic inadunates (with the possible exception of the highly specialized disparid genus *Chiropinna*), was acquired by cladid inadunates only at a late phylogenetical stage. One may therefore question whether all crinoid appendages designated as pinnules

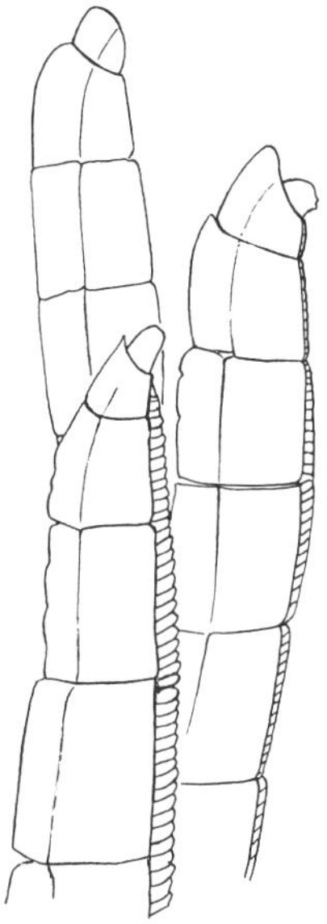

FIG. 130. *Carpocrinus ornatus* (ANGELIN), U.Sil., Gotl.; Stockholm Museum Ec. 11265, distal ends of three pinnules, ×20 (Ubaghs, n).

have the same origin and morphological significance.

ADORAL GROOVE

As seen above, the inner or adoral face of the brachials and pinnulars typically carries a furrow that in life contains extensions of the ambulacral grooves (food grooves), coelom, and nervous, water-vascular, and (at least in recent crinoids) reproductive systems. This furrow (often erroneously called ambulacral groove) is here termed **adoral groove.** It is variously deep and wide, V- or U-shaped in transverse section. Ordinarily it preserves no trace of the soft structures it housed, though, in some fossils, a special groove on its floor or an underlying axial canal perforating the plates indicate the former presence of an entoneural nerve cord (Fig. 107, p. *T*137). In some camerates, the arm ossicles have two ridges, one along each side of the adoral groove; these ridges have been interpreted as indicating the former location of the hyponeural system (HAUGH, 1937). In the floor of the groove of some inadunates pits disposed at regular intervals may be observed. Thus, between successive

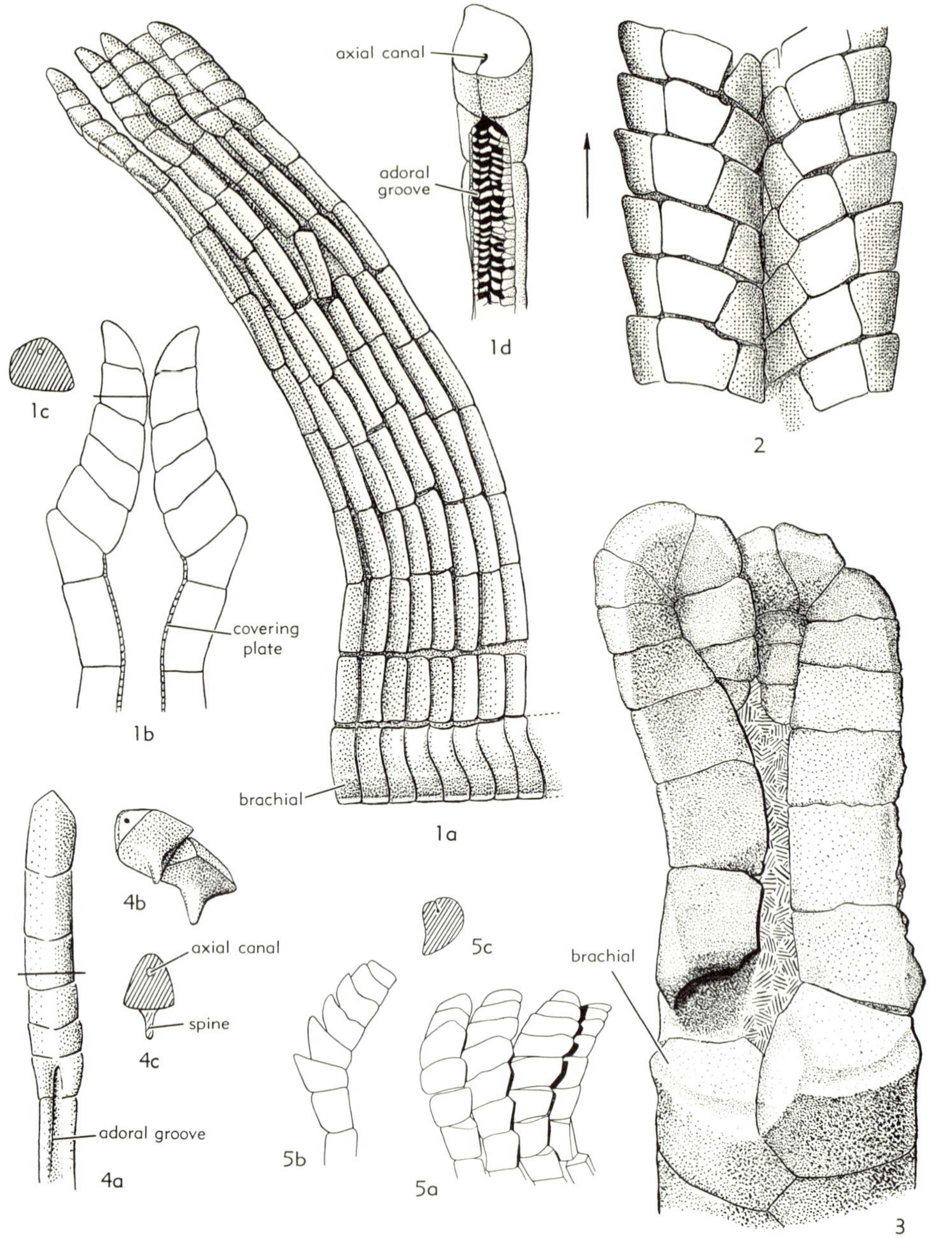

FIG. 131. Modified distal pinnulars in camerates.

1. *Clonocrinus* sp., U.Sil., Gotl.; Stockholm Museum Ec. 6156; *1a*, part of arm with complete pinnules, aboral side, ×10; *1b*, lat. aspect of distal portions of two pinnules in adoral contact, with section *(1c)* of distal pinnular, ×20; *1d*, distal end of adoral groove at base of modified distal portion of a pinnule, ×25 (Ubaghs, n).

2. *Abacocrinus* sp., U.Sil., Gotl.; Stockholm Mus. Ec. 6153, aboral aspect of modified distal pinnulars folded inward so as to protect soft structures; arrow indicates distal direction, ×20 (Ubaghs, n).

3. *Abacocrinus tasseracontadactylus* (GOLDFUSS), U. Sil., Gotl.; Stockholm Museum Ec. 10855, im-

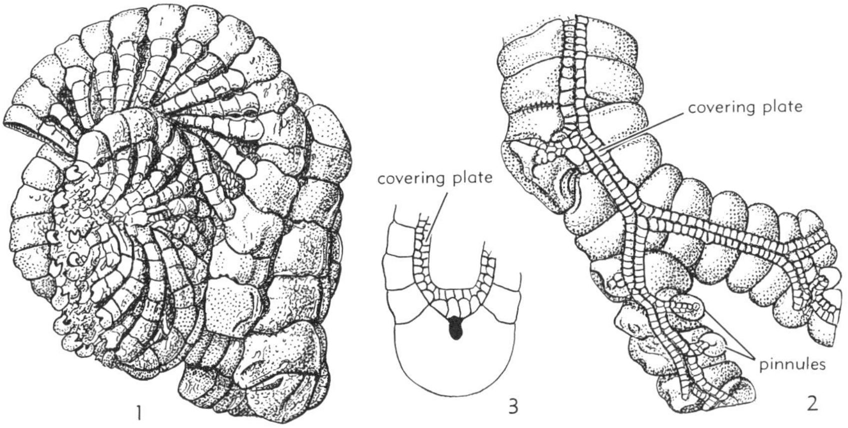

FIG. 132. *Cyliocrinus scolopendra* (BATHER), U.Sil., Gotl. (redrawn from Bather, 1893a).——*1*. Lat. view of arm, ×3.——*2*. Adoral aspect of part of arm, in which only the bases of the pinnules are preserved, ×3.——*3*. Transverse section of a brachial with two pinnules, ×3.

brachials of the crotalocrinitid *Pernerocrinus discus,* three canals link the groove with the axial canal (Fig. 133). In several poteriocrininine inadunates with biserial arms (e.g., *Cromyocrinus simplex, Erisocrinus* sp., *Proallosocrinus glenisteri*) pits occur at the level of each suture between the two series of brachials; they end blindly at some distance from the bottom of the adoral groove and do not appear to connect with any canal (*see* Fig. 138, p. *T*167). The exact function of these pits is unknown, but they probably served as passageways for nutrient fluids and aboral nerve branches.

BRACHIAL AND PINNULAR ARTICULATIONS

The mode of union between crinoid plates has been systematically treated by various authors (CARPENTER, 1884a; BATHER, 1900a; MINCKERT, 1905a; REICHENSPERGER, 1912; A. H. CLARK, 1915a; SPRINGER, 1920; GISLÉN, 1924; VAN SANT in VAN SANT and LANE, 1964). We have already described the articula uniting the columnals (p. *T*74) and those connecting the calycinal plates, including the fixed brachials (p. *T*130). Following this will be considered only the arm-bearing articular facets, the articulations between free brachials, the articulations for attachment of pinnules on brachials (**pinnule socket** or **facet**), and the articulation between free pinnulars.

These articulations, as any others in crinoids, fall into one of two categories, the **muscular articulations** and the **ligamentary articulations.** In the first type, the connection between ossicles is effected by a combination of ligament and muscle fibers. In the second one, ligaments alone intervene in union of skeletal pieces.

The ligamentary and muscular tissues are clearly distinct histologically, but not so much physiologically. The muscles, it is well known, have active contracting power. Ligaments, on the other hand, which are primarily composed of collagenous fibers, are reputed to have no contractile power, but elastic properties; they serve to bind the plates together, to oppose the action of the muscles and to provide cirri, arms, and pinnules with the stiffening required for maintaining the crinoid in position and

(*Continued from facing page.*)

mature brachials and pinnules, distal portions strongly folded, ×30 (Ubaghs, n).

4. Acanthocrinus sp., L.Dev., Belg.; *4a,* adoral side of distal portion of a pinnule; *4b,* two modified distal pinnulars; *4c,* cross section of distal pinnular, ×16 (mod. from Ubaghs, 1947).

5. Polypeltes granulatus ANGELIN, U.Sil., Gotl.; *5a-c,* aboral, lat., and cross section of distal pinnulars, ×14.5 (Ubaghs, 1956).

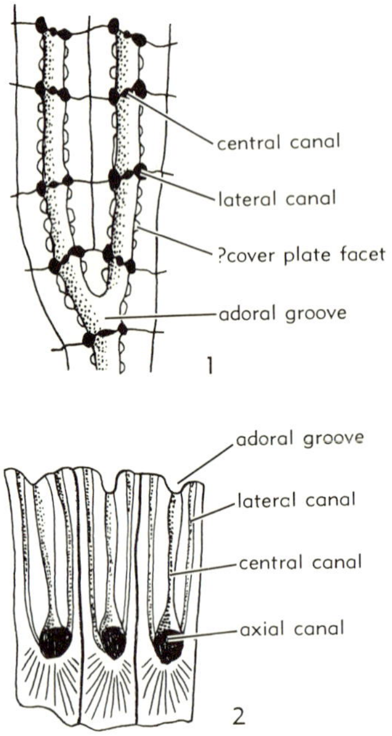

FIG. 133. *Pernerocrinus discus* BATES, L.Dev., Aus. (Bates, 1972).——*1.* Portion of adoral surface of adjacent arms, showing sutural pits in adoral grooves, ×6.——*2.* Articula of three contiguous brachials, ×6.

feeding in moving waters (MEYER, 1971). It appears, however, that some parts of the body, such as the stalk and the cirri, which have so-called ligamentary articulations only, may also accomplish generally slow but active movements, as if the ligaments, activated by the nervous system, had some contractile power, too. [For a review on this subject, see BROWER, 1973, p. 286-287.]

The microstructure of the crinoid endoskeleton reflects the type of the tissues that are bending the ossicles. The meshwork of the stereom in the ligament fields is markedly coarse and galleried, allowing deep penetration of the ligament fibers into the skeleton. In sharp contrast, the sites of muscle attachment are fine meshed, for the muscle fibers do not extend into the stereom. Thus, the stereomic microstructure furnishes a reliable criterion for the paleobiological interpretation of the articulations in fossil crinoids (MACURDA & MEYER, 1975; LANE & MACURDA, 1975).

MUSCULAR ARTICULATIONS

The surface of a typical muscular articulation bears a more or less straight ridge **(transverse ridge)** and five depressions or fossae (Fig. 134). The transverse ridge is a prominent, generally sharp-crested elevation that extends from near one outer lateral edge to the other. It may be marked by minute teeth, or denticles, normal to its axis. Slightly adorally to the middle of the transverse ridge, on a raised calcareous platform and on a level with it, is located the axial canal through which passes the axial cord of the aboral nervous system. The transverse ridge acts as a fulcrum upon which the motion of the articulation is accommodated. For this reason, it may also be called the **fulcral ridge.**

On the aboral (outer) side of the transverse ridge lies a single narrowly semicircular depression serving for attachment of ligament fibers. It is termed the **dorsal** or **aboral ligament fossa.** It usually contains a (generally) well-marked excavation **(ligament pit)** that adjoins the midportion of the transverse ridge. The dorsal ligament probably serves to counteract the muscles on the opposite side of the transverse ridge and, therefore, serves as extensor of the arms and pinnules, to which, as described above, it may also provide rigidity over more or less long periods.

On the adoral (inner) side of the transverse ridge, lies a pair of more or less triangular depressions **(interarticular ligament fossae),** one on either side of the axial canal. These fossae lodge ligaments of uncertain function. According to authors, they may serve merely to connect adjacent arm segments (CARPENTER, 1866), or to serve as antagonists of the muscles (BOSSHARD, 1900; VAN SANT in LANE and VAN SANT, 1964), or to oppose the dorsal ligament (GISLÉN, 1924). These fossae are bounded adorally by strong, oblique ridges that separate them from the two **muscular fossae,** in which are inserted the paired flexor muscles of the arms and pinnules. Between the muscular fossae runs a narrow ridge or a groove (**intermuscular ridge** or **groove**) from the platform about the axial canal to the inner (adoral), generally concave, margin of the facet.

Muscular articulations are of two types.

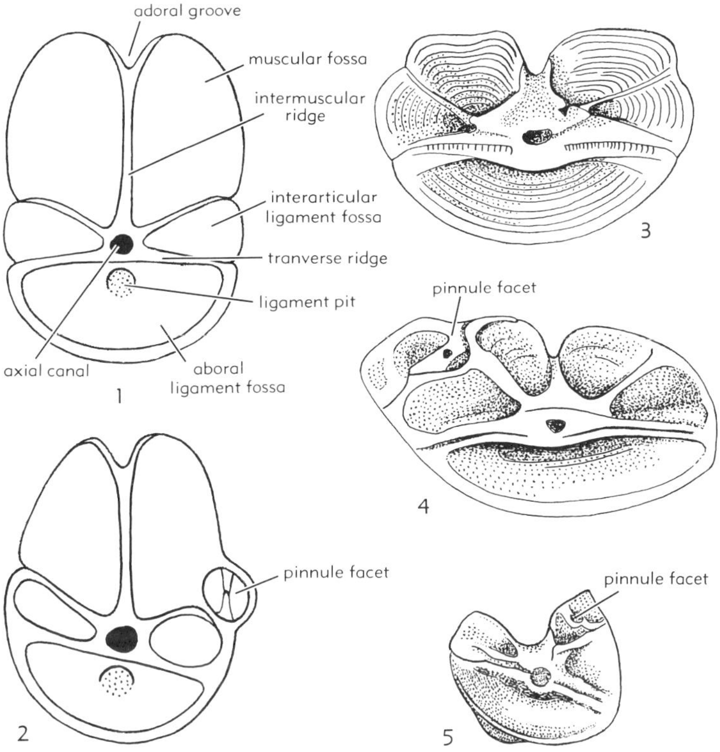

FIG. 134. Muscular articulations.——*1.* Diagram. sketch of a straight muscular articulation (Clark, 1915a). ——*2.* Diagrammatic sketch of an oblique muscular articulation (Clark, 1915).——*3-5. Marsupites testudinarius* (SCHLOTHEIM), U.Cret., Eu.; *3,* radial facet; *4,* second secundibrach; *5,* 25th secundibrach, all ×6 (redrawn from Sieverts, 1927).

In one type, known as the **straight muscular articulation,** the transverse ridge is perpendicular to the adoral-aboral axis of the facet, the right and left halves of which are alike and symmetrical (Fig. 134,*1,3,4*). The plane of such a facet is normal to the longitudinal axis of the arm, and its edge, viewed from the outer (aboral) or inner (adoral) side of the ossicle, cuts the arm transversely.

In the second type, termed **oblique muscular articulation,** the transverse ridge and the adoral-aboral axis meet at an angle so that the two parts of the facet on either side of the intermuscular ridge or groove are unequally developed (Fig. 134,*2,5*). In that case, the plane of the facet, and its edges in outer (aboral) and inner (adoral) views are more or less oblique to the longitudinal axis of the arm. The slope of succeeding articulations is alternately to the right and to the left, so that each brachial, as seen from its aboral or adoral sides, offers a high lateral side and a low lateral side. Usually (but not invariably; see reversion in chapter on recent crinoids, p. *T37*), the socket for attachment of the pinnule is located on the high lateral side.

In arms provided with muscular articulations, the distal face of axillary brachials bears two muscular articulations separated by a projecting median ridge. The apposed sides of these articulations, which are usually intermediate between straight and oblique muscular articulations, are more or less reduced.

Typical muscular articulations such as those just described, are well developed between radials and primibrachs and between most brachials of recent and fossil articu-

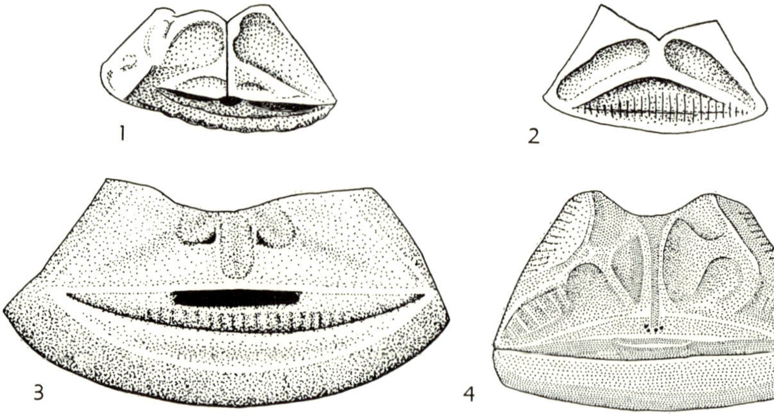

Fig. 135. Muscular articular radial facet of various inadunates.——*1. Synbathocrinus texasensis* Moore & Ewers, Miss., USA(Texas); a disparid inadunate, ×10 (Moore & Ewers, 1942).——*2. Parapisocrinus ollula* (Angelin), U.Sil., Eu., N.Am.; a disparid inadunate, ×6 (Bather, 1893a).——*3. Goleocrinus mangeri* Burdick & Strimple, Miss., USA(Ark.), a poteriocrinine cladid inadunate, ×10 (Burdick & Strimple, 1973b).——*4. Graphiocrinus timoricus* Wanner, Perm., Timor; a poteriocrinine cladid inadunate, ×6 (Van Sant, in Van Sant & Lane, 1964).

lates. They are also present, but slightly modified, between the brachials and first pinnulars and between the two first pinnulars of these crinoids. In the remaining portion of their pinnules, the articulations are generally characterized by the presence of a small adoral muscular fossa and a large aboral ligamentary depression on each side of a median fulcral ridge running in an adoral-aboral direction.

Muscular articulations seem to have evolved independently in several groups and between different parts of the crinoid body. We have seen above (p. *T*132) that in Calceocrinidae the basals and the radials were probably connected during life by ligaments and muscles (Fig. 104, p. *T*134). The radial facets of disparid inadunates like *Allagecrinus, Pisocrinus,* and *Synbathocrinus* have a straight transverse ridge or a sharp-crested angulation separating a single, rather small, ligamentary depression on the outer (aboral) side from a much larger inner part composed of a pair of flat or gently concave symmetrical (?) muscular areas (Fig. 135,*1,2*). It is, however, within the Poteriocrinina that the muscular articulation appears to have more generally evolved during late Paleozoic times. In many Pennsylvanian and Permian representatives of these advanced inadunates, the radial and, but perhaps in a lesser degree, the brachial facets show a prominent, generally sharp-crested denticulate transverse ridge and three large depressions, an outer (aboral) ligament area (commonly with a distinct ligament pit), and two inner (adoral) large fields separated by a narrow groove (intermuscular furrow) normal to the transverse ridge and leading into a V-shaped indentation (intermuscular notch) of the inner edge of the facet (Fig. 135, *3,4*). To these basic elements are usually added small ridges, furrows and pits, but interarticular ligament fossae are rarely clearly defined. It seems highly probable that in that sort of articulation muscles, which could flex the arms inward, occupied at least a part of the inner (adoral) fields, while extensor aboral ligaments served to pull them outward.

The socket serving for attachment of pinnules on brachials in most pinnulate Paleozoic crinoids seems also to represent an antagonistic system of articulation. Although our knowledge of such a small and delicate articulum is very limited, it appears that, in spite of a certain amount of diversity, its basic structure remains approximately the same. In both camerates and inadunates, it consists of two or three unequal fossae separated by a ridge or an

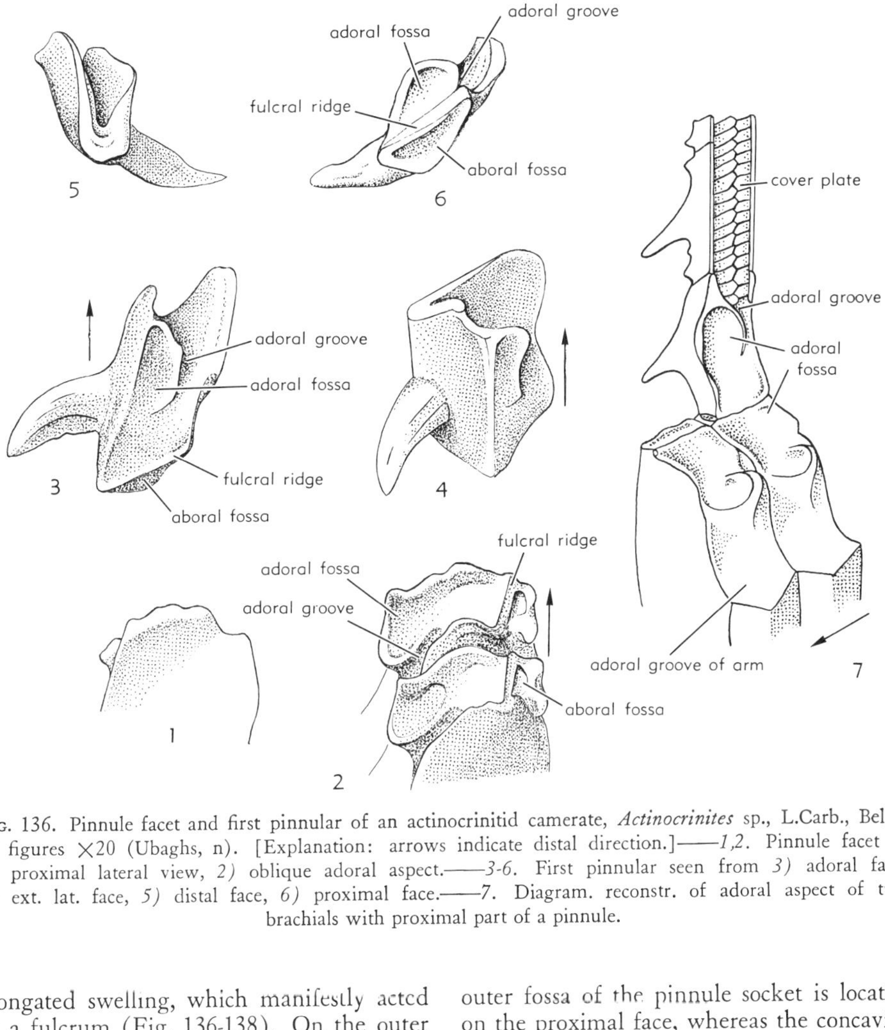

FIG. 136. Pinnule facet and first pinnular of an actinocrinitid camerate, *Actinocrinites* sp., L.Carb., Belg.; all figures ×20 (Ubaghs, n). [Explanation: arrows indicate distal direction.]——*1,2.* Pinnule facet in *1)* proximal lateral view, *2)* oblique adoral aspect.——*3-6.* First pinnular seen from *3)* adoral face, *4)* ext. lat. face, *5)* distal face, *6)* proximal face.——*7.* Diagram. reconstr. of adoral aspect of two brachials with proximal part of a pinnule.

elongated swelling, which manifestly acted as a fulcrum (Fig. 136-138). On the outer (aboral) side of the latter, there is usually one (in some cases two) shallow ligamentary fossa, provided in many inadunates with a distinct ligament pit. On the other side of the fulcrum, there is generally a large and deep, probably muscular fossae that ordinarily slopes rather strongly toward the oral groove of the arm. In the inner (adoral) edge of the pinnule facet, a notch prolonged downward by a straight or contorted furrow may occur, indicating the position of the proximal end of the ambulacral tract of the pinnule. All these features find their counterparts on the most proximal pinnular (Fig. 136, 137). In this ossicle, the depression corresponding to the outer fossa of the pinnule socket is located on the proximal face, whereas the concavity corresponding to the inner fossa covers a more or less large part of the adoral side; as to the edge that separates these two faces, it rested against, and had the same function as the fulcral ridge of the pinnule facet.

LIGAMENTARY ARTICULATIONS

Ligamentary articulations (called non-muscular articulations by some authors) occur in brachia of both living and fossil crinoids. In Paleozoic forms they are the commonest mode of union between radials and arms, and between arm plates. Some of them allow a certain amount of differential movement between apposed joint faces. They are termed **movable** ligamen-

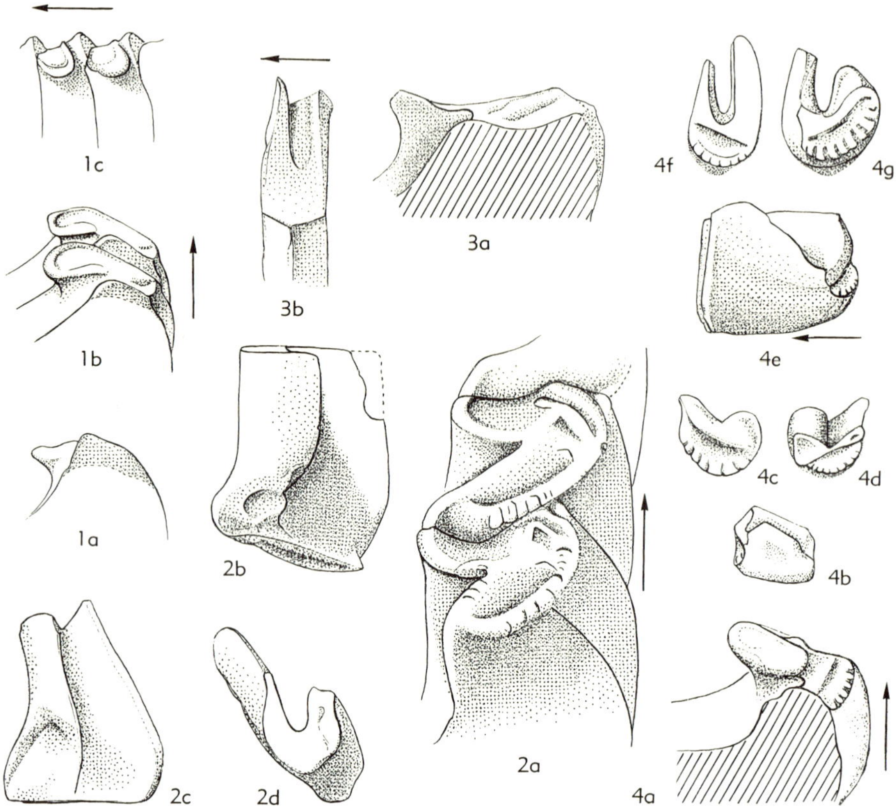

Fig. 137. Pinnule facet and first pinnular in various camerates (Ubaghs, n). [Explanation: arrows indicate distal direction.]

1. *Abacocrinus tesseracontadactylus* (Goldfuss), U.Sil., Gotl.; *1a,* pinnule facet seen from proximal side; *1b,* in oblique adoral view; *1c,* from inner side facing ventral groove of arm, ×20.
2. *Platycrinites* sp., L.Carb., Belg.; *2a,* pinnule facets in oblique adoral view; *2b-d,* first pinnular showing int. lat. face, adoral face, and distal face, ×20.
3. *Eucalyptocrinites* sp., U.Sil., Gotl.; pinnule facet showing *(3a)* proximal side, and *(3b)* inner side facing ventral groove of arm, ×20.
4. *Scyphocrinites* sp., U.Sil., N.Afr.; *4a,* pinnule facet in oblique adoral view; *4b-d,* first pinnular showing int. lat., distal, and proximal faces; *4e-g,* second pinnular showing int. lat., distal, and proximal faces, ×20.

tary articulations. Others, that permit only very slight movement or none at all, are said to be **immovable.**

MOVABLE LIGAMENTARY ARTICULATIONS

Several kinds of such articulations are known, between which many intermediates exist, so that it is not always easy to determine the type of articulation.

Trifascial Articulations.—The ligamentary articulation nearest the muscular one, to which it may have been antecedent is termed **trifascial.** It is characterized by the occurrence on each apposed joint face of

Fig. 138. Pinnule facets of some inadunates (arrows indicate distal direction).

1. ?*Pentaramicrinus* cf. sp. *bimagnaramus* Burdick & Strimple, Miss., USA(Ill.); *1a,* adoral view of part of arm, ×6; *1b,* sketch of two pinnule facets, ×10 (Ubaghs, n).
2. *Erisocrinus typus* Meek & Worthen, Penn., N.Am.; sketch of a pinnule facet, ×10 (Ubaghs, n).
3. *Erisocrinus* sp., U.Penn., USA(Kans.); *3a,* adoral

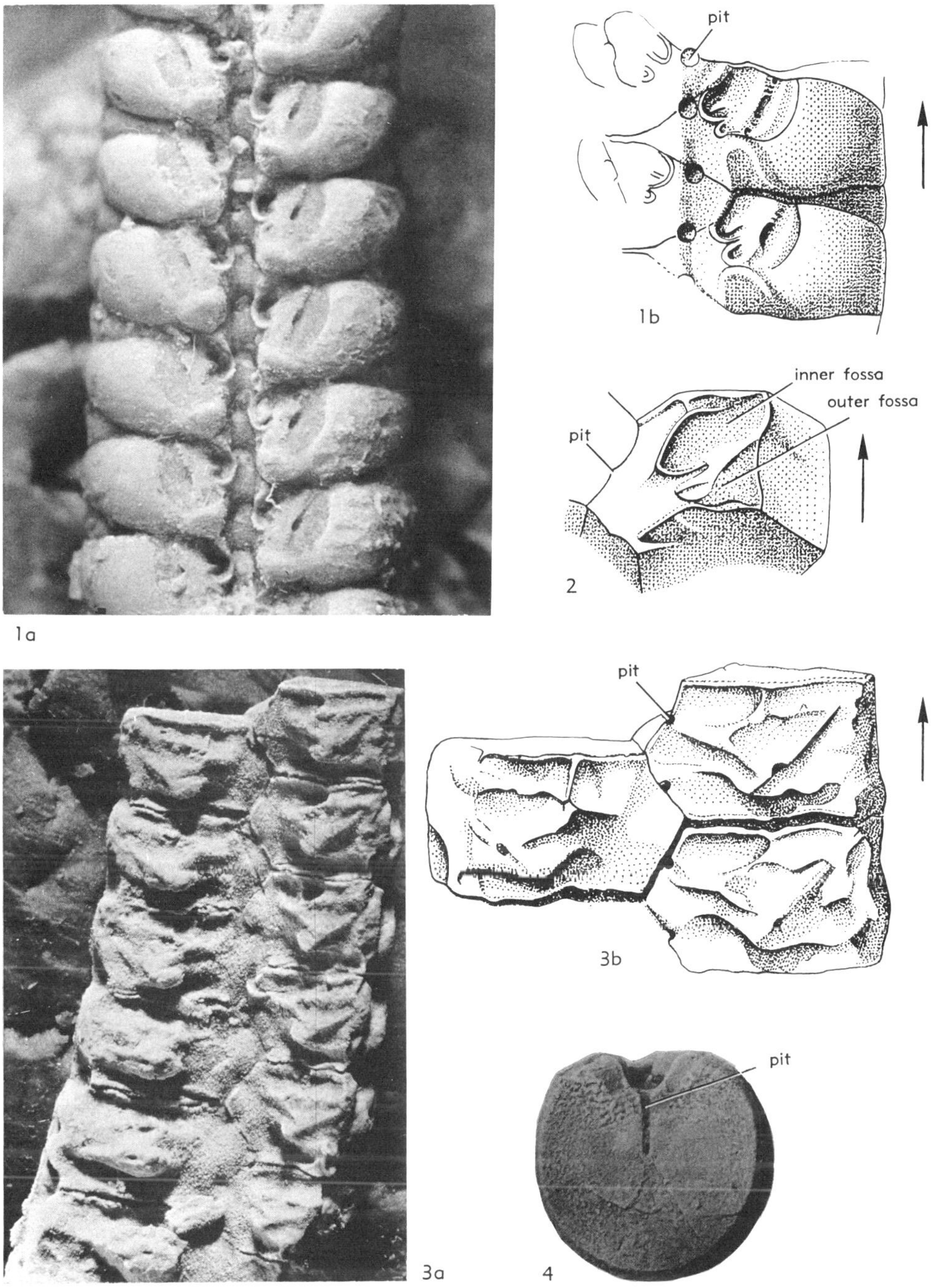

Fig. 138. *(Explanation continued from facing page.)*

view of part of arm, ×6; *3b,* sketch of three pinnule facets, ×10 (Ubaghs, n).

4. Proallosocrinus glenisteri Moore & Strimple, L.Penn., USA (Okla.); articular surface of secundibrach with sutural pit, ×4.5 (Moore & Strimple, 1973).

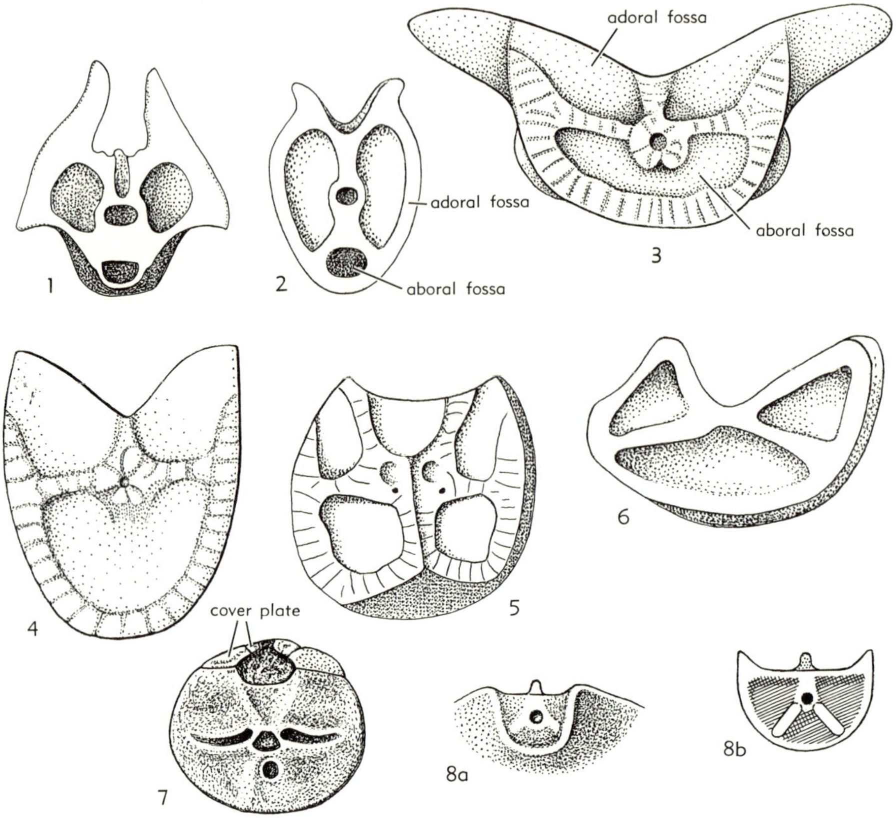

Fig. 139. Trifascial articulations.

1. *Bathycrinus australis* Clark, recent, Ind.O.; proximal face of fifth brachial, ×20 (Carpenter, 1884a).
2. *Bathycrinus aldrichianus* Thomson, recent, Atl.; proximal part of a brachial, ×33 (Carpenter, 1884a).
3-5. *Paradichocrinus polydactylus* (Casseday & Lyon), Miss., USA(Ind.); radial, first secundibrach and primaxil distal facets, ×6 (Van Sant, in Van Sant & Lane, 1964).
6. *Pellecrinus hexadactylus* (Lyon & Casseday), Miss., USA(Ind.); distal facet of a radial, ×4 (Van Sant, in Van Sant & Lane, 1964).
7. *Botryocrinus ramosissimus* Angelin, U.Sil., Gotl.; proximal facet of an axillary, ×4 (Bather, 1892b).
8a,b. *Petalocrinus visbycensis* Bather, U.Sil., Gotl.; *8a*, arm facet, ×4; *8b*, diagram of same (Bather, 1898a).

three depressions for attachment of ligaments. These depressions are a single median outer (aboral) fossa, which corresponds to the outer fossa of a muscular articulation, and two inner (adoral) fossae. They are separated by more or less well-defined elevations, which do not act as a fulcrum, but allow distinct though moderate mobility in two directions. The median outer fossa may consist of a small and rather deep depression, as shown by some recent crinoids (Fig. 139,*1*,*2*), but, in Paleozoic genera, it is usually larger and shallower, and may have distinct markings, like crenulae and even a median ligament pit. Such articulations occur in some advanced camerates (Fig. 139,*3-5*) and in some flexibles (e.g., Lecanocrinidae) between radial and brachial plates or between certain brachials. It is probably among Devonian and Mississippian cyathocrinid and dendrocrinitid inadunates (Fig. 139,*6-8*) that they were commonest, however, though precise and systematic information on that

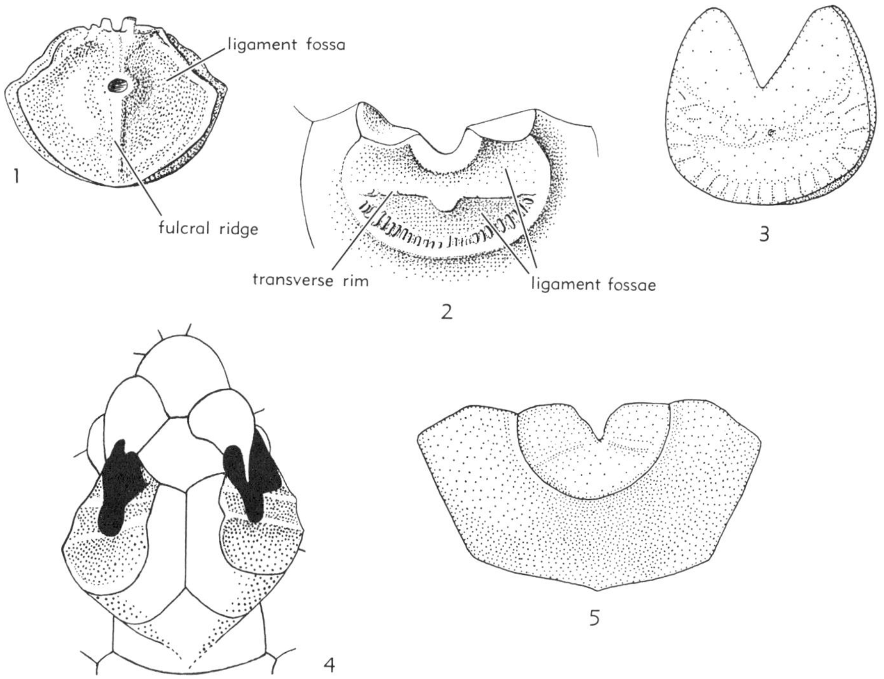

FIG. 140. Bifascial articulations.——*1. Himerometra bassleri* GISLÉN, Eoc., USA(S.Car.); first primibrach, ×6 (Gislén, 1934).——*2. Platycrinites austinianus* DE KONINCK & LE HON, L.Carb., Belg.; arm facet, ×5 (Ubaghs, n; unnumbered specimen from collection of Univ. Liège).——*3. Platycrinites hemisphericus* (MEEK & WORTHEN), Miss., USA(Ind.); distal facet of third tertibrach, ×13.3 (Van Sant in Van Sant & Lane, 1964).——*4. Aacocrinus nododorsatus* BOWSHER, Miss., USA(Mo.); arm facet on first secundibrach, ×1.5 (Bowsher, 1955b).——*5. Barycrinus hoveyi* (HALL), Miss., USA(Ind.); radial facet, ×4 (Van Sant in Van Sant & Lane, 1964).

matter are badly lacking. Besides, it is usually difficult, judging from the shape of depressions and the importance of ridges, to determine the exact nature of the tissues uniting the opposed joint faces in such articulations. For instance, the radial facets of such disparids as *Allagecrinus, Pisocrinus,* or *Synbathocrinus* might be classified almost as well as ligamentary trifascial articulations as muscular articulations.

Bifascial Articulations **(Synarthry).**—A synarthry or **bifascial** articulation is a ligamentary junction in which each apposed joint face bears two fossae separated by a fulcral or articular ridge. In typical brachial synarthry, the two fossae are equal and deep and the ridge is strong and medially located, running in an aboral-adoral direction (Fig. 140,*1*). Articulations of this sort occur between some brachials of living and fossil articulates. They are similar to those found between the cirrals and between the columnals of such crinoids as the Platycrinitidae of the Bathycrinidae (see p. *T77*), and they allow differential movements in two directions normal to the median ridge.

A synarthry that is modified by a general flattening of the joint faces, together with a reduction in size of the ligament fossae and in distinctiveness of the median ridge, is known as a **cryptosynarthry** (hidden synarthry). Such articulation is found between certain brachials of some post-Paleozoic crinoids. It allows a very small amount of differential movement between opposed ossicles and approaches synostosis (*see* below). It is practically an immovable articulation.

Articulations more or less similar, at least functionally, to synarthries have evolved independently in many Paleozoic crinoids between brachials or between arm-bearing thecal plates and free arms. Such facets commonly have two unequal shallow fossae, with or without crenulated margin, and a more or less distinct transverse elevation that allowed slight motion in an aboral-adoral direction (Fig. 140,*2-5*). The elevation, which may extend from one lateral side to the other, or be shorter, or be reduced to two symmetrical knobs, and which may be straight or incurved, smooth or denticulate, resembles a transverse rim rather than a typical fulcral ridge.

A modified articulation of this type occurs in brachials of many flexibles that possess distinctly arcuate transverse sutures (Fig. 141). In these crinoids the proximal margin of the aboral portion of each ossicle bends downward and forms a rounded process or lip that fits into a corresponding socket in the distal aboral margin of the ossicle next below. Both lip and socket have crenulated marginal contact surfaces, which restricted lateral movement of the plates when the structures were fitted together. Between their apposed faces there is a space, which during life was filled by the aboral ligament. The remaining and main part of the articulation is occupied by a single ligament fossa or by paired adoral fossae, separated from the aboral fossa by a transverse crenulated rim. Such articulations could only permit limited folding of the arms inward, but considerably restricted their extension outward. The projecting processes are thin and commonly fractured along their upper margin, so that they take the appearance of separate plates, called "patelloid plates" by HALL (1858). SPRINGER (1920) was the first to recognize their true nature. This type of structure is unknown in other crinoids; however, a few inadunates and camerates, as well as some recent pentacrinids, have similar arcuate sutures (but not necessarily similar articulations) in some parts of their arms.

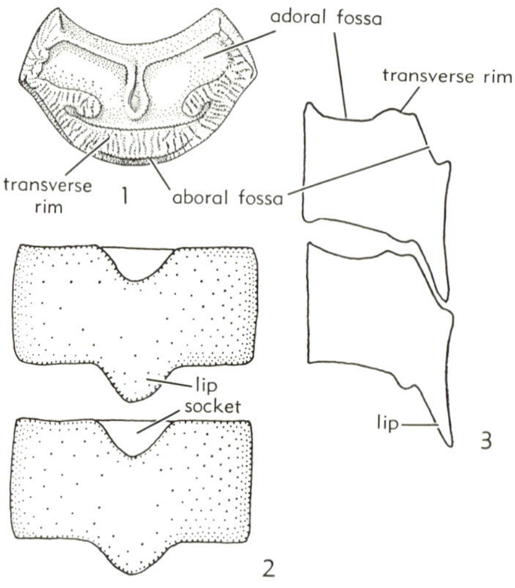

FIG. 141. *Forbesiocrinus nobilis* DE KONINCK & LE HON, L.Carb., Belg. (Springer, 1920).——*1*. Distal facet of second primibrach, ×2.8.——*2,3*. Diagrammatic representation of a pair of brachials, from aboral side and in long. median section.

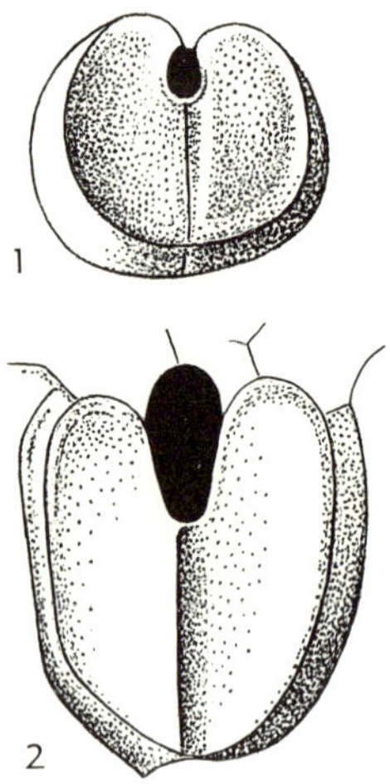

FIG. 142. Compound arm facet (pseudosynarthry) (Ubaghs, n).——*1*. *Alloprosallocrinus conicus* CASSEDAY & LYON, Miss., N.Am.; ×4.——*2*. *Agaricocrinus inflatus* (HALL), Miss., N.Am.; ×4.

Different in origin and nature from the above described articulations, though still somewhat synarthrial in function, are the so-called compound facets occurring between fixed and free brachials in some camerates provided with biserial arms (Fig. 142). Each one consists of two slightly concave surfaces borne by two adjacent ossicles, and, as these two surfaces are not on the same level, the lateral admedian edge of the highest one could probably act in some way as a rudimentary fulcrum, allowing

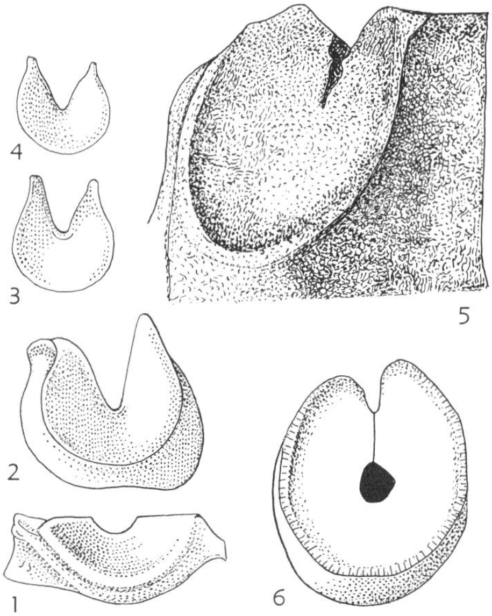

FIG. 143. Synostosial articulations.——*1-4. Aethocrinus moorei* UBAGHS, L.Ord., France; distal facets of second fixed primibrach, of a free primibrach, and of two secundibrachs, ×10 (Ubaghs, 1969). ——*5. Thetidicrinus piriformis* WANNER, Perm., Timor; a radial facet in oblique view, ×3 (Wanner, 1937).——*6. Barycrinus hoveyi* (HALL), Miss., USA(Ind.); distal facet of third primibrach, ×3 (Van Sant in Van Sant & Lane, 1964).

restricted movements in a lateral direction. Such synarthrial type of articulations, which allowed movements in two directions as in true synarthry, but in which there is no distinct median ridge, has been named **pseudosynarthry** (false synarthry) by VAN SANT (in VAN SANT & LANE, 1964).

Synostosial Articulations.—A **synostosis** is a ligamentary junction in which the apposed joint faces are more or less concave. The surfaces of the facets are smooth or barely rugose and may be surrounded by a low, crenulated rim (Fig. 143). Articulations of this type allow a slight flexibility in all directions. They are common in the distal portions of the arms in many recent and fossil crinoids. They may also occur between all the brachials, as in the early dendrocrinid *Aethocrinus* (Fig. 143,*1-4*) or the recent *Calamocrinus* and *Ptilocrinus*. The term synostosis has been used with different meanings. It is here considered as synonym of "loose synostosis" of authors, particularly GISLÉN (1924).

IMMOVABLE ARTICULATIONS

Many crinoid ossicles are united in such a manner that no differential movements or only very slight ones occur between them. Such articulations have been commonly referred to as "close sutures." The following types may be distinguished.

Zygosynostosis.—In the sort of union termed **zygosynostosis** the apposed joint faces are nearly flat and closely fitted together by short ligament fibers (Fig. 144). They commonly have moderate amounts of calcareous deposits on their faces. They may be smooth or have supplementary structures that strengthen the union of the ossicles, such as radial crenulae in a part of the apposed facets, or irregular vermiculation; there may also be a faint ridge on one facet that fits into a corresponding groove on the opposed facet (Fig. 144,*1,3*). This type of union between arm plates is certainly the most common among Paleozoic crinoids. It allows very slight differential movements between joined ossicles. However, even in such cases, a limited flexibility of the arms could be achieved merely by multiplication of the ossicles. Commonly, in biserial arms, one may count 20 or 25 transverse sutures per cm. on each side. This number reaches 60 to 70 in some species of *Desmidocrinus,* and this probably is a maximum. The mode of preservation of certain specimens of this genus suggests, indeed, that their arms were remarkably flexible.

Symplexy.—A **symplexy** is an articulation in which culmina and crenellae of apposed joint faces interlock, so that the corresponding suture is crenulate (Fig. 145, *1-3*). This type of joint is very common in the stem of crinoids, particularly Paleozoic forms. It may also occur in the arms of camerates and between thecal plates. The extremely slight mobility of each articulation may only be partially compensated in arms and stems by reduction in length of the component ossicles and consequent multiplication of the joints. An ill-defined symplectial articulation is known as a **cryptosymplexy.**

Syzygy.—A **syzygy** is an articulation in

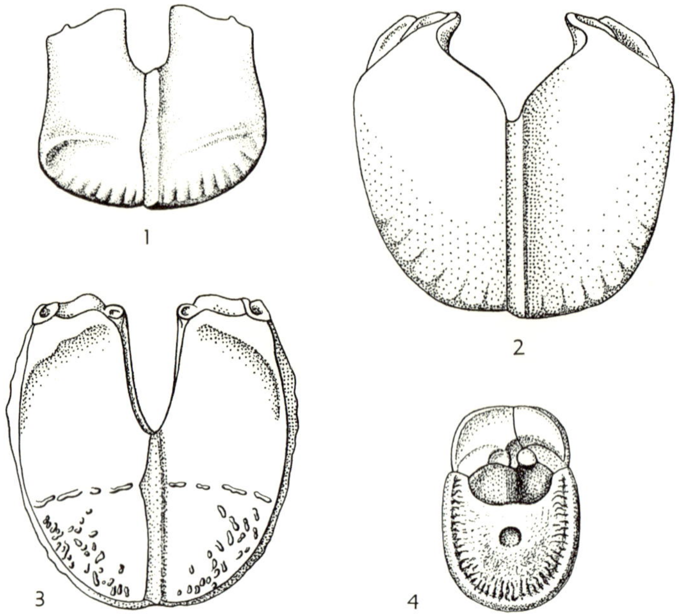

FIG. 144. Zygosynostosial articulations.——*1. Clonocrinus polydactylus* (McCoy), M.Sil.-U.Sil., Eng., Gotl.; articular facet of two biserial brachials, ×20 (Ubaghs, n).——*2. Abacocrinus* sp., U.Sil., Gotl.; articular facet of two biserial brachials, ×20 (Ubaghs, n).——*3. Actinocrinites* sp., L.Carb., Belg.; articular facet of two biserial brachials, ×20 (Ubaghs, n).——*4. Cyathocrinites acinotubus* (Angelin), M.Sil., Eng.; articular facet of a brachial, with covering plates in position, ×10 (Bather, 1892d).

which the joint faces are flat and covered with radiating ridges and furrows, but instead of interlocking as in symplexies, the ridges of one articulum meet the ridges of the other articulum and the furrows are opposite the furrows (Fig. 145,*4,5*). Connective tissue fills the furrows. Externally a syzygy is usually readily recognizable; it appears as a fine dotted line that crosses the arm at a right angle to the longitudinal axis. Two brachials united by syzygy form a syzygial pair. In such a pair the proximal ossicle is called **hypozygal,** the distal one **epizygal.** The latter alone bears a pinnule, the former never. It is clear from all this that a syzygy is a very specialized type of ligamentary articulation that occurs only in the arms (never in the stem) of the Articulata, where it represents a preformed place for autotomy (a function performed in some cases by zygosynostosial articulations). The existence of true syzygies in Paleozoic crinoids is extremely questionable. In most cases what is called syzygy in these crinoids is synostosis or symplexy or another type of ligamentary articulation. Moreover, the term has been used with several meanings, sometimes serving to designate the articulation, sometimes the pair of ossicles united by syzygy, or even all the brachials located between two succeeding syzygies. But, as stressed by Bather (1896a), the term "syzygy" must invariably be used in accordance with its original definition by Müller (1843), as a particular type of ligamentary articulation. Articulations of syzygial type, but with very short ridges that may be replaced by rows of tubercles or granules, with tendency toward irregular arrangement and disappearance, are distinguished as **cryptosyzygies.** They occur in the arms of Articulata (e.g., some Isocrinidae) and certainly permit but a very low capacity for movement.

Symmorphy.—The term **symmorphy** designates a ligamentary junction in which one or two prominent ridges or toothlike elevations on one face fit into corresponding

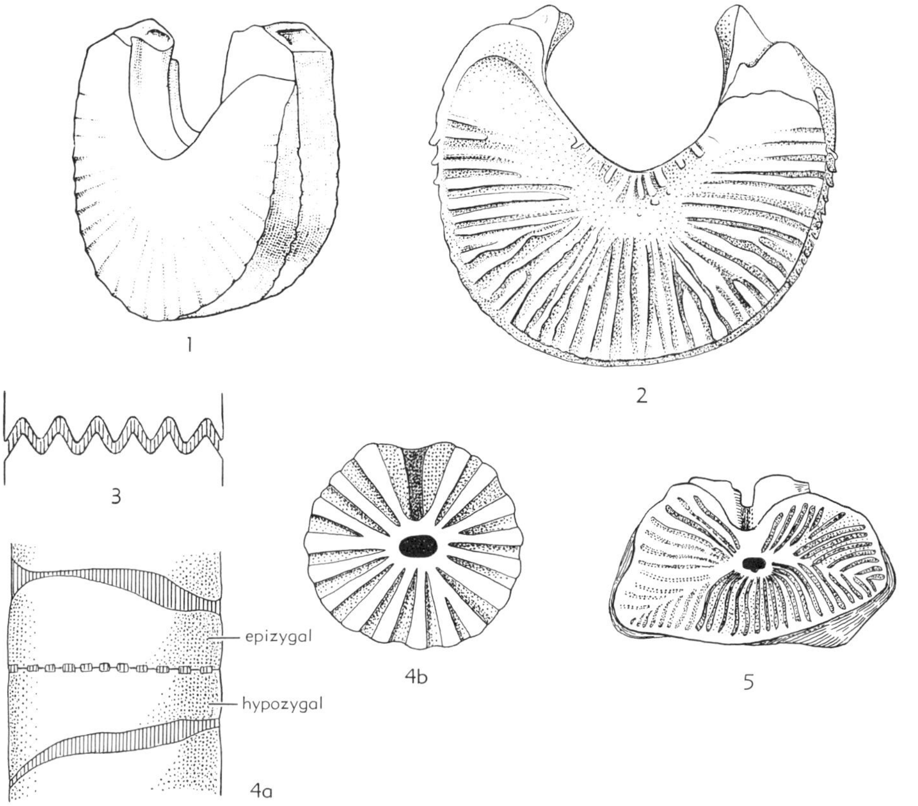

FIG. 145. Symplectial and syzygial articulations compared.

1. *Diamenocrinus stellatus* JAEKEL, L.Dev., W.Eu.; symplectial articulation, ×18 (Ubaghs, 1947).
2. *Scyphocrinites* sp., U.Sil., N.Afr.; symplectial articulation, ×20 (Ubaghs, n).
3. Diagrammatic representation of two brachials united by symplectial articulation (Ubaghs, n).

4a,b. Diagrammatic representation of *(4a)* two brachials united by syzygy, and *(4b)* a syzygial articulation (Minckert, 1905a).

5. Marsupites testudinarius (SCHLOTHEIM), U.Cret., Eu.; syzygial articulation, ×6 (redrawn from Sieverts, 1927).

grooves or sockets on the opposed faces accompanied in some cases by peripheral crenulae (Fig. 146,*1,2*). With such articulations only very slight differential movement of joined ossicles in two directions seems possible. Symmorphial articulations occur in some recent (e.g., *Neocrinus, Hypalocrinus*) and fossil articulates. The presence of similar types of joints in Paleozoic crinoids has also been recorded in a few cases, and even between pinnulars (Fig. 146,*3,4*).

Ankylosis.—An **ankylosis** (also spelled anchylosis) is an immovable union of ossicles cemented by a deposit of stereom on their apposed joint faces. It is commonly accompanied by a partial or complete obliteration of the sutures (Fig. 124, p. *T*152). Lateral fusion of the arms in some inadunates (e.g., *Crotalocrinus, Petalocrinus*) and the formation of hyperpinnulated brachials through fusion of ossicles are well-known examples of this type of union. Ankylosis may also occur between certain brachials normally in some species, or as the result of an adaptive development in other ones. For instance, the support of pendent arms in some camerate species was furnished by fusion of proximal free brachials.

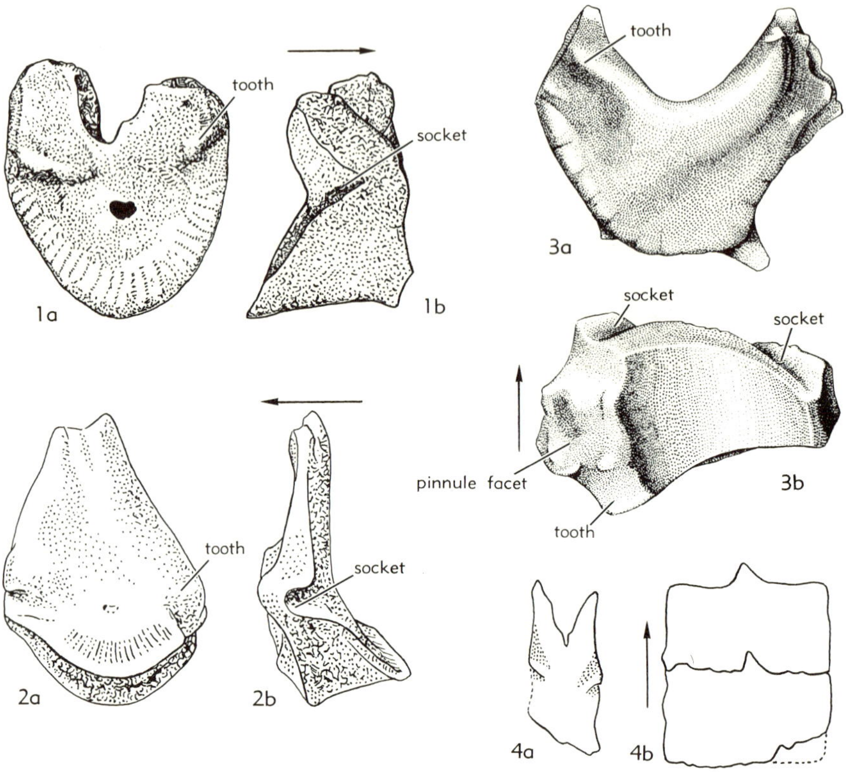

FIG. 146. Symmorphial articulation (arrows indicate distal direction).

1. *Cainocrinus tintinnabulum* FORBES, Eoc., Eng.; *1a,b,* distal facet and lat. side of second tertibrach, ×6 (Rasmussen, 1972a).
2. *Isocrinus cingulatus* (MÜNSTER), Jur., Eu.; *2a,b,* distal facet and lateral side of third secundibrach, ×10 (Hess, 1972b).
3. *Ctenocrinus gottlandicus* (PANDER), U.Sil., Gotl.; *3a,b,* proximal and adoral view of a ramule ossicle, ×70 (Ubaghs, 1958b).
4. *Barrandeocrinus sceptrum* ANGELIN, U.Sil., Gotl.; *4a,b,* distal facet of a pinnular and lat. side of two adjoining pinnulars, ×35 (Ubaghs, 1956).

ADORAL SKELETON

The **adoral skeleton,** also called **perisomic skeleton,** includes every skeletal element that may be present in the teguments that 1) surround or cover the mouth, 2) support the ambulacral tracts, and extend along the adoral side of the free arms from their bases to the tips of their smallest branches and pinnules, and 3) in some crinoids pass down the fixed parts of the rays and their divisions, and cover the interbrachial areas of the calyx. The central part of these teguments extending over the visceral mass and forming the roof of the thecal cavity is distinguished as the **tegmen** (also called "disc" or "vault").

The adoral skeleton consists of plates, ossicles, or spicules of various sizes and shapes. Almost completely lacking in some crinoids, it is fairly well developed in others, forming a thick protective covering over the viscera, and in some cases over the adoral groove of the free brachials. As a rule, it is rather less definite and characteristic than the aboral skeleton. It may even vary in different parts of the same individual. Morphologically it includes the in-

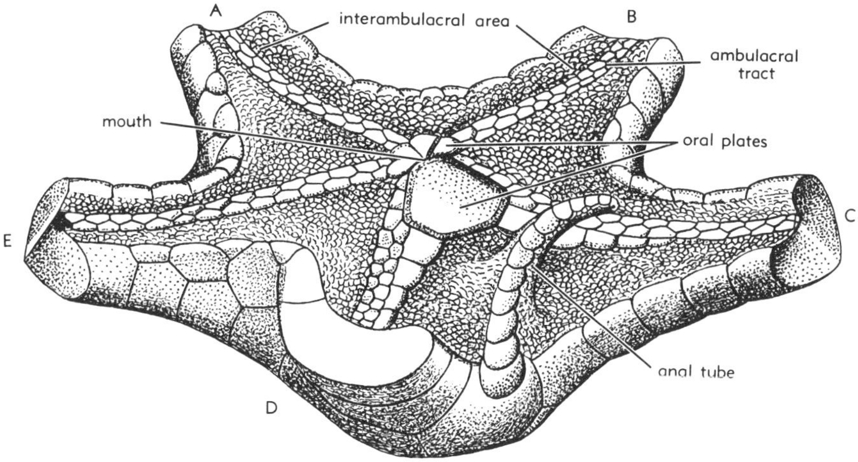

FIG. 147. *Onychocrinus ulrichi* MILLER & GURLEY, Miss., USA(Ind.); reconstr. of calyx, tegmen, and proximal part of arms of a flexible crinoid, oblique adoral view, enl. (Springer, 1920). [Explanation: *A, B, C, D, E,* letter symbols designate rays according to CARPENTER system.]

terbrachials, but as these plates in the crinoids where they are best developed (Camerata, Flexibilia) occupy areas that under normal conditions are on the aboral side of the body, they have been treated along with elements belonging to the aboral skeleton (see p. *T*120).

TEGMEN

GENERAL FEATURES

The tegmen is the adoral portion of the theca. It differs markedly from the aboral portion or cup (calyx), a feature that is most distinctive of the crinoids.

The shape and relative size of the tegmen are quite diversified. This structure may be low, medium, or high; flat, gently rounded, strongly convex, conical, flask-or mushroom-shaped, or have other form. Its surface, generally simple and regular, may comprise ridges that run from near the center to the arm bases. In a few inadunates (e.g., *Tenagocrinus*) and some camerates (e.g., *Paragazacrinus, Methabocrinus, Eucalyptocrinites*), outgrowths of some of its plates form vertical partitions that produce alcoves for reception and protection of free arms. On the tegmen of many crinoids stands a variously developed conical to cylindrical structure (**anal tube, pyramid,** or **cone**) that bears the anal opening; in some cladid inadunates this structure acquires an enormous size and forms the so-called "ventral sac" **(anal sac).** The tegminal plates are smooth or ornamented with pustules, granules, vermiculations, nodes, spines, and so forth.

A typical tegmen is divided into five interambulacral areas by five narrow ambulacral tracts (Fig. 147). The ambulacral tracts radiate from the edges of the central mouth and pass over the tegmen to the bases of the free arms. They are bisected by the radii, and then correspond to the rays of the crinoid. The interambulacral areas correspond to the interrays and they are bisected by the interradii. As the rays may bifurcate within the cup or calyx, so the ambulacral tracts entering the tegmen and prolonging the food grooves of the arms over its surface may divide into as many branches as free arms (and pinnules, if such organs are incorporated in the calyx).

In most crinoids, the main orifices of the body are located on the adoral surface of the body, and this is another characteristic feature of these echinoderms. The mouth is found at the point of convergence of the ambulacral tracts. It is ordinarily central or subcentral, but it may be secondarily dis-

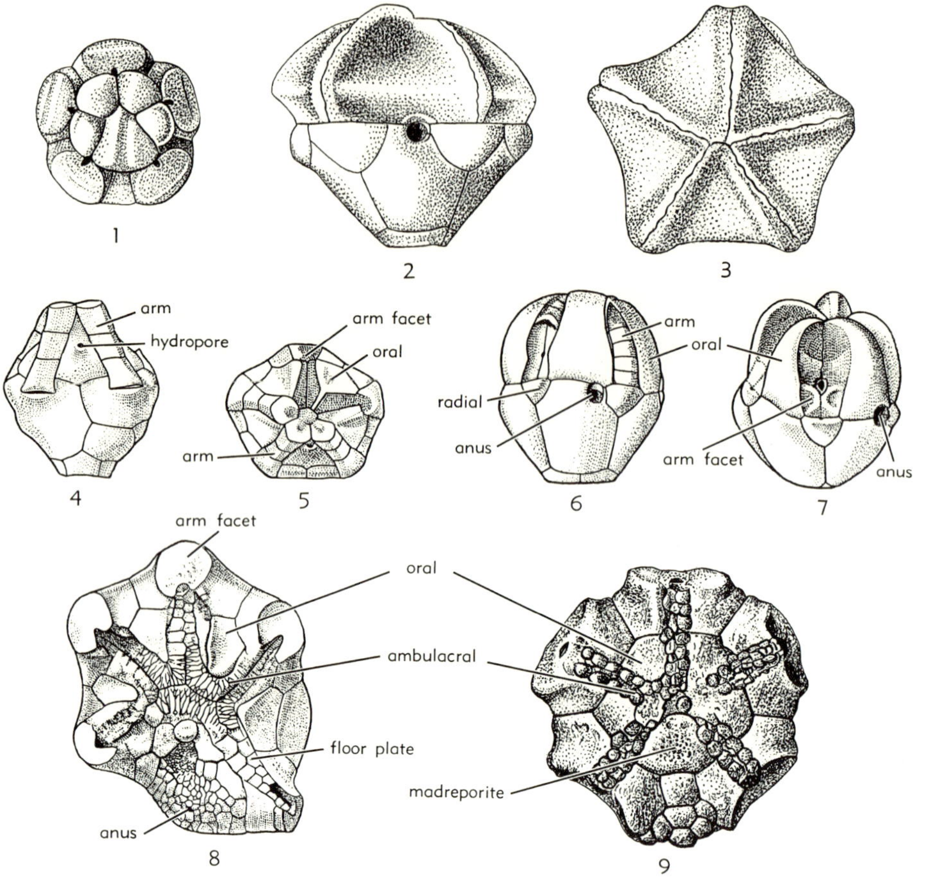

Fig. 148. Tegmens with orals only *(1-7)*, or with orals and ambulacrals *(8-9)*.

1. Neodichocrinus nanus Wanner (compsocrinine camerate), Perm., Timor; adoral view, ×5 (Wanner, 1937).

2,3. Cranocrinus timoricus Wanner (cyathocrinine inadunate), Perm., Timor; *CD*-interray and adoral views, ×8 (Wanner, 1929b).

4,5. Haplocrinites mespiliformis (Goldfuss) (disparid inadunate), M.Dev., Ger.; *CD*-interray and adoral views of a specimen with part of 3 arms fitting closely in grooves on apposed edges of orals, ×5 (Springer, 1926a).

6,7. Tenagocrinus sulcatus Wanner (cyathocrinine inadunate), Perm., Timor; arms forming niches for reception of arms; *6, CD* interray; *7,* oblique adoral view, arms not preserved, ×5 (Wanner, 1929b).

8. Hybocrinus conicus Billings (hybocrinid inadunate), M.Ord., N.Am.; adoral view, ×1 (Springer, 1911b).

9. Gissocrinus incurvatus (Angelin) (cyathocrinine inadunate), Sil., Gotl.; adoral view, ×4 (Bather, 1893a).

placed toward the margin. In a large number of Paleozoic crinoids, it was not exposed on the surface, but it opened beneath the tegmen, so that in order to reach the mouth the food grooves had to be lowered (at least in part) beneath the surface—a position known as **subtegminal.** The anus opens in one of the interambulacral areas designated for that reason as posterior, the opposite ray being anterior. According to Carpenter's nomenclature, this ray is termed *A* ray, and the posterior interray, *CD* interray. Like the mouth, the anus may be secondarily displaced and come to occupy a central or even anterior position. In many forms, it is located on a conical protuberance (anal pyramid, tube, or sac), which may be of considerable height and acquire an extraordinary development. With the mouth and anus, one or several

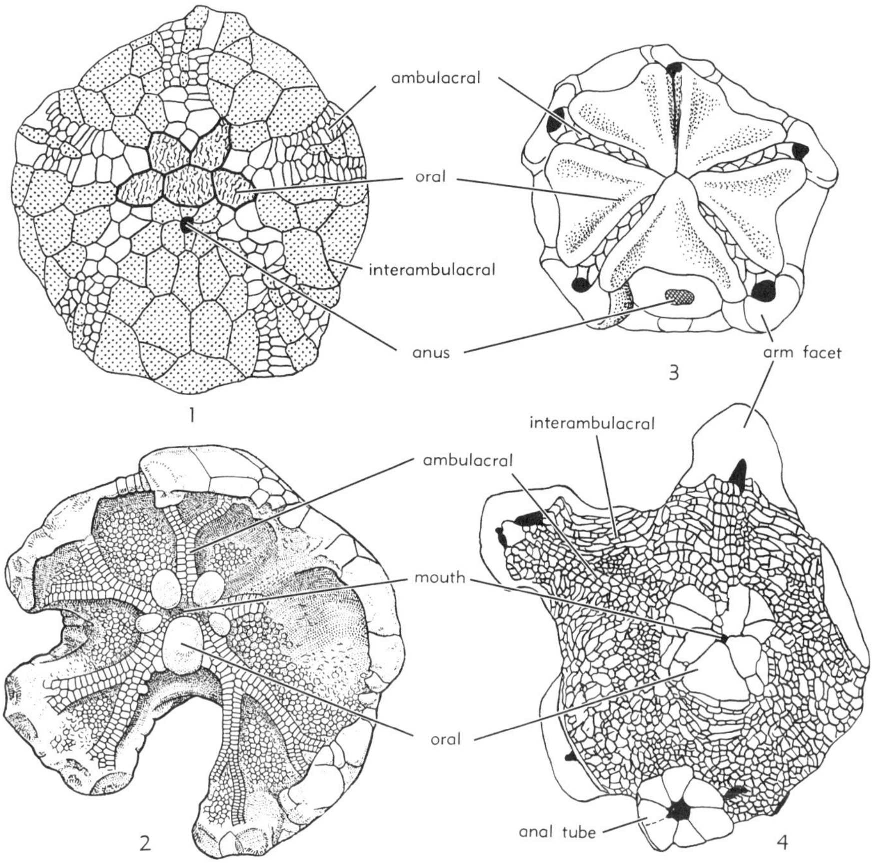

FIG. 149. Tegmens with orals, ambulacrals, and interambulacrals *(1,2,4)*, or with orals and ambulacrals only *(3)*.

1. Marsupiocrinus striatus WACHSMUTH & SPRINGER (glyptocrinid camerate), M.Sil., USA(Tenn.); ×1.5 (Wachsmuth & Springer, 1897).

2. Taxocrinus intermedius WACHSMUTH & SPRINGER (flexible), Miss., USA(Iowa); ×1.5 (Springer, 1920).

3. Cantharocrinus simplex BREIMER (glyptocrinid camerate), M.Dev., Spain; ambulacrals lacking in *A* radius, ×1 (mod. from Breimer, 1962).

4. Ceratocrinus gracilis WANNER (cladid inadunate), Perm., Timor; ×3 (Wanner, 1937).

hydropores may also occur on the tegmen. In extant crinoids these orifices are usually many and scattered over the five interambulacral areas, but in Paleozoic forms, where they exist, they are located in the *CD* interray, between the mouth (or its presumed position, if it is subtegminal) and the anus. A tegminal plate on which the hydropores are concentrated is called **madreporite.**

The tegmen may be formed by a soft integument that contains no skeletal elements except microscopic ones, or may be supported by closely fitted or loosely imbedded plates. Three different types of plates may enter its composition: the **orals,** which form a central, pentameral circlet around or above the mouth; the **ambulacrals,** that protect the ambulacral tracts and their branches (if any); and the **interambulacrals,** which are located between the ambulacral tracts and their branches.

ORALS

The orals (also called deltoids) belong to the primary skeleton. They consist of five

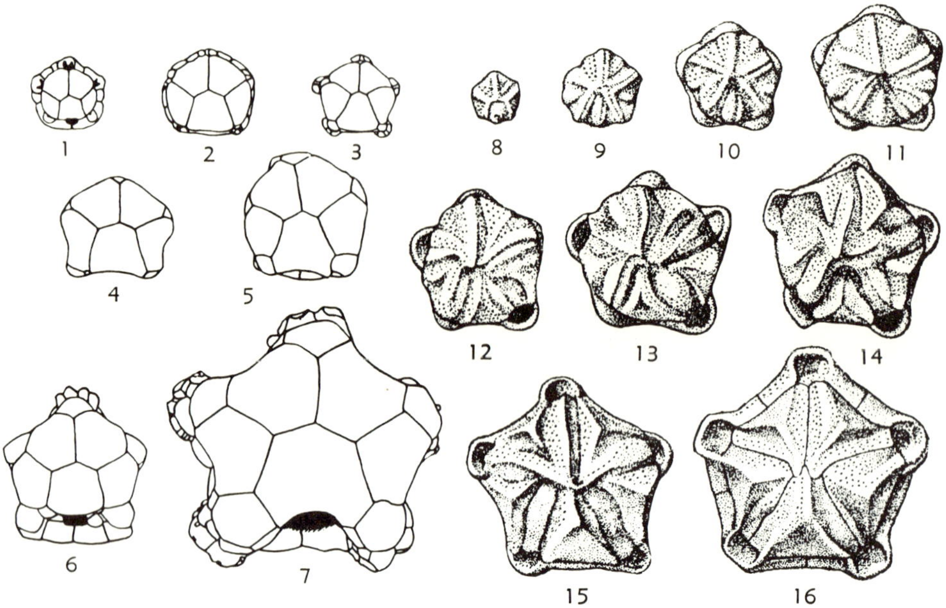

FIG. 150. Growth series of oral plates in a platycrinitid *(1-7)* and a hapalocrinid camerate *(8-16)*.——*1-7. Neoplatycrinus dilatatus* WANNER, Perm., Timor; ×1 (Wanner, 1937).——*8-16. Cyttarocrinus eriensis* (HALL), M.Dev., USA(N.Y.); ×15 (Koenig, 1965).

plates interradially disposed, which in the larvae of comatulids develop at the same time as the basals and long before the radials, but are wholly resorbed before maturity in most species. Five triangular plates that cover the mouth and the proximal parts of the ambulacral tracts in some recent crinoids (e.g., *Holopus, Cyathidium, Hyocrinus, Rhizocrinus*) are judged to be homologous with the larval orals of comatulids.

Plates similar in appearance and position to orals, and for that reason also called orals, occur in the tegmen of many fossil crinoids, including articulates, inadunates, flexibles, and camerates. They may rest on the radials and form the whole tegmen (Fig. 148,*1-7*), but generally they are associated with ambulacrals (Fig. 148, *8,9*) or with ambulacrals and interambulacrals (Fig. 149), or with an undifferentiated pavement of tegminal plates (*see* Fig. 151,*7*). In some streblocrinid inadunates, which have no arms and generally no radial plates, the orals have been shifted to a radial position in advanced genera.

The plates called orals in fossil crinoids are extremely variable in size; they may be very small or remarkably conspicuous. They may carry nodes or even spines, and in some cases appear quite modified. Thus, in Devonian *Cupressocrinites* and *Rhopalocrinus*, they form an annular structure (the so-called "consolidating apparatus") that served at least in part for the attachment of brachial muscles. In the Permian *Tenagocrinus* they had the same function, and also formed niches for the reception of arms (Fig. 148,*6,7*). They may meet centrally (Fig. 148,*1,3*), or be disposed around a central space **(peristome)** that in life remained open and contained the mouth (as in some Flexibilia) (Fig. 149,*2*) or was covered by ambulacrals (see below) (Fig. 148,*8,9*). They may be laterally in contact or separated by furrows. As to the ambulacral tracts, they passed under the orals (Fig. 149,*1*) or between them (Fig. 149,*2*), or they occupied grooves on their apposed edges (Fig. 149,*3*).

Oral plates of fossil crinoids are equal or unequal in size. When they are unequal, the posterior *(CD)* one is the largest. Not infrequently, this large posterior plate occupies a central position between the four others (Fig. 148,*1*), or, as in many inadunates, it is pierced by a single pore or by many pores (madreporite) (Fig. 148,*9*)—a feature suggesting that it performed the same basic function as the perforated orals of the extant *Holopus, Cyathidium,* or *Hyocrinus.*

The homology of the oral plates of fossil crinoids with those of recent ones is indicated by 1) their resemblance to these plates in shape, number, position, and orientation; 2) the relatively common occurrence of a pore or pores in the posterior one; 3) their location over the mouth or in the five angles of the mouth opening; 4) the location of the anus between the posterior one and the adjacent *C* and *D* rays; 5) the discovery of a growth series showing that these plates were already present in the earliest known stages, as if, like the orals of the larval comatulids, they were among the first plates to be developed in ontogeny (Fig. 150); 6) the chronological sequence of genera or species in some lineages, suggesting that early members had more conspicuous orals than later ones.

AMBULACRALS

The portions of the ambulacra that enter the tegmen generally do not have any skeletal floor, unless they rest over the apposed edges of the orals. In *Hybocrinus conicus,* however, a double row of large, alternating **floor plates** lies at the bottom of the groove formed by the lateral extensions of the orals—a quite exceptional feature (Fig. 148,*8*). On the other hand, covering plates or **ambulacrals** are ordinarily present; they protect the food grooves over the surface of the tegmen (Fig. 147; 148,*8;* 149; 151). They are suturally connected with the orals (if present) and (or) with the plates that occupy the interambulacral areas (i.e., interambulacrals), where such plates occur. In their simplest form, they are disposed in a single row on either side of each ambulacral tract, meeting medially along a zigzag line resulting from their alternating arrangement (Fig. 149,*1-3*). To the ambulacrals may be added side plates or **adambulacrals,** which separate them from the edges of the ambulacral tracts. Ordinarily, the side plates form a single row on each side of the ambulacrals (Fig. 151, *2,3*), but they may be arranged in a double interlocking series of outer and inner side plates (Fig. 151,*4*). The ambulacrals may be very distinct (Fig. 151,*1,2*). In some crinoids, however, the plating of the tegminal portion of the ambulacra is less regular, and it tends to merge into the plating of the interambulacral areas (Fig. 151,*5,6*). Eventually, ambulacrals and interambulacrals cease to be distinguishable, and the whole tegmen appears to be composed of an undifferentiated pavement (Fig. 151,*7*).

In flexible, articulate, and inadunate crinoids the plated covering of the ambulacral tracts traverses the perisome from the arm bases to the mouth, and passes between, or on apposed edges of the orals, if such plates are present (Fig. 148,*8,9;* 149, *2,3*; 151,*2,3*). Such condition is referred to as **suprategminal** (Fig. 148,*8,9*). In camerates, the ambulacrals may also be incorporated in the surface (Fig. 149,*1,3;* 151, *1,5*), but more generally a more or less considerable part of them is lowered beneath the surface. Traces of these subtegminal portions of ambulacral tracts in camerates may consist of 1) impressions of covering plates against the inner surface of the tegmen, 2) tubes composed of several rows of plates that connect the arm openings to the edge of a special inner structure, the so-called convoluted organ (see p. *T*201) (Fig. 152); these tubes probably contained the food grooves and associated structures, transformed into tubular passageways leading to the subtegminal mouth, or 3) grooves on the inner surface of the tegmen or canals within the tegminal plates, which have been interpreted (Haugh, 1973) as indicating the location and plan of the water-vascular system (see p. *T*195).

Some ambulacrals may be considerably modified. In the coronatid *Stephanocrinus,* each ambulacral tract in the tegmen is covered by a single pair of elongate plates, which possibly resulted from fusion of several ambulacrals (Fig. 153,*1*). The mouth of the same genus and of some cyathocrinine inadunates is covered by plates that are regarded as proximal ambulacrals. These plates—here termed **peristomials**—are larger than ordinary ambulacrals; they tend to assume, and in some cases effectively have

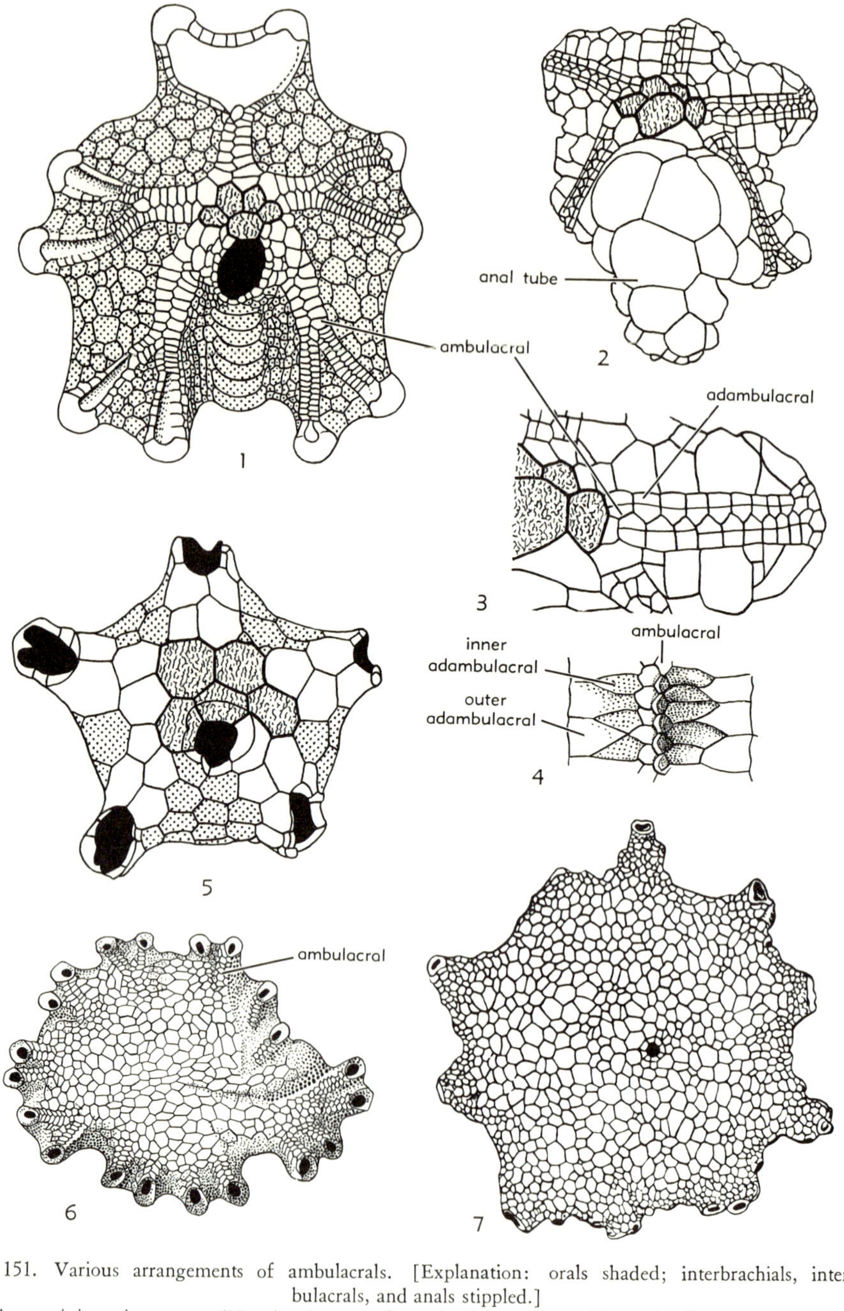

FIG. 151. Various arrangements of ambulacrals. [Explanation: orals shaded; interbrachials, interambulacrals, and anals stippled.]

1. Dimerocrinites inornatus (HALL) (camerate), M.Sil., USA(Ind.); ×3 (Wachsmuth & Springer, 1897).

2,3. Ceratocrinus exornatus WANNER (cladid inadunate), Perm., Timor; *2*, tegmen, ×1; *3*, part of same (*B* ray), ×2 (Wanner, 1937).

4. Hybocystites eldonensis (PARKS) (hybocrinid), Ord., N.Am.; portion of ambulacrum with two sets of adambulacrals, and small arched ambulacrals forming median ridge, ×15 (Springer, 1911b).

5. Steganocrinus pentagonus HALL (camerate),

assumed, a pentameral arrangement and an interradial position so that they simulate oral plates, with which they have been sometimes confounded (Fig. 153,*1*,*2*). At the same time, in many forms, the true orals are reduced, though the posterior one, commonly recognizable by the fact that it is pierced by hydropores, may remain relatively large (Fig. 153,*3*).

In a number of camerates, the ambulacrals located at the points of bifurcation of ambulacra (i.e., axillary ambulacrals) are conspicuous (Fig. 154). Known as **radial dome plates,** they commonly form prominent bosses on the tegmen. In some cases, they are hypertrophied and develop (particularly in the genus *Pterotocrinus*) into an extraordinary variety of forms; according to the species, they may be thick, rounded, club-shaped, or spatulate; they may bifurcate; they may produce knife-like blades, wing-like processes, or gigantic spines (Fig. 154,*3*,*5*,*6*). Radially disposed between the orals (if present) and the arm bases, the radial dome plates may be single (Fig. 154,*1-3*), or several in each ray (Fig. 154,*4*); they may be the only ambulacrals incorporated into the tegmen (Fig. 154,*1*), or be accompanied by ordinary ambulacrals (Fig. 154,*2*); in many cases, they are isolated among undifferentiated tegminal plates, presumably because they remained exposed on the surface, after the sinking of the ambulacral tracts below the tegmen (Fig. 154,*4*), or simply because other ambulacrals are not distinct from adjacent interambulacrals.

INTERAMBULACRALS

Plates other than orals that occur in the tegmen between ambulacral tracts and between branches of ambulacral tracts are termed **interambulacrals.** They correspond in position and nature to the interbrachials of the calyx. In a large number of crinoids, they are essential skeletal components of the tegmen. In the Flexibilia (Fig. 149,*2*), some inadunates (Fig. 149,*4*), and articulates they consist of small plates in a pliant

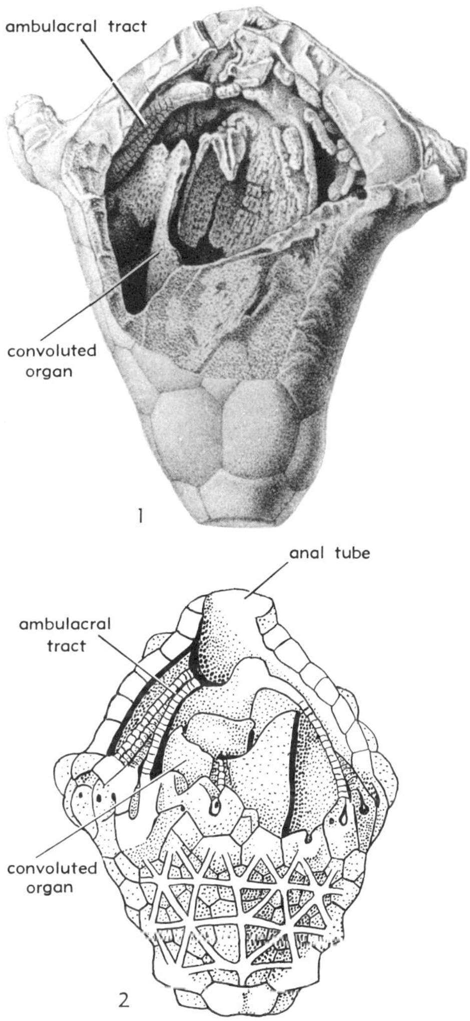

FIG. 152. Subtegminal ambulacral tracts in camerates. Specimens with one side of tegmen broken away, showing partially preserved inner structures.——*1*. *Cactocrinus glans* (HALL), Miss., USA (Iowa); ×2 (Wachsmuth & Springer, 1897).——*2*. *Cactocrinus proboscidalis* (HALL), Miss., USA (Ill.); ×2 (redrawn from Meek & Worthen, 1873).

integument, which supports the ambulacral tracts, and extends downward and outward

(Continued from facing page.)

Miss., USA; ambulacrals similar to interambulacrals, ×1.5 (Brower, 1965).
6. *Periechocrinus whitfieldi* (HALL) (camerate), Sil., USA; ambulacrals still recognizable at periphery, ×1 (Wachsmuth & Springer, 1897).
7. *Strotocrinus glyptus* (HALL) (camerate), Miss., USA (Iowa-Mo.); ambulacrals unrecognizable, ×1 (redrawn from Wachsmuth & Springer, 1897).

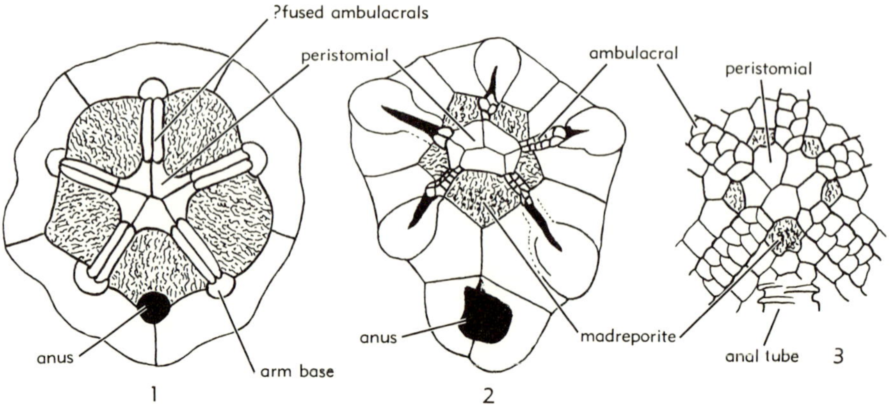

FIG. 153. Modified ambulacrals. [Explanation: orals shaded.]——*1. Stephanocrinus angulatus* CONRAD, M.Sil., USA(N.Y.); diagram of tegmen, X4 (adapted from Fay, 1961).——*2. Thetidicrinus piriformis* WANNER, Perm., Timor; X2.5 (Wanner, 1937).——*3. Gissocrinus arthriticus* (PHILLIPS), M.Sil., Eng.; X3 (Bather, 1900a).

between the rays and their divisions where it may connect the interbrachial areas. Ordinarily the latter are covered with larger and more definite plates, and the distinction between tegminal and interbrachial skeletal elements is plain, but in some forms a gradual passage between the two may be observed. In either case, however, both interambulacrals and interbrachials are judged to have the same secondary origin, and to represent different modifications of a common skeletal component.

Interambulacrals are usually present in Camerata, where with other tegminal plates they form a continuous vault, which, according to its thickness and number of component elements, may be extremely rigid or rather yielding. This vault may even be stratified, as in *Trybliocrinus,* where near the arm bases it is composed of several layers of small plates, a structure called "*Macadam-struktur*" by W. E. SCHMIDT (1937). In some camerates, a clear distinction exists between orals, ambulacrals, and interambulacrals (Fig. 151,*1*), but in others the whole tegmen appears as made of the same type of plates (Fig. 151,*7*). The latter condition may result either from the reduction and disappearance of the orals and the lowering of the ambulacral tracts below the surface, with the consequent extension of the interambulacrals over them, or from a loss of differentiation of the various component skeletal elements. Whereas some camerates (e.g., Hapalocrinidae) have no interambulacrals at all or but a few, others have many such plates. During ontogeny, as more ray and interray plates enter the cup, their number tends to grow. In primitive camerates, they merge gradually into the interbrachial pavement between the arms, but in some more advanced forms they are separated from them by the ray plates in lateral contact. Rather generally indefinite, they may acquire a high degree of specialization; thus, in such genera as *Callicrinus* and *Eucalyptocrinites* they bear vertical partitions to varying heights, which served to protect the free arms.

In most Inadunata, the interambulacrals do not appear to have reached such great development as in Flexibilia and Camerata. They are generally absent or few, and largely confined to the *CD* interray, where they participate in the formation of the anal structures. In some cases, however, they are numerous, covering the largest part of the interambulacral areas (Fig. 149,*4*).

ANAL TUBE

In recent crinoids, the anus is generally located in the *CD* interray at the summit of a fleshy conical elevation. This structure, known as the **anal tube** or **cone** or **pyramid,** according to its shape and height, may be of considerable size. A similar organ occurs in many fossil crinoids, though in some forms it is considerably modified

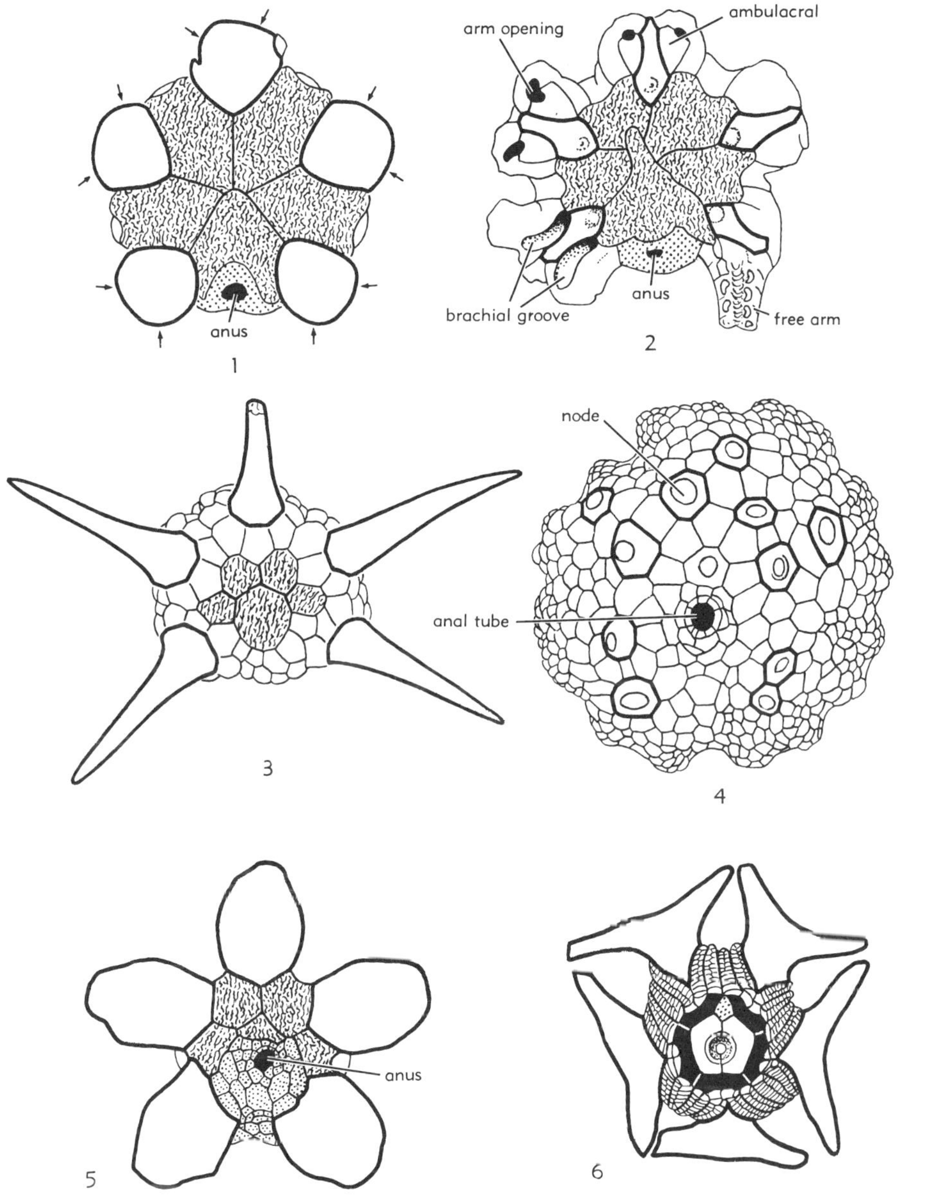

FIG. 154. Axillary ambulacrals and radial dome plates in camerates. [Explanation: radials black; anals stippled; orals shaded; axillary ambulacrals and radial dome plates with heavy outlines.]

1. *Oenochoacrinus pileatus* BREIMER, M.Dev., Spain; arrows indicate positions of arm facets, ×3 (Breimer, 1962).
2. *Oenochoacrinus princeps* BREIMER, Dev., Spain; ×3 (Breimer, 1962).
3. *Dorycrinus missouriensis* (SHUMARD), Miss., USA(Iowa-Mo.); ×1 (redrawn from Wachsmuth & Springer, 1897).
4. *Megistocrinus nodosus* BARRIS, M.Dev., USA (Iowa-Mich.); ×1 (redrawn from Wachsmuth & Springer, 1897).
5. *Pterotocrinus coronarius* (LYON), Miss., USA (Ky.); adoral view, ×1 (Springer, 1926b).
6. *Pterotocrinus bifurcatus* WETHERBY, Miss., USA (Ky.); aboral view, ×1 (redrawn from Springer, 1926b).

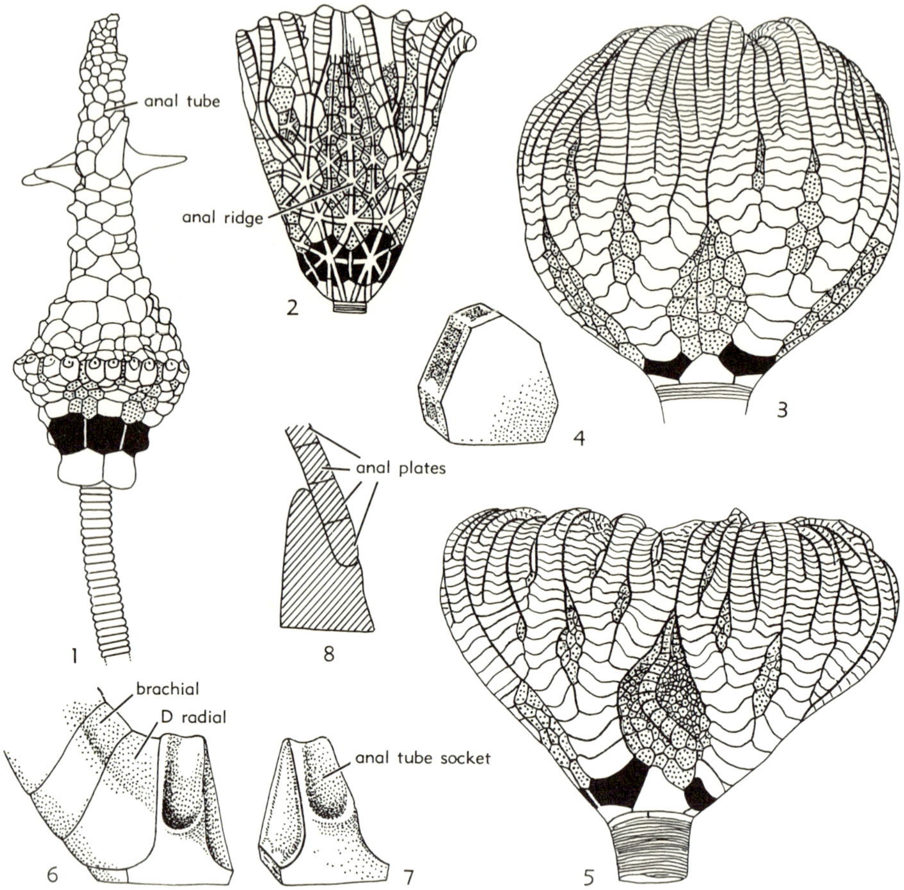

FIG. 155. Morphology and relations of anal tube in camerate *(1,2)* and flexible crinoids *(3-7)*. [Explanation: radials black; interbrachials and anals stippled.]

1. Uperocrinus nashvillae (TROOST), Miss., USA; part of stem, distal end of tube and arms lacking, ×0.45 (Wachsmuth & Springer, 1897).

2. Glyptocrinus decadactylus HALL, Ord., USA; theca seen from *CD* interray with clearly differentiated anal ridge, ×1.5 (Wachsmuth & Springer, 1897).

3. Forbesiocrinus multibrachiatus LYON & CASSEDAY, Miss., USA(Ind.); crown in *CD*-interray view, ×1 (Springer, 1920).

4. Forbesiocrinus sp.; *CD* basal, showing angular distal face, and mode of union (Springer, 1920).

5. Taxocrinus ungula MILLER & GURLEY, Miss., USA(Ind.); crown in *CD*-interray view, ×1 (Springer, 1920).

6-8. Taxocrinus sp.; *CD* basal showing rounded distal margin and deep socket for anal tube, with adjoining *D* radial and brachials *(6)*, isolated and exposed articular facets *(7)*, and in vertical section *(8)* (Springer, 1920).

and acquires an extraordinary development.

An anal tube (sometimes called proboscis) is present in various camerates. It is conical or subcylindrical, and of quite variable size. In some genera it is relatively short, whereas in others it rises well above the distal end of the arms (Fig. 155,*1*). It is generally composed of many irregular, commonly nodose or spinose, plates. In various early camerates, such as *Reteocrinus* and *Glyptocrinus,* it is supported by a line of ridged plates continuous with the anal plates of the calyx (Fig. 155,*2*). This ridge (anal ridge) is connected with the ridges

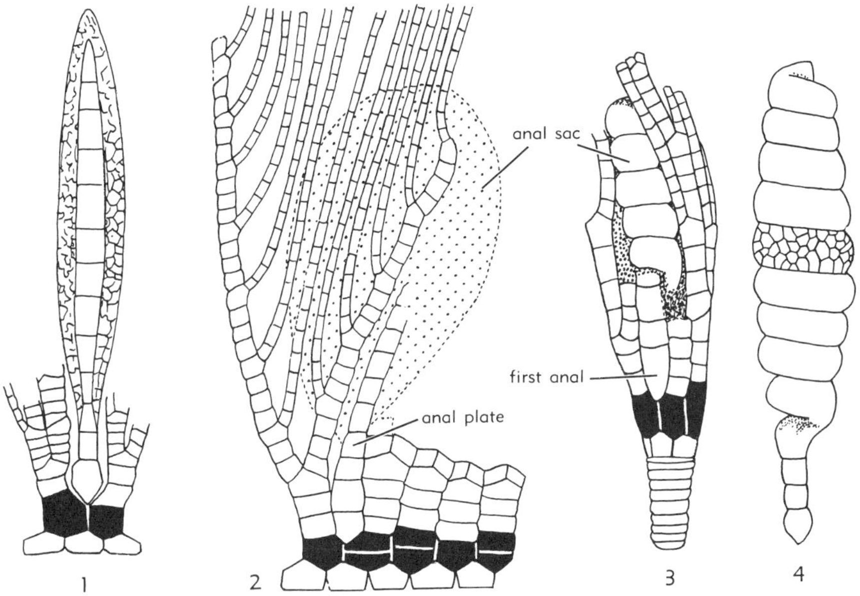

FIG. 156. Morphology of anal tube or sac in disparid inadunates. [Explanation: radials and biradials black.]——*1. Dystactocrinus constrictus* (HALL), U.Ord., USA(Ohio); diagram of part of crown and anal sac (adapted from Ulrich, 1925).——*2. Daedalocrinus kirki* ULRICH, Ord., Can.(Ont.); diagram of part of crown and anal sac (Ulrich, 1925).——*3. Ohiocrinus brauni* ULRICH, U.Ord., USA(Ind.); *CD*-interray view of crown, ×4.5 (redrawn from Ulrich, 1925).——*4. Ohiocrinus laxus* (HALL), U.Ord., USA(Ohio); anal series and spiral anal sac, enl. (Ulrich, 1925).

that unite the *CD* basals with adjacent radials, and this suggests that a nerve cord passed up to govern the motion of the tube. In later forms, where the tegmen is more rigid and the plates of the tube are firmly joined together, the anal ridge disappears (BATHER, 1900a). The anal tube of camerates may be regarded as an outgrowth of the *CD* interambulacrum. Secondarily it may occupy a central or even an anterior position. Its presence is ordinarily considered to be a generic characteristic, rarely (as in *Platycrinites*) a specific one only. It has been suggested by HAUGH (1973) that it could have contained a respiratory structure in addition to the hind-gut.

In flexible crinoids, as in articulates, the anal tube never acquires gigantic size or marked differentiation (Fig. 147). It is a mere conical protuberance from the *CD* interambulacrum, but may have different relations with the calyx. In some genera like *Forbesiocrinus* its plating prolongs upward the rather solid pavement that fills the *CD* interbrachial area of the calyx, so that the tube itself originates at some height above the posterior *(CD)* basal, which is not markedly differentiated (Fig. 155,*3,4*). On the contrary, in such genera as *Taxocrinus* or *Onychocrinus* the anal tube rests directly on the posterior basal, in a socket and a groove that serve for attachment of the proximal ossicle of a series of strong articulated anal plates; these plates support a portion of the pliant tegmen and raise it into an inverted funnel-shaped structure containing the rectum and anal opening (Figs. 147, 155,*5-8*).

It is among inadunates that the anal tube reaches its highest degree of diversity and specialization. While in some groups like the Hybocrinida (Fig. 148,*8*), and Coronata (Fig. 153,*1*), and in such genera as the disparid *Haplocrinites* (Fig. 148,*4,5*) and *Zophocrinus,* or the Cyathocrinitidae, *Carabocrinus* and *Porocrinus,* it is entirely lacking or reduced to a slight protuberance made of small plates, in most members it is a well-developed and commonly complex structure. In disparid inadunates, it is

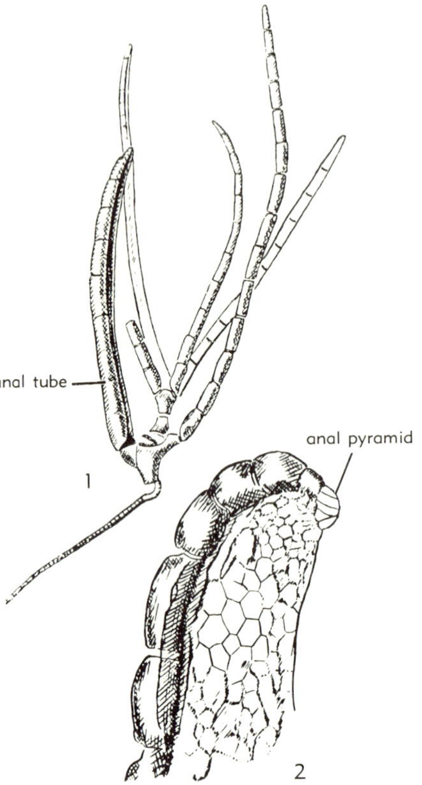

FIG. 157. *Senariocrinus maucheri* W. E. SCHMIDT, L.Dev., Ger.; *1*, crown, ×2; *2*, distal part of anal tube, ×3 (W. E. Schmidt, 1934).

closely related to the adjacent *C* ray (Fig. 156). The series of plates that support it may rest directly on the *C* radial or biradial, or on the *B* and *C* superradials, which have fused in a single plate, or on left shoulder of the second or the third *C*-ray plate, taking the place and, in some cases, the appearance of an arm or an arm branch (see anal plates, p. *T*121). Thus, in *Eustenocrinus* (and probably also in *Ramseyocrinus*) it rests on the whole distal articular facet of the *C* radial or superradial and consists of a single series of brachial-like plates ending in a small anal pyramid (Fig. 93, p. *T*122). In Calceocrinidae also, the anal tube is the sole appendage borne by the *C* superradial or by the fused *B* + *C* superradials, and its uniserially arranged plates closely resemble brachials: they are massive and crescentic in cross section, with their hollowed side enclosed by a saclike covering of small, irregular plates (Fig. 87,*1,5*, p. *T*114); Fig. 157). In other disparids, the anal tube generally consists of an elongate, segmented, armlike appendage that supports a variously expanded wall of small plates, the whole structure being tubelike, saclike, balloon-shaped, or other form, according to the genus (Fig. 156,*1-4*). In consideration of these anatomical relations and general appearance, the anal tube of the disparids has been commonly interpreted as representing an evolutionary modification either of the *C*-arm itself or of a branch of this arm.

The anal tube of the cladid inadunates appears to be somewhat different. Primitively, it does not rest on the radianal, which

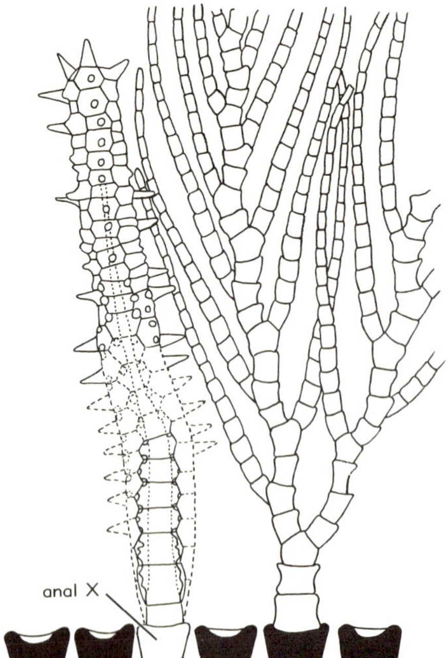

FIG. 158. *Botryocrinus bellensis* KESLING, M.Dev., USA(Mich.); plate diagram of cup, anal sac and *B* ray, showing strong median row of sac plates resting on anal *X* (McIntosh in Kesling, 1973). [Explanation: radials black.]

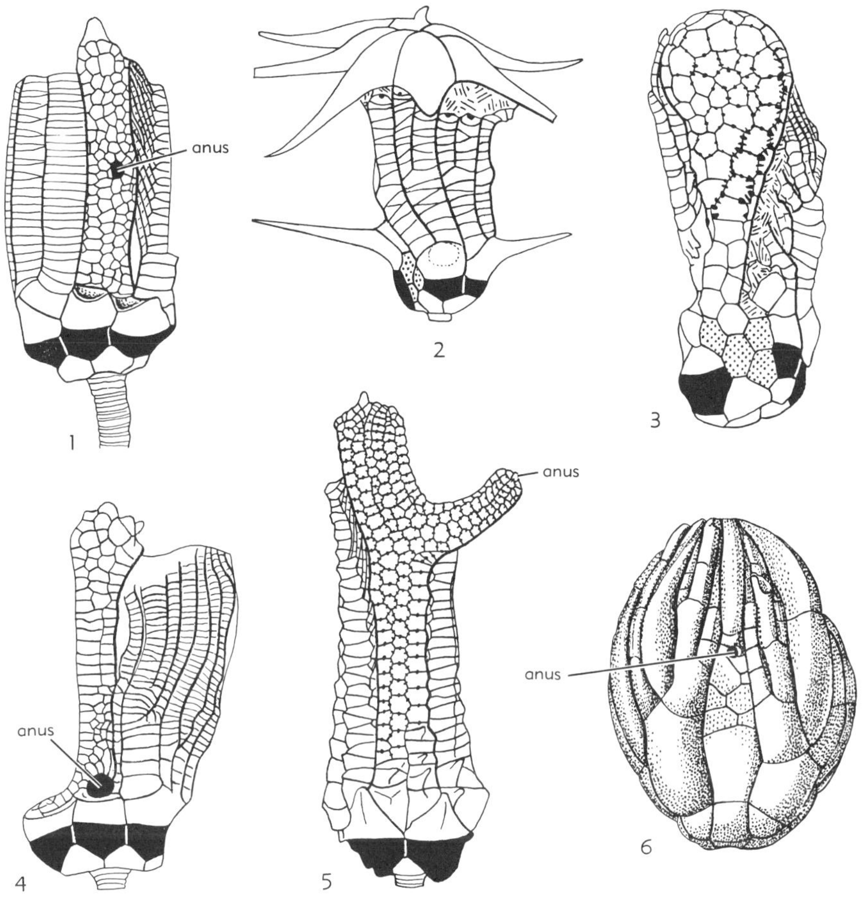

FIG. 159. Morphology of anal sac in cladid (poteriocrinine) inadunates. [Explanation: radials black; anals stippled.]

1. *Scytalocrinus validus* WACHSMUTH & SPRINGER, Miss., USA(Ind.); anterior view; arms of *A* and *E* rays broken off at base, ×1 (Springer, 1926b).
2. *Tholocrinus wetherbyi* (WACHSMUTH & SPRINGER), Miss., USA(Ky.); *C*-ray view; mushroom-shaped anal sac, except distal spiniferous canopy, concealed by arms, ×1.5 (Springer, 1926b).
3. *Linocrinus arboreus* (WORTHEN), Miss., USA (Ala.); *CD*-interray view, ×1.5 (Springer, 1900a).
4. *Eratocrinus commaticus* (MILLER), Miss., USA (Mo.); ant. view; arms of *A* and *B* rays broken off at base, ×1.5 (Springer, 1926b).
5. *Aulocrinus agassizi* WACHSMUTH & SPRINGER, Miss., USA(Ind.); lat. view; arms in front broken off, ×1 (Springer, 1926b).
6. *Timorechinus mirabilis* (WANNER), Perm., Timor; *CD*-interray view, ×1.5 (Wanner, 1916a).

is the sole anal plate that originates as a *C*-ray plate, but it is supported by the interradially located anal *X,* where the strong median row of plates begins when present (Fig. 94,*1-3,* p. *T*123; Fig. 158). Its proximal plates enter into contact with the radianal, but only secondarily, after that element has been shifted from its original position at the base of the *C* ray toward the posterior *(CD)* interray (Fig. 94,*4,* p. *T*123). These facts suggest that the anal tube of the cladid inadunates was not supported, as in disparid crinoids, by a modified process from the *C* ray, but was merely

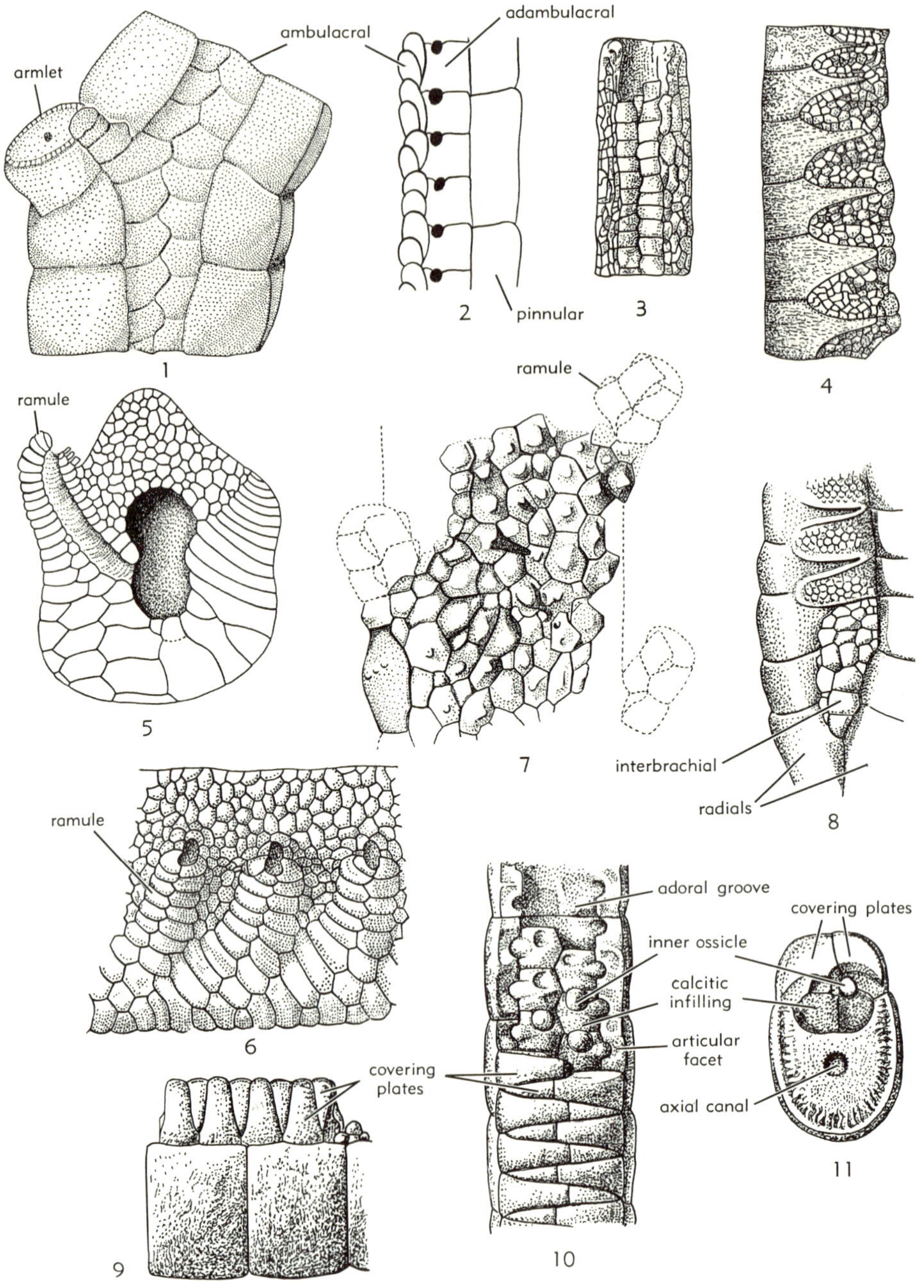

Fig. 160. Perisomic skeleton of the arms.

1. Barycrinus hoveyi (Hall), Miss., USA(Ind.); note single row of ambulacrals over groove of armlet opposed to double row over groove of arm, ×6 (Van Sant, in Van Sant & Lane, 1964).

2. Neometra spinosissima (A. H. Clark), recent, Ind.O.; lat. view of portion of pinnule, enl. (A. H. Clark, 1923).

3,4. Carabocrinus huronensis Foerste, M.Ord., Can.(Ont.); adoral and lat. views of portion an extension from the posterior interambulacrum, as are the anal tubes of camerate, flexible, and articulate crinoids.

It is among cladid crinoids that this organ reaches its greatest diversity in size and shape (Fig. 159). In size it varies from a small protuberance to a huge inflated structure, the so-called "ventral sac," that may exceed the calyx in size and the arms in height, and may involve almost the entire tegmen. It may be cylindrical, pyramidal, club-shaped, balloon-shaped, or mushroom-shaped; it may taper to a pointed end, or enlarge distally to a more or less rounded, commonly nodose or spiniferous terminal surface (Fig. 159,*2*); it may be longitudinally ridged, folded, coiled upon itself, and provided with a lateral branchlike process (Fig. 159,*5*); or it may form niches for the protection of arms, as in the Permian *Timorechinus* (Fig. 159,*6*). It is generally composed of longitudinal rows of plates, commonly provided with strong folds and deep pits, and ornamented with radial ridges. In some species numerous round or slitlike pores, presumably respiratory, occur on the suture lines between the plates (Fig. 159,*3*) (see p. *T*208). The anal opening may be located at the summit (probably the primitive position), but it may also appear at various places along the anterior side: at the base of the sac (Fig. 159,*4*), midway (Fig. 159,*1*) toward its summit, or at the end of a lateral spoutlike process (Fig. 159,*5*). Such anterior position of this opening may be explained by the recurving and more or less complete redoubling of the gut and of its enveloping tube upon themselves, so that the anus may emerge through the wall at any point between the distal end and the base. This deflection and reversal of the gut and tube, the walls of which coalesce where they are brought into contact (as may be observed when ridged rows of plates occur continuously from one end to the other), accounts for the distal expansion and, at least to some extent, the large size of the saclike tube. But this does not exclude the possibility that the latter contained organs other than the gut itself and performed functions other than to protect this part of the digestive tube. But which organs and which functions are unknown.

ADORAL SKELETON OF ARMS AND PINNULES

The perisomic teguments of the arms and pinnules are probably never entirely free from any skeletal elements. These vary considerably in size, from tiny spicules almost completely limited to small epidermal folds or lappets bordering the ambulacral grooves on either side, to well-defined plates, the so-called covering-plates or **ambulacrals.** In their simplest form, they are arranged in two alternating rows (rarely in a single row), one on each side of the ambulacral grooves. They are movable, and could be everted or closed down over the grooves. When closed, the two rows meet in the median line by a zigzag suture (Fig. 160,*1*). Between the ambulacrals and the adoral edges of the brachials and pinnulars may be added small, usually squarish or oblong, side-plates or **adambulacrals,** so that four rows of plates protect the ambulacral grooves (Fig. 160,*2*).

But the perisomic skeleton of the arms may be much more complex. Thus, in some camerates, especially those that possess powerful brachial trunks, the tegminal pavement extends far up the main arms, which appear to have formed tubular structures (Fig. 160,*5-7*); in *Trybliocrinus* the adoral covering is even stratified, and it forms a highly arched roof under which the ramules

(Continued from facing page.)

of arm, ×6 (Foerste, 1925).

5,6. Trybliocrinus flatheanus (Geinitz), L.Dev., Spain; *5,* transv. section through proximal part of brachial trunk with stratified ambulacral covering; *6,* portion of brachial trunk in lat. view, ramules broken off, *ca.* ×5 (Breimer, 1962).

7. Ctenocrinus sp., L.Dev., Belg.; portion of brachial trunk in adoral view with tesselated ambulacral covering, ×12 (Ubaghs, 1945b).

8. Cupulocrinus jewetti (Billings), M.Ord., Can. (Ont.); proximal part of *E* ray in lateral view, ×6 (Springer, 1911b).

9-11. Cyathocrinites acinotubus (Angelin), Sil., Gotl.; *9,10,* lat. and adoral views of brachials and covering plates, ×10; *11,* articular face of brachial, with covering plates in position, ×10 (Bather, 1893a).

penetrate rather deeply (Fig. 160,6). In the diplobathrid *Traskocrinus,* a plated weblike structure between tegmen and arms extends on sides of the ambulacral tracts apparently throughout all arms.

In the cyathocrinine inadunate *Carabocrinus huronensis,* the adoral groove of the arms is protected by two median rows of quadrangular plates, accompanied on each side by a stripe of small irregular plates, which extends laterally as well over depressions developed between the successive brachials (Fig. 160,*3,4*). A similar arrangement seems to have existed in at least part of the arms of the dendrocrinine inadunate *Cupulocrinus jewetti* (Fig. 160,*8*). In some species of the cyathocrinine genera *Cyathocrinites* and *Gissocrinus* heavy plates divided by one or more transverse sutures cover the adoral groove of the arms, to the edges of which they are united by articular facets; they look like small pinnules, and as such have been interpreted by several authors (particularly Sieverts-Doreck, 1952); beneath these plates and alternating with them, minute, inner, rounded ossicles may also occur (Fig. 160,*9-11*).

VARIOUS SKELETAL STRUCTURES

In this section are described skeletal structures that generally are not located in a single part of the body such as the column, the theca, or the arms, but extend into several body regions, or are related to both the aboral and the adoral skeletons (e.g., the cup and the tegmen). Most of these structures are associated with organic systems (i.e., nervous, digestive, water-vascular, or reproductive systems), or are composed of elements combined for the accomplishment of a particular function, like respiration or digestion.

STRUCTURES ASSOCIATED WITH NERVOUS SYSTEM

The crinoids have three nervous systems related functionally and morphologically: the ectoneural, hyponeural, and entoneural systems. The first one, being entirely within the soft tissues, leaves no trace on the skeleton and consequently is unknown in fossil crinoids. According to Haugh (1975), the hyponeural system was represented by ridges and grooves on the inner tegminal surface of many camerates; it has not been detected in fossil representatives of the other subclasses. As to the entoneural or aboral nervous system, which is the main motor system of the crinoids, it lies within the aboral skeleton or near its inner surface, and therefore its former presence is frequently indicated in fossil crinoids by canals within the plates or by grooves or ridges on the interior thecal surfaces, or on internal molds.

The entoneural system is centered in a ganglionic nervous mass surrounding a coelomic cavity divided into five chambers. Typically this **chambered organ,** as it is called, is lodged within the ring of basals in monocyclic and pseudomonocyclic forms, or of infrabasals in dicyclic ones. If the proximal or the two proximal circlets of plates are reduced it may rise up to the next circlet (Fig. 161,*1*); if they are entirely lacking, or are overgrown by the radials, as in roveacrinid and cyrtocrinid articulates, the whole organ occupies the bottom of the relatively wide cavity enclosed by these plates (Fig. 162). In comatulid articulates it is included in the centrodorsal cavity below the basal plates reduced in size or fused into a rosette (see p. *T*108).

Generally no trace of the chambered organ is preserved in fossil crinoids. In some camerates, inadunates, and flexibles, however, the proximal or the next to proximal circlet of cup plates carries an inner funnel-shaped structure which probably housed this organ (Fig. 161,*1,2*). This calcareous structure is trigonal or subpentagonal in outline, and it is divided by short septa into lobed depressions continuous with the stem lumen.

Coelomic vessels, each with a sheath of nervous tissue, proceed from the five lobes of the chambered organ and penetrate directly into the cirri in comatulids, or in stalked crinoids into the axial canal of the column. These coelomic and nervous extensions into the stem may give off branches that go into the cirri, pseudocirri, and

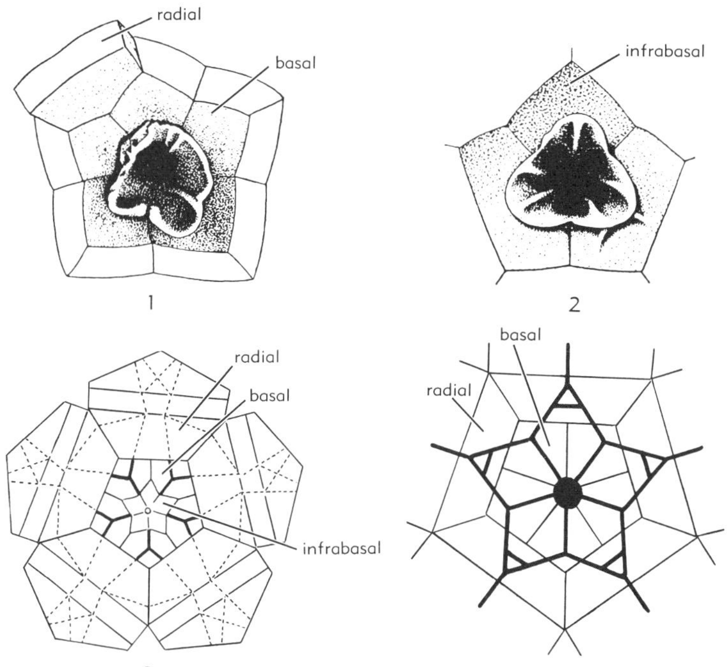

FIG. 161. Morphology of entoneural system.

1. *Icthyocrinus* sp., Sil.-Dev.; funnel-shaped structure lodging chambered organ; infrabasals too minute to support funnel, which rests upon basals, ×8 (Springer, 1920).
2. *Forbesiocrinus saffordi* HALL, Miss., USA(Ind.-Tenn.); funnel resting upon infrabasals, ×5 (Springer, 1920).
3. *Encrinus* sp., Trias., Ger.; course of axial canals within calyx and primibrachs; canals represented by dotted lines when penetrating plates, and by heavy lines when exposed on inner surface of plates (Beyrich, 1858).
4. *Rhizocrinus lofotensis* M. SARS, recent, Atl.; distribution of entoneural canals within the calyx (Carpenter, 1884a).

radicular cirri, which, like the column, are all provided with an axial canal. A description of these longitudinal passageways through the column and its appendages has already been given (p. *T79*).

Relatively little is known about the entoneural system in the crown of fossil crinoids. Its general pattern, however, is clearly shown in fossil articulates and some dicyclic inadunates, the brachials and pinnulars of which are perforated by a single, or rarely a duplicate (as in the Pennsylvanian *Aesiocrinus* or the Triassic *Encrinus*), canal **(axial canal)** that contained the main brachial nerve (Fig. 161,*3*). Proximally the axial canal of each ray penetrates into a radial where it divides into two branches, one of which proceeds to the basal on the right, the other to the basal on the left. If it is duplicate, the two canals simply diverge and, as in the preceding case, extend downward to two separate basals. Within the radials, the axial canals of the five rays are connected by a commissural canal that generally forms a pentagonal ring all around the cup. Proximal to this pentagonal ring interradial commissures may also occur at the level of the basals. Ultimately, as shown by the recent forms, the cords that are housed in this canal system unite into five primary trunks or pass directly into the nervous sheath of the chambered organ.

Various modifications of this scheme may occur. Thus, in some bathycrinid articulates like *Bathycrinus* or *Democrinus* the five primary trunks rise through half the height of the radials before they fork (Fig.

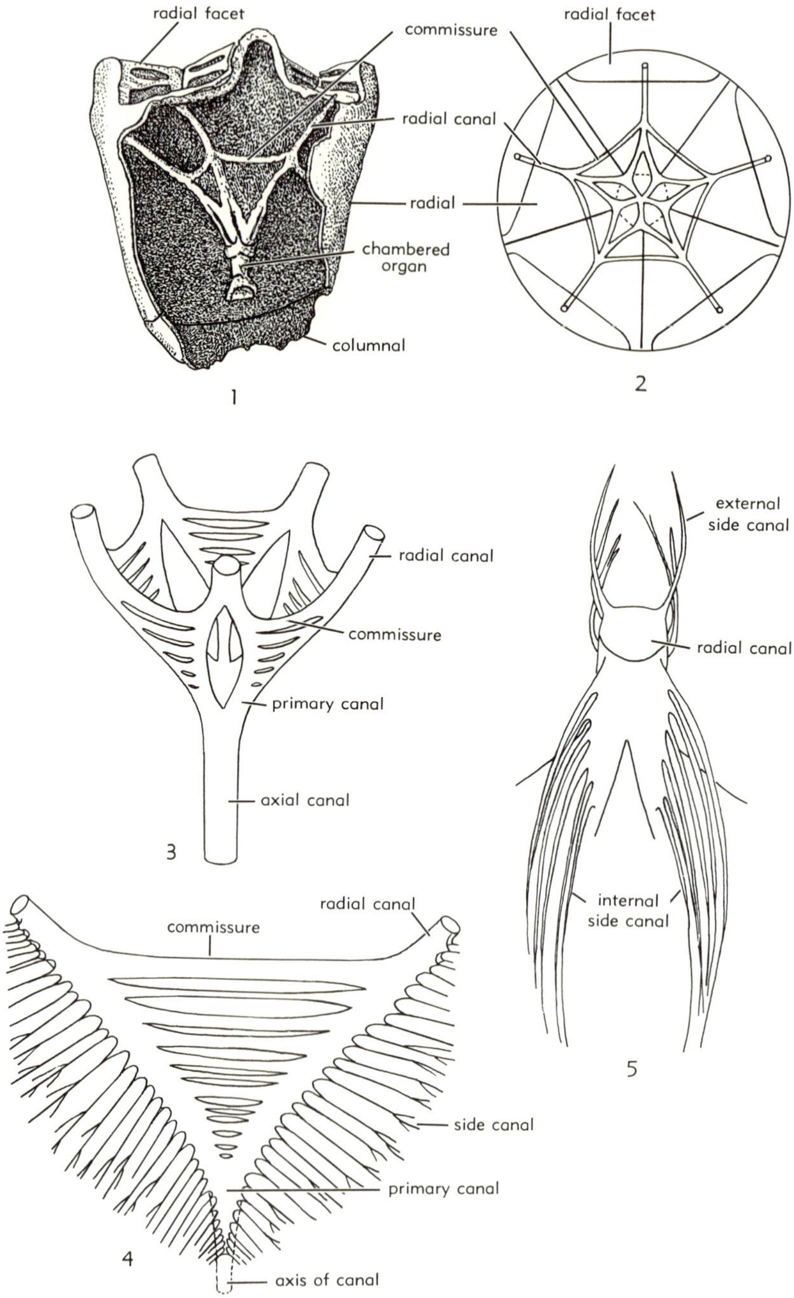

FIG. 162. Course of axial canals within the calyx in cyrtocrinid articulates.——*1,2. Eugeniacrinites caryophyllatus* SCHLOTHEIM, Jur., Eu.; *1,* calyx broken open to show silicified axial canals, ×4; *2,* distribution of axial canals within calyx (Jaekel, 1891a).——*3-5. Gammarocrinites strambergensis* JAEKEL, Jur., Ger.; *3,* reconstruction of entoneural canal system, side canals omitted; *4,* detail of canal system between two adjacent radii; *5,* detail of radial canal with side canals in ventrolateral view (Zitt, 1973).

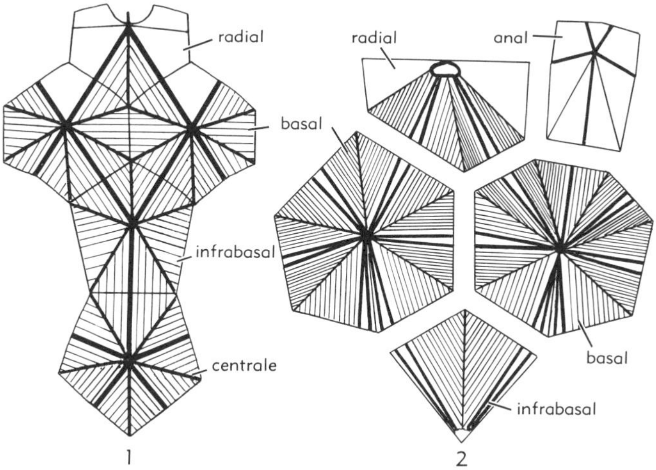

FIG. 163. Diagram of complex entoneural canal system within the calyx. [Explanation: heavy lines indicate primary canals, intermediate lines secondary canals, weak lines tertiary canals.]——*1. Marsupites testudinarius* (SCHLOTHEIM) (diagramm.), Cret., worldwide (Sieverts, 1927).——*2. Calceolispongia* (diagramm.), Perm., Australia, Timor, India (Teichert, 1949).

161,*4*). In cyrtocrinid articulates that have the cup composed of radials only, the axial canal from the stem, at a short distance above its entrance into the cup, widens and takes a form that suggests that the chambered organ was located there (Fig. 162). Five interradial primary canals proceed from this region, each one dividing into two diverging branches. At various levels of the radials, the neighboring branches meet to form radial canals that pass directly up to the radial facets and the arms. In addition, the branches are connected with each other by a single or, as in *Cammarocrinites strambergensis* (Fig. 162,*3-5*), multiple commissures. They may also give off side canals toward the periphery of the cup and toward its central cavity (Fig. 162,*5*).

In the Permian inadunate *Calceolispongia* and in the Cretaceous articulate *Marsupites* a complicated system of canals is present inside the plates near their inner surface (Fig. 163). In the ornamented varieties of *Marsupites* the position of these canals corresponds to the ornament of the calyx. Three sets of canals may be distinguished: 1) main canals, which enter the plates near the middle of the facets and join in the center of the plates, 2) secondary canals that connect the center of the plates with the corners, and 3) tertiary canals that branch off from the secondary canals and cross the facets at right angles. The secondary and tertiary canals are absent from some plates or sectors of some plates in both *Marsupites* and *Calceolispongia,* and they are entirely lacking in the smooth varieties of *Marsupites.*

An unusual branched (?) canal system has also been observed (PAUL, 1970) in a few calycal plates of a specimen of the Silurian camerate genus *Masupiocrinus.* It consists of a delicate branching and anastomosing structure, which has been [incorrectly according to HAUGH, 1975] interpreted by PAUL (1970) as located within the stereom. Each branch has a median keel and is crenulate with short processes that alternate on either side.

Such complex canal systems as those just mentioned have not been found in any recent crinoids. If they housed branches of the entoneural system, the function of these branches is not clear, for the aboral nervous system serves primarily for innervating muscles, and there are no muscles within the calycal plates.

In flexible, camerate, and most inadunate crinoids the entoneural system was not enclosed within the plates, except very rarely. Nevertheless, in some of these crinoids, particularly camerates, the course of the aboral

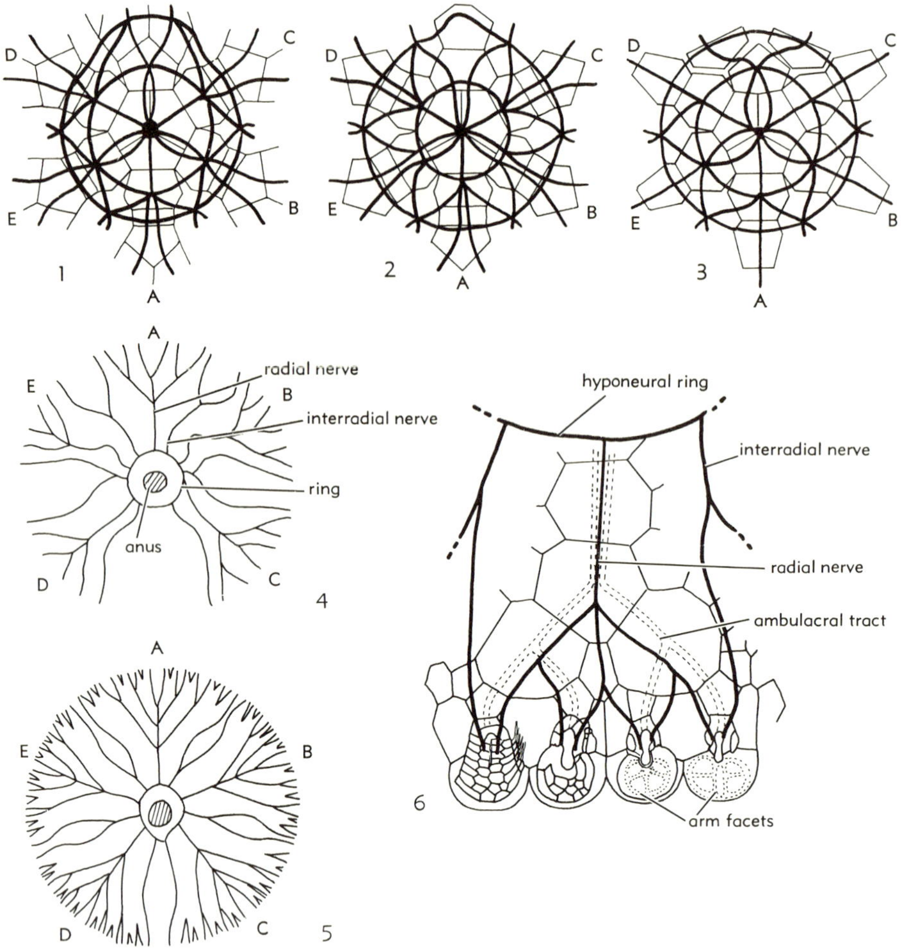

FIG. 164. Diagrammatic plan of the entoneural *(1-3)* and hyponeural *(4-6)* systems in some camerate crinoids (all figures from Haugh, 1975b, slightly mod.). [Explanation: *A, B, C, D, E,* letters designate rays according to CARPENTER.]——*1,4. Batocrinus icosidactylus* CASSEDAY, Miss., USA.——*2. Dorycrinus quinquelobus* (HALL), Miss., USA.——*3. Steganocrinus pentagonus* (HALL), Miss., USA.——*5. Teleiocrinus rudis* (HALL), Miss., USA.——*6.* View from exterior of a part of a tegmen as if it were transparent in order to show the course of the hyponeural nerves in a single ray and adjacent interrays. (From *Paleobiology,* v. 1, no. 3, Summer, 1975, courtesy of The Paleontological Society and Bruce N. Haugh.)

nerve cords within the calyx appears to be indicated by grooves or ridges on the inside of the calycal plates, or on the surface of internal molds (BROWER, 1973; HAUGH, 1975). It has been frequently supposed also that the external ornamentation of the calyx of fossil crinoids reproduced the plan of the entoneural system, but to what extent is not clear.

The evidence at hand suggests that the entoneural network in camerates radiated from the basal region of the calyx at the point of the stem attachment where the chambered organ was presumably located (Fig. 164,*1-3*). From this point, six primary nerve trunks proceeded, one to each ray and one to the anal interray. These trunks were single or double, and they divided or separated at the primaxil of each ray, and at all succeeding higher order axillaries. In addition, branches were present in each of the interrays with, however,

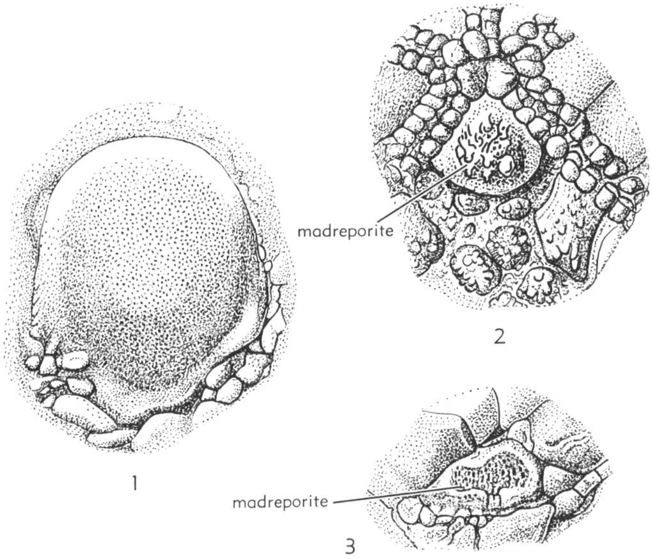

FIG. 165. Morphology of the madreporite.——*1. Aexitrophocrinus jaekeli* MOORE & PLUMMER, L.Carb., Scot., USSR; ×6 (Springer, 1920).——*2. Gissocrinus incurvatus* (ANGELIN), U.Sil., Gotl.; ×6 (Bather, 1893a).——*3. Euspirocrinus spiralis* ANGELIN, U.Sil., Gotl.; ×6 (Bather, 1893a).

in *CD* interray a pattern different from that in the other interrays. Probably two ring commissures interconnected the nerve trunks, but their former presence is commonly vague or even lacking in actual specimens. The resulting general plan of this network closely resembles the plan of the aboral nervous system of the recent crinoids, though the presence of the radial-like *CD* interray trunk produces a dual, trimeral symmetry as opposed to the pentameral symmetry of modern forms (HAUGH, 1975).

Another network of ridges and grooves occurs on the tegminal inner surface of many camerates. According to HAUGH (1975), it would correspond to the hyponeural system of the living crinoids because its plan is very similar to that of this system in the latter. It appears to have consisted of a ring commissure, which was located near the upper part of the tegmen, and from which originated radial and interradial branches. These branches divided so that each free arm received two hyponeural cords, as in modern crinoids (Fig. 164,*4-6*). If this interpretation is correct, the camerates possessed at least two (and probably three) nervous systems homologous to those of the recent crinoids.

STRUCTURES ASSOCIATED WITH WATER-VASCULAR SYSTEM

In most fossil crinoids, the water-vascular system has left no trace, or only scanty ones, and one may only suppose it was essentially similar to that of recent forms. A single (more rarely two or three) external pores or a narrow slit or a perforated small tubercle occurs on the posterior *(CD)* oral plate of some disparid, many hybocrinid and cyathocrinine inadunates; this structure has been commonly interpreted as a hydropore (Fig. 148,*4*). In many flexible genera in which the tegmen is known, an enormously enlarged posterior oral is perforated by numerous pores, and certainly performed the function of a madreporite (Fig. 165,*1*). Similarly, in various cyathocrinine inadunates, the posterior oral had the structure, and most likely the role, of a typical madreporite (Fig. 165,*2,3*).

The presence of a madreporite in dendrocrinine and poteriocrinine inadunates has been rarely reported. However, one such plate has been observed in the dendrocrinine *Parisocrinus* by LANE (1975) and in the poteriocrinines *Stellarocrinus* and *Clathro-*

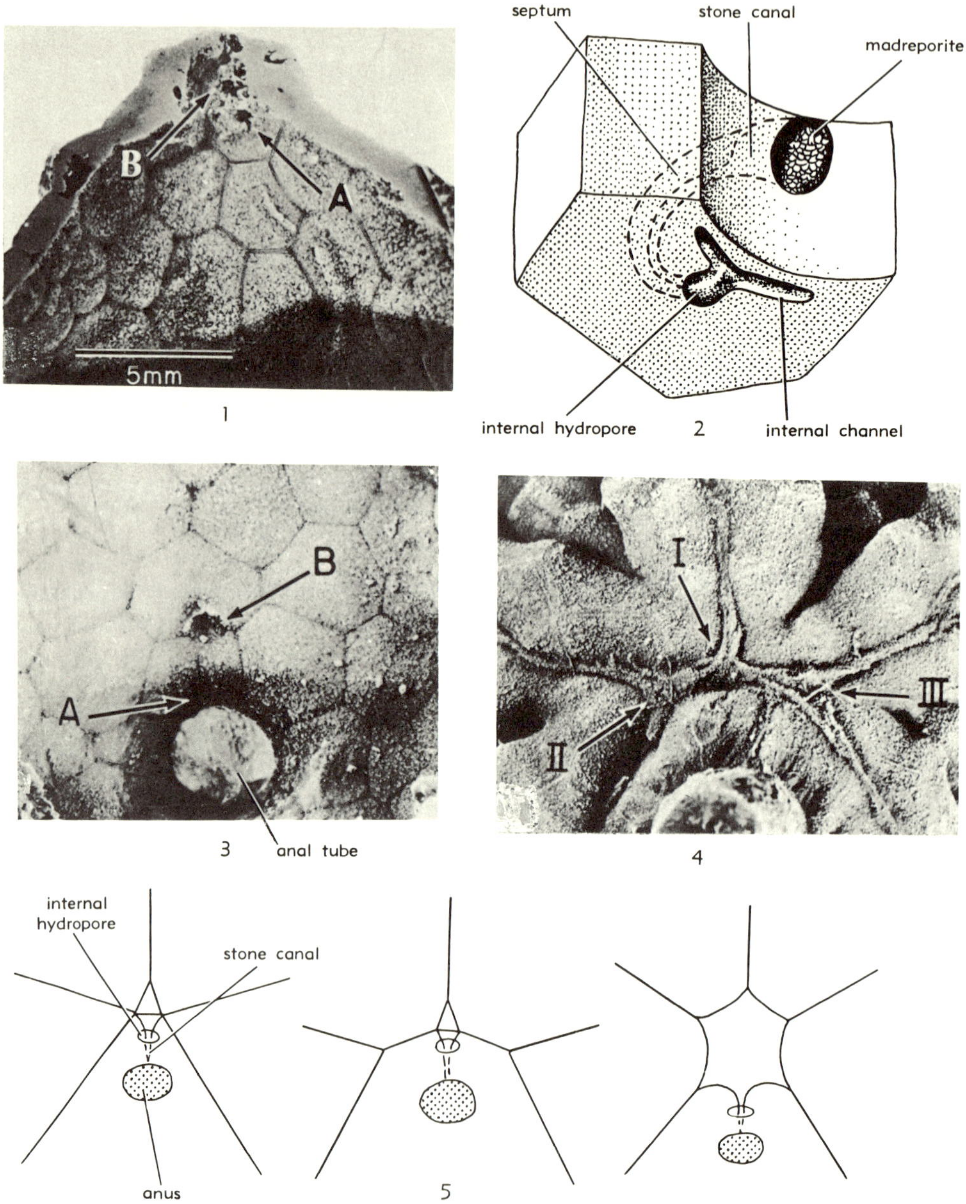

FIG. 166. Structures related to water-vascular system in camerates.

1,2. Batocrinus icosidactylus CASSEDAY, Miss., USA (Ind.); *1*, vertically cut hollow specimen showing location of internal hydropore *(A)* and madreporite *(B)* at base of anal tube, ×4.4; *2*, posterior oral plate with entrance of water-vascular system, diagram. (Haugh, 1973).

3. Gilbertsocrinus tuberculosus (HALL), Miss., USA (Iowa); part of inner face of tegmen with internal madreporite *(A)* and hydropore *(B)*, ×3.75 (Haugh, 1973).

4. Actinocrinites lowei (HALL), Miss., USA (Mo.); part of chert mold with three-fold *(I, II, III)* convergence of radial grooves (here appearing as ridges), ×3.75 (Haugh, 1973).

5. Various patterns of water-vascular system in camerates according to HAUGH (1973).

crinus by STRIMPLE (1973a). According to LANE (1975), a complex tubule system in the anal sac of the Pennsylvanian poteriocrinine inadunate *Aesiocrinus magnificus* might possibly represent a hypertrophied madreporitic system that would be an intermediate evolutionary step between the single madreporite plate of more primitive inadunates and the multiple-element ciliated funnel system of articulates.

Unlike flexible and inadunate crinoids, an external hydropore or madreporite has never been recorded for any camerate crinoid. It is true that the function of this structure was previously assigned (BATHER, 1900a) to the so-called respiratory pores, which in some camerates are located near arm bases (see below, p. *T*207), but this interpretation is made unlikely by the discovery of a structure which probably is an internal hydropore (HAUGH, 1973). This structure has been particularly well observed in *Batocrinus icosidactylus* (Fig. 166), but it has been found also in other camerates and may have existed in many. It is situated in the posterior oral plate at the base of the anal tube. The side of this plate facing the cavity of the tube contains a pore, which may be protected by a calcareous sieve. This pore is followed by an incurved canal within the plate, and this canal, commonly divided by an incomplete longitudinal septum, leads to an internal hydropore with two lateral channels on the inner surface of the same plate. According to HAUGH, this canal would be the stone canal and the two channels, the marks of two branches connecting the stone canal with the water ring or equivalent structure. If so, then the entrance to the water-vascular system in camerates would have been close to the distal part of the gut and completely internal, as in living holothurians.

The other parts of the water-vascular system in fossil crinoids are practically unknown, except possibly in some camerates where radial grooves on the inner surface of the tegmen probably indicate (at least to some extent) the course of the ambulacra and the morphologic plan of the water-vascular system within the theca, but what exactly contained these grooves remains uncertain (Fig. 166; 167,*2,3*). Their diameter is much larger than that of the water vessels of living echinoderms. In individuals of *Strotocrinus* they penetrate into the plates and are completely within the stereom in some places. If generally each groove is single, it may also be multiple and anastomosing, and may even be provided with many small lateral branches (Fig. 166,*4*). The grooves may be preserved peripherally and be lacking centrally, as if the structures they were associated with rested against the tegmen near the arm bases and became subtegminal as they approached the center. They may also be entirely lacking, suggesting that the structures in question were largely located below the tegmen. In forms where the grooves meet centrally, their connection may be quite complex, and the grooves may converge to one, three, or five points (Fig. 166,*4,5*). HAUGH (1973) detected four morphologic patterns in the arrangement of these structures. If this arrangement is a real and reliable expression of the morphologic plan of the water-vascular system, it suggests that the latter differed in organization and symmetry from that of recent crinoids. Thus, according to this view, the radial ambulacral canals in most camerates would have been supplied by a crescent-shaped rather than by a complete ring canal.

STRUCTURES ASSOCIATED WITH DIGESTIVE SYSTEM

The mouth is situated at the focus of the ambulacral tracts. It is generally exposed on the tegmen in articulate and flexible crinoids, but in inadunates and camerates, it opened typically beneath the tegmen—a condition known as subtegminal. Ordinarily the mouth is located approximately at the center of the body. Secondarily it may be displaced along the *A* radius or *AB* interradius; in some cases, as among many recent Comasteridae or in the Cretaceous articulate *Uintacrinus*, it is quite marginal (Fig. 168,*1*). Such shifting is accompanied by an increase in size of the anal area, which tends to occupy a large surface of the tegmen.

In most articulates the mouth consists of a round, oval, or crescentic opening in the plated or naked integument that forms the tegmen. In some articulates (e.g., *Cyathi-*

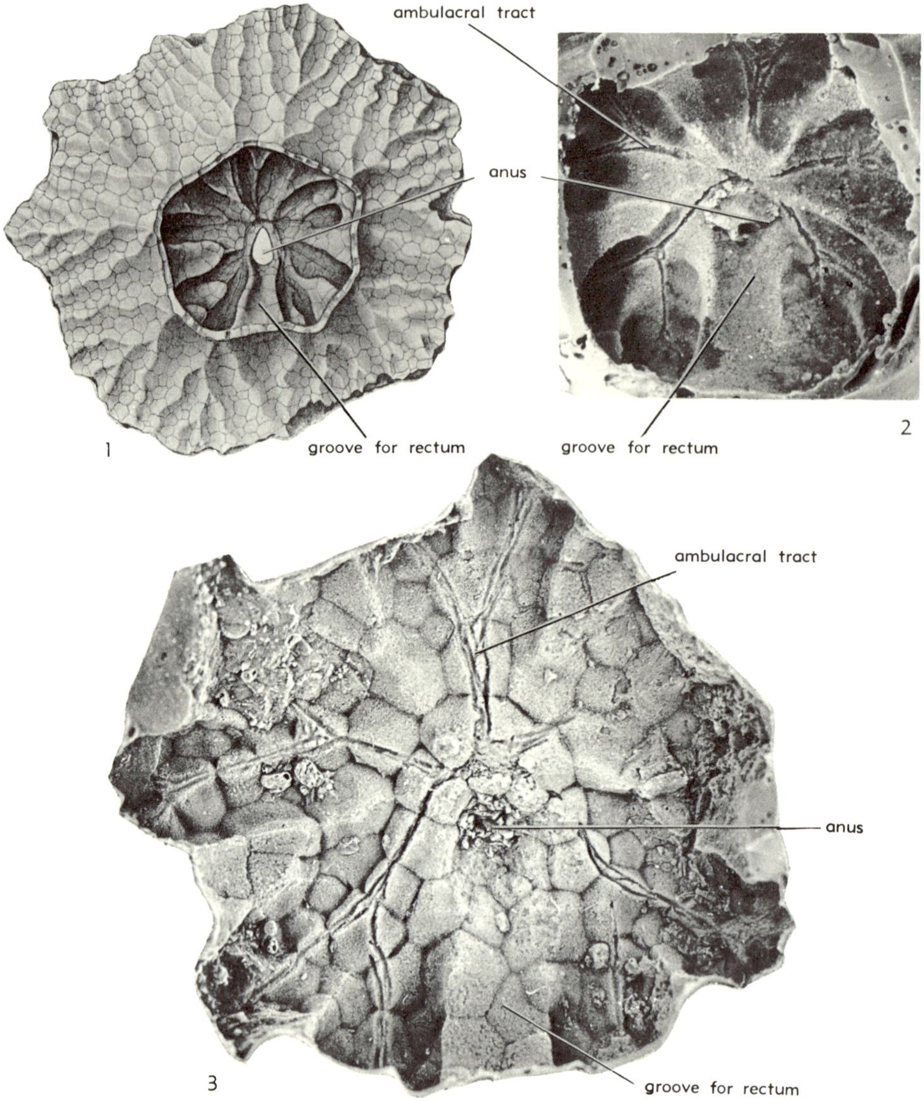

FIG. 167. Inner tegminal structures in camerates.——*1. Strotocrinus regalis* (HALL), Miss., USA(Iowa-Mo.); theca in aboral view, with part of aboral cup broken away, exposing subtegminal galleries and grooves for rectum and ambulacral tracts, ×0.67 (Wachsmuth & Springer, 1897).——*2. Physetocrinus ornatus* (HALL), Miss., USA(Mo.); inner view of tegmen, latex cast, USNM S 1259, ×2 (Ubaghs, n).——*3. Strotocrinus regalis* (HALL), Miss., USA(Mo.); inner view of tegmen, latex cast, USNM S 1275, ×3 (Ubaghs, n).

dium, Holopus, Hyocrinus) and in those flexibles the tegmen of which is known, it is surrounded or covered by five oral plates, which occupy the apices of the interambulacral areas, the ambulacral tracts running between these plates or beneath their apposed edges.

The mouth of all Paleozoic crinoids except flexibles was covered either by orals, proximal ambulacrals (peristomials) (Fig. 153), or undifferentiated perisomic plates. So far as is known, this plated covering could not be opened in life. Some microcrinoids, however, that lacked arms and in

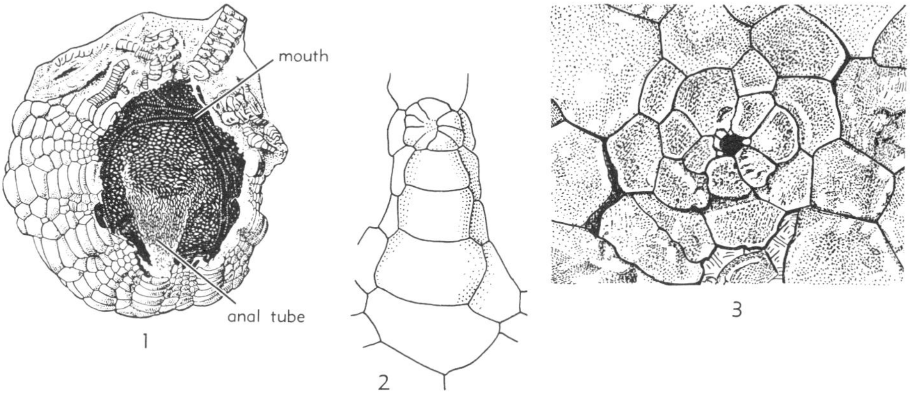

FIG. 168. Skeletal structures associated with digestive tube.——*1. Uintacrinus socialis* GRINNELL, U.Cret., N.Am.; specimen vertically compressed, showing black tegminal integument, ×0.67 (redrawn from Springer, 1901).——*2. Stelidiocrinus laevis* ANGELIN, U.Sil., Gotl.; Stockholm Museum, no. 11774, series of anal plates with terminal anal pyramid, ×7.5 (Ubaghs, n).——*3. Eucalyptocrinites tuberculatus* (MILLER & GURLEY), M.Sil., USA(Ind.); anal cover plates, surrounded by accessory plates, enl. (Macurda, n, mod.).

some cases even an external vent were probably able to open their oral plates and expose their soft structures in order to receive food and expel feces.

An external anal opening occurs invariably in crinoids, though, as just said, none has been recorded in some microcrinoids (e.g., *Atremacrinus, Pentececrinus*). Typically located in the *CD* interambulacrum, it pierces the tegmen directly or (more generally) is raised on an anal pyramid, tube, or sac. Usually in such cases it is situated at the summit or at the posterior side of this structure, but in cladid inadunates provided with an inflated sac it may occur at the anterior side of the sac, either at the base, or part way up, or at the end of a lateral spoutlike process (Fig. 159,*1,4,5*). The vent may also occupy various other positions. In extant comasterids, in the Cretaceous *Uintacrinus* (Fig. 168,*1*), in the Triassic *Encrinus,* and in some camerates with an anal tube it becomes central or subcentral. In the camerate *Siphonocrinus* it may open anteriorly at or even beneath the arm-bases. It has migrated downward into the posterior side of the cup to a position just above the posterior *(CD)* basal in the cyathocrinine inadunate Gasterocomidae, and in most Codiacrinacea; furthermore, some members of the latter superfamily have it in perradial position or between a radius and an interradius. It is laterally located also in the hybocrinid genus *Cornucrinus,* probably as the result of hypertrophy of the radial plates.

The anus of crinoids is commonly surrounded by small plates and protected by a valvular pyramid. In many fossil forms it was located in a well-defined thecal orifice, the **periproct,** which in life was covered by a small cone of triangular plates **(periproctals)** or by a plated periproctal membrane. Both these structures are rarely preserved (Fig. 168,*2,3*).

In most recent and, probably, fossil crinoids the gut makes a single dextral coil, with the mouth in the axis of the coil, a condition known as **endocyclic.** In comasterid articulates it makes several coils, and they do not wind around the axis of the mouth, which becomes marginal, but around the axis of the central anus. Such a gut is said to be **exocyclic.** Some fossil crinoids, like *Uintacrinus,* which had a marginal mouth and a central anus, may possibly have had an exocyclic gut, but this remains to be proved.

The digestive tube of most fossil crinoids has generally left no traces or but scanty ones. In camerates, however, hollow silicified specimens commonly contain highly

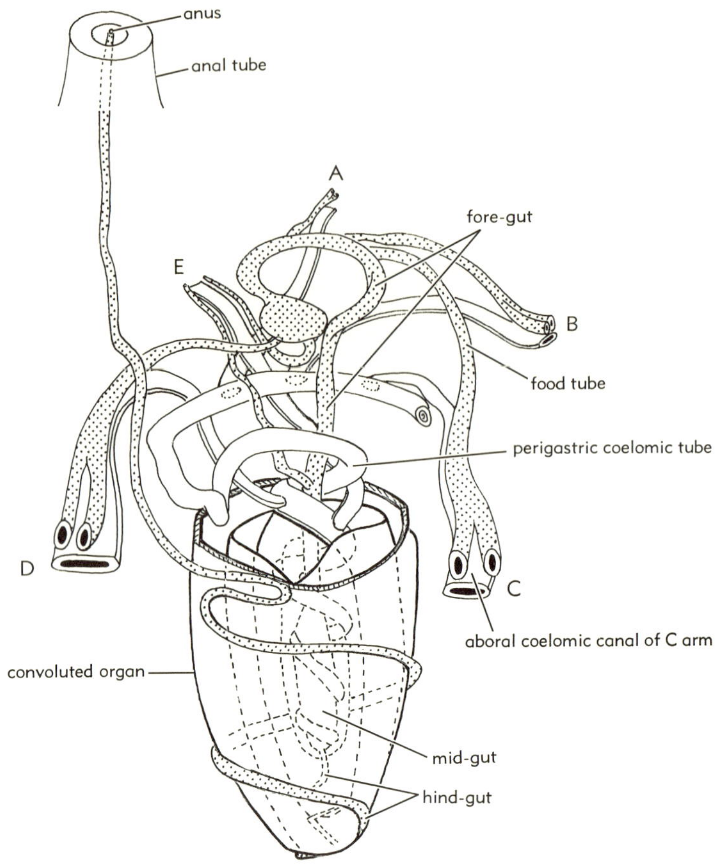

FIG. 169. Diagrammatic reconstruction of the digestive system and part of the coelomic system of camerate crinoids. [Explanation: food tubes and digestive tube stippled. Rays identified by symbol letters of CARPENTER system] (mod. from Haugh, 1975a). (From Journal of Paleontology, v. 49, no. 3, May, 1975, courtesy of The Paleontological Society and Bruce N. Haugh.)

distinctive internal structures that have been interpreted as remnants of the digestive and other visceral organs (HAUGH, 1975). It is probable that in life calcitic spicules included in the walls of these organs formed a supportive framework that has been preserved after death by secondary calcification or silicification. We have already noted (p. T133) that the food grooves of the arms were transformed into tubular conduits as they entered the theca. In some cases these conduits appear to comprise an alternating series of floor-plates, probably accompanied by covering plates. They converged into five or six ducts that led, not into an open mouth as do the ambulacral tracts of the flexible and articulate crinoids, but directly into the fore-gut, each duct reaching this structure at a separate point (Fig. 169). The lack of an open mouth and its replacement by a bulbous closed terminal chamber were apparently required to prevent the food from penetrating into the visceral cavity.

The fore-gut, after a loop, bent abruptly downward, descended vertically, and then expanded in an elongate mid-gut, from the bottom of which started a constricted tube, the hind-gut. The latter, after one or two turns inside the so-called "convoluted organ" (see below), emerged from the base of this structure, wound around its exterior generally three or four times, and finally entered the anal tube or ended directly in the vent. Ordinarily near the upper part of the convoluted organ the coiling direction reversed—a change that took place in the *CD* interray.

STRUCTURE PROBABLY ASSOCIATED WITH COELOMS

The theca of many camerates from the Ordovician to the Mississippian contains a coiled axial body, known as the **convoluted organ** (Fig. 170). This body extends from beneath the origin of the food tubes down to the basal circlet. Open at both ends, it may be fusiform, cylindrical, or trochoid, and appears to have consisted of a helically coiled, vertical wall. This wall in the usual state of preservation is solid and thickly incrusted with inorganic precipitates, probably deposited after burial, but in life it seems to have consisted of a mesentery strengthened by a spicular calcitic meshwork. In many instances, it includes a dark median layer of degraded organic matter from the original tissue.

The convoluted organ, a name introduced by WACHSMUTH & SPRINGER (1879), was first taken by WACHSMUTH (1877) for the alimentary canal itself. Later on WACHSMUTH & SPRINGER (1881, p. 35 [209]) suggested that it could represent "an extensive plexus of blood vessels surrounding the ambulacral canals" and be called the "oesophageal network." BATHER (1900) considered it as a spicular calcification of the connective tissue around the axial sinus, apparently serving to support the digestive tube. As seen above, it certainly surrounded the fore-gut and the mid-gut passing down its hollow axis and carried the hind-gut coiling around it. There is little doubt that its spicular framework belonged to the visceral skeleton, and that the body cavity was divided by this organ into an inner "perigastric subcoelom" and an outer "peripheral subcoelom." HAUGH (1957), who proposed these designations, calls this structure the **"perigastric coelomic organ."**

Within the theca of some camerates one may observe an inner calcareous lining that varies greatly in appearance (Fig. 171). It may consist of simple pillars, or star-shaped protuberances with many rays and deep depressions between the rays, or stereomic extensions that spread out and join similar extensions at their edges, thus forming a perforated plated structure held away from the thecal plates by pillars. This lining extends under the tegmen and over most of the thecal interior but does not cover the basal plates (Fig. 172). Between it and the thecal wall, a space is left that certainly belonged to the coelom, and, because this structure surrounded most of the viscera, it has been termed **perivisceral coelomic organ** by HAUGH (1975).

At the lower (proximal) edge of the lining, passageways formed openings for the aboral nervous cords, which, coming from the chambered organ and leading to the arms, were located between the lining and the inner thecal plate surfaces (Fig. 131). At the upper (distal) end of the lining, there are six extensions: two in the *CD* interray and one in each of the other four interrays (Fig. 173). Those four latter extensions connect directly with a ring commissure **(perivisceral coelomic ring)** pressed against the upper, inner surface of the tegmen. The two extensions in the *CD* interray are attached to an auxiliary ring, which in its turn is connected with the main ring commissure (HAUGH, 1975).

These complex perivisceral structures in camerates have no certain equivalent in other crinoids. They may possibly correspond to the perihemal coelomic system of the extant forms, though they differ from it by their great expansion in the body cavity, and their apparent lack of tubular branches into the arms. Their discovery under the tegmen of many specimens led some authors (WACHSMUTH, 1877; WACHSMUTH & SPRINGER, 1891; P. H. CARPENTER, 1884a) to consider that in camerates a second integument equivalent to the disk or tegmen of the recent forms occurred below the outer pavement, a subtegminal "disk," to which the term "vault" or "dome" was applied and which was regarded as a structure peculiar to Paleozoic crinoids, but lacking in their successors. It was supposed that the "vault" represented a part of the aboral system of plates that has grown to cover and protect the originally flexible tegmen. Such distinction between "vault" and "disk" was rejected long ago (NEU-

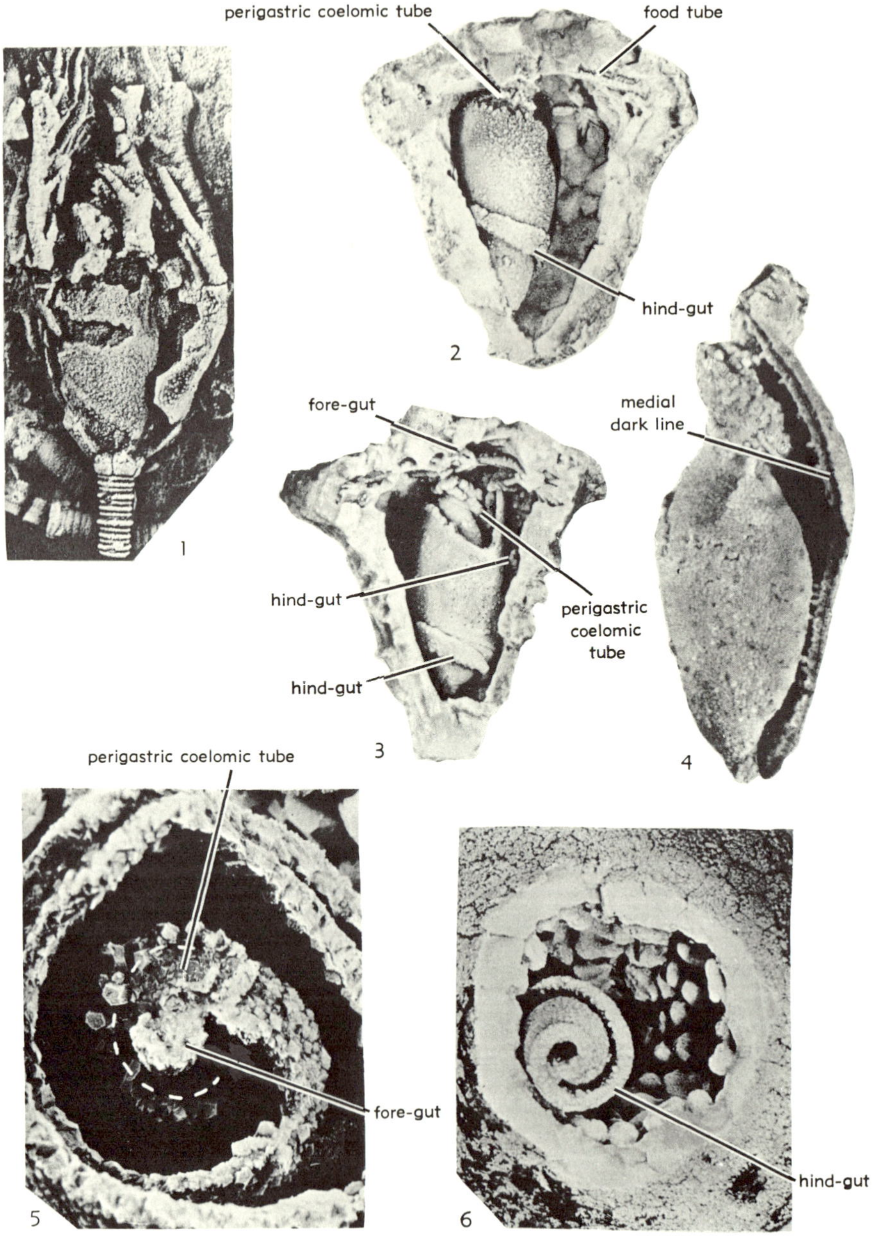

FIG. 170. Convoluted organ of camerate crinoids (*2-6,* from Journal of Paleontology, v. 49, no. 3, May, 1975, courtesy of The Paleontological Society and Bruce N. Haugh).

1. Ptychocrinus fimbriatus (SHUMARD), M.Ord., USA(Mo.); theca open showing convoluted organ (lat. view), ×4 (Brower, 1973).

2. Teleiocrinus umbrosus (HALL), L.Miss., USA (Iowa); theca open with whole convoluted organ, ×1.1 (Haugh, 1975a).

3. Teleiocrinus liratus (HALL), L.Miss., USA (Iowa); theca open with convoluted organ partly broken, ×0.8 (Haugh, 1975a).

4. Actinocrinites verrucosus (HALL), Miss., USA (Ind.); central part of convoluted organ (Haugh, 1975a).

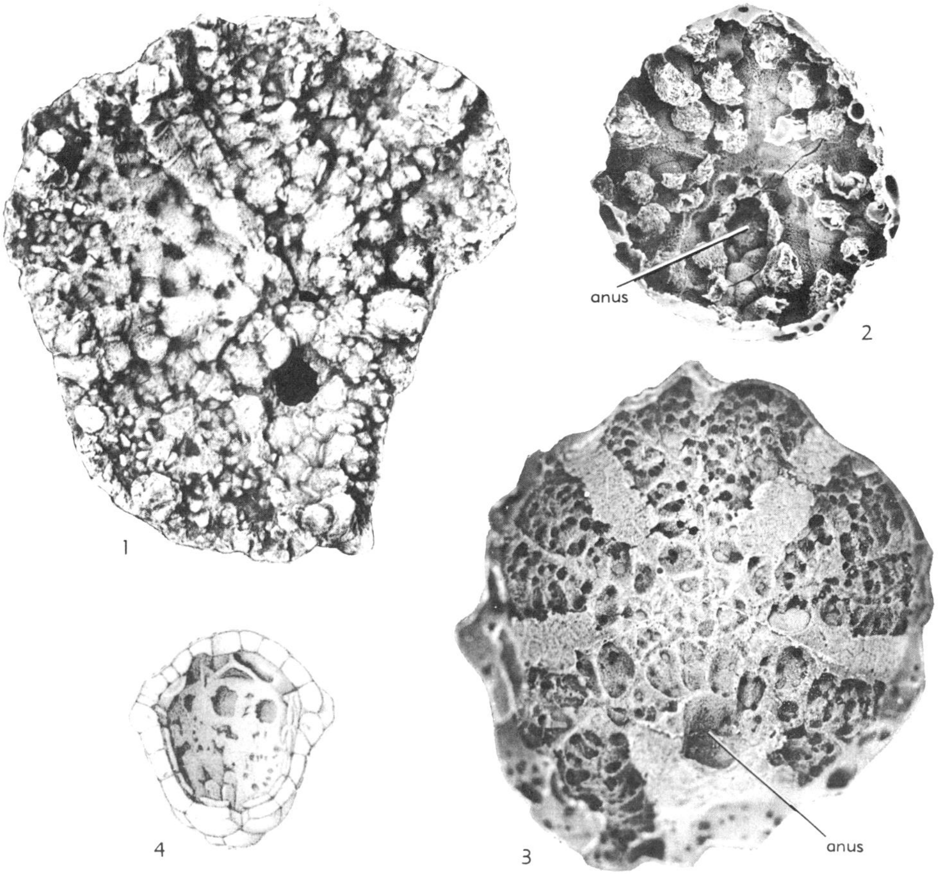

FIG. 171. Perivisceral coelomic lining of camerate crinoids.——*1. Physetocrinus ventricosus* (HALL), Miss., USA(Iowa); inner aspect of portion of tegmen, U.S. Natl. Museum S 1262, ×2 (T. Phelan, n). *2. Eutrochocrinus christyi* (SHUMARD), Miss., USA; inner view of tegmen, latex cast, U.S. Natl. Museum S 885, ×2 (Ubaghs, n).——*3. Teleiocrinus* sp., Miss., Iowa; inner view of tegmen, latex cast, U.S. Natl. Museum SPRINGER Coll., ×2 (Ubaghs, n).——*4. Uperocrinus longirostris* (HALL), Miss., USA (Iowa); broken theca showing perivisceral lining within aboral cup and tegmen, ×1 (Wachsmuth & Springer, 1897).

MAYR, 1889; BATHER, 1890a). It now appears to be groundless.

According to HAUGH (1975), from whom all these proposals have come, tubular coelomic structures were also present in the arms and in the theca of the camerate crinoids in addition to the major body coelom (Fig. 169, p. *T*200). Best observed at their entrance in the thecal cavity, these structures appear as flattened "tubes" that divide the keyhole-shaped arm openings into upper and lower halves. Each free arm has such a tube, and these tubes merge into five (one per ray) primary canals where they enter the body cavity. Just above the convoluted organ, these canals connect with a

(Continued from facing page.)

5. Strotocrinus glyptus (HALL), L.Miss., USA (Iowa); convoluted organ with top broken away revealing fore-gut and perigastric coelomic tube, ×4.5 (Haugh, 1975a).

6. Eutrochocrinus christyi (SHUMARD), L.Miss., USA(Iowa); hind-gut emerging from base of convoluted organ, ×2 (Haugh, 1975a).

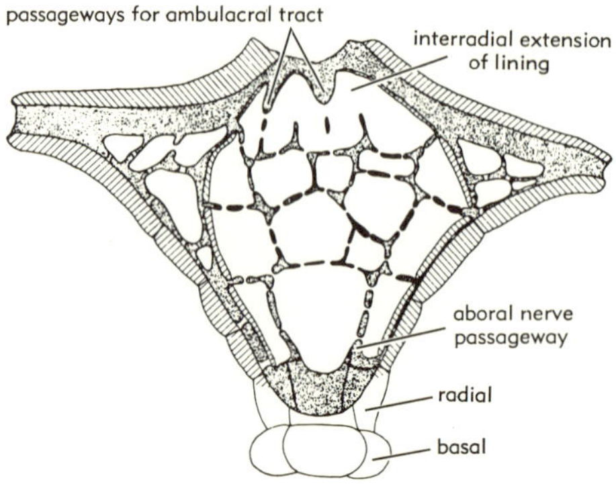

Fig. 172. Diagrammatic reconstruction of the perivisceral coelomic lining of a generalized camerate crinoid (lateral view). The lining is held away from the thecal plates by pillars (Haugh, 1975a). (From Journal of Paleontology, v. 49, no. 3, May, 1975, courtesy The Paleontological Society and Bruce N. Haugh.)

loop-shaped portion of a contorted tube **(perigastric coelomic tube),** the two ends of which do not appear to join any other structures. This tube penetrates the convoluted organ, wraps tightly around the fore-gut and mid-gut, and ends near the base of the mid-gut. The flattened tubes of the arms may be homologous to the aboral coelomic canals of the arms of recent crinoids, though in modern forms these canals do not extend into the thecal cavity but connect directly to the perivisceral compartment of the coelom.

Evidence for the existence of another coelomic system in the arms of camerates corresponding to the subambulacral (subtentacular) canals of the extant crinoids is not conclusive, but this is not proof that such a system was lacking.

STRUCTURES POSSIBLY ASSOCIATED WITH REPRODUCTIVE SYSTEM

In recent crinoids the gonads are usually located in the genital pinnules, more rarely in the arms, but never in the theca. This peripheral location of the gonads is one of the most distinguishing features of the crinoids, which in this respect stand alone among the stalked echinoderms. Each gonad of a crinoid is connected with a strand

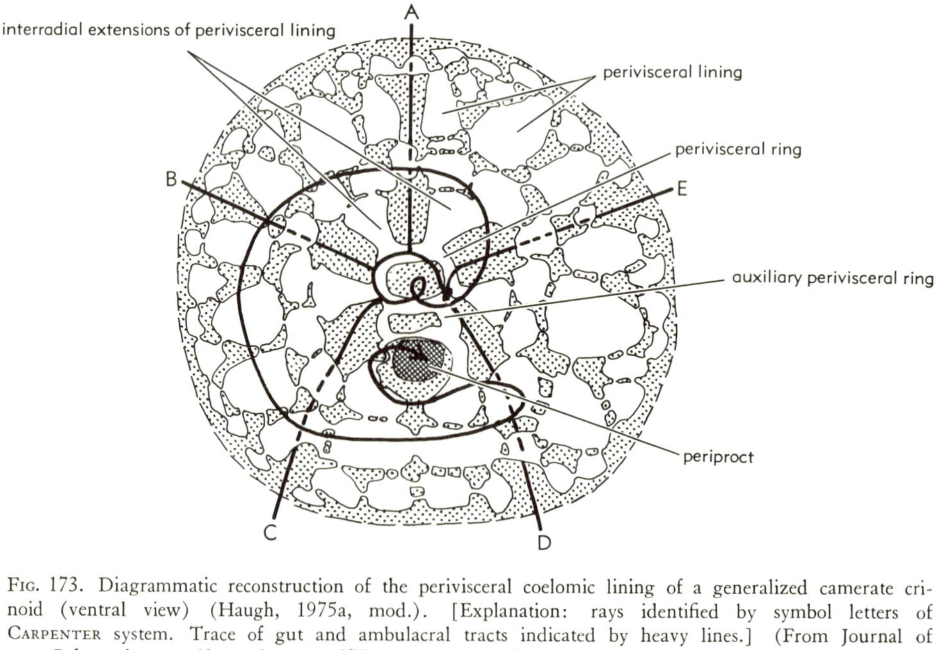

Fig. 173. Diagrammatic reconstruction of the perivisceral coelomic lining of a generalized camerate crinoid (ventral view) (Haugh, 1975a, mod.). [Explanation: rays identified by symbol letters of Carpenter system. Trace of gut and ambulacral tracts indicated by heavy lines.] (From Journal of Paleontology, v. 49, no. 3, May, 1975, courtesy The Paleontological Society and Bruce N. Haugh.)

of cells (genital cord) included in a coelomic canal (genital canal) that transverses the arm and penetrates the theca where it is gradually lost in the meshes of the subtegminal lacunar plexus.

There is no direct evidence of the presence of gonads in the arms or pinnules of fossil crinoids. JAEKEL (1895), it is true, interpreted the distal widening of the pinnules of *Acanthocrinus* and a velum-like extension along the pinnules of *Hapalocrinus elegans* as possibly connected with the former existence of gonads in these two Devonian camerates. But in both cases detailed studies of the structures under question have shown that they were not fit for such a function (W. E. SCHMIDT, 1934; UBAGHS, 1947). Neither is there proof that the gonads were located within the theca and still less within the column as sometimes supposed (YAKOVLEV, 1922). A pore in the posterior *(CD)* oral plate of several inadunates such as *Hybocystites, Hobocrinus, Haplocrinites, Porocrinus,* or *Lageniocrinus* might well be a hydropore.[1] Pores or slits in the oral plates of *Cupressocrinites* have been regarded as gonopores by SIEVERTS-DORECK (SIEVERTS, 1934), who attributed the same meaning to the pores in the anal sac of many cladid inadunates. A similar supposition was made by GISLÉN (1924), who thought that a part of the gonad, in addition to a part of the intestine, may have been enclosed in the anal sac, and that such intrathecal location of the gonad would explain the extension of the adoral thecal perisome high up between the arms, as seen in many flexibles and inadunates. One or two marginal extensions of the radials of the allagecrinid genus *Trophocrinus* have been conjectured by KIRK (1930) as possibly representing "brood pouches" of a sexually dimorphic type, but this interpretation has been judged not entirely conclusive by PECK (1936) and rejected by STRIMPLE & KOENIG (1956).

STRUCTURES ASSOCIATED WITH RESPIRATION

Skeletal structures presumed to have had respiratory functions are far less common in crinoids than in other pelmatozoan echinoderms, very likely because the arms of the former offer a much more extensive surface for gaseous exchanges with the sea water than the food-gathering appendages of the latter. Nevertheless, stereomic structures generally regarded as having had a respiratory function occur in some Paleozoic (mainly Lower Paleozoic) species, which for some reason probably needed a supplementary supply of oxygen. In these structures, the gaseous exchanges are supposed to have taken place either at the outer surface of the body wall or to have been effected at the inner surface of the skeletal armor, so that functionally two main types of organs may be distinguished: **exospires,** which were evaginations of the body wall, and **endospires,** which consisted of invaginations of the body wall (HUDSON, 1915).[2]

EXOSPIRES

Apposed articular faces of thecal plates (rarely of columnals) of some crinoids may enclose simple or branched canals **(interarticular canals).** Each of these canals opens either on the inner or outer surface of the skeleton and so terminates in an internal and an external pore, or it has an internal opening only and no connection with the exterior. These pores are placed either along the sutures, or at the corners of the plates. They may, therefore, be called **sutural pores** and **goniopores,** respectively, though they probably differ in no essential way. In both cases the interarticular canals are supposed to have housed small fleshy projections, similar to the gills or papulae of asteroids, and, like these organs, retractile and provided with a lumen continuous with the general body cavity, so that coelomic fluids could be carried into them and have gaseous exchanges with the sea water.

Several types of these pore-canal structures presumably associated with such branchial vesicles are known. Most of them are remarkably like the sutural pores, or

[1] In *Lageniocrinus* and other microcrinoids without arms, the gonad must have been located within the theca. In such case, genital products may have been emptied outward through a common hydropore-gonopore orifice.

[2] According to HAUGH (1973), a third type of respiratory organ was perhaps present in camerates. He suggested that the distal section of the hind-gut of these crinoids was possibly associated with a passive respiratory sac or a fluid-filled coelomic cavity acting as a cloacal pump. Evidence for the existence of such structures does not appear to be conclusive.

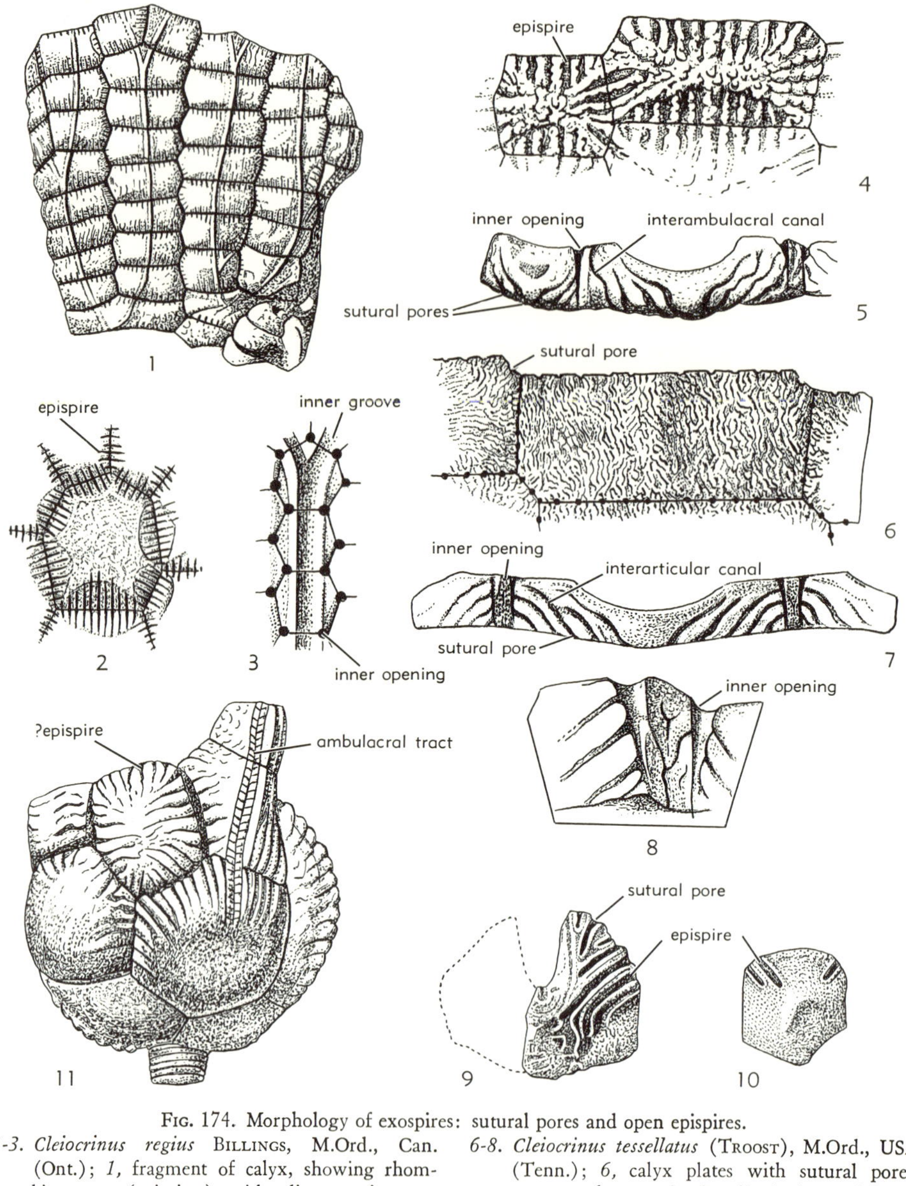

FIG. 174. Morphology of exospires: sutural pores and open epispires.

1-3. Cleiocrinus regius BILLINGS, M.Ord., Can. (Ont.); *1,* fragment of calyx, showing rhombic areas (epispires) with slits crossing sutures, ×3; *2,* axillary plate, outer surface, ×4.5; *3,* calyx plates, inner surface, with groove and large pores at plate corners, ×6 (Springer, 1911c).

4,5. Cleiocrinus sculptus SPRINGER, M.Ord., USA (Ky.); *4,* calyx plates, outer surface, with epispires forming rhombic areas across sutures, ×6; *5,* distal face of cup plates, showing interarticular canals running from sutural pores on outer surface to funnel-shaped depressions with inner opening at plate corners, ×6 (Springer, 1911b).

6-8. Cleiocrinus tessellatus (TROOST), M.Ord., USA (Tenn.); *6,* calyx plates with sutural pores, outer surface, ×9; *7,* distal face of calyx plates, with interarticular canals discharging into funnel-shaped depressions at plate corners, ×9; *8,* funnel-shaped depressions and interarticular canals, ×12 (Springer, 1911b).

9,10. Carabocrinus slocomi costatus FOERSTE, U. Ord., USA (Iowa); radial and anal plates, with bordered epispires, ×3 (Foerste, 1925).

11. Hybocystites eldonensis (PARKS), M.Ord., Can. (Ont.); *CD*-interray view, with grooves (? epispires) crossing sutures, ×3 (Springer, 1911b).

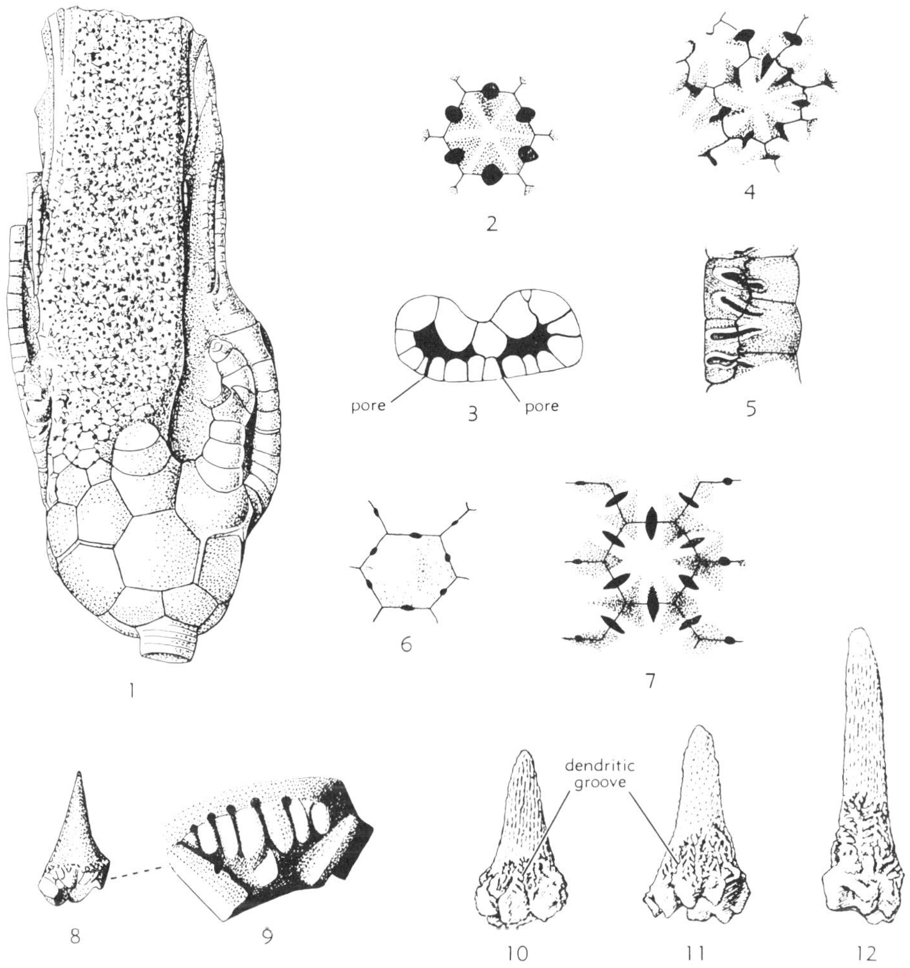

FIG. 175. Pore-canal structures in anal sac of poteriocrinine inadunates.

1-3. Parisocrinus crawfordsvillensis (MILLER), Miss., USA(Ind.); *1,* specimen showing part of perforated anal sac, ×2; *2,* plate of sac, ×6; *3,* transverse section of sac showing pore-canal structure through sac wall, ×4 (Springer, 1900a).

4. Blothrocrinus swallovi (MEEK & WORTHEN), Miss., USA(Iowa); plates of sac, ×6 (Springer, 1900a).

5. Abrotocrinus unicus (HALL), Miss., USA(Ind.); plates of sac, with slitlike pores, ×6 (Springer, 1900a).

6. Aulocrinus agassizi WACHSMUTH & SPRINGER, Miss., USA(Ind.); plate with tiny pores, ×6 (Springer, 1900a).

7. Decadocrinus tumidulus (MILLER & GURLEY), Miss., USA(Ind.); plates of sac, with pores of various shapes and sizes, lying on ridges across sutures, ×6 (Springer, 1900a).

8,9. Hydreionocrinus sp., Perm., Timor; *8,* isolated distal plate of anal sac, lat. view, ×1; *9,* articular facet of same, with pore-canal structures, ×4 (Wanner, 1916a).

10-12. Hydreionocrinus sp., L.Carb., Scot.; isolated spiniferous distal plates of anal sac, lat. view, ×2 (Bather, 1912).

epispires, of eocrinoids, and may be so termed. They consist of interarticular canals and sutural pores or gonipores only, or of those structures coupled with epithecal extensions (grooves or canals) that lay on the outer surface of the thecal plates. Thus, in the camerate genus *Cleiocrinus* (Fig. 174,*1-8*), external pores are present along the sutures and at the angles of the calycal plates. They are extremely numerous; probably several thousands of them are present in a mature individual. They may remain simple or be accompanied by narrow grooves that extend across the adjacent plates and form rhombic striated areas bisected by the sutures. The interarticular canals corresponding to the pores do not generally pass directly through the test, but lead to large, funnel-shaped depressions that open into the thecal cavity at the corners of the plates (HUDSON, 1911; SPRINGER, 1911b). At the junction of the orals, radials, and anal *X* in some species of the cyathocrinine inadunate *Carabocrinus* (e.g., *C. esthonus, C. radiatus, C. slocomi, C. treadwelli*), there is a series of narrow grooves that cross the common junctures of these plates and extend for some length on their outer surface (Fig. 174,*9,10*); these grooves may be bordered by slightly raised rims. They vary greatly in number and size even within a single species, and are connected with sutural pores and vertical canals penetrating to the interior; thus, they form complex structures very much like the epispires of the eocrinoid genus *Lichenoides*. According to SPRINGER (1911b), similar structures are present on the anal and adjacent plates of the hybocrinid *Hybocystites eldonensis* (Fig. 174,*11*), and possibly on the anal plate of *Hybocrinus tumidus*.

In some inadunates (e.g., *Culmicrinus regularis, Hallocrinus ? inagilis, Lophocrinus minutus, Poteriocrinites ? broadheadi*) small perforations occur at most angles of cup plates. The exact function of these so-called goniopores is unknown, but it may have been similar to that of the just-described sutural pores.

Sutural pores may also occur in the adoral skeleton. This is particularly true of many poteriocrinine inadunates, the anal sacs of which have countless perforations (Fig. 175). These small, round or slitlike openings are typically located at the middle of the sutures, generally in depressions between axial folds (Fig. 175,*2,4*), more rarely on folds that cross the sutures at right angles (Fig. 175,*7*). In some cases, these pore-canal structures become very complex; branching repeatedly as they approach the outer edges of the plates, they form a dendritic system of shallow grooves that may extend on the outer plate surface and the base of spines (Fig. 175,*11,12*).

In all of the preceding examples, the interarticular canals are thought to have been open to the exterior, but those of some crinoids were not so. Thus, in the anal sac of the poteriocrinine inadunate *Aesiocrinus magnificus* grooves forming a complex tubule system on lateral articular faces of plates terminate on the outer surface in rounded, commonly expanded tips that are sealed off from the exterior (LANE, 1975). The cup plates of the cyathocrinine genus *Palaeocrinus* are ornamented with parallel, hollow ridges that cross the sutures at right angles and have no opening to the exterior (Fig. 176). In *P. striatus* the lumen of each ridge communicates directly with a large sutural canal, which opens into the interior of the calyx through a vertical canal (HUDSON, 1911). In *P. planobasalis,* the ridges are grouped into raised bundles, and their canals open into a rhomb-shaped central chamber, which in turn opens into the calyx interior (Fig. 176,*4*). Series of transsutural hollow ridges occur also on the outer surface of the cup of the monocyclic inadunate genus *Tiaracrinus* (Fig. 176,*5-7*). But here each ridge contains a canal, which arises from a pore on the inner surface of one radial, crosses the adjacent interradial suture, and ends in a pore located on the internal surface of another radial. Such a set of pore-canal structures is quite similar to a simple humatirhomb as found in some rhombiferan cystoids. Function of these structures in crinoids has been discussed by HUDSON (1911), BROWER & VEINUS (1974), and LANE (1975), and in cystoids by PAUL (1972). It is supposed that body fluids flowed through these interarticular and tangential canals in life, and that gaseous exchanges took place through their thin stereomic outer covering. Possibly the hollow ridges of the perittocrinid *Tetracionocrinus* and of some other early crinoids provided with very thin plicated cup plates served the same role, for they represent extension of the thecal cavity close to the outer body surface, and increase the surface area.

Probably similar in function, though structurally different, are the interarticular pits called **gonioporoids** by W. E. SCHMIDT

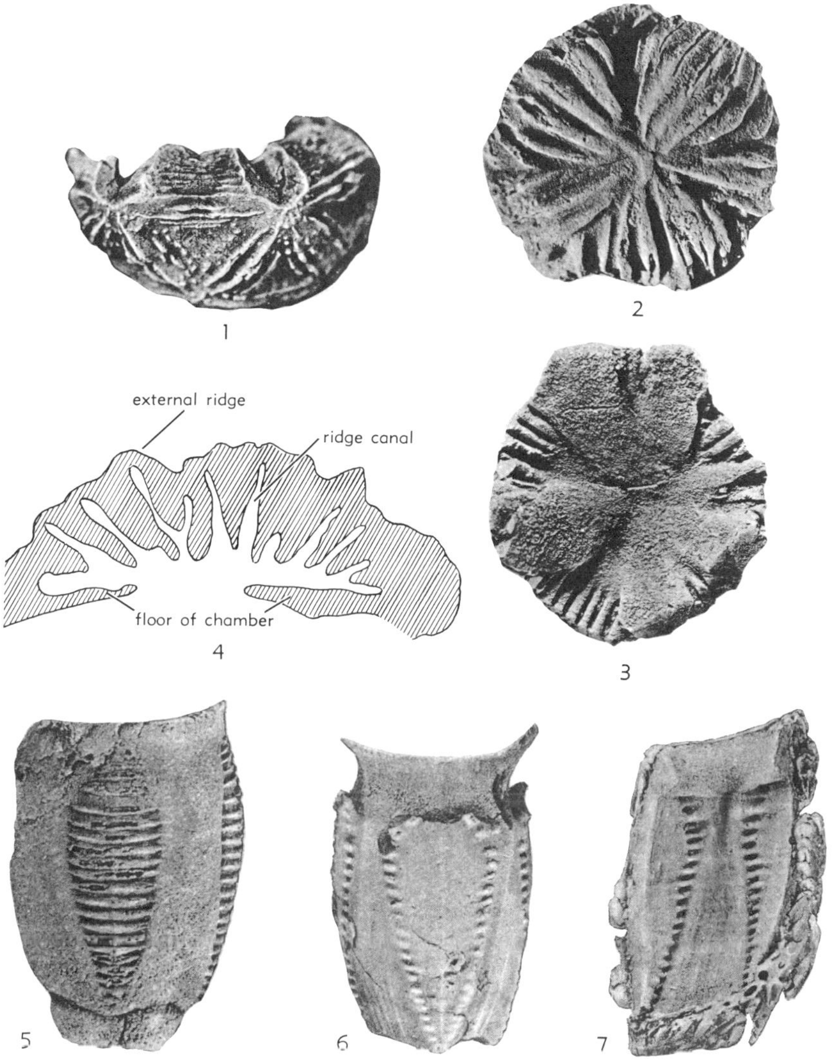

FIG. 176. Morphology of covered epispires.

1-4. Palaeocrinus planobasalis BROWER & VEINUS, M.Ord., USA(Tenn.-Va.); *1*, cup in *BC*-interray view, ×1.7; *2*, isolated basal, external face, with canals of internal ridges exposed by weathering at some places, ×2.7; *3*, same, internal face, ×2.7; *4*, cross section of isolated plate, ×10 (Brower & Veinus, 1974).

5-7. Tiaracrinus moravicus UBAGHS & BOUČEK, L. Dev., Czech.; *5*, cup in lat. view, ×3; *6*, internal mold of thecal cavity, with infilling of inner pores partially preserved, ×3; *7*, internal surface of isolated radial plate, ×3 (Ubaghs & Bouček, 1962).

(1931) (Fig. 177). These pits or depressions open into the thecal cavity, but do not reach the outer surface of the skeleton. They are located at the corner of the plates in the calyx and also in the tegmen (including the anal sac) and possibly (though somewhat modified) in the column. They may be simple and cylindrical, or be provided with lateral ampulla-like cavities, which extend on the adjacent articular facets (Fig. 177,*1-3,5*). All these cavities form a complex system in open connection

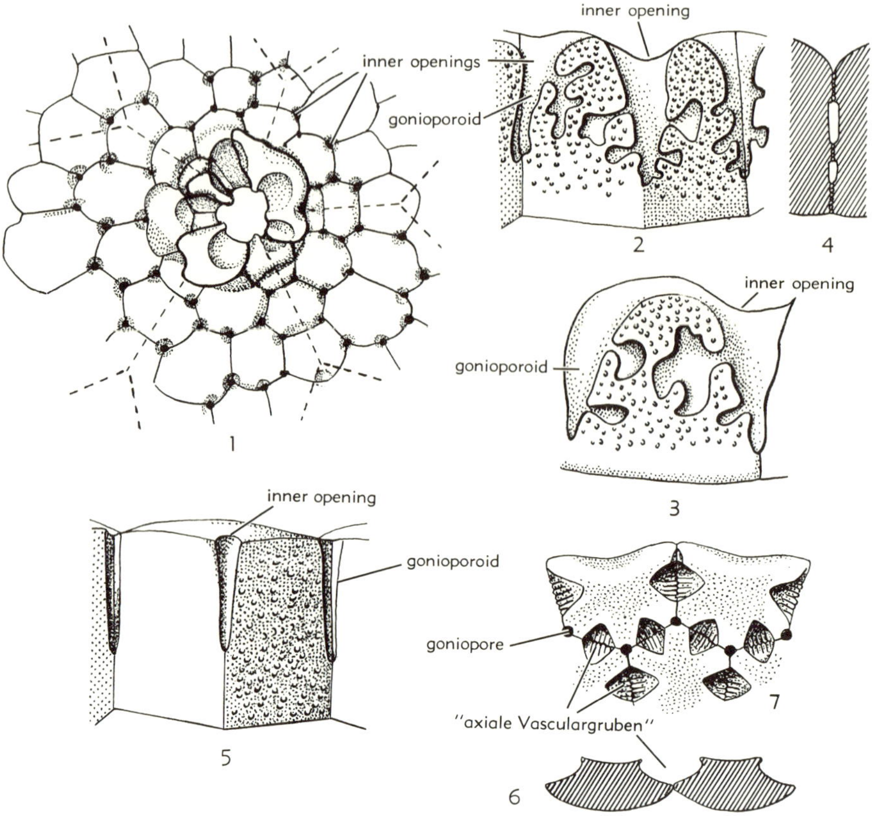

FIG. 177. Morphology of gonioporoids and gonioporoid-like structures.

1-5. Trybliocrinus flatheanus (GEINITZ), L.Dev., Spain; *1*, proximal part of cup, internal side; rays indicated by interrupted lines, ×1.5 (W. E. Schmidt, 1932); *2,3*, articular faces of cup plates with branched gonioporoids, ×4 (W. E. Schmidt, 1932); *4*, diagram. section through juncture of two cup plates with two gonioporoids (W. E. Schmidt, 1932); *5*, articular faces of cup plate with simple gonioporoids, diagram., ×4 (Ubaghs, n).

6,7. Culmicrinus regularis (VON MEYER), L.Carb., Ger.; *6*, transverse section of two radials, inner side above, ×6; *7*, inner side of two radials and parts of three basals, ×6 (redrawn from W. E. Schmidt, 1930).

with the thecal cavity, and thus probably represent an extension of the coelom into the skeletal armor. The occurrence of gonioporoids has been recorded mainly from camerate species, but similar structures seem to have developed also in some inadunates. For instance, deep sutural pits that do not open to the exterior mark the inner surface of the anal sac of the poteriocrinine species *Synyphocrinus magnus* (Fig. 178). Other inadunates (i.e., *Culmicrinus regularis, Hallocrinus ? inagilis, Lophocrinus minutus*) show diamond-shaped depressions bisected by the plate sutures at the inner side of the cup; the very thin and fragile outer wall of these depressions may be both reinforced and divided into narrow compartments by small ridges running transversely across the plate junctions (Fig. 177,*6,7*); though they are placed on plate sides rather than at plate corners, these *"axialen Vasculargruben"* as W. E. SCHMIDT (1930) called them, are best regarded as only one particular type of gonioporoids.

Another type of gonioporoid is exemplified by the irregular spaces between the

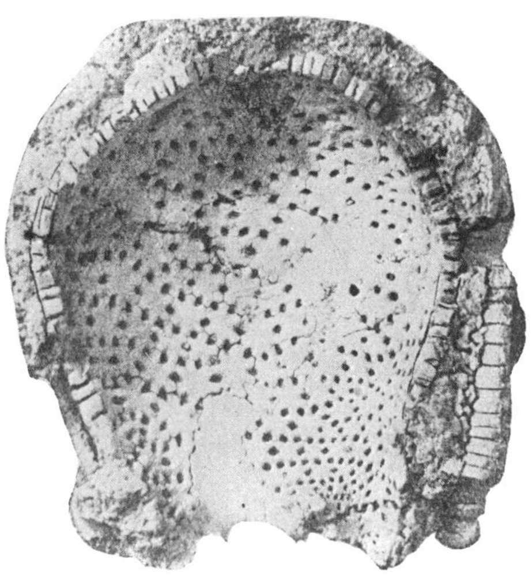

FIG. 178. *Synyphocrinus magnus* YAKOVLEV & IVANOV, M.Carb., USSR; distal part of anal sac, broken open to show gonioporoids on inner face and section, ×1 (Yakovlev & Ivanov, 1956).

tegminal plates of the camerate species *Trybliocrinus flatheanus;* in open connection with the coelomic cavity, these spaces could have provided circulation of coelomic fluids among the plates of the tegmen (Fig. 179,2). Also, in the stem of this crinoid, on the articular face of the columnals, radiating grooves open into the axial canal through vertical slits but terminate near the periphery without communicating with the exterior. They may have had a role similar to that of the gonioporoids of the theca (Fig. 54,*11,* p. *T75*).

Somewhat different in nature but possibly not in function was the system of narrow canals that in some crinoids are found within (not between) the plates. Thus, the tegminal plates of the just-mentioned *Trybliocrinus flatheanus* (Fig. 179, *1,3*), and the undivided basis of the strange species *Lodanella mira* encloses a complicated network of branching canals that open into the thecal cavity but end without reaching the exterior; even if in life those canals were filled with uncalcified tissues, they must have facilitated the circulation of the body fluids in the skeleton armor, and thus helped respiration.

ENDOSPIRES

The endospires are respiratory organs in which the gaseous exchanges take place below or near the inner surface of the thecal

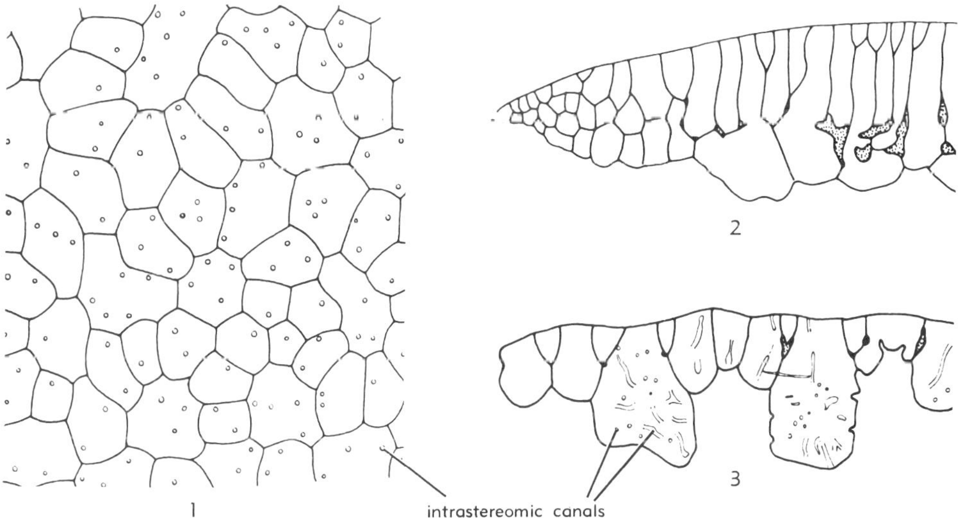

FIG. 179. *Trybliocrinus flatheanus* (GEINITZ), L.Dev., Spain; morphology of tegmen, *ca.* ×3 (Breimer, 1962).——*1.* External view of slightly weathered surface, showing canal system within plates.——*2.* Section through tegmen, with stratified margin at left, and lacunae between plates at right (stippled).——*3.* Section through tegmen, with inner expansions and canal system within plates.

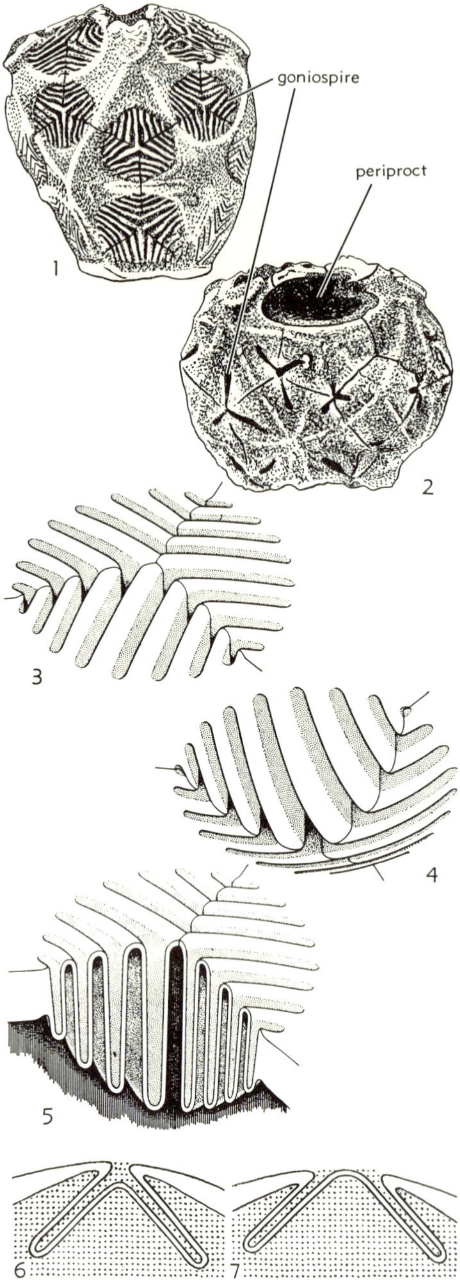

FIG. 180. Morphology of goniospires.——*1. Porocrinus shawi* SCHUCHERT, M.Ord., Can.; *A*-ray view, ×4 (Kesling & Paul, 1968).——*2. Triboloporus cryptoplicatus* KESLING & PAUL, M.Ord., USA (Va.); *CD*-interray view, ×5 (Kesling & Paul, 1968).——*3-5.* Diagram. representation of goniospire in *Porocrinus* as viewed obliquely from *(3)* ext., *(4)* int., and *(5)* with one plate removed to expose folds (Kesling & Paul, 1968).——*6,7.* Diagrammatic sections through plate corners to show structure of goniospires; *6, Triboloporus cryptoplicatus* KESLING & PAUL, M.Ord., USA(Va.); *7, T. xystratus* KESLING & PAUL, M.Ord., USA(Pa.) (Kesling & Paul, 1968).

armor. The calyx of some fossil crinoids is provided with folds, or deep depressions, into which a fresh supply of sea water is presumed to have been introduced continuously by ciliary action to replace that from which the oxygen had been extracted. Neither the folds nor the depressions open into the calyx cavity. It is only through extremely thin sheets of stereom or possibly sac walls that respiration could be effected.

A primary type of endospire structure is represented by the **goniospires** (HUDSON, 1915) found in the calyx of the cyathocrinine inadunate *Porocrinus* (Fig. 180,*1,3-5*). Each goniospire consists of three sets of parallel folds that meet the plate sutures at an angle. Folds of two half-sets on any one plate correspond to those of the adjacent half-sets on the adjacent plates; at the plate juncture they bend at an angle of about 120°. The longest fold was the first formed; it lies nearest the plate corners. The others were added in succession; their length decreases toward the middle of the plate sides. The folds had extremely thin walls, so that respiratory exchanges could be effected through them. In some species the goniospires are relatively small, in others, large; some were protected by being depressed into the plates, others by being adjacent to highly developed ridges. Those of the related genus *Triboloporus* are very simple; they appear externally as three clefts located at meeting points of plate corners, each cleft being the confluent of two lateral folds (Fig. 180,*2,6,7*). The function of goniospires has been discussed in detail by KESLING & PAUL (1968).

A second type of endospire is exemplified by the perittocrinid inadunate *Perittocrinus*. It differs from the first type by the fact that the folds, instead of being located at plate corners, are placed on small triangular accessory plates that are inserted between the large calyx plates (Fig. 181). Each accessory plate has three groups of folds, one

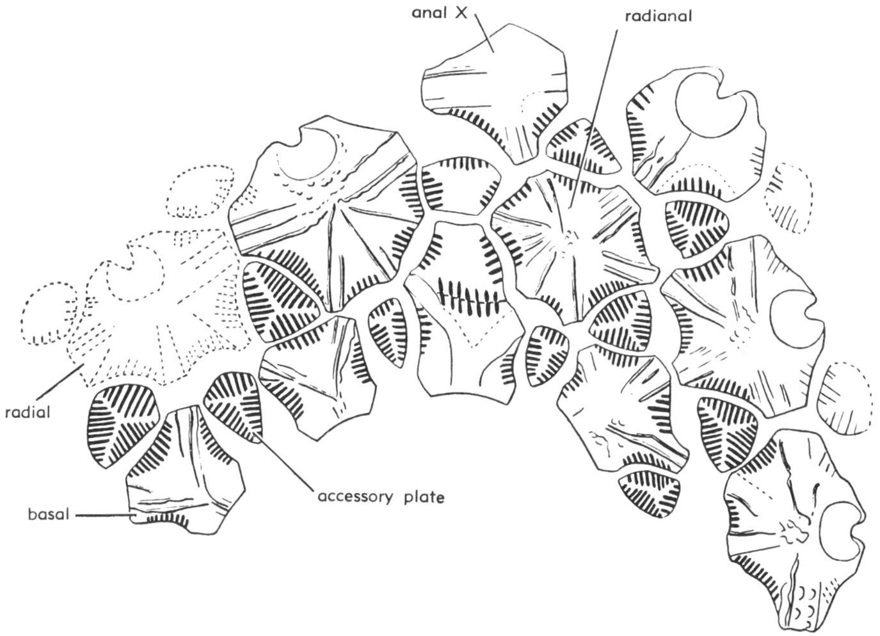

FIG. 181. *Perittocrinus radiatus* (BEYRICH), ?L.Ord., Baltic regions; diagr. of cup, showing location of goniospire structures on accessory plates (Ubaghs, 1971a).

group on each of its curved margins. The folds cross the sutures without bending, and extend into the adjacent large plates. The longest (oldest) folds are at the middle of the accessory plate sides, the shortest (youngest) ones near the accessory plate corners. Technically these structures should be called **craspedospires** (according to HUDSON's nomenclature, 1915), because the folds are at plate margins. In fact, they do not differ essentially from the goniospires of *Porocrinus,* for the accessory plates with which they are associated could only develop by truncating the corners of three large plates.

Although these structures and the pectinirhombs of the rhombiferan cystoids were structurally similar and probably had similar functions, they differed in position and ontogeny (HUDSON, 1915; KESLING & PAUL, 1968; UBAGHS, 1971a). The pectinirhombs are confined to the margins of the plates, their folds are perpendicular to these margins, the last-formed folds lie nearest the plate corners, and during plate growth the folds elongated by simple addition of calcareous material on each side of the suture (Fig. 182,*3*). The goniospires are situated at plate corners (or on accessory plates developed at the expense of the corners of three large plates), the folds meet the sutures at an acute angle, the last-formed folds lie near the margin centers, and during growth adjustment was constantly needed to keep the direction of the folds unchanged, since the growth lines of the plates are angled or arcuate (Fig. 182,*1,2*). These differences suggest that the goniospires and the pectinirhombs evolved independently.

A third type of endospires characterizes the calyx of the Permian poteriocrinine inadunates *Indocrinus* and related genera (Fig. 183). They consist of rounded or subtriangular depressions occurring at every junction of three plate corners. The bottom of these depressions is so thin as to be translucent, but it may be strengthened by small ridges running from edge to edge across the sutures (Fig. 183,*2,3*). In *Indocrinus rimosus* and the genus *Metaindocrinus* the depressions at plate corners are re-

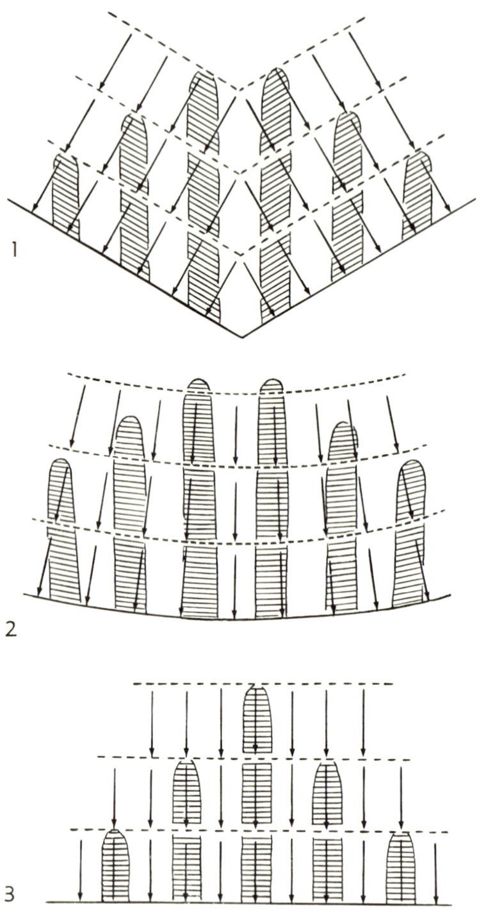

FIG. 182. Differences in mode of growth and development between goniospire of *Porocrinus* (*1*), goniospire-like structure of *Perittocrinus* (*2*), and pectinirhomb of cystoid (*3*) (Kesling & Paul, 1968; Ubaghs, 1971a). [Explanation: heavy lines, plate edges; dotted lines, growth lines; arrows, direction of growth.]

placed by slitlike troughs that cross the sutures at right angles; the longest (oldest) troughs are at the middle of the plate sides, the shortest (youngest), near the corners (Fig. 183,*5*). YAKOVLEV (1950) is the first to have recognized the probable respiratory function of these structures, for which the term **bothrospires** is proposed herein. They differ from goniospires by the fact that they are not composed of folds but of mere pit-like or slitlike depressions on the calyx surface.

A fourth type of endospires is found in the disparid *Acolocrinus* (Fig. 183,*6,7*). According to description and interpretation by KESLING & PAUL (1971), the respiratory system of this genus is divided into five unconnected parts, which are alike (except for the area around the anal plate). Each part develops along an interradius, with extensions along the inferradial-superradial joint faces and along the basal-inferradial joint faces. Serrate portions of sutures located around each point of meeting of two inferradials with two superradials and around each point of junction of one basal with two inferradials may have acted as incurrent slits. Parts of these slits that are farthest from these points of junction lead to canals within plates (each canal being shared by two plates), whereas the parts adjacent to junctions lead to channels (possibly provided with membranous sacs) inside of cup plates. These channels extend most of the length of the sutures between adjacent inferradials; those from above and those from below lead through short excurrent canals to a common pore at the middle of the suture. It is supposed that water entered through the serrate sutures and was expelled through the pores, and that gas exchange took place through membranous sacs inside the channels. A somewhat similar apparatus may have existed in another disparid genus, *Agostocrinus,* but it has not been worked out in sufficient detail.

PINNULE OPENINGS IN THECA AND ARM TRUNKS

In many camerates belonging to various families, such as the Batocrinidae, Actinocrinitidae, Dolatocrinidae, or Platycrinitidae, pores occur at the limit of the cup and the tegmen, between arm bases. They have been called "respiratory pores," "ovarian apertures," or "thecal pinnular openings," according to their presumed function or morphological significance. They also have been designated as "interbrachial pores" and "thecal pores."

These pores do not penetrate the plates, but lie at their corners or within sutures. They may stand in close relation with the arm openings (Fig. 184,*3*), or be located at some distance from them (Fig. 184,*1*).

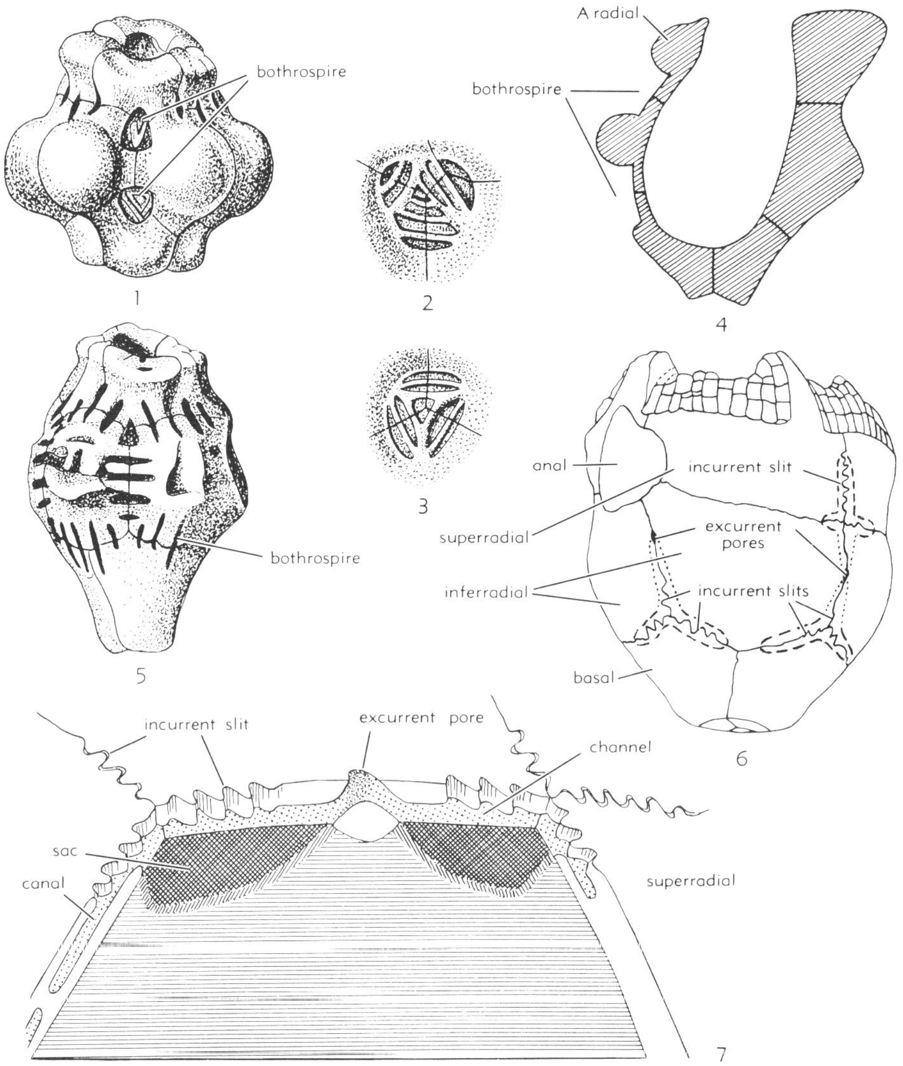

FIG. 183. Morphology of some endospire structures.

1-4. Indocrinus elegans WANNER, Perm., Timor; *1,* cup in *A*-ray view, ×2 (Wanner, 1924); *2,3,* detail of bothrospires, ×4 (Wanner, 1923); *4,* vert. sec. through cup passing through *A* ray and *CD* interray, ×3 (Wanner, 1916a).

5. Indocrinus rimosus WANNER, Perm., Timor; cup in *A*-ray view, ×3 (Wanner, 1916a).

6,7. Acolocrinus hydraulicus KESLING & PAUL, M. Ord., USA(Tenn.-Va.); *6, C*-ray view of cup and arms [Explanation: interrupted lines, outlines of inner canals; dotted lines, outlines of inner channels], ×5 (Kesling & Paul, 1971, mod.); *7,* diagram. view of part of cup with one inferradial plate removed to expose respiratory system; hypothetical respiratory sac indicated (Kesling & Paul, 1971, mod.).

In many cases they are obviously the outer apertures of passageways leading to the main ambulacral tracts. In other cases, such connections are not evident. Their number varies from one to eight per interray. They generally are round and small, but they may have the shape of narrow vertical slits (Fig. 184,*4*). They may be present in im-

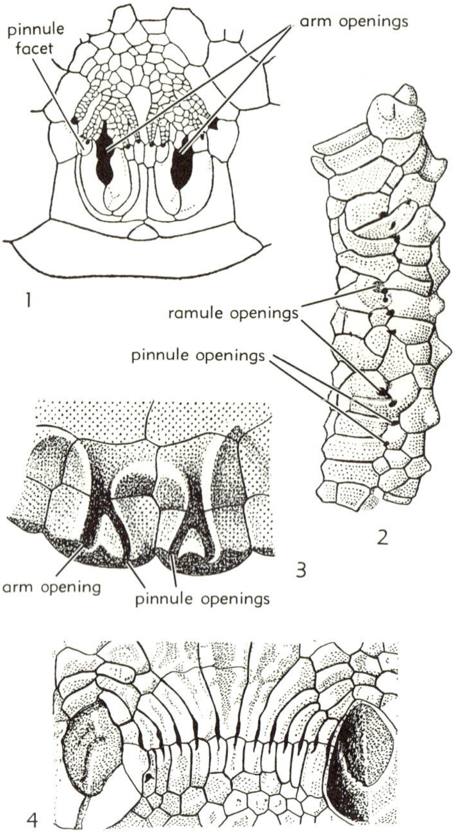

FIG. 184. Pinnule openings in camerates.——*1. Marsupiocrinus inflatus* (TROOST), Sil., USA; part of ray showing pinnule openings and ambulacral tracts leading to them, ×1.5 (Springer, 1917).——*2. Eucladocrinus millebrachiatus* WACHSMUTH & SPRINGER, Miss., USA(Iowa-Ill.); part of arm trunk, lat. view, ×1.5 (Wachsmuth & Springer, 1897).——*3. Abatocrinus clypeatus* (HALL), Miss., USA; transv. sec. of theca at arm-opening level, plates separated at sutures, ×4 (Springer, 1917).——*4. Dolatocrinus grandis* MILLER & GURLEY, M.Dev., USA(Ky.); part of interray with 12 slitlike pinnule openings, ×2 (Springer, 1921a).

mature individuals and permanently closed in adults. Similar pores occur on both sides of arm trunks of *Eucladocrinus* (Fig. 184,*2*) and *Steganocrinus* at the limit of brachials and covering plates.

The internal passageways that correspond to these pores may be complex. Thus, in *Batocrinus icosidactylus* the lower part of the passageways is divided into a blind chamber and a canal that connects with the ambulacral tract of the adjacent arm. The external pore was protected in life by a covering, apparently movable, flap that is composed of the two proximal plates of the proximal pinnule, the hinge socket of which may be observed on the theca below the pore; adjacent to this flap is the first normal, free pinnule (HAUGH, 1973) (Fig. 129,*3*, p. *T*158).

As suggested by WACHSMUTH & SPRINGER as early as 1881 and conclusively demonstrated by SPRINGER in 1917, these pores are the outer openings of pinnules. This interpretation is supported by the following evidence: 1) in very well-preserved specimens pinnules are still attached at the emplacement of these openings; 2) hinge sockets of pinnules may be present below them (Fig. 184,*1*); 3) pores contiguous to arm openings are connected directly with the main arm groove by a narrow diagonal canal (Fig. 184,*3*); 4) if located farther from the arm openings, their connection with the main ambulacral tracts may be marked by the tracing of covering plates on the tegmen (Fig. 184,*1*). These pores therefore belong to the proximal pinnules, which generally became partially incorporated in the calyx during growth. These pinnules might have been somewhat different from the others, as just seen in *Batocrinus icosidactylus,* and like those of recent comatulids might have served special functions. Possibly also in some species the free parts of these pinnules were lost in life, while the pores remained open. But the role of the pores in such a case is unknown. One thing seems almost certain: they were not replacements for the hydropore or madreporite of some inadunates and flexibles, as supposed by BATHER (1900a), for they coexist in some camerates with a structure that has all the appearance of an internal hydropore or madreporite (see p. *T*197) (HAUGH, 1973).

THE CRINOID STEREOM

By D. B. Macurda, Jr., D. L. Meyer, and Michel Roux

[Museum of Paleontology, The University of Michigan; University of Cincinnati; Laboratoire de Paléontologie, Université Paris-Sud]

CONTENTS

INTRODUCTION

The endoskeleton of a crinoid is a porous lattice composed of high-magnesium calcite, an attribute apparently characteristic of all echinoderms. The stereom may be highly differentiated in crinoids, as shown by the study of the skeletal microstructure of modern crinoids (Macurda & Meyer, 1975), and well preserved in fossil crinoids as shown by the study of columnals of Mesozoic-Cenozoic Articulata (Roux, 1970, 1971, 1974, 1975), and of Paleozoic inadunates (Lane & Macurda, 1975; Moore, Jeffords, & Miller, 1968).

Direct observational investigations of modern comatulid and other articulate crinoids (Macurda, 1973; Meyer, 1973a, b; Macurda & Meyer, 1974) have permitted the correlation of life habits with the microstructure of the skeleton, enhancing the paleobiological study of crinoids.

MAIN MICROSTRUCTURAL FEATURES

The crinoid stereom shows two main microstructural features in relation to the organic stroma (Roux, 1970, 1971). The first one is an irregular and labyrinthic lattice of calcite with a thin meshwork of collagen microfibers (Fig. 185,*4;* type β of Roux). The second is a regular galleried stereom with paraxial galleries where collagenous fibers penetrate through the ossicle (Fig. 185,*1;* type *a* of Roux). The articular ridges of the crenularia have various microstructural features: galleried stereom with short paraxial collagenous fibers or dense labyrinthic stereom without paraxial organic stroma (Fig. 185,*4*). The calcite meshwork of the ligamentary areola is always galleried. Some muscular articulations have a muscle fossa with small needlelike calcite projections to which muscle fibers adhere (Fig. 185,*2; see* 191,*1,2*). The microstructural organization of an articular facet is highly differentiated in stems, cirri, brachials, and pinnules.

STEM

Stem plates of crinoids are ubiquitous in Paleozoic sediments, locally bulking large in volume. They are much less important in the Mesozoic-Cenozoic and are confined to sediments at depths below 100 meters in the recent. The stems of modern crinoids are represented by different types of columnals, which can be compared with fossil specimens.

The heteromorphic column of the isocrinids is composed of stellate, pentagonal, or cylindrical plates, which are divided into five petaloid areolae that are bordered by a symplexial crenularium (Fig. 186,*1*). The stereom of the areola is characteristically galleried (Fig. 186,*3,4*). Each culmen is

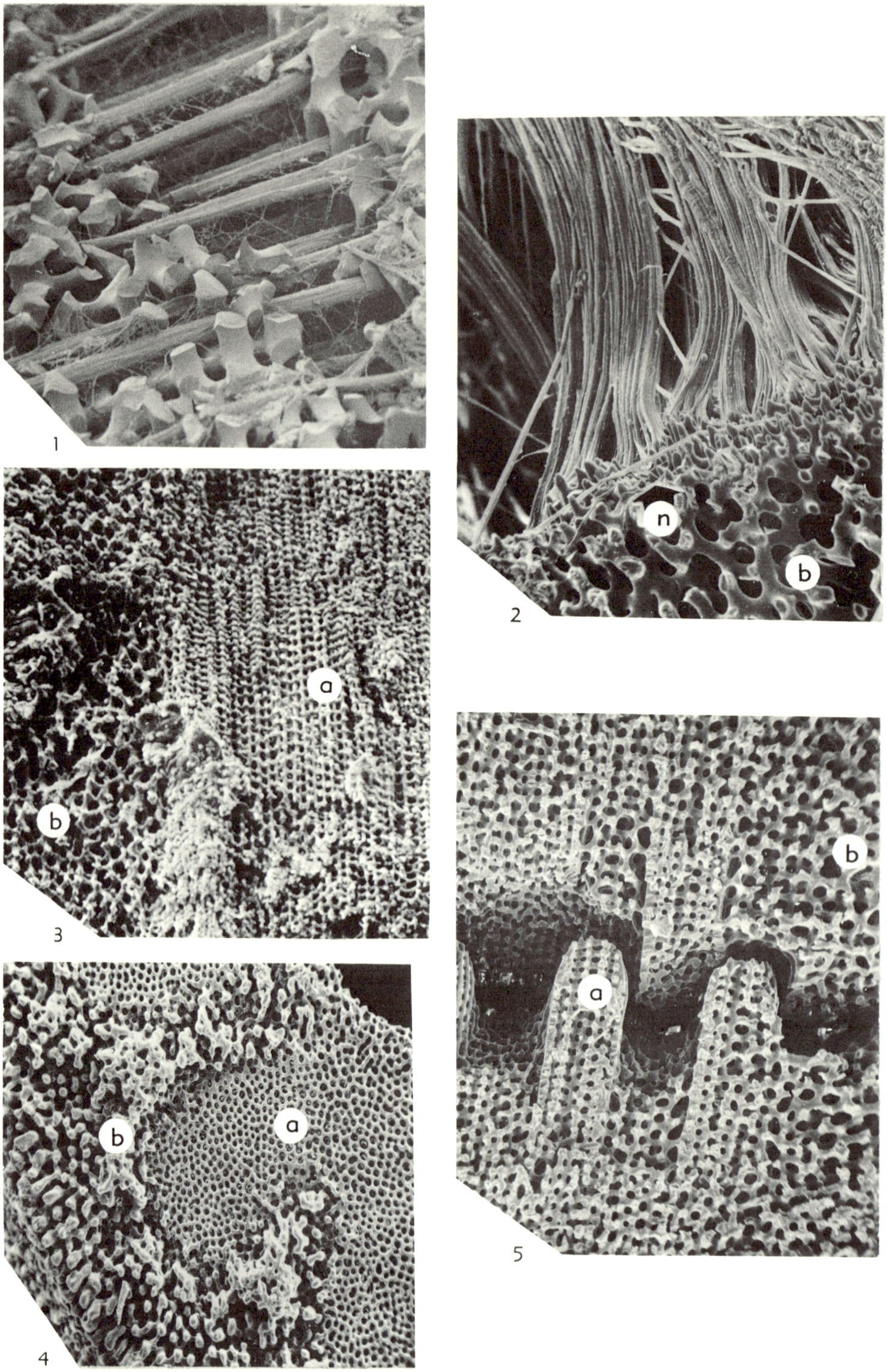
1
n
b
2
a
b
3
b
a
5
a
b
4

composed of galleried stereom developed on the labyrinthic stereom of the interpetaloid area. Short and long paraxial collagenous fibers that occupy the galleried pores (Fig. 185,*1*) provide a strong bond between two internodal plates (Fig. 185,*1,5*). The tensile strength of these bonds account for the ability of a recent crinoid such as *Cenocrinus* to elevate the crown almost one meter above the sea floor in currents in excess of 0.5 knots (Macurda & Meyer, 1974).

The lower (distal) surface of each nodal is a cryptosymplexy to almost smooth synostosial suture (Fig. 186,*2*), which can result from the secondary development of stereom filling in interarticular space of a symplexy and anchylosing it. The facets appear smooth because of the greatly reduced crenularium and the smaller pores of the synostosial stereom. These may be labyrinthic as in the stem of *Annacrinus*, or galleried as in the stem of *Endoxocrinus*. Proceeding distally on a stem, the synostosial stereom becomes more highly developed and can fill in the lumen. A large mass of labyrinthic stereom may develop below the distal synostosis of an isocrinid stem; with cirri it serves as an anchorage in mud or sand (e.g., *Annacrinus*), or as a terminal attachment disc it may encrust a hard substrate (e.g., *Hypalocrinus*).

The homeomorphic stem of the Hyocrinidae is instructive because this type is very frequently found in fossils. Agassiz's work (1892) on *Calamocrinus* is the most detailed study of such a stem. The proximal part of the stem has very primitive articulations that probably lack galleried stereom and paraxial collagenous fibers. These are present in the middle part of the column where articulations have a symplectic crenularium. Around the lumen, a large annular medulla (like the claustrum of some Paleozoic columnals) is composed of labyrinthic (*β*) stereom characterized by large pores. In the distal part of the stem, the articular facets are plain with a multiradiate crenularium. The main part of the columnal is composed of a labyrinthic stereom, which has circular growth lines.

The recent Bathycrinidae have a xenomorphic column. The proxistele is well developed in *Monachocrinus* and *Bathycrinus;* the first stem plates immediately below the calyx have labyrinthic stereom and synostosial articulations. Quickly, articulations are developed with thick labyrinthic stereom for the crenularium and galleried stereom for the areola (Fig. 185,*4*). A new secondary columnal grows into the interarticular space; it has only galleried stereom. The mesistele of the Bathycrinidae is characterized by elliptical bifascial synarthries (Fig. 186,*5*). Each synarthry is bisected by a large, massive, fulcral ridge (Fig. 186,*6*), which may show abrasive wear. In *Democrinus,* the ridge is bordered on either side by a series of knobs and large pores, which are offset on either side of the ridge. The knobs are the high points of the articulation and apparently fit into corresponding large pores on the opposing plate. This prevents slippage in any direction. This type of fulcral ridge with its bordering knobs and pores also developed in the brachial articulation of some Paleozoic inadunate crinoids. Most of the surface of the synarthrial articulation in the bathycrinids is deeply concave and characterized by a well-developed galleried stereom penetrated by ligament fibers. Since the fulcral ridges are set from 30 to 90 degrees to one another on the opposite ends of each columnal, as seen in a stem of one specimen of *Monachocrinus,* this permits flexure in any direction. The tensile strength of

(See facing page.)

Fig. 185. Ligament fibers and articular surfaces (Roux, n).——*1.* Paraxial collagenous fibers through a symplexy on the stem of *Metacrinus nobilis* Carpenter, recent; with galleried stereom to left and interarticular space to its right, ×900. *2.* Muscular articulation of brachial ossicle of *Annacrinus wyville-thomsoni* (Jeffreys), recent. Muscular fibers on needlelike projecting surface *(n)* of muscle fossa; labyrinthic stereom of interarticular ligament fossa *(b),* lower right, ×350.——*3.* Paraxial section of a columnal of *Isselicrinus subasaltiformis* (Miller), Eocene, London Clay, showing well-preserved galleried stereom *(a)* on the right and labyrinthic stereom *(b)* on the left, ×60.——*4.* Crenularium of a proximal columnal of *Monachocrinus recuperatus* (Perrier), recent; ×150.——*5.* Symplexial crenularium of a columnal of *Annacrinus wyville-thomsoni* (recent); paraxial section through the articulation with organic stroma removed, ×100. [*a,* galleried stereom (reseau *α*); *b,* labyrinthic stereom (reseau *β*); *n,* needle-like projections.]

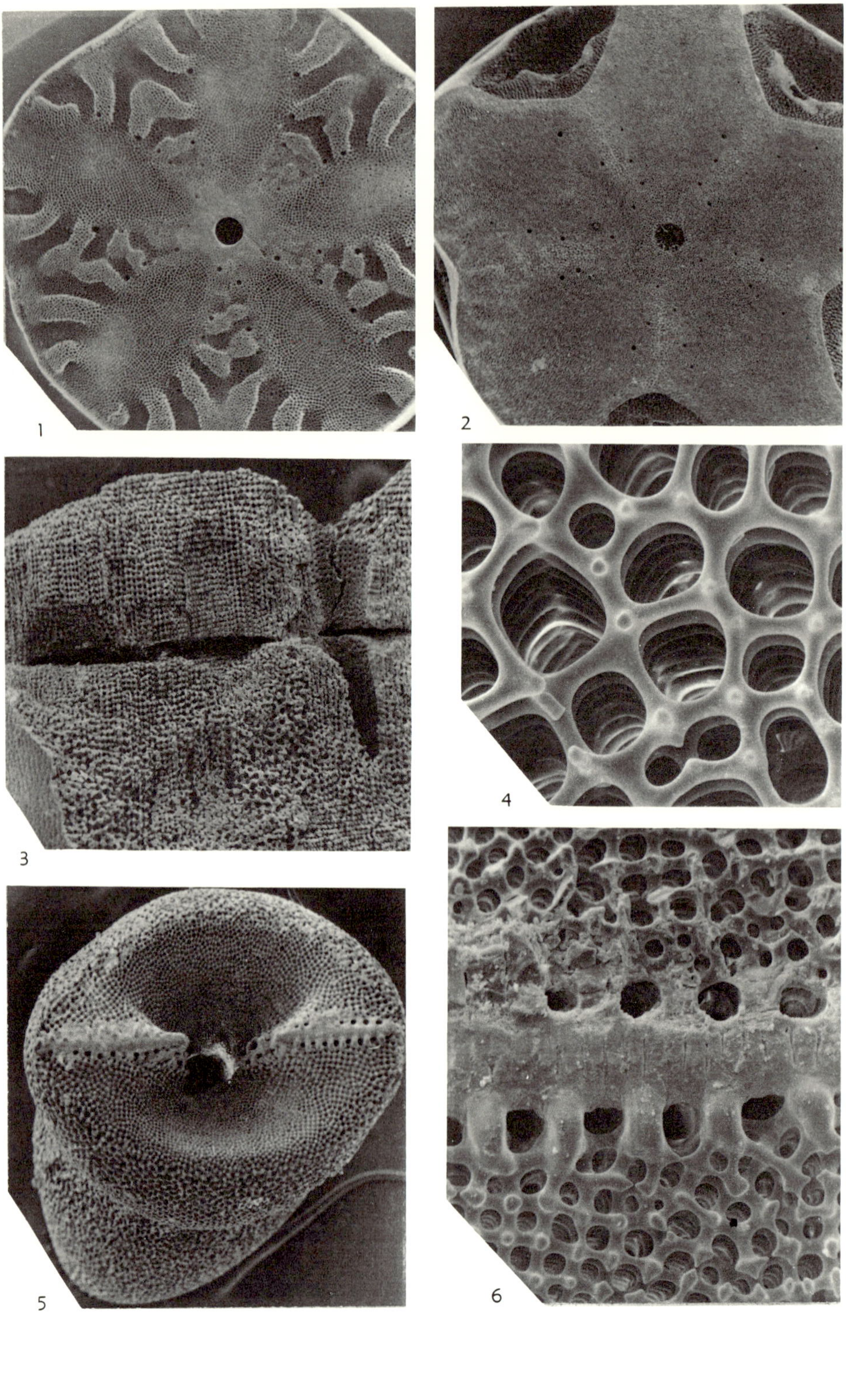
1
2
3
4
5
6

these articulations is strong enough to hold the calyx upright in moderate currents, only the roots being in contact with the substrate (MACURDA & MEYER, 1974).

The most detailed investigation of the stereom of fossil crinoids has involved the stem plates of Mesozoic and Cenozoic articulates (ROUX, 1970, 1971, 1974, 1975). During diagenesis, the magnesium that is randomly distributed within the crystal lattice of a crinoid plate is lost. Calcite may secondarily encrust the surface of the plate or penetrate throughout the stereom, being deposited in optical continuity with the original calcite. The original structural detail may be preserved and visible in a thin section, or brought into relief on a surface by natural or artificial etching (Fig. 185,*3*). Further recrystallization may completely destroy the fabric of the stereom. The most fortuitous conditions for stereom preservation appear to be when the plates are preserved in marl or clay. Sometimes pyrite precipitated within the stereom will make an internal cast if the calcite is leached away.

Biometric studies of the stereom of fossil articulate columnals demonstrate that α and β stereom are easily recognizable and that the distributional pattern of these within the stem is of ontogenetic, functional, and taxonomic significance (ROUX, 1970, 1971, 1974, 1975). Figure 187,*1* illustrates the growth of the stereom and the gradual increase in pore diameter, the subsequent formation of two smaller pores, and their gradual increase in diameter. The resulting biometrical graph diagrams this process (Fig. 187,*2*). Preparation of a graph for the β stereom of the Jurassic millericrinid *Apiocrinites* (Fig. 187,*4*) shows the ontogenetic development of this plate; the arrows in Figure 187,*4* indicate the directions of growth. The patterns of the stereom may differ between proximal and distal columnals. Thus, in Triassic *Encrinus,* proximal columnals have a highly complex pentalobate crenularium; in median columnals it is radially organized and restricted to near the periphery and in the distal columnals it has become more extensive from the border to near the center (Fig. 188). Thin transverse sections through the columnals show the ontogenetic evolution of the facet morphology during columnal growth and the main stages of microstructural organization. When fossil columnals are dissociated in the sediments, such a study shows that the three types of columnals belong to the stem of one species (Fig. 188,*B,C*). Due to these differences and the pentalobate structure of isocrinid stems, they are best represented by having separate radial biometrical plots of the dimensions of the α stereom of the areola and β stereom of the petalodium in the proximal, median, and distal columnals, as in the Jurassic isocrinid *Balanocrinus subteres* (Fig. 187,*5*). Such graphs are highly distinctive for taxonomic differentiation, summarize the ontogenetic development, reveal functional changes (e.g., Fig. 187,*3a-c*), and can be used for evolutionary comparisons.

CIRRI

The cirri of modern crinoids grow outward from the nodal columnals of the isocrinids or from the centrodorsal of comatulids. In the former, the distal cirri function as anchors or props for the distal part of the stem; in the latter they are used to grasp the substrate (e.g., rocks, sponges, alcyonarians, corals, etc.). They are very tenacious and must be carefully detached. If removed from the substrate they will reattach fairly rapidly if provided an opportunity.

Cirri have basically the same structure in the isocrinids and comatulids. They are cylindrical and pierced by a lumen. The stereom of the isocrinid nodicirral articulation may be organized into two distinct

(See facing page.)

FIG. 186. Stem plates (Macurda & Meyer, 1975).——*1.* Plan view of symplexy on internodal stem plate of *Isocrinus blakei* (CARPENTER), recent, W.Indies; ×28.——*2.* Plan view of synostosis on nodal stem plate of *Isocrinus blakei,* recent, W.Indies, ×28.——*3.* Lateral view of broken interior of internodal stem plates along axis of areola of *Isocrinus blakei,* recent, W.Indies, lumen at right, ×56.——*4.* Galleried stereom in center of areola on internodal stem plate, *Endoxocrinus parrae* GERVAIS, recent, W.Indies; ×700.——*5,6.* Synarthrial articulation on stem plate of *Democrinus* sp., and enlargement of fulcral ridge, recent, W.Indies; ×56 and ×350.

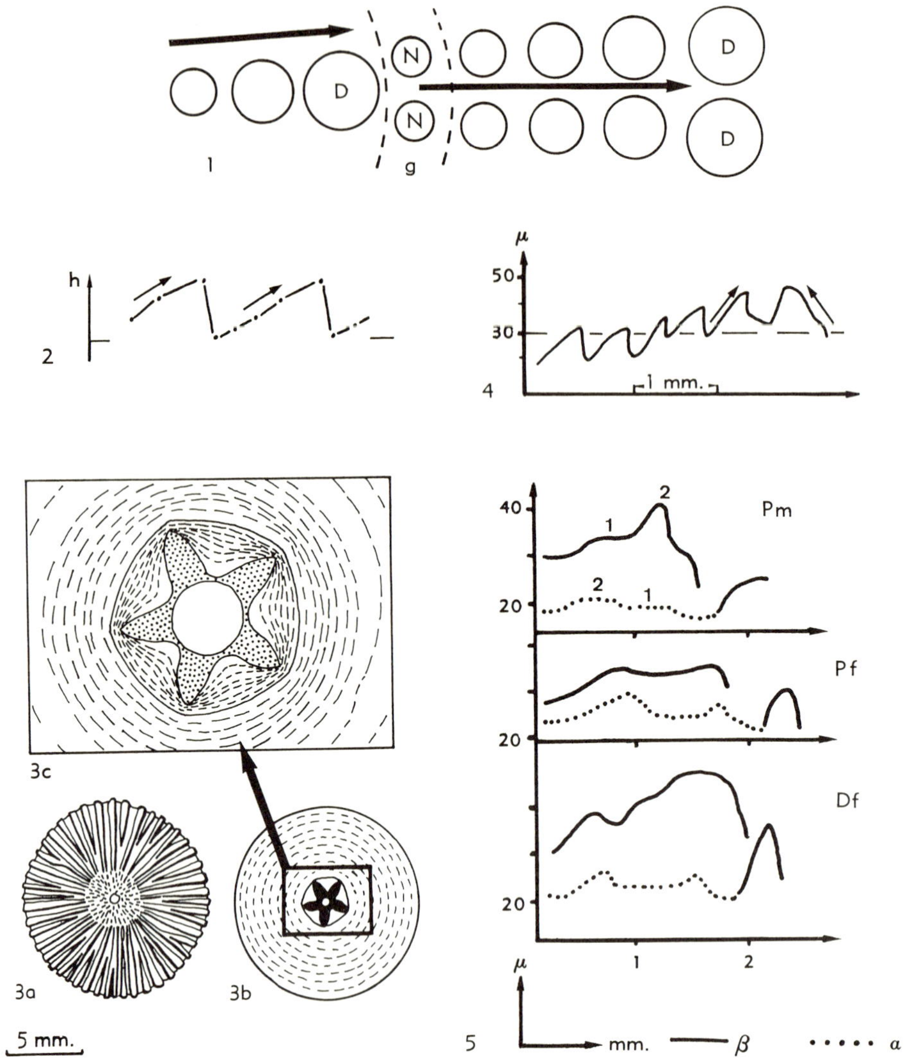

FIG. 187. Biometric graphs of stem growth and ontogenetic modification (Roux, 1975).——*1.* Radial growth of β stereom in stem, subdivision of mesh *(D)*, formation of new mesh *(N)* at growth line *(g)* and continued growth.——*2.* Radial biometric graph of *1* with vertical distance *(h)* equaling dimension of stereom and lateral distance radial growth; arrows indicate directions of growth.——*3a.* Articular facet of a distal columnal of *Millericrinus milleri* (SCHLOTHEIM), Jur., France.——*3b.* Transverse section showing the pentalobate organization of the young columnal *(3c)*.——*4.* Radial biometric graph of β stereom of an *Apiocrinites*, Jur., columnal.——*5.* Radial biometric graphs of α and β stereom of proximal *(P)* and distal *(D)* columnals of *Balanocrinus subteres* (MÜNSTER), Jur. (*m,* near the median of the thin section; *f,* on the articular facet). Numbers indicate growth stages.

fields, one proximal and one distal as on some centrodorsals; others are undifferentiated. The articulum of a cirral is typically organized into a larger distal field, which has galleried stereom, and a smaller proximal field, also galleried, with some type of articular surface(s) between them (Fig. 189,*1a,b*). The mean pore diameter is smaller in the comatulids than in the isocrinids. The articular faces are of varied structure; they can be two knobs lateral to the lumen with corresponding sockets on the opposing face, a bar or horseshoe-shaped area above the lumen, or flattened elevated surfaces. All have denser stereom; the projecting articular surfaces are on the distal ends of cirrals, the corresponding depressions on proximal surfaces. The greater development of the ligament field in the lower part of the articulum enhances the clasping function of the cirri since the contraction of these pulls the cirrus in toward the stem or centrodorsal, firmly affixing the crinoid. The upper ligament field allows the cirrus to be detached and raised. The most distal cirral is a small, sharp hook; its stereom is usually almost solid, transparent calcite. If short spines are developed along the lower surface of the distal cirrals, these and the hook provide even firmer anchorage for the crinoid.

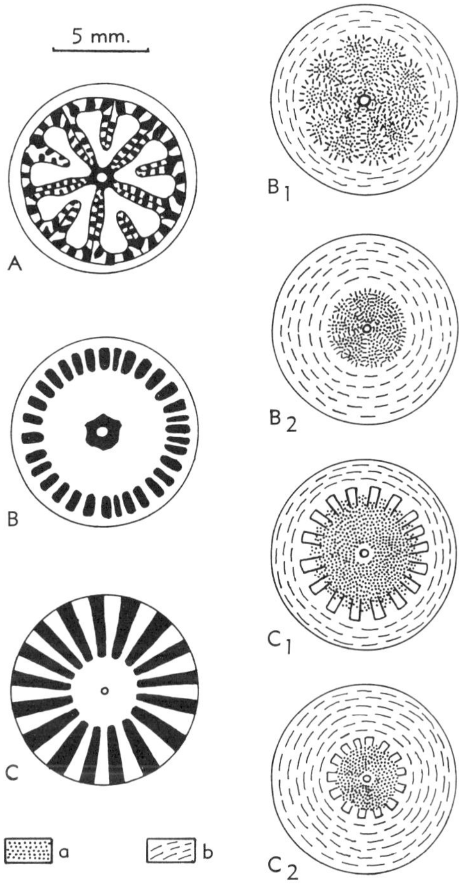

Fig. 188. Microstructural evolution of columnals of *Encrinus liliiformis* Lambert, Trias. (Roux, 1975). On the left, the morphology of the facets externally; on the right, microstructural organization as seen within a transverse thin section. [*A*, proximal columnal; *B*, median; *C*, distal; *1*, section near the facet; *2*, section near the median plane, *a*, α stereom; *b*, β stereom.]

CALYX PLATES

The calyx plates of the articulates (e.g., basals, radials, centrodorsal) surround and protect the lower part of the crinoid soft body, which includes the chambered organ and main nervous center, located aborally. The nerves, which extend outward from the latter, are richly expressed in the numerous nerve canals (Fig. 189,*2,3*) penetrating the plates and extending into the arms, stem, and cirri (Macurda & Meyer, 1975). In the isocrinids, accessory nerve canals may penetrate the basals to extend well down into the stem (Fig. 186,*1,2*). The calyx plates have a porous stereom, except on the exterior, where it is denser as in the arms. The surface of contact between the calyx plates is by definition a synostosis.

BRACHIAL ARTICULATIONS

In crinoids, articulations reach their greatest diversity and complexity in the arms. Four types of articulations are known: a muscular articulation, a syzygy, a symmorphy, and a synarthry. The most common is the muscular articulation that articulates the arm with radials, branching arms with their axillaries, and most brachials with one another. Syzygies are a type of stiffener, particularly just proximal to an axillary. Synarthries are relatively uncommon; the fulcral ridge is vertical and thus accentuates the lateral motions of the arms. Symmorphies are quite rare and their functional significance not fully understood.

The muscular articulation (Fig. 190,*1a, b,3a,b*) has a lower (dorsal) ligament fossa for the extensor ligament, a fulcral ridge

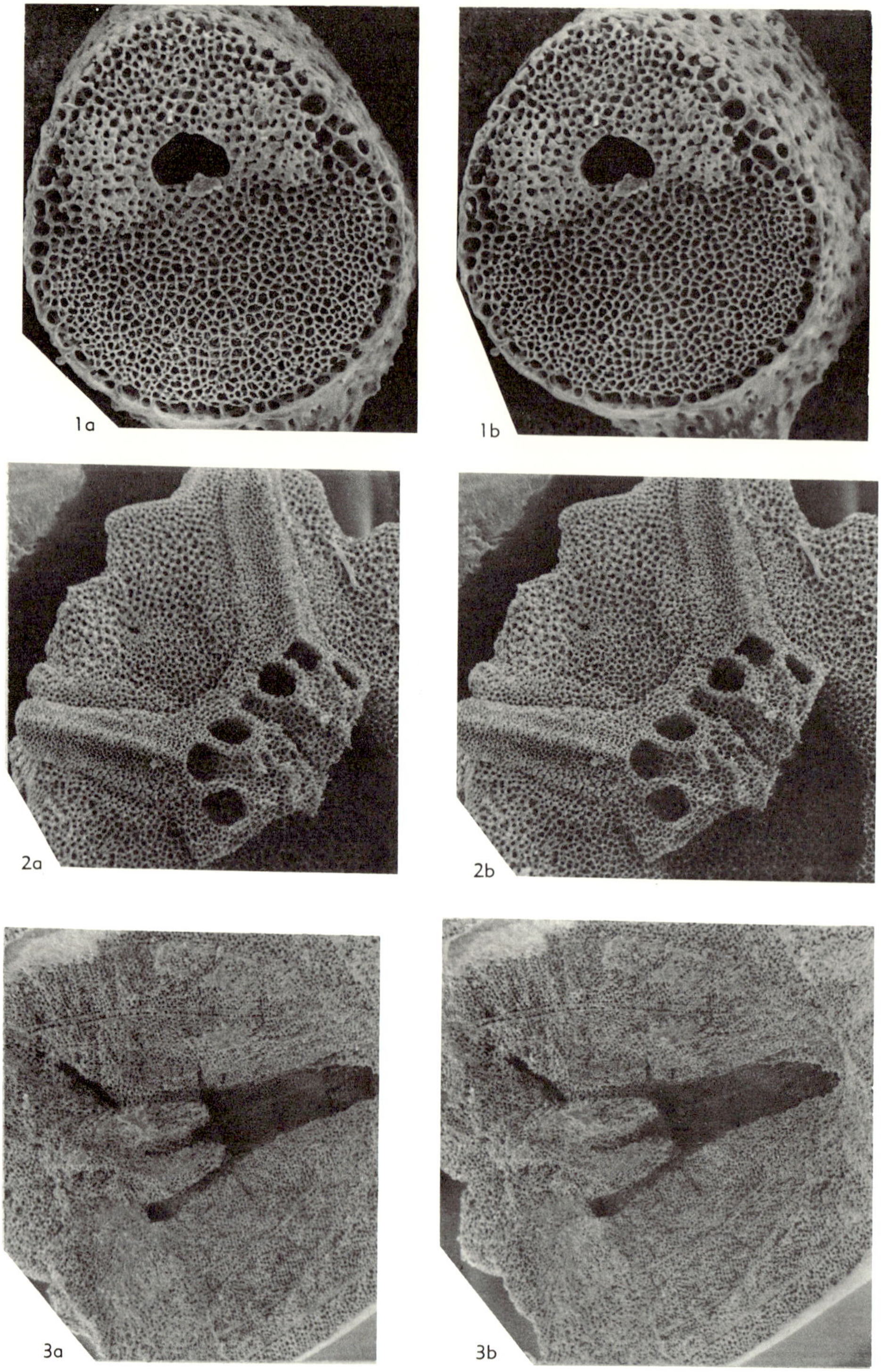
1a
1b
2a
2b
3a
3b

that may be horizontal or oblique, a subcentral lumen for the brachial nerve above the fulcral ridge, two fossae lateral to the lumen for interarticular ligaments, two fossae above these for flexor muscles, and a gap between the muscle fossae, the intermuscular furrow, which contains an extension of the aboral coelomic canal. Crinoid arms can move rapidly, up to four strokes per five seconds in swimmers as *Analcidometra,* can be used for locomotion by slowly creeping across surfaces by pushing and pulling, or can be held erect in moderate currents to form filtration fans (MAGNUS, 1967; MEYER, 1973a; MACURDA & MEYER, 1974). Arm extension is effected by the dorsal extensor ligament. The stereom of the extensor ligament fossa is porous and galleried (Fig. 190,*3a,b*) and there is commonly a ligament pit in the upper center. The fulcral ridge is composed of rather massive calcite and may show signs of abrasive wear. The stereom of the interarticular ligament fossae is also galleried (Fig. 190,*2a,b*). Long ligament fibers penetrate the stereom in both the extensor and interarticular ligament fossae; these fibers are in part collagenous (MEYER, 1971). The function of the extensor ligaments is presumably to act in conjunction with the muscles to enroll the arm but they might conceivably act in opposition to the interarticular ligaments to stiffen the arm when it is extended for feeding.

The microstructure of the stereom of the muscle fossae stands in strong contrast to that of the ligament fossae (Figs. 190,*2a,b, 3a,b*; 191,*1a,b, 2a,b*). In the comatulids it usually has a very irregular mesh; the pores are not aligned and this has been called labyrinthic (MACURDA & MEYER, 1975). The pore diameters are similar to those of the ligament fossae but are more variable and can be larger or smaller. The surface pore area varies from 15 to 40 percent. In the isocrinids, the surface of the muscle fossae has a clotted appearance produced by many small needlelike projections projecting outward from the fossa (Fig. 191,*2a,b*). Labyrinthic stereom is present just beneath these needles. The labyrinthic stereom of the muscle fossae (and the needles of the isocrinids) is due to the fact that muscles attach to the surface of the stereom and do not penetrate into the skeleton as do ligament fibers. This clear difference in stereomic microstructure has been used by LANE & MACURDA (1975) to distinguish muscles and ligaments in brachial and pinnular articulations of the Pennsylvanian inadunate *Aesiocrinus.*

A syzygy is a ligamentary articulation in which the culmina (ridges) of one articulum are opposed to the corresponding elevations of the other articulum (Fig. 191, *3,4*). The culmina radiate from the central lumen and the ligaments are located primarily in the crenellae between them. The syzygy, typically developed only in the comatulids, allows limited mobility in all directions. The tops of the culmina are composed of knobby, massive stereom with small pores; abrasion marks may be visible on the top. The stereom of the crenellae is porous (50 percent pore area) but may or may not be galleried.

A synarthry is a ligamentary articulation in which the opposed articula each bear a medial fulcral ridge all of which are aligned with one another. Each articulum has relatively broad bifascial ligament fields next to the fulcral ridge, which permit mobility in a direction normal to the axis of the fulcral ridge (Fig. 192,*1a,b*). The fulcral ridges may be somewhat porous and show evidence of abrasion. The stereom of the bifascial ligament fields is porous, but may or may not be galleried. In the bathycrinids and some isocrinids, the lower (dorsal) fulcral ridge may be replaced by two divergent ridges radiating dorsally to form a trifascial articulation.

(*See facing page.*)

FIG. 189. Cirral and calyx plates (Macurda & Meyer, 1975).——*1a,b.* Stereo pair of distal cirral articulation of *Analcidometra armata* (POURTALÈS), recent, W.Indies; lower edge of plate in lower half of view, ×210.——*2a,b.* Stereo view of oral surface of centrodorsal of *Comactinia echinoptera* var. *meridionalis* (AGASSIZ & AGASSIZ), recent, W.Indies, with two basal rays still in position. Latter form cover for concavity of main aboral nerve center in centrodorsal. Outer openings of each basal ray contain nerve canals, which lead to adjacent overlying radials, ×56.——*3a,b.* Stereo view of fractured interior of calyx of *Democrinus* sp. showing internal space for chambered organ and main aboral nerve center and continuation of nerve canals outward (left) toward radial facets, ×35.

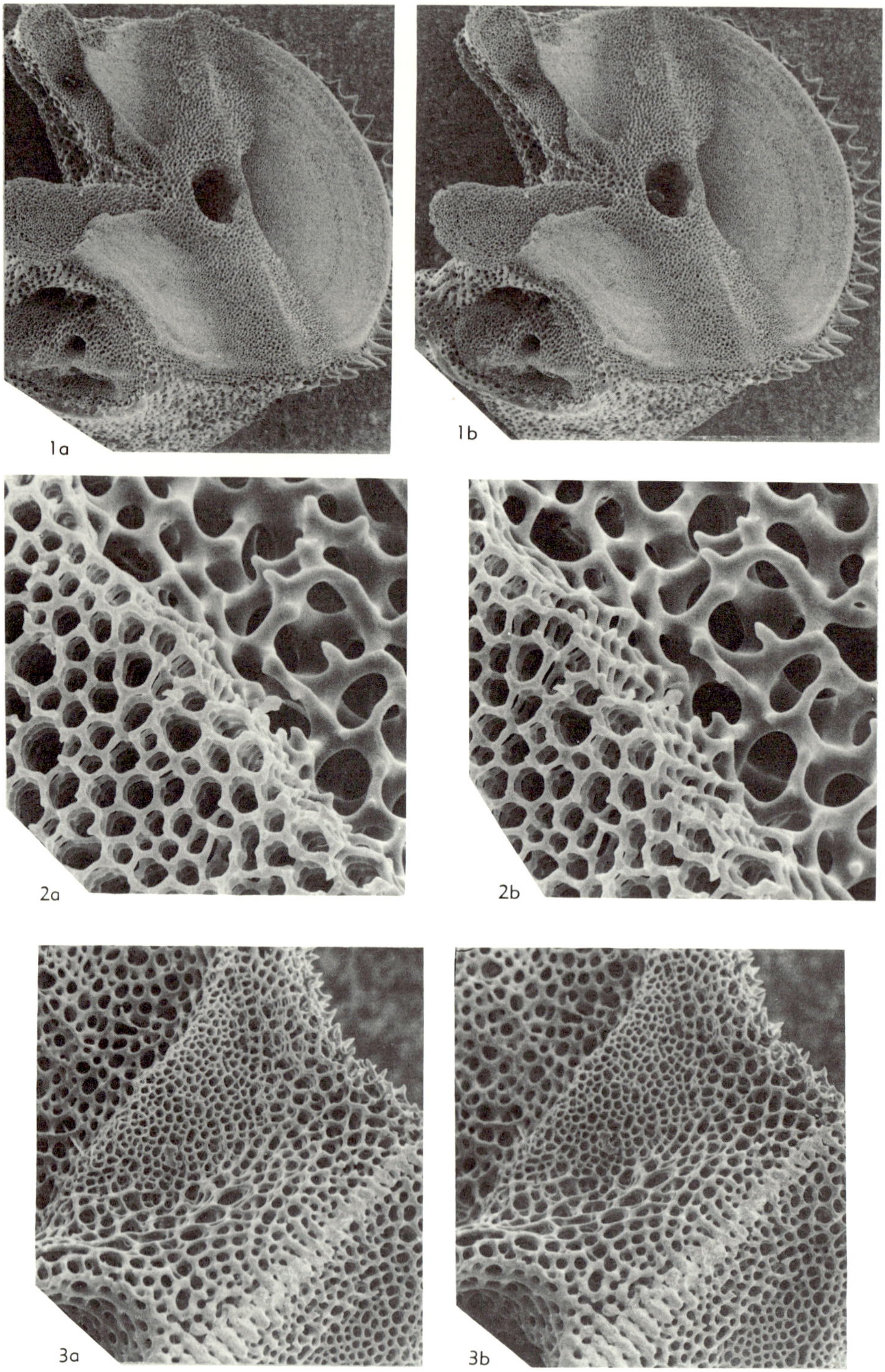
1a
1b
2a
2b
3a
3b

A symmorphy is a ligamentary articulation in which a prominent transverse culmination of the epizygal brachial fits into a corresponding depression of the hypozygal (Fig. 192,*2a,b*). It is not a common articulation, but is found in the brachials of the isocrinid *Isocrinus,* for example. Sharp, dense culmina border a central flat area of the articulum, which has galleried pores.

PINNULAR ARTICULATIONS

The brachials of the modern articulates bear uniserial pinnules. The pinnules extend outward at a sharp angle from the arm and when the tube feet are extended from the adjacent pinnules to almost touch, a dense filtration fan is formed. Pinnules can stand stiffly in a prevailing current, reverse in a bidirectional wave surge in a few species, extend to increase surface area while swimming as in *Analcidometra,* and then retract to lie parallel to the arm on the upstroke, or twitch and writhe convulsively for unfathomed reasons.

The pinnule articulates with a brachial on its upper lateral surface (Fig. 190,*1a,b;* 191,*1a,b*). The articulation is a muscular articulation similar to that between the brachials except that the muscle fossae have been consolidated in a single, deep muscle concavity (Fig. 192,*3a,b*). This is a result of the rotation of the articulation relevant to the axis of the brachial and the consequent asymmetry and reduction of area available for the inner fossa. The pinnulars themselves have a variety of shapes, from trough-shaped in the isocrinids (Fig. 193,*1*) to rodlike in many comatulids (Fig. 193,*2*). The trough-shaped isocrinid pinnular has cover plates that articulate along the upper edge and the tube feet can thus be protectively retracted within. In many comatulids, the pinnular acts as a stiffening rod for the tissue and the tube feet stand exposed on the surface. Peculiar pores penetrate the upper surface (Fig. 193,*2*).

The first two or three pinnulars articulate along a muscular articulation, which has the same transverse ridge, extensor ligament fossa, interarticular ligament fossae, and muscle fossae as the brachials. The transverse ridge between the first and second pinnulars is strongly skewed so that the pinnule extends outward at more of a right angle to the brachial. Most subsequent pinnulars have a modified muscular articulation. The transverse ridge is replaced by two dorsally diverging ridges with thickened stereom (Fig. 193,*3a,b*); a vertical trough with a corresponding ridge on an opposite articulum may be present ventrally (Fig. 193,*4a,b*). Although the form of the articular surface is modified (probably to allow greater flexibility), the stereom is clearly differentiated into galleried pores for ligament fibers and labyrinthic microstructure for muscles.

CONCLUSIONS

The stereom of the crinoid endoskeleton displays a wide variety of structures that reflect the functional morphology and ontogenetic development of the animal. Investigation and interpretation of these by scanning electron microscopy is a recent development (Macurda & Meyer, 1975; Roux, 1974), but the study and preservation of similar microstructures in fossil crinoids will enhance the paleobiological interpretation of these animals (Roux, 1970, 1971, 1975; Lane & Macurda, 1975).

(See facing page.)

Fig. 190. Muscular articulations (Macurda & Meyer, 1975).——*1a,b.* Inclined oral stereo view of distal muscular articulum of brachial of *Nemaster rubiginosa* (Pourtalès), recent, W.Indies, with pinnular articulum in lower left, ×28.——*2a,b.* Stereo view of transition from muscle fossa (upper right) to interarticular ligament fossa (lower left) on brachial of *Nemaster rubiginosa,* ×1,050.——*3a,b.* Stereo view of transition from muscle fossa (upper left) to interarticular ligament fossa (upper center); dorsal ligament fossa in lower right, *Analcidometra armata,* recent, W.Indies, ×245.

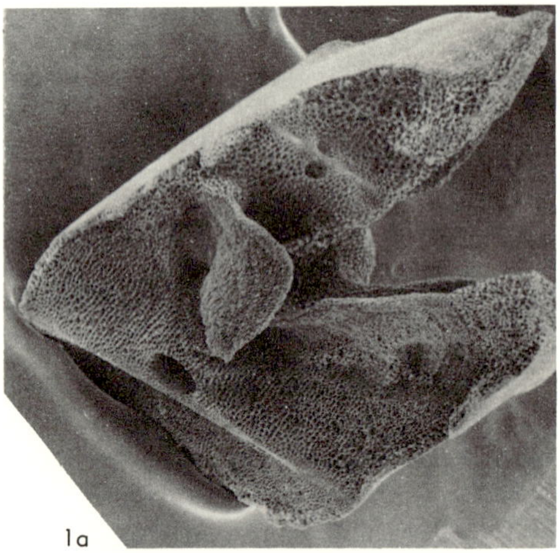

1a

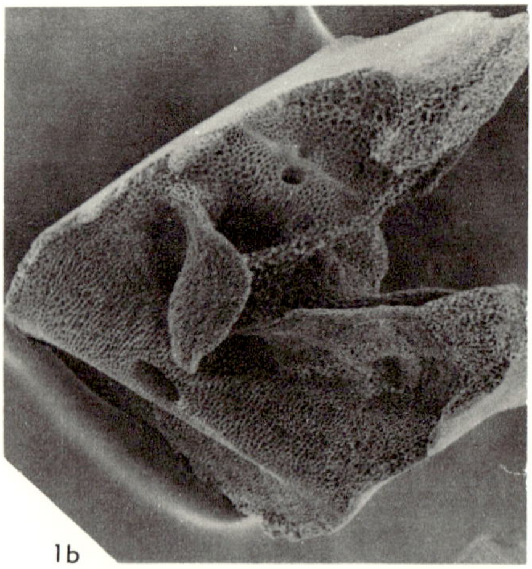

1b

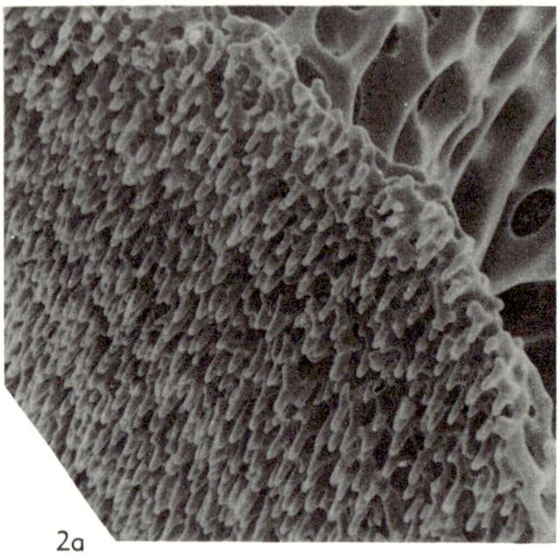

2a

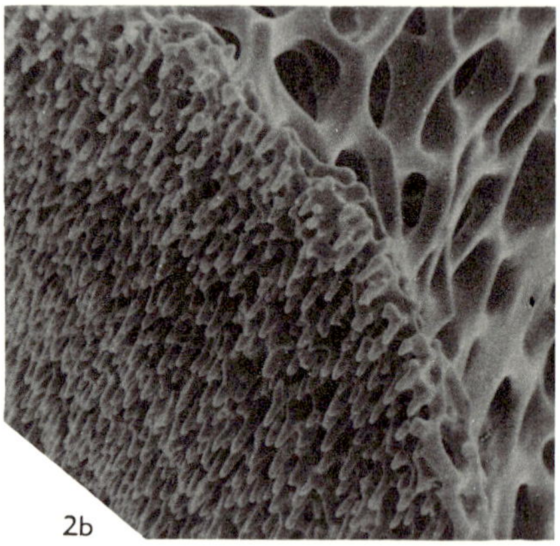

2b

3

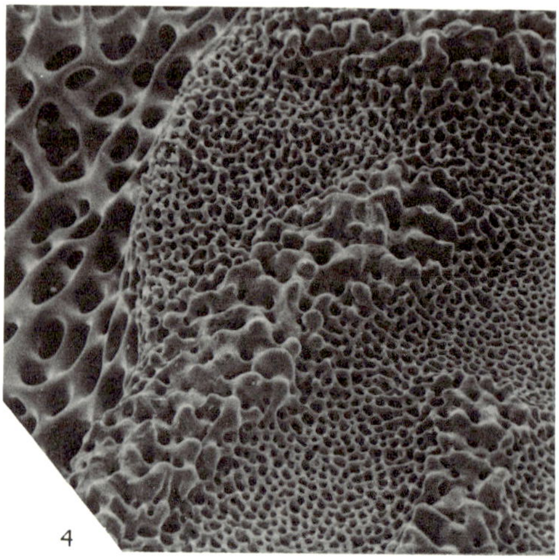

4

GLOSSARY OF CRINOID MORPHOLOGICAL TERMS

By Raymond C. Moore, with additions by Georges Ubaghs, H. Wienberg Rasmussen, Albert Breimer, and N. Gary Lane

[Obsolete terms are printed in italic type, and if synonyms, also enclosed in brackets]

A ray. Ray (anterior) located opposite interray designated *CD* (posterior), generally wider than others and containing anus or anal plates; includes radial and succeeding brachitaxes with intervening plates as well as pinnules. Other rays are designated *B, C, D,* and *E* in clockwise order from *A,* viewing adoral side of theca.

AB interray. Interray (anterior right) next adjoining *A* ray in clockwise direction when crinoid is viewed from adoral side; between *A* and *B* rays. Other interrays are designated as *BC, CD, DE,* and *EA* in clockwise order from *AB* viewing adoral side of theca.

abactinal. Applied to aboral side of theca or plate (syn., aboral, dorsal); preferred term is aboral.

abmedial. Away from median line of ray (syn., abradial).

aboral. Applied to surface of body opposite adoral surface (syn., abactinal, dorsal); directed away from mouth.

aboral cup. Part of aboral side of body between origins of free arms and column (syn., dorsal cup, calyx).

aboral ligament fossa. Narrowly semicircular depression on aboral side of articular face of any ray ossicle, serving for attachment of ligament fibers that function in opposed pairs of ossicles.

aboral skeleton. Part of calcareous framework located on aboral side of body; includes columnals, holdfast structures, centrodorsal, cirrals, centrale, infrabasals, basals, radials, intercalaries, brachials, and pinnulars.

abradial. Away from median line of ray (syn., abmedial).

abthecal. Applied to side of pinnule or pinnular directed away from theca.

actinal. Side of theca or arm directed toward mouth (syn., adoral, ventral); preferred term is adoral.

adambulacral (*adamb,* pl., *adambb*). Small plate of tegmen or arm between ambulacrals and ventral edges of brachials and pinnulars (syn., side plates).

adapical. See aboral, dorsal (syn., abactinal).

adcentral crenulae (of petalodium). Adradial crenulae of columnar articulum located near lumen; may merge with central area of columnal.

admedial. Toward median line of ray (syn., adradial).

adoral. Applied to surface of body that contains mouth and ambulacra (syn., actinal, ventral); direction toward mouth.

adoral groove. Furrow along adoral surface of free brachials and pinnulars; in life contains food groove and radial extensions from coeloms and water-vascular, hemal, nervous, and reproductive systems (syn., ventral groove).

adoral skeleton. Part of calcareous framework located on adoral side of body; includes orals, ambulacrals, adambulacrals, interambulacrals, interbrachials, and some anals (syn., perisomic skeleton).

adradial. Toward median line of ray (syn., admedial).

adradial crenulae (of petaloid columnal). Crenulae of columnal articulum located along margin of petal adjacent to interpetal radii inside periphery, disposed obliquely or nearly normal to margin of petal.

adthecal. Applied to side of pinnule or pinnular directed toward theca.

alphabrach. Brach of proximal brachitaxis of any axil-arm (typically developed in Calceocrinidae); succeeding brachitaxes not belonging to ramules are designated as betabrachs, gammabrachs, etc.

alpha-ramule. Branchlet borne by axillary alphabrach of calceocrinids, invariably directed abanally; succeeding ramules of a given axil-arm are designated as beta-ramules, gamma-ramules, etc.

ambulacral (*amb,* pl., *ambb*). Small plate of tegmen or arm covering part of food groove; may be separated from brachials or pinnulars by adambulacrals (syn., covering plate).

ambulacral groove. Simple or branched furrow in adoral surface of tegmen, arms, and pinnules, underlain by ambulacral epidermis; serves to convey food to mouth (syn., food groove).

ambulacral lappet. Small epidermal fold bordering ambulacral groove.

ambulacrum. Simple or branched, elongate area on adoral surface of body, extending radially from mouth onto tegmen, arms, and pinnules, formed

(*See facing page.*)

Fig. 191. Muscular articulation and syzygy (Macurda & Meyer, 1975).——*1a,b.* Inclined oral stereo view of distal muscular articulum of brachial of *Isocrinus blakei,* recent, W.Indies, with pinnular articulum in upper center, ×35.——*2a,b.* Inclined view of muscle fossa of brachial *Isocrinus blakei,* recent, W.Indies, such as that of left center of *1a,b,* ×315.——*3,4.* Syzygy on epizygal of *Nemaster rubiginosa,* recent, W.Indies, and enlarged view of culmina at outer edge of syzygy, ×28 and ×280.

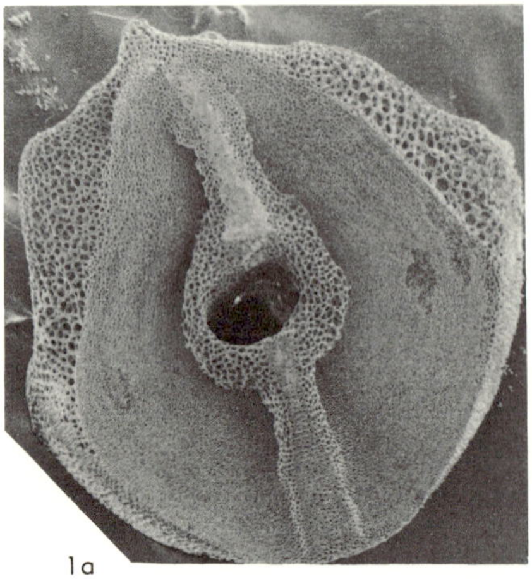

1a

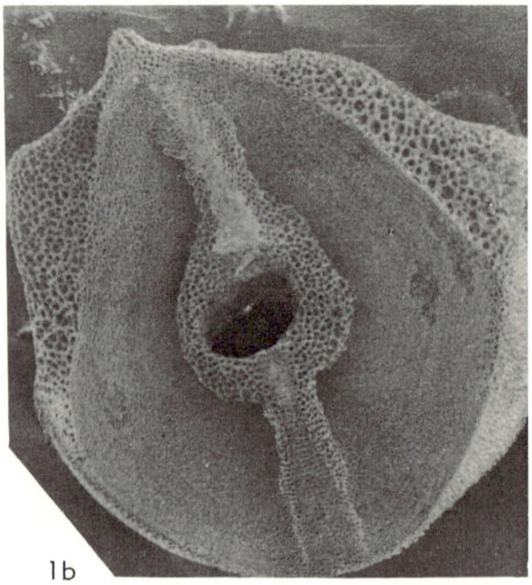

1b

2a

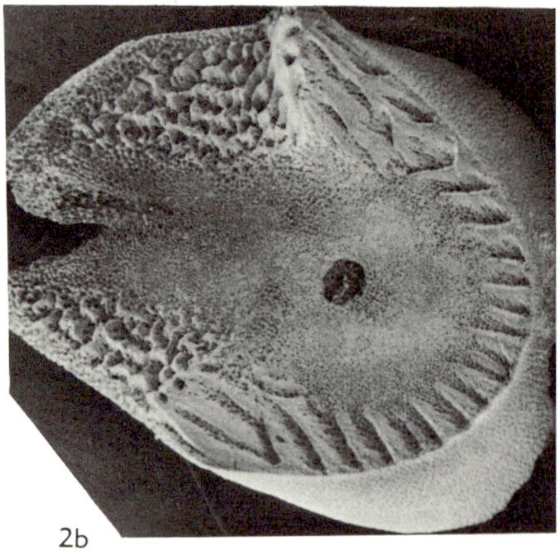

2b

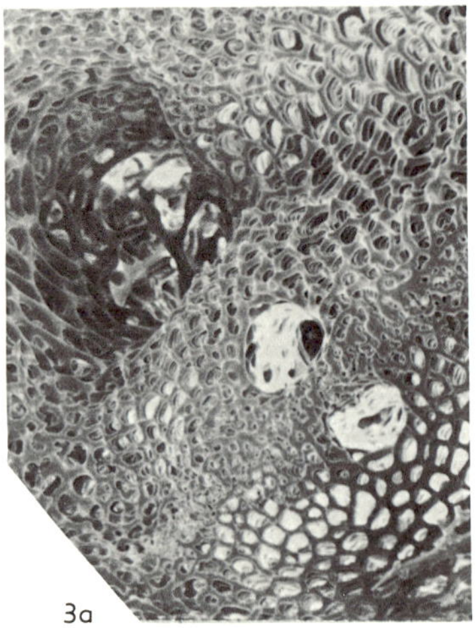

3a

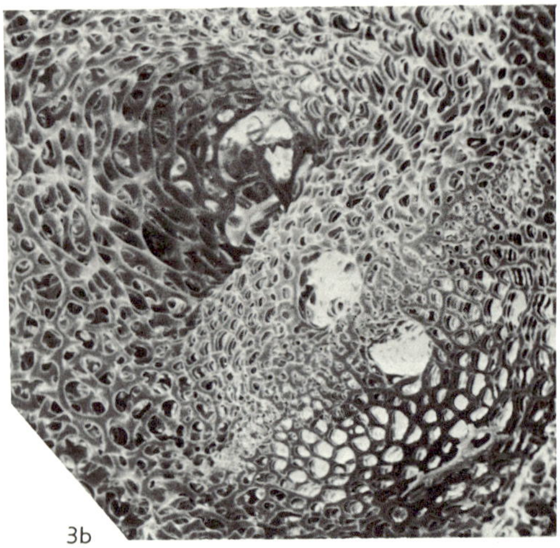

3b

by ambulacral groove and its associated structures such as ambulacral lappets and tube feet (syn., ambulacral tract).

anal. Any plate of posterior *(CD)* interray, mostly confined to cup but excluding interbrachials and fixed pinnulars and among inadunate and flexible crinoids may include plates of right posterior *(C)* ray other than radial, brachials, and pinnulars.

anal pyramid. Low, generally conical elevation of tegmen around anus.

anal sac. Variously shaped, generally inflated and strongly elevated part of tegmen as developed among inadunate crinoids (syn., ventral sac).

anal tube. Conical to cylindrical structure, usually of considerable height, bearing anal opening at its summit, typically developed in many camerate, flexible, and articulate crinoids (syn., proboscis).

anal *X (X)*. Special anal plate in inadunate and flexible crinoids, typically located between posterior *(CD)* radials, distal to and at left of radianal if present.

angustary. Radial articular facet very much narrower than width of plate, generally horseshoe-shaped.

anibrachial. Axillary brachial plate with left shoulder bearing proximal plate *(X)* of anitaxis and on right shoulder proximal plate of brachitaxis (e.g., *Iocrinus, Merocrinus*). [This plate was undesirably designated as brachianal by MOORE, (1962a, page 28).]

aniradial. Radial (or superradial) plate with left shoulder bearing proximal plate *(X)* of anitaxis or anal sac and on right shoulder proximal plate of brachial series (in some catillocrinine allagecrinids bearing proximal plates of as many as four arms [e.g., *Eucatillocrinus, Isocatillocrinus, Xenocatillocrinus*]). Aniradials are common in monocyclic (disparid) crinoid genera (e.g., *Synbathocrinus, Heterocrinus, Allagecrinus*) and they prevail in most dicyclic (cladid) inadunates.

anisuperradial. *C* superradial supporting proximal anal plate on its left shoulder (e.g., *Ectenocrinus*).

anitaxis. Lineal succession of anal plates; commonly raised above laterally adjacent plates of posterior interray (e.g., *Reteocrinus, Xenocrinus*).

ankylosis. Fusion of ossicles effected by calcareous deposits at their interfaces, with or without disappearance of suture lines.

anterior. Referring to direction or ray designated as *A*, located opposite *CD* (posterior) interray, which contains the anus.

anterior left. Equivalent to *EA* interray.

anterior right. Equivalent to *AB* interray.

apical. See aboral, dorsal (syn., abactinal).

areola (pl., **areolae**) *(A)*. Area of columnal articulum between lumen (or perilumen if present) and inner margin of crenularium, generally smooth and featureless but may be granulose or marked by fine vermicular furrows and ridges.

areolar index. Ratio of total width (diameter) of areola to that of columnal articulum multiplied by 100 to avoid fractional numbers (see columnal indices). Areolar index combined with other articular indices has value of 100.

arm. Radial evagination of body above radial, normally extending upward or outward from theca, containing coelomic canals, and composed of pinnulate or nonpinnulate brachitaxes. Restricted by some authors (CLARK) to undivided distal branches.

arm facet. Articulate surface serving for attachment of free arm to theca.

arm trunk. Powerful ramule-bearing arm in some camerates.

armlet. *See* ramule.

articular face (or facet) *(F)*. Smooth or sculptured surface of columnal, cirral, or ray ossicle serving for ligamentary or muscular articulation with contiguous one, also (by some authors) joint face of thecal plate or arm ossicle toward adjacent skeletal element; see articulum.

articular index. Ratio of width (diameter) of columnal articulum to that of entire columnal multiplied by 100, directly measurable in most circular columnals but computed as mean of maximum and minimum values in pentagonal and elliptical columnals.

articular rim. Raised border of articulum.

articulation. Flexible to nearly immovable union of adjoined ossicles effected by ligaments or ligaments and muscles attached to articular faces.

articulum (pl., **articula**). Same as articular face (or facet). All columnal and cirral articula are divisible into lumen and surrounding area designated as zygum.

atomous. Type of ray characterized by lack of branching.

attachment disc. Terminal disc in column of comatulid larva and in some Bourgueticrinida, serving as holdfast.

augmentative regeneration. Replacement of lost part of arm by an axillary (instead of ordinary) arm plate and two branches, common among multibrachiate comatulids.

axial canal. Longitudinal passageway for axial cord penetrating columnals, cirrals, thecal plates, arms,

---

(See facing page.)

FIG. 192. Synarthry, symmorphy, and pinnular articulum (Macurda & Meyer, 1975).——*1a,b.* Stereo view of synarthry on brachial of *Atelecrinus balanoides* CARPENTER, recent, W.Indies, ×42.——*2a,b.* Stereo view of symmorphy on epizygal brachial of *Isocrinus blakei,* recent, W.Indies, ×35.——*3a,b.* Stereo view of pinnular articulum on brachial of *Endoxocrinus parrae* GERVAIS, recent, W.Indies, lower right is adoral direction of brachial, ×112.

1

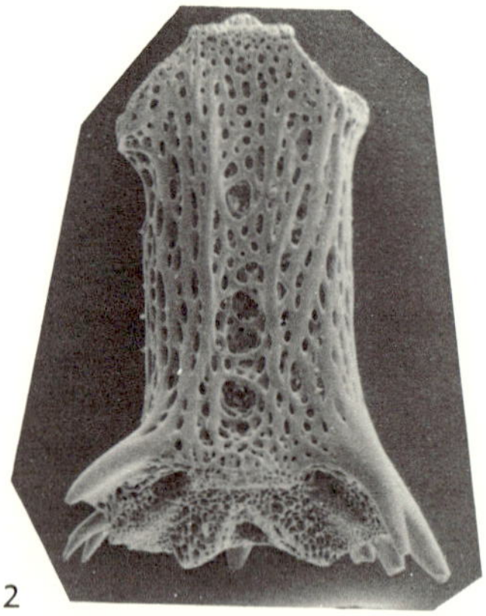

2

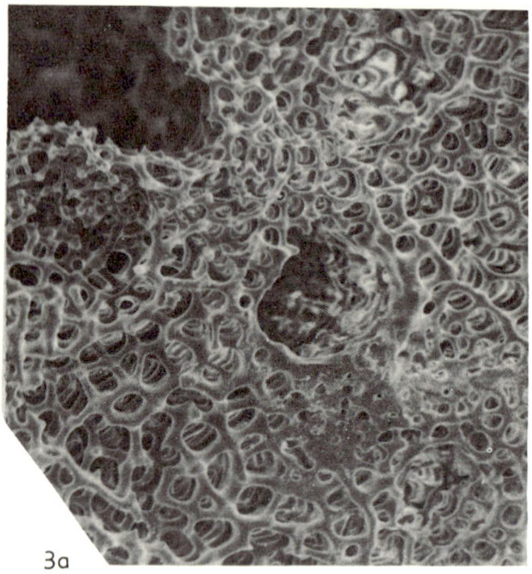

3a

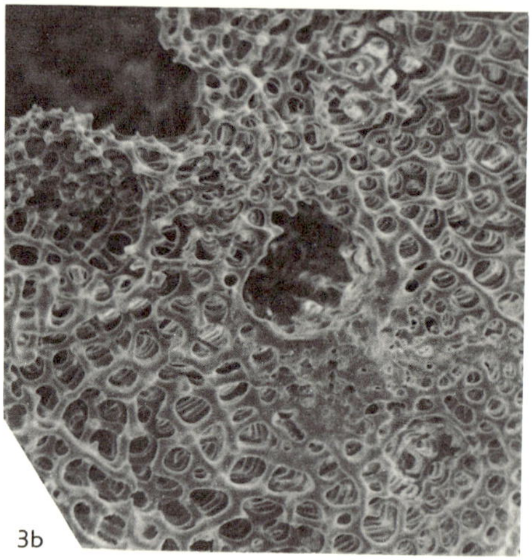

3b

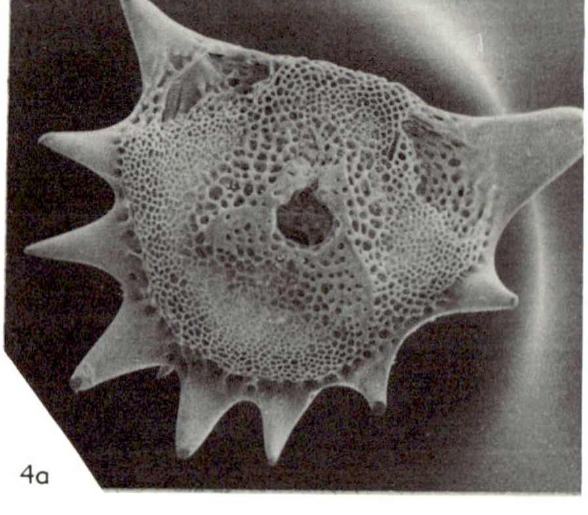

4a

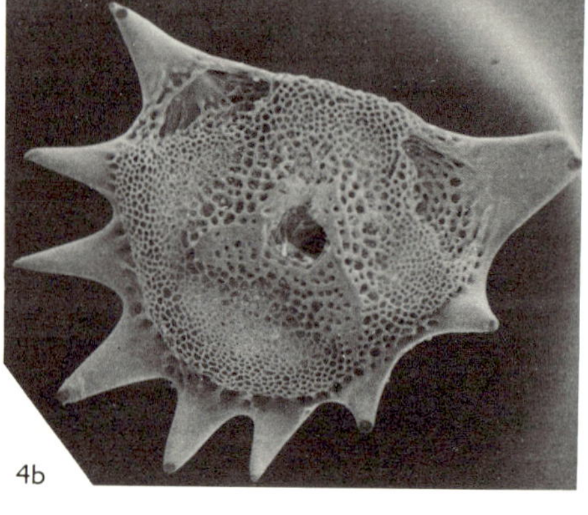

4b

and pinnules, generally but not invariably located centrally; may be simple or multiple with main canal accompanied by smaller accessory ones.

axial tube. Thin-walled, straight-sided cylindrical passageway within axial canal (found in exceptionally well-preserved pluricolumnals).

axil-arm. Arm including its branches borne by any brach of main-axil as seen in calceocrinids but excluding terminal branchlet termed omega-ramule; may be differentiated as primaxil-arm, secundaxil-arm, etc.

axillary *(AX)*. Brachial supporting two arm branches.

axis. Transverse or longitudinal median line defined in theca or longitudinal medial line of ray or stem.

azygous basal. Small plate in 3-basal circlet.

B ray. Right anterior ray (*see A* ray).

BC interray. Lateral interray on right posterior side of cup or calyx (*see AB* interray).

basal (*B*, pl., *BB*). Any plate of circlet next proximal to radials, each basal typically in interradial position.

basal ray. Any of five rodlike structures lying on adoral surface of centrodorsal of some comatulids in interradial position.

base. Part of cup or calyx between radials and stem, normally composed of basals or of basals and infrabasals but in a few articulates and inadunates may include centrale or comprise proximal surface of radial pentagon (Comatulida).

basilarid. One of small number of most proximal columnals which seem to be permanently fixed to base of theca, so that newly added columnals are introduced below them, rather than above most proximal columnal (STRIMPLE, 1963a).

betabrach. *See* alphabrach.

beta-ramule. *See* alpha-ramule.

bifascial articulation. *See* synarthry.

bifascial field. Generally broad smooth area on either side of fulcral ridge of articulum in elliptical columnals, may be bordered on outer side by articular rim, developed also on some cirrals (e.g., *Isocrinus, Austinocrinus, Nielsenicrinus*), may be bordered on outer side by articular rim.

bilateral heterotomy. Type of arm branching characterized by occurrence of ramuli on either side of main arm.

binodal. Paired nodal columnals which share equally in supporting cirri, with distinct apposed articula or with columnals fused together (e.g., *Camptocrinus*); called paired nodals by SPRINGER (1926a).

biradial. Proximal pair of plates in any ray, exclusive of brachials; components of pair designated as inferradial and superradial but in *C* ray may comprise radianal and radial or inferradial and anisuperradial (syn., compound radial).

biserial arm or brachitaxis. Composed of brachs arranged in double row with interlocking sutures along junction of rows (majority of cladid inadunates and camerates, but lacking in disparid inadunates and flexibles).

bivium. Differentiated *C* and *D* (posterior) rays, generally shorter than rays of opposed trivium *(A, B, E)* (e.g., many *Comasterida, Holopodidae*).

bothrospire. Pitlike or slitlike depression on calyx outer surface, presumably serving for respiration (e.g., *Indocrinus*).

brachial (brach) (*Br*, pl., *Brr*). Any ray plate above radial or biradial, exclusive of pinnulars, ambulacrals, and adambulacrals; may be axillary or nonaxillary and incorporated in theca (fixed brachials) or not (free brachials).

brachianal. Fixed brachial of *C* ray supporting proximal and anal plate on its left shoulder (e.g., *Peniculocrinus*).

brachitaxis (pl., **brachitaxes**) *(BrT)*. Series of brachials (brachs) extending from radial or biradial to and including primaxil or in arms lacking axillaries to distal extremity of arm; likewise brach series extending from any axillary to and including next one or to distal extremity of arm or branch.

brachium (pl., *brachia*). *See* arm.

C ray. *See A* ray.

calyx. Part of theca excluding tegmen (equivalent to aboral cup) (employed by some authors as equivalent to theca).

canal. *See* axial canal, interarticular radial canal.

canalicula (pl., **canaliculae**). Subhorizontal radially disposed tubular passageway in body of nodal columnal leading from axial canal of stem to axial canal of cirrus, typically grouped in fives with angle of 72 degrees between adjacent canaliculae, but with one or more of these suppressed in many columnals; also, canaliculae may terminate at outer surface of columnal in pimple-like node (aborted cirrus) or small open pore.

cavannulus (pl., **cavannuli**). Low hollow ring in inner medulla of some columnals girdling axial canal and sloping somewhat inward toward midplane of columnal, paired with another on opposite side of mid-plane; may contain delicate,

(*See facing page.*)

FIG. 193. Pinnulars and pinnular articulations (Macurda & Meyer, 1975).——*1.* Inclined oral view of pinnular of *Endoxocrinus parrae,* recent, W.Indies, ×70.——*2.* Oral view of pinnular of *Nemaster rubiginosa,* recent, W.Indies, ×77.——*3a,b.* Stereo view of muscular articulum on pinnular of *Endoxocrinus parrae,* recent, W.Indies, with upper surface in upper left (compare with *1*), ×210.——*4a,b.* Stereo view of muscular articulum on distal surface of a pinnular of *Nemaster rubiginosa,* recent, W.Indies, upper surface at top, ×140 (compare with *2*).

rather open vesicular tissue (function unknown).

CD interray. *See AB* interray.

[*central area* (Bather). *See* perilumen, areola.]

central axis. Line connecting aboral and adoral poles or centers (syn., polar axis).

[*central canal. See* axial canal.]

central cavity. Central lumen of cup surrounded by radial and basal circlets.

central nodicirral articulum. Cirrus attachment scar socket located at or very near midheight of latus of nodal, generally facing straight outward (= *central cirrus facet,* Bather, 1909a).

central plug. Large or small, generally spongy calcareous deposit, on adoral surface of radial pentagon (e.g., some comatulids).

centrale. Noncirriferous thecal plate typically occurring inside infrabasal or basal (if infrabasals are vestigial) circlet of some inadunates and articulates (e.g., *Marsupites, Uintacrinus*).

centrodorsal. Commonly cirriferous columnal or semifused to fused columnals attached to theca of Comatulida (Thiolliericrinidae).

centrodorsal cavity. Depression on adoral surface of centrodorsal containing chambered organ and accessory structures.

centrum (pl., **centra**). Substance of columnal or cirral including luminal septa, if present; may be divisible into parts distinguished by differences in microstructure, such as outer zone (between proximal and distal crenularia), intermediate zone (between proximal and distal areolae), and inner zone (corresponding to perilumina on columnal articular facet).

chiasma. Figure formed by division of aboral nerve trunks within axillary.

cirral. Single cirrus segment.

cirrinodal. Columnal bearing cirrus or cirri; other nodal columnals are termed nudinodals.

cirripore. Small circular opening of canalicula approximately at midheight of columnal latus, commonly accompanied by four others distributed with even spacing around columnal, interpreted as rudimentary type of cirrus structure but of unknown function (e.g., *Mooreanteris*), cirripore-bearing columnal classed as nodal.

cirrus (pl., **cirri**). Unbranched, jointed appendage of crinoid stem of centrodorsal, exclusive of radix.

[*cirrus facet, scar, socket. See* nodicirral articulum.]

cirrus root. *See* radicular cirrus, radix.

cirrus socket. Articular face on nodal or centrodorsal for articulation of cirrus (syn., nodicirral articulum).

claustrum (pl., **claustra**). Thick or thin inward projection of columnal medulla constricting axial canal, inner extremity acuminate to bluntly rounded, truncate with rabbeted edges, or clavate, composed of dense stereom or showing microstructure of fine annular lamellae subparallel to midplane of columnal, with or without intersecting longitudinally disposed lamellae which form microscopic cribwork. Transverse sections of claustra may show pentastellate indentations which are extensions of jugulum and between such indentations inner parts of claustrum may be thickened to form jugular ramparts.

clinate. Distinctly sloping, with inclination gentle, moderate or steep, invariably applied to attitude longitudinally and understood to be essentially straight, unless modified by adjective (e.g., curved, with possible addition of descriptive designation such as gently, strongly, evenly, and the like); refers to attitude of plates in indicated circlet, disposition of mean surface of radial articular facet, or orientation of interbrachial facets and sutures in relation to arm axis.

[*close suture. See* zygosynostosis.]

column. Series of segments composing stem; excludes cirri and anchorage structures (holdfasts).

columnal. Individual ossicle of crinoid stem (exclusive of cirri and holdfast structures).

columnal diameter. Dimension transverse to longitudinal axis, may be uniform in all longitudinal planes or notably dissimilar in different ones.

columnal height. Dimension in longitudinal plane, generally any such plane, but in comparatively rare columnals with opposite articula inclined to one another, plane of measurement needs to be specified for definition of minimum, maximum, and mean height.

columnal indices. See different types: areolar, crenular, epifacetal, facetal, height, jugular, luminal, periluminal, septal, shape, zygal. Variations in outline of facetal elements, as well as facets or columnals in whole introduce difficulties in determining total width factors needed for computation of indices; normally, the sum of measurements along two opposite radii (whether these are identical or different in length) provides the measurement of total width, but in elliptical facets or columnals maximum and minimum radii (disposed at right angles to one another) are chosen.

comb. Peculiar comblike modification of distal part of lower pinnules (in Comasteridae).

commissural canal (ring canal). Passageway within cup plate mainly in transverse direction for entoneural branch (ring-nerve or commissure) connecting neighboring entoneural cords.

complex axial canal. Medial perforation of crinoid column characterized by successive alternating constrictions (jugula) produced by adaxial annular projections (claustra) of columnals and intercolumnal expansions (spatia).

compound basal. Ossicle resulting from union of a basal ray with an interradial process of rosette in some comatulids.

compound nodal. Two or more columnals that share in bearing cirrus or cirri.

[*compound radial. See* biradial.]

concavodeclinate. Downwardly and outwardly sloping circlet of plates or proximal portions of plates located within basal concavity of aboral cup (e.g., *Delocrinus*).

concavoplanate. Horizontal or subhorizontal circlet of plates located within basal concavity of aboral cup.

consolidating apparatus. Peculiarly specialized oral plates (e.g., *Cupressocrinus*).

convoluted organ. Loose calcareous coiled body within theca of many camerates (syn., perigastric coelomic organ).

cortex. Peripheral substance (stereom) of columnal or cirral next to latux, rather clearly distinct from medulla or grading into it; substance of columnals with undifferentiated cortex and medulla referred to simply as stereom.

[*costal. See* primibrach(ial).]

[*covering plate. See* ambulacral.]

craspedospire. Infold at plate margin, presumably serving for respiration.

crenella (pl., **crenellae**). Narrow furrow between culmina of columnal articulum (commonly misapplied to culmina by many authors).

crenula (pl., **crenulae**). Ridge (culmen) combined with adjacent furrow (crenella) of columnal articulum.

crenularial index. Ratio of total width of crenularium to that of columnal articular facet (C/F), multiplied by 100 to avoid fractional numbers.

crenularium. Entire area of columnal articular facet bearing crenulae.

crenulate suture. Externally visible wavy line of contact between symplectically united columnals.

crinoidal plane. Plane of bilateral symmetry passing through *A* ray and *CD* interray.

crown. Whole crinoid exclusive of stem.

cryptodicyclic. Characterized by concealment of infrabasals or by occurrence of these plates only in immature growth stages (syn., pseudomonocyclic).

[*cryptosymplectic articulation. See* cryptosymplexy.]

cryptosymplexy. Weak, ill-defined articulation of symplectial type, as developed between nodals and infranodals of many crinoid columns (e.g., Isocrinidae).

[*cryptosynarthrial articulation. See* cryptosynarthry.]

cryptosynarthry. Weak, ill-defined articulation of synarthrial type, as developed between pairs of brachials of some Articulata.

cryptosyzygy. Weak, ill-defined articulation of syzygial type, as developed between pairs of brachials of Articulata (e.g., some Isocrinidae) in which opposed culmina of articular faces are reduced to discontinuous short ridges or granules, with tendency toward irregular arrangement and disappearance.

culmen (pl., **culmina**). Narrow ridge between adjoining crenellae of columnal articulum (= crenella of many authors).

cup. See aboral cup.

[*cup-brachial. See* fixed brachial.]

[*cup-pinnular. See* fixed pinnular.]

D ray. *See A* ray.

DE interray. Left anterior lateral interray. *See AB* interray.

declivate. Sloping downward and outward (e.g., infrabasal plates of *Delocrinus,* radial articular facets of *Zeacrinites*).

defective pinnulation. Characterized by lack of pinnules in places where they should normally be present.

deltoid. *See* oral.

dichotomous. Characterized by division of arm into two branches (dichotomy), which may be equal (isotomy) or unequal (heterotomy).

dichotomy. Division of arm into two branches.

dicyclic. Having two thecal circlets of plates proximal to radials or (in some inadunates that lack radials) proximal to orals.

disc. Tegminal surface between arm bases or visceral mass resting on aboral cup.

discoid holdfast. Subcircular, depressed, upwardly convex to crateriform plated structure with interior supported by radial walls, base plane or somewhat concave for cementation to foreign object such as shell, central articulum on upper surface for attachment of most distal columnal of crinoid stalk (e.g., *Lichenocrinus, Aspidocrinus*).

distal. Referring to direction or position away from polar or central axis.

distal pinnule. In crinoids provided with oral and genital pinnules, any pinnule beyond the latter.

[*distichal. See* tertibrach.]

dististele. Distal region of crinoid column.

divergence of fulcral ridges. Azimuthal angular difference in orientation of fulcral ridges on opposite articula of synarthrially joined columnals.

[*division series. See* taxis.]

dorsal. Referring to direction or side away from mouth, normally downward and outward; preferred term is aboral.

dorsal cup. Theca exclusive of tegmen; preferred term is aboral cup or cup.

dorsal ligament fossa. See aboral ligament fossa.

dorsal star. Stellate hollow around aboral pole of centrodorsal in some comatulids, often fused with depression.

E ray. Left anterior ray; *see A* ray.

encrinoidal. Characterizing perfect pentamerous cup in which each ray coincides with a plane of bilateral symmetry.

endocyclic. Characterized by central location of mouth with respect to coiled digestive tube, hence at or near center of tegmen.

endospire. Invagination of body wall, serving for respiration.

endotomous. Arm structure characterized by bifurcation in two main arms which give off branches only on their adradial side.

epifacet. Extrafacetal part of crinoid columnal articulum.

epifacetal index. Ratio of total width of epifacet to that of entire columnal multiplied by 100 to avoid fractional numbers (*see* columnal indices). [Combined epifacetal and facetal indices have value of 100.]

epispire. Epithecal structure, open or covered, associated with sutural pore or goniopore, and presumably serving for respiration.

epizygal. Proximal segment of syzygial pair. [Considered by BATHER (1909a) to be equivalent to nodal in crinoid columns, but misnomer, since distal articulum of nodals is not a surface of syzygial articulation.]

eustenocrinoid (eustenocrinoidal) plane. Plane of bilateral symmetry passing through *C* ray and *EA* interray; prominent in some disparid inadunates.

even synarthry. Articulation of synarthrial type characterized by general evenness of opposed faces of two brachials joined by synarthry.

exocyclic. Characterized by location of mouth near or beyond border of coiled digestive tube, hence near margin of tegmen.

exospire. Evagination of body wall, serving for respiration.

exotomous. Arm structure characterized by bifurcation in two main arms which give off branches only on their abradial sides.

facet (articulum). Differentiated part of proximal and distal surfaces of crinoid columnals serving for articulation with contiguous columnals (includes lumen); similarly, in thecal plates and ray ossicles any surface that functions for connection with adjoined skeletal element.

facetal index. In crinoid columnals, ratio of total width of facet to that of entire columnal multiplied by 100 to avoid fractional numbers. [Combined facetal and epifacetal indices have value of 100.]

facetal rim. Raised border of columnal or cirral articular facet.

finial. Ossicle of most distal brachitaxis which does not fork again.

fixed brachial (fixed brach). Ray plate above radial or biradial (exclusive of pinnulars) comprising part of calyx, joined more or less firmly to neighboring plates (e.g., *Ohiocrinus, Dystactocrinus* among inadunates; *Amphicrinus, Artichthyocrinus* among flexibles; *Periechocrinus* and most camerates; *Uintacrinus* among articulates).

fixed pinnular. Plate forming part of pinnule incorporated in calyx (e.g., *Scyphocrinites*).

floor (of columnar articular surface). Generally smooth ligament area in median part of petal, mostly flush with articular surface of columnal but may be depressed or (rarely) slightly elevated, bordered by short crenulae; equivalent to areola.

food groove. Simple or branched furrow running in adoral surface of tegmen, arms, and pinnules; serves to convey food to mouth (syn., ambulacral groove).

fossa (pl., **fossae**). Elongate depression on articular face of ossicle for attachment of muscles or ligaments; less localized than pit.

fossula (pl., **fossulae**). Equivalent of half-canalicula in intercolumnal location, thus traversing articular surfaces of two contiguous columnals and leading peripherally to axial canal of cirrus, distinguished from crenellae in this way (as well as by size, connection with lumen of columnal, and other characters).

free arm. Part of ray exclusive of pinnules not incorporated in crinoid theca, freely mobile.

free brachial (free brach). Ray plate above radial or biradial (exclusive of pinnulars) not incorporated in theca (e.g., nearly all crinoids).

free face. Surface of any ossicle not joined to another.

free pinnular. Plate forming part of aboral skeleton of pinnule not incorporated in theca.

free pinnule. Arm branchlet not incorporated in crinoid theca (essentially synonymous with pinnule).

fulcral ridge. Linear elevation in articular face forming hinge line in synarthrial and muscular articulations of ray plates, synarthry of columnals, and ligamentary articulation of some cirri; facilitates differential movement in directions normal to ridge. Fulcral ridge may be interrupted by lumen or continuous around it and may be simple or variously modified. In cirrus sockets and cirrals, often reduced to pair of lateral tubercles or missing.

gammabrach. *See* alphabrach.

gamma-ramule. *See* alpha-ramule.

genital pinnule. Pinnule used for storage of gametes.

goniopore. Opening at plate corner, presumably serving for respiration.

gonioporoid. Pit or depression generally at plate corner, opening into thecal cavity but not reaching outer surface of skeleton; presumably respiratory.

goniospire. Infold or group of infolds at plate corner, presumably serving for respiration (e.g., *Porocrinus, Triboloporus*).

growth-index line. Graphic plot of selected dimensions of crinoid skeletal elements in graded series of specimens differing in size or number of elements considered.

height. Dimension of columnal, thecal plate, or arm ossicle measured in proximal-distal direction.

height index. For columnals, ratio of height to total width of columnal; for thecal and arm plates, ratio of height to width, multiplied by 100 to avoid fractional numbers (see columnal indices).

heterocrinoid (heterocrinoidal) plane. Plane of bilateral symmetry passing through *D* ray and *AB* interray; prominent in some disparid inadunates.

heteromorphic (column on pluricolumnal). Crinoid column composed of dissimilar columnals, which commonly are classifiable as nodals (with or without cirrus scars) in one group and internodals of first-, second-, or higher-order subgroups in the other.

heterotomous. Arm characterized by division into unequal branches.

heterotomy. Division of arm into unequal branches.

holdfast. Any structure at distal extremity of column serving for fixation.

holotomous. Arm characterized by division on every successive brachial.

homeomorphic (columnal or pluricolumnal). Crinoid column composed of similar columnals throughout observed length but admitting gradual change of characteristics from proximal to distal parts of stem. Pluricolumnals from different parts of xenomorphic stems may be homeomorphic within themselves but dissimilar when compared with one another.

homocrinoid (homocrinoidal) plane. Plane of bilateral symmetry passing through *E* ray and *BC* interray; prominent in some disparid inadunates.

hyperpinnulation. Development of two or more pinnules on a single brachial.

hypozygal (columnal). Distal segment of syzygial pair. [Internodal adjoining distal extremity of nodal (BATHER, 1909); considered inappropriate term because articulation of nodal and this internodal is not syzygial.]

inferradial (*IR*, pl., *IRR*). Proximal plate of biradial; includes radianal.

inferradianal (*IRA*). Plate other than basal on proximal side of radianal, directly in series with it or disposed somewhat obliquely.

infrabasal (*IB*, pl., *IBB*). Any plate of proximal circlet in aboral cup having two circlets of plates (dicyclic) below radials.

infracentral cirrinodal articulum (or facet). Cirrus attachment scar located below midheight of nodal latus, generally directed obliquely downward and outward. In dissociated pluricolumnals (excepting those derived from a few genera of stalked Articulata) infra- and supracentral nodicirral articula are not distinguishable because proximal and distal extremities are indeterminate (= infracentral cirrus-facet, BATHER, 1909a).

infranodal. Internodal adjoining distal face of any nodal (CARPENTER, 1884a).

inner ligament area. Confluent interarticular ligament fossae, located on ventral (inner) side of transverse ridge of articular face.

interambulacral (*iamb*, pl., *iambb*). Any plate of tegmen lying between ambulacrals.

interarticular canal. Simple or branched duct between apposed articular faces of thecal plates (rarely columnals) with outer and inner openings or inner opening only, these openings being invariably located on sutures (sutural pores) or at plate corners (goniopores).

interarticular ligament fossa. Elongate depression on articular face of ray ossicle located on either side of axial canal and on adoral side of transverse ridge.

[*interarticular pore* (CARPENTER). *See* radial pore.]

interarticular radial canal. Small radially directed passageway formed by apposed radial grooves on petaloid articula of some columnals (e.g., *Isocrinus*); not identical to canal produced by apposed fossulae on binodal articula.

interbrach(ial) (*iBr*, pl., *iBrr*). Ossicle of calyx above basals (except anals and fixed pinnulars) between rays and branches of any single ray.

intercalary. One of many calyx plates occurring between radial and basal circlets of some crinoids (e.g., *Acrocrinus*).

[*intercostal. See* interbrachial, interprimibrach.]

[*interdistichal. See* interbrachial, intersecundibrach.]

intermuscular furrow. Linear depression separating muscular fossae of articular face of brachials joined by muscular articulation.

intermuscular ridge. Linear elevation separating muscular fossae of articular face of brachials joined by muscular articulation.

internal suture. Line of contact of apposed columnals or cirrals inside of their latera, as seen in longitudinally cut or in weathered specimens.

internodal (*iN*, pl., *iNN*). Columnal intercalated between pair of neighboring nodals, commonly lacking cirri. Internodals are classifiable according to their different successive generations as first-order, second-order, third-order, etc., each generally distinguished by diagnostic width and height measurements of their own.

internodal index. Ratio of height of internode to height of noditaxis containing it, multiplied by 100 to avoid fractional numbers. [Combined internodal and nodal indices have value of 100.]

internode (*IN*). Section of stem between any two successive nodals.

interpinnular (*iP*, pl., *iPP*). Ossicle between any fixed-pinnulars or fixed-pinnulars and fixed-brachials of same ray.

interprimibrach(ial) (*iIBr*, pl., *iIBrr*). Plate of calyx (except anals and fixed pinnulars) above basals located between rays.

interradial (adj.). Indicative of location between any two adjacent radii.

[*interradial* (noun). Any interray plate above basals (except anals, fixed pinnulars, and tegminal plates). Although used by many authors, this term is rejected here, because the plates referred to are identical in kind and origin to interbrach(ial)s. *See* interbrach(ial), interprimibrach(ial).]

interradial impression. Faint petaloid impression on aboral side of centrodorsal in some comatulids.

interradius. One of five planes, regularly alternating with radii (*see* definition of radius).

interray. Part of calyx between any two adjacent rays.

intersecundibrach(ial) (*iIIBr*, pl., *iIIBrr*). Any ossicle (except pinnulars) between neighboring secundibrach series (brachitaxes) of same ray, including succeeding higher plates.

intertertibrach(ial) (*iIIIBr*, pl., *iIIIBrr*). Any ossicle (except pinnulars) between neighboring tertibrach series (brachitaxes) of same ray, including succeeding higher plates.

isotomous. Characterized by division of arm in equal branches.

isotomy. Division of arm into equal branches.

joint. Connection between any pair of contiguous ossicles.

[*joint face* (BATHER). *See* articulum.]

jugular index. Ratio of total width of jugulum to that of columnal articular facet, multiplied by 100 to avoid fractional numbers (*see* columnal indices).

jugular rampart. Localized thickening of flange in columnal canal at margin of jugulum, typically developed as five petaloid bulges between radiating jugular slits.

jugulum (pl., **jugula**). Localized constriction of axial canal, commonly at midheight of columnals; may be circular transversely or extended by very narrow slits radiating into septa in form of lineate star.

latus (pl., **latera**). Surface of crinoid columnal or cirral exclusive of articular facets, equivalent to epifacet.

left anterior. Ray or radius (designated by letter "*E*") next to anterior (*A*) ray or radius in counterclockwise direction when crinoid is viewed from adoral side.

left posterior. Ray or radius (designated by letter "*D*") next to *E* ray or radius in counterclockwise direction when crinoid is viewed from adoral side.

ligament field. Concave or flat part of articular face for attachment of ligaments.

[*ligament fossa. See* ligament field.]

ligament pit. Generally steep-sided small depression in aboral ligament fossa adjoining center of transverse ridge.

ligamentary articulation. Union of ossicles effected solely by ligaments, lacking striated muscle fibers but in some articulations of this type supplemented by more or less calcareous deposition.

longitudinal oblique articulation. Union of ray ossicles characterized by obliquity of articular facet plane in relation to longitudinal axis of ossicles; defined externally by oblique sutures.

loose suture. Externally visible line of contact between movably united ossicles.

lumen (pl., **lumina**). Open space approximately in plane of columnal or cirral articulum comprising intercept of axial canal, generally located centrally, highly variable in size and shape, surrounded on all sides by zygum.

luminal index. Ratio of total width of lumen to that of columnal articular facet, multiplied by 100 to avoid fractional numbers (*see* columnal indices). [Combined luminal and zygal indices have value of 100.]

main axil. Brachial supporting two pinnulate branches on equal or subequal shoulders; in calceocrinids conjoined proximal and following series of contiguous axillary brachials forming a distinctive lowermost part of rays next above lateral radials (e.g., *Synchirocrinus*).

manosynostosis. Ligamentary articulation characterized by relatively large and deep fossae and well-developed peripheral rim that provides contact with adjacent plate; differs from typical synostosis in depth of fossae and presence of rimmed edges; common in flexible crinoids.

medulla. Part of columnal girdled on outward side by cortex, from which it is distinguished by differences in microstructure if discernible at all, may be divided into well-defined or indistinctly bounded inner (proximal) and outer (distal) portions and may consist solely of substance of claustrum, remainder of columnal then being classed as thick cortex.

mesistele. Intermediate part of crinoid column between proxistele and dististele regions, doubtfully distinguishable in pluricolumnals.

monocyclic. Having only single circlet of plates proximal to radials.

multibrachiate. Characterizing comatulids which have more than ten free arms.

muscle field. Concave or flat area on adoral side of articular faces of muscularly articulated ray plates, serving for attachment of muscle fibers.

[*muscle fossa. See* muscle field.]

muscular articulation. Union of ossicles effected by muscle fibers in addition to ligaments, opposed articular faces being characterized typically by presence of aboral ligament fossa, ligament pit, transverse ridge perforated by axial canal, two interarticular ligament fossae, and two muscle fields.

nodal (*N*, pl., *NN*). Columnal generally distinguished by maximum width and height in succession of heteromorphic columnals differentiated as noditaxis. In columns with cirri, these are articulated to cirrus sockets on the nodals (cirrinodals), but in some crinoids apparently lacking canaliculae and cirrus scars entirely. In columns without cirri, nodals are distinguished only by maximum size (nudinodals). Columnals that share in bearing a cirrus, with fossulae on their articular surfaces leading to the intercolumnal cirrus, are defined as compound nodals. In a few crinoids (e.g., *Camptocrinus*) characterized by cirri borne by pairs of fused columnals, having cirrus scars along the line of the vanished intercolumnal suture, cirrus-bearing nodals (separated by internodals) are termed binodals.

nodal index. Ratio of height of nodal to that of noditaxis containing it, multiplied by 100 to avoid fractional numbers. [Combined nodal and internodal indices have value of 100.]

nodicirral articulum (or facet). Scar or socket on latus of nodal for articulated attachment of cirrus.

noditaxis (pl., **noditaxes**). Nodal and contiguous succession of internodals (in most fossil crinoid stems arbitrarily chosen in either direction from nodal, because proximal and distal directions are undeterminable, but in extant stem-bearing crinoids and some fossil forms, such as *Isocrinus,* internodals on proximal side of nodal are joined with it in defining noditaxis, because each nodal precedes in origin next nodal above it and internodal series between pairs of nodals became intercalated after appearance of subjacent nodal).

nonaxil-brach. Plate of any brachitaxis adjoined on its distal side by single brach.

noncrenulate suture. Externally visible straight or curved nonwavy line of articular contact between ossicles.

nudinodal. Nodal lacking nodicirral articula on its latus.

oblique articulation. Union of ray ossicles by ligaments or by muscles and ligaments characterized by obliquity of the articular facet plane of its surface, or both; includes longitudinal oblique articulation and transverse oblique articulation.

oblique ridge. Linear elevation separating interarticular ligament fossa and muscle field on articular face of muscularly articulated ossicles.

oblique suture. Externally visible line of articular contact between contiguous ossicles of brachitaxes and columnals not perpendicular to longitudinal axis of united ossicles (e.g., distal suture of enlarged columnal in abruptly bent stalk of *Nevadacrinus* and *Lampidocrinus*).

omega-ramule. Branchlet borne by terminal axial of main-axil in calceocrinids, invariably directed adanally.

oral (*O,* pl., *OO*). Any of five interradially disposed plates forming circlet on tegmen surrounding or covering mouth.

oral pinnule. Any proximal pinnule differentiated from distal ones in function or structure, or both.

ossicle. Any single calcareous segment of crinoid skeleton (e.g., columnal, cirral, thecal plate, brachial, pinnular).

[*outer ligament area. See* aboral ligament fossa.]

[*palmar. See* tertibrach(ial).]

pararadial. Supplementary arm-bearing plate in radial circlet.

patina. Essential, primitive part of cup or calyx comprising radials, basals, and in some crinoids, infrabasals.

pelma. Stalk and holdfast of crinoid beneath crown.

peneplanate. Almost but not quite same as planate (q.v.), deviating very slightly by upward or downward slope, or with extremely gentle convexity or concavity.

peneplenary. Type of radial articular facet occupying most but not all of distal extremity of plate, leaving nonarticular surfaces (generally narrow) next to sutures at plate margins (e.g., *Zeacrinites, Stellarocrinus*).

pentamere. Fifth part of columnal or column; may be discrete or laterally ankylosed.

[*perigastric coelomic organ. See* convoluted organ.]

perigastric coelomic tube. Contorted conduit above and within convoluted organ in some camerates; possibly part of aboral coelomic system.

perilumen (pl., **perilumina**). Raised inner border of columnal articular zygum, surrounding lumen as rim of tabular field with smooth, granulose, tuberculate, or vermiculate surface. Internally, perilumen of some columnals corresponds to dense inner medulla, which is very distinct from reticulate to spongy outer medulla between areolae and possibly part of crenularia of opposite articula.

periluminal index. Ratio of total width of perilumen to that of columnal articular facet, multiplied by 100 to avoid fractional numbers (*see* columnal indices). [Combined periluminal, areolar, and crenular indices are equal to zygal index.]

peripheral crenulae (of petalodium). Crenulae along abaxial border of petal, generally reaching margin of articulum next to columnal latus.

periproct. Opening in theca for anus, covered in life by membrane or pyramid of small plates.

periproctal. Any plate covering thecal opening for anus, generally part of small pyramid.

perisomic skeleton. Part of skeleton located on adoral side of crown; includes all tegminal plates, interbrachials, and skeletal covering of food grooves in arms and pinnules (syn., adoral skeleton).

peristome. Opening in theca for mouth, may be open or covered by plates.

peristomial. Any plate, probably ambulacral in origin, covering peristome.

perivisceral coelomic organ. Perforate calcareous meshwork separating narrow coelomic compartment adjacent to inner thecal surface from main thecal coelomic cavity of some camerates.

perivisceral coelomic ring. Double-ring commissure connected with perivisceral coelomic organ at summit of thecal cavity in some camerates.

perradial. Precisely in position of some one of crinoid radii.

perradial crenulae (of petalodium). Coalesced or inosculating adradial crenulae near central area of some columnals (e.g., *Balanocrinus*).

petal. One of five main lobate divisions of petaloid columnal articulum (petalodium).

petalodium. Pentalobate, petal-shaped arrangement of short crenulae typically developed on articular faces of some Articulata (e.g., Isocrinidae, Pentacrinidae); equivalent to rosette of BATHER but not of authors referring to Comatulida.

pinnular. Plate forming part of aboral skeleton of pinnule; may be incorporated in theca (fixed pinnular) but almost universally forming part of crown above theca (free pinnular).

pinnular arm unit. Any arm segment giving off one pinnule; may be composed of a single brachial or more.

pinnulation. Arrangement of pinnules.

pinnule. Generally slender, unbifurcated, uniserial branchlet of arm, typically borne on alternate sides of successive brachials except hypozygals and axillaries.

pinnule socket. Articular facet on brachial for attachment of pinnule.

planate. Level or horizontal; refers to shape of aboral-cup base, attitude of basal or infrabasal circlet, or disposition of general plane of radial articular facet.

plate. Tabular ossicle, mostly having articular facets at edges.

plenary. Type of articular facet completely occupying distal extremity of plate and extending laterally to sutures at its margins (e.g., *Delocrinus*).

pluricirral. Two or more cirrals attached to one another.

pluricolumnal. Two or more columnals attached to one another.

posterior. Referring to interray *(CD)* containing anus or anal plates, generally wider than other interrays.

posterior left. Interray or interradius next to left posterior *(D)* ray in clockwise direction when crinoid is viewed with adoral side directed upward.

posterior right. Interray or interradius next to right posterior *(C)* ray in counterclockwise direction when crinoid is viewed with adoral side directed upward.

[*postpalmar.* Any brachial distal from *SBr* series (*see TBrr, QBrr,* etc.).]

primanal. Most proximal anal plate in camerates.

primary skeleton. Part of calcareous framework composed of first ossicles developed in ontogeny; includes columnals, cirrals, infrabasals, basals, radials, brachials, pinnulars, orals, and ambulacrals.

primaxil *(IBrax).* Axillary primibrach(ial).

primaxillary (primaxil). Axillary plate of proximal brachitaxis.

primibrachial (primibrach) (*IBr,* pl., *IBrr*). Plate of proximal brachitaxis; may be axillary or nonaxillary and fixed or free.

priminternodal. First-order internodal (may be indicated by symbol *iN-1*).

[*proboscis. See* anal tube.]

proximal. Referring to direction or position toward polar or central axis.

proximale. Noncirriferous topmost columnal or fused topmost columnals, distinguished typically by enlargement and permanent attachment to aboral cup.

proxistele. Proximal region of crinoid column near theca, generally not clearly delimited from mesistele. Dissociated columnals and pluricolumnals are rarely identifiable as belonging to proxistele.

pseudocirrus. Unsegmented sideward projection from columnal resembling cirrus in having axial canal but very irregular in form and distribution.

pseudohomeomorphic. Crinoid column with perfect or near-perfect homeomorphic appearance externally possessing internodals which do not reach stem periphery (e.g., *Dianthicoeloma*).

pseudomonocyclic. Characterized by occurrence of infrabasals in immature growth stage only (syn., cryptodicyclic).

[*pseudosyzygy. See* cryptosyzygy.]

quartibrach(ial) (*IVBr,* pl., *IVBrr*). Any ray plate of fourth brachitaxis.

quartinternodal. Fourth-order internodal (may be indicated by symbol *iN-4*).

rabbet. Channel or groove along suture between adjoined skeletal elements formed by beveling of their edges.

radial (noun) (*R,* pl., *RR*). Undivided proximal plate or any ray, may be simple or compound (biradial) and bearing proximal anal plate (aniradial).

radial (adj.). Pertaining to a radius.

radial canal (of petaloid columnal). Radially disposed tubular passageway in apposed articula of contiguous petaloid columnals formed by matched radial grooves between petals, extending to periphery but not reaching lumen (e.g., *Isocrinus*) (syn., interarticular radial canal).

radial cavity. Lumen inside radial circlet.

radial circlet. Same as radial pentagon.

radial dome plate. Prominent ambulacral, especially axillary in nature, on tegmen of some camerate crinoids.

radial facet. Distal face of radial, smooth or sculptured, bearing marks of ligamentary or muscular articulation with first primibrach; facet lacking in radials that bear no arms.

radial groove. Half of radial canal located on either of apposed petaloid articula of adjoining isocrinid or pentacrinitid columnals. Narrow space between adjacent petals of petaloid columnal articulum.

radial pentagon. Subpentagonal ring formed of mutually adherent radials after removal of all other structures.

radial pit. Radially disposed depression in ventral surface of centrodorsal in some comatulids.

radial pore (of petaloid columnals). Small opening in radial position visible at periphery of two attached columnals formed by coincident radial grooves of apposed columnals.

radial ridge groups (of petaloid columnal). Various types of perradial crenulae ranging from alternating, gable-shaped or rectilinear joined crenulae of adjacent petals.

radial space (of petaloid columnals). Area between contiguous petals distinguished by absence of crenulae; may be broadly triangular, with base on rim of columnal and apex at or near central area (e.g., *Pentacrinus*), or very narrow, with apex not reaching central area (e.g., *Isocrinus*), or restricted to narrow radial groove.

radianal *(RA)*. Inferradial of *C* ray adjoining anal *X* or anal opening; may be shifted in position leftward and upward to lie even with *C* radial or rarely somewhat above it.

radicle. Individual rootlike branch of radix.

[*radicular cirrus. See* radicle.]

radius. One of five reference planes passing through polar or central axis and median line of any ray.

radix. Rootlike distal anchorage of column.

ramule. Bifurcating or nonbifurcating minor branch of arms, differing from pinnule in less regular occurrence and in some crinoids by presence of pinnules on it.

ramus. Main arm branch.

ray. Radial plate, together with all structures borne by it.

[*rebate* (BATHER). *See* rabbet.]

regular dichotomy. Bifurcation of arm, constantly repeated in regular manner.

reproductive regeneration. Reproduction of lost part of arm without any increase in number of branches.

right anterior. Ray or radius (designated by letter "*B*") next to anterior *(A)* ray or radius in clockwise direction when crinoid is viewed from adoral side.

right posterior. Ray or radius (designated by letter "*C*") next to *B* ray or radius in clockwise direction when crinoid is viewed from adoral side.

right tube plate *(RX)*. A special anal plate in inadunate crinoids commonly located above the radianal and to the right of anal *X*.

rim (of columnal articulum). *See* articular rim.

rosette. Delicate calcareous plate formed of metamorphosed basals, centrally located within radial pentagon in comatulids.

secondary skeleton. Part of calcareous framework composed of ossicles which are intercalated between primary pieces; includes interbrachials, interambulacrals, intercalaries, and some anals.

secundaxil *(IIBrax)*. Axillary secundibrach(ial).

secundibrach(ial) (*IIBr*, pl., *IIBrr*). Any ray plate of second brachitaxis.

secundinternodal. Second-order internodal (may be indicated by symbol *iN-2*).

septal index. Ratio of total width of septum to that of columnal articular facet, multiplied by 100 to avoid fractional numbers (*see* columnal indices).

septum (pl., **septa**). Thick or thin inward projection of columnal centrum which locally constricts lumen, generally comprised of horizontal microlamellae with or without being crossed by vertically disposed annular microlamellae, thus producing a microscopic cribwork. In some crinoids central septa are indented by very narrow radially placed slits extending outward from jugula and opposite septal surfaces may be swollen between neighboring slits in manner that produced petaloid bulges (jugular ramparts).

[*side-plate. See* adambulacral.]

spatium (pl., **spatia**). Localized widening of columnal axial canal opposite interarticular sutures.

stem. Stalklike attachment of many crinoids; column.

straight articulation. Union of ray ossicles by ligaments or by muscles and ligaments characterized by perpendicular orientation of transverse elements of articular surface with respect to longitudinal axis of joined brachials.

straight suture. Externally visible line of articular contact perpendicular to longitudinal axis of adjoined ossicles.

subaxil. Subordinate type of axillary brach bearing pinnule or ramule on one of its distal facets.

subradial cleft. Deep narrow furrow between dorsal surface of radials and opposed surface of centrodorsal in comatulids.

subtegminal. Beneath adoral integument of theca.

superradial (*SR*, pl., *SRR*). Distal plate of biradial; includes *C* radial in crinoids having radianal and anisuperradial.

supracentral nodicirral articulum (or facet). Cirrus attachment scar located above midheight of nodal latus, generally directed obliquely outward and upward.

supranodal (CARPENTER). Columnal adjoining proximal articulum of any nodal.

sursumate. Surface sloping outward-upward, generally refers to radial articular facet.

sursumclinate. Sloping upward and outward (e.g., radial, basal, and distal part of infrabasal plates of *Poteriocrinites*, radial articular facets of *Aesiocrinus* (longitudinal axis of all crinoid facets defined as extending from inner to outer margins at mid-width).

sutural pore. Opening, presumably respiratory, located on line of juncture between ossicles.

suture. Externally visible line of articular contact between adjoined ossicles, employed by many authors as equivalent to joint.

symmorphy. Ligamentary articulation in which one or two prominent ridges or toothlike prominences on one face interlock with corresponding grooves or sockets on opposed face (e.g., *Chladocrinus*, *Hypalocrinus*); peripheral crenulae may occur on parts of united joint faces; slight differential movement of joined ossicles is possible in two directions.

symplectic articulation. *See* symplexy.

symplexy. Ligamentary articulation in which culmina on one joint face interlock with crenellar grooves on opposite joint face, marked externally by crenulate suture; joined ossicles almost entirely immobile.

synarthrial articulation. *See* synarthry.

synarthry. Ligamentary articulation in which each opposed joint face bears transverse ridge separating two fossae for attachment of ligament bundles; allows moderately flexible differential movement of joined ossicles in two directions. Articular face generally rather flat, but if embayed, articular face is strongly curved, the convexity downward.

synostosial articulation. *See* synostosis.

synostosis. Ligamentary articulation in which opposed joint faces are gently and broadly concave, comprising attachment surfaces for ligament bundles which allow moderately flexible differential movement of joined ossicles in all directions; corresponds to "loose synostosis" of authors.

syzygial articulation. *See* syzygy.

syzygial pair. Two ossicles joined by syzygy, the distal one termed hypozygal, the proximal one epizygal.

syzygy. Ligamentary articulation of brachials and some thecal plates but not observed in columnal or cirrals in which culmina or less well-defined ridges of one articulum are apposed to corresponding elevations of other articulum and crenellae or ill-defined depressions of apposed articula to each other. This term has been misapplied by many previous authors to crinoid columnals, especially articulation of nodals with next-distal internodals; Also used by previous authors for all more or less immovable (close) ligamentary articulations (except symplexy), including cryptosyzygy, cryptosynarthry, symmorphy, synostosis, and even cryptosymplexy (symbol: +).

taxis (pl., **taxes**). Definite linear series of plates in any part of crown; superfluous for general application to stem but useful for differentiation of successive nodal-internodal groups of columnals. (*See* brachitaxis, anitaxis, noditaxis.)

tegmen. Adoral part of theca above origin of free arms or occupying space between them; may be calcified or not; may bear anal vent, pyramid, sac, or tube.

[*tergal.* Proximal anal plate in Camerata; *see* primanal.]

[*terminal stem plate.* Most distal columnal in comatulid larva. *See* holdfast.]

tertaxil *(IIIBrax)*. Axillary tertibrach.

tertibrach(ial) (*IIIBr*, pl., *IIIBrr*). Any ray plate of third brachitaxis.

tertinternodal *(iN-3)*. Third-order internodal.

theca. Crinoid skeleton exclusive of pelma and free arms.

transverse oblique articulation. Union of ossicles characterized by oblique position of transverse elements of articular face in relation to longitudinal axis of joined ossicles.

transverse ridge. Generally denticulate fulcral elevation on articular face of ray ossicles, disposed perpendicularly or slightly oblique to greatest width of facet.

trifascial articulation. Union of ray ossicles characterized by presence on articular faces of three gently concave areas for attachment of ligaments; allows distinct mobility in two directions.

trivium. Differentiated *A, B, E* rays, generally longer than rays of opposed bivium (*C, D* rays) in posterior position (e.g., many Comasterida, Holopodidae).

unilateral heterotomy. Type of arm branching characterized by occurrence of ramuli on one side of main arm only; includes endotomous and exotomous arm branching.

uniserial arm or brachitaxis. Composed of brachs arranged in single row, with or without subparallel sutures (e.g., *Barycrinus, Allagecrinus,* inadunates; *Icthyocrinus, Lecanocrinus,* flexibles; *Allocrinus, Lampterocrinus,* camerates; *Saccocoma, Marsupites,* articulates).

united compound basal. Ossicle formed by union of two basal rays with two interradial processes of rosette in some comatulids.

[*vault. See* tegmen.]

ventral. Referring to adoral side of theca and rays, in living crinoids normally directed upward; preferred term is adoral.

ventral groove. Longitudinal trough on adoral side of ray plates (radials, brachials, pinnulars); preferred term is adoral groove.

[*ventral sac. See* anal sac.]

visceral skeleton. Spicules or calcareous network developed within crinoid body, especially in connective tissues surrounding visceral mass and walls of digestive tube.

xenomorphic column. Crinoid stalk containing dissimilar sorts of columnals in proxistele, mesistele, and dististele regions, but dissimilarity excluding contrast between homeomorphic and heteromorphic pluricolumnals, either or both of which may be represented in the xenomorphic differentiates of the stalk.

zygal index. Ratio of total width of zygum to that of columnal articular facet, multiplied by 100 to avoid fractional numbers. *See* columnal indices.

zygocirral. Most proximal segment of cirrus, articulating with cirrus scar on nodal columnal (cirrinodal).

zygosynostosis. Ligamentary articulation in which opposed joint faces are nearly flat areas for attachment of short ligament fibers combined with moderate calcareous deposits; may allow extremely slight differential movement of joined ossicles in all directions but generally forms immovable union; corresponds to "close synostosis" of authors.

zygous basal. One of two large plates in tripartite basal circlet.

zygum *(Z)*. Part of columnal articular facet between borders of lumen and facet; may be divisible into parts (crenularium, areola, perilumen, facetal rim, bifascial fields, fulcral ridge) or contain crenularium alone.

ABBREVIATIONS AND SYMBOLS USED IN CRINOID DESCRIPTIONS

By N. Gary Lane
[University of Indiana]

Beginning with crinoid studies by F. A. Bather (1900), various schemes of abbreviations and symbols have been used to denote certain plates, parts, articulations, or directions of orientation of the crinoid skeleton. There have been considerable revision and changes in usage of this terminology over the years, resulting in a needlessly complex jargon of terms that is surely confusing to the beginning student. Compiled below are selected lists of symbols used in crinoid descriptions. Although not complete, they contain all of the more commonly used abbreviations. The first list contains symbols that begin with a letter and is arranged alphabetically, followed by a list of symbols beginning with a numeral. Miscellaneous symbols that are neither alphabetical nor numerical are listed last. These lists were compiled from the references cited below, and letters in parentheses in the lists indicate the reference from which the symbol or explanation was obtained: Bather, 1900a (B); Clark, 1915a (C); Gislén, 1924 (G); Moore & Laudon, 1943a (ML); Moore, Lalicker, & Fischer, 1952 (MLF); and Moore, Jeffords, & Miller, 1968 (MJM). For a more elaborate list of symbols, see Moore & Laudon, 1941.

ALPHABETICAL SYMBOLS

A	anterior (ML); areola (MJM), anterior ray or radius of authors
A-B	interray between *A* ray and *B* ray (C)
A_1, A_2	left and right brachials, respectively, of *A*-ray arm (C)
Amb	ambulacral (B,ML,MLF); pl., *AmbAmb* (ML), *Ambb* (MLF)
Ant	anterior (MLF)
AntL	antero-left (MLF)
AntR	antero-right (MLF)
ant.R.	anterior radius (B)
AR	anterior radial (ML)
Ax	axillary (ML,MLF); pl., *AxAx* (ML), *Axx* (MLF)
B	basal (B,ML,MLF); pl., *BB* (ML,MLF); *B* ray or radius of authors
B-C	interray between *B* ray and *C* ray (C)
B_1-B_2	left and right brachials, respectively, of *B*-ray arm (C)
Br	brachial (ML,MLF); free brachial (B); pl., *BrBr* (ML), *Brr* (MLF)
Br_1	first brachial (C)
Br_2	second brachial (C)
$\overline{Br}$	fixed brachial (B)
C	columnal (MLF); pl., *CC;* crenularium (MJM); *C* ray or radius of authors
C-D	interray between *C* ray and *D* ray, posterior position (C)
C_1-C_2	left and right brachials, respectively, of *C*-ray arm (C)
CBr	cup brachial (MLF); pl., *CBrr*
CIBr	cup primibrachial (MLF); pl., *CIBrr*
CIIBr	cup secundibrachial (MLF); pl., *CIIBrr*
Cd	centrodorsal plate (G)
Ci	cirral plate (MLF); pl., *Cii*
D	*D* ray or radius of authors
D-E	interray between *D* ray and *E* ray (C)
D_1-D_2	left and right brachials, respectively, of *D*-ray arm (C)
E	*E* ray or radius of authors; epifacet (MJM)
E-A	interray between *E* ray and *A* ray (C)
E_1-E_2	left and right brachials, respectively, of *E*-ray arm (C)
F	finials or brachials of final arm brachs (B)
fBr	fixed brachial (ML); pl., *fBrBr*
IAmb	interambulacral (MLF); pl., *IAmbAmb*
IB	infrabasal (B,C,ML,MLF); pl., *IBB*
IBr	interbrachial (ML); pl., *IBrBr*
IN	internode or internodal (MJM)
IR	interradius (B)
ISBr	intersecundibrachial (ML); pl., *ISBrBr*
ITBr	intertertibrachial (ML); pl., *ITBrBr*
iAmb	interambulacral (B,MLF); pl., *iAmbb*
iIAmb	interprimambulacral (B)
iIIAmb	intersecundambulacral (B)
iBr	interbrachial (B,C,MLF); pl., *iBrr*
iIIBr	intersecundibrachial (B)
iR	inferradial (ML); interradial (MLF); pl., *iRR*
iRA	inferradianal (ML)
L	lumen (MJM)
N	nodal (MJM)
NT	noditaxis (MJM)
P	perilumen (MJM)
P_a	first inner pinnule (C)
P_d	distichal pinnule (C)
P_p	palmar pinnule (C)
P_1	first pinnule (C)
P_2-P_b	second pair of proximal pinnules (C)
P_3-P_c	third pair of proximal pinnules (C)

R radial (B, ML, MLF)
RA radianal (B, ML, MLF)
T taxis (MJM)
Z zygum (MJM)

NUMERICAL SYMBOLS

IAx primaxil (B); pl., *IAxx* (MLF)
IAmb primambulacral (B)
IBr free primibrach (B); primibrach (C, MLF); costal (C)
IBr_1, IBr_2 free first primibrach, free second primibrach (B); first primibrach, second primibrach (MLF)
IBr_2 costal axillary (C)
$\overline{IBr}$ fixed primibrach (B)
IIAx secundaxillary (B); pl., *IIAxx* (MLF)
IIBr free secundibrachial (B); secundibrachial (C,MLF); distichal (C); pl., *IIBrr* (MLF)
$\overline{IIBr}$ fixed secundibrachial (B)
$IIBr_1$, $IIBr_2$ first secundibrachial, second secundibrachial (B,MLF)
IIIBr tertibrachial (B, C); palmar brachial (C)
$IIIBr_{ax}$ third postradial axillary (C)
IVBr first post-palmar brachial (C); tetrabrachials (C)
VBr second post-palmar brachial (C)
VIBr third post-palmar brachial (C)

MISCELLANEOUS SYMBOLS

+ syzygy (G)
— synarthry (G)
□ anchylosis (G)
△ deltoid plate (B)
/_ armlet or ramule (G)
$\underline{1},\overline{2}$ bar over or under numeral that stands for brachial indicates position of pinnule (G)

POSTLARVAL ONTOGENY OF FOSSIL CRINOIDS

By J. C. Brower, N. Gary Lane, and H. Wienberg Rasmussen
[Syracuse University; University of Indiana; Geologisk Museum, Universitet København]

CONTENTS

CAMERATES

By J. C. Brower

INTRODUCTION

A recent crinoid crown consists of an internal skeleton composed of a series of adjacent calcite plates. The endoskeleton is mesodermal and occupies the major part of this layer. The outer plate surfaces are covered by a thin epidermal layer, which is not completely developed in some specimens. Plates of the calyx comprise most of the body wall, which encloses the viscera of the crinoid (see Hyman, 1955, p. 47-61, for summary). In simple statement, three processes are seen in recent and fossil crinoid ontogeny: 1) development of new plates; 2) calcite accretion and increase in size of previously formed plates; and 3) complete or partial resorption of plates.

This paper represents a summary of my more detailed discussions of camerate crinoid ontogeny (BROWER, 1973; 1974a; 1974b).

ACKNOWLEDGMENTS

The initial stages of the work were supported by a Grant-in-aid from the Illinois Geological Survey; later phases were financed by NSF Grant GB-4115 and Syracuse University.

PLATE GROWTH SEQUENCES

Among recent crinoids complete plate growth sequences can be observed directly. In camerates, this applies only to the later developmental phases, since earliest sequences must be inferred. In general, the plate growth sequences of camerates and comatulids are quite similar, especially with respect to the arms and column (Fig. 194). The basic calyx plate development sequence of all taxa, listed from first to last appearance, comprises: basals, orals, infrabasals (if present), radials, primibrach *1,* primibrach *2* (primaxil), secundibrach *1,* and so forth.

All brachials and pinnulars first appear at the distal tips of the arms or pinnules. In modern crinoids, pinnules do not form until after the arms are well developed. The first pinnule of *Hathrometra sarsii* is seen when about secundibrach *14* is present; in *Antedon bifida,* pinnules occur along with secundibrach *12.* The first pinnules are formed at the arm tips; these are followed by the proximal (oral) pinnules on secundibrach *2.* After these, the intervening ones appear. Pinnules developed earlier in camerates. The smallest known arm-bearing individuals have a calyx height of about 1.0 to 1.3 mm., possess four to nine brachials in an arm, and show a full complement of pinnules.

The development sequence of the stem and arm plates of living species and the camerates follows the same order. Columnals typically are introduced immediately distal to the calyx, below the centrodorsal in pentacrinid growth stages of comatulids and below the infrabasals or basals of camerates. Also, new columnals are generally intercalated between previously formed stem plates. In many living comatulids, the animal discards its column at the end of the pentacrinid growth stage and assumes a more or less free-living existence. This does not occur in camerates and recent isocrinids, in which (barring traumatic accidents) all or part of the column is retained until death.

The main differences between camerate and modern crinoids in sequences of plate development are related to the presence or absence of fixed brachials. During the growth of most camerates, free brachials were incorporated in the calyx, although at variable rates. This does not occur in modern crinoids and inadunates and flexibles, in which the arms remained free above the radials throughout life. Typically, the proximal interprimibrachs of recent crinoids develop late in ontogeny. For example, they are first seen after the primibrach *2* (primaxil) is well defined in *Comactinia meridionalis* (Fig. 194), but the interprimibrachs of *Promachocrinus kerguelensis* do not appear until about five secundibrachs are present. Generally, the living crinoid radianal forms earlier than interprimibrach *1,* ranging from just prior to soon after appearance of the radials. I believe that the radianal and proximal interprimibrachs of living crinoids are homologous with the primanal and proximal interprimibrachs of camerates, as evidenced by similarities of topographic position, ontogeny, and phylogeny (BROWER, 1973, p. 301-304; 1974b). Most likely, the interprimibrach *1* and the primanal of camerates were probably initiated along with or immediately after the radials (BROWER, 1973, p. 301-308; 1974b). Relative to modern crinoids, the camerate interprimibrach *1* is postulated to have appeared earlier, whereas the proximal *CD* interray plates of both camerates and recent forms began at roughly the same time. In turn, this allowed the formation of extensive interbrachials throughout subsequent ontogeny. This plate sequence was probably retained by all camerates including forms both with

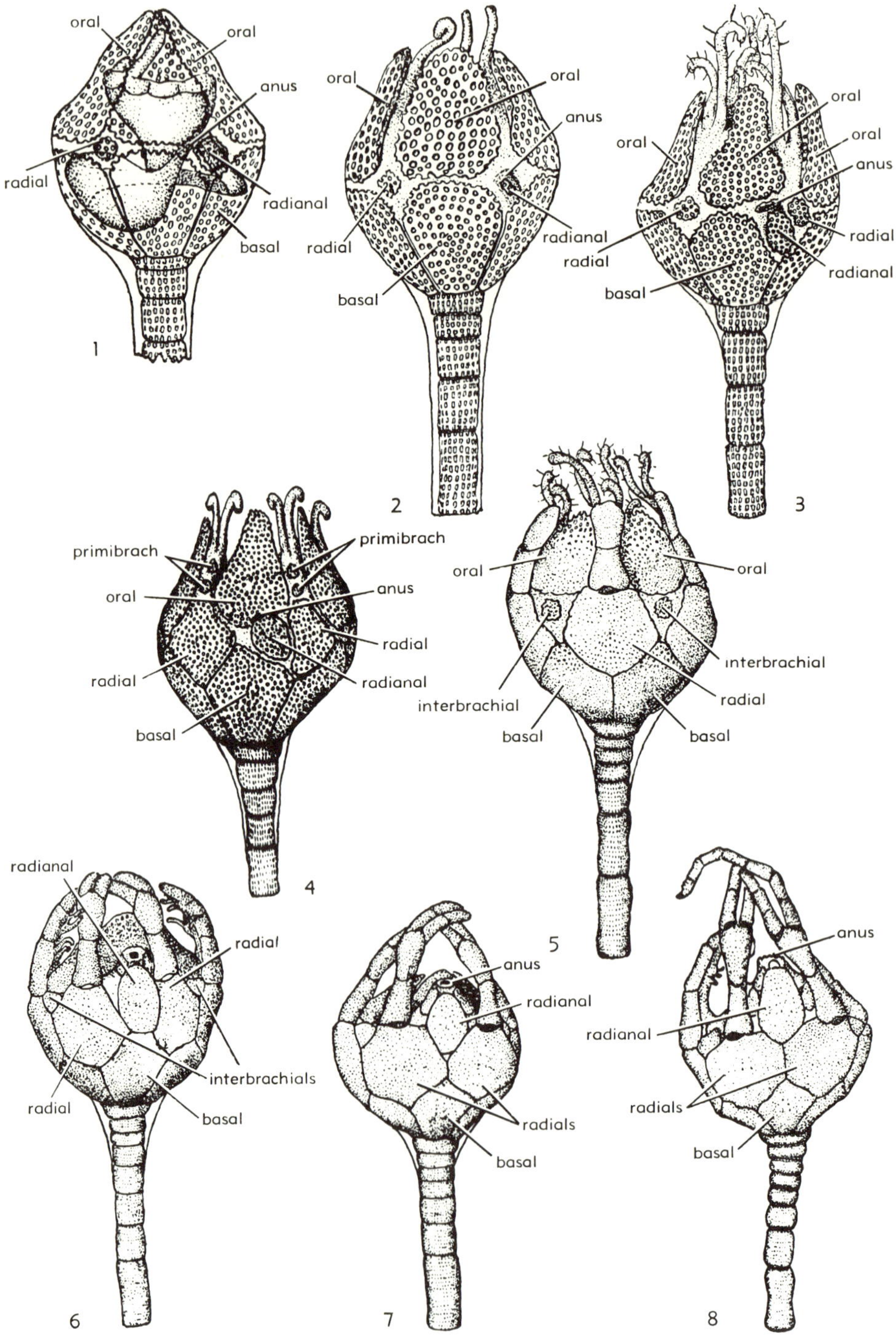

FIG. 194. Growth stages of *Comactinia meridionalis,* recent, from Yucatan, Mexico (Springer, 1920).

numerous and with few fixed brachials.

The few available data suggest that growth rates of size relative to time in modern crinoids fit a "slow-fast-slow" model. "Slow" growth rates seem to occur throughout the microcrinoid growth phase and well into the pentacrinid stage. Apparently, the "fast" growth is initiated just prior to loss of the column. When these growth rates are terminated is uncertain, but probably these end at or immediately prior to maturity. Older crinoids are characterized by "slow" time-size growth rates. These growth rates are evidenced by direct observation on some living crinoids and by the spacing of growth lines in fossil crinoids.

INTEGRATION AND COORDINATION

In the earliest growth stages of extant crinoids, the calyx plates are not fully in lateral contact. They are isolated from one another by the soft tissues (Fig. 194). At this time, the plates commonly are arranged irregularly and vary widely in size and shape. This shows that the growth of a certain plate is poorly coordinated and integrated with other plates during this stage. Due to calcite accretion along the plate sides, the plates eventually join, except for certain "strategic holes," such as the arm openings, anus, and columnal axial canal. This is probably dictated by the functions of the internal plated calcite skeleton. Basically, the plates serve to support and protect the fragile organs housed on the inside of the skeleton. In extremely small and immature individuals of living species, the water-vascular system and other coelomic cavities probably provide the necessary support by hydrostatic means. However, with increasing size, mass, and volume, the (coelomic cavity volume)/(total mass or volume) ratio probably becomes too low to ensure the necessary support, and this function is taken over by the skeleton. Once the plates join, the support problem requires that this configuration be maintained throughout subsequent ontogeny. This probably simplifies growth of the plate mosaic, and it presents an approach to delineation of ontogeny. In general, if one plate increases in size, its neighbors must do the same. Otherwise, gaps would appear between the adjacent plates and the skeleton could not support the animal. In other words, the development of all plates must be integrated and coordinated to prevent the formation of gaps. Matrices of correlation coefficients yield information on this type of integration and coordination. Progressive increase of plate and calyx size is denoted by positive correlation coefficients. Resorption would result in decreased size and would be shown by negative correlations for the plate or plates affected. However, resorption was not common in Paleozoic crinoids except for the stem facet, arm openings, and possibly the anal opening.

Integration and coordination gradients exist, and the dimensions of a given plate correlate best with those of the directly adjoining plates and less well with plates farther away. In general, the ontogeny of calyx plates is slightly better integrated and coordinated than that of free brachials and pinnulars (data are not available for the stem). This is related to basic geometry. In most calyx plates all margins are in contact with adjoining plates and only the inner and outer sides are not opposed by other plates. Conversely, brachials and pinnulars possess one or more free lateral margins not bounded by other plates (Fig. 194). In general, correlations are higher for plates with rapid growth rates than for those with lower developmental vectors.

The mechanism of integration and coor-

(Continued from facing page.)

1. Prebrachiate stage theca consisting only of basals and orals. [Note highly porous plates not fully joined together.]

2. Slightly older specimen with embryonic radials and **radianal.**

3. Largest prebrachiate stage with radianal and all radials (radianal below and left of *C* radial).

4. Specimen with embryonic primibrachs (proximal plates fully joined but gaps between distal ones).

5. Specimen with well-developed primibrachs and embryonic interbrachials *1*.

6-8. Growth sequence of progressively older individuals. [Note distalward migration of radianal and interbrachials *1*.]

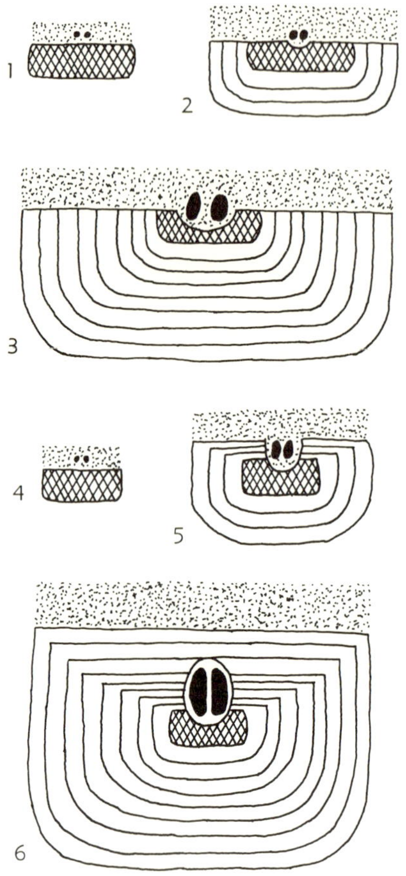

FIG. 195. Schematic cross sections showing growth of crinoid calyx and cup plates by growth lines (Brower, 1974b). [Explanation: soft parts on plate interiors stippled; axial nerve cords black; embryonic plates cross ruled].——*1-3.* Camerate crinoid, plates arranged from youngest to oldest. ——*4-6.* Recent comatulid, plates arranged from youngest to oldest.

dination is conjectural, although a few observations can be presented. Growth of the plate mosaic is adjusted to cause the minimum possible amount of interference with the aboral nervous system. This is the dominant nerve level in crinoids and it controls such basic reactions as flexing the arms, bending the stem, and other movements (A. R. MOORE, 1924). The youngest crinoids lack an aboral nervous system. Consequently, I conclude that the plate geometry and growth of very young crinoids is determined by the physical relationship of the mesenchyme to the ectoderm (RAUP, 1966, p. 393). In older crinoids the aboral nervous system is present and it may integrate indirectly and coordinate growth of the arms and many calyx plates. The control probably operates through the physical relationship of the aboral nervous system and the surrounding mesenchyme. The transition from ectodermal to mesenchyme and aboral nervous system control would be expected prior to the development of secundibrachs.

CALYX PLATES

In Paleozoic crinoids, new calcite was deposited on the margins and exterior surface of calyx plates, but no new calcite was added to the inside of plates (FIG. 195,*1-3*). This is evidenced by the locations of growth lines (BROWER, 1973, p. 293-294; 1974b; MACURDA, 1968; MEYER, 1965b; LANE, 1963b). When present, growth lines are always observed on the interior of calyx plates; this shows that calcite was not deposited on the previously formed plate interior. Growth lines are not seen on the outside of crinoid plates. As noted by LANE (1963b), this is related to placement of the aboral nerve cords. These are located along the plate interior, some simply lying on the flat plate surface, others being housed in grooves on the interior of the plate. Deposition of calcite on the plate interior could have interfered with or disrupted the critical and fragile aboral nerve cords. Growth of Paleozoic crinoid plates simply extended the previously established aboral nervous system. Most tegminal plates seem to have followed the same pattern, although in some cases calcite may have been deposited on the interiors of the plates. All calyx plates are in contact with the aboral nervous system, but this is not true for tegmen plates; however, some tegmen plates were in contact with the hyponeural nervous system (HAUGH, 1973, p. 86).

During the early development of living comatulids such as *Antedon,* the aboral nerve cords lie on the inner sides of the radials (FIG. 195,*4-6*). Subsequent internal calcite deposition on the plates gradually buries the nerve cords within the theca

(W. B. CARPENTER, 1866, p. 738, 739; 1876, p. 454). Resorption of calyx plates is more common in recent crinoids than in Paleozoic camerates (A. H. CLARK, 1915a, p. 322-340). In general, plate growth of living forms is far more complex than that of most Paleozoic species.

BRACHIAL GROWTH

INTRODUCTION

Development of brachials follows the same pattern seen in the radials of living and fossil crinoids. The radials are the first ray plates formed during ontogeny and they establish the growth pattern of all subsequent ray plates, both fixed and free brachials. The role played by the radials during growth of the rays establishes the paramount importance of these plates in crinoid taxonomy.

Generally, axes of the axial nerve cords constitute constant topographic reference points. Brachial accretion simply extends these throughout ontogeny. Lateral migration of the nerve cords and resorption of the brachials is either absent or minimized as much as possible.

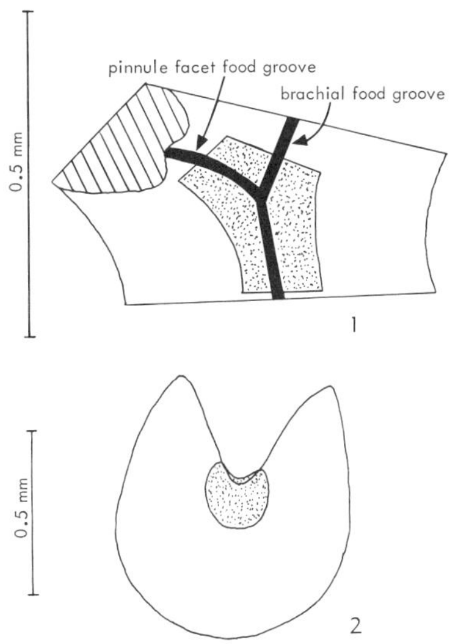

FIG. 196. Growth of uniserial pinnulate brachials in camerates (Brower, 1974b). [Explanation: axial nerve cords cross ruled; food groove axes black; pinnule facet oblique ruled.]——*1.* Ventral views of two superposed brachials of *Eopatelliocrinus latibrachiatus,* Upper Ordovician, Girardeau Limestone, Illinois and Missouri.——*2.* Superposed outlines of two idealized growth stages in articular surface view.

UNISERIAL NONPINNULATE BRACHIALS AND PINNULARS

All flexible and some inadunate crinoids had uniserial nonpinnulate brachials. The axial nerve cord and food-gathering system, and associated organs, were housed in a single V-shaped ventral food groove. By analogy with extant crinoids, the axial nerve cord lay at the base of the ventral groove. The food groove was deepened by calcite deposition along its ventral margins (FIG. 196,*2,* which depicts the ontogeny of a uniserial pinnulate brachial, nonpinnulate brachials developing in the same way). Calcite accretion also extended the other dimensions. In living and, presumably, fossil crinoids, the axial nerve cord diameter increases during ontogeny. This probably causes some resorption along the base and associated sides of the ventral groove.

UNISERIAL BRACHIALS OF CAMERATES

All uniserial brachials of camerate crinoids bear pinnules. Throughout development, these plates retained the same basic geometry in which all ontogenetic changes were gradual and incremental (Fig. 196). The brachial ontogeny plan minimized calcite accretion across the articular surfaces as much as possible within basic geometrical limits. Minimizing these growth vectors caused the least possible interference with ligaments and muscles that hinged and flexed the contiguous brachials. Studied relative growth rates toward the articular surfaces, listed from greatest to least, are: pinnule facet, distal brachial, and proximal brachial. The calcite accretion differentials closely correlate with mechanics for support of the developing arms. The rapidly developing vectors of width provided increased supporting area for the growing arms and pinnules. The lowest accretion vector (proximal height) was directed toward the facet which supported the higher brachials. The largest growth rate of an articular surface was toward the pinnule

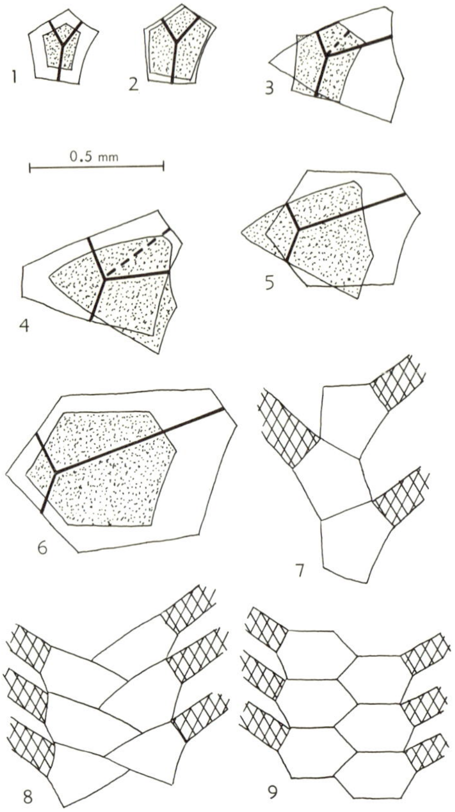

FIG. 197. Ventral view outlines of biserial brachials in free arms of *Macrostylocrinus pristinus,* Upper Ordovician, Girardeau Limestone, Illinois and Missouri (Brower, 1973, 1974b). [Explanation: food-groove axes in heavy black, smaller brachial stippled; if resorption and migration of the pinnule facet food groove was involved, that of the smaller brachial is shown in dashed line.]——*1.* Early uniserial stages.——*2.* Late uniserial stages.——*3,4.* Transitional uniserial to immature biserial stages.——*5.* Immature to mature biserial stages.——*6.* Mature biserial stages.——*7-9.* Schematic sketches of arm segments composed of uniserial, immature biserial, and mature biserial brachials, respectively (pinnules cross ruled; not to any scale).

facet that held only a corresponding pinnule.

The immature brachials present a comparatively high and angular appearance with prominent, roughly horizontal pinnule facets. Mature plates are relatively wider and possess smoother outlines with less prominent and more steeply inclined pinnule facets than juvenile brachials. The convergence angle between proximal and distal faces of brachials decreased throughout ontogeny. In young crinoids and juvenile brachials of adults (near distal arm tips), the height axes of succeeding plates are not parallel, imparting a crooked or bent appearance to the arms. This configuration was produced by unequal angular convergence of the proximal and distal brachial faces. During later growth, the height axes became more or less parallel because equal or nearly equal proximal and distal brachial-face convergence angles were formed by adjustments in the various height growth rates (compare mature and immature brachials in Fig. 196,*1*). This ontogenetic pattern is related to problems of supporting the growing arms (BROWER, 1973, p. 314-315; 1974b).

The ventral calcite accretion rates probably were small or nil at the food groove axes (Fig. 196,*2*). Deepening of the food grooves was produced principally by ventral calcite deposition on the areas flanking the food grooves. Some resorption probably occurred along the base and sides of the food grooves, resulting in increased diameter of the axial nerve cords. The average dorsal developmental vectors probably exceeded the ventral ones.

The most immature brachials, which are located at the distal arms tips, lack pinnules and these closely resemble pinnulars. These terminal brachials exhibit parallel proximal and distal faces. The least mature ones have rounded distal margins, but slightly older terminal plates are characterized by angular distal margins. The highest pinnulate brachials occur immediately below the terminal brachials. Small pinnule facets are present with a short pinnule consisting of one or two pinnulars. The various parts of the brachial did not become differentiated until the pinnule facet and its pinnule appeared.

BRACHIALS OF MODERN CRINOIDS

Except for the aboral nerve cords, the main growth patterns of living crinoids are like those of camerates. Like camerates, in most immature plates, the axial nerve cord and food-gathering apparatus lie in a single ventral groove. During later growth, the food-gathering tissue is displaced ventrally by calcite deposition. Eventually, the axial nerve cords are fully separated from the food-gathering tissue, and the axial nerve cords are housed within the brachial as in cyathocrinids. The most important parts of the axial nerve cords comprise constant topographic reference points throughout ontogeny (BROWER, 1973, p. 294-301; 1974b).

BISERIAL BRACHIALS

The food-groove configurations of biserial-armed crinoids resembled those of uniserial forms except that the "pinnule-facet food groove" was straight, rather than curved as in uniserial brachials (Fig. 197). Throughout the early growth stages, the brachials retained a uniserial configuration during which the previously established food groove axes comprised constant topographic reference points (Fig. 197,*1,2,7*). The convergence angle between the proximal and distal faces of the brachials increased, unlike the invariable convergence-angle decrease of uniserial-armed crinoids. This is related to contrasts in the growth rates of brachial height and width of uniserial and biserial-armed crinoids.

As the biserial brachials developed from the uniserial to biserial configuration, major changes took place (Fig. 197,*3-5,8,9*). The convergence angle continued to increase and eventually resorption of the inner sides (opposite the pinnule facet) of the brachials began as the brachials started to interlock on opposite sides of the arm. In some forms, the pinnule-facet food-groove axis underwent resorption and migrated laterally (Fig. 197,*3-5*).

Within the last growth stages, resorption ceased and growth became adjusted so that the brachial faces remained parallel to one another or nearly so (Fig. 197,*6,9*).

COMPARISON OF UNISERIAL AND BISERIAL BRACHIALS

The nonresorption development of uniserial-type brachials is relatively simple and straightforward. Conversely, resorption is a major factor in the development of biserial brachials so that ontogeny is more complex and a higher degree of integration and coordination of the various dimensions is required to prevent gaps from developing

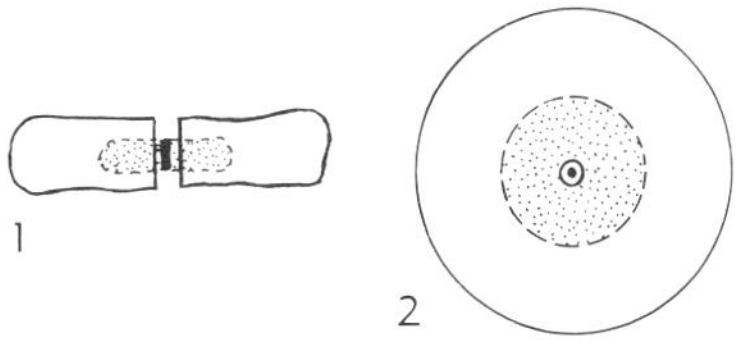

FIG. 198. Diagrams of superposed camerate columnals showing growth stages (Brower, 1973, 1974b). [Explanation: axial canal, black; smaller plate stippled with dashed outline; larger columnal shown by solid line.]——*1*. Side view.——*2*. Axial plane view.

between adjacent wedge-shaped brachials. In many crinoids pinnules are more densely packed along biserial arms than uniserial ones, owing to the presence of two rows of pinnulate brachials that alternate from side to side of the arm. The amount of curvature observed in many arm segments indicates that biserial arms are more flexible than uniserial arms, at least in species with ill-developed ligamental articulations between the brachials.

In Paleozoic camerates and pinnulate inadunates, the usual evolutionary sequence was from uniserial to biserial arms, indicating that the advantages of greater food-gathering capacity and more arm flexibility outweighed the disadvantage of ontogenetic complexity. All living crinoids have uniserial arms with well-developed muscular arm articulations, which allow much flexibility. Apparently, muscular arm articulations dictate uniserial arms (BROWER, 1973, p. 317, 318; 1974b).

COLUMNALS

The growth of a columnal follows the same pattern in both living and fossil crinoids (Fig. 198). The axial nerve cord penetrates the center of the columnals and passes through the axial canal. As in other plates, the axis of this structure forms a constant point of topographic reference. During ontogeny, the axial canal diameter is augmented by marginal resorption. The principal direction of calcite accretion is outward or peripheral. The much smaller vectors of growth in proximal and distal height are symmetrical for columnals with and without cirri. Typically, the articular

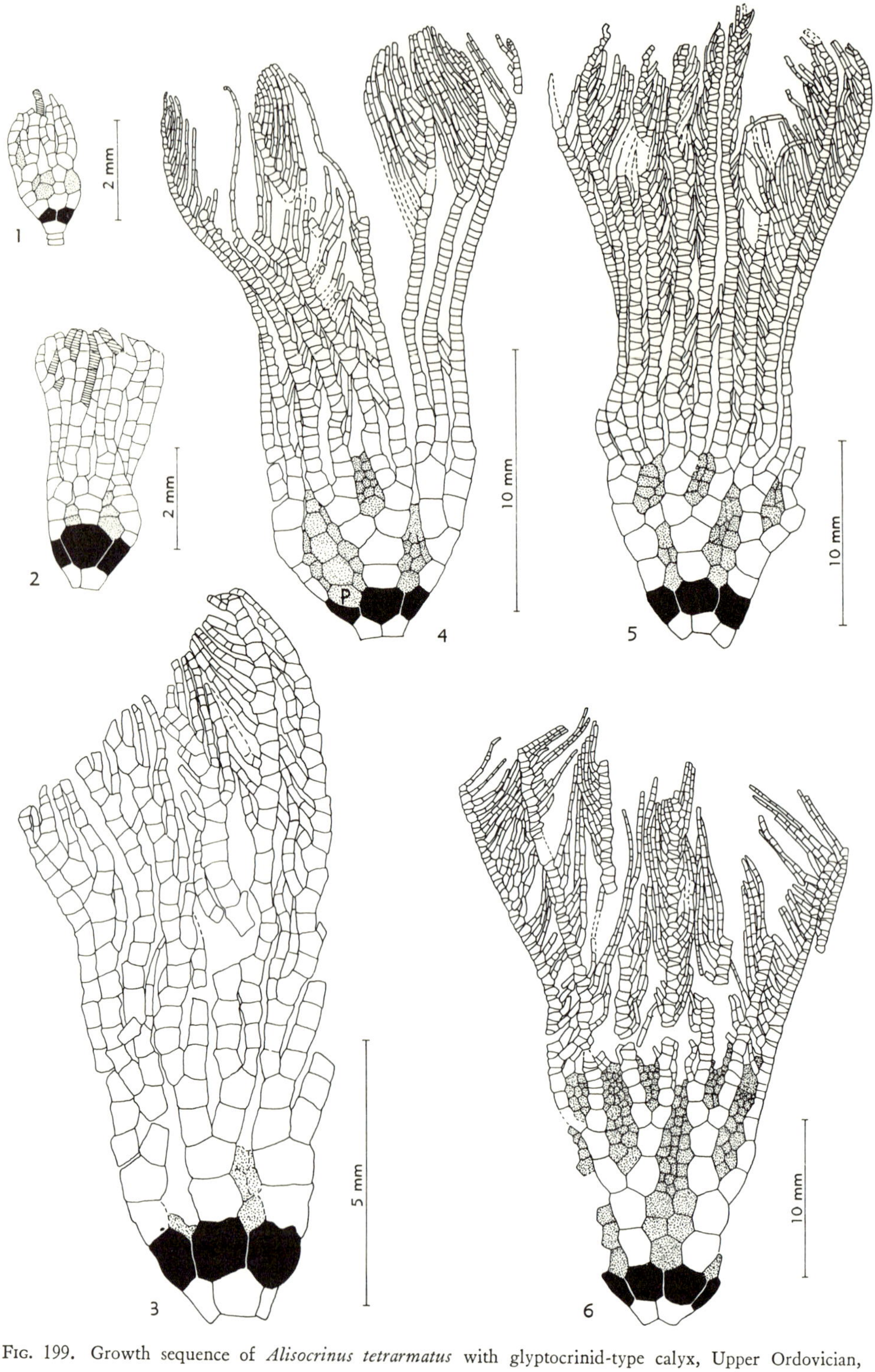

Fig. 199. Growth sequence of *Alisocrinus tetrarmatus* with glyptocrinid-type calyx, Upper Ordovician,

surfaces have culmina or ridges arranged in a radiating pattern (e.g., JEFFORDS & MILLER, 1968, pl. 3). As height and width of the columnals increase, widths of the culmina are augmented and new ones are intercalated between older ones.

ONTOGENY OF THE CALYX

Camerate crinoid calices are divided into five types depending on the extent of fixed brachials, number and nature of interbrachials, and to some degree on number of arms (BROWER, 1973, p. 310-313; 1974b). Regardless of calyx type, the youngest camerates are quite similar. For example, crinoids with calyx heights of about 1 mm. typically have primibrach *1* or *2* as the distal fixed brachial, a small number of interbrachials, a calyx which is dominated by the radials, and four to nine free brachials in a single arm (Fig. 199-203). Throughout subsequent ontogeny, the crinoids diverged and adults of the various calyx types are usually easily separated.

GLYPTOCRINID-TYPE CALICES

These crinoids are characterized by many fixed brachials, numerous large and regular interbrachials, and a small number of fixed arms in each ray (Fig. 199). The basic growth pattern of these crinoids was the incorporation of many fixed brachials in the calyx during ontogeny. This caused a series of correlated developmental changes. Young crinoids which lack intersecundibrachs have closely spaced arms within a single ray and the different rays are separated by wide interray areas. Older crinoids formed intersecundibrachs and, in some cases, intertertibrachs. Consequently, the adjacent arms within a single ray were spread apart so that the arms became spaced evenly around the calyx. High rates of fixed-brachial incorporation resulted in rapid distal extension of the calyx and viscera. A lobate tegmen was formed because the intersecundibrachs and intertertibrachs generally reached higher than the lateral and *CD* interray interprimibrachs. Young crowns have calices in which radials are the dominant ray plates. The subsequent growth rates of the radials and primibrachs relative to calyx height were roughly equal. This resulted in an adult calyx with primibrachs and radials roughly equal in height.

XENOCRINID-TYPE CALICES

The morphology and development of an xenocrinid calyx is similar to the glyptocrinid type except for the interbrachials (Fig. 200). The interbrachials of all camerates probably comprise tegmen interambulacrals, which shifted downward to become fixed in the calyx. In glyptocrinids, the number of interbrachials at any one level is small and was stabilized after the fixed brachials were incorporated in the calyx. The interbrachials show large height and width growth rates relative to calyx height and the interbrachial areas were probably rigid during life of the animal. Conversely, xenocrinid interbrachials remained small throughout ontogeny and the interbrachial areas became wider in larger individuals mainly owing to the intercalation of new interbrachials between older plates. Xenocrinid-type calyces were derived from ancestors with large and regular interbrachials (BROWER, 1974a). The growth pattern of the xenocrinid interbrachials is unique among camerates and probably represents a specialized adaptation, perhaps in response to respiration.

ACTINOCRINITID-TYPE CALICES

The basic morphological and developmental features of crinoids having actinocrinitid-type calices parallel the more primitive glyptocrinid group with one exception.

(Continued from facing page.)

Girardeau Limestone, Illinois and Missouri (Brower, 1973, 1974b). [Explanation: radials black; interbrachials stippled; pinnules oblique ruled in fig. *1* and *2;* specimens arranged in order of increasing calyx height.]

1. Side or *CD*-interray view of smallest crinoid showing few fixed brachials and short arms composed of few brachials.

2. D-ray view of slightly larger young specimen with better developed arms.

3. Side view of juvenile crinoid.

4. *C*-ray view of young adult showing development of proximal intersecundibrachs.

5. Side view of submature adult.

6. Side view of largest crinoid with numerous interprimibrachs, intersecundibrachs, intertertibrachs, and fixed brachials.

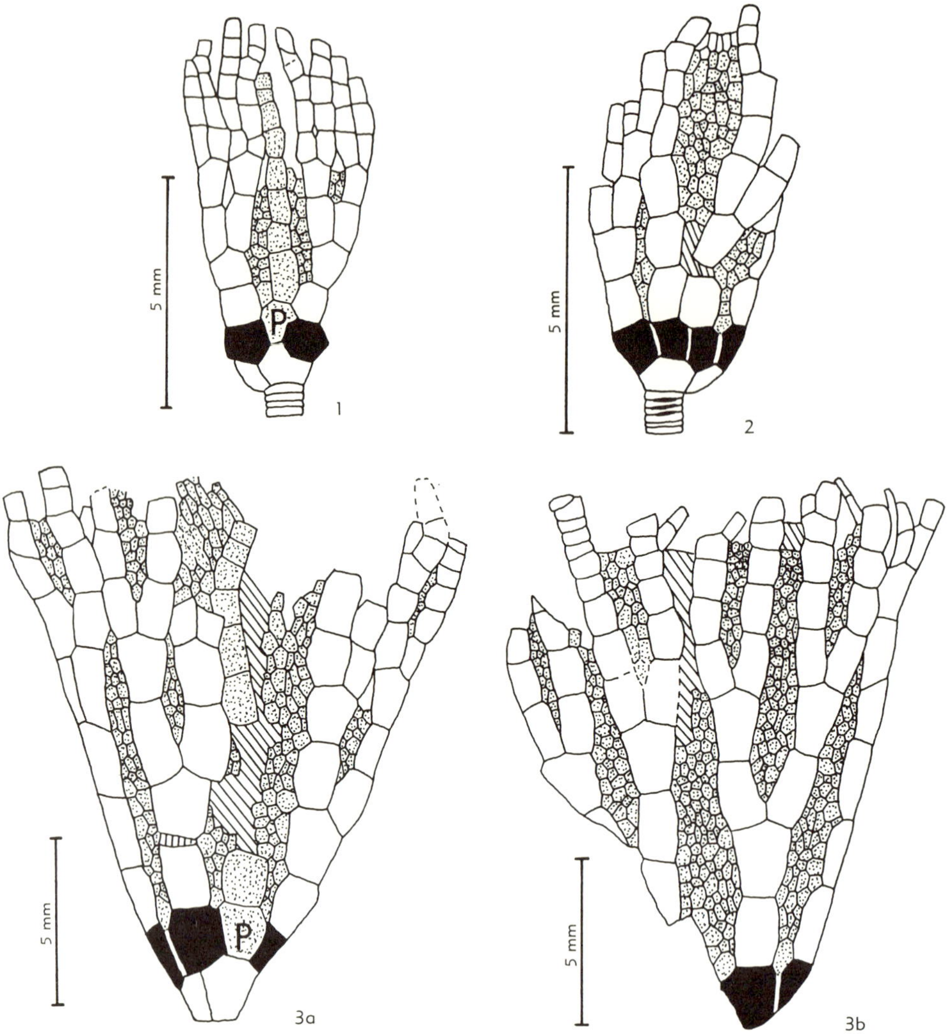

FIG. 200. Growth sequence of *Xenocrinus multiramus,* Upper Ordovician, Scotland, with xenocrinid-type calyx (Brower, 1974a, b). [Explanation: *P,* primanal; radials black; interbrachials stippled; damaged parts of specimens oblique ruled; specimens arranged in order of increasing calyx height.]

1. CD-interray view of young specimen with relatively few interbrachials.

2. A-ray view of another young specimen with more interbrachials; the *A* ray is abnormal and fails to branch.

3a,b. CD-interray and *A*-ray view of adult individual with numerous small and irregular interbrachials.

The glyptocrinids have few fixed arms in a single ray (usually two or four), whereas six or more fixed arms generally occur in rays of actinocrinitid calices. The development of numerous arms and associated axillaries fixed in the calyx tended to spread the arms apart, because the axillaries are roughly pentagonal and expanded distally (Fig. 201). The large number of axillary fixed brachials appears to be correlated with fewness of intersecundibrachs, intertertibrachs, and higher interbrachs, which are decidedly less well developed than in the more primitive glyptocrinid camerates (compare Fig. 199, 201).

PATELLIOCRINID-TYPE CALICES

These crinoids show a small number of fixed brachials which are joined into the calyx by large and regular interbrachials. Each ray bears a small number of arms, usually two or four (Fig. 202). Adults of all species examined have the proximal secundibrachs as the distal fixed brachials. The youngest crinoids lack intersecundibrachs and have closely spaced arms within a single ray that are separated by wide interray areas. Adults retained the juvenile condition because of the low rates of fixed brachial fixation in the intersecundibrach and intertertibrach areas. The lower rate of development of fixed brachials dictated slower distal extension of the calyx compared to glyptocrinid-type calices with the same shape. The radials are the largest ray plates in the youngest crinoids. The subsequent growth rates of the ray plates are adjusted so that mature patelliocrinids either retained or accentuated this juvenile radial and primibrachial arrangement.

PLATYCRINITID-TYPE CALICES

These crinoids are most similar to and doubtless were derived from patelliocrinid-type calices from which they diverged in two respects. The arms remained free above the fixed primaxil throughout platycrinitid ontogeny (Fig. 203). The patelliocrinids joined fixed brachials (both primibrachs and secundibrachs) in the calyx throughout development and proximal secundibrachs usually comprise the most distal fixed brachials in mature individuals. Primitive platycrinitid-type calices exhibit two primibrachs, but more advanced types tended to lose the primibrach *1,* and primibrach *2* (primaxil) rests on the radial. The primibrach width and height growth rates are much smaller in platycrinitids than in patelliocrinids. Thus, the platycrinitid primibrachs become progressively smaller in relation to size of the calyx throughout growth. For example, in young *Platycrinites bozemanensis* (Fig. 203), a pentagonal primaxil completely supports the secundibrachs *1.* Mature individuals have a triangular primaxil, which only partially serves this function; the outer secundibrach *1* margins rest on the radials.

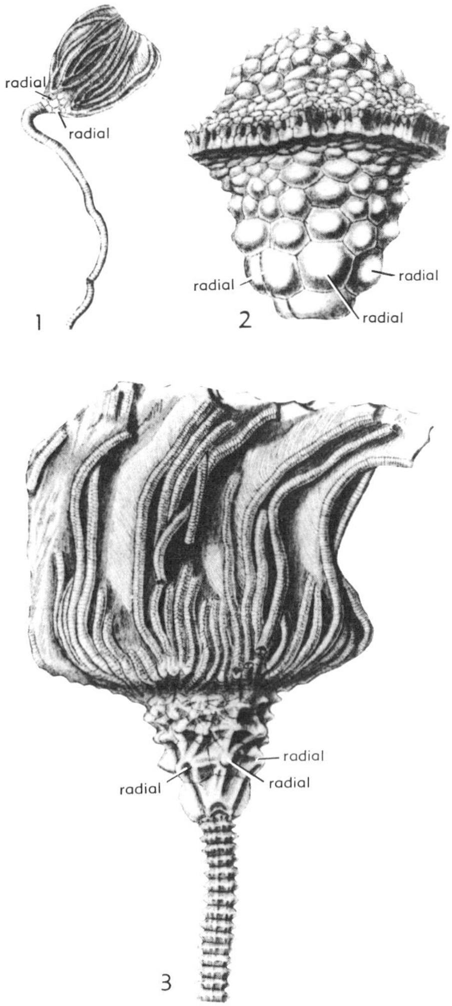

FIG. 201. Growth sequence of *Teleiocrinus umbrosus,* Lower Mississippian, Midcontinent, with actinocrinitid-type calyx (Wachsmuth & Springer, 1897). ——*1.* Side view of a young crinoid with arms free above the secundibrachs, ×1.——*2.* *A*-ray view of adult with many fixed brachials and fully developed wheel-like rim of free-arm bases, ×1. ——*3.* Side view of submature adult with moderately numerous fixed brachials and incipient wheel-like rim at free-arm bases, ×1.

GEOLOGICAL HISTORY OF CAMERATE CALYX TYPES

Consideration of lineages and the overall distribution of calyx types suggests a se-

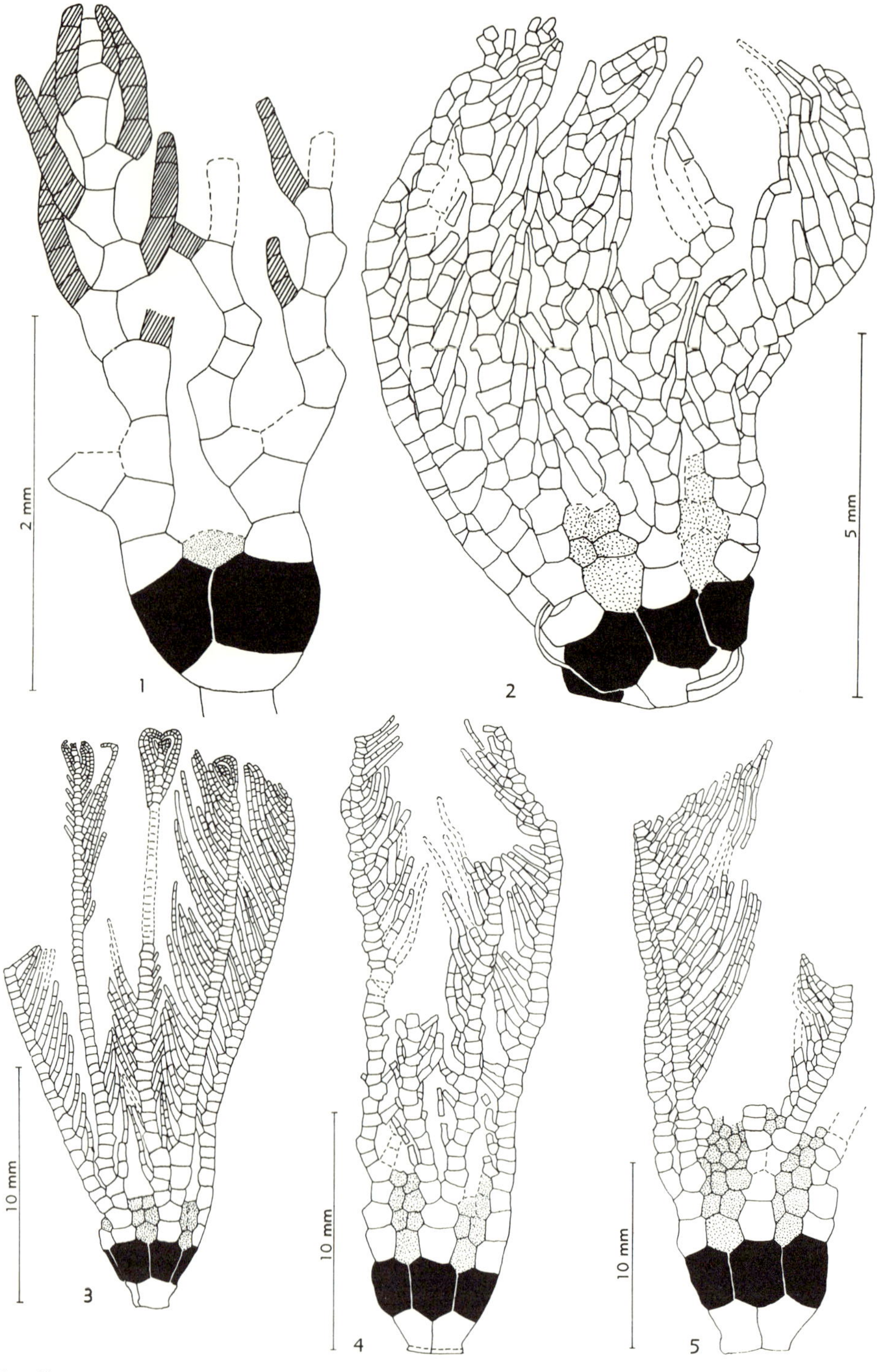

FIG. 202. Growth sequence of *Eopatelliocrinus scyphogracilis,* a patelliocrinid-type calyx, Upper Ordo-

ries of successions and replacements (see Brower, 1973, p. 401-407; 1974b). During the early and middle Paleozoic, xenocrinid-type calices evolved from glyptocrinid types but because these were not successful, they were replaced probably by glyptocrinid types (Brower, 1974a); extinct glyptocrinid groups were succeeded by actinocrinitid and patelliocrinid types; and platycrinitids supplanted patelliocrinids and perhaps some others. Camerates became extinct in the Permian. During the late Paleozoic, camerates probably were replaced by inadunates and flexibles (Lane, 1972) that almost exclusively had arms free above the radials. This indicates that crinoids lacking fixed brachials were most successful in the long run.

ONTOGENY OF THE FOOD-GATHERING SYSTEM

Crinoids face the same problems of growth as most filter-feeding marine invertebrates. The food-gathering capacity is, at best, an area or (linear dimension)^2 function, but the tissue to be supplied with food is a volume or (linear dimension)^3 function. Throughout ontogeny, the length of the food-gathering system is augmented by formation of new plates and increase in height of older ones. Addition of new plates, both in the arms and pinnules, results in the development of more food-catching tube feet. All new plates are initiated at the distal tips of the arms and pinnules. The following variables are the most critical (see Brower, 1974b, for more details): 1) the total volume of tissue is approximated by external volume of the calyx since total amount of tissue cannot be determined because of complex geometry of the crown and unknown complete stems. Both the entire tissue volume and calyx volume are (linear dimension)^3 functions and should scale similarly during ontogeny; 2) length of the entire food-gathering system including the arms and pinnules is substituted for the area of water covered by the food-gathering system because it is impractical to measure the area due to its geometrical complexity; 3) number of food-catching tube feet. The latter two variables are highly correlated and both scale the same way during ontogeny. Later discussion will emphasize the length of the food-gathering system which is easier to measure.

The basic statistics derive from the simple power function or allometric equation $Y = b \cdot X^k$ (see Huxley, 1932, and Gould, 1966, for discussion of the equation) in which X and Y are the independent and dependent variables, respectively. The intercept b gives the value of Y when X equals unity. The exponent k is the ratio of the specific growth rates of Y and X. "Ideal cases" may be defined for the development of food-gathering system length (Y) relative to calyx volume (X). For a crinoid without pinnules, the length of the food-gathering system is a simple linear dimension, i.e., the length of all free arms. In this case, the ideal exponent is 0.33. All camerate crinoids bear pinnulate arms and the length of the food-gathering system is:

$$\left[\begin{pmatrix}\text{average}\\ \text{pinnule}\\ \text{length}\end{pmatrix} \times \begin{pmatrix}\text{number of}\\ \text{pinnule-bearing}\\ \text{brachials}\end{pmatrix}\right] + \begin{pmatrix}\text{length}\\ \text{of all}\\ \text{free arms}\end{pmatrix}$$

The length of the food-gathering system is a linear dimension that is mainly a product of two other linear dimensions, namely, number of pinnule-bearing brachials and average pinnule length. The length of the arms is negligible relative to the food-gathering system length represented in the pinnules. For a crinoid with pinnulate arms, the approximate ideal exponent comprises 0.67.

Table 1 presents the statistical data for several Ordovician camerates from the

(Continued from facing page.)

vician, Girardeau Limestone, Illinois and Missouri (Brower, 1973, 1974b). [Explanation: radials black; interbrachials stippled; pinnules oblique ruled in smallest specimens; crinoids arranged in order of increasing calyx height.]

1. Side view of young specimen with few fixed brachials and one interbrachial in each interray.

2. *E*-ray view of juvenile with more numerous fixed brachials and interbrachials.

3. Side view of small adult.

4. Side view of mature crinoid.

5. Side view of largest crown, showing moderately numerous interbrachials and fixed secundibrachs.

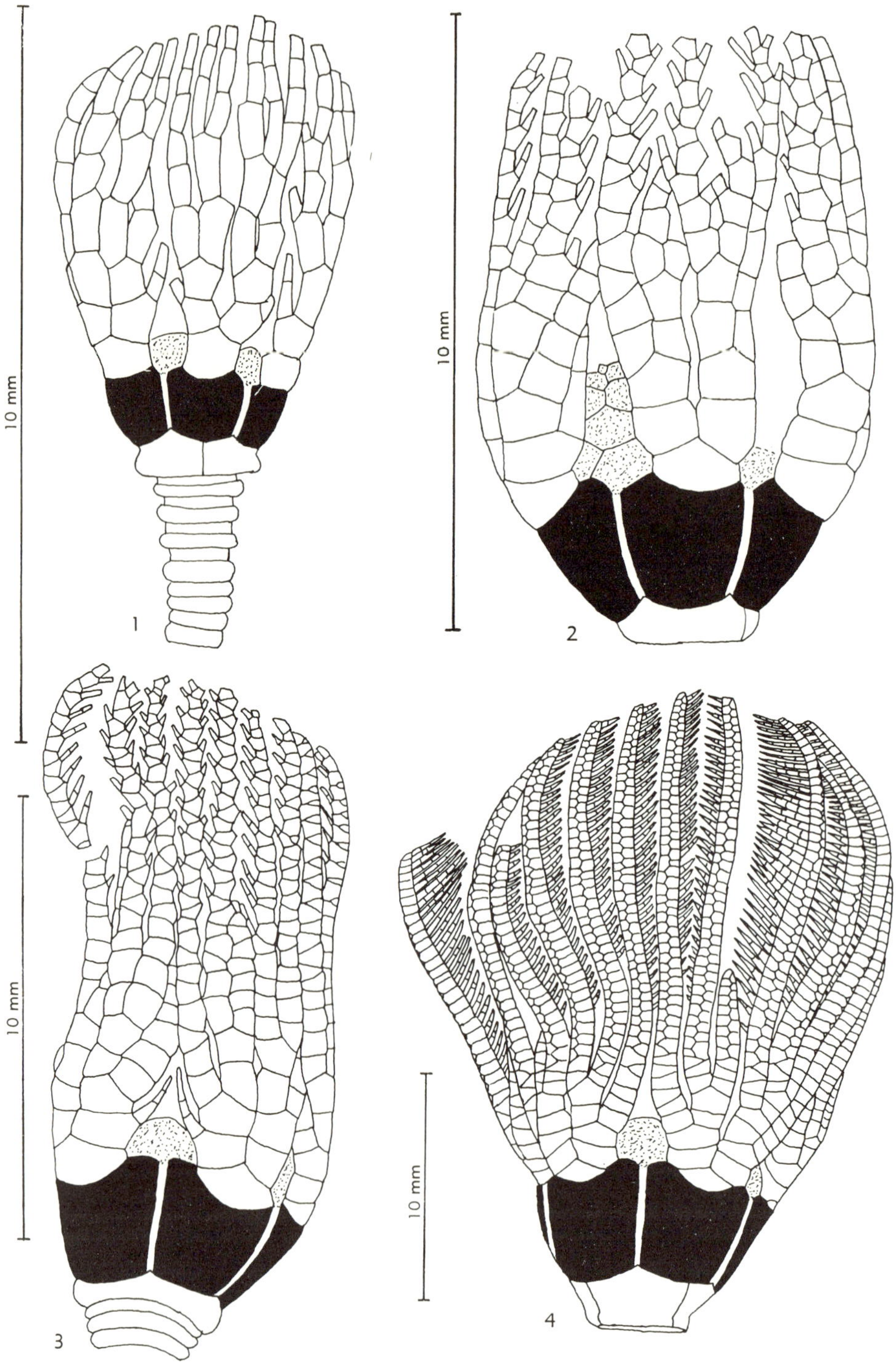

Fig. 203. Growth sequence of *Platycrinites bozemansis,* Lower Mississippian, Montana, wth platycrinitid-

TABLE 1. *Growth of Calyx Volume and Length of Food-gathering System in Some Girardeau Limestone Camerates (Ordovician).*

[X = calyx volume in cc.; Y = length of food-gathering system in cm.]

Species	Initial Intercept	Exponent	Minimum X	Predicted Minimum Y	Maximum X	Predicted Maximum Y	(Minimum Predicted Y)/(Minimum Observed X)	(Maximum Predicted Y)/(Maximum Observed X)	Correlation Coefficient	Number of Specimens
Eopatelliocrinus scyphogracilis	2,134	0.807	0.0007	6.07	0.220	629	8,670	2,860	0.976	16
E. latibrachiatus	1,337	0.717	0.0006	6.56	0.132	313	10,900	2,370	0.937	10
Macrostylocrinus pristinus	2,900	0.872	0.0003	2.46	0.237	826	8,200	3,480	0.988	10
Alisocrinus tetrarmatus	4,360	0.920	0.001	7.57	0.216	1,060	7,570	4,910	0.996	12
Ptychocrinus splendens	1,390	0.729	0.012	55.4	0.865	1,250	4,620	1,370	0.912	16
P. fimbriatus	482	0.452	0.006	47.8	0.184	224	7,970	1,220	0.885	6

Girardeau Limestone (BROWER, 1974b). The youngest crinoids all have similar food-gathering systems (Fig. 199, 202), but divergence occurred throughout subsequent development. Ontogeny of length of food-gathering system relative to calyx volume was curvilinear in all Girardeau camerates in which the growth rates of length of food-gathering system per 0.01 cc. of calyx volume increment dropped with increasing volume and age, i.e., negative allometry (Table 1). Except for *Ptychocrinus fimbriatus,* the food-gathering system grew more rapidly than one would predict based on the ideal case. In *P. fimbriatus,* the food-gathering system developed more slowly than in the ideal case. In general, crinoids with pinnulate arms seem to have maximized the growth rates of length of the food-gathering system with respect to calyx volume as much as possible within the inherent geometrical limits of the food-gathering system. This was accomplished by several basic mechanisms:

1) Development of pinnules must greatly accelerate the rate of development of the food-gathering system. Once pinnules are present, further increases in the rate of growth are achieved through augmenting the number and length of pinnulars in each pinnule.

2) Crinoids have several ways of accelerating the rate of formation of new brachials in a ray. This increases the number of pinnules because usually all plates within a free arm bear pinnules except for one or two distal free brachials, free axillaries, and free brachials located immediately above the free axillaries. Development of biserial brachials augments the number of brachials per unit free arm length compared to crinoids with uniserial brachials of roughly

(Continued from facing page.)

type calyx (mod. from Laudon, 1967). [Explanation: radials black; interbrachials stippled; crinoids arranged in order of increasing calyx height.]

1. Anterior view of young specimen with relatively large axillary primibrach *1* and uniserial free arms consisting of relatively high brachials.

2. Side view of older crown with relatively wider and shorter uniserial brachials.

3. Juvenile crown with strongly cuneiform brachials in upper arms.

4. Mature specimen with biserial arms, small primaxil and outer edges of secundibrachs *1* resting on radials, proximal part of free arms above axillary secundibrach *2* uniserial but regularly biserial higher.

Table 2. *Ontogeny of Feeding Tube Feet in* Alisocrinus tetrarmatus, *Ordovician, Girardeau Limestone.*

"Size" (mm.)	Calyx volume (cc.)	Feeding tube feet per arm	Feeding tube feet in pinnules of single arm	Total number of feeding tube feet	Total number of feeding tube feet/ calyx volume
1.1	0.001	150	60	4,200	4,200,000
1.4	0.002	180	150	6,600	3,300,000
4.2	0.20	1,710	5,700	148,000	740,000
5.5	0.20	1,950	4,550	130,000	650,000

the same height (compare Fig. 202, 203). The number of fixed or free arms can be increased. For example, *Alisocrinus tetrarmatus* has four unbranched uniserial free arms per ray, whereas eopatelliocrinids have only two free arms in a single ray (Table 1, compare Fig. 199, 202). *Ptychocrinus splendens* has about eight uniserial free arms in each ray, which results in very rapid formation of new brachials as the arms are extended distally. Development of hypertrophied arms, such as in melocrinitids, probably also accelerates the rate of formation of new brachials.

3) Length of the arms can also be increased through the growth of higher brachials, although this does not affect the number of food-gathering tube feet.

4) Many crinoids combine two or more of the above mechanisms. For example some actinocrinitids have numerous branched free arms composed of biserial brachials that bear long pinnules with many pinnulars.

The "food-gathering ratio" equals (food-gathering-system length)/(calyx volume). Despite the fact that camerate crinoids seem to have maximized the rate of growth of the food-gathering system as much as possible, the food-gathering ratio invariably declined during progressive development of a single species (Table 1).

In an attempt to evaluate ontogeny of the food-gathering system further, the approximate number of food-catching tube feet was computed for several individuals of *Alisocrinus tetrarmatus* (Table 2; see Brower, 1973, p. 323; 1974b, for details). The capacity of the food-gathering system, as measured by the number of feeding tube feet in relation to calyx volume, decreased throughout ontogeny (Table 2). The initial downtrend of the ratio was most rapid, for more than half of the total decline occurred over roughly five to 10 percent of the calyx volume interval. It is notable that (food-gathering-system length)/(calyx volume) and (number of food-catching tube feet)/(calyx volume) behaved similarly during development. Both ratios declined as the crinoids became larger, indicating that the food-gathering capacity relative to volume dropped throughout growth. Continuation of this ontogenetic pattern could produce a "hypothetical crinoid" in which the number of food-catching tube-feet is too low and the food-gathering system too short to provide the volume of soft tissues with food. Obviously no crinoid reached this critical limit, although some forms may have approached it. Perhaps some ratio of (food-gathering-system length)/(tissue volume) or (number of food-catching tube feet)/(tissue volume) serves as a limit beyond which further increase of tissue cannot take place.

GROWTH OF SUPPORTING STRUCTURES

A crinoid endoskeleton is a complex of supporting relationships. Essentially, each plate bears all higher ones. For example, the proximal columnal supports the crown, basals underlie the radials and all higher plates, distal fixed brachials elevate the arms, and pinnule facets hold the pinnules. Not all supporting relationships have been evaluated and attention is focused on the proximal columnal area and the area of the distal fixed brachial (see data on Girardeau camerates in Brower, 1973; 1974b). In both cases, the supporting ratio (area of supporting plate)/(weight of structure supported) declines throughout ontogeny and mature crinoids are "less well supported" than youngsters. Seemingly, continuation of this decrease eventually could have produced a stem too small to have elevated and supported the crown or arms which were too long and heavy for the crinoid to support, or both. Perhaps critical ratios provide limits, beyond which further size growth cannot proceed. Whether or not any camerate crinoid approached or reached this limit is not known.

GROWTH OF THE STEM

During development of most living and fossil crinoids, new plates form in the stem. Typically some of these are initiated immediately below the calyx (nodals), whereas others grow between previously formed columnals (intercalation of internodals). In most camerate crinoids which I have examined, more than half of the columnals formed by intercalation. In a few crinoids (e.g., some *Pisocrinus*) all new columnals are developed just below the calyx.

The function of a crinoid stem is to elevate the crown above the substrate. Elevation is progressive and continues throughout life of a stalked crinoid. The rate of elevation is dictated by two growth vectors —the rate of initiation of new columnals and the height growth rates of old ones. During growth, crinoid stems must achieve a balance between two factors—suitable rate of lengthening and maintenance of an appropriate amount of flexibility in the column. Flexibility allows the crinoid to bend the stem so as to attain advantageous feeding orientation and to avoid undue stress on the column that might break the stem (Brower, 1973, p. 283-290, 298-299; 1974b). Most camerates show the usual flexibility gradient of stalked crinoids in which maximum flexibility generally occurs in the middle 67 to 75 percent of the column (Seilacher, *et al.*, 1968). Relatively rigid stem segments are associated with the rooting device and immediately below the calyx. This allows the crinoid to flex the stem against a rigid holdfast, thus achieving maximum mechanical leverage. With a certain articular type, column flexibility is directly proportional to the ligament volume/external columnal volume ratio.

In most crinoids, growth of width in the columnals tends to decrease stem flexibility because the ligament volume/external columnal volume ratio of mature plates is lower than that of young columnals. Consequently, a continuous supply of intercalated plates is required to maintain a flexible stem. Some crinoids develop more strongly depressed articular surfaces in larger columnals and in such crinoids the ligament volume/external columnal volume ratio increases throughout ontogeny. Such forms need not depend on intercalation to maintain a flexible stem.

COMPARATIVE ONTOGENY OF CAMERATES AND LIVING CRINOIDS

Camerates and recent crinoids show the same plate-growth sequence with respect to the arms, column, and most calyx plates. In many living crinoids (comatulids), the animal discards the column at the end of the pentacrinid growth stage and assumes a more or less free-living existence. This does not occur in camerates and recent isocrinids, where, barring traumatic accidents, all or part of the stem is retained until death. Nevertheless, growth of the comatulid larval column and the camerate and isocrinid stem follows the same pattern. The sequences of calyx-plate development of camerates and comatulids are the same except for proximal interbrachials. During the ontogeny of most camerates, free brachials became incorporated in the calyx, although at rates varying with calyx type. This does not occur in modern crinoids, among which the arms remain free above the radials throughout life. Typically the interprimibrachs *1* of recent crinoids develop late in ontogeny, usually after primibrach *2* is present or still later. In camerates, these proximal interbrachials probably were initiated along with or soon after the radials. Camerate crinoids were probably also characterized by early interprimibrach

1 development. It is notable that the contrasts between the plate-growth sequences of living crinoids and camerates can be correlated with the presence or absence of fixed brachials.

Growth of the camerate and modern crinoid calyx plates follows different paths. Most camerate crinoids probably had a calcitic convoluted organ located inside the calyx. I believe that this structure both supported the gut and separated it from the chambered organ and aboral nerve cords. If so, the camerate convoluted organ served roughly the same purpose as the basal rosette in comatulids. The basic camerate calyx functions were to enclose, protect, and support the viscera, and to bear the free arms. These free arms were supported by distal fixed brachials, which in turn rested on more proximal plates. All camerate calyx plates performed more or less the same functions. Consequent on this functional morphology, major resorption was rare in calyx plates of camerates. I believe that resorption in camerate calyx plates was largely confined to plates located near the openings for free arms, anus, and axial canal of the column.

As noted by A. H. Clark (1915a, p. 344-348), comatulids depart greatly from the camerate plan (see Brower, 1974b, fig. 4). The principal specializations of comatulids are related to reduction of the cup plates and increase in cup strength. Different parts of the comatulid cup served different functions. The comatulid infrabasals, if present, are reduced. Early in ontogeny these fuse with the distal columnal to form the centrodorsal with cirri which the crinoid uses to grasp the substrate or foreign objects after the column is discarded. The comatulid basals are also insignificant cup elements that are metamorphosed and shifted upward to form the basal rosette during growth. The mature chambered organ is almost completely enclosed within the centrodorsal on its base and sides, by radials on its sides, and the basal rosette at its top. The viscera proper (i.e., inner and outer coelom, gut, etc.) rests on the distal margin of the basal rosette and radials. Thus, the basal rosette serves as a platform that largely separates the chambered organ from the viscera proper. The principal function of the comatulid radials is probably to support the arms, which are free above the radials. Lateral protection and support for the viscera proper are provided by the primibrachs. Thus, in comatulids and to a lesser extent in isocrinids, the infrabasals, basals, and radials do not serve to support and protect the sides of the viscera as in camerates. Consequently, the cup-plate growth patterns of modern crinoids and less complex camerates diverge. As mentioned above, resorption was rare in camerate calyx plates. Conversely, in comatulids (isocrinid ontogeny is not well known), resorption during cup-plate growth is common and can be documented for basals, centrodorsals, radianal, orals, and other tegminal plates (A. H. Clark, 1915a, p. 322-340).

The initially formed comatulid radials have flat inner and outer surfaces with axial nerve cords located inside of the plates (Fig. 195,*4-6*). Throughout ontogeny, calcite is deposited on the inside of the comatulid radials, and the axial nerve cords are completely buried within the radials during the last development stage (W. B. Carpenter, 1866, p. 738-741). Growth of comatulid brachials follows the pattern established for radials. In camerates, calcite was not deposited on the interior of the calyx plates and the axial nerve cords were housed on the plate interiors throughout life (Fig. 195,*1-3*). Growth of the brachials produced a distal extension of the calyx plate type (Fig. 196) and the axial nerve cord lay at the base of the brachial food grooves regardless of age.

The camerate type of uniserial brachial growth predominated in the Paleozoic, whereas living forms are characterized by the comatulid type. This suggests progressive evolution and adaptation. Obviously, the recent crinoid-axial nerve cords are better protected. Damage to the food-groove tissue need not harm the axial nerves. This is important because the aboral nervous system, including axial nerve cords of the arms, is the major level that controls posture and other basic reactions. Conversely among camerates, many injuries to the food-gathering structures probably also damaged the axial nerve cords. Geometrically, the comatulid growth pattern is far more complex than the camerate one. This implies that comatulid brachials require a higher degree of developmental integration and coordination than camerate plates. Perhaps this partially explains why camerates retained their type of growth, despite vulnerability of their axial nerve cords.

Compared to cup or calyx plates and brachials, the main growth patterns of columnals are highly conservative. As far as can be determined, these are uniform in living crinoids and all fossil forms.

INADUNATES

By N. Gary Lane

Ontogenetic studies of inadunate crinoids have been confined to two groups of these fossils, each of which is classified informally as microcrinoids. These small crinoids are included in the monocyclic, disparid superfamily Allagecrinacea, and the dicyclic suborder Cyathocrinina, superfamily Codiacrinacea. The ontogeny of larger, and more diverse, macroscopic inadunates is as yet unstudied, although immature or very small specimens of some species are known.

The three dicyclic inadunates for which ontogenetic development is best known are three species from the Artinskian of the U.S.S.R., described by Arendt (1970a): *Monobrachiocrinus oviformis* Yakovlev, *Cranocrinus praestans* Arendt, and *Hemistreptacron abrachiatum* Yakovlev (Fig. 204).

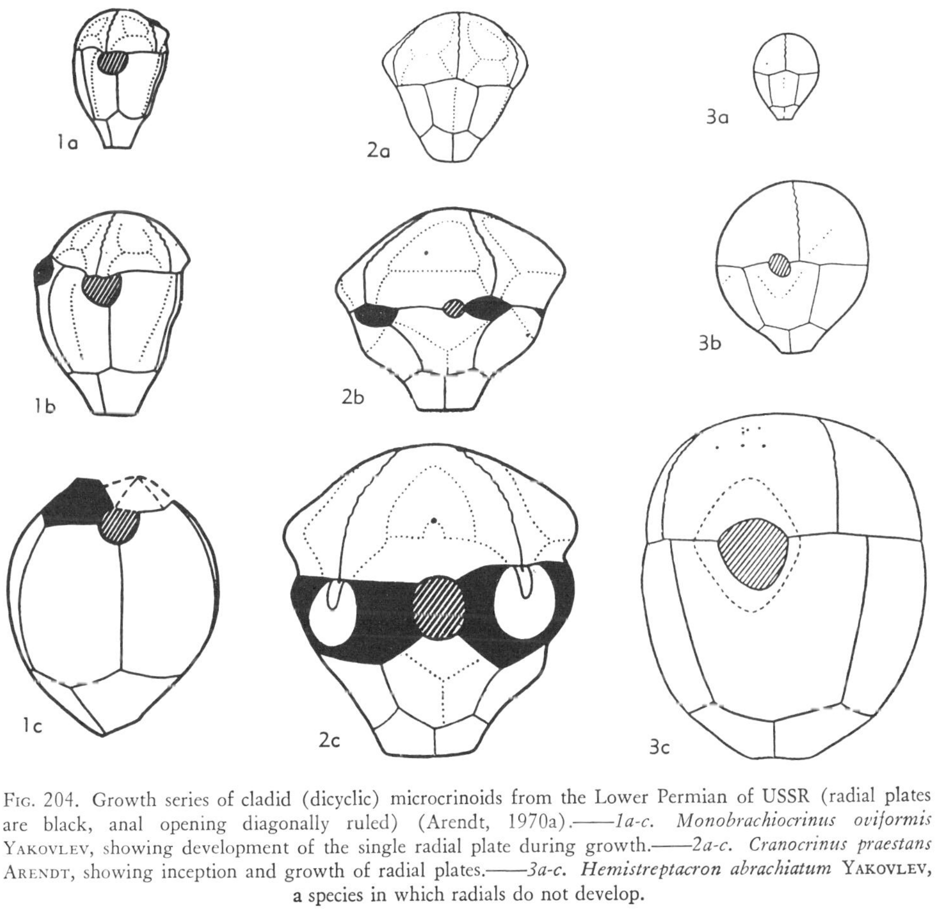

Fig. 204. Growth series of cladid (dicyclic) microcrinoids from the Lower Permian of USSR (radial plates are black, anal opening diagonally ruled) (Arendt, 1970a).——*1a-c. Monobrachiocrinus oviformis* Yakovlev, showing development of the single radial plate during growth.——*2a-c. Cranocrinus praestans* Arendt, showing inception and growth of radial plates.——*3a-c. Hemistreptacron abrachiatum* Yakovlev, a species in which radials do not develop.

The growth series for *Monobrachiocrinus oviformis* begins with a high, pyriform theca about 4.5 mm. high, consisting of three infrabasals, five basals, and five oral plates. Radials are completely absent and a large opening, presumably an anal opening, exists between the upper lateral edges of two of the basals. In the next stage, a single radial plate appears as a small, transversely lens-shaped element, directly atop one of the basals, bounded above by two oral plates. This radial plate, the only one that develops in this crinoid, increases abruptly in size, and a radial facet for the single arm appears on thecae that are about 6.9 mm. high. Subsequent changes during growth include change in thecal shape from high and pyriform to a globular shape, occasioned by enlargement and bulging out of the basal plates. The oral plates seemingly cease growth at an early stage and form a small cap-shaped pyramid perched on top of the theca in mature stages, maintaining their same relative position to the single radial and basals.

The earliest known growth stage in *Cranocrinus praestans* consists of a small theca composed of five large orals, five basals, and an uncertain number of infrabasals. In the next, slightly larger, stage, an anal opening appears above one of the basals, followed shortly by appearance of a small, diamond-shaped radial adjacent to and just to the right of the anal opening. This radial is judged to be the *C* radial. Next, three additional radials appear virtually simultaneously, those of the *B, D,* and *E* rays. The *A* radial is the last to appear. All radials develop at the top of interbasal sutures; they are small, diamond or lens-shaped, where they first appear, and none of them bears an arm facet until they have become conspicuously larger. An arm facet appears first on the *C* radial, followed by a facet on the *D* radial and very small facets on the *B* and *E* radials. The arm facet of the *A* radial is the last to appear. The facets of the *C* and *D* rays gradually enlarge until they are conspicuously larger than the facets of the other three rays.

Differential growth of the lateral edges of the oral plates results in gradual shift in orientation of these plates relative to the basals during growth. In smallest known specimens interoral sutures are directly above the midlines of the basals. As growth continues the interoral sutures are gradually displaced to the left, until in largest known specimens the orals are directly in line with the basals, and are, therefore, in normal interradial position.

In *Hemistreptacron abrachiatum* radial plates are lacking and the principal changes during growth include the appearance of an anal opening after smallest known stages, and rotation of the oral plates relative to the basals. Where the anal opening first appears, it is directly above the midline of a basal, and one of the interoral sutures is directly above the anal opening. During growth the basal plate that bounds the anal opening exhibits slightly greater growth along its right lateral edge than along the left side, so that in large specimens the anal opening is shifted slightly to the left of center of the basal. The oral plates alternate with the basals in early stages, so that interoral sutures are in line with the midlines of basals, as in *Cranocrinus*. The interoral sutures become progressively shifted to the right during growth, indicated especially by rightward shift of the interoral suture that was directly above the anal opening in early stages. The orals stop short of being aligned directly above the basals, however, their sutures being slightly offset to the left of interbasal sutures in largest specimens.

The growth series of these three species indicate that the anal opening appears after basal and oral plates are well formed and in close contact with each other. The radial plates appear quite late in ontogeny, and arm facets are relatively slow to develop on these plates. The oral plates undergo rotation during growth, perhaps in response to internal shifts in the main visceral components.

Reasonably complete growth series are known for several disparid crinoids, one of the most complete and most interesting being that of *Allocatillocrinus rotundus* Moore, described by Moore (1940) (Fig. 205). The smallest specimens of this species have a single arm facet on each of three radials, the other two radials lacking facets. These first three facets are on the *B, C,* and *E* rays. In the next stage there

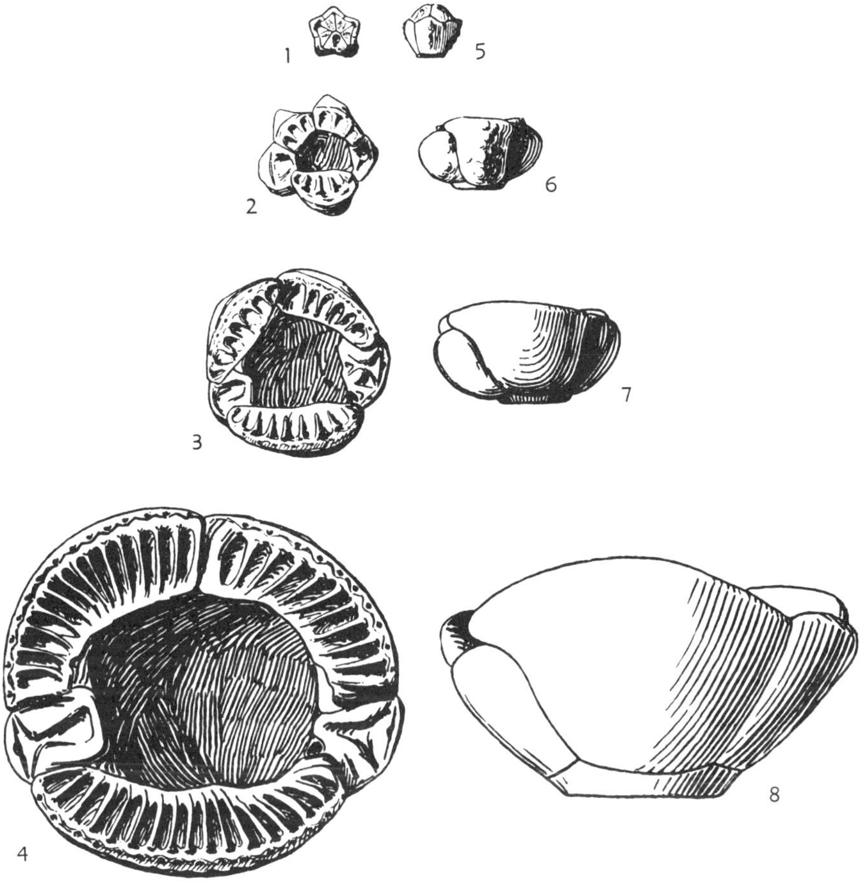

FIG. 205. Growth series of a monocyclic, disparid crinoid, *Allocatillocrinus rotundus* MOORE, all ×5 (Moore, 1940a).——*1-4.* Oral views of cup showing increase in number of arm facets on the *A, B,* and *D* radials, *D* radial at the bottom, and retention of a single facet on the *E* and *C* radials.——*5-8.* Lateral views of same specimens showing allometric lateral growth of *D* radial.

is a single facet on the *A, E,* and *C* radials, and two facets, side by side, on each of the other two radials. As growth continues, additional facets continue to be added to the *A, B,* and *D* radials; the *C* and *E* radials continue to have a single radial facet. Each radial facet supports a single, uniserial, unbranched arm. As the number of arm facets increases, there is a corresponding increase in width of the three multifaceted radials, so that they become conspicuously larger than the two small single-facet-bearing radials. Largest individuals with greatest number of facets have 10 facets on the *B* radial, 14 on the *A* radial, and 16 on the *D* radial. During growth, new arms and arm facets appear either at one end of a radial plate, so that the oldest facet is at one end and the most newly formed facet at the other, or the oldest facet is in the center of the radial, with new facets, and arms, added at both extremities of the plate. In *Allocatillocrinus,* the oldest facet is at the anterior edge of the *B* radial, and the newest facet at the posterior end. On the *A* radial, new facets are added at each end.

Specimens of *Allagecrinus* show a reasonably complete growth series that resembles the early growth stages in *Allocatillocrinus.* In *Allagecrinus pecki* MOORE (1940), small specimens, about 1 mm.

wide, have a cap of five conspicuous oral plates and one arm facet on the *C* radial only. A facet develops next on the *E* radial, then on the *B* and *D* radials almost simultaneously. The facet for an arm on the *A* ray develops last, in specimens that are about 2 mm. wide. In fully mature individuals, the oral cap of plates is typically lost, and two facets are developed on the *A, B,* and *D* radials, with single facets on the other two radials. Other species of *Allagecrinus* may have three or more facets on the three multifaceted radials. Presumably the oral plates are firmly fixed to the radials in early growth stages, but become progressively more loosely bound to the radials as growth continues, so that they are rarely preserved in larger specimens.

A growth series for *Synbathocrinus texanus* has been described by Moore & Ewers (1942). Smallest specimens of this species have a very low basal circlet, five conspicuous radials and five orals; the radials completely lack arm facets. The first arm facet appears on the *C* radial, closely followed by facets on first the *E* and then either the *D* or *B* radials. The facet on the *A* radial is the last to be developed. The arm facets are initially small, rounded, and smooth. With continued growth the facets become progressively wider until they occupy the entire distal faces of the radials. The facets also become progressively more prominently sculptured, with a strong transverse ridge, outer ligament pit, and other features typical of advanced arm facets in inadunate crinoids. The basal circlet grows in height, becoming a more conspicuous part of the lateral wall of the cup. The five oral plates are found in position in all of the small specimens, but mature specimens invariably have the oral plates missing, perhaps indicating a change in the strength of oral-radial articulations with growth.

FLEXIBLES

By N. Gary Lane

Knowledge of morphological change consequent upon individual growth in flexible crinoids is scanty, and has been little studied since Springer's (1920) monograph on these crinoids. The smallest known specimens of flexible crinoids, less than one centimeter in crown height, are probably best judged to be young adult, rather than immature, individuals. The plates of the cup—infrabasals, basals, radials, and plates of the posterior interray—undergo little change in number or relative proportion of plates from smallest to largest individuals. These plates increase in size with growth, but maintain constant relations to each other and to the arms and stem, resulting in a cup that maintains the same outline and shape during growth. The only observed change in these plates was reported by Springer (1920) in *Homalocrinus parabasalis* Angelin, in which the infrabasal plates expanded differentially during growth so that in large individuals the infrabasals cover most of the proximal parts of the basals and the radianal, plates that are mostly exposed in small specimens.

Two growth features that affect the stem have been reported in flexible crinoids. The proximal part of the stem, the proxistele of Moore and Jeffords (1968), is sharply differentiated from more distal stem parts in many flexible crinoids. The proxistele is composed of wide, very thin columnals that are firmly united and commonly preserved attached to the cup. In at least one flexible, *Eutaxocrinus curtus* (Williams), discussed by Goldring (1923), small individuals have very few columnals distinguishable as a proxistele, whereas large specimens of this species have many columnals in the proxistele, indicating a progressive increase in number of plates in the proxistele during growth. In other flexible crinoids the number of plates in the proxistele changes very little, if at all, from quite small to large individuals. A growth series of specimens of *Taxocrinus colletti* White, illustrated by Springer (1920), indicates that the number of plates in the proxistele in this species remains relatively constant from small to large individuals. Increase in height of the proxistele with growth is accomplished by slight increase in thickness of individual columnals, four per mm. in small specimens

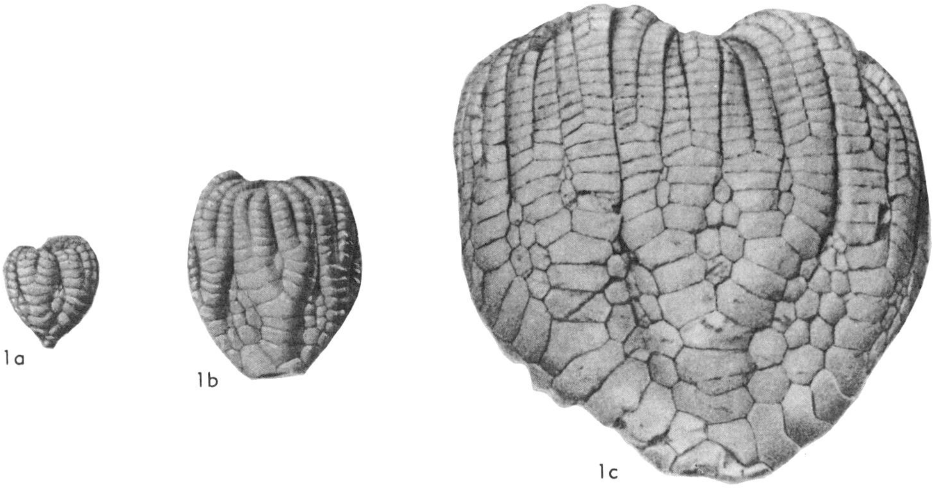

FIG. 206. Growth series of *Forbesiocrinus multibrachiatus* LYON & CASSEDAY, Miss.; *1a-c*, ×1 (Springer, 1920).

and 2.5 to three columnals per mm. in large specimens, according to SPRINGER. Consequently, the slight evidence that is presently available concerning growth of the proxistele is equivocal and deserves further study.

The more distal parts of the stem, the mesistele and dististele, also have been observed to undergo conspicuous changes during growth in a number of flexible crinoid species. In young, or small, individuals this part of the stem consists of highly elongate, barrel-shaped columnals, the degree of elongation increasing distally. Large specimens of species that show this character have relatively short, wide columnals with lateral faces evenly rounded. Presumably, these columnals attain their full height quite early in growth, with subsequent additions of stereom serving to increase the width of the column. This mode of growth, with emphasis first on increasing the overall length of the stem, and then on making the stem wider and sturdier, may be related to an adaptive strategy on the part of the crinoid to first raise itself up off the sea floor a given distance as rapidly as possible, and once this minimum height is attained, to concentrate on providing as firm and strong a support as possible for the increasingly bulky crown. SPRINGER illustrated small specimens with highly elongate distal columnals for *Taxocrinus colletti* WHITE, *Eutaxocrinus alpha* (WILLIAMS), *Icthyocrinus laevus* CONRAD, and *Asaphocrinus ornatus* (HALL). Specimens of *Mespilocrinus* retain these stem characteristics, typical of small specimens in other flexibles, in fully adult specimens.

One of the most conspicuous changes in the crown of flexible crinoids during growth is the increase in number of interradial plates in the four interrays other than the posterior interradius. This growth trend is best seen in specimens of *Forbesiocrinus multibrachiatus* LYON and CASSEDAY, in which small specimens have one, two, or three interray plates per interray, whereas large specimens have 19 or 20 such plates (Fig. 206). SPRINGER (1920) observed this increase in number of interradial plates in virtually every flexible crinoid for which he had both small and large specimens available. The interray plates served to bind flexibly together the proximal parts of the arms, thus as the interray plates increased in number with growth, an increasingly higher part of the arm was laterally bounded by such plates. This increasing support of proximal arms may have added needed strength and support for the presumably highly flexible arms during arm movement.

An increased height of interradial plates also resulted in a proportionately greater volume to the visceral cavity of the crinoid. In small individuals with only one or two interradial plates per interray the tegmen was situated very little above the bases of the arms, and the main visceral organs were confined mostly to the cup. With increase in number of interradial plates the tegmen was situated progressively higher on the crown and space internal to the arm bases would have been a significant part of the visceral cavity. Presumably the internal organs grew sufficiently large that they could not be accommodated within the cup in large individuals, at least partly because cup plates did not increase in size proportionate to the increase in volume of vital organs. This adult configuration would have provided an additional reason for restricting movement of proximal parts of the arms by laterally bounding interray plates—excessive movement of the proximal parts of the arms could conceivably have jostled adjacent internal organs sufficiently to be disadvantageous to the crinoid.

With increase in size, most flexible crinoids also exhibit an increase in the number of arm branches per ray. Small specimens, on the order of 1 cm. or less in crown height, typically have either secundibrachs or tertibrachs as the highest arm plates in the crown. Large specimens of the same species, as for instance *Taxocrinus ungula* MILLER and GURLEY as illustrated by SPRINGER (1920), have quartibrachs and additional higher arm divisions preserved. As new series of brachials are added to the tips of the arms during growth, proximal brachials constitute a progressively smaller portion of the total crown height. The mechanism by which additional arm branches are added at the ends of the arms is not known. Living multibrachiate comatulids add new divisions to the arms by autotomically casting off a series of distal arm plates and regenerating a new axillary plate and higher brachials on the brachial above which the rupture took place. Whether flexible crinoids utilized the same, or a similar, mode of increasing the number of arm divisions per ray is not known.

ARTICULATES

By H. WIENBERG RASMUSSEN

INTRODUCTION

Larval ontogeny and early development have been studied in a few recent crinoids, mainly comatulids, and are described in the *Treatise* section on recent crinoids. No such information on fossil crinoids exists, but the influence of larval ontogeny in recent comatulids on phylogenetic interpretation of some Articulata is mentioned in the section on evolution of Articulata, especially in connection with Comatulida and Bourgueticrinida.

Postlarval growth and change of skeletal form is very important in the determination of fossil and recent Articulata. Several morphological features undergo considerable change during growth of the crinoid skeleton and single skeletal elements. Nominal species based on specimens differing only in age and growth stages has resulted in a large number of synonyms. Thus, 17 species have been based on growth stages of the Cretaceous *Glenotremites paradoxus,* according to RASMUSSEN (1961, p. 284). Growth series of fossil Articulata have been described by SPRINGER (1901) and RASMUSSEN (1961). MOORE (1967), RASMUSSEN (1972), and PECK & WATKINS (1972) have given information on growth in fossil Articulata. Studies on growth have been based on correlation of size with change of form, and must be closely connected with the species concept, but also direct evidence of growth seen as growth rings in sections of skeletal elements has been mentioned by GISLÉN (1927) for recent *Democrinus,* by RASMUSSEN (1961) for *Phyllocrinus,* and by MOORE (1967) for *Dunnicrinus.*

In different parts of the crinoid skeleton the form changes during growth according to a few general patterns common to all Articulata or to large groups of the subclass. Identification of fossil Articulata is not possible without familiarity with these general patterns of postlarval ontogeny, which are therefore also to some extent included in the systematic description of Articulata in the *Treatise.*

During growth of a crinoid, the single skeletal elements increase in size and new columnals are added at the top of the stalk, and new brachials in the tip of the arms. Therefore, distal columnals and proximal brachials are older than proximal columnals and distal brachials of the same specimen. However, the new skeletal elements, especially columnals, introduced in an old individual are different from those introduced in a young specimen. This complicates the growth pattern and makes a concise terminology necessary in the description of ontogenetic features, so as not to confuse age of specimen with age of skeletal element.

A general rule for all Articulata is that during growth of a brachial its width increases more than its height. Young brachials are high and slender, but during growth they become wider and more robust. This change is found from juvenile to adult specimen as well as from the distal to the proximal part of an arm. Division of arms may take place either by two branches growing out from an axillary in the tip of the arm, or as augmentative regeneration, the distal part of an arm being shed from a nonmuscular articulation and replaced by an axillary with two branches. Except in damaged specimens, augmentative regeneration is found only from an axillary following a nonmuscular articulation. It has been observed in recent *Endoxocrinus, Teliocrinus,* and after the first division in most or all multibrachiate comatulids. Also, modification of pinnules may lead to a multibrachiate structure in some Isocrinida. Crinoids with arms divided at primibrachs *1* are extremely few, but have been observed among Cyrtocrinida and a few Comatulida. They have often been explained away by suggesting fusion of two primibrachials, but there has been no observation supporting the premise that brachials may fuse during growth.

The radials follow the same pattern of growth as the brachials, increasing in width, and in many groups also leaning progressively outward. The growth of basals is more variable. In Millericrinida the growth of basals and radials is almost equal, the cup becoming wider, but not changing much in form during growth. In most Isocrinida and Comatulida the basals increase less in size than the radials, or they may be reduced in size during growth. In some Bathycrinidae, especially *Democrinus,* the height of basals is very greatly increased.

Growth of the column and columnals is more complicated. An articulated column is developed together with other skeletal elements early in the larval embryology of recent comatulids, and the crinoid attaches itself by a terminal disc of the column. The juvenile column of stalked crinoids is small and slender, and consists of single columnals, which are small in diameter, but may well be rather high. During growth the columnals increase in size, and new columnals are added in the proximal end of the column. These new proximal columnals are concealed between adjacent proximal columnals or between the uppermost columnal and the cup until they have reached the columnal diameter and show up as extremely low ("thin") discs. By continued introduction of new proximal columnals, the previously formed columnals gradually move to a more and more distal position in the column, and at the same time they rapidly increase to their final height, and may also change in outline and ornament. It follows from this that diameter of the column and of single columnals is a function of the size and age of the individual, but height of the columnals is mainly correlated with its proximal or more distal position within the column. Very slender columnals, whether low or high, are found only in juvenile specimens, but low columnals are formed proximally throughout the lifetime of the specimen or, rather, its growing period. Some recent and fossil crinoids have very few or no low (new) proximal columnals in large specimens, and this indicates a final decrease or cessation of columnal formation (Rasmussen, 1961, pl. 24, fig. 16; pl. 27, fig. 14; pl. 30, fig. 1; pl. 32, fig. 3; A. M. Clark, 1973b, fig. 5).

The diameter of the column, although increasing during growth of the individual, is often the same or almost the same throughout its entire length, but a slight increase toward the distal end is not unusual, and a slight decrease distally is found in some Isocrinida. Also, proximal columnals may be enlarged and modified, especially in Apiocrinitidae, some Millericrinidae, and in *Bourgueticrinus*.

New columnals in the proximal end of the column may be identified by their very low height. They may all be introduced at the boundary between cup and the preceding uppermost columnal, which is generally considered the most common pattern of growth in crinoid columns, such as in Bathycrinidae, but it is uncertain how common it is among other Articulata. Differences in height of proximal columnals indicate other patterns in the succession of new columnals in several groups of Articulata.

ISOCRINIDA

In Isocrinida, the proximal part of the column with low columnals shows strong variation in height and generally in diameter as well. In this zone the new columnals are introduced. The largest columnals, provided with cirri, are called nodals or cirrinodals. Immediately below the cup only cirrinodals occur, generally, and then nodals separated by a single internodal succeeded by parts with an increasing number of internodals regularly alternating in size. This arrangement leads to the conclusion that all nodals are introduced at the boundary between cup and column, and the first and largest internodals introduced between two nodals are succeeded by several generations of internodals intercalated in alternation with previous columnals in the proximal zone until the full number of internodals is reached. This number may vary even within the single column, but variation within a species is often rather small. These proximal columnals display a considerable difference in height and commonly some variation in diameter according to succession of generations, but during further growth they gradually reach the same diameter and almost the same height except for the larger cirrinodals.

New proximal columnals are always strongly stellate or deeply pentalobate, with radial canals (radial pores) between. During growth, stellate columnals become pentagonal and pentalobate columnals become more rounded subpentalobate, in some specimens even circular. Also, different kinds of ornamentation such as ridges, tubercles, or spines are gradually formed during growth. The symplectial articular face of columnals in Isocrinida is characterized by five interradial petals (areolae) surrounded by crenulae. The number of crenulae increases during growth by the addition of new culminae in the interradial points. The change in outline of columnals from pentalobate to subcircular may influence the pattern of petals and crenulae so that petals grow wider and crenulae may be more or less restricted to the marginal zone. This is most distinct in *Isselicrinus,* where the pattern of small or proximal columnals is similar to *Isocrinus,* but during growth these are transformed to large and circular columnals similar to those of *Balanocrinus.* Many species of fossil Isocrinida are characterized by size, form, and ornamentation of the columnals, but the affinities of proximal columnals where these features are not yet developed are in most cases indeterminable.

The column of Isocrinida generally has almost the same diameter in its entire length, but a few species show a slightly increasing diameter toward the distal end. This is most distinct in *Proisocrinus,* but the same is presumably the case in the column of *Austinocrinus.* The articular face of columnals in *Austinocrinus* shows a petaloid central pattern of crenulae, which cover the entire articular face of proximal columnals, but apparently never increases in size. In the distal part of the column, formed when the crinoid was young, this petaloid pattern covers only a narrow central area of the articular face, but during growth, a marginal zone with radiating crenellae is formed outside the petaloid area, and this marginal zone increases in width toward the distal end.

In the very short stem of recent *Endoxocrinus maclearanus* the diameter does not change, and the entire column maintains its juvenile features with low, pentalobate, strongly alternating columnals and very short internodes.

In a few species early columnals do not keep up with diameter of later columnals during growth, so that diameter of the column decreases toward the distal end, and also other juvenile features may be maintained in the distal part of the column, such as alternating size, pentalobate outline, and rather short internodes. This is most evident in the very long stem or Jurassic *Seirocrinus*, but to some degree similar features or at least a distal decrease of diameter and of internodal length may occur in *Pentacrinites* and in young specimens of some recent *Annacrinus, Chladocrinus, Metacrinus*, and *Teliocrinus*. The reason this is not seen in most old specimens may well be that all Isocrinida from time to time break off distal parts of the column.

The cup and brachials of Isocrinida increase in diameter more than in height, and the radials constitute an increasing part of the cup and become more outwardly directed. The basals are rather variable in size, quite often meeting as a contiguous circlet on the surface of small specimens, but in larger specimens separated by the lower point of the radials.

MILLERICRINIDA

In Millericrinida the column is generally very long and columnals rather uniform. The diameter of the column may remain almost the same in its entire length or may increase a little toward the distal end. In all or most species a permanently uppermost columnal, the proximale, different in form and often rather large, is included as a five-sided plate in the dorsal side of the thecal structure inside the basal circlet. Introduction of new columnals during growth must either be restricted to juvenile specimens or take place in the upper part of the column below the proximale. These new, low columnals occurring in the proximal zone may be rather uniform or, in several species, distinctly alternating. So it seems that at least in some species small, new columnals are introduced in succeeding generations between previous columnals in a zone below the proximale. In other specimens no small, new columnals are seen, so it is quite possible that introduction of new columnals ceases early in the adult. In some species there are a few enlarged, proximal columnals below the cup and a zone with smaller columnals below this proximal cone. It is uncertain whether new columnals are introduced within the conical part or only below it, or possibly the conical part is not formed until formation of new columnals has ceased. Some variation in height of columnals, less frequent in diameter, may be maintained in more distal parts of the column but is generally inconspicuous. Form of columnals generally does not change much during growth, although in some species a nearly five-sided section may occur in the proximal part of the column.

The articular face of columnals in Millericrinida is more or less completely covered by radiating crenulae, the number of which increases during growth by division or by insertion of new culminae.

Some species generally referred to Apiocrinitidae are very similar to Millericrinidae and show similar indications of columnal growth with a zone of low columnals below a proximale.

In typical *Apiocrinites*, a conical, enlarged, uppermost part of the column forms a gradual transition to the large, wide cup, and the columnals within this cone are generally higher and more uniform than in the zone below, thus indicating that new columnals are probably not formed within the conical part.

Growth and development of cup and arms in Millericrinida have never been studied. In Apiocrinitidae the cup and conical uppermost part of the column seem to increase in width during growth. The relative size of basals to radials is rather variable, but no distinct change in this relation during growth has been observed.

The radial and proximal brachial articulations of Apiocrinitidae are more or less modified by reduction of the fulcral ridge and of ventral and interarticular fossae and by the greatly enlarged and modified dorsal ligament fossae, which may have a feeble ornament of radiating crenellae or irregular rugosity. These modifications, which may also be indicated in some Millericrinidae, are no doubt introduced or increased during growth, although this has not been studied in detail.

BOURGUETICRINIDA

Columnals of small Bourgueticrinida, including juvenile *Bourgueticrinus* and most species of Bathycrinidae, are very slender, rather high, reel shaped or hourglass shaped, and twisted with elliptical, synarthrial articulations and with a funnel-shaped axial canal enlarged toward the articular face. They may be similar in size or slowly increasing toward the distal end.

During growth the columnals increase in diameter more than in height, as shown in recent *Rhizocrinus,* and in large specimens (e.g., *Bourgueticrinus*), may become very stout, cylindrical to barrel shaped and with a narrow axial canal not widening toward the articular face.

New columnals are introduced at the proximal end of the column, presumably immediately below the cup, since height of proximal columnals gradually decreases toward the cup without alternation. These proximal columnals are low discs with smooth articular faces, but gradually change to the size, form and synarthrial articulation found in more distal columnals. The number of low disc-shaped proximal columnals has been used by GISLÉN (1938a) for discrimination between recent genera and species, but this must be taken with some reservation since the number decreases in adult specimens and large specimens devoid of any low disc-shaped proximal columnals may be found. This has been demonstrated by A. M. CLARK (1973b) for recent Phrynocrinidae. It is obvious that introduction of new columnals decreases and may stop in adult specimens.

Bourgueticrinus has a large, cylindrical, conical or pyriform proximale composed of a few fused columnals, generally without trace of sutures. The articular face of the proximale toward the free columnals may vary within a species, either smooth and flat synostosial with circular outline or synarthrial and with elliptical outline. It is presumed, therefore, that if new columnals are introduced in *Bourgueticrinus,* it must be below a proximale with smooth, circular articular face, and when introduction of new columnals has ceased, the articulation between proximale and column may become synarthrial as in more distal parts of the column. This interpretation is supported by the proximal decrease of columnal height in some specimens and not in others (RASMUSSEN, 1961, pl. 27, fig. 12 and 14; compare with pl. 24, fig. 16 and pl. 30, fig. 1).

A considerable variation is seen in the cup of *Bourgueticrinus,* but no ontogenetical change or correlation between size and form of cup has been observed except that the proximale, at least in some species, increases in size more than the radials and basals during growth.

A juvenile specimen of *Dunnicrinus* shows radials and first brachial to be as high as wide, while in larger specimens width is greater than height.

In *Democrinus* growth of the cup mainly affects the basal circlet, height increasing much more than width. Adult specimens thus attain a very high, slender, more cylindrical form. GISLÉN (1927) demonstrated change in growth lines in a recent *Democrinus* from a low, conical to a high, slender, almost cylindrical cup. Juvenile *Conocrinus* is very similar to *Democrinus,* but during growth the basals and radials fuse, and there is an excessive growth of basals and maybe radials, so that in large specimens no suture is seen, and the arms are separated by ventral projections from the cup. Also, the uppermost columnals may be overgrown by the basals.

Studies of recent *Rhizocrinus* show that height of cup increases during growth more than diameter, but in *Zeuctocrinus* it seems that diameter of cup and brachials increase more than height. Considering the few specimens of *Zeuctocrinus* known, and the large individual variation, this must be taken with some reservation, however.

Development of cup and arms in *Bathycrinus* has not been studied, but GISLÉN (1938a) suggested that the small species *B. gracilis* with its keeled arms may well be a juvenile specimen. Figures by CARPENTER (1888) of *B. aldrichianus* and its synonyms indicate almost no change of form during growth.

CYRTOCRINIDA

Cyrtocrinida live attached to a hard substrate by a short column or directly by the cup, including a nonarticulated dorsal element. Growth and change of the column are unknown.

Irregular or excessive growth, increasing compactness, and fusion of skeletal elements are found during growth of most Cyrtocrinida. In Sclerocrinidae, width and stoutness of the cup increase. In *Phyllocrinus,* and presumably in *Eugeniacrinites,* the ventral projections of the cup separating the arms increase in size during growth, and a study of growth lines in *Phyllocrinus* shows the cup changes from stellate to a more rounded outline. In *Hemicrinus* fusion of cup and proximal columnals take place, and compactness, irregularity, and possibly angular growth increase. Cups and dorsal elements generally fuse in the adults of Hemibrachiocrinidae. In *Cyathidium* small juvenile specimens are very low and shield shaped, but during growth the height is greatly increased and the form is variable, dependent on the space available. Thus, specimens from the Danian coral limestone at Fakse in Denmark may occur in narrow pipes from dissolved coral branches and are extremely slender, following the form of the pipe until they reach the surface of the limestone, where they expand as a cone, but other specimens found on the ceiling of larger crustacean burrows or on the underside of overhanging walls are wide, conical or columnar, and rather low. A juvenile specimen of recent *Holopus* is low and shield shaped, and similar juvenile specimens of recent *Cyathidium* have been recorded from the mid-Atlantic. More detailed studies of brachial growth and apparent change in arm division or fusion of primibrachials of recent *Holopus* and *Cyathidium* are needed.

COMATULIDA

Larval Comatulida are the only Articulata known to agree with Bourgueticrinida in having a column with synarthrial articulations, a radix, and no true cirri until late in the pentacrinid stage. The possibility that Bourgueticrinida may be derived in the early Cretaceous by proterogenesis from the Comatulida seems not to have been previously considered.

The juvenile centrodorsal of most comatulids is conical to hemispherical, and this form is maintained in many genera. In most comatulids, however, the diameter increases during growth more than the height, and the form develops from conical through hemispherical to truncated or discoidal, and by obliteration of apical cirri, and presumably resorption in the apical area, they may obtain a large, flat or concave, cirrus-free dorsal area. Further reduction of the centrodorsal to a low disc or plate with few or no cirri is found in some Comasteracea.

The centrodorsal cavity, lodging the chambered organ and its surrounding nerve capsule, is large in small and juvenile specimens, often more than half of centrodorsal diameter, but is reduced in relative size during growth, generally to about a third or a quarter of the centrodorsal diameter. Exceptions are found in the Atelecrinidae and some Antedonacea, especially Zenometrinae and Pentametrocrinidae.

The first five cirri of the larval centrodorsal are radially placed. New sockets grow out at the ventral edge of the centrodorsal and are commonly seen as a ridge on the ventral face. The sockets at the edge of the centrodorsal are at first very small, but rapidly increase to the same size as preceding sockets or larger, so that only one or two small, juvenile sockets may be seen in each radial side. Generally the first sockets are placed alternately right and left of the midradial line, thus forming ten vertical columns on the centrodorsal, two in each radial side. This form is maintained in the adult stage of many conical centrodorsals. In other species new sockets are intercalated during further growth also in the radial areas between the first ten columns, the number of columns in each radial side thus increasing to three or four, and the arrangement may become irregular. The first sockets near the dorsal pole are often obliterated during growth, so that in low, disc-shaped centrodorsals the sockets remaining form a single or a few irregular, marginal circles surrounding a large, cirrus-free dorsal area. Further reduction of sockets is found in some Comasteracea and Thiolliericrinidae.

The most enlightening studies of growth series of fossil comatulids have been described by Rasmussen (1961) for *Glenotremites,* and by Peck & Watkins (1972) for *Decameros.*

The cup and arms of comatulids follow a pattern of growth similar to that of the Isocrinida. Basals are well developed in the larval comatulids, but during growth radials increase in size and basals are rapidly reduced. Generally, the basals of adult comatulids are reduced to narrow rods exposed only in the interradial point, or in many groups are further reduced to a tiny, perforate, central plate, the rosette. Even in Atelecrinidae, where basals are large plates exposed on the surface, a reduction takes place during growth.

Radials and brachials increase in width more than in height, and many comatulids have a radial circlet that is exposed and contiguous on the surface of juvenile specimens, but with a very low, free surface in larger specimens, and often exposed only near the interradial edge.

UINTACRINIDA

In small specimens of *Marsupites,* the basals are the largest plates of the cup, but during growth the centrale and infrabasals increase to the same size as basals and radials or slightly larger, and the greatest diameter of the cup moves from the upper end of the basal circlet in small specimens to the lower part of basals or upper part of infrabasals in large specimens. In the basal plates the upper and lower angle and the height of the lateral edges decrease during growth. The surface ornament of ridges and granules on thecal plates, although extremely variable, seems to weaken during growth. It thus seems that both change in relative size of the plates and in ornament of these planktonic crinoids is opposite to that of most other crinoids.

In *Uintacrinus* thecal size cannot be determined exactly, the specimens generally being flattened by fossilization and the boundary between thecal structure and free arms being more or less indistinct. Still, it appears that width of theca increases more than height during growth. The number of interbrachial plates, although quite variable, increases during growth by intercalation of small, new plates at arbitrary points of the interbrachial areas. The presence or absence of infrabasals are not correlated with growth and size of the specimens. Growth of brachials follows the usual pattern, increasing in width more than in height, and the same change is found from distal to proximal part of an arm.

ROVEACRINIDA

Growth in Roveacrinida has not been studied, but it was demonstrated by RASMUSSEN (1961, 1971) that basals are overgrown by downward prolongations of the radials, and this undoubtedly took place during early growth. Also, spines, ridges, and flanges serving as floats in Roveacrinida increase during growth; thus, several species and subspecies of *Plotocrinus* and *Poecilocrinus* are interpreted by RASMUSSEN (1961) as probable synonyms based on different ontogenetical stages.

EVOLUTION

By GEORGES UBAGHS, N. GARY LANE, H. WIENBERG RASMUSSEN, and H. L. STRIMPLE

[Université de Liège; University of Indiana; Geologisk Museum, Universitet København; The University of Iowa]

CONTENTS

ORIGIN OF CRINOIDS

By GEORGES UBAGHS

INTRODUCTION

The earliest known undisputed crinoids are Early Ordovician in age. They are *Ramseyocrinus cambriensis* (HICKS) from the Lower Arenigian of Wales (BATES, 1968), *Aethocrinus moorei* UBAGHS from the Lower Arenigian or possibly the Upper Tremadocian of France (UBAGHS, 1969), *Hybocrinus* sp. and two undetermined inadunate species from near the Tremadoc-Arenig boundary of Utah (LANE, 1970), and *Proexenocrinus inyoensis* STRIMPLE & MCGINNIS, from the Al Rose Formation (upper Arenigian) of California (STRIMPLE & MCGINNIS, 1972). To these perhaps may be added *Perittocrinus radiatus* (BEYRICH) and *Tetracionocrinus transitor* (JAEKEL) of uncertain provenience but reputed to be from the Kunda Formation (Arenig-Llanvirn) of the Baltic regions (UBAGHS, 1971a). Among these identified Lower Ordovician species, one *(Aethocrinus moorei)* is a dicyclic inadunate, one *(Proexenocrinus inyoensis)* is a monocyclic camerate, and all the others are monocyclic inadunates belonging to two or possibly three different orders. Such considerable taxonomic diversity indicates that crinoid ancestry must extend far back in pre-Ordovician time—a conclusion fully supported by the fact that the just-mentioned forms already have all the distinguishing features of normal crinoids.

The brachiole-bearing echinoderms—particularly the Cambrian eocrinoids—often have been suggested as being the ancestral stock of the crinoids. This seems quite improbable, considering the differences that exist between a brachiole and a crinoid arm. A brachiole is a purely exothecal appendage that has no direct connection with the interior of the theca and, consequently, could not have carried extensions from the main coelomic cavity, the hemal, genital, and entoneural systems. An arm or brachium of a crinoid is an evagination of the central body mass, carrying extensions from the various coeloms, as well as from the hemal, genital and entoneural systems. Arms and brachioles, therefore, are structures of a different nature and are not homologous.

Unlike crinoids, intimate correlation between brachia and theca does not exist in the Paracrinoidea, which have asymmetrical food-gathering appendages, provided with an ambulacral tract on one side and a row of uniserial side branches along the other side. A canal, connecting with the thecal cavity at its proximal end, runs internally through the main appendages, but does not penetrate into the branches (Parsley & Mintz, 1975); it is unlikely that such a canal could have contained the same extensions (certainly lacking in the side branches) as those present in the arms of crinoids. Here again fundamental differences appear to exist between crinoid arms and paracrinoid armlike processes.

If the presence of true brachia seems to be very distinctive of the crinoids, other features serve also to distinguish them from the brachiole-bearing and paracrinoid classes. Examples of these are: 1) the clear division of the theca into an aboral cup and an adoral tegmen; 2) the invariable presence of a radial plate at the origin of each ray; 3) the mainly radial growth pattern, which finds its expression in the powerful development of the arms, directed away from the theca; 4) the fact that the aboral nervous system is the main motor system; 5) the occurrence of the gonads outside the theca, in the arms or pinnules (at least in recent members). Because of these many and considerable differences, derivation of the crinoids from the brachiole-bearing and paracrinoid echinoderms may be regarded as very unlikely.

However, certain similarities in the arrangement of thecal plates in some Eocrinoidea and Rhombifera, the presence of respiratory structures comparable to epispires or pectinirhombs in some early crinoids, and continuation of the ambulacral tracts over the thecal plates in the hybocrinid inadunate *Hybocystites* have led some authors to look for the origin of crinoids among Rhombifera (Yakovlev, 1918, 1927; Moore, 1954), hypothetical minute Cystoidea (Kirk, 1911), Eocrinoidea (Fell, 1962, 1963b; Jaekel, 1918; Moore, 1954; Nichols, 1969), or proto-blastoids (Bather, 1900a). But such analogies are no proof of close affinities. They may merely reflect the fundamental unity of the stem-bearing echinoderms, and they may have arisen independently in each class. After all, the crinoids and the brachiole-bearing echinoderms must have met the same sort of problems, having the same attached mode of life and probably highly similar food-catching methods.

Some earliest known crinoids have features judged to be very primitive, and thence possibly significant for the problem of the origin of the class. These features are: 1) the merging of the calyx into the column and of the thecal cavity into the wide lumen of the stem (e.g., *Aethocrinus, Ramseyocrinus*); 2) the partition of the column into pentameres, accompanied by irregular interlocking of the columnal plates and passage to a distal mass of tiny skeletal elements (e.g., *Aethocrinus*); 3) the extension of this columnal interlocking into the theca, the proximal circlets of which are not arranged in quite regular circlets (e.g., *Aethocrinus*); and 4) the occurrence of supplementary plates (anals, interbrachials, accessory plates) in the cup (e.g., *Aethocrinus, Perittocrinus, Tetracionocrinus*). These characters suggest derivation of the crinoids from an elongate ancestor, provided with an irregular many-plated calyx that graded downward into an irregularly plated hollow holdfast and extended upward into radial outgrowths of the central body mass.

Similar views were first advocated by Jaekel (1918), who stated that in the beginning there was no simplicity, but lack of regularity. He thought that originally the stem and root were the rear extension of the body, and that some forms like the Middle Cambrian eocrinoid *Acanthocystites* with its many and irregularly arranged plates might well represent the ancestral type of the crinoids. Consequently he strongly opposed the opinion, generally accepted in his time, that the crinoids originated from small and simple forms, the Inadunata Larviformia of Wachsmuth &

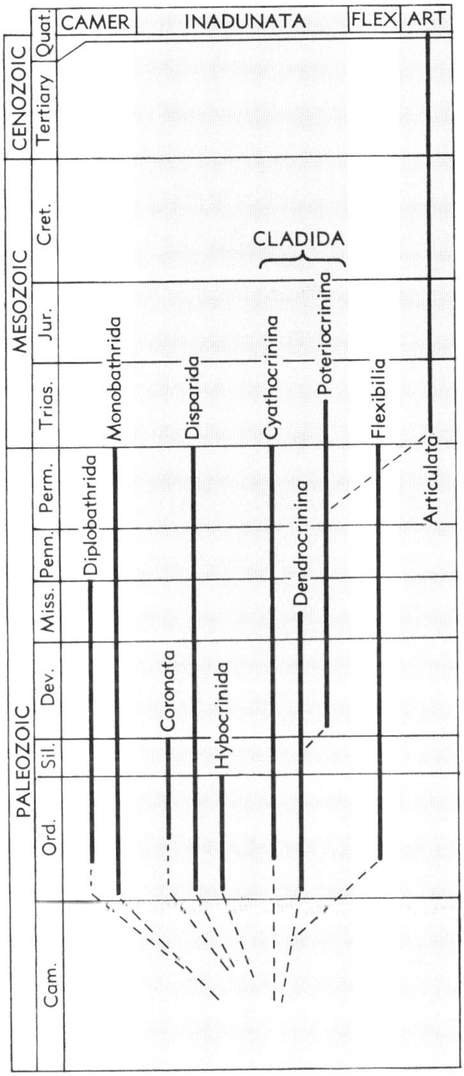

FIG. 207. Stratigraphical distribution and presumed phylogenetic relationships of major groups of crinoids (Ubaghs, n).

SPRINGER (1885), or passed through a larviform stage, as imagined by BATHER (1900a). The discovery of new Lower Ordovician crinoids and revision of those already known have brought partial support to JAEKEL's judgment, except that the eocrinoids or other brachiole-bearing pelmatozoans are probably not the forerunners of the crinoids.

As yet no Cambrian echinoderm has been recorded that could be accepted with certainty as the ancestor of the crinoids. However, in this connection one form deserves much attention. It is *Echmatocrinus brachiatus* SPRINKLE, a very primitive attached echinoderm from the Middle Cambrian (Burgess Shale) of western Canada (SPRINKLE, 1973a). With its stalk barely differentiated from the theca, and its many irregularly arranged plates, it does recall some very archaic eocrinoids. It has armlike uniserial processes with apparently soft appendages (? tube feet, ? uncalcified armlets) given off alternately (?) left and right. Because of these processes, it was placed by SPRINKLE in an indeterminate crinoid subclass and order, though in most other respects it does not fit the definition of a crinoid. Besides, it is not certain that its armlike processes are really homologous with the crinoid brachia, that is to say, that they are outgrowths from the central body mass. At any rate, the regular arrangement of the plates of these processes contrasts with the poorly organized plating of the theca. Even if *Echmatocrinus* seems to be separated from any unquestioned crinoid by a wide gap, at least it furnishes evidence that pelmatozoan echinoderms provided with arms similar in outer appearance to those of crinoids were living in the Cambrian seas. From such forms the crinoids could have arisen.[1]

ORIGIN OF MAJOR GROUPS

From the outset of their known geological record the Camerata and the Inadunata, which seemingly represent the two main branches of the Crinoidea, were already clearly differentiated (Fig. 207). In many respects, the Camerata stand apart from the other crinoids. They are morphologically fairly distinct, and never gave rise to any other subclass. The Inadunata, on the other hand, are the basic stock from which both the Flexibilia and the Articulata proceeded. There seems, therefore, to be some justification in the classification advocated by JAEKEL (1894, 1901, 1918), accord-

[1] In the systematic part of this *Treatise*, *Echmatocrinus* is considered by SPRINKLE and MOORE to represent a new order Echmatocrinida and new subclass Echmatocrinea, both by SPRINKLE and MOORE (see p. T405).—CURT TEICHERT.

ing to whom the Camerata, which he named Cladocrinoidea, should be separated from the rest, named Pentacrinoidea.

It has sometimes been suggested that the Camerata and the Inadunata could have originated separately, and also that each of them was polyphyletic (BROWER, 1969). JAEKEL (1894, 1902, 1918) went even so far as to suppose that the arms, pinnules, and most thecal plates of the Camerata were different in origin and morphological significance from the corresponding plates and structures of the other crinoids. Such views do not appear consistent with the fact that all crinoids share the same basic organization and are composed of highly similar elements, which are arranged in the same manner and obviously had the same functions. So many and far-reaching resemblances can hardly be explained as the result of convergent evolution. Besides, some early inadunates (e.g., *Aethocrinus, Cupulocrinus, Ottawacrinus*) have fixed brachials and interbrachials like camerates, a condition that rather favors judgment that both Inadunata and Camerata are derived from a common source.

The division of the Crinoidea into two subclasses, the Monocyclica and the Dicyclica, as proposed by BATHER (1899b, 1900a), would signify at least an early dichotomous branching of the class and a large number of parallel modifications, since each main branch would have produced a camerate and an inadunate type of theca and arms. It is true that the structure of the base has been a fundamental, highly distinctive and generally constant feature of the major groups of crinoids throughout their history. Nevertheless, the monocyclic camerates have more characters in common with the dicyclic camerates than with the monocyclic inadunates, and the same is true for the monocyclic and dicyclic inadunates. Among early camerates, *Reteocrinus,* which is dicyclic, and *Xenocrinus,* which is monocyclic, are so essentially alike, except for characters of the base, that it is improbable that they could have been derived from widely separated ancestors. Similarly among inadunates, the dicyclic *Merocrinus* is practically like the monocyclic *Iocrinus* but for the presence of infrabasals. Such examples tend to indicate that the differentiation of the monocyclic and dicyclic types of base took place separately within the camerates and the inadunates though it must be agreed that we are completely ignorant as to how they originated and what were their mutual relationships.

As described previously, the camerates form a group fairly well defined on a morphological basis. Yet they appear to have been divided into monocyclic and dicyclic stocks from their earliest known appearance in Lower or Middle Ordovician time, and we do not know which of these stocks is ancestral to the other. But once they had differentiated they seem to have kept the fundamental monocyclic or dicyclic character of their base unchanged throughout their entire existence, allowing recognition of two orders within the subclass; the Diplobathrida, which include all dicyclic camerates, and the Monobathrida, which comprise all monocyclic camerates.

Such sharp cleavage into two main branches does not seem to characterize the evolution of the inadunate crinoids. Comparison of their representatives rather indicates either an early division into six clearly marked groups, or a polyphyletic origin of the subclass. These six groups are the Disparida, Hybocrinida, Perittocrinacea, Coronata, Cyathocrinina, and Dendrocrinina. The first four include monocyclic inadunates, the last two, dicyclic inadunates. Inasmuch as important differences existed between them throughout their whole geological record and as no intermediate forms are known which could have reduced the gaps between them, their origin and interrelationships remain problematical.

The Disparida are the most important and diversified group of monocyclic inadunates. They largely correspond to the Inadunata Larviformia of WACHSMUTH & SPRINGER (1885)—a term referring to their generally small size and simple structure, and to the belief formerly held that they were the most primitive crinoids and that they represented the ancestral type of the class. In fact, they disclose evolutionary trends very different from those shown by the other crinoids. Thus they show a common tendency to develop bilateral symmetries in planes different from the so-called crinoidal plane *(A-CD)* prevalent in most other crinoids; they commonly have compound radial plates in rays other than the *C* ray; their proximal *C*-ray plate never takes the appearance or function of a radianal; the median line of ossicles supporting the anal tube is ordinarily so intimately related to the adjacent *C* ray that it may replace it or its proximal left ramus. Such highly distinctive features probably evolved within the stock and were not inherited from an ancestor common to all crinoids. Apparently the disparid inadunates form a separate group, which became extinct in Upper Permian time, and did not give rise to any other group.

Hybocystites and other genera now included in the order Hybocrinida were thought by BATHER (1900a) as possibly derived from early forms of Protoblastoidea. Consequently, and because of their simple structure, they were placed at the starting point of his classification. As clearly shown, however, by JAEKEL (1902, 1918), and largely confirmed by REGNÉLL (1948a), the peculiarities of these genera and similarities with blastoids are purely secondary. They resulted from the progressive (or degenerative) evolution from an ordinary five-rayed crinoid, and were largely brought about through the progressive reduction, and eventually the loss, of the arms and replacement of these processes by recumbent ambulacral tracts. The hybocrinid inadunates differ markedly from the disparid inadunates by lack of compound radials, and presence of a radianal situated obliquely below the *C* radial and that supports the anal *X*. On the other hand, in such other features as the shape of the cup, the narrowness of the arm facets, and absence of an anal tube, they resemble primitive Cyathocrinina (e.g., Palaeocrinidae), except that they are monocyclic instead of being dicyclic. These similarities are not sufficient, however, to indicate any close relationships between these crinoids. The hybocrinids are among the earliest recorded crinoids. So far as we know, they disappeared in Late Ordovician time, without giving rise to any other group.

The Perittocrinacea are represented by two monotypic genera of probably Early Ordovician age. Like the Disparida and Hybocrinida, they are monocyclic inadunates. They differ from both in having numerous triangular accessory plates and only four basals. Also, they are clearly separated from the Disparida by the presence of a radianal and lack of compound radials. They are less distinct from the Hybocrinida, for, like them, they have simple radials, a typical radianal and anal *X*, a *C* radial smaller than the other four and located above their level, and relatively narrow and rounded arm facets. In other words, the pattern of the main cup plates in Perittocrinacea is similar to that in Hybocrinida and in primitive Cyathocrinina. Of course, such resemblances may result from an evolutionary convergence, but they may also indicate inheritance from a common ancestry (UBAGHS, 1971a). On the whole, however, their peculiarities suggest tendencies of their own, so that they appear to form a rather aberrant offshoot of the inadunate stock, which apparently left no descendants. They certainly are not transitional between the camerates and the other crinoids, as supposed by JAEKEL (1902, 1918).

Perhaps still more puzzling are the few Ordovician and Silurian genera known as the Coronata. Because they look very much like blastoids in general appearance and arrangement of thecal plates, they have been considered as such by some workers, including JAEKEL (1918, 1927), WANNER (1924), and REGNÉLL (1945). However, they lack lancet plates, hydrospires, gonopore and brachioles, all features typical for blastoids. On the other hand, they have biserial arms and armlets comparable to those of crinoids (SPRINGER, 1926a; BATHER, 1900a; FAY, 1961). For that reason they were accepted as crinoids first by WACHSMUTH & SPRINGER (1885), and then by BATHER (1900a), SPRINGER (1926a), FAY (1961), and others. They certainly are quite different from any typical crinoid. They may be considered, at least for the time being, as representing a particular evolutionary development from an inadunate stock, though their origin and relationships remain equally enigmatical.

The Cyathocrinina and the Dendrocrinina comprise the earliest known cladid, or dicyclic, inadunates. The former first appear in the Middle Ordovician, the latter in the Lower Ordovician. Both were clearly differentiated from the outset and were subject to an early diversification—two conditions that point to a long previous history. There is no evidence at present for a common origin or derivation of one group from the other. Relationship with monocyclic inadunates is also obscure. As stated above, the thecal pattern of some Ordovician Cyathocrinina resembles that of the Hybocrinida, except for characters of the base. But such similarities are no proof of phylogenetic relationships. As to the Dendrocrinina, they display a number of features and evolutionary tendencies quite different from those exhibited by the monocyclic inadunates. If one compares, for instance, their earliest known representatives, the Lower Ordovician *Aethocrinus moorei,* with the disparid *Ramseyocrinus cambriensis* of about the same age, one is struck by the many differences which exist between them. Most of their characters are contrasting and certainly none indicates unquestionable origination from a common source.

Although the Cyathocrinina appear not to have given rise to any other group, the Dendrocrinina are reputed to include the direct ancestors of the Poteriocrinina and the Flexibilia. The former are the most advanced dicyclic inadunates. They probably arose during Devonian time from such dendrocrinine families as the Botryocrinidae or the Mastigocrinidae, which, in several respects, approximate primitive Poteriocrinina very closely. The passage from one type to the other assuredly was quite progressive and possibly repeated several times. It was especially characterized by the development of the pinnulate condition from a strongly heterotomous condition, the appearance of muscular articulation between the radial and first primibrach, and the more and more prevailing occurrence of three (or more) anal plates in the cup.

The earliest known representatives of the Flexibilia are two species of the genus *Protaxocrinus* from the lower Trenton rocks (Middle Ordovician) of Canada. They closely resemble species of the dendrocrinine genus *Cupulocrinus,* which occur in the same beds, in that they have erect infrabasals, a radianal in primitive position, a series of anal plates bordered by finely plated integument, interbrachial areas largely occupied by numerous small and irregular plates, arcuate brachial sutures, and a pliant tegmen. The main differences lie in number of infrabasals, number and distribution of primibrachs, and presence of an anal sac. *Protaxocrinus* has three unequal infrabasals, two primibrachs per ray, and no distinct anal sac. *Cupulocrinus* possesses five equal infrabasals, three or more primibrachs, which are unequally distributed among the rays, and a prominent anal sac. These divergences are not so important as to prevent accepted judgment that both genera derive from a common Ordovician or slightly pre-Ordovician ancestor.

The exact origin of the Articulata is still obscure. Their earliest known representatives are Middle Triassic in age. They belong to three distinct families (Dadocrinidae, Holocrinidae, and Roveacrinidae) that cannot be derived with certainty from any known Paleozoic genera. Yet the poteriocrinine inadunates may display features or evolutionary tendencies which approach and in some cases reach the articulate condition. Thus, all poteriocrinid inadunates are dicyclic (all articulates are dicyclic or cryptodicyclic), and several have the infrabasals reduced in number and size. Some may have all anal plates eliminated from the cup, as is invariably the case in articulates. Their arms are pinnulate like those of articulates, and free above the radials, as in most articulates. Many of them have uniserial arms, as do all articulates. Typically, their radial facets and proximal arm plates bear marks of well-developed muscular articulations similar to those of articulates. These articulations may be oblique and perforate, and distributed in exactly the same manner as those of the living comatulids (LANE & MACURDA, 1975). Finally, the mouth and the ambulacral furrows were probably exposed in some advanced poteriocrinine members such as the Encrinidae, which have lost the large anal sac so typical of most poteriocrinine inadunates and acquired a flattened tegmen quite similar to the disc of articulates. For these reasons it is generally agreed that the Poteriocrinina comprise the stock from which the Articulata originated.

There is no evidence that the Articulata or even some of the members of the subclass descended from the Flexibilia. It is true that the Flexibilia also had a pliant tegmen, with exposed mouth and food grooves, and that certain Permian forms, such as *Palaeoholopus* and *Permobrachypus,* were outwardly remarkably like the articulate Holopodidae. But the Flexibilia as a whole are characterized by a combination of structural peculiarities very different from those of the Articulata. They never have pinnulate arms, and their arm-branching never really tended toward the pinnulate condition. Their radial facets and brachial articulations do not resemble the muscular articulations of the recent crinoids, but are of a rather special type of movable ligamentary juncture. Peculiar also are the articulations between the calyx plates. The brachials usually have undulate sutures, resulting from the fact that the lower edge of each has a projection that fits into a depression of the plate below. Almost invariably there are three infrabasals, two large plates and a small one, which always occur in the *C* radius, whereas in articulates the infrabasals generally number five. In short, the Flexibilia practically do not show any feature that could suggest an articulate descent. They rather appear as a specialized group that became extinct in Permian time.

Even if all the Articulata were derived from the Poteriocrinina, as indicated by their essential unity of structure, they did not necessarily originate from the same poteriocrinine ancestor. Their early diversification and large diversity speak rather in favor of a moderate polyphyletism. In any case, the origin and affinity of their various orders remain largely conjectural, if not entirely unknown. This problem will be discussed in the chapter devoted to the phylogeny of these crinoids.

EVOLUTION OF CAMERATE CRINOIDS

By Georges Ubaghs

As for the other crinoid subclasses, the phylogeny of the Camerata is essentially based on comparative morphology of genera. Practically no lines of ancestry and descent are known at the species level. The most comprehensive treatments of the evolutionary development of these crinoids are purely qualitative (Wachsmuth & Springer, 1897; Bather, 1900a; Jaekel, 1918; Moore & Laudon, 1943a; Ubaghs, 1953). Few quantitative phyletic studies have been undertaken, and they all concern restricted groups (Brower, 1973, 1974a; Lane, 1963b; Macurda, 1974). For the present, the study of the phylogeny of Camerata appears very subjective and still in its infancy.

The earliest known camerate is *Proexenocrinus inyoensis* Strimple & McGinnis from the Lower Ordovician Al Rose Formation of California. It was classified as belonging to the family Xenocrinidae by Strimple and McGinnis (1972). Uncertainty concerning the structure of the proximal part of its calyx up to the radial circlet makes its systematic position somewhat uncertain, however. Nevertheless, it demonstrates that the Camerata had already acquired all their distinguishing features by Early Ordovician time, suggesting a more remote, probably Cambrian, origin for these crinoids.

This premise is strongly supported by the early diversification of the subclass. At least six distinct camerate families occurred in the Middle Ordovician, to which four or possibly five more were added in the Late Ordovician. At the end of this period, all orders and suborders, and no less than six superfamilies of the twelve recognized in this *Treatise* were differentiated.

Which of the camerate orders and suborders are the most primitive is unknown, for all of them were clearly separated since their first appearance in the geological record. In particular there is no convincing evidence suggesting derivation of the monocyclic camerates from dicyclic camerates, or descent of the Glyptocrinina from the Compsocrinina. Since all these major groups are distinguished throughout their history by the structure of the proximal part of the theca, it appears that when once evolved this structure stabilized (except for minor changes), so that the modifications occurring afterwards mainly affected the distal part of the theca and (or) the free arms, more rarely the column.

EVOLUTIONARY TRENDS IN CAMERATES

The geological succession of the main calyx types (as recognized by BROWER, 1973) and the comparison of the earliest known stages with more advanced members in some lineages allow recognition of some definite trends in the evolutionary development of the camerates. Some of these trends were restricted to a few genera, others were widespread, but none seem to have involved the whole subclass. They probably affected the young crinoids at various growth stages and caused divergences during the later ontogeny (BROWER, 1973).

EVOLUTION OF THE CALYX

One of the most remarkable trends was the elimination of fixed brachials, interbrachials, and anal plates from the calyx, so that eventually the arms became entirely free. This condition is partly filled by the Gazacrinidae, Carpocrinidae, Coelocrinidae, Eucalyptocrinitidae, Stelidiocrinidae, Patelliocrinidae, some Dimerocrinitidae, and a few Hapalocrinidae; members of these families have few fixed brachials and few but large interbrachials. The trend is still more advanced in the Nyctocrinidae, Hexacrinitacea, and Platycrinitacea, which generally have no brachials and no interbrachials incorporated in the calyx. The genera included in the two last superfamilies were referred by BATHER (1899b, 1900a) to a monocyclic order, the Adunata, which he considered to have been derived from the Inadunata and modified after the fashion of the Camerata. This view is not accepted in the present *Treatise*. The earliest known representatives of these so-called Adunata already had the arms fully pinnulated, that is to say, long before Inadunata acquired pinnules. Their many-plated tegmen generally was of a camerate type. The Silurian Marsupiocrinidae and some Hexacrinitidae and Hapalocrinidae had fixed brachials and large interbrachials. These crinoids, therefore, are not morphologically distinct from typical Camerata. They are regarded here as highly specialized members of this subclass, which became superficially similar to some Inadunata.

In camerate crinoids, as in other animal groups, the different parts of the body evolved more or less independently and at various rates, so that many genera share primitive and advanced features. For instance, *Melocrinites* combines an archaic type of calyx with a most specialized kind of arms, as do also *Thamnocrinus, Manillocrinus,* and several other genera.

The main changes observed in the base of the calyx are related to size, shape, and number of component elements, and already have been discussed in the chapter devoted to the skeletal morphology of the crinoids (p. *T*107). Here it will be merely noted that primitively the base was probably erect and composed of one or two circlets of five elements each. Bases flattened or concave, or with proximal circlet reduced and concealed by stem, or having less than five infrabasals or five basals are judged to be specialized. All or at least some of these modifications may have benefited the crinoid. Thus the reduction of the number of plates in the proximal circlet probably increased the mechanical strength of the calyx, better protected the chambered organ, and simplified the problem of integration and coordination of calyx growth (BROWER, 1973). The number of basals—the number of infrabasals less so—is an important diagnostic feature on the family or even the suprafamilial level. It was generally established very early and commonly persisted throughout the history of the families. For instance, the Periechocrinacea and the Carpocrinacea, which flourished abundantly from the Late Silurian to the Early Carboniferous, invariably have three basals, which most generally are equal. All the Melocrinitacea have four basals. Two is the characteristic number of these plates for the Dichocrinidae and Acrocrinidae, and other similar examples could be given.

The fixed ray plates are usually more numerous in Ordovician and Early Silurian genera than in later forms. This suggests, as seen above, a rather general tendency toward reduction of the number of these plates up to their complete elimination (except radials) from the calyx—a stage represented by the dicyclic Nyctocrinidae and the most advanced representatives of the monocyclic Hexacrinitacea and Platycrinitacea. In some lineages, however, an opposite trend seems to be recognizable. The Polypeltidae and some genera like *Scyphocrinites, Strotocrinus,* and *Teleiocrinus* have, for example, such a high number of fixed ray plates there is little doubt that in these crinoids a progressive incorporation of the arms into the calyx took place.

Another feature of the fixed part of the ray which tends to disappear with time is the presence of a median ridge on each ray of the calyx. These ridges, strongly marked in various Ordovician genera, were ordinarily replaced by faint ridges in later forms, and ultimately vanished completely.

Accompanying these changes, modifications in number, size, and arrangement of interbrachials occurred in many lineages. Generally, in the Ordovician genera, they were numerous, small or moderately large, and more or less irregular. Also, they were continuous with the interambulacrals. The Reteocrinidae, Archaeocrinidae, Xenocrinidae, and Glyptocrinidae are good examples of this condition, which is regarded as primitive. In most later camerates, the interbrachials tended to diminish in number, to increase in size, to become more definitely arranged, and, in some lineages, to be separated from the tegminal plates by fixed brachials and pinnulars in lateral contact. It must be noted, however, that by the Middle Ordovician the interbrachials were lacking in the Cleiocrinidae and separated from the interambulacrals by fixed ray plates in the Anthracocrinidae. On the other hand, the extremely numerous, small and irregular interbrachials of the Ordovician genera *Reteocrinus* and *Xenocrinus* probably represent an advanced rather than a primitive evolutionary character. Indeed, while the rays and interrays of *Xenocrinus* (and doubtless of *Reteocrinus*) expanded continuously during ontogeny, the interbrachials showed small growth rates and a complex intercalatory type of development. This mode of growth contrasted with that observed in other camerates, such as the Glyptocrinidae. In those, the growth rates for dimensions of the interbrachials were large relative to size, and all new interbrachials developed at distal borders of the interbrachial areas. Multivariate, statistical information about phylogeny suggest this condition to be more primitive than that shown by *Xenocrinus* and *Reteocrinus* (BROWER, 1974a).

A consequence of the progressive enlargement of the interprimibrach in several lineages consists of a change in the shape of the adjacent primibrachs. In primitive genera or in lineages which retain a primitive type of calyx, as in the Melocrinitidae or the Periechocrinidae, the interprimibrachs are ordinarily small or of moderate size, and the first and second primibrach are, respectively, hexagonal and heptagonal in outline. But in more advanced or specialized forms (e.g., the Paragaricocrinidae, Batocrinidae or Eucalyptocrinitidae, which have well-developed interprimibrachs), the first primibrach becomes quadrangular and the primaxil usually pentagonal in outer appearance. This apparently unimportant difference serves as a significant diagnostic feature for separating most Periechocrinacea from the Carpocrinacea.

In dicyclic Rhodocrinitacea, the proximal interprimibrachs lie between adjacent radials and rest on basals. The basic change to the organization seen in the Dimerocrinitacea seems to be the displacement of these first interprimibrachs to a position above the radial circlet. Whether such displacement really took place in evolution remains conjectural. It must be noted, however, that both types of structures occur in some specimens of *Dimerocrinites icosidactylus, Lyriocrinus melissa, L. dactylus,* and *Griphocrinus nodulosus.* Also, in the ontogeny of some recent comatulids, interradial plates having the same location as the proximal interray plates of camerates develop between adjacent radials and then migrate to above the radial circlet. Possibly a similar process occurred in the evolution of the dimerocrinitid from the rhodocrinitid camerates. Such elimination of plates from the radial circlet was perhaps advantageous, for it must have increased the strength of the proximal part of the radial circlet (BROWER, 1973).

The structure of the posterior *(CD)* interray of the calyx seems to have evolved in many lineages. Primitively this interray was wider than others, contained more plates, and was commonly marked by a prominent sagittal and generally ridged series of extra plates. Subsequently, this ridge disappeared, the plates were reduced in number, and the posterior interray became similar to the other interrays. As a consequence, an almost perfect pentamerous symmetry was acquired by the calyx, contrasting with the strongly marked bilateral symmetry of many early forms. The similarity of the median anal series of plates in some archaic camerates with a series of fixed brachials has led some authors (MOORE & LAUDON, 1943a; SPRENG & PARKS, 1953) to assume that the anal series might have had its origin in a sixth ray. Until now no fossil has been recorded to support this hypothesis. Neither do we have proof that the primanal migrated effectively from a primitive position between *C* and *D* radials to a more advanced position above these plates, or, on the contrary, was secondarily interpolated into the radial circlet. Both explanations have been advocated, but no definitive solution to this problem has ever been found.

EVOLUTION OF THE TEGMEN

The phylogeny of the tegmen of camerate crinoids has not been studied in detail. It appears that in many Ordovician genera judged to be primitive in many respects the tegmen consists of an incompetent and many-plated structure, in which the orals, ambulacrals, and interambulacrals are not clearly differentiated. Rather surprisingly, in many lineages, it is in advanced members that these plates become really distinct. But an opposite tendency may also occur. For instance, in the Platycrinitidae, there seems to have been a definite trend toward loss of differentiation of orals and ambulacrals through incorporation of a large number of interambulacrals. On the whole, the tegmen of camerates has been modified in various ways, forming, for instance, protective devices for the free arms, such as grooves, niches, bladelike processes, wing plates, and gigantic spines, or being provided with an anal tube that in some species was elevated well above the summit of the arms. Most of these changes developed at the generic level, probably as a response to some particular need.

EVOLUTION OF THE FREE ARMS

During and after evolution of the main calyx structures, the arms generally evolved toward greater complexity and efficiency. Several trends may be recognized in their evolution. One of them consists of the change from a uniserial to a biserial arrangement of the arm plates. Species with uniserial free arms were prevalent in the Ordovician, progressively less frequent in the Silurian, and rare after the Early Devonian. Members of some lineages went a step further toward the biserial condition; they acquired compound brachials that carried one or more pinnules on each side of every brachial. These changes are interpreted as adaptative, for they allowed a considerable increase in the number of pinnules and thence in the food-gathering capacity of the brachial system.

Increase in the number of free arms is another dominant trend in many camerate lineages. All camerates, including the earliest known, had two arms per ray at least. It is true that a few camerate genera have only one arm per ray, but this is because the two arms of each ray fused together (as in *Melocrinites*) or developed so unequally that one of them took the appearance of a mere armlet (as in *Cytidocrinus*). Two arms per ray is certainly a primitive stage in camerates, which persisted in some lineages (e.g., *Scyphocrinites, Stelidiocrinus, Patelliocrinus*) but generally was replaced by a more advanced one characterized by four or six arms per ray, more rarely three, five, seven or even more (large specimens of *Strotocrinus* may have up to 30 arms per ray). This increase in number of free arms resulted from either the transformation of a proximal pinnule into a pinnulate arm or the incorporation of one or several bifurcations of arms into the calyx. An example of the first case seems to be furnished by the transition from the Middle Ordovician *Pycnocrinus ornatus,* with two arms per ray, to the Late Ordovician

Glyptocrinus decadactylus, with four arms per ray (BROWER, 1973). The second case may be illustrated by batocrinid and actinocrinitid genera which have many brachitaxes incorporated in the calyx and consequently many free arms given off by the theca.

Primitively the free arms were probably undivided. At least they are so in several archaic genera, such as *Cleiocrinus, Rhaphanocrinus, Proexenocrinus, Xenocrinus,* and *Compsocrinus.* The next step was marked by isotomous bifurcations repeated at long intervals. Examples of this stage include Ordovician genera like *Archaeocrinus* or *Canistrocrinus.* Subsequently other types of arm branching (heterotomy, endotomy, exotomy) evolved in various lineages, forming diagnostic features for many genera and sometimes species. At the same time there was a widespread tendency to concentrate ray division within the calyx or immediately above it—a probably advantageous modification, for any accidental loss of an arm involved only a small part of the whole system. This tendency was frequently accompanied by the grouping of the arms, thus producing lobation of the calyx at arm regions (as in Actinocrinitidae or Platycrinitidae), or was manifested by their arrangement in a continuous belt around the calyx (as in many Batocrinidae).

A special type of arm structure peculiar to camerate crinoids resulted from the hypertrophy and very unequal, heterotomous branching of arms. It arose independently in various families (i.e., Rhodocrinitidae, Lampterocrinidae, Actinocrinitidae, Melocrinitidae, Polypeltidae, and Platycrinitidae), and consisted of the formation of one or two powerful arm trunks in each ray. These trunks carried relatively slender ramuli that were pinnulate, whereas the trunks themselves usually (if not invariably) lacked pinnules. In advanced melocrinitids, the two adradial arms of each ray coalesced and produced a compound ramule-bearing trunk. These structures probably originated in response to a need for enlargement and strengthening of the food-gathering apparatus.

Another specialization which appeared independently in several camerate families (i.e., Rhodocrinitidae, Carpocrinidae, Dichocrinidae, Aorocrinidae, Patelliocrinidae, and Platycrinitidae), but very rarely in Inadunata (and never in Flexibilia and Articulata), lies in the downward growth of the arms. That in such cases the pendent or recumbent attitude was permanent and not accidental is shown by its association with various structural devices which prohibited, or at least hindered, any motion in upward direction. This curious peculiarity developed at the specific level, except in the desmidocrinid genus *Barrandeocrinus,* the whole crown of which is transformed in connection with a fixed recumbent position of the arms.

These were the dominant trends that affected the evolutionary development of the camerate crinoids. Now, guided by this knowledge, we may endeavor to trace the phylogeny of this subclass, taking the Diplobathrida first and then the Monobathrida (Fig. 208).

PHYLOGENY OF CAMERATES

DIPLOBATHRIDA

Reteocrinus has been commonly accepted as the most primitive known camerate crinoid (MOORE & LAUDON, 1943), and certainly it has many archaic features. However, as explained above (p. *T*283), its pliant interbrachial areas studded with numerous small and irregular plates seem to represent a specialized and advanced feature, suggesting that this genus derived from a form with larger and more regular interbrachials. Such an ancestor may perhaps be visualized as an archaeocrinid which, like *Rhaphanocrinus,* would have an erect calyx, prominent ray-ridges, moderately large and regular interbrachials, a primanal supporting an anal median series of plates, and uniserial unbranched free arms, but, in contrast to *Rhaphanocrinus,* should be provided with infrabasals visible from the side. These characters are widespread among Ordovician genera belonging to various lineages, and consequently can be considered as primitive.

Rhaphanocrinus is a member of a widely diversified assemblage of dicyclic camerates, all of which have the radials separated by interbrachials in contact with basals, and are conveniently grouped in the superfamily

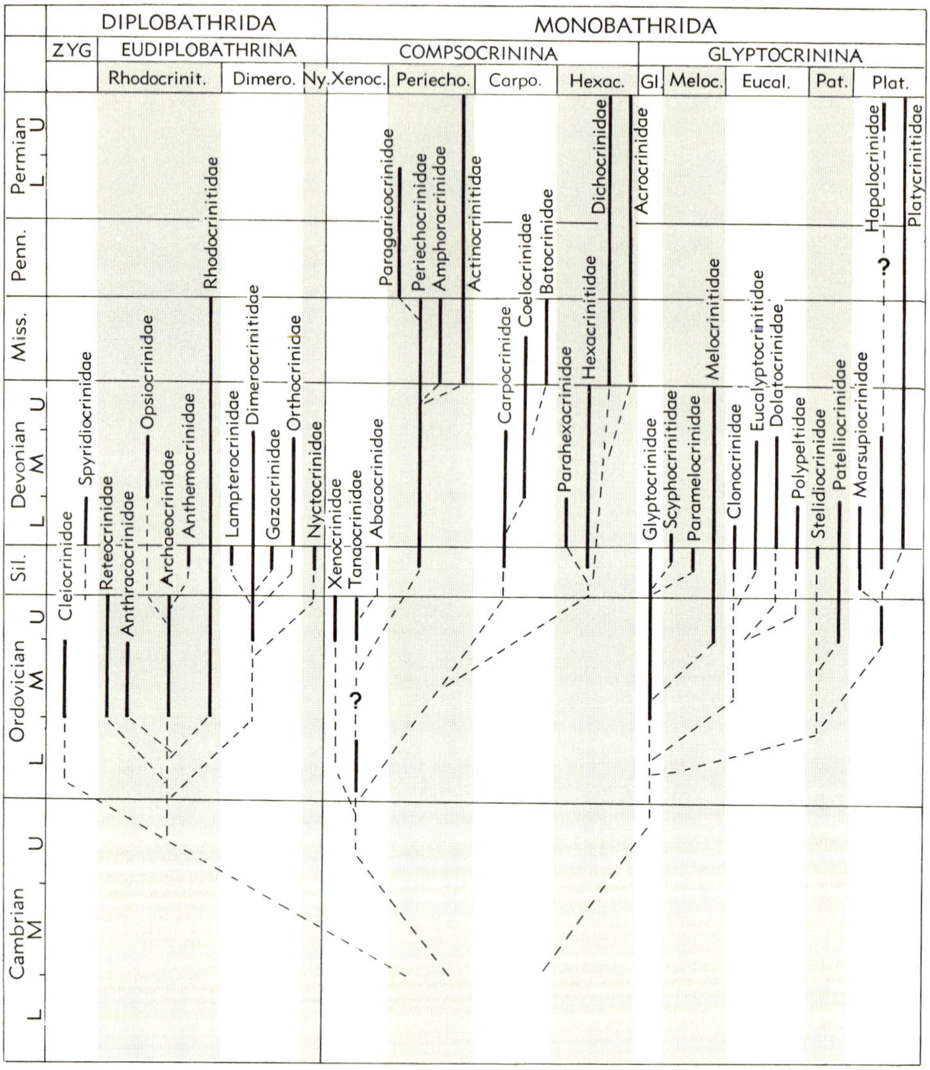

FIG. 208. Diagram showing evolutionary trends of the Camerata (Ubaghs, n).

Rhodocrinitacea. This classification is admittedly artificial, for the phyletic relations of these forms are practically unknown. By middle Ordovician time they already had reached a high degree of diversification and, in some cases (e.g., Anthracocrinidae), of specialization, which suggests a long previous history and possibly a polyphyletic origin.

The relations of the Rhodocrinitacea with the Cleiocrinidae and Spyridiocrinidae (here placed in the suborder Zygodiplobathrina) are obscure. *Cleiocrinus* is one of the most puzzling crinoids. It was referred successively, and generally with doubt, to the Crotalocrinidae (ZITTEL, 1879), Ichthyocrinidae (WACHSMUTH & SPRINGER, 1885), and the Flexibilia Impinnata (BATHER, 1900a); it was also regarded as an intermediate form between the Camerata and the Flexibilia (SPRINGER, 1905), an *incertae sedis* echinoderm (SPRINGER, 1911b, 1920), and an aberrant crinoid-like offshoot from the cystoids (Moore & Laudon, 1943a). Because of its pinnulate arms and subtegminal mouth, it was classified as a camerate closely related to *Reteocrinus* by Springer (1913), as a camerate by Bassler (1938) and Bassler & Moodey (1943), a monocyclic camerate by Jaekel (1918), and a dicyclic camerate by Ubaghs (1950, 1953). It differs, however, from all other dicyclic camerates in having the infrabasals overlapped by the basals and radials, which form a circlet of ten plates that project downward over the column like a collar. Furthermore, in contrast to all other camerates provided with many fixed brachials, it lacks interbrachials completely. However, the location of the basals between the radials (zygodiplobathrid type of base) does not seem to be fundamentally distinct from the eudiplobathrid type in which the basals and radials are arranged in two different circlets, for in some species (e.g., *Dimerocrinites pentlandicus, Paulocrinus biturbinatus, Rhipidocrinus crenatus*) some or even all radials may be inserted between the basals and thus be in contact with infrabasals. According to Brower (1975) the zygodiplobathrid base can be derived from the eudiplobathrid type by reducing the growth rates of the basals. *Cleiocrinus,* however, appears in many respects as an aberrant camerate, which presumably separated early from the ancestral diplobathrid stock. As to *Spyridiocrinus,* whether it is related to *Cleiocrinus* or not remains conjectural. If its basals, like those of *Cleiocrinus,* alternate with the radials in a circlet of ten plates, they rest on the distal edge of the infrabasals but do not overlap them, and with them and other calycal plates participate in the formation of a deep basal concavity which, as generally in crinoids, results from an invagination of the thecal wall. Moreover, contrary to the preceding genus, *Spyridiocrinus* has well-developed interbrachials and no anal plates. These and other differences, as well as the chronological gap (Middle Ordovician-Lower Devonian) that separates these two forms make a direct phylogenetic connection between them very doubtful. A rhodocrinitid ancestry is postulated for *Spyridiocrinus* by Brower (1975), who considers the zygodiplobathrids to be probably polyphyletic, and proposes to drop the suborder Zygodiplobathrina and to group the two genera *Cleiocrinus* and *Spyridiocrinus* within the Eudiplobathrina along with the most closely related families.

The Dimerocrinitacea originated probably from one or several archaeocrinid ancestors through upward displacement of the first interprimibrachs, so that the radials were brought into lateral contact except at the posterior side, where a single anal plate (primanal) remained. This line of ancestry is strongly suggested by: a) specimens of several species (e.g., *Atactocrinus wilmingtonensis, Dimerocrinites icosidactylus, Griphocrinus nodulosus, Lyriocrinus melissa, L. dactylus*) in which the proximal interprimibrachs separate the radials and rest on basals as in archaeocrinids, or occur above the radials in lateral contact as in dimerocrinitids; b) the similarities that exist between an archaeocrinid such as *Rhaphanocrinus* and the dimerocrinitid *Ptychocrinus.* Both genera have the same type of theca, except for the presence or absence of interbrachials between the radials, and very similar arms. The primitive dimerocrinitids had two or four unbranched uniserial arms per ray, as does *Rhaphanocrinus.* They were followed by forms provided with more or less numerous arm branches and biserial arms, although the uniserial arrangement of brachials persisted in at least one genus *(Macarocrinus)* until Devonian time. Other evolutionary trends in dimerocrinitids include formation of arm trunks in *Lampterocrinus,* development of tegminal ridges forming grooves for protection of arms in *Gazacrinus,* and allometric growth of base relative to height of calyx in *Orthocrinus*—three genera widely divergent from most other dimerocrinitids.

The Silurian genus *Nyctocrinus* was interpreted, probably correctly, by Moore & Laudon (1943a), as representing the most advanced evolutionary stage of the dicyclic camerate crinoids, inasmuch as the fixed brachials, interbrachials, and anals have been eliminated from the cup, thus establishing perfect pentamerous symmetry in this part of the theca. Unfortunately, the ancestors of *Nyctocrinus,* which may be regarded as a dicyclic homeomorph of *Platycrinites,* are unknown.

MONOBATHRIDA

Whether the monobathrid camerates derived from the diplobathrid camerates is uncertain. Both groups coexisted in the Ordovician, and no intermediate forms between them have ever been recorded. The earliest known (presumably) monobathrid species is *Proexenocrinus inyoensis* from the Lower Ordovician of California. It is a small crinoid with ten uniserial, unbranched free arms, few relatively large and regular interbrachials, and a well-defined median series of anal plates. Its systematic position remains somewhat obscure, because the structure of its depressed base has not been elucidated. Nevertheless, comparison of this species with later monobathrid camerates points to the probable line of development of these crinoids.

The monobathrid camerates comprise a large and diversified group which, in this *Treatise,* is divided into 28 families (against 13 for the diplobathrid camerates). They include two main stocks characterized by persistent structures of the calyx, namely, the occurrence of the primanal in the radial circlet in the Compsocrinina and its absence from the radial circlet in the Glyptocrinina.

COMPSOCRININA

So far as we know, all Ordovician, and probably primitive, members of the compsocrinine stock had a quadripartite base of hexagonal outline. [*Tanaocrinus,* erroneously reported as having five basals, has only four such plates; it is here considered to be a subjective synonym of *Canistrocrinus.*] This type of base was maintained in the Upper Silurian *Abacocrinus,* but in most other compsocrinine camerates it was replaced by an equally or subequally tripartite type that persisted throughout the history of the group, except in some terminal members, which evolved an equally bipartite type of base.

The Upper Ordovician genera *Canistrocrinus* and *Compsocrinus* stand closest to *Proexenocrinus,* inasmuch as they have retained its (probably primitive) kind of interbrachial structure. At the same time, *Xenocrinus* developed very small and irregular interbrachials and a complex mode of ontogeny of these plates. Like its homeomorph dicyclic *Reteocrinus,* this genus is therefore considered as representing a specialized offshoot, which apparently died out at the end of the Ordovician period.

Proexenocrinus had only two arms in each ray. This arm number was maintained in some species of *Xenocrinus,* but other xenocrinid species and most tanaocrinids acquired three or, more commonly, four arms per ray, presumably through conversion of the first pinnule (prominent in *Proexenocrinus*) of each arm into a pinnulate arm (Brower, 1974a). The free arms of *Proexenocrinus* were unbranched, as were also those of *Xenocrinus* and *Compsocrinus.* Those of *Canistrocrinus* became isotomously forking, but only once or twice, and at long intervals. The same type of arm branching occurs in the Upper Silurian *Abacocrinus*—possibly a descendant of *Canistrocrinus.*

During the Silurian period, three large branches of compsocrinine camerates appeared, namely the Periechocrinacea, Carpocrinacea, and Hexacrinitacea. Their precise origin and interrelations are not known, but they are supposed to be derived from tanaocrinid ancestors.

The Periechocrinacea contain the families Periechocrinidae, Paragaricocrinidae, Amphoracrinidae, and Actinocrinitidae. All these crinoids are characterized by having an equally tripartite base, a generally large number of fixed brachials, large and regular interbrachials, a first primibrach with hexagonal outline becoming quadrangular only in advanced members, and branching free arms, commonly. The Upper Silurian *Periechocrinus* is the earliest known representative of this group. Its primitiveness is indicated by its high conical calyx, its upflaring basal plates, its numerous fixed brachials, interbrachials and anal plates, its interbrachials connected with interambulacrals, its median ray and anal ridges, its many-plated tegmen, and its relatively small number of free arms per ray. From this or a similar form, various genera, some of them highly specialized (e.g., *Gennaeocrinus, Megistocrinus, Thamnocrinus*), evolved within the Periechocrinidae during the Devonian.

In Early Carboniferous time, or probably somewhat earlier, the Periechocrinidae gave rise to the Actinocrinitidae and to the closely related Amphoracrinidae, which both (mainly the former) flourished abundantly in the Mississippian. The less advanced actinocrinitids (e.g., *Maligneocrinus*) closely resemble periechocrinids. The principal difference between them lies in the number of plates in the second range of the *CD* interray: three in the periechocrinids, two in the actinocrinitids and amphoracrinids. However, many actinocrinitid specimens still have the primanal followed directly by three plates. Another difference appears in the number of free arms. The periechocrinids usually have two to four free arms per ray (exceptionally, as in *Gennaeocrinus,* six or eight), whereas six or more free arms in the ray are common among actinocrinitids, reaching a maximum of approximately 16 per ray in *Teleiocrinus* and 30 per ray in *Strotocrinus.* In the two latter genera, incorporation of many brachitaxes (up to the twelfth order in *Strotocrinus*) resulted in the formation of a prominent calycal flange from the edge of which the unbranched free arms arose in a belt around the theca. Such genera are among the most specialized camerate crinoids.

Another, rather late, offshoot probably derived from the periechocrinids is the Paragaricocrinidae, a small group restricted to Pennsylvanian and Lower Permian deposits. These crinoids are very advanced camerates, as indicated by the low, rounded shape of their theca, the occurrence of a basal concavity commonly including the radials, the quadrangular outline of their first primibrachs, and the large size and small number of their interray plates, which are separated from the tegmen by the arm bases except generally on the posterior side.

The second branch that possibly originated from tanaocrinid ancestors comprises the superfamily Carpocrinacea, which includes the Carpocrinidae, Batocrinidae, and Coelocrinidae. Like the Periechocrinacea they have three, generally equal, basals, but their first primibrach is typically quadrangular instead of hexagonal in outline, their interbrachials are generally fewer, larger and more regular, and they have a lesser number of fixed brachials, and ordinarily fewer free arms, and those mostly unbranched. These features and their early appearance in the Silurian indicate that the Carpocrinacea are more progressive than the Periechocrinacea. They coexisted with the latter, but disappeared earlier, at the end of the Mississippian period, during which they showed an extraordinary development followed by a rapid decline.

Among the Carpocrinacea, the Carpocrinidae are believed to be the ancestral family. They are practically restricted to rocks of Silurian age, while the Coelocrinidae, which are not known before Middle Devonian time, are found mainly, and the Batocrinidae exclusively, in Mississippian deposits. On the other hand, the Carpocrinidae contain genera which seem well fitted to represent the source from which the Coelocrinidae could be derived. *Desmidocrinus,* for instance, combines primitive features, such as uniserial free arms, interbrachials connected with interambulacrals, a median series of extra plates in the *CD* interray, and a variable number of free arms, with typical carpocrinacean characters. Between *Desmidocrinus* and the oldest known coelocrinid *Aorocrinus* there is not a wide morphological gap. In its turn, *Aorocrinus* does not seem to be very far from the less advanced batocrinids. Those differ essentially from the coelocrinids in having an anal tube, the interrays commonly separated from the tegmen by fixed brachials, and no median series of anal plates; in addition, they are generally provided with cup-pinnule openings, which are lacking in coelocrinid genera. But all batocrinids do not have all these distinguishing features. For instance, in *Uperocrinus,* which is classified among the Batocrinidae because it has an anal tube, the interbrachials are connected with the interambulacrals and there are no cup-pinnule openings. Such a form constitutes a morphological transition between the two families and suggests that the Batocrinidae originated from the Coelocrinidae.

The third branch with a possible tanaocrinid ancestry is composed of the Hexacrinitidae, Parahexacrinidae, Dichocrinidae, and Acrocrinidae, which together form the superfamily Hexacrinitacea. As here interpreted, these crinoids are the most advanced and specialized compsocrinine camerates. Known from Upper Silurian to Upper Permian, they are characterized by a cup that typically comprises only two circlets of plates: a basal circlet and a radial circlet, the latter including a primanal in line with the radials. This simple structure may not be considered primitive, but probably results from elimination from the cup of all plates above the radial circlet and derives from a more complex stage. Like the Periechocrinacea and Carpocrinacea, the Hexacrinitidae have an equally tripartite hexagonal base, and the primanal located between the *C* and *D* radials. They seem to be nearer the Carpocrinacea than to the Periechocrinacea, for the former have fewer fixed brachials and interbrachials, and generally larger radials than the latter. A form like *Prohexacrinus,* considered by Yakovlev (1946) as a primitive hexacrinitid, might well be regarded as a carpocrinid insofar as concerns its cup (the only part of this crinoid known). This concept of a carpocrinid origin of the Hexacrinitidae is opposed by those who think that these crinoids are closely related to the Platycrinitidae or their hypothetical ancestors (Bather, 1900a; Wilson, 1916; Strimple, 1969; Strimple & Watkins, 1969).

Several hexacrinitacean camerates (e.g., some *Hexacrinites, Talarocrinus, Pterotocrinus*) show a marked tendency toward reduction of the primibrachs: they have only one such plate left in each ray, and this plate, which is axillary, is commonly not visible and may even be lacking, so that the secundibrachs become supported partly or completely by the radials. One would expect a similar origin for the multiple-arm bearing radials of Parahexacrinidae. However, the known stratigraphic succession of the three genera now included in this family is not consistent with this theory. According to Shevchenko (1967), it is in the oldest genus that the free arms are directly borne by the radials, whereas they are carried by short, cuneate, irregular proximal brachials in the youngest.

The Dichocrinidae are commonly regarded as directly related to the Hexacrinitidae, from which they differ mainly in the structure of their tegmen and in having two, instead of three, equal basals, but in fact their origin is unknown. This prolific family, restricted to Upper Paleozoic deposits (Lower Mississippian-Upper Permian), contains genera which, like *Camptocrinus* or *Pterotocrinus,* are among the most unusual camerates. In these, as well as in other members, of this lineage, the tegmen, arms, or column were modified, in some cases very markedly, but the cup showed no significant change. The Acrocrinidae, which are distinguished from all other crinoids by the presence of supplemental calyx plates (intercalaries) between the basal and radial circlets, exhibited an explosive evolution in Late Mississippian, followed by a decline in the Pennsylvanian. Although they share possession of an equally bipartite base with the Dichocrinidae and are generally regarded as a specialized offshoot from *Dichocrinus,* their precise origin remains obscure (Moore & Strimple, 1969).

GLYPTOCRININA

The Glyptocrinina are the other main stock of monobathrid camerates. Their earliest known representatives, the Glyptocrinidae, occur in Middle Ordovician deposits. Their great antiquity combined with primitiveness of features designates them as the main source of the monocyclic camerates which, like them, have no anal plate in the radial circlet and, consequently, have a pentagonal base. Whether this type of base is primitive or not is a much debated but still unsolved problem (see discussion in section on skeletal morphology of crinoids, p. *T*102). In fact, no transition form between compsocrinine and glyptocrinine camerates has ever been recorded. The origin of both groups is unknown.

The Glyptocrinidae are regarded as primitive Glyptocrinina because they have a conical calyx with five upflaring basals, prominent ray and anal ridges, numerous fixed brachials, medium-sized and regular interbrachials connected with interambulacrals, a many-plated tegmen, and few arms (2 to 4 per ray), which are unbranched or fork only once or twice, and are ordinarily uniserial.

A first branch probably given off from the Glyptocrinidae is represented by the superfamily Melocrinitacea, which includes the Scyphocrinitidae, Paramelocrinidae, and

Melocrinitidae. Those camerates have retained the primitive glyptocrinid type of calyx, except that the number of basals was reduced from five to four. Also, in the Scyphocrinitidae, loose articulations evolved between cup plates, and very many brachials and pinnulars were incorporated in the calyx, separating the interbrachial areas from the tegmen. But the major evolution concentrated on the free arms, increasing the food gathering system considerably. This was accomplished by a) transition from the uniserial to biserial arrangement of brachials, b) by branching of the arms: those of the Scyphocrinitidae and Paramelocrinidae divided isotomously several times, whereas those of the Melocrinitidae were transformed into powerful arm trunks bearing a large number of pinnulate ramuli.

The principal steps in the evolution of these arm trunks, as elucidated by Jaekel (1895, 1902, 1918), Olsson (1912), Kirk (1929c), Ubaghs (1958b), and Brower (1973), include: 1) development of four arms per ray from two arms per ray in the Upper Ordovician—a change already accomplished at the glyptocrinid level, probably by converting the first pinnule of each arm into a pinnulate arm; 2) exotomous branching of the inner arms of each ray, followed or accompanied by loss of the pinnules proximal to the arm branches, in the Silurian; 3) fusion together of the inner arms, starting proximally and extending progressively distalward, thus forming arm trunks, in the Upper Silurian and Lower Devonian; 4) disappearance of the unbranched outer arms, in the Middle Devonian; 5) division of the arm trunks themselves into ramule-bearing branches, in the Upper Devonian. Worth mentioning also about the Melocrinitacea is the unique type of root that evolved in the genus *Scyphocrinites:* a large, hollow, spheroid body that probably served as a float.

A second main section presumed to be derived from glyptocrinid ancestors consists of the families Clonocrinidae, Eucalyptocrinitidae, Dolatocrinidae, and Polypeltidae. For convenience they are placed in the superfamily Eucalyptocrinitacea, although their origin and interrelations are admittedly obscure. They have few distinctive features in common. In most of them, the calyx is low and wide, and the tegmen is stoutly plated. The base comprises four (or three) unequal basals, which may be fused together. The posterior side of the cup is barely, if at all, differentiated from the other interrays. The free arms are typically biserial. But all these characters could have evolved independently, and they do not serve to designate any precise forerunner. The Eucalyptocrinitidae and Clonocrinidae are probably related, for they have four basals, a relatively small number of fixed brachials and interbrachials, and ordinarily few free arms per ray. Different in appearance are the Dolatocrinidae, which have three, commonly fused, basals and comprise a well-diversified, mainly North American, group of Lower and Middle Devonian species. Still more distinct are the Polypeltidae, with their huge theca generally deeply modified proximally and including an extremely large number (up to about 700 in *Himerocrinus*) of fixed brachials. Their relatively short free arms may be as few as two per ray *(Hadrocrinus)* but as many as 16 or 17 per ray *(Himerocrinus)*. This type of theca is certainly not primitive, but rather suggests the existence of a peculiar tendency to incorporate a very large number of ray plates into the calyx.

The superfamily Patelliocrinacea, which includes the Stelidiocrinidae and Patelliocrinidae, constitutes a third main branch from the glyptocrinid stock. Compared to their presumed Middle Ordovician ancestors, they appear to be very progressive, and, with their few fixed brachials and interbrachials, morphologically intermediate between the Glyptocrinidae and Platycrinitidae. Although the Stelidiocrinidae retained the primitive number of five basals, in most other respects (particularly in the stoutly plated structure of their tegmen) they are highly specialized, and probably represent an isolated offshoot of unknown origin and descent.

The ancestry and evolution of the Patelliocrinidae is better documented, thanks to the discovery of an early representative of this family in Upper Ordovician rocks, *Eopatelliocrinus,* which differs only from Middle Ordovician glyptocrinids in having three, instead of five, basals, a lesser number of fixed brachials, interbrachials and anal plates, and larger radials but relatively smaller primibrachs; in addition, its first primibrach is commonly quadrate instead of being hexagonal. On the other hand, *Eopatelliocrinus* resembles the Glyptocrinidae in its general habitus, in showing traces of median ray and anal ridges, in the structure of its many-plated tegmen, and in having few unbranched and uniserial free arms. In the evolution of the family, which became extinct in the Lower Devonian, the following trends are recognizable: 1) reduction of the number of fixed brachials and interbrachials; 2) change in shape of the first primibrach and primaxil, which by Late Silurian time became generally quadrangular and pentagonal, respectively; 3) gradual reduction of the heights of these two plates relative to calyx size; 4) progressive disparition of the median ridge and extra plates in *CD* interray; 5) formation of biserial arms (Brower, 1973).

The fourth and last branch from the glyptocrinid stock comprises the Marsupiocrinidae, Hapalocrinidae, and Platycrinitidae, which together form the superfamily Platycrinitacea. As interpreted here, they represent the end products of one of the most dominant evolutionary trends of camerates, namely the elimination of the fixed brachials, interbrachials and anal plates from the calyx, which thus tends, and eventually is, confined to the sole basal and radial circlets. They are presumed to have a common ancestor with the Patelliocrinidae or to be descended from them, and to represent a more advanced stage along a similar line of evolution. Their precise origin is unknown, and their occurrence in rocks older than Upper Silurian doubtful. The Hapalocrinidae are antecedents to the Platycrinitidae, and in some respects they seem to be more primitive. As in these two families the main characters of the cup were stabilized very early, it is in the structure of the tegmen, the arms and the column that the principal changes occurred. Whereas the hapalocrinid stem remained cylindrical, the platycrinitid column became elliptical in cross section, twisted, and provided with synarthrial articulations, except proximally. The Hapalocrinidae had two primibrachs in each ray, generally more than two secundibrachs in each half ray, and, the arm bases not being in close contact with the calyx, the arms never formed proximal trunks; on the contrary, the arms of the Platycrinitidae ordinarily comprised only one, commonly very small, primibrach followed in each half ray by two secundibrachs; those arm plates therefore tended to be in close contact with interambulacrals and commonly were covered by tegminal ambulacrals, so that the arm bases became connected. The hapalocrinid tegmen, mainly composed of five orals, did not incorporate a large number of ambulacral and interambulacral plates, nor show a loss of differentiation of its component elements. Very different was the platycrinitid tegmen, which primitively had ambulacra protected by alternating orals, with axillar ambulacrals, but exhibited a strong tendency toward incorporation of many interambulacrals, sinking of ambulacra under those plates, and loss of differentiation of its plates. As to the Marsupiocrinidae, they do not seem to have shared the same tendencies, and probably were distantly related to the hapalocrinid-platycrinitid line of evolution.

EVOLUTION OF INADUNATE CRINOIDS

By N. Gary Lane and H. L. Strimple

DISPARIDA AND HYBOCRINIDA

By N. Gary Lane

Disparid and hybocrinid monocyclic inadunates are highly diverse and common Ordovician crinoids. Of the nine superfamilies of disparids and hybocrinids recognized in this *Treatise,* six are first represented in the Ordovician. They appeared very early in the Ordovician and had undergone an extensive adaptive radiation before Silurian time. Most of these inadunates are characterized especially by having a plane of bilateral symmetry through the crown situated other than through the anterior ray and posterior interray, the bilateral plane typical of most crinoids. In Ordovician disparids these other symmetry planes are expressed especially by presence of compound radial plates in certain rays and undivided ones in other rays.

Seemingly, the most primitive and archaic disparid crinoids belong to the Eustenocrinidae of the superfamily Myelodactylacea. *Ramseyocrinus* is one of the oldest known crinoids and possesses features such as infer- and superradials in all rays and *C*-ray supporting an anal sac, but not an arm, which are judged to be very primitive. *Peniculocrinus* is only slightly advanced beyond and younger than *Ramseyocrinus.* These crinoids, which have a *C-EA* symmetry plane, provide evidence that the distinctive homocrinoidal and heterocrinoidal symmetry planes of other Ordovician disparids evolved within this stock and were not inherited from a pre-disparid ancestor. The Eustenocrinidae are thought to have evolved directly into Ordovician *Iocrinus,* which lacks compound radials in all rays, but is distinguished by the presence above the cup rim of an anibrachial in the *C*-ray. *Iocrinus,* in turn, evolved in the Silurian into the Myelodactylidae, characterized by an *Iocrinus*-like crown enclosed within a coiled bilateral stem.

The Eustenocrinidae also gave rise, by Middle Ordovician time, to two stocks of crinoids that are characterized by retention of a high, narrow primitive cup and advanced, complexly heterotomous arms. The Heterocrinacea, exclusively Ordovician, with two compound radials, and the Homocrinacea, dominantly Ordovician, with three, record experimentation with various patterns of arm-branching, differing from each other mainly in presence or absence of compound radials in the *B*-ray. The heterocrinoids died out by the end of Ordovician time without known descendants in the Silurian, but, before they became extinct, they may have given rise to one small superfamily of Ordovician disparids, the Anomalocrinacea. These crinoids display a heterocrinoid symmetry plane and heterotomous arms, but have a globose cup and other features that may point to an origin independent of the Heterocrinacea.

The Homocrinacea surely were the ancestral source for the unique Calceocrinacea, the "bent-crown" crinoids which appear to have lived with the stem prostrate and the crown bent upward just above the bottom. These disparids first appeared in the Middle Ordovician and were reasonably successful Paleozoic crinoids, achieving maximum generic diversity in the Silurian and persisting into the Early Permian. The calceocrinids have a homocrinoidal symmetry plane *(E-BC)* and heterotomous arms that point clearly to an origin within the Homocrinacea.

The superfamilies discussed above constitute the bulk of known Ordovician disparids and groups remaining to be treated below, including the majority of post-Ordovician disparids. Three distinct, and presumably separate, evolutionary trends are evident among these younger disparids. One trend was toward fusion of all infer- and superradial plates to produce a cup with five equal, large radials. This was accompanied, in some, by evolution of five simple atomous arms, one to a ray. This trend is seen especially well-developed in the Belemnocrinacea with a crinoidal *(A-CD)* symmetry plane, which had achieved this much simplified crown by Silurian time with appearance of the Pygmaeocrinidae, Zophocrinidae and oldest representatives of the Synbathocrinidae.

A second trend was development of a small bowl-shaped cup in which compound radials were retained in one or more rays. Conspicuous differences in size of radials became evident, especially by hypertrophy of the *A* and *D* radials. The arms tended to be simple and atomous. This trend is present in the Pisocrinacea, Allagecrinacea, and the Perissocrinidae and Holynocrinidae of the Belemnocrinacea, and is especially well-represented among Silurian and Devonian disparids.

The third trend was for multiple simple atomous arms to be developed on broad radial facets, an adaptive alternative to branched arms in effecting an increased ambulacral surface. This innovation appeared first in the Middle Devonian Anamesocrinidae and is extensively developed in the dominantly late Paleozoic Catillocrinidae, and Allagecrinidae, all families of the Allagecrinacea.

These three trends point up one of the most conspicuous differences between Ordovician and post-Ordovician disparids in that the great majority of the former have heterotomous arms and most of the latter have simple atomous arms. Although evolutionary simplification of the arms may have occurred once, or several times, it seems unlikely that any of the Ordovician disparids with elaborate arm-branching led directly to younger disparids with atomous arms. All post-Ordovician disparids have either the homocrinoidal *(E-BC),* eustenocrinoidal *(C-EA),* or crinoidal *(A-CD)* symmetry plane. Therefore, a likely ancestor for at least some of these younger crinoids is an Ordovician form that had a rather generalized cup exhibiting homocrinoidal symmetry and simple isotomous arms that might have evolved directly into atomous arms. Based on these qualifications, the most likely ancestor for many of the post-Ordovician disparids is *Tunguskocrinus,* known from the Ordovician of Russia. Specimens of *Tunguskocrinus* have homocrinoidal bilateral symmetry, and simple isotomous arms that could readily have evolved into the atomous arms typical of so many younger disparids. The steeply conical cup of this genus is closely similar in construction to that of the oldest known synbathocrinid, *Abyssocrinus,* which has a crinoidal *(A-CD)* bilateral symmetry. It seems likely that *Tunguskocrinus* stands closest morphologically to the ancestral type for many of the Belemnocrinacea.

The Pisocrinacea, which display a homocrinoidal *(E-BC)* type of symmetry, appeared abruptly in Upper Silurian rocks and their origin is obscure. The only Ordovician disparids with a bowl-shaped cup are the Anomalocrinacea, which can be ruled out as precursors of the pisocrinoids because they possess the wrong kind *(D-AB)* of bilateral symmetry, have specialized heterotomous arms and other features which indicate that they are a specialized sterile offshoot of the Heterocrinidae.

The origin of the multi-armed radial plates of the Allagecrinacea is also unknown. The Anamesocrinidae are known only from the Middle Devonian and are contemporaneous with oldest known Catillocrinidae. The development of multiple atomous arms was not confined to this superfamily, but also occurred in very different fashion, and clearly polphyletically, in some genera of the Pisocrinacea.

The prevailingly globose form of the theca observed in the Hybocrinida differs from that of disparid superfamilies, as does the tendency toward suppression of arms accompanied by placement of ambulacra on the surface of cup plates. Also, the Hybocrinida are unlike Disparida in having a radianal plate directly or obliquely beneath the *C* radial. The hybocrinid *Baerocrinus* and disparid *Ramseyocrinus* are the only monocyclic inadunate genera recorded from lower Ordovician rocks. Beyond doubt, one type was not derived from the other; rather both descended from unknown pre-Ordovician ancestors.

CLADIDA

CYATHOCRININA AND DENDROCRININA

By N. Gary Lane

[University of Indiana]

The two most primitive groups of cladid, or dicyclic, inadunates are the suborders Cyathocrinina and Dendrocrinina. The latter is first represented by specimens in the Lower, the former in the Middle Ordovician. The Cyathocrinina are dominantly a lower and middle Paleozoic group of crinoids, being represented in the late Paleozoic mainly by a group of microcrinoids. The Dendrocrinina, on the other hand, gave rise in the Middle Ordovician to the subclass Flexibilia, and in the Devonian, evolved into the suborder Poteriocrinina, the most successful and diverse group of late Paleozoic crinoids. The nature of Ordovician genera of the cyathocrinids and dendrocrinids is so disparate, that it is virtually impossible to postulate a common ancestor for these two early groups of dicyclic inadunates, and it seems quite likely, therefore, that the Cladida arose polyphyletically from two or more pre-Ordovician crinoids or other pelmatozoans.

TABLE 3. *Comparison of Morphological Features of Ordovician Genera of Cyathocrinina and Dendrocrinina.*

CYATHOCRININA	DENDROCRININA
Cup large, bowl- or egg-shaped	Cup small, low, or conical
Infrabasals large, conspicuous	Infrabasals generally small and low
Tegmen low or with a small, short sac	Tegmen produced into a conspicuous anal sac
Goniospires or trace of them present	Goniospires absent
Posterior oral a madreporite	Posterior oral not known
Arms narrow, rounded, less than one-half width of radials	Arms mostly wide, one-half to two-thirds width of radials
Arms branch isotomously once or a few times	Arms richly branching, isotomous or heterotomous

In the Ordovician, the cyathocrinids are represented by six genera, which are placed in five families assigned to two superfamilies in the *Treatise* classification. The contemporaneous dendrocrinids include 11 genera assigned to ten families in three superfamilies. Consequently, even though the number of genera and specimens known is small, these crinoids do show a wide range of morphological features which point to an extensive evolutionary history that is as yet unknown.

A summary of conspicuous differences between Ordovician cyathocrinids and dendrocrinids is provided in Table 3. Comparison of the lists points up the difficulty of postulating a common origin for the two suborders. The presence of goniospire slits or traces of these structures in several cyathocrinid genera may well represent a hold-over from some kind of slit-bearing noncrinoid pelmatozoan such as the cystoids. If this should prove to be true, then the major adaptive innovation of the cyathocrinids was the evolution of true crinoid arms from brachioles. The inflated egg-shaped theca of several Ordovician cyathocrinids superficially resembles the theca of the monocyclic Hybocrinida inadunates, but this similarity is judged here to be convergent and not indicative of phyletic relationship. The quite large, prominent infrabasals of early cyathocrinids seemingly precludes any close relationship with monocyclic inadunates, the disparids or hybocrinids.

The Dendrocrinina, on the other hand, exhibit several features that point toward a much closer relationship between these dicyclic crinoids and monocyclic disparids. Early dendrocrinids tend to have quite small, low infrabasals that could have disappeared in pre-Middle Ordovician time, resulting in monocyclic inadunates, or conversely, the small infrabasals could have been a new feature, evolved in dendrocrinids from a disparid ancestor. Both the dendrocrinids and disparids have infer- and superradial plates in most primitive, and oldest, representatives. Both groups include types with a small, high, conical cup, have arms that are densely branched, and a conspicuous anal sac. It seems likely, therefore, that disparids and dendrocrinids may have had a common ancestor, or one group may have evolved from the other. Only discovery of new Early Ordovician, or better, Cambrian, crinoids will resolve these phyletic problems.

Ordovician genera of the Dendrocrinina can be arranged into four morphological grades that may approximate the principal early evolutionary pathways within this group. At the base of the series stands *Aethocrinus,* the oldest dicyclic inadunate known, which possesses several very primitive features, especially the arrangement of cup or calyx plates. The next grade contains *Ottawacrinus,* which has five compound radials in the cup, that of the *C*-ray consisting of a superradial classed as a radial and an inferradial identified as the radianal plate. The third stage, as represented by *Cupulocrinus* and *Dendrocrinus,* consists of crinoids that retain an inferradial, identified as a radianal, in the *C* ray only. A fourth grade is defined by *Merocrinus* in which the *C* radial is an undivided plate, exactly like the four other radials, but unlike them is followed above the cup rim by an axillary anibrachial, which on its left articular facet supports an anal *X* plate followed by other plates of the anal sac and on its right facet a series of normal primibrachs.

Four nominal Ordovician genera assigned to the Dendrocrinina are omitted from this ranking because they are poorly known. All were described by JAEKEL (1918) and may be very important crinoids for understanding the early history of this group of crinoids. JAEKEL's descriptions are so incomplete and his illustrations so unreliable, as recently demonstrated by UBAGHS in redescription of JAEKEL's 1918 genus *Perittocrinus,* that it would be foolhardy to discuss these four genera *(Esthonocrinus, Pandoracrinus, Polycrinus, Metabolocrinus)* as if we really understood their significance.

The Dendrocrinina constitute a reasonably homogeneous group of crinoids displaying few striking modifications of the basic morphological types established in the Ordovician. The main interest in post-Ordovician dendrocrinoids concerns their phylogenetic relationship to the advanced dicyclic cladids assigned to the Poteriocrinina. The two advanced groups of dendrocrinids are the Botryocrinidae, a family of the Dendrocrinacea, and the Mastigocrinidae of the Mastigocrinacea. Either, or both, of these families may have served as the phyletic source of the Poteriocrinina, which assuredly evolved from the Dendrocrinina during Devonian time. The two key characters used for judging phyletic derivation of the poteriocrinids are the appearance of muscular articulation between radials and first primibrachs, and the development of pinnules from nonpinnulate heterotomous arms with closely spaced ramules. Two other features of secondary value are relative width of the radial articular facets, generally narrow in more primitive, and wide in advanced forms, and the number of anal plates in the cup. Width of radial facets as a criterion for discriminating early Poteriocrinina is a feature of dubious value. Many earliest known pinnulate cladids have relatively narrower radial facets than even very generalized Ordovician dendrocrinids, like *Cupulocrinus,* for instance. Advanced genera of the Mastigocrinidae have wider radial facets than do many primitive poteriocrinids.

The evolution of pinnules during the transition from dendrocrinids to poteriocrinids was taking place on a broad front within more than one lineage, resulting in a mosaic of crinoids that independently achieved or closely approximated a pinnulate condition. Advanced genera of both the Botryocrinidae and Mastigocrinidae exhibit a sequence of arm types in which two main isotomous branches occur above the primaxils. In less advanced genera, ramules are developed at irregular intervals on alternate sides of brachials, spaced from four to nine brachials apart, as in *Lasiocrinus* and *Cradeocrinus.* In more advanced genera, the ramules are attached to alternate sides of every second brachial, as in Devonian *Pagecrinus* and *Iteacrinus.* Only one further step is necessary, having alternate ramules on alternate sides of each successive brachial in a series, to achieve a truly pinnulate condition. It is quite clear that true pinnules did evolve within crinoids classed as members of the Dendrocrinina. Both *Imitatocrinus* and *Dictenocrinus,* of Devonian age, have pinnulate arms, each brachial having a side branch. These two genera are placed within the Mastigocrinidae and neither could have been the direct ancestor of the Poteriocrinina, because they possess only two anal plates in the cup, like most other genera of this family with less advanced arm structure. One Silurian dendrocrinid, *Cyliocrinus,* records a precocious, independent origin of pinnules in which proximal brachials have pinnules on alternate sides of each brachial, and distal brachs each bear a pinnule on both sides of each plate, a hyperpinnulate condition that did not evolve in the Poteriocrinina until Mississippian time.

The pinnulate dendrocrinids discussed above cannot reasonably be accorded importance as potential ancestors of the Poteriocrinina, but one genus, *Quantoxocrinus,* stands as close to such an ancestral type as is now known. This crinoid has well-developed pinnules, and three anal plates in the cup, but is placed in the Dendrocrinina because its anal sac is closely comparable to that of other genera grouped in the Mastigocrinidae.

The relationship between these two suborders is still more complicated by the fact that two genera of Devonian poteriocrinids, *Rhenocrinus* and *Charientocrinus* of the Rhenocrinidae, are not pinnulate but have ramules on every second brach above the primaxils. Consequently, the arms of these genera are less advanced than those of some dendrocrinids. They are placed in the Poteriocrinina because three anal plates occur in the cup and because features of the sac and configuration of the cup ally them with several other Devonian genera, all of which have pinnules and otherwise qualify as belonging in the Poteriocrinina.

The pinnulate condition in dicyclic inadunates clearly evolved gradually and separately in more than one distinctive stock of advanced dendrocrinids and primitive poteriocrinids and cannot be utilized as an all-or-nothing "key" character to differentiate the two suborders.

A second feature used to distinguish dendrocrinids from poteriocrinids is the presence or absence of muscular articulation in the arms, especially at the base of the arm between the radials and first primibrachs. The condition of muscular articulation is recognized traditionally in fossil skeletal morphology of crinoids by presence of a transverse articular ridge on the facet. Adjacent to the ridge externally is a pit that housed ligaments and two or more internal fossae that contained ligaments and muscles. Strictly ligamentary articulation is judged to have been present on radial and other facets that lack a prominent ridge and associated fossae, although other less conspicuous structures may be present (VAN SANT, 1964). The radial articular facets of most primitive poteriocrinids and advanced dendrocrinids are not known in sufficient detail to allow direct observations concerning morphological steps leading to this important advance. Not all of the crinoids now placed in the Poteriocrinina had muscular articulations, for genera such as *Poteriocrinites* and *Springericrinus* lack necessary fossae, and none of the Dendrocrinina attain this condition. Consequently, muscular arm articulations evolved within the Poteriocrinina and appeared somewhat later in the phyletic history of inadunate crinoids than did the pinnulate condition, which is present in both suborders. The time-sequential relationship between these adaptations may provide a clue to the evolutionary pathway by which muscular articulations evolved in the Poteriocrinina.

If one examines the articular facets between brachials and first pinnular plates in many crinoids that possess pinnules, whether camerate, inadunate, or articulate, these facets have a closely similar construction. There is a distinct, short transverse ridge with rather large fossae on either side. In living crinoids the external fossa houses ligaments and the internal one muscle fibers, acting in opposition to each other. Contraction of the muscle pulls the pinnule up over the arm into a closed position, and relaxation of the muscle allows the ligament to contract, extending the pinnule into a feeding posture. Because of the close skeletal similarity of these facets in living crinoids and fossils it seems reasonable to postulate that many camerates and advanced inadunates, all of which have pinnules, possessed muscle fibers that helped to move the pinnules. With very rare exceptions *(Planacrocrinus),* camerates never evolved beyond this stage to one wherein muscles developed between brachials or between radials and arms.

Ramules of dendrocrinids generally have smooth or only faintly sculptured articular surfaces between plates and presumably were united to brachials by ligaments. Pinnular facets of advanced poteriocrinids clearly had both muscles and ligaments. The evolution of ligament-muscle from strictly ligamentary articulations is therefore thought to have occurred in cladid inadunates contemporaneously with or shortly after attainment of the pinnulate condition. The presence of muscle-activated pinnules was a necessary pre-condition for the evolution of muscular articulations in the main part of the arms. Instead of stopping at the point reached by camerates early in their development, with muscles confined to helping move pinnules, the cladids went a step further and sites of muscle insertion spread from pinnular facets to the main part of arms, first between radials and arms, then on either side of axillary plates, and later, in articulates, between most of the brachials. If this interpretation of evolution is correct, the development of pinnules or "near-pinnules" was a necessary first step toward incorporation of muscle fibers into the main part of the arms. The sequential appearance of, first, pinnules, and then muscular articulation on the radial facets, in the phyletic transition from dendrocrinids to poteriocrinids is thus explained.

EVOLUTIONARY TRENDS AMONG POTERIOCRININA

By H. L. Strimple

The general trend in evolution of the Poteriocrinina is toward increase in overall size and moderate thickening of cup plates. As an example, the cup of *Ulocrinus elongatus* Strimple (1961d), Middle Pennsylvanian, is known to have attained a height of 39.5 mm. and width of 44.7 mm., as compared to the similarly shaped *Glossocrinus naplesensis* Goldring (1923), Late Devonian, with a height of 3.5 mm. and width of 6.0 mm., an increase in size of about 85 percent.

Evolution of the primitive cup outline (high conical) may have proceeded along different paths such as 1) retention of primitive straight-sided form modified by reduction in height, 2) longitudinal curvature of sides with constriction of the summit producing globose shapes, and 3) broadening of the cup base accompanied by development of a shallow to deep basal concavity.

The mechanics for changing cup shape appear to be fairly simple: 1) infrabasals developed a flattened surface about the stem attachment with only their distal tips flexed upward, producing narrow, flat-based cups as in *Decadocrinus* Wachsmuth & Springer (1880), Middle Devonian-Lower Mississippian; 2) subhorizontal distal tips of infrabasals joined upwardly curved proximal ends of basals providing a broad flat base as in *Laudonocrinus* Moore & Plummer (1940), Middle Pennsylvanian-Upper Pennsylvanian, 3) horizontal or downflared infrabasals were surrounded by downflared proximal parts of basals or entire basals, which was characteristic in many genera of Mississippian to Permian age.

Although most Poteriocrinina became stabilized with a moderately deep, broad, bowl-shaped cup by Chesterian or Pennsylvanian time, some continued to lower cup height by extending proximal ends of the radials into the basal plane and eventually most of these plates became horizontal so that the cup became little more than a platform beneath the visceral mass. Examples of the latter condition are *Galateacrinus* Moore (1940a), *Oklahomacrinus* Moore (1939), and *Sciadiocrinus* Moore & Plummer (1938), all Lower or Middle Pennsylvanian to Upper Pennsylvanian. Among the Pirasocrinacea (Lower Mississippian to Lower Permian), loss of body space in the cup was partially compensated by extension of the steep, outward-downward, radial articular facets; however, contributing factors were probably the prominent musculature of the arm facets and large number of arms found in these genera. A similar condition is found in Chesterian species of *Zeacrinites* Troost in Hall (1858) in which the arms are numerous, the cup is very shallow, and large, outward-downward sloping, radial articular facets appear. Reduction in capacity of the cup may be compensated also by enlargement of the anal sac.

Primitive Poteriocrinina had ridges extending across sutures from plate to plate in the cup, strengthening the thin plates. Evolution normally trended toward thicker plates with reduction or disappearance of the ridges as in *Exoriocrinus* Strimple & Moore (1971a), Upper Pennsylvanian; *Elibatocrinus* Moore (1940a), Middle to Upper Pennsylvanian; and *Indocrinidae* Strimple (1961d), Lower to Upper Permian. Thin plates and surficial ridges persisted in various post-Devonian Poteriocrinina (e.g., the Stellarocrinidae Strimple, 1961, Upper Mississippian-Lower Permian), which retained the ridges or remnants of them even though plates of the cup were considerably thickened.

Primitive (normal) anal plates in the cup of Poteriocrinina numbered three, consisting of radianal placed obliquely left below the *C* radial and above the *BC* basal, followed obliquely left by anal *X* resting on the truncate distal edge of the posterior *(CD)* basal, and directly or obliquely above by a right tube plate (= *RX,* of older usage). No representative of the Poteriocrinina possessed a radianal in its most primitive position, directly beneath the *C* radial, as found in some Dendrocrinina (e.g., *Dendrocrinus* HALL, 1852, Middle Ordovician-Upper Silurian, and *Ottawacrinus* BILLINGS, 1887, Middle Ordovician).

Evolution of anal plates in cups of crinoids belonging to this suborder is readily demonstrable to have produced a step-by-step reduction in size and number until they became eliminated completely. First, the right tube plate was squeezed upward out of the cup, leaving the first two anals in their normal positions. Then the anal *X* was expelled and the radianal shifted leftward and upward so as to replace the anal *X*. Very exceptionally the radianal was resorbed (e.g., *Zeacrinites*[1] MOORE & PLUMMER, 1940, text-fig. 43a,c), leaving the anal *X* next above the posterior basal to take over as radianal. Many Pennsylvanian and Permian genera of the Poteriocrinina had only one anal plate (radianal) in their cup. In such crinoids as *Erisocrinus* MEEK & WORTHEN (1865b), Lower Pennsylvanian-Lower Permian, *Arkacrinus* KNAPP (1969), Lower Pennsylvanian, and *Paradelocrinus* MOORE & PLUMMER (1940), Lower to Upper Pennsylvanian, this anal plate generally was not visible externally, for it was reduced in size and confined to a notch on the inner edges of the posterior radial articular facets. Bilateral symmetry had nearly vanished. Finally, perfect pentameral symmetry was attained in *Encrinus,* Middle Triassic, which genus is usually considered to belong to the Articulata. Normally the arms were pinnulate and in later genera with uniserial arms the arms tended to become hyperpinnulate (more than one pinnule to each brachial). Uncommonly, syzgial pairs of brachials developed for increased flexibility of the arms, although with loss of pinnules borne by infrazygals. The latter structure is known only in the families Ampelocrinidae and Cymbiocrinidae and the genus *Araeocrinus* (Rhenocrinidae).

In some lineages the number of arms was reduced to five, which generally but not invariably became very long.

The delicate arms of primitive forms diverged upward with no apparent tendency ever to be closed. Addition of arms led to crowding so that they abutted one another when closed. Most Middle Mississippian and later genera, even those with relatively few arms, tended to close them in a compact manner and if their structure did not allow them to abut (e.g., Stellarocrinidae) spaces between proximal parts of the arms might be occupied by interlocked pinnules. Most pre-Pennsylvanian Ampelocrinidae and Cymbiocrinidae did not attain the ability to close their arms and some genera, never.

A tendency for brachials of adjacent arms to interlock when closed, particularly in proximal sections, is found in the Pirasocrinacea, many genera of the Erisocrinacea, in the Laudonocrinidae of the Lophocrinacea, and Ampelocrinidae of the Agassizocrinacea (e.g., *Arroyocrinus*).

The arms of oldest Poteriocrinina are mostly very numerous as result of isotomous branching in the middle and upper parts of the crown (e.g., Poteriocrinitidae BASSLER, 1938, L.Dev.-U.Perm.; Proctothylacocrinidae KIER, 1952, M.Dev.; Blothrocrinidae MOORE & LAUDON, 1943a, L.Miss.-L.Perm.; Bursacrinidae KIRK, 1947, L.Miss.), or endotomous heterotomous branching (e.g., Zeacrinitidae BASSLER, 1938, L.Miss.-U. Perm.). Generally, the course of evolution led to crinoids having only ten arms, as seen in the host of genera grouped in the families Scytalocrinidae MOORE & LAUDON, 1943a (M.Dev.-U.Perm.); Apographiocrinidae MOORE & LAUDON, 1943 (L.Penn.-U. Perm.); Erisocrinidae S. A. MILLER, 1890 (L.Penn.-L.Perm.); Decadocrinidae BATHER, 1890 (M.Dev.-U.Penn.); Graphiocrinidae WACHSMUTH & SPRINGER, 1886 (L.Carb.-U. Perm.); Diphuicrinidae STRIMPLE & KNAPP, 1966 (L.Penn.-M.Penn.); Paradelocrinidae KNAPP, 1969 (L.Penn.-U.Perm.); Protencrinidae KNAPP, 1969 (M.Penn.-L.Perm.); Catacrinidae KNAPP, 1969 (M.Penn.-U. Perm.); Stachyocrinidae MOORE & STRIMPLE, 1973 (U.Perm.); and others.

Some small crinoids belonging to the Poteriocrinina appear to have been very successful and thus survived through considerable lengths of time. For example, the Ampelocrinidae and Cymbiocrinidae, with long slender arms, were common throughout the Chesterian and by Pennsylvanian

[1] Misidentification by MOORE & PLUMMER.

time increased in size and general resemblance to associated forms of other families. The ampelocrinids and cymbiocrinids retained many of their unique characters, however, and some genera (*Allosocrinus, Halogetocrinus*) reduced the number of their arms to five. The Decadocrinidae remained small throughout their history but were never very abundant, the last survivors (*Ramulocrinus, Glaukosocrinus*) apparently dying out in the Middle or Late Pennsylvanian.

Most of the Phanocrinidae are small but dominant crinoids in Chesterian deposits. Their affinities with Pennsylvanian derivatives have not been determined, although nearly identical cup shape and small, tube-like anal sac of *Delocrinus* (Catacrinidae) suggest that this widely distributed, very common genus may be one of them. The ten arms of *Delocrinus* are biserial, whereas the arms of *Phanocrinus* are typically uniserial. One large species (*P. imoensis* BURDICK & STRIMPLE, 1973) from the uppermost Chesterian of northern Arkansas has distinctly biserial arms, however.

Crinoids with ten arms, of small or moderate size and with relatively small anal sacs, were very successful in Pennsylvanian time, as indicated by their abundance and variety. The most common genera are *Delocrinus* and *Erisocrinus* with biserial arms and *Apographiocrinus* with uniserial arms. *Apographiocrinus* reached its peak in the Upper Pennsylvanian (Missourian). All ranged into the Permian.

At least moderate mobility on or slightly above shallow sea floors is judged to characterize such late Paleozoic stemless genera as *Agassizocrinus, Paragassizocrinus, Cryphiocrinus, Staphylocrinus,* and *Exochocrinus.* The same habitat is thought to have characterized some stalked crinoids also (e.g., *Calceolispongia, Jimbacrinus, Utharocrinus, Metutharocrinus, Lasanocrinus*), as indicated by pronounced downward projections of their basal or radial plates or both, which served to lift the crinoids slightly above bottom sediment. The stems of these crinoids were relatively slender and probably acted as tethers or sea anchors rather than as upright stalks.

Some lineages display a tendency toward reduction in width of the *B* and *E* radials and ultimately disappearance of arms in these rays. Also, like *B* and *E,* the *C* and *D* rays are mirror-image pairs; changes in the *B* ray simultaneously or eventually appear also in the *E* ray and those in the *C* ray are duplicated in the *D* ray or vice versa. In one of the few available studies of crinoid ontogeny, STRIMPLE (1938, p. 5, 7) observed that primibrachs *1* in the *B* and *E* rays of *Graphiocrinus carbonarius* (= *Apographiocrinus typicalis*) and *Erisocrinus typus* are shorter (weaker) than in the other three rays. It thus follows that any tendency toward suppression of arms should appear first in the *B* and *E* rays, enhancing bilateral symmetry of the cup oriented in the crinoidal (*A-BC*) plane. Among Permian-age Indocrinidae suppression of one arm took place in the *B* ray of *Metaindocrinus* and in both *B* and *E* rays in *Indocrinus* and *Proindocrinus. Sundacrinus* WANNER (1916a) and *Tribrachyocrinus* M'COY (1847), of Permian age, have two armless radials located in the *B* and *E* rays. STRIMPLE (1951b, p. 200) reported a specimen of *Delocrinus* sp. of Pennsylvanian age in which the *B* radial lacks an arm. *Hosieocrinus* WRIGHT (1952, p. 137) from the Visean of Scotland is reported to have *B* and *E* radials "on which is fused a small triangular *PBr* [primibrach *1*] curved over the top of the cup," or these radials bear no arms. Thus, the evolutionary tendency discussed is recognized in widely different crinoid stocks of Poteriocrinina.

Primitive columns were quinquelobate or quinquestellate and moderately large. Evolution of them was toward pentagonal and then subpentagonal outlines, ultimately to transversely circular, accompanied by reduction in diameter to the point of complete atrophy in some genera. Large quinquestellate columnals are known as late as Morrowan (Early Pennsylvanian) in species of *Heliosocrinus* and quinquelobate stems of *Chlidonocrinus* in Missourian (Upper Pennsylvanian) rocks. Pentagonal or subpentagonal columnals are known in at least proximal portions of the stems of several genera of the Ampelocrinidae and Cymbiocrinidae as late as Virgilian (Late Pennsylvanian) and in *Hydriocrinus* from Missourian (Upper Pennsylvanian) formations. The vast majority of stems belonging to

Mississippian, Pennsylvanian, and Permian Poteriocrinina are circular in cross section.

Search for a possible Paleozoic ancestor of the Jurassic *Pentacrinites* and *Isocrinus* points to Pennsylvanian Ampelocrinidae because several of its genera possess axillary primibrachs 2, syzygial pairs of brachials, and have highly cirriferous pentagonal or quinquelobate stems. *Chlidonocrinus* from the Upper Pennsylvanian has such a stem and therefore may belong to the lineage containing the Mesozoic crinoids mentioned.

EVOLUTION OF FLEXIBLE CRINOIDS

By N. Gary Lane

The origin of flexible crinoids is known much more precisely than for any of the other three subclasses of crinoids. The only and earliest known flexible crinoid from Ordovician rocks is *Protaxocrinus,* which closely resembles associated *Cupulocrinus,* classed as a dicyclic inadunate. The principal differences between these two genera are that *Protaxocrinus* has three unequal infrabasals, rather than the five of *Cupulocrinus,* possesses a distinctive series of anal plates in the posterior interray, and displays a different number and arrangement of primibrachial plates. *Cupulocrinus* has arcuate sutures between brachials which somewhat resemble those associated with the patelloid processes of brachials in many flexible crinoids. This close relationship between flexibles and dicyclic inadunates has been accepted by crinoid workers since first pointed out by Bather in 1900. The probable evolution of *Protaxocrinus* from *Cupulocrinus* provides one of the closest evolutionary links known between major groups of crinoids.

After the Ordovician appearance of *Protaxocrinus,* a great gap exists in our knowledge of flexible crinoids until Late Silurian time. No crinoids of this type are known from Upper Ordovician rocks which yield a variety of camerate and inadunate crinoids, and information is lacking concerning this interval of geologic time when flexible crinoids must have undergone relatively rapid diversification. A single genus of flexibles *(Clidochirus)* is known from the Early Silurian and when next younger Late Silurian flexibles appeared, five families are represented. The order Sagenocrinida is dominant, represented by 46 genera, and the family Homalocrinidae is especially characteristic. Silurian representatives of the Taxocrinida, presumably derived directly from *Protaxocrinus,* include only three genera, in addition to *Protaxocrinus.* It is assumed that the Sagenocrinida evolved from the Taxocrinida, but there is little or no direct fossil evidence for this evolutionary step, which mainly involved changes in the posterior interray.

Phylogenetic relationships among the families of the Sagenocrinida are poorly known. The Icthyocrinidae are recorded in Early Silurian deposits. Five of the 12 families in this order appeared more or less simultaneously in the Late Silurian. Whether the pre-Devonian families each evolved independently from the Taxocrinida, or whether post-Silurian families radiated after the Sagenocrinida had become established, is not known.

Several broad evolutionary trends are evident within the flexible crinoids. The superfamily Lecanocrinacea and the Homalocrinidae among Sagenocrinitacea developed an arrangement of cup and lower arms that is closely similar to and essentially homeomorphic with the cup and arms of inadunate crinoids. Careful study of articular facets and individual brachials has been necessary in order to distinguish some of these flexibles from advanced poteriocrinitid inadunates. The Pennsylvanian and Permian flexible *Cibolocrinus* is a case in point as this genus was classified as an inadunate crinoid for many years before its affinity with other Flexibilia was documented. Specialized end products of flexible evolution, shortly before they became extinct near or at the end of the Permian Period, have been described from the Upper Permian of Timor. These include inadunate-like genera of the Lecanocrinacea, as well as small compact sessile forms adapted for rough-water reef habitats which resemble superficially the modern articulate *Holopus.* Several of these last-surviving flexibles have very short, unbranched arms, that are tightly folded over the top of the theca. Two unusual Permian genera have been reported that have a "bent-stem," analogous with, but not related to, the "bent-crown" of the disparid inadunate family Calceocrinidae. This adaptation permitted lowering of the crown to a position just above the sea floor and may have been an adaptation for rough-water environments. Early flexibles have simple, isotomous arms, but in various groups in both orders of flexibles, heterotomous arm branching developed at various times during the Paleozoic. This trend reached a peak with the late Paleozoic taxocrinoid *Onychocrinus,* which has closely spaced much-branched heterotomous ramules.

Other evolutionary trends include the following: 1) Change in lateral outline of the cup or calyx from steeply conical to a low bowl shape and, in some genera, development of a basal concavity. 2) Infrabasals evolved from high plates that form a conspicuous part of the lateral wall of the cup or calyx to low plates that are completely hidden beneath the proximal stem columnal. In some late Paleozoic flexibles the infrabasals may become fused into a single plate. 3) The primanal shifted from directly below the *C* radial in *Protaxocrinus* to an oblique position. The radianal, or the anal *X* plate, or both, may be eliminated from the cup. 4) Evolutionary trends affecting the development of interbrachial plates between rays are not clearly understood. Many early flexible crinoids have small areas of interbrachial plates, and Springer (1920) believed that the general trend was toward increase in size and number of plates in the interbrachial areas. However, the Sagenocrinitidae have many interbrachials between the rays when they first appear in the Late Silurian. Presumably there were trends both to increase and decrease the importance of these plates in the crown, because some late Paleozoic lecanocrinacean genera lack interbrachials altogether. These plates served to bind together in a flexible way the lower parts of the arms. Perhaps the development of interbrachials was related to the relative strength of the arms and to the strength of waves or currents in various habitats.

EVOLUTION OF ARTICULATE CRINOIDS

By H. Wienberg Rasmussen

The subclass Articulata includes all Mesozoic to Holocene crinoids except Encrinidae, which are now generally referred to Inadunata. No Paleozoic member has been recorded.

Since the time of Jaekel (1892), it has been generally thought that the Articulata evolved from the dicyclic Inadunata, and more specifically from the Poteriocrinina, by complete elimination of all anal plates from the cup, reduction in size of the five infrabasals as well as the height of tegmen, exposure of mouth and ambulacral groove on the surface of tegmen, enclosure of the axial nerves in canals penetrating basals, radials, and brachials, by formation of well-developed muscular articulations in radials and arms, and by the arms being invariably uniserial and pinnulate.

In the Poteriocrinina the infrabasals are commonly reduced in number to three, or are fused. The radial articular face has a fulcral ridge and ligament fossae, and may have muscular fossae. Anal plates are generally present in the cup, but may disappear. The tegmen generally forms a large ventral sac, but may be flattened in Triassic members. The arms are pinnulate, generally branched, and uniserial or biserial; the column may or may not bear cirri on the nodals. Several of these characters show a trend toward the Articulata.

The Encrinidae, often referred to Articulata, have five small infrabasals, a low tegmen similar to that of the Articulata with exposed mouth and ambulacral groove, and there is no anal plate in the cup. They were considered by Bather (1896b, 1897a) to be the ancestors of the Isocrinida and the Millericrinida, but the arms are biserial and brachial articulations less advanced. According to JAEKEL (1892) and HILDEBRAND (1926), the Encrinidae are closely related to *Erisocrinus* [= *Stemmatocrinus*] (Erisocrinidae), now referred to the Poteriocrinina. The Triassic Articulata, the Dadocrinidae and Holocrinidae, according to HILDEBRAND, are more similar to *Poteriocrinites* in structure of the cup than to the Encrinidae.

It has not yet been demonstrated whether the Articulata had a monophyletic or polyphyletic origin, nor which of the many families of the Poteriocrinina may have been ancestral.

The Triassic fauna of the Articulata is poor, yet it contains the first representatives of three orders important in the Mesozoic fauna (Fig. 209).

1) The Millericrinida, represented in the Triassic by the Dadocrinidae and perhaps by incompletely known specimens of the Millericrinidae, include the Articulata in which the column has simple symplectial articulations with radiating crenulae and no cirri. Nonmuscular articulations are synarthrial in Millericrinina and synostosial in the Hyocrinina. This order may be ancestral to the Cyrtocrinida with their more or less reduced column and basal circlet and generally without nonmuscular articulations in the arms, and maybe to the Bourgueticrinida, which have synarthrial articulations in the column and synostosial or synarthrial articulations in the arms.

2) The Isocrinida, represented in the Triassic by the Holocrinidae and by various incompletely known Isocrinidae, includes Articulata with a generally five-sided column with symplectial articulations, the crenellae forming a petaloid pattern, and with nodals with a verticil of five radially directed cirri. Nonmuscular brachial articulations include synarthry as well as cryptosyzgy. This order is ancestral to the large order of Comatulida, in which the larval column is discarded and proximal nodals are fused to a centrodorsal with cirri. Nonmuscular articulations include synarthry as well as typical syzygy.

It is possible that the Bourgueticrinida may have been derived by proterogenetic evolution from the Comatulida.

3) The Roveacrinida, represented in the Triassic by the subfamily Somphocrininae, include planktonic microcrinoids with a small cup of radials, more or less reduced basals, and with or without a dorsal plate or rod-shaped element, but no true column. The nonmuscular articulations are synarthrial.

The order Uintacrinida, including only the two Upper Cretaceous genera *Uintacrinus* and *Marsupites,* are large planktonic species with a cup formed by infrabasals, basals, radials and a centrale, but no trace of a column. The arms are connected by a variable number of interbrachial plates. The nonmuscular brachial articulations are syzygial. The affinity of the Uintacrinida with other orders of Articulata is unknown.

It is remarkable that, with very few exceptions, the ramification and articulations of the arms follow the same general pattern in all Articulata. Almost all Articulata have arms divided at primibrachs *2* and have nonmuscular articulations at primibrachs *1-2* and secundibrachs *1-2,* and with variable intervals more distal in the arms. The number of further divisions is variable, and the kind of nonmuscular articulations is different in different groups of Articulata.

The presence or absence of nodals with cirri, and the radiate or petaloid pattern of crenellae in symplectial articulations of the column are the two most distinct and permanent differences between the two large orders of sessile Articulata, the Millericrinida and the Isocrinida, but the origin of these two groups may not be far apart. There is a considerable resemblance between the Triassic Dadocrinidae and Holocrinidae, and nodals with cirri are found in several Poteriocrinina. Also, a petaloid pattern of crenellae may be approached in pentalobate proximal columnals of the Dadocrinidae and several Millericrinidae, and simple radiating crenulae are found in the almost noncirriferous column of the recent *Proisocrinus.* We cannot exclude, therefore, that Isocrinida may be derived from Millericrinida near the beginning of the Mesozoic, or that the two groups may have a common origin from the Poteriocrinina.

The origins of the Roveacrinida and Uin-

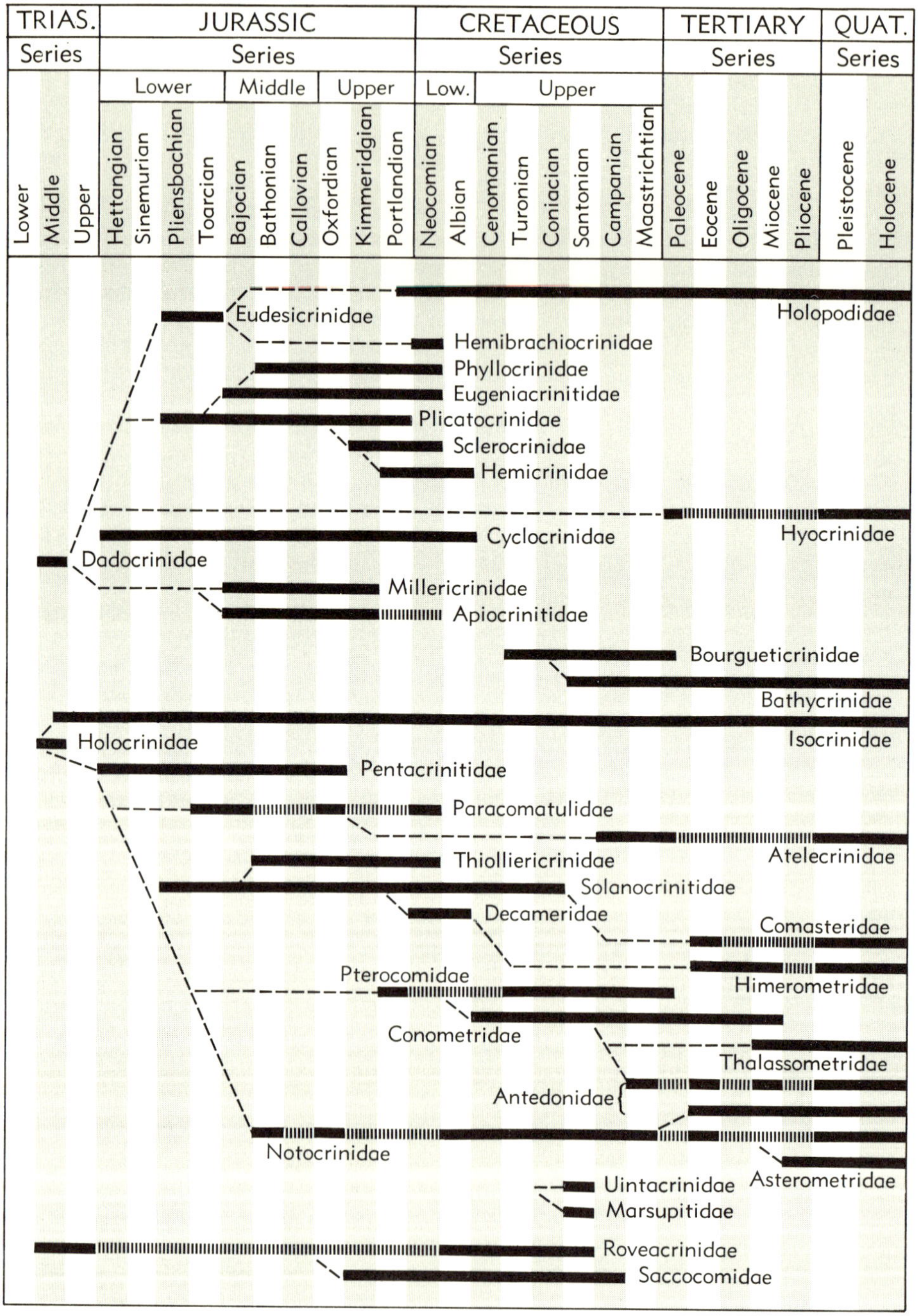

FIG. 209. Diagram showing evolutionary trends of the Articulata (Rasmussen, n). (The lower Paleocene now includes the Danian, which was established in 1846 as uppermost stage of the Cretaceous.)

tacrinida are entirely unknown, although the structure and articulation of the arms of Roveacrinida may have features in common with the Millericrinida, and those of Uintacrinida with the Comatulida.

MILLERICRINIDA

The Dadocrinidae, containing the single genus *Dadocrinus* from the Middle Triassic are the earliest known representatives of the order Millericrinida. According to von Koenen (1895a), *Dadocrinus* is related to *Erisocrinus* (Inadunata). Specimens of *Dadocrinus* show considerable individual variation. The column is circular in section, but may be more or less pentagonal in its proximal part as in most of the Millericrinidae. The diameter of the column may be uniform, or it may increase conically in its upper part as in *Apiocrinites* and in some species generally referred to the Millericrinidae. Some specimens of *Dadocrinus* show columnals apparently elliptical in section as in *Bourgueticrinus* although without a fulcral ridge. According to Bather (1897a), small and low columnals below the conical top of the column may indicate the growth zone below the conical part, thus anticipating the evolution of the uppermost columnals as a proximale similar to that of *Bourgueticrinus.* All of these variations in *Dadocrinus* thus demonstrate features found in the Millericrinidae, Apiocrinitidae, and possibly the Bourgueticrinidae.

The genera of the Millericrinidae are distinguished mainly on form of cup and transition between cup and column. *Liliocrinus* and *Angulocrinus,* with their conical cup, show the greatest resemblance to *Dadocrinus,* and also the gradual transition from column to cup in *Liliocrinus* may be an approach to *Apiocrinites. Millericrinus* is more unique in its low, wide cup and flat base, and *Orbignycrinus* with its hemispherical cup may be an approach toward *Pomatocrinus.*

The Apiocrinitidae, separated by Jaekel (1918), on the family level from the Millericrinidae, include advanced forms derived from the Millericrinidae, but it is uncertain whether the two genera, *Apiocrinites* and *Guettardicrinus,* which are included by most authors, constitute a natural group. They are both advanced in having modified articular faces in radials and proximal brachials, connected with reduced mobility in the proximal parts of the arms, and this is generally considered a decisive feature of the Apiocrinitidae, but the degree of modification in the articular face is rather variable and to some degree dependent on growth and age of the specimen. We may also consider differences in the transition between cup and column, where the Millericrinidae have a modified uppermost columnal or proximale included in the cup, but *Apiocrinites* a very gradual transition from the cup to the conical upper part of the column. In this respect *Apiocrinites* may well be an extreme end member of evolution from *Liliocrinus* or directly from *Dadocrinus,* while *Guettardicrinus* appears much more similar to *Pomatocrinus* and to certain species similar to *Pomatocrinus* but generally referred to *Apiocrinites* (e.g., *A. insignis, A. beltremieuxi, A. magnificus, A. crassus, A. changarnieri,* and *A. murchisonianus*).

The family Cyclocrinidae, based on very large, cylindrical columnals of *Cyclocrinus,* shows some resemblance to the Millericrinidae and Apiocrinitidae in absence of nodals and cirri, in size, and to some degree in the articular face of columnals, and has been placed in the Millericrinida in the absence of further information.

CYRTOCRINIDA

The Cyrtocrinida were considered separately as a group, Coadunata, by Miller (1821), as suborder Compacta by Jaekel (1918), and as order Cyrtocrinida by Sieverts-Doreck (in Ubaghs, 1953).

The Cyrtocrinida have a stout cup consisting of more or less fused radials, without distinct indication of basals, attached directly to a short column or to a dorsal element variously interpreted as fused basals or as a proximale. The column, when present, is more or less cylindrical without nodals or cirri, the articular face with radiating marginal crenulae. Synostosis may occur at primibrachs *1-2,* but generally all other articulations are muscular.

Wachsmuth & Springer (1889) referred

Holopus (together with *Hyocrinus* and *Bathycrinus*) to monocyclic Inadunata (Larviformia), due to its large oral plates and absence of infrabasals. WANNER (1916a, 1929) compared *Holopus* with Permian *Palaeoholopus* and *Brachypus* [*Permobrachypus*] among the dicyclic Flexibilia (Lecanocrinidae) having a sessile cup of similar structure, but without muscular articulations and pinnules.

The Cyrtocrinida, generally referred to Articulata, were derived in the Early Jurassic from the Millericrinidae by reduction of the column and basal circlet and by various modifications in relation to currents and hard substrate, mainly in reef sediments.

The least specialization is found in the Plicatocrinidae, which maintain cylindrical columnals and a conical, upright cup, although unique in fusion of pinnules. From the Plicatocrinidae we may have an evolution through Eugeniacrinitidae to Phyllocrinidae. *Proholopus* is still rather similar to *Plicatocrinus,* but with increasingly compact structure of the cup and beginning separation of the arms it may have evolved to *Pilocrinus* and by further development of interradial projections separating the arms to *Remisovicrinus, Eugeniacrinites, Psalidocrinus* and ultimately to *Phyllocrinus.*

Another line of evolution in Cyrtocrinida is characterized by the cup forming an angle with the column and by increasing fusion within cup and column, which can be followed through Sclerocrinidae to Hemicrinidae.

A third group within the evolution of the Cyrtocrinida is formed by the Holopodina, already present in the Lower Jurassic among the earliest Cyrtocrinida known, and also the only one found in the recent. They have a large and wide cup with radials more or less fused with the dorsal element attached to the substrate. It is not really possible to see any sign of the origin of this group, except that they agree with other Cyrtocrinida in the structure of the cup, the stout arms protecting the ventral side, the muscular articulations, and the presence of pinnules. In Eudesicrinidae and juveniles of Hemibrachiocrinidae the radials are still separated by distinct sutures from the dorsal element and there is an increasing difference in form and size of the radials (and arms). In Holopodidae the elements of the cup are completely fused, but the cups are less oblique.

AFFINITIES OF HYOCRINIDAE

The Hyocrinidae have a long column without nodals or cirri, articular face of columnals with radiating crenulae. Cup thin-walled, formed by large radials and by basals, which are often reduced in number or fused. The arms are narrower than the radials and with several synostosial articulations.

Hyocrinus was placed by ZITTEL (1882) in Plicatocrinidae due to resemblance in column, high conical and thin-walled cup, and presumed absence of basal sutures. CARPENTER (1884a) found *Hyocrinus* to differ from *Plicatocrinus* by the presence of distinct basals and undivided arms, and established the Hyocrinidae as a family. JAEKEL (1892) followed ZITTEL considering *Hyocrinus* closely related to *Plicatocrinus,* but also to *Saccocoma,* and considered this group derived from the early dicyclic Inadunata (Cyathocrinidae).

The genus *Calamocrinus* described by AGASSIZ (1890; 1892) was first considered closely related to *Apiocrinites, Millericrinus,* and *Hyocrinus,* and was referred by BATHER (1900a) to the Apiocrinidae, and by JAEKEL (1918) to the Millericrinidae, far removed from *Hyocrinus,* which was placed among monocyclic Inadunata.

GISLÉN (1939) gave a review of the Hyocrinidae and a discussion of their affinities. He clearly demonstrated *Calamocrinus* with its five distinctly separated basals, high tegmen, and inconspicuous orals as the most primitive member of the family and also the one least similar to the Inadunata. He considered the Hyocrinidae related to the Plicatocrinidae and other Cyrtocrinida but with distinct and well-preserved basals. Moreover, these recent deep-sea crinoids have none of the specializations or adaptations characteristic to the reef-dwelling Cyrtocrinida. SIEVERTS-DORECK (in UBAGHS, 1953) separated the Cyrtocrinida as an order from the Millericrinida without comment on the classification of the Hyocrinidae. So, Hyocrinidae are maintained herein within the Millericrinida, believed to be derived probably from the same or a closely related form as the Plicatocrinidae but following an entirely different line of evolution.

BOURGUETICRINIDA

The Bourgueticrinidae and the Bathycrinidae have generally been classified as belonging to Millericrinida, and agree with that order in possessing a long column without nodals or cirri. Furthermore, in the earliest and most common fossil genus, *Bourgueticrinus,* the uppermost columnals fuse to a large proximale, often compared to the conical proximal part of the column in *Apiocrinites* and a few other Millericrinida. The cup is simple, with exposed circlets of basals and radials, but no trace of infrabasals. The frequent nonmuscular brachial articulations are either synostosial or synarthrial to trifascial.

The Bourgueticrinida, however, did not appear until the Upper Cretaceous after the Millericrinidae and the Apiocrinitidae had disappeared, and they do not share the advanced characters in the modified articular face of radials and proximal brachials found in the Apiocrinitidae, nor is there any close resemblance in the column.

The column of all Bourgueticrinida typically consists of elongate, twisted columnals with elliptical section and synarthrial articulations. This character is also found in the Platycrinitidae among Paleozoic Camerata, but is unique among Articulata except for the larval column of comatulids, discarded by the adults of all comatulids except the Thiolliericrinidae. Synarthrial articulations may also be anticipated in proximal columnals of very small, juvenile specimens of some Isocrinidae (CLARK, 1908d, p. 88).

Synarthrial columnals are unknown in the Millericrinida, the only approach being a few, apparently elliptical columnals found in some specimens of *Dadocrinus,* and no intermediate forms between Triassic *Dadocrinus* and Upper Cretaceous *Bourgueticrinus* have been found. On the other hand, it should be mentioned that columnals with simple radiating crenulae similar to the Millericrinida are found in recent *Proisocrinus* and in a few distal columnal articulations of *Porphyrocrinus.*

The identical form of the column in the Bourgueticrinida and in larval Comatulida may indicate a possible relationship between these two orders. The possibility that *Bourgueticrinus* may be ancestral to the Thiolliericrinidae and maybe other comatulids, as proposed by KIRK (1911), can hardly be accepted, since there was a rich evolution of comatulids, including the Thiolliericrinidae, before the first arrival of the Bourgueticrinidae, and furthermore, all comatulids, including the Thiolliericrinidae, have a centrodorsal with cirri derived by evolution from the Isocrinida.

GISLÉN (1924) considered Bourgueticrinida to be closely related to the Thiolliericrinidae and derived, together with Comatulida, from the Isocrinida (Pentacrinitidae), but in 1937 he postulated Bourgueticrinida to be derived from the Apiocrinitidae or their predecessor, the Dadocrinidae. Although synarthrial columnals are unknown in the Millericrinida, they might, according to GISLÉN, have been characteristic of juvenile specimens of the Apiocrinitidae as they are of juveniles of comatulids and to some degree isocrinids, thus explaining evolution of the Bourgueticrinida by "neoteny." The larval growth and early ontogeny of stalked crinoids, except *Rhizocrinus,* is, however, virtually unknown.

The "pentacrinoid" larval stage of comatulids, before the development of cirri, has a remarkable resemblance to the Bourgueticrinida, not only in the high and slender, twisted, synarthrial columnals, but also in the disc-shaped, generally enlarged, proximal columnals and the cup with large basal circlet. This may strongly indicate a possible proterogenetic evolution of Bourgueticrinida from the comatulids. The absence of syzygial articulations in the arms is shared not only with the Millericrinida, but also with many early comatulids.

Among late species of the Bourgueticrinida, near the end of the Cretaceous, we find specimens with high basals and a small proximale, indicating a transition from *Bourgueticrinus* toward *Democrinus* and its more specialized aberrant, *Conocrinus,* and also occasionally specimens without a proximale, showing affinity to other Bathycrinidae.

Considering the distinct and unique common features of the column and the ambiguous origin of the group, RASMUSSEN

(1969) removed the Bourgueticrinidae from the Millericrinida and classified the Bourgueticrinida as an order.

ISOCRINIDA

The origin of the Isocrinida from the Triassic Millericrinida or from unknown Inadunata has never been traced. The Middle Triassic Holocrinidae are considered the most primitive members of this order, and are the only Articulata with exposed infrabasal circlet, but we also find columnals of typical Isocrinidae in the Middle Triassic and uncertain records from the Lower Triassic.

The most obvious feature common to all Isocrinida is the column with nodals and cirri and a petaloid pattern of crenulae.

It is not likely that the Holocrinidae with their exposed infrabasals are closely related to monocyclic Inadunata (Belemnocrinidae) as proposed by WACHSMUTH and SPRINGER (1886), but columns with nodals and cirri are found among several dicyclic Inadunata (Poteriocrinina) such as the Ampelocrinidae and some Botryocrinidae with five exposed or concealed infrabasals and with uniserial, pinnulate arms undivided or divided at primibrachs *2*. It may be possible, perhaps, that the Holocrinidae have evolved from similar forms by reduction of the anal plates and the large anal sac.

HILDEBRAND (1926) considered the Holocrinidae closely related to or even derived from *Dadocrinus* by evolution of cirri and increasement of infrabasals and basals. The greatest resemblance, however, in the column with cirri as well as in the crown is found between the Isocrinidae and the Middle Triassic *Encrinus pentactinus* BRONN, 1837, except for the biserial structure of the arms.

According to VON KOENEN (1895a), the mouth of *Holocrinus wagneri* appears excentric as in the Comasteridae and *Uintacrinus*.

Considerable variation among the Isocrinidae is seen in the articular pattern of columnals, arrangement of cirrus sockets, ramification of arms, and in kind and distribution of nonmuscular brachial articulations. All of these features have been used in separation of the many genera. In fossil identifications, this is inconvenient since numerous distinct species are known from columnals only, and their assignment to genus, therefore, uncertain.

A subdivision of the large family Isocrinidae above the genus-level has never been made, because we cannot demonstrate which characters will unite natural phylogenetic groups. The old division into three genera, *Pentacrinus, Isocrinus,* and *Balanocrinus* (partly under other names), was based on a pattern of columnal articulations. The three fossil genera *Austinocrinus, Isselicrinus,* and *Doreckicrinus* are distinguished by columnal articulations and arrangement of cirrus sockets, but their mutual relationship or relation to other Isocrinidae is unknown, and a similar pattern of columnal articulations may occur among Triassic Isocrinidae as well as in recent *Annacrinus*.

SIEVERTS-DORECK (1944) considered *Balanocrinus* (including *Isselicrinus*) to be derived from early *Isocrinus* through change of columnal articular pattern and by development of cryptosyzygial instead of synarthrial articulation at primibrachs *1-2*, and *Austinocrinus* to be derived from *Balanocrinus* by evolution of a marginal zone of crenellae in the columnals.

CARPENTER (1884a) divided the recent Isocrinidae into what he considered "two very natural groups," those with synarthrial articulation and those with cryptosyzygial articulation at primibrachs *1-2*. If the many genera of Isocrinidae, now accepted, are divided according to this criterion, we find in the first group: *Isocrinus, Chladocrinus, Chariocrinus, Hypalocrinus,* and apparently *Balanocrinus* (with different columnal articulation), and in the second group: *Nielsenicrinus, Cainocrinus, Teliocrinus, Cenocrinus,* and *Endoxocrinus* as well as *Metacrinus* (with greater number of primibrachs) and the four genera *Austinocrinus, Isselicrinus, Doreckicrinus,* and *Annacrinus* (with different columnal articulations). According to CARPENTER (1879a) and CLARK (1908e), the articulations primibrachs *1-2* and secundibrachs *1-2* are homologous and identical, but any synarthrial articulation may be replaced by other kinds of nonmuscular articulation ("syzygy"). In *Cainocrinus, Nielsenicrinus,* and *Teliocrinus* we find primibrachs *1-2* cryptosyzygial and secundibrachs *1-2* synarthrial. So it is uncertain in which group these three genera belong. Also, it is clear that subdivision based on columnal articulations differs from subdivision based on brachial articulation.

The very few Lower Triassic columnals referred to Isocrinidae are dubious and indeterminable, but in the Middle Triassic there is a rich fauna of columnals showing remarkable variation. Although described by BATHER as *Isocrinus,* there are also columnals with an articular face similar to *Balanocrinus* as well as some with very long crenellae in the radial areas as in recent *Annacrinus,* and cylindrical columnals rather similar to *Austinocrinus.* The first species with some resemblance to *Pentacrinites* seem to be *"Isocrinus" gravinae* from the Upper Triassic of Alaska and *P. versistellatus* from the Alpine Upper Triassic. The Pentacrinitidae, or at least certain species, are very different from the Isocrinidae in columnal articulations, basal circlet, and endotomous branching of arms after the second division. They may be derived from early Isocrinidae or more likely from their predecessors.

COMATULIDA

By the beginning of this century about 188 species of recent comatulids had been described and were referred to six generally accepted genera and about 127 fossil species referred to nine fossil and some of the recent genera. The distinction of modern genera was based mainly on the central or excentric position of the mouth, the size of basals and the arms undivided or divided; further informal grouping of species was based on pattern of division and articulation in arms.

Since A. H. CLARK (1907) began his studies of recent comatulids, the number has increased to 530 accepted species, 138 genera, and 17 families, synonyms and fossils excluded. Of these new forms, CLARK himself established 310 species, 124 genera, and 14 families, most of them during the years 1908 to 1915. He developed an entirely new scheme of classification on the basis of recent forms alone, and this was done explicitly and quite regardless of their fossil relatives: "The palaeontological succession of crinoid types is not at all to be trusted in matters of crinoid phylogeny, except possibly on the basis of broad averages. Phylogenetic facts must be acquired through the study of comparative anatomy of the group, combined with the study of the embryology and later development; later they may be tested in the light of the palaeontological record if one so desires" (CLARK, 1915a, p. 183). A discussion of the geological history of recent crinoids and relationship to their fossil representatives was, according to CLARK (1915a, p. 15), planned to end his monograph on the existing crinoids. The first volume of this large and valuable monograph, including introduction and comatulids, was started in 1915, and the fifth and last part of the volume, although more or less completed before 1924, was published in 1967 in collaboration with A. M. CLARK after the death of A. H. CLARK. A section on the phylogeny of crinoids was never written, except for a small note by CLARK (1910b) and scattered remarks on primitive or advanced and specialized groups or features in his monograph do not reveal any phylogenetic pattern of comatulid groups except that the two suborders, Oligophreata and Macrophreata, into which he divided the comatulids are considered as "two rather independent groups."

The main features used by CLARK for identification and classification of comatulids are size and structure of proximal pinnules, size, structure and number of cirri and cirrals, and division of arms. Such features are, with few exceptions, unknown in fossil comatulids. Other features such as form of centrodorsal, arrangement of cirrus sockets, ventral face of centrodorsal, form and reduction of basals, and form of articular face of radials, which are seen in most fossil comatulids, are less considered in the classification by CLARK, and to some degree undescribed in many modern species and genera. To bring fossil and recent comatulids together in a common classification is therefore at present a most difficult task. Moreover, it is not certain that present subdivision of recent comatulids reflects a natural classification in agreement with phylogeny. GISLÉN (1922) expressed his reservation to the validity of species and genera based on very small differences in length of pinnules and other features.

E. KIRK (1911) made the first important contribution to the origin of comatulids. He considered the comatulids to be a probable polyphyletic group representing a number of convergent lines of evolution, derived at different times in the Mesozoic from one or more groups of crinoids by a reduction of the columnal and fusion of columnals to form a centrodorsal. The dicyclic origin according to the law of WACHSMUTH & SPRINGER is demonstrated by the radial orientation of the first verticil of cirri in larval comatulids. In several groups of stalked Mesozoic crinoids, we find the tendency to stop the permanent formation of new columnals immediately below the cup, and thus a permanently uppermost columnal, which may be enlarged and differentiated, is obtained, forming a proximale. This may be indicated in *Millericrinus charpyi* and *M. beaumonti* and is characteristic of *Bourgueticrinus,* where the proximale is intimately connected with the thecal structure and infrabasals have disappeared or fused with the proximale. The column in such forms has a tendency to break immediately below the proximale, where the small and new columnals are placed. Such a break and detachment, at first only in the adult, may in later forms take place earlier in the ontogeny. By acquisition of cirri on the proximale, a true centrodorsal is formed and the comatulid stage is reached. The centrodorsal in such comatulids consists of a single enlarged columnal and is low, with few and irregularly arranged cirrus sockets. This transition from *Bourgueticrinus* or forms related to the comatulids is illustrated, according to KIRK, by the Lower Cretaceous (U.Jur.-L. Cret.) *Thiolliericrinus,* which maintains a column similar to *Bourgueticrinus* in the adult stage, and a centrodorsal with only few and small, irregularly placed cirrus sockets in some species *(T. heberti),* but larger sockets in other presumed later species. *Thiolliericrinus* was therefore considered as the start of one line of comatulid evolution, leading to comatulids with a low centrodorsal and few, irregularly placed cirrus sockets.

Other comatulids, according to KIRK, have a centrodorsal formed by fusion of several columnals, as may be indicated by larval development of some species *(Koehlermetra porrecta)* in which a number of proximal columnals are enlarged and the first cirri are formed from a single verticil of radially placed sockets.

Contemporary with *Thiolliericrinus,* we find "highly specialized" comatulids which may be traced far back in the Jurassic, mainly forms with large centrodorsal and stout arms. The extremely variable degree of specialization among contemporary comatulids was taken by KIRK to indicate different lines of comatulids developed at different times from stemmed crinoids, and the ancestors may be looked for among the Apiocrinitidae and the Bourgueticrinidae as well as the Pentacrinitidae.

The pentagonal section of the centrodorsal in many comatulids, the stellate scar from the larval column in juvenile centrodorsals of many comatulids, and the petaloid, crenulate basals in some Jurassic comatulids *(Archaeometra scrobiculata)* extremely similar to *Pentacrinites dargniesi,* indicate comatulids derived not from *Bourgueticrinus* but from forms with a pentagonal column, i.e., the Pentacrinitidae (Isocrinida). Also, the tendency to abbreviation and modification of the column, resembling an articulated centrodorsal in *Pentacrinites collenoti* and *P. sorlinensis,* indicates that some comatulid lines may have evolved from Pentacrinitidae.

Among the Millericrinida we may also find species with columns resembling the Pentacrinitidae in a more or less petaloid articular pattern such as *Millericrinus charpyi* and species with an abbreviated column such as *Millericrinus partti,* although without cirri. KIRK did not exclude an evolution also from such forms to the comatulid stage.

CLARK (1915a, p. 16, 17, 222) followed the concept of KIRK with some modification, deriving not only comatulids but also Isocrinida by evolution in opposite directions from *Thiolliericrinus* or closely related Thiolliericrinidae. In comatulids the formation of columnals stops after the formation of the first and only nodal, and this nodal remains attached to the cup as a centrodorsal when the larval column is discarded. In the Isocrinida, the column is strongly elongated by a continuous formation of proximal columnals and by repeated formation of cirriferous nodals, each of which corresponds to a centrodorsal.

GISLÉN (1924) slightly modified the arrangement of recent comatulids given by CLARK to agree better with their fossil representatives. He based the groups mainly on features in the centrodorsal and radials that are present also in fossil species. From the study of early (Jurassic) comatulids he concluded, as did KIRK, that there are such considerable differences between Jurassic comatulids that a polyphyletic origin is most likely. He found that most primitive fossil comatulids have a conical or columnar centrodorsal with cirrus sockets arranged in two columns on each radial side and with a distinct fulcral ridge on the cirrus sockets as in the Isocrinida. He therefore rejected an evolution from species with a noncirriferous column and proximale by subsequent evolution of cirri in forms like *Thiollericrinus,* and he accepted only the Pentacrinitidae (Isocrinida) as ancestors to the different lines of comatulid evolution. The more or less five-sided centrodorsal and the presence in some recent and fossil specimens of five interradial impressions in the dorsal side of the centrodorsal, resembling the petaloid pattern of columnals in *Pentacrinites,* and formed presumably as remains of a petaloid articulation in the uppermost larval columnals, were taken as indication that comatulids evolved from species with a pentagonal column. He concluded that the origin of comatulids should be searched among stalked Triassic or Jurassic crinoids with a pentagonal column and with verticils of five cirri attached to nodal columnals. The greatest resemblance to the primitive centrodorsal he found in Pentacrinitidae with closely placed nodals and almost no internodals in the proximal part of the column. In species with very low nodals such as *Pentacrinites dargniesi* the cirrus sockets of successive nodals may be displaced alternating right and left of the midradial line, thus forming two columns of sockets on each radial side of the column. He also found that the retarded evolution of the most proximal pinnules in comatulids indicated that comatulids evolved from crinoids with reduced proximal pinnules due to extension of the disc up between the arms as described for *Seirocrinus* among the Pentacrinitidae, but also found in several other groups of the Articulata. He excluded the Thiolliericrinidae from the comatulids and rejected this group as ancestral to other comatulids, but considered the Thiolliericrinidae as a separate line of evolution, in most of their characters so similar to the Solanocrinitidae that most likely they came from the same ancestral form among Pentacrinitidae.

One line of evolution among comatulids, according to GISLÉN, is formed by the Comasteridae, to which he referred the fossil *Palaeocomaster,* which is the oldest comatulid known. It has a large, discoidal centrodorsal with rather few and small, closely placed, almost smooth cirrus sockets arranged in a few irregular circles. The centrodorsal cavity is small. The basals are rod shaped. The articular face of radials is steep to vertical, generally with small and low ventral muscular fossae except in the oldest forms, and with a large radial cavity. This group, according to GISLÉN, shows resemblance to *Seirocrinus* among the Pentacrinitidae in the commonly reduced cirri.

Archaeometra and other Solanocrinitidae with a large discoidal to columnar centrodorsal, large cirrus sockets with distinct fulcral ridge, generally arranged in distinct columns and with very stout basals at least in the oldest forms *(Archaeometra)* constitute another line of comatulid evolution, probably derived (together with Thiollieri crinidae) a little later in the Jurassic from forms more similar to *Pentacrinites.* The radial articular face is steep to vertical with very low muscular fossae, and the radial cavity is large. The presence of coelomic furrows in the ventral face of the centrodorsal, found in many Solanocrinitacea, indicates evolution toward recent Mariametracea by reduction of basals and of sculpture on cirrus sockets.

A third group comprising two lines of evolution are, according to GISLÉN, the Notocrinacea (including Asterometridae) and the Conometridae (succeeded by the Thalassometridae), both lines characterized by a conical to discoidal centrodorsal and often somewhat sculptured cirrus sockets more or less columnar in arrangement. The centrodorsal cavity is moderate. Basals are rod shaped. Articular face of radials display rather large muscular fossae. The radial cavity is moderate to small. Pinnulars are generally carinate. The Notocrinacea are represented in the Cretaceous by *Loriolometra* and *Glenotremites,* which have large, crenulate cirrus sockets in 10 to 20 more or less distinct columns and deep coelomic canals forming radial pits in the centrodorsal. Later forms such as *Semiometra* and the recent *Notocrinus* have smooth cirrus sockets, which are in *Semiometra* small and numerous. The Conometridae represented in the Cretaceous by *Amphorometra* and in the Tertiary by *Conometra* also include, according to GISLÉN, the Cretaceous genera *Placometra* and *Jaekelometra.* Although there are great similarities to the Jurassic *Archaeometra,* GISLÉN found that the Conometridae and the recent Thalassometridae may be related to the Upper Jurassic *Pterocoma,* and together with Notocrinacea these groups may have evolved from the Pentacrinitidae.

The Macrophreata (Antedonacea) also, according to GISLÉN, are a polyphyletic group. One line of evolution may lead from the Notocrinacea through fossil Paleantedonidae to the Antedoninae, Thysanometrinae and perhaps other Antedonacea with more or less hemispherical centrodorsal, moderate centrodorsal cavity, numerous small, closely alternating cirrus sockets, reduced basals, moderately embayed synarthrial articulations, and in some cases with coelomic radial pits in the centrodorsal. A second line of evolution may lead from the Conometridae to the Thalassometridae and to groups of the Antedonacea including the Zenometrinae, Bathymetrinae, Heliometrinae, Isometrinae, and also to the Atelecrinidae and Pentametrocrinidae, all characterized more or less by a conical or columnar centrodorsal, large to very large centrodorsal cavity, cirrus sockets arranged in columns, rod-shaped basals commonly present, and synarthrial articulations generally deeply embayed. The Perometrinae may belong to the same line of evolution as the Antedoninae or may have evolved from the Colobometridae (Mariametracea).

Few of the modifications made by GISLÉN to the classification by A. H. CLARK were accepted by CLARK in his great monograph on all recent comatulids, and no attempts at new classifications of comatulids have been published since.

Apart from the description of several recent and fossil comatulids, the most important new contribution to the evolution of comatulids is the discovery by H. HESS (1951) of the Middle Jurassic *Paracomatula.* *Paracomatula* has an articulated "centrodorsal" consisting of a few nodals and resembling a very short column of *Pentacrinites.* The stellate basal ring has a petaloid pattern of crenellae toward the centrodorsal. Primibrachs *1-2* and secundibrachs *1-2* are synarthrial and rather flat. The arms divide at primibrachs *2* only. Secundibrachs *3-4* are syzygial, more distal syzygies with very few, radiating ridges. This species shows the most perfect transition from the Jurassic Isocrinida such as the Middle Jurassic *Pentacrinites dargniesi* to early comatulids such as the Middle to Upper Jurassic *Archaeometra,* and supports the theory of KIRK and GISLÉN on evolution of comatulids from Pentacrinitidae. However, the long, slender arms and the large, muscular fossae of the radials are not similar to Comasteridae but rather to *Archaeometra* or to the Atelecrinidae.

Since this discovery another two genera and three species of the Paracomatulidae with articulated "centrodorsal" have been found, although not yet described. These new forms show a beautifully preserved, conical, articulated centrodorsal, and one of them shows a most remarkable resemblance to the Atelecrinidae in form of the centrodorsal and cirrus sockets but still with a stellate ring of basals with a petaloid pattern of crenellae. The variation in the centrodorsal cavity in *Jaekelometra* and in the Antedonacea indicates that the division of comatulids into Oligophreata and Macrophreata is more problematic than considered by A. H. CLARK.

Evolutionary tendencies among comatulids are now seen as a gradual modification of the primitive ancestral or juvenile features in different parts of the crinoid skeleton. The conical or columnar centrodorsal dominating in the Jurassic and expected from presumed ancestors among the Isocrinida may modify to hemispherical or discoidal. The arrangement of cirrus sockets in 10 columns is modified ontogenetically and presumably phylogenetically by intercalation of new sockets in the midradial areas, the number thus increasing to 15 or 20 less distinct columns, and may proceed to a more irregular pattern of verticils or "circles." The cirrus sockets are commonly large and with more or less distinct fulcral ridge or tubercles in most Mesozoic comatulids, but are almost smooth in most Tertiary and recent species. The centrodorsal cavity, which is small and narrow in most early comatulids, becomes extremely small and shallow in the Decameridae, very variable in *Jaekelometra,* and in *Microcrinus,* and large to very large and deep in most Atelecrinidae and some Antedonidae. The basals, which form a rather stout, stellate ring of plates similar to the Isocrinida in *Paracomatula* and *Archaeometra,* become narrow and rod shaped with few or no crenellae and are centrally almost separated in most comatulids and may be further reduced to a delicate, perforate central plate, the rosette, in the Mariametracea and most Antedonacea. A special evolution of large polygonal basals surrounded by the radial ring is found in Decameridae and a circlet of large basal plates is maintained from the larval stage by a proterogenetic evolution in most Atelecrinidae. Coelomic furrows or depressions in the ventral face of the centrodorsal are characteristic of the Decameridae and most Mariametracea. Coelomic canals are seen as radial pits in the centrodorsal of most Notocrinacea, Asterometridae, and in some *Jaekelometra, Microcrinus,* and *Antedon.* The radial articular face is rather high and steep in the Paracomatulacea and *Archaeometra,* but may change in different directions, being very high with large muscular fossae on thin and steep ventral flanges and with a narrow radial cavity in most Atelecrinidae, Antedonidae, and Thalassometridae, whereas the radials are low and compact with small muscular fossae and a large radial cavity in most Solanocrinitacea, Comasteracea, and Mariametracea. The arms are almost always divided at primibrachs 2 and often further divided, and there is a synarthrial articulation at primibrachs *1-2* and generally also in succeeding brachials *1-2.* A syzygial articulation at primibrachs *1-2* is found in the Zygometridae and *Eudiocrinus* among the Mariametracea and in a few Comasteracea. Arms divided at primibrachs *1* are found in some Solanocrinitacea, and undivided arms are found in some Solanocrinitacea, in *Atopocrinus, Eudiocrinus,* and in the Pentametrocrinidae. The synarthrial articulations are rather flat in the Solanocrinitacea, Comasteracea, Mariametracea, and Notocrinacea, but may be more or less embayed in some Atelecrinidae, Tropiometracea, and Antedonacea. Syzygial articulations are almost always present at brachials *3-4* and with variable intervals more distal, but are missing in some Solanocrinitacea. Syzygy with very few radiating ridges are found in some Paracomatulidae, Atelecrinidae, and Solanocrinitacea. Very special features in the Comasteridae are the excentric mouth and the pinnular comb, although the mouth may also be slightly excentric in *Atelecrinus.*

A phylogenetic classification of comatulids should be based on further studies of the evolution in several features such as described above, and of the distribution of such characters especially in fossil species, as well as on more detailed studies of early comatulids and especially the structure and affinities of Paracomatulacea.

The new, partly unpublished observations on fossil crinoids support that comatulids evolved from the Isocrinida through the Paracomatulidae, and we find within this family a considerable variation including species similar to the Atelecrinidae and *Archaeometra* in several features. A polyphyletic origin of comatulids is no longer certain. The interpretation by A. H. Clark of the centrodorsal as a modified single columnal is not supported by the study of fossils, nor is the interpretation of the basal rays (rod-shaped basals) as secondary calcifications independent of the basals. The continuous growth and formation of new cirrus sockets and cirri at the ventral edge of the centrodorsal does not support the general opinion that columnal growth is stopped by a fusion of one columnal with the infrabasals to form a centrodorsal. It is possible, therefore, that only columnals are included in the centrodorsal, and that infrabasals, if present in early larval stage, are resorbed. The Thiollierıcrinidae appear to be a proterogenetic evolution from the Solanocrinitacea. The origin of the Comasteridae is uncertain. The excentric mouth often associated with reduction in posterior arms, the specialized proximal pinnules, and the common reduction of centrodorsal and cirri indicate that this is a specialized group, hardly ancestral to other groups of comatulids. Most of these special features are not seen in fossils, and there is no certain indication of the Mesozoic Comasteridae. The origin of the Comasteridae is therefore uncertain. A more or less excentric mouth may be found in some Atelecrinidae, and cannot be excluded in some Paracomatulacea resembling the Atelecrinidae in other characters. The centrodorsal and the radial ring of the Comasteridae are more similar to those of *Palaeocomaster* and other Solanocrinitacea.

The classification of comatulids used in the *Treatise* follows, with very few modifications, the classification used by A. H. Clark in his great monograph on the existing crinoids (1915-67), not because this classification is considered final or satisfactory, but in order to facilitate comparison between fossil and recent comatulids and thus encourage future reconsideration of comatulid evolution. For this reason, also, the text includes a record of all genera, also when unknown as fossil, and a description of all groups above the genus level with special reference to features relevant in the study of fossils, even if these characters have not been used in classification of modern species, and are described only for some of the species or genera of each group. The use of superfamilies by A. H. Clark has been followed with change of endings in order to agree with other sections of this *Treatise* volume.

UINTACRINIDA

The two families included, each with a single genus restricted to the Upper Cretaceous (Santon.), are both characterized by a large, ovate cup and thin plates including a centrale without trace of any column or attachment. The arms are long, pinnulate, divided at primibrachs *2* only, and with muscular and syzygial articulations. There is a variable number of fixed brachials, fixed pinnulars, and interbrachial plates.

In *Uintacrinus,* the fixed proximal brachials are thin and flat like plates of the cup, and there is a considerable number of fixed brachials, pinnulars, and interbrachial plates. The structure of the base in *Uintacrinus* is very confusing and has been much discussed. Apparently about half of the specimens are dicyclic with small but distinct infrabasals and half are without trace of infrabasals. Furthermore, there is always a centrale, variously interpreted as a rudiment of the column or a secondary additional plate like the interbrachials. Since the infrabasals are by no means reduced or vestigial when present, this appears to be a unique species of crinoid with both monocyclic and dicyclic specimens. It has never occurred to the many specialists discussing this paradox that if the centrale is a secondary additional plate filling up the space inside the basal circlet, then the same explanation may as well be given to the presumed infrabasals. The latter explanation is consistent also with the presence of a few specimens with different composition of the dorsal area (see Fig. 609,*1f-j*) including two or three plates of similar size, or one large and three to five smaller plates of different size. It is a general rule to the secondary plates that a stellate outline or a re-entrant angle or a meeting of more than three plates in a point never occur. Therefore, a centrale is never stellate; if the dorsal area inside the basal circlet is stellate, then a smaller plate is formed in each re-entrant angle, and these are the plates generally interpreted as infrabasals.

The flexible tegmen of *Uintacrinus* shows a central anal tube and excentric mouth as in most Comasteridae and, according to von Koenen (1895a), maybe also in *Holocrinus*.

Marsupites has five large infrabasals, basals, radials, and a large centrale. The arms are narrow, widely separated and not flattened as plates of the cup, although the arms and most proximal pinnules may be connected by a few thin interbrachial plates. Provided the infrabasals are true primary plates of the cup, this is the only genus of the Articulata with large exposed infrabasals, except for the Triassic Holocrinidae (and maybe *Uintacrinus*).

Several authors have classified *Uintacrinus* with the Flexibilia or compared it with *Forbesiocrinus* and the Icthyocrinidae (SCHLÜTER, 1878; NEUMAYR, 1889; ZITTEL, 1895), many of which show interbrachial plates, a flexible tegmen, and uniserial arms, but have anal plates in the cup, three infrabasals, a column, no syzygy, and no pinnules.

JAEKEL (1894) placed *Marsupites* in dicyclic Inadunata and *Uintacrinus* in dicyclic Camerata.

BATHER (1896b) considered all resemblance between *Marsupites* and *Uintacrinus* secondary as results of adaptation to a pelagic life. He found that primary structures such as absence of anal plates in the cup, uniserial pinnulate arms with axial canal, muscular and syzygial articulations, arms divided at primibrachs *2*, and primibrachials connected by interbrachial plates, indicate affinity with Articulata and especially with *Dadocrinus,* although differing in all those characters interpreted as adaptation to the pelagic habit.

SPRINGER (1901) observed the structure of tegmen with excentric mouth and exposed ambulacra. He emphasized the resemblance to the Comasteridae not only in tegmen and mouth, but also in syzygial articulations.

CLARK (1909h) included the Uintacrinidae and the Marsupitidae as pelagic comatulids with unmetamorphosed basals and infrabasals under the name "Comatulides Innatantes." Also, GISLÉN (1924) considered *Uintacrinus* to be probably derived from the same ancestral form as the comatulids.

KIRK (1911) considered the flexible tegmen and excentric mouth of *Uintacrinus* as secondary characters, and that *Uintacrinus* was derived from the same dicyclic Inadunata as *Dadocrinus* and other Articulata.

ROVEACRINIDA

The small, planktonic, stemless Roveacrinida, including the Roveacrinidae and the Saccocomidae, have a generally thin-walled cup of large, thin radials, with or without small basals, and in the Saccocomidae a small centrale and in the Somphocrininae a large, conical or rod-shaped dorsal element variously interpreted as a centrale, a centrodorsal, or a fused basal circlet. Arms generally divided at primibrachs *2* and with a proximal pinnule gap. Muscular brachial articulations alternate with cryptosynarthrial or synostosial articulations.

GOLDFUSS (1831) described species of *Saccocoma* under the genus *Comatula.*

NEUMAYR (1889) considered *Saccocoma* to be juvenile specimens of an unknown stemless crinoid similar to *Plicatocrinus.*

JAEKEL (1893) compared *Saccocoma* with *Plicatocrinus* and *Hyocrinus,* all having a thin-walled cup with large, thin radials, reduced or fused basals and narrow arms, in *Saccocoma* and *Hyocrinus* with many nonmuscular articulations and a proximal pinnule gap, but in *Hyocrinus* with undivided arms. He considered *Saccocoma* closely related to *Hyocrinus* and *Plicatocrinus,* and these groups derived not from the Articulata or the Poteriocrinitidae but from primitive, dicyclic Inadunata, the Cyathocrinoidea.

DOUGLAS (1908) described *Roveacrinus* and considered it related to *"Extracrinus"* (*= Pentacrinites*) due to the downward prolongation of the radials.

In 1918 JAEKEL placed the Saccocomidae and the Hyocrinidae, together with the Plicatocrinidae and the Eudesicrinidae, in his monocyclic order Costata, also including families now referred to the Inadunata and the Camerata, but *Roveacrinus* (under the name of *Drepanocrinus*) was placed in the Rhizocrinidae (= Bathycrinidae).

BATHER (1928) considered *Roveacrinus* allied to *Phyllocrinus,* due to ventral projections from the cup separating the arms.

SIEVERTS (1932a, 1933b), following JAEKEL, considered *Roveacrinus* probably related to the Bathycrinidae due to similar structure of arms, but in 1943 she included the genus in Saccocomidae.

PECK (1943) established the family Roveacrinidae as a group of monocyclic Inadunata, but at the same time as derived from Jurassic or Cretaceous comatulids. In 1948, after a description of Triassic Roveacrinidae with a spine-formed dorsal element, he declared all previously suggested relationships improbable, and the derivation of the Roveacrinidae unclear. In 1955, he classified the family in the Articulata.

Patrulius (1956) considered the thick-walled *Pseudosaccocoma* related to the Thiolliericrinidae, but Bachmayer (1958) found it related to *Saccocoma*.

ECOLOGY AND PALEOECOLOGY

By Albert Breimer and N. Gary Lane

[Instituut voor Aardwetenschappen der Vrije Universiteit, Amsterdam; University of Indiana]

CONTENTS

ECOLOGY OF RECENT CRINOIDS

By Albert Breimer

GENERAL STATEMENT

In this chapter on ecology the crinoid's function and behavior is described as it reflects interrelations between the organism as a functional system on the one hand, and the surrounding medium on the other. Thereby, the organism is not considered as a "closed" functional system. From a point of view of physiology probably no such "closed" functional system exists. All metabolic processes (respiration, digestion, and excretion) require interchange of the animal with its environment. For this reason, not only the usual biological factors influencing the crinoid's mode of life are included, but also physiological factors are briefly dealt with in this chapter.

A detailed monographic treatment of the physiology of echinoderms has been published by a team of echinoderm zoologists under editorship of R. A. Boolootian (1966a). The book gives pertinent descriptions and compilations on the present stage of our knowledge on the subject. The result of this work again clearly demonstrates to the student of crinoids that virtually nothing is known about the basic physiological processes and mechanisms in crinoids. This lack of knowledge partly may be explained by the difficulties met with in finding and maintaining suitable crinoid material for physiological experiments.

Thus, a discussion of crinoid physiology is necessarily short. It should provide pertinent information at least on fundamental physiological processes which are vital to the crinoid organism as a functional system: the metabolic processes (respiration, digestion, and excretion) and their possible regulatory mechanisms, either nervous or glandular. The paucity of available observations is reflected in the brevity of descriptions given here, derived mainly from the above-cited book.

FELL (1966) has reviewed the ecology of recent crinoids, with attention given to several interesting aspects of the subject. The physicochemical interrelations of the crinoid and its environment, as well as its distributional patterns are described in separate sections of the chapter on Ecology.

PHYSIOLOGICAL FACTORS

RESPIRATION

No special respiratory organs are found in crinoids, for it is only the podia or tube feet that provide a respiratory surface as they do in echinoderms generally. All cells, however, are permeable to oxygen and it may be expected that oxygen penetrates both specialized and nonspecialized surfaces. Hence, the body wall presumably obtains oxygen directly from sea water. Probably about 60 percent of the oxygen intake of crinoids is utilized mainly by metabolism of the body wall.

The digestive tract is sometimes said to be involved in respiratory exchange inasmuch as water is frequently observed to be taken in by the anal cone and again ejected. This is undoubtedly a minor supply of oxygen, since histologically and physiologically the tissues of the digestive tract are not adapted to serve for respiration.

A specialized respiratory surface of crinoids is found in terminal branches of the water-vascular system (podia or tube feet). In other echinoderm classes it is known that oxygen is transported in the lumina of the podia by ciliary currents to the respective ampullae, where it diffuses across the ampullar membrane into the perivisceral coelomic fluid. Neither the water vessels nor the hemal lacunae are directly involved in a vascular transport of oxygen. The transmittal of oxygen to the internal organs is effected by the perivisceral coelomic fluids, which are kept in motion by flagellae of the coelomic endothelial cells.

Since crinoids have a primitive sort of water-vascular system, lacking ampullae and fully open to the external medium, the supply of ambient oxygen to the perivisceral coelomic fluid can be furnished only by direct intake of water through the hydropores and ciliated funnels, which for this purpose probably beat in one direction. In this way ambient oxygen is brought directly to the viscera. Supply of oxygen to the subambulacral coelomic fluid is inferred to be provided by diffusion through the horizontal membrane.

The fluid present in the hemal lacunae is often referred to as blood. However, any differences in the chemical constituents of perivisceral coelomic fluid and hemal lacunar fluid remain to be demonstrated. Probably it is the perivisceral rather than hemal fluid that transports the oxygen. In any case, a free exchange between the two fluid bodies is possible, since coelomocytes are present in both. The presence of specialized hemocytes containing pigments closely allied to the hemoglobins of vertebrates are unknown in crinoids. Such cells have been clearly demonstrated to function in transporting or storing oxygen, or both. They occur in some genera of holothuroids and in one species of ophiuroid.

Crinoids thus are devoid of effective respiratory organs, effective vascular circulation, and respiratory pigments. This probably makes them unable to maintain a constant consumption of oxygen. Rather, this is thought to vary with changes in oxygen of the environment.

DIGESTION

The nature of the food taken by the crinoids has been described previously. The processes and mechanisms of digestion in crinoids and the manner of transporting and storing nutrients required by them are little known. Normally in echinoderms digestive enzymes are secreted by secretory cells in the epithelia of digestive caeca dependent from the foregut. The functional nature of the diverticula in crinoids appears to be unknown, although supposedly it is digestive. The source and nature of digestive enzymes in crinoids remains to be investigated. Similarly, nothing now is known about the mechanisms of uptake and the transport and storage of nutrients. Generally in echinoderms coelomocytes play a role in ingesting sufficient nutrients to satisfy metabolic requirements. In crinoids the activity of certain cells is known to be phagocytic; some also may be adapted for transport and storage of nutrients. The perivisceral coelomic fluid, however, in many echinoderms has been proved to contain products of digestion, with traces of protein and amino nitrogen, as well as a small quantity of reducing sugar. The perivisceral coelomic fluid seems to be the most important medium of nutrient transport. An exchange of substances between the coelomic fluid and tissues of the body seems to be continuous in crinoids.

EXCRETION

Crinoids lack well-defined excretory organs. In older literature the sacculi and the globular brown bodies known to live in the digestive tract of *Antedon* (Fig. 1) have been interpreted from time to time as excretory bodies. The structure of the sacculi and brown bodies is well known. Each sacculus is a special spherical body enclosed within a thin membrane and containing a number of pyriform sacs filled with refractive spherules. The sacculi of the crinoid disc, arms, and pinnules periodically rupture and extrude their granular contents freely into the surrounding medium. Those in the wall of the gut (especially at its posterior end) mostly discharge into the gut. No mucus is associated with the extruded matter and the nature of excretory products and manner of their elimination is physiologically unknown.

Recent literature, however, takes into account the fact that phagocytic coelomocytes may convey ingested particles from the coelomic fluid to the exterior and it has been assumed that the coelomocytes could be responsible for the elimination of metabolic waste, at least partly. Excretory products, principally ammonia and urea, occur in the coelomic fluid. Crinoids have a special type of coelomocyte with short pseudopods possessing a large nucleus and a cytoplasm which commonly contains granules, some of which may be colored. The phagocytes with short pseudopods are known to be actively phagocytic, ingesting colored particles. No information is available as to whether ingested particles are conveyed to the exterior by the crinoid phagocytes or are deposited at definite sites in bodies of these animals. The coelomic epithelia could contribute in this way to excretion since they are supposed to have phagocytic powers and to be able to accumulate substances. In other echinoderms coelomocytes carrying ingested material may aggregate in the axial gland or may be eliminated through gills or respiratory trees, or through the stone canal and madreporite.

BIOLOGICAL FACTORS

In dealing with crinoid behavior and habits, most attention needs to be directed to feeding and reproducing, the two most essential functions of the organism. Recently the feeding behavior of extant crinoids in their natural habitats has received much attention. Magnus (1963, 1964, 1967), Fishelson (1974), and Rutman & Fishelson (1969) have reported on the feeding behavior of shallow-water comatulids from the Red Sea; Meyer (1973a,b) and Macurda (1973) on shallow-water comatulids from the Caribbean Sea; Pérès (1958, 1959) on deep-water comatulids from the Pacific; and Macurda & Meyer (1974) on deep-water isocrinids from the Caribbean. Less attention is given to the reproductive behavior. This subject lately has been excellently reviewed by Boolootian (1966b).

PERCEPTION

A key question in crinoid ecology relates to possibilities for the organism to perceive its environment, react to stimuli received from it, and to translate these into actions. Crinoids have only an unspecialized sensory system, mainly located in the sensory papillae of the podia and tube feet and in free nerve endings distributed over the body surface and connected to the subepidermal nerve plexus. Crinoids probably are capable of discriminating chemical, photic, tactile, and thermal stimuli—particularly their presence or absence and probably also changes in their intensity. The motor system is located in a well-developed aboral nerve system, which allows the animal to make and coordinate independent movements of arms and pinnules in respect to one another and to the tegmen. Among stalked crinoids both the stalk and cirri are capable of movement.

POSITION, RIGHTING, LOCOMOTION

The normal position of both stalked and stalkless crinoids is determined by fixation to the bottom or any substrate in such a way that the mouth is directed upward. Crinoids are the only surviving echinoderms that retain this primitive condition. Stalkless comatulids normally are attached by their cirri to a bottom object. They will not move about as long as ecological conditions at the spot of their attachment remain satisfactory. They move actively by swimming or creeping only in case environmental factors become unsatisfactory. Seemingly, the stemless crinoids have only gained the vagile capacity of active movement in order to gain efficiency as sedentary animals. Movements are aimed at finding better attachment sites. CLARK (1921) has described the desperate need of comatulids to attach themselves to suitable bottom objects. He wrote:

> If a dozen specimens of *Antedon* were thrown at night into a large basin of water and were left without any means of attachment they were all found dead in the morning, conglomerated at the bottom of the basin, clinging to each other with their cirri and having their arms intertwined in such a manner as to suggest the idea that they had died of the asphyxia produced by overcrowding after exhausting themselves in efforts to find a suitable attachment; while if, in a basin of the same size and containing the same quantity of water, there were placed with a like assemblage of specimens a sufficient number of rough stones to afford them all a basis for attachment, they would be all found in the morning in a state of full expansion, with every appearance of health and vigor!

Suitable objects for clinging and attachment are rock, coral, arborescent growths such as algae, coelenterates, seaweeds, and any object available which projects slightly above the bottom surface.

If comatulids happen to turn to an upside down position with the mouth below, they are able to right themselves by means of a pull and push mechanism operated by the arms.

Movements of comatulids are twofold. Antedontids are specialized for swimming and have a slender construction suited to this, whereas comasterids only creep around and have a more sluggish appearance. Both arms and cirri play an active role in locomotion.

The well-known swimming movements of *Antedon* begin by release from an attached position in which its five bifurcated arms are spread out horizontally. The arms then act as two groups of five, one comprising the left half-rays and the other the right half-rays. Arms of one group are first raised upward with the pinnules flexed inward against the arm. Next these arms lash rapidly downward but now with extended pinnules to give maximum strength to their stroke. The same movements are performed by the other group of arms repeated again and again. The crinoid can swim in any direction. A differential horizontal component is given to the motion in swimming by variation in the strength of strokes with arms in one position or another. The swimming of *Antedon* is graceful; however, it is induced by unfavorable conditions in the environment, such as oxygen deficiency, intense illumination, and thermal changes. Swimming is restricted to short distances, several meters at most.

Comasterid movements are best described as creeping in which they mostly utilize their cirri, although the arms may assist. These crinoids persistently decline to swim, even when diligently stimulated to do so in experiments. The creeping movements of comasterids are based on a pull-and-push mechanism carried out by the cirri. They may crawl about for hours, with arms aiding their movement. The anterior arms, extended forward during locomotion, serve as tactile organs and assist in pulling, while the posterior arms assist in pushing. Fell (1966) has stated that the longer arms lie on one side of the disc, to which the mouth is displaced, and these longer arms invariably occupy an anterior position during locomotion. He stressed the theoretical importance of this phenomenon for the relationship between mouth and anus as markers of the anteroposterior axis of echinoderms.

AUTOTOMY AND REGENERATION

It is well known that crinoids may shed their arms and then are able to regenerate them. Some authors have doubted whether the shedding of arms constitutes true autotomy, for a majority of regenerated arms are found to have originated from a syzygial contact or from another sort of nonmuscular contact. Autotomy may easily be brought about by degeneration of the very short and delicate ligament fibers connecting brachials in syzygial contacts. Of course, true autotomy has to be distinguished from the breaking of arms in response to mechanical external pressure. Arms of crinoids are indeed liable to break at syzygial contacts by reason of inflexibility at such contacts. Shedding of arms mostly takes place under unfavorable physical conditions in the environment (e.g., high temperature, oxygen deficiency, etc.).

Regenerative powers are very strong in crinoids. If arms are lost or shed, they are readily regenerated, and if two or more are lost at the same time, all are regenerated. Crinoids with all arms missing at the same time are reported to die, but the cast-off arms remain alive for a long time (up to several weeks) and continue to feed, without ability to regenerate an entire animal, however. The tegmen, if lost, regenerates, as do crinoids which have been eviscerated. Cirri, if all are removed at the same time, do not regenerate. Regeneration is only successful as long as the aboral nerve center remains intact. Therefore, any crinoid part containing the aboral nerve center plus one arm generally will regenerate an entire animal.

Minckert (1905a) has observed the connection between regeneration and growth of crinoids mainly in the later stages. Autotomy of arms occurs in juvenile ten-armed comatulids as a means of growing out to a multibrachiate condition. In such a case, each shed arm is not simply replaced but reduplicated, for two arms replace one that is lost. This is termed **augmentative regeneration.**

Fell (1966) has considered regeneration in connection with reproduction. He envisaged the possibility of dividing a crinoid in such manner as to regenerate two entire specimens from the original one. This would mean asexual reproduction. He concluded that spontaneous or induced division of the body does not lead to the production of two individuals from one. Crinoids lack asexual reproductive powers.

REPRODUCTION

Sexual reproduction is the only natural means of self-duplication among crinoids. Most have distinct reproductive periods (breeding seasons) which are marked by periodical outgrowth of the gonads and the spawning of gametes. The crinoid gonads are lodged in specialized pinnules which are interpreted to serve mainly as storage organs. Therefore, periodic growth of the gonads is inferred to succeed gametogenesis, growth being due to gamete accumulation.

The first, and probably only, report of crinoid gamete shedding behavior in its natural habitat is by Fishelson (1968) for *Lamprometra* from Eilat, Red Sea. The shedding behavior of this crinoid is described as follows:

> . . . at 17.45 h. individuals were observed climbing out of their daytime hiding places on an isolated coral block at a depth of 2 m. At 18.00 h. twelve were found attached by their cirri along a deep notch in this block with fully extended arms and swollen genital pinnules. One of these individuals started violent undulations of its arms, and after several seconds the remaining individuals were whipping their arms vigorously. Simultaneously, a dense greenish-coloured cloud of gametes arose around them, partly covering them. Arm-whipping continued for 25 s. and then stopped abruptly. Laboratory observations revealed that the individual which began the gamete shedding was a male, and that in this group there were females and males.

In spawning, all gonads in all genital pinnules of every arm release their gametes at the same time. Male individuals spawn first and their products probably stimulate females to release their eggs. This is a very effective mechanism for immediate fertilization of the just-shed eggs. Release of the gametes is by rupture of the pinnular body wall, either in preformed thin spots or elsewhere. In many cases and typically in *Antedon,* eggs are stuck to the pinnules by secretion of an adhesive from gland cells.

Boolootian (1966b) tabulated data supplied by Clark and Mortensen on the reproductive periods of 21 species of comatulids. A majority of these have distinct and mostly short reproductive periods of one or two months. The only species which seems to spawn the year round is *Antedon mediterranea.* Breeding seasons of the different species of crinoids are scattered throughout the year except in November and December. No data are available on the spawning and breeding behavior of stalked crinoids. Attention already has been given (see Morphogenesis) to the fact that several Antarctic species of comatulids breed their young in special brood-pouches or marsupia (Fig. 36).

No information can be given on the physiological mechanisms which regulate the course of events in reproductive cycles of crinoids. Crinoids are known to spawn at very specific moments of the day or night, and their spawning activities have been observed to correlate with several external factors such as presence of moonlight. *Comanthus japonicus* spawns on a single afternoon in October at 3 o'clock, the date depending on lunar periodicity (Dan & Dan, 1941; Dan & Kubota, 1960).

FEEDING

The importance of feeding for crinoids is indicated by the estimation that they probably spend at least half of their entire life time in obtaining food. They are exclusively suspension-feeders that depend on small planktonic and nektonic organisms such as algae, dinoflagellates, diatoms, radiolarians, foraminifers, small crustaceans (e.g., amphipods, copepods, larval stages of malacostracans), and larvae of other organisms.

Rutman & Fishelson (1969) have studied the food composition of comatulids in the Red Sea. Food selectivity concerns the size of the food organisms. Nearly 85 percent of the ingested organisms are within the size range of less than 1 micron to 300 microns. There seem to be two size-limiting factors; one is the width of the food-conveying ambulacral tract, and the other is the locomotory strength of the organisms captured in the mucus net. Larger organisms are capable of forcing their way out of the net. The food of these crinoids consists of organisms belonging mainly to the finer fractions of microplankton and to the nannoplankton. Also benthonic micro-organisms, such as brown algae, swept up from the bottom by currents, are ingested as food. The average diet of these crinoids contains about 10 percent phytoplankton, 50 percent protozoans and about 40 percent crustaceans and mollusks. Phytoplankton is always found to be the minor constituent in the food, the most frequent relation between phytoplankton and zooplankton being one to nine in favor of the latter.

The crinoid feeding mechanism has been studied by Gislén (1924) and Nichols (1960) on specimens in aquaria. The crinoid tube feet are adapted to collect food particles, for this is their main use besides respiration and subsidiary sensory functions. The tube feet possess papillae with muscle-operated glands which produce strings of mucus for trapping food particles. Holland (1969) found that the papillar muscle cell is actually filled with microtubules. The activities of the tube feet in a feeding *Antedon* have been described by Nichols (1960, p. 106, 107, 115) as follows:

> The largest tube feet of each group project laterally almost at right angles to the long axis of the pinnule. The medium-sized tube feet of each group project upwards and outwards at an angle of about 45°, while the smallest of each group projects almost straight upwards. Bordering the food grooves of arms and pinnules are the groove lids or lappets. When the animal is feeding these project upwards at an angle, and each long tube foot projects laterally through the valley between the two adjacent lappets. The other two tube feet of each group lie against the inside wall of each lappet with the outer side of the proximal part of the tube feet fused to this wall, so that when the tube feet bend in

towards the mid-line of the food groove they move the lappets in this direction too. If these tube feet contract, the lappet closes over the groove, so that only the largest tube feet remain protruding, though usually they too bend inwards to lie across the oral surface of the pinnule.

When the animal is feeding, the tube feet make sudden and rapid bending movements every few seconds. This apparently happens whenever food material touches the tube feet and stimulates the sensory cells, since crushed fragments of food dropped onto a pinnule cause the tube feet immediately to bend and flick in an attempt to throw the food into the groove. The rapid movement of the long tube feet is almost entirely at right angles and towards the food groove; that of the medium-sized tube feet is almost always at right angles to the groove, but both towards and away from it; the small tube feet can twitch in any direction. The tube feet of the arms can move in any direction, though most of their movements are at right angles to the food groove. After a twitch the tube feet return to the "waiting-position" more slowly. The main food-entrapping organs are the largest tube feet of the pinnules and the medium-sized ones of the arm. The middle and small tube feet undoubtedly act mainly to transfer the particles into the groove. The median tube feet are able to move outwards as well as inwards to pick up the food-laden mucus strands, and the small feet can twist in any direction to catch the strands from the medium-sized feet and drop them into the groove.

GISLÉN (1924, p. 274, 275) has described the activities of ambulacral grooves and the mouth in feeding as follows:

> . . . the margins [of the ambulacral furrows] . . . generally lie pressed tightly together. [The furrow] opens as soon as any grains fall upon it, with a wavy movement proceeding towards or away from the mouth. The mouth, only a narrow slit before, opens to its widest extent, and becomes rounded. . . . [Food grains] are cast into the ambulacral groove, where the ciliary current takes hold of them and carries them to the mouth. The ambulacral furrow itself is slightly concave and ciliates strongly towards the mouth. The mucus is secreted in the ambulacral groove . . . and the grains are glued together into larger lumps by it. These are transported to the mouth, the edges of which are often pressed together like a pair of lips, during movements similar to swallowing.

Systematic observations on the feeding behavior of crinoids in their natural environment have resulted in important new insights into their feeding habits and postures. Crinoids living in very shallow water tend to be very sensitive to illumination. During daylight they hide in shadowy places, resting with the arms rolled spirally inward over the tegmen. The start of feeding activities is regulated by the sunset, feeding continues during all of the night and stops shortly before sunrise. Such very shallow-water crinoids are truly nocturnal animals. Crinoids living in somewhat deeper water seem to lack such pronounced diurnal rhythms, and are sometimes supposed to feed almost continually.

Among reef-dwelling comatulids two feeding postures are known. The first posture has been called a **"brachial filtration fan"** by MAGNUS (1967) and BREIMER (1969). This feeding posture is illustrated in Figure 210, and was described by MAGNUS (1967, p. 649-650) as follows:

> Feeding obviously takes place only by filtration from the current. To catch food the animals try to climb higher places such as the tips of seagrass-leaves or rock ledges, and at first unroll only few of their arms, but later unroll all of them. Each arm takes a position in which the aboral side faces the current. The pinnules, pressed against the oral side of the arm when inactive, become erect and bend towards the aboral side so that all of them form a grating on each arm, opened against the current. The tube feet bordering the pinnular food grooves are erected so far that they contact the neighbor pinnules. In this position they form a fine network. With the arms extended against the current they orient themselves in a single plane to form a filtration fan across the axis of the current. In order to do this the arms on the side of the body that is not facing the current have to turn up to 180° around their long axis. If two animals are found side by side, they together form a unique fan with 40 arms, because each arm tries to get into a position with undisturbed current. The filtration fan, thus built, is usually held vertically. It can be pressed down and finally be resolved by a stemming pressure of the current. Generally, however, the animals seem to settle down at those places where there is a more or less constant and slight current (2-5 cm. per sec.). In this current strength the fans could remain such. They could not withstand stronger currents of even brief duration. When the current direction changes, the fans are first untied and then the arms are held randomly upward or to the side, but they never form a cup or funnel. The arms turn against the new current direction and form a new uniform filtration fan.

FIG. 210. Undetermined stalkless crinoid, probably comasteroid, on reef off Dominica (Caribbean), clinging to sponges with help of four arms; specimens form filtration fans and probably show their feeding posture (Breimer, n, by courtesy of Dr. Porter M. Kier, Washington).

Comatulids employing a brachial filtration fan actively search for favorable places to feed. Such places are mostly on top of the reef structure. They extend the arms above the surrounding topography so that the arms are exposed to horizontal or unidirectional currents and wave oscillation. The filtration fan is typically held normal to the direction of water movement, which ensures a maximum exposure of the tube feet to food-carrying currents. The arms of such crinoids are featherlike, with the pinnules placed in two rows, but lying in a single plane.

The brachial filtration fan probably is widely in use by current-seeking or rheophilic crinoids. It is now known to occur in reef-dwelling comatulids from the Mediterranean (MAGNUS, 1963, 1964, 1967; FISHELSON, 1974; RUTMAN & FISHELSON, 1969), from the Caribbean (MEYER, 1973a,b; MACURDA, 1973), and also from the Indian and Pacific Oceans (MACURDA & MEYER, pers. commun.).

The second feeding posture among reef-dwelling comatulids is called **radial feeding posture** by MEYER (1973a,b). Crinoids employing this feeding posture live in crevices and other restricted places deeper within the reef infrastructure. Their arms do not extend above the surrounding topography. Water movement within the reef infrastructure is more turbulent and of reduced velocity compared to that just above the irregular reef topography. In response to the complex flow regime of the

FIG. 211. *Cenocrinus asterias* (LAMARCK), 180-240 meters, off Discovery Bay, Jamaica. (Photograph by W. K. Sacco; Breimer, n, through courtesy of D. L. Meyer.)

reef infrastructure, these crinoids utilize the radial feeding posture. In this posture, the arms are extended in many directions and the pinnules are held in a four-row, radial arrangement, providing maximum exposure of the tube feet to multidirectional water movement prevailing in their microhabitats.

The radial feeding posture also seems to be in general use. Although it is not yet known from the Mediterranean, it seems to occur also in Indian and Pacific reef crinoids (MACURDA & MEYER, pers. commun.).

Deep-water comatulids (PÉRÈS, 1958, 1959) use still another feeding posture, known as the **collecting bowl.** These crinoids seem to feed on the settling plankton falling upon them from the "plankton rain," and passively await the arrival of food. The mouth in such crinoids is directed upward, and the arms are spread out so as to form a collecting bowl or funnel to capture food. This feeding posture may be interpreted as being in use by current-avoiding or rheophobic crinoids, relying for their food on plankton rains, rather than water currents.

Traditionally (HYMAN, 1955; NICHOLS, 1960), all crinoids were believed to form a "collecting bowl or funnel" with the mouth at the center. This opinion was based on observation of *Antedon* held in aquaria and fed by means of artificial plankton rains, produced by grinding plankton samples and dropping them in aquaria with motionless water. This traditional opinion has now been abandoned. *Antedon*'s feeding habit in aquaria must be held to represent a special adaptation to living under artificial aquarium conditions in motionless water.

The feeding posture of deep-water isocrinids from the Caribbean has recently been described by MACURDA & MEYER (1974). Isocrinids also seem to use a brachial filtration fan (Fig. 211). During feeding the stalk is slightly bent over, and the crown is held in a more or less vertical position, with arms and pinnules all stretched out.

The arms are held with the aboral side facing the current, and the tips of the arms are recurved. In this posture a parabolic filtration fan is built. In some multibrachiate forms this may form an almost continuous filter. Caribbean isocrinids are moderately rheophilic crinoids, relying for their food on slight water movements.

Recent investigation has shown that many crinoids prefer to live in areas with sufficient water currents so that they can rely on these currents for feeding. Such crinoids are **rheophilic.** Their preferred feeding posture seems to be the brachial filtration fan, either planar (as in comatulids) or parabolic (as in isocrinids). The widespread use of this feeding posture in modern rheophilic crinoids has led BREIMER (1969) to infer that many stalked Paleozoic crinoids may be interpreted as using a brachial filtration fan for feeding, and thus be rheophilic crinoids. (See also section on Autecology of fossil crinoids, this *Treatise* volume.)

Feeding postures discussed so far occur in adult crinoids. LANE & BREIMER (1974) have distinguished a sequence of feeding habits during ontogeny of modern crinoids. For crinoids in the armless cystid growth-stage they inferred that they relied on epidermal feeding, utilizing dissolved exogenous nutrients. Crinoids in the pentacrinid growth-stage use their long, tentacular oral podia for capture of smaller planktonic particles, a mechanism called tentacular feeding. The juvenile, nonpinnulate crinoid growth-stage would use tube feet and ciliary feeding, whereas the mucus net feeding is only to be expected in adult crinoids with pinnulated arms.

PREDATION, PARASITISM, AND COMMENSALISM

No animals are known to feed regularly upon crinoids. Some authors believe that the supposed toxic excretions in mucus produced by the crinoids is unattractive to other animals. Crinoids thus are judged to lack enemies. Many animals, however, live in association with crinoids either as casual or permanent commensals or parasites. CLARK (1921) and FISHELSON (1974) have listed many such organisms, among which are crustaceans (mostly casual commensals), ophiuroids (semiparasitic commensals), gastropods (ectoparasites), and especially polychaete worms (mostly permanent semiparasitic commensals). The polychaetes all belong to different species of *Myzostomum,* which are highly specialized and aberrant. They mostly live on the outer side of the crinoids, arms or pinnules or make burrows, which induce malformations such as cysts, etc. Myzostomes have adopted the habit of sucking up food from the mucus streams coming down along the ambulacral grooves of their crinoid host. Crustacean commensals are interesting in that their color patterns provide camouflage for their life among crinoid arms and pinnules. These have been described by POTTS (1915).

MAGNUS (1963) has observed fishes feeding on the food-laden mucus strands in the ambulacra of *Heterometra savignyi.* CLARK (1921) has discussed the apparent immunity of crinoids to attack by fishes.

An ectoparasitic gastropod and commensal gorgonian octocorals on bourgueticrinids are illustrated in Figure 212. Both are rare and little known cases of parasitism and commensalism in stalked crinoids.

PHYSICAL FACTORS

Attention here is drawn to physiochemical interrelations between crinoids and their environment. It is evident that such physical factors as light, salinity, bottom conditions, water movements, temperature and pressure influence the acceptability of various habitats to crinoids. Some factors may influence or even govern both bathymetrical and geographical distribution of these echinoderms.

LIGHT

The diurnal rhythm of crinoids may depend entirely on environmental light conditions. Very shallow-water reef-dwelling comatulids are known to be night-active, whereas some other tropical reef-inhabiting comatulids are inferred to be day-active. Night- and day-active crinoids are supposed to have quite different reactions to light conditions in their environment.

Few precise data on light responses of crinoids are known, and these only for a small number of species. Generally crinoids

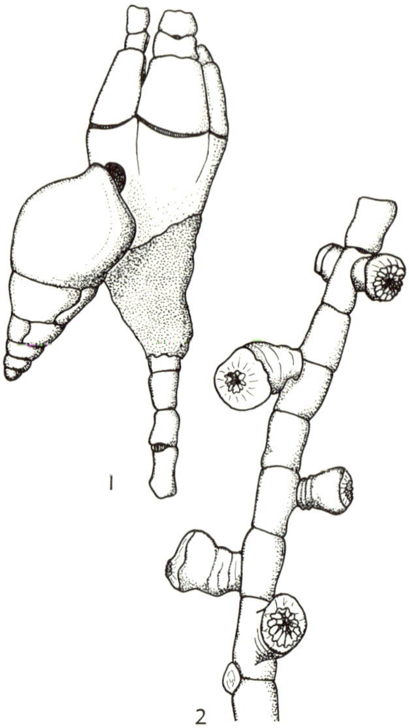

FIG. 212. Parasitic and commensal organisms associated with recent crinoids (Breimer, n).——*1.* Ectoparasitic gastropod attached to calyx of specimen of *Rhizocrinus lofotensis* next to hole drilled in its side.——*2.* Commensal gorgonian octocorals attached to column of specimen of *Democrinus rawsoni.*

appear to avoid direct sunlight and withdraw into shady places. This behavior is also known to characterize littoral species of *Antedon.* Another littoral species, *Tropiometra picta,* which lives in tropical waters, is reported to show no response to intense illumination for it does not avoid direct sunlight and will not seek it when in shadowy places.

The spawning activity of *Comanthus japonicus* is known to be correlated with occurrence of moonlight. Other crinoids have also shown an attraction to weak light during darkness.

The nature of the protoreceptive ability of crinoid sensory systems needs further investigation. HOLLAND (1967) proposed that each crinoid saccule constitutes a photoreceptive lens.

SALINITY

Some crinoids of littoral tropical habitats have shown relative indifference to experimental changes in salinity, including both increase and decrease. Probably they can withstand temporary changes in salinity that might occur in their natural environment. On the whole, crinoids are not markedly tolerant to a definite decrease of salinity, in view of the fact that they are absent from brackish and freshwater environments.

SUBSTRATE

Soft sandy or muddy bottoms are inhabited by stalked crinoids attached by radicular cirri. The radices may penetrate the bottom to provide a holdfast. A majority of stalked crinoids use this mode of attachment. A small minority of comatulids also live on sandy or muddy bottoms, for temporary anchorage to which they have developed very long, slender and straight cirri. These spread out radially to prevent the animal from sinking into the mud.

Most crinoids living on firm substrates such as rocky and shelly bottoms and on arborescent growths such as coelenterates, weeds, and other available substrates commonly possess holdfasts formed by the excretion of lime carbonate. Cementation in exceptional cases is effected by the base (e.g., *Holopus*), or by radicular cirri (e.g., *Democrinus rawsoni*), but usually by an attachment disc that in some stalked crinoids is permanent or that in all comatulid pentacrinoid juvenile stages is transitory. Therefore, pentacrinoids or comatulids may live on substrates other than those adopted in adult stages. Adult comatulids living on rocky or shelly substrates have developed long, but stout and rigid curved cirri. Those which adhere to arborescent growths have short, stout, curved cirri for grasping such means of anchorage.

WATER MOVEMENT

FELL (1966) has noted the relation between crinoids and movements of water masses surrounding them. In this context crinoids may be classed either as rheophobes (current avoiders) or rheophiles (current seekers). FELL concluded that most crinoids are moderate rheophiles. Several good rea-

TABLE 4. *Chemical Composition of Crinoid Skeletal Parts in Relation to Temperature of Environment.*

[Compiled by A. BREIMER from data by F. W. CLARKE in A. H. CLARK, 1921 (*A Monograph of the Existing Crinoids,* v. 1, pt. 2, p. 296-301).]

	Latitude Longitude	Depth (in m.)	Temperature (degrees C.)	$MgCO_3$ (percent)	$CaCO_3$ (percent)
Promachocrinus kerguelensis Antarctic	67°S., 90°E.	375	−1.85	7.86	91.55
Anthometra adriani Antarctic	67°S., 90°E.	375	−1.85	8.23	91.05
Psathyrometra fragilis Japan	44°N., 145°E.	750	1.61	9.25	87.77
Florometra asperrima Washington Coast	47°N., 125°W.	1,145	3.28	9.44	89.45
Pentametrocrinus japonicus Japan	34°N., 137°E.	1,123	3.39	10.15	87.34
Hypalocrinus naresianus Philippines	9°N., 121°E.	612	10.22	10.16	89.66
Parametra granulata Philippines	9°N., 123°E.	502	11.95	11.08	87.86
Crinometra concinna Cuba	23°N., 82°W.	59	26.17	11.69	87.96
Zygometra microdiscus Aru Arch.	6°S., 134°E.	13	–	13.37	85.48
Tropiometra picta Tobago	12°N., 61°W.	littoral	28.–	13.74	83.13

sons favor such an interpretation. Many recent observations on feeding habits of crinoids have shown that these animals rely at least partly on water currents for their supply of food particles. Also for respiration the animals need well aerated water movements. Excessive currents and other turbulence, however, could be disadvantageous to crinoids since they may disturb bottom sediments, with danger of fouling the minute hydropores, if ciliary movements of their funnels are unable to keep polluted water out of the animal. Too strong water movements have been known to prevent individuals of *Heterometra savignyi* from erecting their filtration fans for feeding. Crinoids are also liable to break in strong currents.

TEMPERATURE

Crinoids inhabit waters of temperatures ranging from those of warm tropical littoral seas to cold polar seas and frigid waters of oceanic depths.

Crinoids show several interesting morphological features that are probably correlated with temperature. First, it appears that multibrachiate comatulids (with more than ten arms) preferably inhabit the warmer shallow waters of tropical and subtropical seas, whereas forms living in cold waters of polar seas and abyssal depths predominantly have five arms or ten arms. Notable exceptions are found, but the overall simplification holds true. Next, a correlation between temperature and chemical composition of the crinoid skeleton seems discernible, for the proportion of magnesium carbonate in crinoid hard parts appears to be a function of temperature. Crinoids living in warmer habitats tend to be richer in $MgCO_3$ than cold-water forms (Table 4).

The literature provides some indications that the length of crinoid arms varies with temperature. Representatives of one species living in colder waters tend to have longer arms than those of the same species living in relatively warmer waters.

The nature of response in crinoids to

temperature changes in their habitat, either natural or experimental, is not well known. It is to be expected that warm-water crinoids of littoral tropical seas should be able to tolerate temperature fluctuations. CLARK (1921) has reported that *Tropiometra picta* at Tobago can tolerate temperature fluctuations of at least 15 degrees C., but probably this is an extreme case of temperature tolerance.

On the whole, crinoids probably include both eurythermal and stenothermal types.

POPULATION DENSITY, SPECIES DIVERSITY, COMMUNITY FORMATION

Long known is the fact that crinoids may occur together in very large assemblages. A classic example of such an occurrence is found in a littoral community near Roscoff, France, where multitudinous individuals of *Antedon bifida* cling to laminarians. The crinoid population at this place reportedly is not stable, for specimens vary greatly in number from year to year. *Antedon* may be the dominant member in the community, which otherwise is composed of sponges, bryozoans, and some other benthonic invertebrates. Other dense littoral communities in which crinoids are subdominant have been described by MAGNUS (1963) from the vicinity of Ghardaqa on the Red Sea, and by FISHELSON (1974) from near Eilat, Red Sea.

Trawl samples taken by zoological expeditions have sometimes yielded such massive numbers of crinoids that one can only explain them by postulating the existence of dense crinoid populations on the sea bottom. The record for one such haul is some 10,000 specimens of *Hathrometra tenella* from a depth of 240 meters off Massachusetts.

In recent years sea-floor photography has contributed to our knowledge on this topic. The preliminary results published seem to confirm that crinoids occur in aggregations, although FELL (1966), who has been especially interested in the matter, doubts that crinoids may figure as dominant or even subdominant members of benthonic communities.

Exact figures for crinoid population densities are scarce. The number of specimens of *Heterometra savignyi* near Ghardaqa, estimated from data given by MAGNUS (1963), is about two or three specimens per square meter over an area of 250 to 300 square meters. FELL has reported the density of an undetermined, probably antedontid species, living at a depth of 650 meters on Galicia Bank off northern Spain, as 65 individuals per square meter over an area of about 100 square meters.

FISHELSON (1974) estimated the total number of crinoid specimens counted during one night along a 200 meters long coral table near Eilat as 12,000. In this crinoid community *Lamprometra klunzingeri* constituted about 70 percent, *Heterometra savignii* about 25 percent, and *Capillaster multiradiatus* about 5 percent. The population density of *L. klunzingeri* may be as high as 50 to 75 specimens per square meter.

Dense populations of crinoids seem to be restricted to areas with optimum ecological conditions. The formation of crinoid aggregations is interpreted by HYMAN to reflect the slight amount of dispersal afforded by the feeble swimming powers of larvae. If these settle close to parent animals and if the conditions are favorable they will grow to maturity near the latter, and a large population may result, especially as adults are not much inclined to leave a good environment.

MEYER (1973a,b) found species diversity and abundance of shallow-water comatulids in Colombia and Panama to be greater than in the islands of the Caribbean Sea. He tentatively proposed that this may be related to more favorable food conditions provided by increased or more diversified primary productivity close to the larger, nutrient-shedding land masses.

DISTRIBUTION

The overall distribution of crinoids is governed by several ecological factors, among which temperature and pressure presumably are the most influential in determining large-scale bathymetric and geographic distributions. As a class, crinoids inhabit every part of the marine environment, having adapted themselves to all natural temperatures and pressures and living at all latitudes and depths. On a smaller scale, however, ecologic parameters importantly influence the acceptability of a given habitat for the crinoids. FELL (1966) has defined as two principal parameters 1) clearness of surrounding sea water resulting from absence of turbulence which might introduce sediment into suspension and 2) availability of suitable microplankton and micronekton. To these should be added 3) presence of proper bottom requirements.

For the purpose of clarity, the large-scale bathymetric and geographic distributions of crinoids are discussed below in separate sections. Some emphasis is placed on the distributional patterns of the separate higher systematic crinoid categories.

BATHYMETRIC DISTRIBUTION

The bathymetric distribution of crinoids is mainly governed by the ecologic parameters of temperature and pressure. At the specific level, however, crinoids include stenobathic or eurybathic and stenothermic or eurythermic forms, with every possible interrelation and intergradation between them. Generalizing, crinoid species are prevailingly stenobathic rather than eurybathic. Following observations by ZENKEVITCH (1959), FELL (1966) has listed 12 eurybathic crinoid species, some of which may even tolerate differences in depth up to about 1,000 meters and a factorial pressure increase of up to 100 times. ZENKEVITCH has analyzed the known bathymetric ranges of all extant crinoid species known to him (615 in number). As a result one may state that comatulids generally are stenobathic, showing less tolerance to difference in depth and pressure than stalked crinoids. Eurybathic forms are found chiefly among the isocrinids.

Also generalizing, it is possible to say that the vast majority of comatulids inhabit shallow waters (to 200 m.) in such environments as found on continental shelves, surrounding reefs, and on fringes of island archipelagos. Only a minority of forms occur on the continental slopes. The isocrinids are moderately deep-water forms, mostly occurring between 200 and 1,000 meters in depth, thus being typical inhabitants of upper parts of the continental slopes. Only a few *(Neocrinus decorus, Cenocrinus asteria)* are known to occur in the shallow shelf waters and adjacent to island archipelagos. Their known maximum depth is at about 2,500 meters. Bourgueticrinids are also relatively deep-water forms, entirely absent from waters shallower than 200 meters. They typically inhabit the full bathymetric range of the continental slope and even descend to abyssal depths on the ocean floor, their known maximum depth being 6,000 meters. The few species of cyrtocrinids, like those of the isocrinids and bourgueticrinids again are deep-water crinoids, known to occur between 500 and 5,000 meters. *Holopus* is the only stalked crinoid restricted to shallow waters (10 to 250 m.). As a reef-dweller it is very specialized in its way of direct cementation to the bottom.

GEOGRAPHIC DISTRIBUTION

When geographic distribution of crinoids as a class is studied, it is found that they are cosmopolitan, inhabiting every marine environment of the globe. However, if the distribution of the several different crinoid orders is studied separately, the influence of restrictive ecological factors becomes evident.

The crinoid orders of millericrinids, bourgueticrinids, and cyrtocrinids are worldwide in distribution. This is not surprising, inasmuch as they thrive in deep-water habitats of the lower continental slopes and the oceanic floor where conditions of pressure and temperature vary little and thus do not act as limiting factors on distribution of these crinoids.

The crinoid order Isocrinida and to a lesser degree that of the Comatulida show the influence of restrictive ecologic factors in governing their geographical distribution. It is believed that temperature acts as the principal limiting factor. This, too, is not surprising, taking account of the facts that comatulids predominantly inhabit the shallow waters of shelf seas and that isocrinids flourish on the upper parts of continental slopes. In these upper regions the conditions of water pressure and temperature are far more variable than in lower regions. This is particularly true of temperature conditions, which are subject to far more environmental changes (wave and current actions, seasonal changes, etc.) than

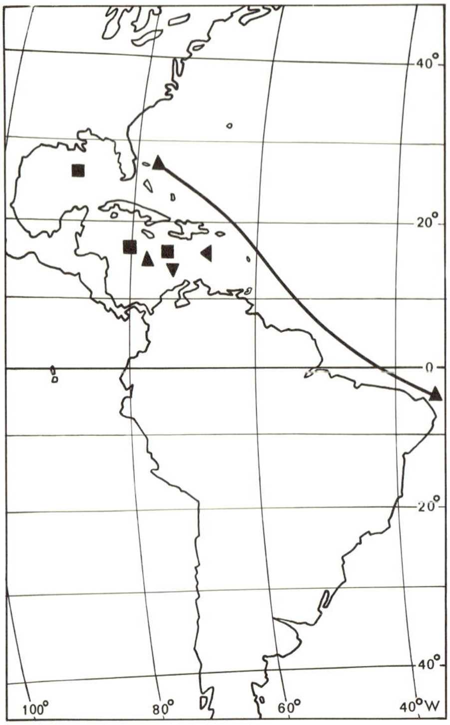

▼ Endoxocrinus
◀ Cenocrinus
■ Neocrinus
▲ Diplocrinus

FIG. 213. Distribution of isocrinid species in central West Atlantic region (Breimer, n). [Each symbol represents one species. A full drawn line between two symbols indicates the total geographic range of one species.]

pressure. Therefore, it is logical to suppose that temperature rather than pressure influences the geographic distribution of crinoids.

Temperature influence is probably to be introduced as a factor to explain the distributional pattern of isocrinids, as given in Figures 213 and 214. They do not occur beyond 40 degrees North and South latitudes. Geographically, the isocrinids have their dominant occurrence in two areas: 1) the central West Atlantic, mainly the Caribbean Sea (Fig. 213), and 2) the Indo-Pacific and West Pacific, dominantly in Banda, Sulu, and Moluccan Seas (Fig. 214). These areas are curiously equivalent in being situated between two larger continental blocks in geologically unstable belts which provide much bottom relief of island archipelagos and deeper trenches; both areas receive water currents from main oceanic streams coming in from the East.

The bulk of comatulids has the same two principal areas of geographic distribution as isocrinids, although they inhabit shallower habitats. Also as an order they are not restricted to these areas, but are worldwide. Their maximum occurrence is in the shallow tropical zones just mentioned, and from there they spread to the polar regions, even reaching the Northern Ice Sea and the Antarctic shelf (MARR, 1963). Probably their number of species and even more so their number of specimens per species decreases in poleward directions.

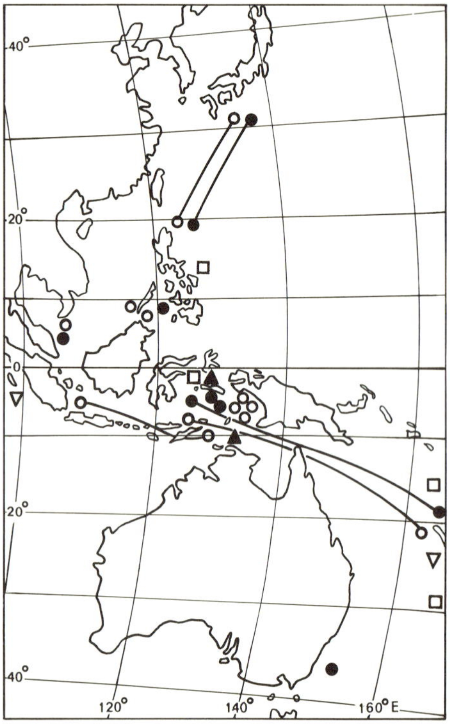

● Metacrinus
○ Saracrinus
▽ Teliocrinus
□ Hypalocrinus
▲ Diplocrinus

FIG. 214. Distribution of isocrinid species in Indo-West Pacific region (Breimer, n). [Symbols and explanations as in Fig. 213.]

PALEOECOLOGY

By Albert Breimer and N. Gary Lane

The morphological diversity of fossil crinoids, expressed taxonomically by the fact that more than 800 genera and 5,500 species have been described, indicates that they have been highly successful animals, especially during the Paleozoic Era. More genera and species of fossil crinoids have been described than of all other fossil echinoderms taken together, a clear indication of their prominent role in ancient marine habitats. The great variety of form that extinct crinoids exhibit surely reflects their presence in many different shallow-water marine environments and is evidence that different forms had somewhat different life styles. These two aspects of study of fossil crinoids, their communal relations and the adaptive significance of their morphology, together constitute the paleoecology of fossil crinoids and provide the content of this chapter. The autecological aspects of crinoids will be discussed first, followed by considerations of the synecology.

AUTECOLOGY

By Albert Breimer

PREVIOUS LITERATURE

The literature on autecology of fossil crinoids, based either on direct analogy with recent forms, or on a functional interpretation of their morphology, is scattered among papers of a predominantly morphologic and systematic nature, and is mostly limited to remarks concerning single crinoid species. In many such cases the crinoids involved are unusual or specialized forms. Such special life histories are dealt with further in this chapter.

Some data on autecology of fossil crinoids may be found in works of a more general ecological nature, e.g., by Bather (1928), Dacqué (1921), and Yakovlev (1964). Older, more specific papers on autecology of fossil crinoids are scarce (Wachsmuth, 1868; Kirk, 1911; Ehrenberg, 1922c, 1928b, 1929, 1930a,b, 1954). Much of the work of Kirk and Ehrenberg was centered on whether fossil crinoids were sessile or vagile animals. Their deductions were almost exclusively based on morphological comparisons, data on ecology of recent crinoids as observed in their own habitat being extremely scarce, if not absent, in their day. Kirk claimed a free-living existence for many fossil crinoid species. Ehrenberg was also strongly attracted to potential free-living forms, some of which (mostly Myelodactylidae) he considered very specialized *"Nebenformen"* or crinoid heteromorphs. One suspects, however, that these crinoids attracted so much attention because of their being curiosities. No attempts were made as yet to explain the mode of life of the more normal or usual crinoid species in relation to important ecological parameters.

More recent literature on autecology of fossil crinoids has been greatly stimulated by two sources. One source is the strongly increased wealth of data and progress in the general fields of ecology and paleoecology. Resultant from this trend are papers (Brower, 1973; Dubatolova, 1973; Halleck, 1973; Lowenstam, 1957; Manten, 1970) seeking to connect the mode of life of fossil crinoids with major physical parameters in the environment (mostly bottom conditions), with special habitats (mostly reefs), and with important ecological factors such as ecological succession. From this type of research many new data are to be expected in the future.

Another stimulus has come from new studies on ecological conditions of modern crinoids in their natural habitats (Fishelson, 1968, 1974; Macurda, 1973; Macurda & Meyer, 1974; Magnus, 1963, 1964, 1967; Meyer, 1973a,b; Pérès, 1958, 1959; Pérès & Picard, 1955; Rutman & Fishelson, 1969). These studies have made possible for the first time more accurate and more valid actualistic deductions for the mode of life of fossil crinoids in relation to major ecological parameters in their environment. Such actualistic deductions were recently made by Breimer (1969), Breimer & Webster (1975), Lane (1968), Lane & Breimer (1974), and Meyer & Lane

(1976). Their papers tried to combine results from actualistic deduction with interpretation from functional morphology. Their work was done in an effort to find paleoecological conditions under which the more "normal" or usual crinoid species could have lived, and to develop a model for their mode of life. A summary of their results is presented below.

GENERAL REMARKS

Autecology of fossil crinoids necessarily is based to a large extent on actualistic extrapolations made by analogy to known habits of recent crinoids. To a certain extent conclusions as to the ecology of fossil crinoids may also be derived from an interpretation of their functional morphology.

The possibility of actualistic interpretations of mode of life and behavior depends to a large extent on direct morphological and anatomical comparability of the major organ systems in crinoids. Key questions are whether organ systems in recent crinoids are organized in a similar manner as in fossil crinoids, to what extent they might differ from one another, and to what the difference in structure might imply for differences in functions of the organs or organ systems. Characteristic differences may indeed be suspected to exist, especially for the fully extinct subclasses Camerata and Flexibilia. Direct comparability of fossil and recent structures indeed determines the degree of accuracy of our actualistic extrapolations. Therefore, the several organ systems of the crinoids will be briefly reviewed.

The ambulacral system, as associated with the water-vascular system (Nichols, 1972), is held to be very uniform and homologous throughout the class Crinoidea. Delpey (1942) and Haugh (1973, 1975a) have described characteristic differences in the arrangement of the digestive tract (including proximal ambulacral tracts) in the theca of the camerates, as compared to recent articulates. Nevertheless, the means for food gathering by arms and pinnules is established to be very similar throughout the entire class. The available evidence suggests that typical crinoids (leaving aberrant armless forms and microcrinoids apart) were always suspension feeders, feeding on planktonic and detrital matter.

The reproductory system is poorly known in fossil crinoids. The genital systems of the several subclasses may have known differences. For instance, the genital system in the non-pinnule-bearing flexible crinoids could have been different from the one known in recent pinnule-bearing articulates, which store the ripe gametes in special genital pinnules. Such differences could have implied differences in reproductory behavior, and even in reproductory advantages or disadvantages influencing their taxonomic diversity. Neotenic, armless crinoids, such as known from the Permian of the Soviet Union and Timor, certainly had a primitive genital system and almost certainly a modified reproductory behavior.

The mode of life of crinoids, and especially the functioning of their arms, depends to a large extent on the organization of the nervous system. This may be specifically true for the main motor system. Throughout the crinoids the aboral nervous system may be considered to be the motor system (see Bather, 1917b; Haugh, 1975b; Paul, 1970; Spreng & Parks, 1953; Yakovlev, 1954b; Žítt, 1973). Any direct indication of the sensory system is lacking, but it is reasonable to suppose that the tube feet exercised a sensory function.

Respiration in crinoids is essentially a function of the entire body wall, but may be located more specifically in specialized respiratory tissues, such as the tube feet, or in specialized respiratory organs. Yakovlev (1944) interpreted the inadunate anal sac as a respiratory organ, by analogy of anal respiration in some recent crinoids. Springer (1900a) has described specialized pores in inadunate crinoids, which Lane (1957) supposed were either respiratory or modified madreporic structures. Among others, Kesling & Paul (1971) have described specialized respiratory structures in the aboral cup of fossil crinoids, which are unknown in recent crinoids.

As the main organ serving the crinoid for support, the skeleton deserves attention here. The microstructure of fossil crinoids has not been studied comprehensively as yet, but from scattered data in the literature it is not expected to differ significantly from that of recent crinoids (see Macurda

& Roux, this *Treatise* volume). The porosity of the skeleton may have been variable (Strimple, 1972a). Also, the microstructure of fossil crinoids may have been adapted to special functions it might have had in relation to associated mesodermal tissues. This is known to be the case for ligaments and muscles in articular surfaces in the arms (Lane & Macurda, 1975).

The chemical composition of the skeleton in fossil crinoids (Bethune & Martin, 1969; Weber, 1968) is known to be comparable to that of recent crinoids, although actually few data are available as yet. Also, the chemical composition may be expected to vary strongly with ecological parameters in the environment, perhaps especially with temperature.

The growth of fossil skeletons is accretionary throughout their forms. The growth of plates characteristically produces growth lines on the internal sides of the plates (Meyer, 1965). Regeneration in fossil crinoids must have been very much the same as it is now (see Hattin, 1958; Strimple & Beane, 1966; Yakovlev, 1952; and many sources scattered in literature). The reactions of fossil crinoids to mechanical influences during growth have been studied by Ehrenberg (1922b) and Yakovlev (1947c,d; 1949a). The functional interpretation of sculpture patterns in fossil skeletons (Yakovlev, 1950) is in need of rethinking, protection being their prominent function postulated so far.

Many authors have described aberrant and anomalous growth phenomena, even leading to deformed growth-structures (Etheridge, 1879; von Graff, 1885; Sieverts-Doreck, 1963; Springer, 1926b; Strimple, 1957; Wanner, 1949a, 1954). Such structures are either related to pathological or mechanical causes, or are to be seen as a response to activities of parasites or commensals. These phenomena seem to be directly comparable to facts known from recent crinoids. A special case of aberrant growth may be represented by the phenomenon of gigantism (Yakovlev, 1954a). The most important deviations from normal growth patterns are to be found in fossil crinoids that underwent regressive trends during their evolution and have developed rudimentary organs (see Arendt, 1968, 1971, 1972; Wanner, 1920; Yakovlev, 1946, 1951).

The evidence available so far seems to indicate that deductions on behavior of fossil crinoids are indeed valid, even within the limitations discussed above.

FUNCTIONAL MORPHOLOGY AND ACTUALISTIC INTERPRETATION

General remarks. Consideration of the mode of life of stalked crinoids has proven to be more fruitful if water movements in the environment which they inhabit are taken into account. Crinoids may be classed either as current-seekers or rheophiles, or current-avoiders or rheophobes. Their way of life was probably strongly influenced by this important ecological parameter.

With regard to water movement in the environment, two different feeding postures are postulated for fossil crinoids as adaptations to either a rheophilic or a rheophobic mode of life. Rheophilic crinoids probably employed a mode of feeding known in recent crinoids as brachial filtration fan feeding (see section on Ecology). In such a feeding posture the arms and pinnules are spread out, so as to form a brachial filtration fan, which is held with the aboral side facing the water current. In this posture crinoids are thought to have fed by means of filtering plankton from a horizontal water current employing a mucus net formed by the erect tube feet in order to intercept planktonic and detrital food particles. Rheophobic crinoids probably employed a mode of feeding described as collecting-bowl feeding. Arms are spread out horizontally, and food is collected from the settling planktonic and detrital matter.

More detailed information on the mode of life of any particular crinoid species may be obtained from an interpretation of their functional morphology. From the point of view of functional morphology, the living position of the stalked crinoids as elevated benthic animals is deeply influenced and even dependent upon 1) the mechanical and functional properties of the stalk; 2) the position of the center of gravity in relation to the bottom; 3) the mobility of the arms and probably also their cirri; and 4) their ability to form an effective food-gathering apparatus.

Most Paleozoic stalked crinoids have the essential part of the body (the food-gathering and food-digesting crown) elevated to a level well above the bottom. Most sessile benthic invertebrates rest directly on the bottom. By elevating themselves many crinoids have given up all the direct advantages that the bottom affords as a supporting surface, and they have had to solve all the mechanical and hydrodynamical problems connected with their elevation from the bottom. Their main problem lies in the fact that by elevating their weighty crown, they have also elevated their center of gravity to a position well above the bottom. Commonly the center of gravity is in the crown or just below it, in the proximal part of the stalk. By elevating themselves from the bottom, crinoids have made themselves top-heavy. Nevertheless, this elevation may be explained as a meaningful adaptation, clearly advantageous to the animals and working to their benefit.

Four basic mechanisms of elevating the crinoid crown from the bottom may be distinguished. The crown is kept elevated 1) on a vertical stalk, serving as a direct support for the crown; 2) by adjusting movements of arms, probably aided by righting movements of the cirri; 3) by obtaining lift from horizontal water currents; and 4) by decreasing its specific gravity, that might result in neutral buoyancy.

Benthic forms elevated on rigid stalks; specialized arm structures; rheophobic adaptations. Elevation of the crown by direct support of a vertical stalk is best performed by a stout and rigid stalk. Any stalk with little or no inherent flexibility and with good dimensions and weight serves this purpose best. A stalk of this nature could bear the functional name column. It is evident that a crinoid in the possession of such a rigid column would have to feed on the settling plankton, because in the absence of flexibility in the stalk its crown would be unable to bend over to form a filtration fan. If it had to rely on settling plankton, forming a collecting bowl with its arms to catch food particles, it is also evident that, theoretically at least, there is no need for the crinoid to be elevated from the bottom, other than avoiding some crowding on the bottom. Feeding in this way can as well be performed by animals directly resting on the bottom.

These considerations may explain why stalks of this columnar type are relatively rare among crinoids. *Euspirocrinus,* a Silurian inadunate crinoid from Gotland; *Calpiocrinus,* a Mississippian flexible crinoid from North America, and *Apiocrinites,* a Jurassic articulate from Western Europe, may be examples. It may also explain why some benthic crinoids are devoid of a stem, or have only rudiments of the stem left, or are even cemented to the bottom.

Reduction of the stalk, leaving only a rudimentary stump, occurs in some species of *Millericrinus,* a Jurassic articulate crinoid from Europe. These forms rested directly on the bottom. Some genera have completely given up the stalk, at least during their adult life stages. A much cited example is to be found in the genus *Agassizocrinus,* a Mississippian inadunate crinoid from North America (see ETTENSOHN, 1975). Such genera have added much weight to the aboral cup, bringing the center of gravity down. They are constructed like "roly-poly" dolls. They rest freely on the bottom and employ collecting bowl feeding. Similar phenomena are known from species of *Edriocrinus,* a Devonian camerate crinoid, and some species of *Pterotocrinus,* a Mississippian camerate crinoid, all from North America. This trend is most pronounced, however, in the genus *Timorocidaris,* a crinoid from the Permian of Timor (LAKEMAN, 1950; WANNER, 1951).

Some forms, apparently also resting directly on the bottom during adult life, have modified their base. Such is known in the genus *Calceolispongea,* a Permian inadunate crinoid from Australia, Timor, and India (TEICHERT, 1949). In this form the basals underwent excessive growth so as to become bulbous or spatulate, to allow the crown to rest directly on the bottom.

Still other genera became cemented to hard bottoms, or hard substrates or objects. This is known to occur in some species of *Edriocrinus,* but also in *Palaeoholopus, Calycocrinus,* and *Permobrachypus,* flexible crinoids from the Permian of Timor, in the Jurassic genus *Cotylederma,* and the Cretaceous genus *Cyathidium* (see JAEKEL, 1907).

It is increasingly evident that crinoids in possession of rigid, columnar stalks, or without stalks, must be looked at as adapted to a rheophobic mode of life. Further evidence to support this view comes from a functional interpretation of the structure and mobility of their arms. Examples may be drawn from both the flexible and camerate crinoids.

As far as flexible crinoids are concerned, it is probably highly significant that their arms do not bear pinnules. Moreover, in some of these forms the free arm endings may be very short, making the surface area of the arms comparatively low if compared to the surface area of the theca. Such short, non-pinnule-bearing arms are not well adapted to filtering plankton from a current. For this reason the crowns of many flexible crinoids could never have functioned as filtration fans. Rather, they are to be regarded as collecting bowls. Theoretically, extreme adaptations could exist in which the whole crown is cup or beaker shaped for better performance of its function. Indeed, such forms did exist among the flexible crinoids. Forms like *Icthyocrinus,* with its expanded theca and relatively short arm endings, may be taken here as an example.

As far as camerate crinoids are concerned, it becomes evident that elevation of the crown on a fully rigid, columnar type stalk is to be found in very specialized forms, probably living in special habitats. The genus *Trybliocrinus,* a camerate crinoid from the Lower Devonian of Spain is a typical example (BREIMER, 1962). In this form the columnar stalk serves to elevate a very weighty, almost gigantic theca, with strongly modified, and nearly immovable arm structures.

The camerate genera *Barrandeocrinus* and *Polypeltes* from the Silurian reefs of Gotland may provide even more extreme examples (UBAGHS, 1956). These crinoids are interpreted as rheophobic crinoids, probably living in the infrastructure of the reef, where currents are absent or non-directional (turbulent). The crown of *Barrandeocrinus* is highly specialized, with arms recurving and forming special arm chambers by close juxtaposition of adjoining pinnules. *Barrandeocrinus* is interpreted by UBAGHS as a crinoid creating its own water currents for feeding and respiration. This current is created by ciliary actions on the ambulacral tracts of the arm chambers, thus drawing water into the chambers. Other examples of rheophobic crinoids creating their own currents for feeding may be found. *Eucalyptocrinites,* a camerate genus from the Silurian of North America and Europe, has a very specialized adoral body wall, which together with the pinnulated arms could have formed suction chambers for creating its own feeding currents (C. E. BRETT, pers. commun.), in a way more or less analogous to *Barrandeocrinus.*

The highly specialized genus *Crotalocrinites,* a Silurian inadunate crinoid from Europe and North America, and probably also a reef dweller, may be taken as an extreme example of a collecting bowl feeder. Arm groups are joined laterally in each ray, and spread out horizontally to collect food. The stalk is of the rigid, columnar type, but very short. Perhaps this genus also may be seen as creating its own water currents.

Also, flexible crinoids adapted to a rheophobic mode of life may be expected to have created, at least partly, their own feeding currents. Such currents could have been produced by pulsatory movements due to contractions and expansions of the crown, brought about by inward and outward movements of the arms.

Crinoids, like the ones discussed above, are adapted to living under conditions of slack water or slight water movement (probably mostly nondirectional or "turbulent"). As such they are interpreted as rheophobic crinoids. These crinoids may be active rheophobes, which are not actively current avoiding crinoids, but actively creating their own water currents for feeding and respiration. As such they are contrasted to passive rheophobes, which do not create their own currents.

Benthic forms elevated on partly flexible stalks; nonmuscular arm structures; rheophobic trends. Usually, however, the crinoid stalk is not fully rigid, but partly flexible. This is probably the most common situation found in fossil crinoids. Such stalks are found in all three Paleozoic subclasses of crinoids. The flexible part of the stalk is generally the upper or proximal part of the stalk, the lower or distal part being rigid. Generally, there is in the stalk a gradual decrease of flexibility away from the cup. This gradual change of flexibility is inherent to the way many crinoid stalks are formed during ontogeny.

When the stalk is not entirely rigid, a certain slight amount of water pressure exercised by horizontal water movements tends to displace laterally the heavy, elevated crown, and with it the elevated center of gravity. If a crinoid wanted to keep its stalk vertical so as to support directly the weight of the crown, keeping its center of gravity in a vertical position above the point of attachment of the stalk, the crinoid should be able to achieve certain functions, either separately, or in any combination of them: 1) control of rigidity in the proximal part of the stalk, 2) righting movements with the cirri, and 3) adjusting movements with the arms.

Rigidity control could be effectuated by stiffening the upper, or proximal, part of the stalk through tightening of the ligaments in between the stem segments. It is completely feasible to think that somewhat delicate and partly flexible stalks could have functioned as columns lending direct support to the crown.

It is believed that arms would be needed to make occasional adjusting movements in order to assist in maintaining balance. LANE (1968) first suggested such actions for the arms. Also, it is feasible that righting movements of the cirri would produce the same effect of elevating, righting, and balancing the crown in slightly flowing water. It certainly would be correct to think that a combination of these three possibilities has the best effect under conditions of slight, horizontal water pressure. Righting and balancing the crown probably was a combined action of the stalk and the arms. It would perhaps be advantageous if most of the efforts were extended by the stalk and its cirri, thus not disrupting the feeding of the arms.

Camerates probably did not have the right sort of arms (biserial arrangement of brachials, absence of articular brachial surfaces) to be used for exercising adjusting movements. In such a case righting movements were probably made by long and flexible cirri at the proximal part of the stalk, as in the genera *Clarkeocrinus* and *Cordylocrinus,* Devonian camerate crinoids from North America. Such movements would create water currents from which the arms could profit in feeding.

Adjusting movements of the arms could be very efficient in case of specially modified arm structures such as the brachial petals of *Petalocrinus,* a Silurian inadunate crinoid from Europe and North America, probably also a reef dweller.

If arms are not specifically modified for such a function, adjusting movements of the arms in general would be possible only for Paleozoic flexible and inadunate crinoids, and in all Mesozoic and Cenozoic articulate crinoids, which have the proper articular connections in the arms.

As an overall conclusion, it seems valid to state that elevation of the crinoid crown by direct support of a rigid, columnar stalk, or by a partly flexible stalk capable of rigidity control, is probably useful only in the presence of slack water or slight, mostly nondirectional, water currents. In such cases there is evidence of modified, or specialized, brachial and cirral structures, further testifying to the probable rheophobic nature of their adaptations.

Benthic forms elevated on partly flexible stalks; pinnulate nonmuscular arm structures; rheophilic trends. It is believed that a vertical position of the crinoid stalk, particularly if its upper part is flexible, is of no use in case of stronger currents, especially not if these currents were unidirectional, or periodically unidirectional (bidirectional, such as tidal currents). It would be unrealistic to think of a crinoid actively fighting any stronger current action in efforts to maintain its balance in a vertical position. Rather, it is believed that the prime reaction of a crinoid toward a persistent horizontal water current is to form a subvertical brachial filtration fan and to initiate feeding from the current. This is known to greatly increase its feeding efficiency.

There is a potential action-reaction system between a crinoid and a persistent, horizontal current in its environment. The hydrodynamic effect of such a current on a crinoid whose stalk is flexible in its upper portion is to passively orient it. A passive orientation of the crown can be obtained without any positive reaction of the crinoid toward the current. The crinoid will lean over with the crown downcurrent from the stem attachment. The stalk is required to support the crown; it has to be partly rigid, at least in its lower or distal half, in order to ensure the elevation of the crown. The stalk typically stands with its convex side facing the current. It must be firmly and definitely anchored in order to prevent the crinoid from breaking or becoming uprooted and being carried away by the current. Anchoring would be either by cementation to hard objects or by rooting in soft bottoms. Passive orientation of the crown by persistent water currents is probably connected with definite anchoring of the crinoid in one place. This may have had unfavorable consequences for the crinoid in case of deterioration of environmental conditions in its habitat.

If the crinoid spreads out its arms and pinnules, a brachial filtration fan is built in essentially the same way as is known in recent crinoids. The brachial fan is oriented perpendicular to the current, the aboral sides of the arms and pinnules facing the current. Filtration fans thus built could be either planar, conical or parabolic, depending on the brachial and thecal structures of the crinoid. In case the crinoid is unable to make active, discriminate, muscular movements with the arms, the crown remains passively oriented, perpendicular to the current. The current will exercise upon it as much upward as downward pressure. As a consequence, no lift is derived from the current to help elevate the weighty crown.

The biological effect of the rheophilic mode of life, outlined above, is considerable. Not only is an enormous body of water being filtered during each active period of the crinoid, but also, the capturing of food is greatly aided. The brachial fan has a baffling effect, and at the lee side of the fan a turbulence may be created, which greatly facilitates the capture of food by the tube feet. Food transport in ambulacral tracts occurs at the sheltered sides of the arms. Fecal wastes are carried away hygienically by the current.

The ecological requirements for a rheophilic mode of life was passive orientation of the crown in the current, but without receiving lift from it, seem to be almost perfectly filled by the batocrinid camerates. The essential morphological characters of camerate crinoids seem further to indicate the rheophilic nature of their adaptations. In this feeding posture the arms are not required to make complicated movements. Arms of camerates lack muscular contacts between brachials. Also, feeding would be greatly aided by the presence of as many pinnules as possible. Camerates have provided for this by making the arms biserial, increasing the number of pinnules per arm two or more times.

Passively oriented crinoids, such as the ones discussed above, may be able to regulate the pressure exercised on their extended brachial fans. This pressure is a function of the total surface area exposed to the current. By folding pinnules in and out, a pressure regulation system may be developed. The advantage of it evidently would be to provide for an equilibrium between the pressure exercised by the current, and the strength of the anchoring device. Such a regulatory system, again, may save the crinoid from being uprooted or carried away by the current.

Some camerate crinoids, like the rhodocrinitids and dimerocrinitids, have the distal parts of the stalk coiled around objects. A well-known example of it is *Acanthocrinus rex* from the Devonian of Europe (Jaekel, 1895). This type of attachment was a definite one, judging from the wedge-shaped columnals in the coiled part of the stalk. Perhaps such type of attachment also was profitable from a point of view of pressure regulation, the anchoring being less rigid and able to give somewhat if necessary in order to prevent too much longitudinal stress in the stalk. Some slip at the attachment site would have prevented the stalk from breaking. This type of crinoid is typically devoid of cirri and has many nodal plates in the stalk (see also Bather, 1912; Brett, 1978; Ehrenberg, 1928b, 1929).

Passive orientation of crinoids in currents

may have had a disadvantage. Partial or differential movements of different arm groups may have caused the crown to rotate about a horizontal axis, thus causing torsion in the flexible, upper part of the stalk. It is believed that many crinoid stalks could have coped with a certain amount of torsion, but this may be doubtful for some others.

Pterotocrinus, a camerate crinoid from the Mississippian of North America, probably was not able to cope with such a situation. In some of its species (e.g., *P. spatulus*) there are large, bladelike processes on the adoral body wall, which could well have functioned as stabilizing fins, in order to prevent the crown from rotating. Other species (e.g., *P. coronarius*) probably gave up the stalk in adult life, adding much weight to the crown, and rested directly on the bottom. Another mechanism preventing the crown from rotating on the stalk, which would resist torsion, could perhaps be found in the spirally coiled stalks of some platycrinitid crinoids. Spirally coiled stalks are supposed to resist torsion.

A rheophilic mode of life is also possible for some flexible crinoids, especially those with densely ramulate arms, like taxocrinids. These flexibles are interpreted by Meyer & Lane (1976) as forming a filtration fan similar to the modern basket star, *Astrophyton muricatum* (Lamarck), which forms the fan at low current speeds.

Benthic forms; elevation involving lift; pinnulated, muscular arm structures; strong rheophilic trends. Rheophilic adaptations for the crinoid's mode of life could be still more advanced. The situation in a persistent current could be entirely different if the crinoid were able to exert some sort of active orientation of its crown in relation to the direction of the current. If the crinoid were able to actively orient its filtration fan at a proper oblique angle to the current direction the current would exercise an upward pressure on the crown. The crinoid crown then receives a certain amount of lift, which helps it to be elevated, and partly compensates for negative buoyancy which it might have.

Active orientation of the crown in the current direction, so as to derive some degree of lift from it, is of great consequence for the mode of life of the crinoid. Because the crown is elevated from the bottom partly because of the lift, the stalk may provide less support, and become more delicate and flexible, and even much longer. Also, the type of attachment of the stalk is influenced. Attachment may be either definite or not, a less definite anchoring affording many advantages.

If lift were about equal to the negative buoyancy of the crown, or even slightly higher, the crown would be suspended in the water, or even have a tendency to ascend. The stalk no longer would be required to support the crown. On the contrary, the crown would be supporting the weight of the stalk, which could merely serve for anchoring. The total organism would then act as a kite on a line. Extreme flexibility of very long crinoid stalks would be in agreement with mechanical requirements for kite lines.

Crinoids also would be able to regulate the degree of lift. The degree of lift is determined by the total surface area of the crown exposed to the current. Regulation of the degree of lift would be possible, 1) as a response to current direction by changing the angle between the current direction and the plane in which the brachial fan is formed, 2) as a response to current strength, by changing the total surface area by flexing pinnules in and out, thus letting water slip freely through the fan if necessary.

Active orientation, understood as an activity of the crinoid itself in response to current direction, is thought to have been possible mostly in rheophilic inadunates and articulates. This type of orientation requires the arms to be pinnulate and to be movable by muscular control. Muscular actions are clearly impossible for the arms of camerates, but advanced Carboniferous and Permian inadunates and all articulates developed the necessary muscular articulations in their arms to allow for active orientation of the crown, and derivation of lift from a current, as an auxiliary device for elevating their crowns from the bottom. The efficiency of active orientation is probably dependent on the number of muscular articulations in the arms. Clearly, the ar-

FIG. 215. Several specimens of *Seirocrinus fasciculosus* (VON SCHLOTHEIM), Lower Jurassic (upper Liassic), Holzmaden, Germany; attached to driftwood heavily encrusted with shells of *Pseudomytiloides dubius* (SOWERBY), ×0.04 (SMF XXIII/115a, Natur-Museum Senckenberg, Frankfurt; photograph by courtesy of Forschungsinstitut Senckenberg).

ticulates could be much more efficient than the advanced inadunates.

Some Jurassic isocrinids, as *Seirocrinus* and *Pentacrinites,* have exceedingly long stalks (up to 15-20 m.), which are flexible throughout. Juvenile stages of some of their species are known to be pseudopelagic and attached to driftwood (Fig. 215). Mature specimens of some species could have lived as benthic species, deriving lift from the current in order to remain elevated, the stalk acting as a kite-line. It is not considered likely that such crinoids actually stood vertically on a stalk that long. Neither is it realistic to think that such heavy crinoids were hanging down from driftwood in all their life stages. [For an alternative interpretation see this volume, p. *T*865—Eds.]

Among flexible and inadunate crinoids with rheophilic life habits special adaptations for active orientation may exist. They are probably to be found in the pivot-joint in the proximal part of the stalk of some taxocrinid flexible crinoids, such as in *Nevadacrinus,* a Permian flexible crinoid from North America (see Fig. 542,*1*); and in the muscular thecal hinge, developed between the triangular basal circlet and the compound *E*-radial of the calceocrinid inadunates. These special adaptations could be interpreted as allowing the crinoid to rest on the bottom, and to obtain an elevated feeding position, deriving lift from the current in order to ascend. Such special adaptations apply only if they were definitely anchored in one place, and if the arm density of the crown is high enough, as in the genera *Chiropinna* and *Chirocrinus.*

Many rheophilic crinoids must be interpreted as no longer being definitely attached to one and the same spot, but to have achieved a less definite type of attachment, allowing the crinoid to be moved about by the current, to establish itself only temporarily in one place or another. Such temporary attachments would be greatly advantageous for the crinoid, in case of deterioration of ecological conditions on one particular spot.

Many adaptations are known in crinoid stalks that may best be interpreted as being structures for temporary anchoring. This may be found in stalks with cirri placed at one side only, and such stalks should be interpreted as recumbent, and only loosely anchored by the cirri. There are tapering stalks with pointed ends, coiled around objects, which could be interpreted as prehensile stalks, such as found in the genus *Eifelocrinus,* a Devonian inadunate crinoid from Germany. Also, special organs have developed such as the anchor in *Ancyrocrinus,* a Devonian inadunate crinoid from North America (Goldring, 1942; Lowenstam, 1942; McIntosh & Schreiber, 1971) (see Fig. 396,*4*). This heavy anchor-like organ apparently served the crinoid as a drag, allowing it to settle temporarily in one place or another.

Another, very interesting case of specialization of the stalk for a limited free-living mode of life is presented by the inadunate genus *Myelodactylus* (Sil., Dev.; N.Am.,Eu.) and its younger camerate homeomorph *Camptocrinus* (Carb., Perm.; N.Am., Eu., Asia). These genera are essentially benthic crinoids, frequently inhabiting reef environments (Ehrenberg, 1922a,b, 1926a, 1930a,b, 1954; Wolburg, 1938). In both genera the distal part of the stalk is greatly developed, and provided with two rows of cirri. The proximal part of the stalk is delicate, as is the entire crown. These crinoids could spirally coil, so as to enclose the crown within the coiled distal part of the stalk, the cirri radially covering the crown at both sides, thus providing protection. The coiled living position is interpreted to be the resting and hiding position, the crinoid lying on the bottom. During activity the crinoid would uncoil, the bulky distal part of the stalk lying flat on the bottom, serving as a drag, and eventually the cirri could grasp some neighboring parts of the bottom to afford some additional, but temporary fixation. The proximal part of the stalk would stand subvertically during feeding. Movement would be achieved by actions of the cirri, either for crawling over the bottom, or for short swims slightly above the bottom, using the cirri for rowing.

A somewhat similar situation may be found in the genus *Ammonicrinus,* a Devonian inadunate crinoid from Europe (Ehrenberg, 1939; Krause, 1927; Ubaghs, 1952; Wanner, 1954; Wolberg, 1937). In this benthic genus the modified stalk is also spirally coiled around the delicate crown, but is devoid of cirri. One of the two known species was free-living. In its coiled resting position it had a globular shape, which allowed it to be rolled passively along over the bottom by current actions.

Considering benthic stalked crinoids as potential rheophiles has brought us to the subject of vagility versus sessility in crinoids. As we have seen, definite attachment, and thus sessility, is thought to be related to rheophobic life conditions or, at most, to be related to rheophilic crinoids only capable of being passively oriented in a current for brachial filtration fan feeding. Vagility in stalked, benthic crinoids is probably connected, and caused by current action of the waters they inhabit.

Elevation involving density-control; free-living forms. An excellent means for elevating the crown of a stalked crinoid above the bottom would be to lower the overall specific gravity (density) of the crown, so as to approach that of the surrounding seawater. Such a crown would be suspended in seawater, or even have slight buoyancy. A stalk would not have to serve for support, but could merely act as a buoy line.

Lowering the specific gravity of the crown could be achieved by modification of 1) the body wall, 2) the body cavity, and 3) specialized organs. Compensation for the heavy weight of the crinoid skeleton, should, in theory at least, come from light materials, such as gases or light oils, stored either within the body wall, within the body cavity, or within organs specially developed for that purpose. The body wall is known to be very porous, in some cases even very much so, but actually gases never have been proved to be stored in skeletal pore volumes. The body cavity, though small in many cases, could have had special compartments to store light materials, as do special organs.

Not much is known about benthic crinoids employing this type of elevation from the bottom. Perhaps the only case in which this ought to be considered is the prominent, thin-plated anal sac of some dendrocrinid and poteriocrinitid inadunate crinoids, which could have stored light materials. This would not conflict too much with a presumed respiratory function of that organ. If the anal sac were not slightly buoyant, these inadunates are among the more top-heavy crinoids known.

Once achieved, the full benefit of a suspended crinoid crown seems to come from a pelagic mode of life. Probably, the only crinoids who ever evolved a pelagic life by modification of the body wall, making the skeleton delicate and porous, and maybe even by employing light materials, are the roveacrinids (e.g., *Saccocoma,* an articulate crinoid from the Jurassic and Cretaceous of Europe). Some uintacrinids, such as *Uintacrinus,* and perhaps also *Marsupites,* articulate crinoids from the Cretaceous of North America and Europe (Sieverts-Doreck, 1927; Springer, 1901) (Fig. 216), might have modified the body cavity by developing special oil or gas compartments in their expanded, thin-walled theca. Both roveacrinids, and uintacrinids are stalkless crinoids. The best example of modification of a special organ for buoyancy is the gas-filled lobolith at the distal extremity of the stalk in the genus *Scyphocrinites,* a Silurian and Devonian camerate genus from North America and Europe (Ehrenberg, 1926; Haude, 1972; Jaekel, 1904b; Sardeson, 1908; Schuchert, 1904; Springer, 1917a; Yakovlev, 1953). This form is considered to be epipelagic.

A pseudopelagic mode of life may be obtained by crinoids settling on floating or drifting objects, or by settling on planktonic or nektonic organisms. Settling on driftwood is known with certainty to be the case in juvenile growth stages of the camerate *Melocrinites* from the Devonian of Ohio (Wells, 1941), and of certain isocrinids, e.g., *Seirocrinus* and *Pentacrinites,* from the Jurassic of Europe (Seilacher *et al.,* 1968). Settlement on the float of a *Scyphocrinites* is known from a species of *Edriocrinus,* a Devonian crinoid from North America (Kirk, 1911, p. 114). Settlement on orthoconic nautiloids (endocerids and orthocerids) is known from an unidentified Ordovician crinoid genus (Ganss, 1937). Settlement on ammonoids is known from the genus *Cyrtocrinus,* a Jurassic articulate crinoid from Europe (Ganss, 1936; Kuss, 1963).

The most efficient, and least vulnerable, crinoids ever developed, are the stalkless comatulid crinoids. During adult life stages they are free-living. They have developed the ability to crawl over the bottom with the help of cirri, and to swim with the help of their arms. They are able to actively select their living sites, can escape from them if necessary, and can hide away whenever necessary for whatever reason. From the point of view of evolution in crinoids, they seem to be in full bloom at the present time, having colonized many different habitats, from extremely shallow water to the deeper parts of the continental slope.

MODIFICATION OF MODE OF FEEDING

Suspension feeding in crinoids must have been diverse with regard to type. Lane & Breimer (1974) have distinguished several modes of suspension feeding in crinoids, based on an interpretation of the different arm structures, and by analogy with the

FIG. 216. Specimens of free-swimming crinoid *Uintacrinus* clustered on a slab of thin limestone found interbedded with chalk of Cretaceous age in western Kansas (about two-fifths natural size) (from *Historical Geology,* by R. C. Moore, copyright 1933, McGraw-Hill Book Company; used with permission of McGraw-Hill Book Company and University of Kansas Museum of Natural History). [For detailed discussion of this occurrence see STRUVE, 1957.—Eds.]

mode of feeding employed in successive growth stages by recent crinoids.

Full mucus-net feeding, with potential rheophilic adaptations using brachial filtration fans, would be possible for pinnulate crinoids, such as camerates, advanced inadunates, and articulates. Many nonpinnulate inadunate and flexible crinoids could have employed only a limited mucus-net feeding. Crinoids with undivided, nonpinnulate arms, such as gasterocomids and pisocrinids, could have used only tube-foot and ciliary feeding.

A strong reduction of number of arms

is known in neotenic crinoids from the Permian of Timor and Ural Mountains (Arendt, 1968, 1971, 1972; Wanner, 1920; Yakovlev, 1946, 1951), and also from many microcrinoids with arm facets. These crinoids must have relied on a type of feeding called tentacular feeding, employed by recent crinoids, shortly after heterotrophic feeding was initiated. These crinoids could have employed epidermal feeding, in part, by absorbing dissolved nutrients from the environment. Certainly, the armless crinoids must have relied entirely on such type of feeding.

SYNECOLOGY

By N. Gary Lane

The single most conspicuous feature of fossil occurrences of crinoids is a strong tendency for specimens to occur in close proximity to each other. This clustering habit is evidenced by numerous slabs that have several to many crinoid crowns preserved on them (Fig. 216). Common references in the literature are to nests, clumps, patches, or colonies. Examples include the well-known occurrences of Mississippian crinoids at LeGrand, Iowa, and Crawfordsville, Indiana, as well as Pennsylvanian crinoids from LaSalle, Illinois (Strimple & Moore, 1971a), and Devonian occurrences of *Clarkeocrinus* figured by Goldring (1923). In some instances the specimens on a slab all may be of a single species, a condition called a "stand" by Lane (1973); in other cases, several different species may be found in proximity. This clustering habit is still evident in living crinoids as well as in many other shallow water marine invertebrates. Fossil crinoids were mainly sessile attached dioecious animals with external fertilization. Thus, an adaptation that would result in mature males and females living close together would be advantageous in helping insure that a reasonable number of released ova would be fertilized. Although some living crinoids retain the fertilized eggs in special brood pouches, there is little evidence that ancient crinoids did so, but, in any event, close proximity of males and females would still be advantageous. Clusters of mature adults could have resulted from sensing of adults by free-swimming larvae and settling of the larvae nearby, or by free-swimming larvae that did not move far from their parents. Alternatively, passive recruitment near adults may have been enhanced by selective survival of the young in suitable microhabitats where adults already lived. The adults of a few fossil crinoids may have been able to swim or crawl close to other individuals of the opposite sex. The great hindrance to study of these aspects of the clustering habit is that the skeletons of fossil and living crinoids do not allow us to distinguish males from females.

There is little evidence that Ordovician and Silurian crinoids had yet specialized into distinctive communities occupying different habitats. The principal exception to this generalization are the distinctive assemblages of reef-dwelling crinoids of the Middle Silurian, found both in western Europe and North America. The structure of crinoid communities that lived during different phases of Silurian reef growth has been documented by Lowenstam (1957), who recognized that a variety of specialized camerate crinoids were conspicuous reef dwellers during rough water stages. During quiet water phases crinoids were not prominent and camerates and inadunates were about equally represented. Some Silurian crinoids are conspicuous in the ubiquity of their distribution. Both *Eucalyptocrinites* and *Pisocrinus* are reef dwellers, but also occur in many areas of nonreef rocks. Other Silurian crinoids, such as the camerate *Siphonocrinus,* are known only from reefal habitats. Although reefs are found throughout the Paleozoic, specialized crinoids that were confined to reef habitats do not reappear again after the Middle Silurian until the Late Permian, when a variety of small, unusual inadunates are found exclusively on or near reefs in West Texas and Timor. Crinoids are found on or near many Devonian through Pennsylvanian reefs but the genera found in these habitats are also known from nonreefal areas.

During the lower and middle Paleozoic, crinoids probably did compete indirectly with other stalked echinoderms, principally blastoids and cystoids, for suitable life sites. Partitioning of microhabitats among cystoids and crinoids has not yet been demonstrated but may well be shown by careful studies in the future. Crinoids and cystoids are found together in many rocks of Ordovician through Devonian age, and crinoids and blastoids occupied similar habitats in Silurian through Mississippian time, after which blastoids became much more restricted in distribution. Throughout the Paleozoic after the Cambrian crinoids were more widely distributed than either cystoids or blastoids. Many localities have yielded specimens of crinoids without a trace of other stemmed echinoderms, whereas there are few cystoid or blastoid localities at which at least a few crinoids have not been found. This difference in availability of habitats has not been worked out in detail, nor have explanations for this gross aspect of Paleozoic echinoderm distribution been forthcoming. The great diversity of Paleozoic crinoids, compared to other stalked echinoderms, is surely related to their ability to occupy habitats not available to these other echinoderms. One possible explanation for this phenomenon may be that crinoids early in their history developed much longer stems than any of the other stalked echinoderms. As adults, crinoids were farther from the sea floor and not in such direct competition with benthonic dwellers as were cystoids and blastoids, most of which were raised only a few centimeters above the bottom. If this explanation has merit, still it poses another problem concerning crinoid paleoecology that has also not been solved. If crinoids were so successful in their "high-rise" life style, as they surely were, they also had to be successful animals during all of their growth stages, before they attained their lofty position above the bottom. Thus, they had to compete with brachiopods, bivalves, and other bottom-dwelling animals during very early growth stages, and with bryozoans, sponges, corals, and short-stalked echinoderms during intermediate growth stages. The adaptive strategies utilized by crinoids to grow up through the lower levels of stratified communities, of which they were the upper level as adults, have not yet been elucidated.

By Devonian time crinoids began to show a clear division into major groups with respect to gross habitat (Lane, 1971). Camerate crinoids predominated in areas of carbonate sedimentation and continued to do so until the Late Mississippian, when they dwindled so much in diversity and abundance that they were no longer a conspicuous element in any marine community. Camerates reached their apogee during Early Mississippian time, when their remains are primarily responsible for the very widespread, thick crinoidal limestones of this time interval. Inadunate crinoids of the Devonian are found in greatest abundance and diversity in rocks composed of fine, terrigenous clastics. They lived on muddy bottoms and in turbid waters from which most, though not all, camerates were presumably excluded. Inadunate crinoids were present on carbonate sea floors, where they tended to be relatively diverse but sparse compared to the great numbers of camerates that contributed much bioclastic debris in these areas. Inadunates, especially the most advanced ones belonging to the suborder Poteriocrinina, continued to be dominant in areas of terrigenous sedimentation until Late Mississippian time when, for reasons still not clear, the camerates underwent an abrupt decline. By Early Pennsylvanian time, the poteriocrinids had taken over virtually all marine environments and occupied carbonate habitats left vacant by camerates. Whether camerates were eventually "forced out" of these areas by the direct competition of inadunates, or whether the latter simply occupied the niches left vacant by camerates is still an unsolved problem of Mississippian paleoecology.

Flexible crinoids, from their origin from dicyclic inadunates in the Ordovician until they became extinct in the Permian, were always a minor component of Paleozoic marine communities. Although specimens of flexibles may be common at some localities, and in a few instances outnumber specimens of other crinoids, they seemingly were limited by a lack of diversity. One or two genera of flexibles are all that are generally found at most good crinoid localities, whereas at the same sites many genera and species of camerates or inadunates, or both, may be known. The flexibles apparently had a distinctive set of adaptive features that allowed them to consistently occupy specific niches. Although they presumably were quite successful in these niches, for very long periods of time, they were unable to radiate out from them. The main exception is found in the lecanocrinid and homalocrinid flexibles that became homeomorphs of inadunates and that are the most diverse of any flexible groups, especially in the late Paleozoic.

Studies of fossil crinoid communities to date have concentrated on localities where reasonably complete crowns or calyces have been collected. Such sites are widely separated stratigraphically and geographically and do not permit evaluation of important paleoecological problems in areas where crinoids are represented mainly by disarticulated ossicles. Future research on the synecology of fossil crinoids may have as one important facet the study of individual crinoid ossicles, especially those of the calyx and arms, obtained from bulk samples. Many such plates can, by reference to more complete specimens, be assigned to nominal genera or at least to a suborder. In this way the gross composition of fossil crinoid communities could be ascertained from many sites where complete calyces or crowns have never been found. Such studies would undoubtedly lead to new hypotheses concerning the evolution and structure of ancient crinoid communities.

MUTUALISTIC RELATIONS OF FOSSIL CRINOIDS

By N. Gary Lane

Many fossil crinoids had quite close and long-lasting relationships with other kinds of invertebrate animals. The exact nature of some of these relationships is still uncertain, hence use of the general term mutualism in the heading for this section, rather than more explicit terms like commensalism, symbiosis, parasitism, or predator prey relations. Even the disarticulated ossicles of dead crinoids that lay on the sea floor were favorite settling sites for the free-swimming larvae of brachiopods, bryozoans, worms, acrothoracic barnacles, corals, and other animals.

The most enduring and best-known mutualistic relationship of crinoids with other animals is with members of the gastropod family Platyceratidae (Clarke, 1908, 1921; Bowsher, 1955). This association began in the Ordovician and continued into the Permian when both these gastropods and the crinoids on which they lived became extinct. Early in this relationship, especially in the Ordovician, the gastropod is commonly found among the arms of crinoids but not in a specific position, as if the gastropod may have simply used the crinoid tegmen as a firm living site and nothing more (Fig. 217,*3*).

By Devonian time the relationship was much more explicit, the snail being invariably found in the posterior interray of the crinoid tegmen, with the aperture over the anal opening of the crinoid (Fig. 217, *2,4;* also see Fig. 40,*2b-d*). These platyceratids are thus interpreted as having been coprophagous, feeding at least partly on the excrement of the host crinoid. Whether this relationship was mutually beneficial to each animal, beneficial for the snail and neutral for the crinoid, or at least slightly harmful for the crinoid, is not known. That the snail lived on the crinoid for most of its life is certain. The growth lines of numerous gastropod conchs match in detail irregular nodes or other ornament on the tegmen of the host crinoid. Lane (1973) has shown that different subgenera of platyceratids were species specific as to the crinoid on which they settled. Thus, either their free-swimming larvae could sense appropriate species on which to settle or an immature snail was able to crawl to a suitable individual. If the life spans of the crinoid and snail were approximately equal, then the snail may have settled as a larva or young adult on the crinoid when the latter was quite young and elevated only a centimeter or so above the sea floor. As the crinoid grew, the snail would have been raised higher and higher above the sea floor as the crinoid stem lengthened. Platyceratids are most commonly found on crinoids that have a solid tegmen with firmly fused plates on which the anal opening is flush with the surrounding surface of the tegmen. A few specimens are known from camerate crinoids with a long anal tube, like *Actinocrinites,* but in these instances the tube is typically broken or the distal end is plugged with small plates and a secondary anal opening was resorbed through the tegminal plates.

An ophiuroid, *Onychaster,* is found among the arms of some Mississippian crinoids (Fig. 217,*1*). It is not known to be associated with the anal vent and is commonly found on specimens with a long, functional anal tube. In such individuals the brittle star is commonly wrapped around the base of the tube. Whether the ophiuroid fed from excrement, from the food tracts of the crinoid, or preyed on the crinoid, or simply used the crinoid as an angling perch, from which to capture its own food supply from passing currents, is not certainly known. Some living ophiuroids live on the stem of stalked crinoids and exhibit close mimicry to the long slender cirri of the stem, in this case surely using the crinoid mainly as a perch.

A variety of organisms caused holes, pits, burrows, or cysts to be developed on various parts of the crinoid skeleton (Warn, 1974; Welch, 1976). Some of these are cysts caused by peculiar ectoparasitic annelids called myzostomes that today infest the arms and pinnules of living crinoids. Insofar as known, myzostome cysts are confined to brachial and pinnular plates in both fossil and living crinoids. Swollen cystlike structures on the stems of Ordovician through Permian crinoids that have been referred to the myzostomes are now known to have been a response on the part of the host crinoid to attached animals with a small, golf tee-shaped, phosphatic skeleton (Fig. 218,*2-4*). The affinities of the attaching animals is not known, but they have been most recently assigned to the Hyolithelminthes (Welch, 1976).

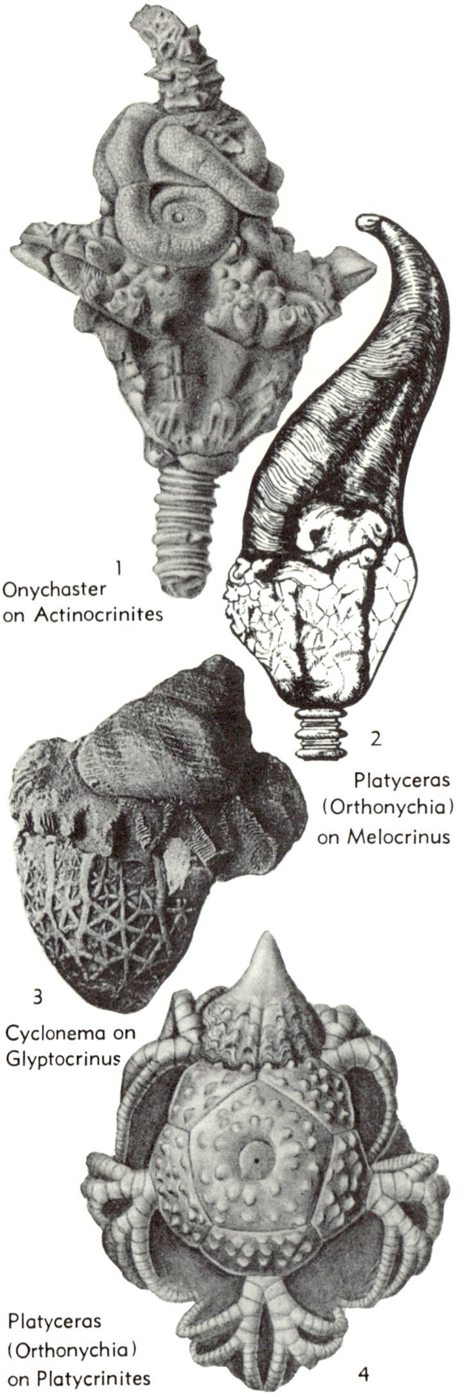

Fig. 217. Mutualistic relationships of fossil crinoids (*1,4,* Wachsmuth & Springer, 1897; *2,* Clarke, 1921; *3,* Bowsher, 1955a).——*1. Actinocrinites multiramosus* Wachsmuth & Springer, with an ophiuroid *Onychaster* fastened to the anal tube; Lower Mississippian (Keokuk), Indiana.——*2. Melocrinites micmac* Clarke with attached shell of *Platyceras (Orthonychia),* Lower Devonian, Gaspé Peninsula.——*3.* Shell of *Cyclonema* sp. attached to tegmen of *Glyptocrinus dyeri* Meek, from Upper Ordovician (Cincinnatian) of Ohio, ×1——*4. Platycrinites hemisphericus* Meek & Worthen, dorsal view, with a *Platyceras (Orthonychia) infundibulum* covering anal opening, from Lower Mississippian (Keokuk), Indiana.

Other induced structures on stem and calyx plates are circular depressions that resemble holes drilled in shells by carnivorous gastropods (Fig. 218,*1*). The great majority of these holes fail to penetrate the plate in which they were started. Whether these are abortive preying attempts on the part of a carnivorous animal or depressions excavated by an organism as a dwelling site is not known.

Very little is known about the predator-prey relationships of fossil crinoids. Several authors have stated that living crinoids apparently have no enemies in the sense that no other animal is known to feed on them (Hyman, 1955). Laudon (1957) has suggested that late Paleozoic bradyodont or hybodont sharks with batteries of flat crushing teeth may have fed on the crowns of crinoids. He envisioned these fishes grazing over extensive crinoid meadows that formed the Burlington Limestone, which also contains numerous shark teeth of this type. The closest living relative to the hybodonts, the Port Jackson shark, does today feed on echinoids *(Strongylocentrous)* and starfishes, which does lend some tenuous support to this hypothesis. Predation of crinoids by sharks would help explain why so many stems and so few heads are found in some rock units, although alternate hypotheses have been proposed (Lane, 1971).

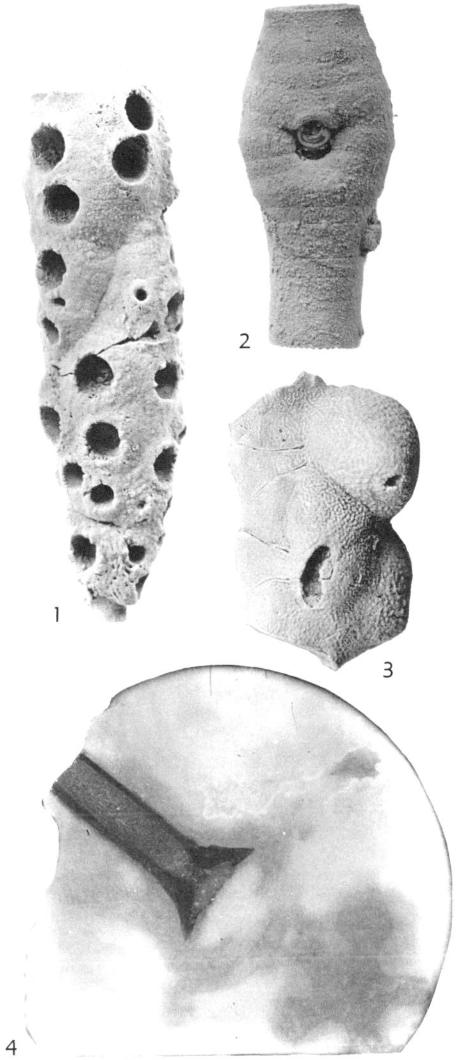

Fig. 218. Mutualistic relationships of fossil crinoids: cysts and other structures (Lane, n; photographs by James R. Welch).——*1*. Crinoid stem that has been penetrated numerous times by borings of unknown affinity, from Upper Pennsylvanian (Seminole F.), Tulsa County, Oklahoma; ×3.2.——*2*. Broken base of a *Phosphannulus* funnel within a cavity in a swollen stem, from Upper Mississippian (Haney Formation), Crawford County, Indiana; ×3.——*3*. Myzostome gall of an arm of an per Pennsylvanian crinoid from the Seminole rmation, Tulsa County, Oklahoma; ×3.7.——*4*. Transverse section of a crinoid stem with a parasitic *Phosphannulus* on a plug of stereom; cavity surrounding funnel is open to the axial canal of the crinoid stem; from Haney Formation, Crawford County, Indiana; ×10.

HISTORICAL REVIEW OF CLASSIFICATION OF CRINOIDEA

By N. Gary Lane

The earliest written account of crinoids was by Agricola in 1546, who proposed four names for the stems of fossil crinoids. *Entrochus* referred to individual round columnals and *Trochites* to a series of such columnals joined together. *Encrinus* was the name given to isolated columnals of pentagonal outline, and *Pentacrinus* to a series of united ossicles of this shape. These four names continued to be used for virtually all known fossil crinoids until J. S. Miller (1821) introduced several generic names that provided the foundation for modern studies of fossil crinoids. Some of the most important advances in knowledge concerning crinoids—such as the conclusion that their fossil remains were really skeletal parts of organisms, that the fossils were of animal rather than of plant origin, that the stem and crown really belonged together and were part of a single individual, that the fossils were related to living echinoderms—were accomplished long before first attempts at classification and hence will not be elaborated here.

Prior to J. S. Miller's clear distinction between stalked crinoids and starfishes, all living and fossil crinoids were commonly associated with the starfishes as one major group of the echinoderms, which were in turn placed within the zoophytes or radiate animals. A distinctive group name for crinoids was lacking until William Martin in 1809 proposed the family Stylastritae for Lower Carboniferous crinoids from Derbyshire. Martin's name was specifically rejected by J. S. Miller because it referred to an assemblage and could not be combined readily with prefixes to form new generic names. Miller therefore replaced Martin's group name with Crinoidea, or lily-shaped animals.

Miller restricted the term Crinoidea to those living and fossil crinoids that had a stem or column, and specifically did not include living stalkless crinoids, called *Comatula,* or the fossil *Marsupites.* He considered these stemless forms to be transitional between starfishes and crinoids. Miller divided the Crinoidea into four divisions:

Classification of Crinoids by Miller (1821)

- Division Articulata. Plates of cup loosely articulated: *Apiocrinites, Pentacrinites, Encrinites.*
- Division Semi-articulata. Plates of cup articulating imperfectly with each other: *Poteriocrinites.*
- Division Inarticulata. Plates of cup adhering by sutures lined by muscular integument: *Cyathocrinites, Actinocrinites, Rhodocrinites, Platycrinites.*
- Division Coadunata. Plates of cup fused to first columnal: *Eugeniacrinites.*

Miller's classification and new group name for these animals were slow to be accepted by other paleontologists. Several German authors continued to use Martin's older name, as Stilastritidae or Stylastritae, rather than Crinoidea, for several decades, and French authors commonly used *Encrinidés* or *Encrines,* based on Agricola's old name, *Encrinus.* A new family-level name, Asterencrinidae, was proposed by de Blainville in 1834 as a synonym of Crinoidea. English authors also did not accept Crinoidea readily. In 1842, Austin and Austin proposed Pinnastella as a replacement name for Crinoidea without attempting to justify the change. They also rejected Forbes' Pinnigrada which was proposed in 1841 as a substitute name for Crinoidea. Neither of these latter names was accepted by subsequent authors. Of Miller's four division names, only Articulata has survived to the present-day classification, where it is used as a subclass for all Mesozoic and Cenozoic crinoids except for the Triassic genus *Encrinus.*

Münster, in 1833, placed the crinoids within the Stellerides or sea-stars. He recognized two major groups of starfishes, the stalked sea-stars or Stilasteritae, using Martin's name for this group, and the Asterites liberi or free sea-stars, within which he placed *Comatula* and *Ophiura.* Within the Stilasteritae, Münster recognized two of Miller's four divisions, distinguishing the articulated crinoids from the nonarticulated ones, which he called Crinoidea inarticulata. He included *Eugeniacrinites, Solanocrinites, Pentacrinites, Encrinites* and *Apiocrinites* in the first group and *Platycrinites, Cyathocrinites, Actinocrinites, Melocrinus* [= *Melocrinites*], *Rhodocrinus* [= *Rhodocrinites*], *Cupressocrinus* [= *Cupressocrinites*], and *Eucalyptocrinus* [= *Eucalyptocrinites*] in the latter group.

The classification of crinoids by F. A. ROEMER (1836) was closely similar to that of MÜNSTER. ROEMER used Stylastritae as a group name equal in rank to the Stellerides, in which he placed *Comatula* with *Ophiura* and *Asterias.* ROEMER accepted three of MILLER's subdivisions based on articulation of cup plates, but included, for the first time, *Marsupites* with the articulated Stylastriten, separated *Poteriocrinus* [= *Poteriocrinites*] as the one genus in the Semi-articulata *(halbeingelenkte Stylastriten),* and included the newly named genera *Caryocrinus* [= *Caryocrinites*] and *Scyphocrinus* [= *Scyphocrinites*] with the inarticulate genera that had been listed by MÜNSTER.

AUSTIN and AUSTIN (1842) proposed a classification of crinoids in which all but one of the then-known genera were placed in the class Pinnastella of the section Echinodermata, substituting this new class name for MILLER's Crinoidea. The class Pinnastella was divided into two orders: Cionacineti for crinoids with a jointed, flexible column, and order Liberidae for crinoids that lacked a stem or were capable of free motion. The latter group included *Marsupites* and *Comatula,* thus placing the living stalkless crinoids with fossil relatives. The stalked crinoids were divided into nine families that contained 25 genera as well as the cystoid *Caryocrinites.* One crinoid, *Sycocrinites,* was placed in the class Adelostella, which was divided into two orders, the first containing the echinoids; the second, order Columnidae, consisting of stem-bearing forms that supposedly lacked arms and included cystoids, blastoids, and *Sycocrinites.*

Most of the aforementioned early workers on fossil crinoids coined new generic names that ended in *crinites.* This procedure was in accord with an early informal practice that generic names of fossils ended in *ites.* This tradition ended abruptly as far as fossil crinoids are concerned when LOUIS AGASSIZ, in 1836, changed the names of all crinoid genera known to him from a *crinites* to a *crinus* ending, for the sake of uniformity. AGASSIZ' authority was accepted tacitly by virtually all later crinoid workers until 1938, when BASSLER published the *Fossilium Catalogus* volume on Paleozoic pelmatozoans. BASSLER returned to the original spellings for these oldest generic names of fossil crinoids, and, in accordance with the rules of nomenclature now in effect, these names are now used with their original, correct endings.

As various groups of stalked, extinct Paleozoic echinoderms, especially cystoids and blastoids, came to be better known and new genera were described, these forms were placed in the Crinoidea. At first, only a few genera were involved (e.g., *Caryocrinites* and *Pentremites*), but the concept of crinoids was expanded gradually until it included all known Paleozoic echinoderms except echinoids and starfishes. This broadening of the idea of crinoids led to an increasingly diverse and morphologically dissimilar group of fossils being included together under this name. For a long time, the tendency was to retain Crinoidea in this broad sense and to propose other new names for crinoids in a strict sense.

In his pioneering studies of the morphology of living stalked crinoids, MÜLLER (1843) divided the Crinoidea into two major groups, *Crinoiden mit Arme* for crinoids, and *Crinoiden ohne Arme* for blastoids and cystoids. The crinoids proper were split into four divisions of very different content. The two large, inclusive groups were the Articulata and Tessellata. The former name was used in the sense of MILLER (1821), but included both stalked and unstalked Mesozoic and recent forms. The latter group of tessellate (plated) crinoids included most Paleozoic genera, as well as *Marsupites.* MÜLLER defined the Articulata as including crinoids with radii free down to the base of the calyx, implying a flexible ventral integument and lack of solid interradials binding the ray plates together. The tessellate crinoids were characterized by a solid plated calyx. The division Costata was named solely for *Saccocoma,* principally because this crinoid was supposed to have opposed "pinnules," and Testacea was erected for *Haplocrinus* [= *Haplocrinites*]. *Holopus* was included but without placement in any of the four groups.

LEUCKART (1848) established a class Pelmatozoa for (transl.) "echinoderms that throughout life, or at least for some period during their youth, are stalked and attached." He divided the class into two orders, which he called *Cystideen* and *Crinoideen.*

BURMEISTER (1856) included in the Crinoidea the Brachiata, formalizing MÜLLER's term, and the Anthodiata, for blastoids and cystoids. ROEMER (1855 in BRONN & ROEMER, 1851-56) included three suborders in the so-called order Crinoidea: Actinoidea for true crinoids, Blastoidea for blastoids, and Cystidea for cystoids. This broad usage of Crinoidea continued until the 1880's (ZITTEL, DE LORIOL), but after that the cystoids and blastoids were generally separated as classes or orders of echinoderms equivalent in rank to the Crinoidea.

D'ORBIGNY (1852) contributed little to the classification of crinoids, simply listing under Order Crinoidea families that included cystoids and blastoids. He informally divided the order into fixed and free crinoids *(crinoïdes fixés et libres).*

ROEMER (1855 in BRONN & ROEMER, 1851-56) attempted the most detailed arrangement of the known kinds of crinoids into hierarchial groups. Within his suborder Actinoidea he devised a dichotomous key based on important morphological features, including presence or absence of a column, and especially on various features of arm development. Only the two primary branches of the key were given names—the Astylida, crinoids without an articulated column, and Stylida, crinoids possessing such a column. The key of ROEMER was formulated as follows:

Classification of Crinoids by Roemer (1855)

Suborder Actinoidea. Crinoids with large pinnule-bearing arms.
- A. Astylida. Crinoids without articulated column.
  - a. Cup adnate: Halopocrinidae. Cyathidiocrinidae.
  - b. Cup free: Astylocrinidae, Marsupitidae, Saccocomidae, Comatulidae.
- B. Stylida. Crinoids with articulated column.
  - a. Arms strongly developed.
    - 1. Ventral side a leathery integument.
      - aa. Arms not normally folded over tegmen: Pentacrinidae.
      - bb. Arms folded into regular pyramid over tegmen: Apiocrinidae [= Apiocrinitidae], Eugeniacrinidae [= Eugeniacrinitidae], Encrinidae, Cyathocrinidae [= Cyathocrinitidae].
    - 2. Ventral side composed of immovable flat plates.
      - aa. Tegmen diffuse between arm bases; arms folded over tegmen.
        - aaa. Arm branches separate: Poteriocrinidae [= Poteriocrinitidae], Rhodocrinidae [= Rhodocrinitidae], Platycrinidae [= Platycrinitidae], Actinocrinidae [= Actinocrinitidae], Melocrinidae [= Melocrinitidae], Ctenocrinidae, Sagenocrinidae [= Sagenocrinitidae].
        - bbb. Arms coalesced into five petal-shaped, reticulate laminae: Anthocrinidae.
      - bb. Tegmen forming apex of crown; arms in resting position embedded into sides of tegmen: Eucalyptocrinidae [= Eucalyptocrinitidae].
  - b. Arms incompletely developed. Haplocrinidae [= Haplocrinitidae], Gasterocomidae.

PICTET (1857) retained the scheme formulated by D'ORBIGNY, merely listing families of crinoids, cystoids, and blastoids under the order Crinoides. GOLDFUSS (1862) continued to use an antiquated classification, placing the crinoids in the Stellerites and dividing them into the Stilasteritae, or stalked sea stars, and the Asterites liberi, or free sea stars. He recognized two divisions of stalked crinoids (Articulata and Inarticulata), retaining these groups in the sense of MILLER (1821). QUENSTEDT (1852) simply listed crinoid genera under the order Crinoideae.

The next important step forward in classification was by WACHSMUTH (1877) who proposed that all Paleozoic crinoids be grouped together in the Paleocrinoidea. He based this grouping primarily on a distinction between the heavily plated tegmen of Paleozoic crinoids, with subtegminal ambulacral tracts, and the flexible tegmen of

Mesozoic and younger crinoids that had an exposed mouth and ambulacral grooves. WACHSMUTH believed that the "visceral disc" of younger, post-Paleozoic crinoids is not homologous with the solidly plated oral part of the theca of Paleozoic crinoids, which he called the "vault." This division into two major groups based on the nature of the oral side of the theca was similar to that of ROEMER (1855, in BRONN & ROEMER, 1851-56) who had recognized a similar, unnamed division of stalked crinoids with well-developed arms. ROEMER, however, placed the Cyathocrinidae [= Cyathocrinitidae] in the group with flexible integuments, whereas WACHSMUTH demonstrated that these Paleozoic crinoids had a solid "vault" similar to that of other Paleozoic crinoids. Within the Paleocrinoidea WACHSMUTH recognized three basic kinds of thecae, which he did not name but informally called the taxocrinid, cyathocrinid, and spheroidea plans. The third name applied to crinoids now called Camerata, and WACHSMUTH set the stage for the eventual recognition of three prime groups of Paleozoic crinoids—camerates, flexibles, and inadunates.

In 1880, WACHSMUTH and SPRINGER elaborated on the Paleocrinoidea, when they changed the spelling to Palaeocrinoidea and named the remaining, exclusively younger, crinoids the Stomatocrinoidea, in reference to an external mouth. They recognized three so-called families within the order Palaeocrinoidea, each corresponding to one of WACHSMUTH's original three plans of organization: the Ichthyocrinidae and Cyathocrinidae, discussed in 1880, and the Sphaeroidocrinidae, named in 1881. They thus formalized the distinctions discussed earlier by WACHSMUTH, while making some changes such as substitution of Ichthyocrinidae [= Icthyocrinidae] for the taxocrinid plan and changing the informal name spheroidea to Sphaeroidocrinidae.

Between these important publications by WACHSMUTH and SPRINGER, the posthumous work of ANGELIN (1878) was issued. The classification adopted there was an extremely artificial grouping of families into four sections based on the number of plates in the proximal circlet of the theca: Trimera, Tetramera, Pentamera, and Polymera. This classification was not used by subsequent workers. Another contemporaneous work was ZITTEL's *Handbuch der Palaeontologie,* which appeared in 1879. Within the class Crinoidea, ZITTEL designated three orders: Eucrinoidea, or true crinoids, Cystoidea, and Blastoidea. Among the Eucrinoidea ZITTEL recognized three suborders: the Tessellata, Articulata, and Costata of MÜLLER.

Beginning with the publications by WACHSMUTH and SPRINGER, one of the most intense periods of discussion concerning the study of both fossil and living crinoids began. This period of several decades was characterized by the description of diverse and well-preserved Paleozoic crinoid faunas from the central United States, and by a substantial increase in knowledge of living, especially stalked, crinoids. Sharp differences of opinion arose as to which morphological features constituted reliable criteria for subdivision of crinoids into major groups. Especially characteristic of this period were the published arguments and rebuttals between English paleontologists, especially P. H. CARPENTER and, later, F. A. BATHER, on the one hand, and WACHSMUTH and SPRINGER, on the other. Each side in these controversies had certain strengths and weaknesses. The American paleontologists generally had much larger, better-preserved collections of Paleozoic crinoids on which to base their arguments than did the British workers. On the other hand, CARPENTER, especially, was a trained biologist and had much experience with living stalked and unstalked crinoids, whereas most of the American scientists had little biological training.

CARPENTER and ETHERIDGE (1881) promptly objected to calling all Mesozoic and younger crinoids the Stomatocrinoidea, principally because they were not convinced that all Paleozoic crinoids had a subtegminal mouth and they predicted that open ambulacral tracts would be found on the tegmen of the Ichthyocrinidae, a discovery that was announced by WACHSMUTH and SPRINGER in 1889. ETHERIDGE and CARPENTER proposed Neocrinoidea as a substitute name and considered other morphological features diagnostic of the neocrinoids, especially lack of anal or interradial plates and perfect five-fold symmetry of the cup as well as consistent division of the arms on the "third radial" (in modern terms, the second primibrach).

De Loriol (1882-84) used Crinoidea in the broad sense of Zittel and others, and under the Eucrinoidea accepted Etheridge and Carpenter's divisions named Palaeocrinoidea and Neocrinoidea. In 1882, S. A. Miller named two new orders of crinoids, each containing a single family, the Lichenocrinoidea and the Myelodactyloidea. The first name applied to plated holdfasts for which the crown was unknown, and the second partly to coiled crinoid stems that Miller believed represented an echinoderm body related to the cyclocystoids, which he also described and placed in the same order.

In 1885 and 1886 the first and second sections of Part 3 of Wachsmuth and Springer's *Revision of the Palaeocrinoidea* were published. Although these authors continued to disagree with the reasons for Etheridge and Carpenter's substitution of Neocrinoidea for Stomatocrinoidea, they accepted the former name, principally because it was euphonious. Wachsmuth and Springer accepted the principle implicit in Zittel's (1879) classification that a relatively large number of families of palaeocrinoids should be recognized, and they both elevated in rank and changed the names of their earlier three subdivisions of the Palaeocrinoidea. The Sphaeroidocrinoidea was changed to suborder Camarata, which was later corrected to Camerata; the Cyathocrinidae was changed to suborder Inadunata, based on the arms being free above the radial plates; the taxocrinid plan or Ichthyocrinidae was changed to Articulata. Wachsmuth and Springer recognized that they were not using this name in the sense of Miller (1821) or Johannes Müller (1843) but considered it such an appropriate name for Paleozoic crinoids referred to the group that no misunderstanding would arise. The use of Articulata in this new sense was not acceptable to many paleontologists, and Zittel in 1895 proposed Flexibilia for these crinoids, Articulata in its original sense applicable to Mesozoic and younger crinoids. Thus, with the completion of Wachsmuth and Springer's *Revision,* the main outlines had been drawn of our present classification of crinoids.

Within the Camerata, Wachsmuth and Springer included both monocyclic and dicyclic crinoids characterized by a boxlike theca of solidly united plates. The Articulata included known crinoids now placed in the Flexibilia, as well as the Crotalocrinitidae. The Inadunata were divided into two branches named Larviformia and Fistulata. The first of these included both monocyclic and dicyclic inadunates characterized by a very simple theca, commonly consisting of only basals, radials, and orals. The simplest genera, *Haplocrinites* and *Allagecrinus,* were regarded as representing true crinoid larvae, "not only of the Inadunata, but of the Palaeocrinoidea generally." The Fistulata were defined as inadunate crinoids in which the visceral disc was partly or completely exposed as a ventral or anal sac that had pores along sutures between some or all of the plates. They divided the suborder into ten families, some monocyclic, others dicyclic, that included, as youngest representatives, the Triassic Encrinidae.

A new classification of echinoderms with several new names for higher categories was published by Neumayr in 1889, who divided the class Crinoidea into two new subclasses, named (in German) the *Hypascocrinen* and the *Epascocrinen.* These were distinguished mainly on the location of ambulacral tracts beneath or upon the tegmen. Thus, these units correspond reasonably closely in definition to Wachsmuth's Paleocrinoidea and Etheridge and Carpenter's Neocrinoidea, although the content of Neumayr's subclasses is quite different. He used three superfamily names within the *Hypascocrinen:* Sphaeroidocrinacea (equivalent to Wachsmuth and Springer's old name for the camerates); Haplocrinacea, corresponding approximately to the larviform crinoids; and Ichthyocrinacea, which included the flexibles, Crotalocrinidae, and Uintacrinidae. The *Epascocrinen* included superfamilies named Cyathocrinacea and Pentacrinacea, thus bearing close resemblance to Roemer's old grouping together of these crinoids. Neumayr's classification, which was far removed from the gradually developing arrangements by British and American workers, did not gain acceptance.

In CARPENTER's (1884a) important work on recent stalked crinoids published in the *Challenger* reports, the long-neglected name Pelmatozoa, introduced by LEUCKART in 1848, was revived. CARPENTER defined the pelmatozoans as a "branch" or subphylum of the phylum Echinodermata, assigning to it all stalked echinoderms. LEUCKART (1848) had proposed two other major divisions of the echinoderms in addition to the Pelmatozoa: Scytodermata for holothurians, and Echinozoa for starfishes and echinoids. CARPENTER's classification was used by WACHSMUTH and SPRINGER in 1885, who published their recommendations based on proofs of the *Challenger* report supplied by CARPENTER. They recognized Pelmatozoa as a class containing two subclasses, BURMEISTER's Anthodiata (for cystoids and blastoids) and Crinoidea (or Brachiata).

In 1886 the final classification adopted by WACHSMUTH and SPRINGER in the last part of their *Revision,* was briefly as follows:

Classification of Crinoids
by Wachsmuth & Springer (1886)

Phylum Echinodermata
- Class Pelmatozoa
  - Subclass Crinoidea (Brachiata)
    - Order Palaeocrinoidea
      - Suborder Inadunata
        - Branch Larviformia
        - Branch Fistulata
      - Suborder Camarata
      - Suborder Articulata (or "Articulosa")
    - Order Neocrinoidea

In 1890, these authors again revised the major divisions of crinoids and abandoned the two main groups, palaeocrinoids and neocrinoids. Instead they recognized three groups, the Camerata; the Inadunata, still divided into larviform and fistulate types; and the Articulata, which they now expanded to include the Ichthyocrinidae, or all flexible crinoids, as well as post-Paleozoic forms. Thus, they returned to a usage of Articulata that was modified and expanded, but more in keeping with the original concepts of MILLER (1821) and MÜLLER (1843). This three-fold division was also used by them in their camerate monograph of 1897, where the following classification was used:

Classification of Crinoids
by Wachsmuth & Springer (1897)

Class Crinoidea
- Order Inadunata
  - Suborder Larviformia
  - Suborder Fistulata
- Order Camerata
- Order Articulata
  - Suborder Impinnata
  - Suborder Pinnata

They recognized division of their broad group Articulata into two suborders, the Impinnata referring to Paleozoic flexible crinoids, and Pinnata to Mesozoic and younger crinoids. These two divisions furnished the basis for BATHER's (1899b) separation of the order Flexibilia into two grades with the same names, although his grade Pinnata was very different in scope from WACHSMUTH and SPRINGER's suborder, including only about one-half of the recognized families of post-Paleozoic crinoids, the remainder being included in dicyclic and monocyclic Inadunata.

In 1890, BATHER published a detailed classification of the fistulate inadunates that foreshadowed his gradually developing ideas concerning the phylogeny and classification of crinoids. BATHER followed WACHSMUTH and SPRINGER's definition of the Inadunata and of the Fistulata, within which he attempted to arrange genera and families of crinoids on a phylogenetic basis. Although no formal names were applied between the suborder Fistulata and the family level, BATHER recognized two divisions which he called group A and group B, and a secondary subdivision of group B, into what he called divisions in the text but omitted from a tabular summary of his classification. Divisions were viewed by BATHER as categories that differed in kind, whereas series represented differences in degree and were stages in evolutionary lineages. He used series for subdivisions of families, equivalent to subfamily rank. Within the Fistulata the family Hybocrinidae was set aside as a separate entity without close relationship to any of the other groups. Group A of BATHER included three families of monocyclic fistulates, the heterocrinids, calceocrinids, and catillocrinids. Group B included all dicyclic forms except for the monocyclic Belemnocrinidae. Within group B, BATHER recognized three main groupings of families, which he informally called divisions in text, but did not name. The first division consisted of the families Dendrocrinidae, ?Carabocrinidae, and Euspirocrinidae. This group was characterized by having three anal plates in the cup and dichotomously branching arms. The Dendrocrinidae contained genera which he judged to be direct progenitors of the other two main groups, *Homocrinus* to the Decadocrinidae, and *Ottawacrinus* to the Cyathocrinidae. The second group, the Decadocrinidae, was characterized especially by having ten arms. Within this later assemblage anal plates were early added to the cup, with a later, secondary loss of anal plates. In addition, there was an evolutionary trend toward development of pinnules. The third group consisted of the Cyathocrinidae, divided into three series.

Implicit in this classification was BATHER's judgment that advanced fistulate crinoids of the late Paleozoic were derived from two different ancestral stocks, genera such as *Poteriocrinites, Scaphiocrinus,* and *Zeacrinites* evolving from a *Dendrocrinus*-type ancestor, whereas scytalocrinids, graphiocrinids, erisocrinids, and cromyocrinids evolved from a *Botryocrinus*-type ancestor. BATHER completely abandoned this scheme three years later in 1893, when, in revising the Silurian inadunates of Gotland, he discarded Larviformia and Fistulata and instead proposed two new suborders: Inadunata Monocyclica and Inadunata Dicyclica, with strict separation based on the presence or absence of infrabasals. BATHER emphasized that in monocyclic inadunates three of the radials were commonly bisected transversely, or two of the radials were conspicuously larger than the other three radials, whereas neither of these conditions was seen in dicyclic inadunates. Monocyclica and Dicyclica were thus initially proposed as subdivisions of the Inadunata.

Six years later, in 1899, BATHER again proposed substantial changes in crinoid classification, but he had so altered his views that the Monocyclica and Dicyclica took on a completely different content. By this time, he had become convinced that crinoids were biphyletic in origin and that monocyclic forms had not evolved into dicyclic crinoids, or vice versa. He thus proposed two subclasses, Monocyclica and Dicyclica, each of which contained different grades of increasing morphologic complexity followed by secondary simplification. These levels of complexity were basically the inadunate, camerate, flexible, and articulate plans distinguished by other crinoid workers. Within each subclass he regarded the inadunate conditions as the ancestral stem from which other major groups arose. The monocyclic inadunates included the larviform crinoids of WACHSMUTH and SPRINGER, as well as monocyclic inadunates (mainly the Hybocrinidae) which previously had been placed in the Fistulata. The Monocyclica Inadunata were supposed to continue to the present, represented by the Hyocrinidae, and to have given rise to two additional Paleozoic orders; the Adunata and the Camerata. The Adunata are composed of three monocyclic families which previously had been classified as camerates: the Platycrinidae [= Platycrinitidae], Hexacrinidae [= Hexacrinitidae], and Acrocrinidae. The Camerata included all other monocyclic "camerates" and were divided into three suborders, named Melocrinoidea, Batocrinoidea, and Actinocrinoidea.

The dicyclic crinoids were also divided into three orders, the Inadunata, Flexibilia, and Camerata. Two informal grades were recognized within the dicyclic inadunates: grade Distincta comprising crinoids with all brachials free above the radials, and grade Articulata characterized by proximal brachials flexibly incorporated into the cup and provided with suprategminal ambulacral tracts. BATHER did not indicate what families or genera should be assigned to these two grades and, instead, divided the dicyclic inadunates into two suborders called Cyathocrinoidea and Dendrocrinoidea. The former included Paleozoic families exclusively, whereas the latter (including *Bathycrinus*) ranged from Paleozoic to recent.

The order Flexibilia was derived from the Dicyclica Inadunata Distincta and encompassed two grades, Impinnata for non-pinnulate forms, and Pinnata for crinoids with pinnules. The Impinnata consisted of all crinoids now called flexibles, and the Pinnata of the great majority of Mesozoic and younger crinoids. The Dicyclica Camerata were crinoids usually classed as camerates that possess an infrabasal circlet.

BATHER's classification was a serious attempt to relate all crinoids within a comprehensive phylogenetic scheme, most earlier classifications being based on morphological similarities and dissimilarities, without discussion of implied origins or relationships. His scheme suffered from two principal defects. There was substantial evidence available that BATHER's basic premise was in error—that monocyclic and dicyclic forms could be more closely related to each other than to other crinoids with the same number of circlets. In addition, the categories proposed by BATHER were awkward and unnecessarily confusing. In order to specify several major groups at the order and suborder level, it was necessary to use binomial or even trinomial terms, such as Dicyclica Inadunata and Dicyclica Inadunata Distincta. BATHER adopted this same classification in LANKESTER's *Treatise on Zoology,* published in 1900. His classification was never used by SPRINGER in numerous subsequent publications on fossil crinoids, or by other American crinoid workers, but was utilized by British paleontologists.

Beginning in 1894, OTTO JAEKEL began a series of papers on fossil crinoids which expressed his views on classification. His initial proposal, in connection with a study of crinoids of the Devonian Hunsrückschiefer, was to divide the Crinoidea into three main groups, Blastoidea, Cladocrinoidea, and Pentacrinoidea; the latter two names were new but were not defined. Only camerate genera were assigned to the Cladocrinoidea and he arranged the Pentacrinoidea in three divisions of unstated rank, but approximately at suborder level, because he named superfamilies within one of these divisions. All pentacrinoid genera were inadunates. JAEKEL proposed the Larvata (a variant of Larviformia), Fistulata, and Costata, the latter used solely for Hapalocrinidae. He divided the Fistulata into two superfamilies named Cyathocrinacea [=Cyathocrinitacea] and Dendrocrinacea, both shortly raised to suborder rank by BATHER (1899b).

JAEKEL gradually defined and expanded his ideas on classification of crinoids and other echinoderms in a series of papers that culminated in his important summation issued in 1918 on the *Phylogenie und System der Pelmatozoen.* The class Crinoidea was divided into three subclasses named Eocrinoidea, Cladocrinoidea, and Pentacrinoidea. He considered the eocrinoids to be the ancestral stock from which the other two groups were derived, the cladocrinoids corresponding to camerates, and the pentacrinoids to all other crinoids. Within the subclass Cladocrinoidea, two orders were recognized, the Monocyclica and Dicyclica, respectively equivalent to BATHER's Monocyclica Camerata and Dicyclica Camerata and to the currently used Monobathrida and Diplobathrida. Within each of these orders several suborders were proposed, each based primarily, but not exclusively, on number of plates in the lowest circlet of plates in the theca.

Within the Pentacrinoidea six orders were recognized, Fistulata, Articulata, Articulosa, Reducta, Turbata, and Costata. The fistulate crinoids included six suborders of Paleozoic, primarily dicyclic, crinoids and correspond approximately to the Cladida of present classification. The Articulosa, which JAEKEL adopted from WACHSMUTH and SPRINGER's provisional name, were the flexible crinoids, which JAEKEL divided into four suborders.

The Articulata were composed of Mesozoic to recent crinoids, including the Triassic Encrinidae, as well as most but not all stalked and stemless crinoids currently assigned to the Articulata. Three suborders were recognized: suborder Typica, stemmed crinoids with basals; suborder Libera, stemless crinoids; and suborder Compacta, comprising crinoids with only radials in the cup. A small group, order Reducta, was composed of secondarily simplified, dicyclic Paleozoic reef-dwellers with fused infrabasals, now placed in the Cyathocrinina. Order Turbata consisted of monocyclic crinoids having radials of unequal size and shape, arranged in five suborders. The Costata, used by JAEKEL in a very different sense from that of MÜLLER, or of JAEKEL's early papers, was also composed of monocyclic forms and included two suborders, Hybocrinites, now considered disparid inadunates, and suborder Hyocrinites, for living and fossil stalked crinoids as well as the stemless Saccocomidae. These were separated from the Articulata because they have large oral plates and, according to JAEKEL, ramules rather than pinnules. A summary of his classification of 1918, above the family level, is as follows:

Classification of Crinoids by Jaekel (1918)

- Class Crinoidea
  - Subclass Eocrinoidea
  - Subclass Cladocrinoidea
    - Order Monocyclica
      - Suborder Tetramera
      - Suborder Disjuncta
      - Suborder Pentamera
      - Suborder Miomera
    - Order Dicyclica
      - Suborder Tetramera
      - Suborder Pentamera
  - Subclass Pentacrinoidea
    - Order Fistulata
      - Suborder Cyathocrinites
      - Suborder Dendrocrinites
      - Suborder Merocrinites
      - Suborder Barycrinites
      - Suborder Rhenocrinites
      - Suborder Poteriocrinites
    - Order Articulata
      - Suborder Typica
      - Suborder Libera
      - Suborder Compacta
    - Order Articulosa
      - Suborder Taxocrinites
      - Suborder Calpiocrinites
      - Suborder Lecanocrinites
      - Suborder Ichthyocrinites
    - Order Reducta
    - Order Turbata
      - Suborder Heterocrinites
      - Suborder Calceocrinites
      - Suborder Triacrinites
      - Suborder Symbathocrinites
    - Order Costata
      - Suborder Hybocrinites
      - Suborder Hyocrinites

Several aspects of JAEKEL's classification deserve amplification. The primary reason for dividing crinoids into two main groups, cladocrinoids and pentacrinoids, was JAEKEL's premise that the arms of these two groups were not homologous. He judged the biserial pinnulate arms of cladocrinoids, which appear very early in the fossil record, to be an intermediate stage between the brachioles of eocrinoids, blastoids, and other primitive echinoderms, and the true arms of pentacrinoids which only gradually developed ramules and later pinnules. He also thought that the arm-bearing calyx plates of the cladocrinoids were not homologous to the radials of pentacrinoids, and he called the former costalia. Another important difference was that the ambulacral tracts branched within the theca of cladocrinoids, hence the name Cladocrinoidea. JAEKEL disagreed vigorously with BATHER's separation of crinoids into two primary divisions, Dicyclica and Monocyclica, maintaining that many monocyclic crinoids represent secondary simplification from a dicyclic condition, hence are not part of a single phylogenetic lineage.

Thus, upon publication of JAEKEL's classification (1918) three very different arrangements of crinoids were offered, each developed more or less independently over a period of years, and each presenting divergent views concerning relationships and phylogeny within the Crinoidea. WACHSMUTH and SPRINGER's (1886) classification, later modified (WACHSMUTH & SPRINGER, 1897) recognized four major groups of crinoids, three major Paleozoic groups, the camerates, flexibles, and inadunates, and one post-Paleozoic group, the articulates. BATHER recognized two throughgoing lineages from early Paleozoic to the present day, based on presence or absence of an infrabasal circlet. JAEKEL set apart camerate crinoids as a distinctive Paleozoic group rather far removed from all other crinoids. Like BATHER, he judged that some crinoids, at the ordinal level, at least, had persisted from Paleozoic to recent time and that living crinoids are polyphyletic.

By 1920, SPRINGER had abandoned Articulata for any Paleozoic crinoids and adopted ZITTEL's (1895) name, Flexibilia, but restricted it to Paleozoic forms, whereas ZITTEL had initially proposed the name to include the Ichthyocrinidae as well as *Marsupites* and *Uintacrinus*. SPRINGER recognized four major groups of crinoids, the Paleozoic Inadunata, Camerata, and Flexibilia, and the post-Paleozoic Articulata.

Since publication of SPRINGER's chapter on crinoids in the ZITTEL-EASTMAN *Textbook of Palaeontology* (VON ZITTEL, 1913) and his monograph on flexible crinoids (1920), virtually all students of fossil and living crinoids have accepted this arrangement, although there has been considerable revision within each of these four groups.

In the English adaptation of VON ZITTEL's textbook, SPRINGER (in VON ZITTEL, 1913) restated the basic classification of crinoids proposed earlier by him and WACHSMUTH, recognizing four main divisions: camerates, flexibles, inadunates, and articulates. He thus accepted VON ZITTEL's name for the Flexibilia and returned to the long-standing usage of Articulata for all post-Paleozoic crinoids, except the Encrinidae. The main new feature of this classification (1920) was division of the Flexibilia into two suborders, named Taxocrinoidea and Sagenocrinoidea. The order Articulata was revised jointly by SPRINGER and A. H. CLARK, who was responsible for treatment of living crinoids in the classification. No divisions between the order Articulata and the family level were recognized, CLARK's earlier order Comatulida, for instance, but the family Pentacrinidae was divided into several sections, tribes, and subtribes to take account of most of the stalked articulates, as well as all known stemless forms except *Saccocoma*.

The classification of living comatulid crinoids was considered by A. H. CLARK in several papers in 1908 and 1909. He proposed the order Comatulida in 1908, and also proposed, in another paper that year, division of living stemless crinoids into two main groups, the Thalassometroida and Antedonoida. The first group comprises crinoids with pinnules of triangular cross section and small eggs; the Antedonoida have round pinnules and large eggs. The two groups were re-named in 1909 as Oligophreata and Macrophreata, respectively, in reference to size of the visceral cavity within the centrodorsal. The two groups also correspond generally to comatulids bearing multibrachiate arms and those having ten or fewer arms, although other important morphological differences are observed. A similar subdivision of the comatulids into ten-armed forms and those with more than ten arms had been proposed as long ago as 1849 by MÜLLER, who recognized two species groups within the broadly defined genus *Comatula*. The concept of comatulids was broadened by SPRINGER and CLARK (in VON ZITTEL & EASTMAN, 1913) to include the fossil *Marsupites* and *Uintacrinus*, classed as tribe Innatantes equivalent in rank to the Oligophreata and Macrophreata.

These major divisions of living comatulids were recognized by CLARK in his exhaustive monograph of the existing crinoids, the comatulid portion of which was started in 1915 but not completed until after CLARK's death. This two-fold division was not accepted by GISLÉN (1924), who proposed four groups of comatulids, splitting the Oligophreata into three tribes. This change was rejected by CLARK in subsequent parts of his monograph. However, GISLÉN's divisions have been accepted by other workers, notably SIEVERTS-DORECK, who utilized GISLÉN's four groups as suborders within the Order Comatulida in UBAGHS' (1953) chapter on Crinoidea in PIVETEAU's *Traité de Paléontologie*. In this classification of comatulids, the Macrophreata stand apart. Living members of the group include the most active crinoids which are characterized by the presence of ten arms, prominent muscular articulations, and a centrodorsal with a large internal cavity. The suborder Comasterina includes comatulids in which cirri are absent or rudimentary and an exocyclic tegmen is developed. The suborder Mariametrina include multibrachiate crawling crinoids with a discoid centrodorsal and the suborder Thallassometrina comprise crinoids which generally bear a pentagonal impression on the base of the centrodorsal.

The remainder of the Articulata, exclusive of comatulids, were classified for many years as miscellaneous families. CLARK (1910b) proposed three orders of living articulates: the Holopodida, containing *Holopus;* the Ptilocrinida, comprising four families of stalked crinoids; and the Comatulida, including stalkless forms and the living pentacrinites. The fossil Articulata were not revised comprehensively until SIEVERTS-DORECK, in MOORE, LALICKER & FISCHER (1952), proposed division of these crinoids into six orders: the Isocrinida, Millericrinida, Cyrtocrinida, Uintacrinida, Roveacrinida, and Comatulida. Her isocrinids encompassed living representatives of the Pentacrinidae as well as several fossil groups. The Millericrinida included CLARK's Ptilocrinida and several exclusively fossil families. The Uintacrinida are Cretaceous stemless forms, and the Roveacrinida are small stalkless Mesozoic crinoids. The Cyrtocrinida include a variety of fossils as well as the living *Holopus*. This classification was repeated by SIEVERTS-DORECK in the *Traité de Paléontologie* (UBAGHS, 1953), with addition of two new suborders within the Millericrinida. This provides the latest authoritative statement on classification of the Articulata.

Between 1920 and 1943 there were few changes in definition of the principal groups of crinoids. Major emphasis in studies was directed toward discovery and description of many late Paleozoic crinoids from the Midcontinent of the United States, especially by MOORE and STRIMPLE, from Russia by YAKOVLEV, and from Timor by WANNER. Prior to this research few crinoids of Pennsylvanian or Permian age were known, and those mainly from western Europe. MOORE and PLUMMER (1940) described many new genera of advanced inadunates and placed them all in the family Poteriocrinitidae.

A comprehensive survey of the classification and evolution of all known Paleozoic crinoids was undertaken by MOORE and LAUDON in 1943. They reviewed the different bases for classification used by WACHSMUTH and SPRINGER, BATHER, and JAEKEL, and concluded that the primary divisions utilized by WACHSMUTH and SPRINGER were in closest accord with the evolution of Paleozoic crinoids as interpreted by them. Therefore, they recognized three subclasses, the Camerata, Flexibilia, and Inadunata, in which they included Triassic Encrinidae. Within the flexibles, they accepted SPRINGER's division into two major groups, the orders Taxocrinoidea and Sagenocrinoidea, but rejected the group called Adunata by BATHER, placing these advanced crinoids with a secondarily simplified cup among the Camerata. They divided the camerates into two major groups, orders Monobathra and Diplobathra, based on presence or absence of infrabasals. WACHSMUTH and SPRINGER had never recognized a primary division within their Camerata, although JAEKEL had divided his Cladocrinoidea into two orders, Monocyclica and Dicyclica, similar in content to MOORE and LAUDON's two orders.

The most significant revisions in classification made by MOORE and LAUDON were within the Inadunata by grouping all monocyclic inadunates in the order Disparata, and within this group recognizing two informal groups called the homo-synbathocrinid stock and the hybocrinid stock. This order was equivalent to BATHER's awkward and potentially confusing Inadunata Monocyclica, and was very different in concept from JAEKEL's separation of these crinoids into two orders, one of which included living crinoids. The dicyclic inadunates were assigned to the new order Cladoidea divided into two suborders, the more primitive Cyathocrinoidea, and the advanced Dendrocrinoidea, terms adopted from BATHER. The coining of new names at the order and suborder level within the camerates and inadunates by MOORE and LAUDON fulfilled several purposes. It effectively set apart their classification from earlier ones, and prevented confusion with earlier concepts of phylogeny and interpretation of morphology which had caused long-established names to be utilized in quite divergent ways by different authors; Larviformia, for instance. They abandoned the old Monocyclica and Dicyclica, substituting for them Monobathra and Diplobathra in the Camerata and Disparata and Cladoidea in the Inadunata. The name Disparata referred to the very unequal size of cup plates, especially radials, in this group of crinoids, Cladoidea was chosen in reference to the richly branched and eventually pinnulate arms of the dicyclic inadunates.

This classification of Paleozoic crinoids has been used with little modification to the present day, and with some elaboration, but without substantive change, is accepted in the present *Treatise* volume. MOORE, in MOORE, LALICKER & FISCHER (1952), changed the endings of some of the new names established in 1943—Disparata altered to Disparida, Cladoidea to Cladida, Monobathra to Monobathrida, and Diplobathra to Diplobathrida. These modifications were made in order to conform with general adoption of the ending -ida for orders and -ina for suborders. MOORE also separated the hybocrinids from the disparids as a new order, the Hybocrinida, based on JAEKEL's (1918) suborder Hybocrinites. The suborders Cyathocrinoidea and Dendrocrinoidea were changed to Cyathocrinina and Dendrocrinina, and the endings for the two flexible suborders were altered in similar manner. The monobathrid camerates were divided into two new suborders, named Tanaocrinina [= Compsocrinina] and Glyptocrinina, thus formalizing the two stocks recognized within this group by MOORE and LAUDON in 1943. Within the cladid inadunates the crinoids that had all been classed in the Dendrocrinoidea in 1943 were divided between the more primitive suborder Dendrocrinina and the more adanced, pinnulate, forms, suborder Poteriocrinina.

The classification of MOORE and LAUDON (1943a) and MOORE (1952a) was adopted by UBAGHS (1953) with some amplification. The principal features of the arrangement by UBAGHS are a number of new categories, especially superfamilies, within established suborders and some new ones. Within the diplobathrid camerates he introduced new suborders named Eudiplobathrina and Zygodiplobathrina, the former divided into three new superfamilies. Each of the two monobathrid suborders was divided into several superfamilies. The disparid inadunates, exclusive of the hybocrinoids, which were recognized as a separate order, were divided into three superfamilies based on different symmetry plans of the theca. No new higher categories were established for the cladid inadunates or flexibles.

The Russian treatise volume on crinoids by GEKKER (1964) adopted UBAGHS' classification, except that superfamilies were not utilized.

CLASSIFICATION OF THE ECHINODERMS

By GEORGES UBAGHS
[Université de Liège]

In Parts S (1968) and U (1966) of the *Treatise* the classical division of the phylum Echinodermata into usually attached forms (Pelmatozoa) and free-living forms (Eleutherozoa) was replaced by a division into four subphyla (Homalozoa, Crinozoa, Asterozoa, and Echinozoa), which had been used for the first time by FELL in 1962. Since the publication of these volumes, the classification of the echinoderms has been the object of varying criticisms, additions, and modifications, of which a short resumé is given below.

Haplozoa. This subphylum, which was not recognized in the *Treatise* (Part S, 1968, p. S364), had been established by WHITEHOUSE (1941) for two enigmatic forms, *Cymbionites* and *Peridionites,* from the Middle Cambrian of Australia. DURHAM (1971), however, believed that the subphylum Haplozoa should be preserved because its two genera appear to be sufficiently distinct from other echinoderms to occupy a place of their own in the classification. The fact remains, nevertheless, that these fossils, which resemble each other only in their apparent simplicity, do not furnish any information as to the organization of the organisms to which they belong. It therefore seems preferable at the present not to assign to them a definite systematic position.

Homalozoa. In 1941 WHITEHOUSE united the classes Machaeridia WITHERS, 1926, and Carpoidea JAEKEL, 1901, in a separate subphylum which he called Homalozoa.

The affinities of the Machaeridia to the echinoderms had first been suggested by BATHER (in WITHERS, 1926), because of the fact that the plates of the genus *Lepidocoleus* show cleavage surfaces judged to be similar to those seen in the plates of fossil echinoderms. In reality, however, as S. BENGSTON (personal communication, 1976) is about to prove, these plates possess a microstructure completely different from that of echinoderm plates. On the contrary, they seem to be of exoskeletal origin, and to have been formed as successive deposits secreted by an epithelium. If this is so, there would be no reason to retain the Machaeridia in the echinoderms, and one can reject with confidence the hypothesis of Pope (1975), according to which *Lepidocoleus* is supposed to be the skeletal envelope of the posterior spines of the stylophoran *Enoploura*.

The carpoids, whose skeleton is typically echinodermal, then remain the only group within the subphylum Homalozoa. They differ from almost all other echinoderms in the complete absence of radial symmetry (at least their skeleton reveals no trace of such) and they deserve to be distinguished taxonomically. Their diversity is such that it was judged necessary by Ubaghs and Caster (1968) to abandon the name Carpoidea and to divide these forms into three classes known as Stylophora, Homostelea, and Homoiostelea. To these a fourth class, Ctenocystoidea, was added by Robison and Sprinkle (1969).

In a series of papers, Jefferies (1967, 1968a, 1968b, 1969, 1973, 1975; Jefferies & Prokop, 1972) has stated the opinion that the Stylophora are not echinoderms but chordates with affinities to the echinoderms, for which he proposed the subphylum Calcichordata Jefferies, 1967, within the phylum Chordata. This thesis, which had already been suggested by Matsumoto (1929) and Gislén (1930), was accepted by Eaton (1970) and favorably considered by Bone (1972). On the other hand, it has been the object of criticism and even rejection by numerous authors (Denisson, 1971; Nichols, 1969; Regnéll, 1975; Ubaghs, 1970, 1971b, 1975; see also discussion in Jefferies, 1967, 1968a). It is true that this theory meets with many difficulties, particularly in view of the fact that the skeleton of the Stylophora is in every way comparable to that of the echinoderms. It is in fact inconceivable that such a skeleton whose characters are unique in the animal kingdom should have been associated with such soft parts as a notochord, muscle blocks, a dorsal nerve cord, a brain and cranial nerves more or less like those of fishes, that is, all features which belong to another phylum, namely that of Chordata. In general, it does not seem to be advisable to substitute for the simplest and most direct interpretation of the observed facts, a construction perhaps ingenious, but nevertheless with a basis whose elements are essentially hypothetical.

Crinozoa. This subphylum was introduced by Matsumoto in 1929. In proposing it for the cystoids, blastoids, and crinoids, Matsumoto only came back to the concept of Leuckart (1848) who, 81 years earlier, had united two of these classes, the cystoids and crinoids, under the name of Pelmatozoa, literally signifying animals possessed of a stalk. The term "Crinozoa," therefore, was unnecessary; however, at the time of Matsumoto's writing the term Pelmatozoa had gradually lost its original meaning, because under this name in addition to the three above-mentioned classes, the edrioasteroids and the carpoids had also been included. It was undoubtedly because of a reaction to such usage that Matsumoto believed that it was necessary to replace the term Pelmatozoa with the new term Crinozoa.

This latter term is found in the classification of Fell (1962) and it was adopted in Parts S and U of the *Treatise* to include the Eocrinoidea, Lepidocystoidea, Paracrinoidea, Cystoidea, Edrioblastoidea, Blastoidea, Parablastoidea and Crinoidea. In other words, all echinoderms with radial symmetry, a stalk, a generally globular theca, and possessing feeding appendages called arms or brachioles were included.

During the last few years, several changes in this classification have been proposed. In 1968, Paul suggested abandoning the class Cystoidea and elevated to class rank the two orders Rhombifera and Diploporita of which the class Cystoidea was usually composed. Durham (1971), on the other hand, retained the class Cystoidea, but added, in addition to the above-mentioned classes, the Edrioasteroidea and the Cyclocystoidea, although these forms have neither stalk, nor arms, nor brachioles. Although he didn't use the term, he thus returned to the older enlarged concept of Pelmatozoa. Sprinkle (1973a), on the other hand, restricted the term Crinozoa solely to the Crinoidea and Paracrinoidea (the latter with some reservation) and proposed a new subphylum, Blastozoa, which comprises the Eocrinoidea, Parablastoidea, Blastoidea, and Rhombifera, that is, all groups whose representatives possess brachioles (not arms like the crinoids). In the absence of sufficient knowledge of their feeding appendages, he did not assign the Diploporita to any subphylum, he assigned Lepidocystoidea to the Eocrinoidea, and he incorporated the Edrioblastoidea in the Echinozoa.

BREIMER & UBAGHS (1974) proposed to retain all stalked echinoderms in a single subphylum and to return to the original concept and term "Pelmatozoa"; however, they recognized the necessity to separate the forms with brachioles or appendices which are morphologically equivalent from those which possess true arms. Inspired by a dualistic classification proposed by BURMEISTER (1856), they distinguished a superclass Cystoidea containing the classes Eocrinoidea, Rhombifera, Diploporita, and Blastoidea (including the Parablastoidea), and a superclass Brachiatoidea which contains only the class Crinoidea, and perhaps the enigmatic Middle Cambrian genus *Echmatocrinus*. The Paracrinoidea, whose origin and phylogenetic relationships are unknown, were left in an indeterminate position.

This classification has been criticized by SPRINKLE (1976), who considers that there is probably no fundamental unity in stalked echinoderms. For him, the known differences between arm-bearing and brachiole-bearing groups greatly outweigh the features in common, which could have arisen through parallel evolution. Therefore, he still favors his former division (1973) into a subphylum Blastozoa and a subphylum Crinozoa. He admits, however, that the class Diploporita probably belongs in the subphylum Blastozoa.

Finally, PARSLEY & MINTZ (1975), in the face of the difficulties of classifying the Paracrinoidea, created a subphylum Paracrinozoa, to be added to the two subphyla proposed by SPRINKLE (1973).

These modifications resulted no doubt from considerable progress made during the last few years in the knowledge of Paleozoic echinoderms and also from a better appreciation of the differences which distinguish the crinoids from other stalked echinoderms. But they also have their drawbacks. With so many and frequent changes, there is a risk of introducing instability into the classification. The term Crinozoa is used with different meanings. The terms Pelmatozoa and Cystoidea, which have a long tradition, have been abandoned without visible advantage. Finally, and most important, the multiplication of major units of equal rank has the consequence of obscuring the degrees of similarity that exist between the different classes of echinoderms, for obviously the stalked echinoderms are more similar to each other than to the echinoids or the asteroids.

Asterozoa. No important modifications have been introduced into the composition of this subphylum since the publication of Parts S and U of the *Treatise*.

Echinozoa. The classification adopted in Part U of the *Treatise* divided the subphylum Echinozoa into seven classes, named Helicoplacoidea, Holothuroidea, Ophiocistioidea, Cyclocystoidea, Edrioasteroidea, Camptostromatoidea, and Echinoidea. WEBBY (1968) and SPRINKLE (1973a) have added the class Edrioblastoidea (not recognized by BREIMER & UBAGHS, 1974), but this addition has been contested by MINTZ (1970), who regarded *Astrocystites* (the only representative of this group presently known) as a probable descendant of the eocrinoids and a member of the subphylum Blastozoa SPRINKLE. On the other hand, the assignment of some of these classes to the Echinozoa has been put in doubt. This applies particularly to the Helicoplacoidea (UBAGHS, 1971b, 1975) and the Edrioasteroidea (BELL, 1976).

As this brief review shows, no consensus exists in regard to the general classification of the echinoderms. On the contrary, profound differences are apparent, but this is only to be expected in matters which are so subjective. For this reason, the present writer, conscious of the necessity to stabilize the classification and not wanting to impose his personal preferences on a collective work such as the *Treatise*, has chosen to maintain in the following synopsis the four subphyla recognized in Parts S and U; however, some minor modifications and additions have been made in the contents of some of the subphyla.

Synopsis of Classification of Echinodermata

Subphylum Homalozoa Whitehouse, 1941.

Echinoderms without trace of radial symmetry; theca depressed, asymmetrical but showing varying degree of secondary bilateralization. *M.Cam.-M.Dev.*

Class Ctenocystoidea Robison & Sprinkle, 1969.

Body composed of a theca, without appendages; thecal outline nearly symmetrical; marginal and central plates usually differentiated; marginal frame two-layered, distinct; mouth in medial inferior face near anterior margin, with highly distinctive ctenoid (probably feeding) apparatus; anus near middle of posterior margin. *M. Cam.*

Class Stylophora Gill & Caster, 1960.

Body composed of a theca, and an armlike appendage (aulacophore), without peduncle (stele); mouth probably intrathecal, at or near proximal end of aulacophore; anus at opposite thecal extremity; aulacophore divided into 3 distinct regions. *M.Cam.-M.Dev.*

Order Cornuta Jaekel, 1901.

Theca with (generally strongly) asymmetrical outline; marginal thecal frame ordinarily well differentiated, commonly braced by a skeletal bar (zygal) on lower face; pores of various types generally present in upper right anterior area; aulacophore typically devoid of aboral spines. *M.Cam.-U.Ord.*

Order Mitrata Jaekel, 1918.

Theca with moderately asymmetrical to bilaterally symmetrical outline; marginal thecal frame slightly distinct from other thecal plates; no zygal; 1 or 2 pairs of pores may be present on lower or anterior thecal face; aulacophore typically with aboral spines or knobs. *L.Ord.-M.Dev.*

Class Homostelea Gill & Caster, 1960.

Body composed of a theca and a stele, without armlike appendage; thecal outline moderately asymmetrical; marginal frame one-layered, strongly differentiated; one large and one small orifice on margin opposite stele insertion; 1 or 2 epithecal marginal (?ambulacral) grooves leading to small orifice; stele not divided into several regions. *M.Cam.*

Class Homoiostelea Gill & Caster, 1960.

Body composed of a theca, a stele and an armlike appendage; theca moderately asymmetrical to almost bilaterally symmetrical; marginal frame usually not differentiated; mouth intrathecal, probably near proximal end of arm; anus commonly near left posterolateral margin; stele differentiated into 3 regions. *U.Cam.-L.Dev.*

Subphylum Crinozoa Matsumoto, 1929[1] (= Pelmatozoa Leuckart, 1848).

Radiate echinoderm typically attached throughout life or in young stage by a stalk inserted on aboral surface; viscera enclosed in a more or less globoid plated test or theca; mouth located at or near free pole of theca, exposed or covered by tegument; anus usually in adoral part of theca, never aboral; ambulacra acting as food grooves, extended distally onto projecting appendages (arms or brachioles). *L.Cam.-Holo.*

Class Eocrinoidea Jaekel, 1918.[2]

Crinozoa with biserial brachioles; theca globular, pyriform or flattened, generally made of numerous irregularly arranged plates, which may imbricate; sutural pores or epispires present or lacking; 2 to 5 ambulacral grooves confined to adoral end of theca and leading to erect brachioles; stem columnal bearing or irregularly multiplated, rarely absent. *L.Cam.-M.Ord., ?Sil.*

[Six orders have been recognized by SPRINKLE (1973), but only one has been named.]

Order Imbricata Sprinkle, 1973 (= Class Lepidocystoidea Durham, 1967).

Aboral part of conical theca and holdfast composed of imbricate plates lacking epispires; adoral part composed of adjacent plates with numerous epispires. *L.Cam.*

Class Rhombifera Zittel, 1879.[2]

Crinozoa with biserial brachioles; globular pyriform or oval theca; exothecal or endothecal pore structures which consist of rhombic sets of thecal canals. *L.Ord.-U.Dev.*

Order Dichoporita Jaekel, 1899.

Rhombifera with endothecal pore structures composed of dichopores and only developed across certain plate sutures; theca comprising a small number of plates arranged in 3 to 5 circlets; well-developed stem throughout life. *L.Ord.-U.Dev.*

Order Fistuliporita Paul, 1968.

Rhombifera with exothecal pore structures composed of fistulipores and developed across all possible plate sutures; theca comprising a large number of randomly arranged plates; stem lost in adult or possibly totally absent in rare examples. *L.Ord.-U.Ord.*

Class Diploporita Müller, 1854.

Crinozoa with uniserial appendages (probably brachioles) very rarely preserved; globular or pyriform theca generally composed of a large

[1] BREIMER and UBAGHS (1974) have shown that the name Crinozoa MATSUMOTO, 1929, is a synonym of Pelmatozoa LEUCKART, 1848, and thus superfluous. However, in order to preserve continuity with usage adopted in *Treatise* Part S, Dr. UBAGHS has agreed to give preference to the junior name in the present volume.—CURT TEICHERT.

[2] SPRINKLE (1973) has erected a subphylum Blastozoa containing the classes Eocrinoidea, Rhombifera, Blastoidea, and Parablastoidea. Dr. UBAGHS feels that the validity of this concept is in need of further testing.—CURT TEICHERT.

number of randomly arranged plates; exothecal pore structures (diplopores) consist of single thecal canal; usually present on all thecal plates; stem present or lost in adult. *L.Ord.-M.Dev.*

Class Blastoidea Say, 1825.[1]

Crinozoa with biserial brachioles bordering ambulacral areas; well-developed pentameral symmetry; 17 major plates arranged in 3 circlets, in addition with 1 to 6 anal deltoids in *CD* interray; ambulacral system with underlying lancet plate; infolded thin-walled calcareous linear structures (hydrospires) on either side of each ambulacrum crossing plate suture. *Sil.-Perm.*

Order Fissiculata Jaekel, 1918.

Theca with exposed hydrospire slits or elongated spiracle (spiracular slit). *Sil.-Perm.*

Order Spiraculata Jaekel, 1918.

Theca with hidden hydrospire slits, spiracles, and hydrospire pores. *Sil.-Perm.*

Class Parablastoidea Hudson, 1907.[1]

Crinozoa with biserial brachioles bordering ambulacral areas; well-developed pentameral symmetry; theca blastoid-like with many regularly arranged plates; ambulacral areas composed of biserially arranged plates, without lancet plates; deltoids particularly prominent; external cataspire slits developed only through deltoids (not across plate suture). *L.Ord.-M.Ord.*

Class Paracrinoidea Regnéll, 1945.[2]

Crinozoa with typically uniserial, free or recumbent food-gathering appendages bearing uniserial side branches arranged in single row; theca boxlike, many-plated, asymmetrical, but tending to become bilaterally (rather than radially) symmetrical; peristome and column off set; with or without internally opening transverse sutural slits that variously extend through thickness of plates but do not open to exterior or connect with neighboring slits. *M.Ord.-U.Ord., ?Sil.*

Class Crinoidea Miller, 1821.

Crinozoa provided with true arms; pentameral symmetry well developed; theca divided into aboral cup and adoral tegmen, comprising 5 radial plates from which invariably the aboral skeleton of the arms starts; radial growth pattern concentrated on arms, which are directed away from theca; column ordinarily well developed, lost in postlarval stage in some forms. *M.Cam.; L.Ord.-Holo.*

Subclass Echmatocrinea Sprinkle & Moore, new subclass.[3]

Primitive Crinoidea with irregularly plated cup; no stem, cup attached to substrate by irregularly plated holdfast; 8 to 10 short, uniserial arms, 10 to 12 heavily plated brachials per arm bearing short, soft appendages; tegmen not known. *M.Cam.*

Order Echmatocrinida Sprinkle & Moore, new order.

Characters of subclass. *M.Cam.*

Subclass Camerata Wachsmuth & Springer, 1885.

Crinoidea with thecal plates typically united by rigid sutures; aboral cup generally including fixed brachials, interbrachials, and anal plates; tegmen usually strong, concealing mouth; arms typically pinnulate. *L.Ord.-U.Perm.*

Order Diplobathrida Moore & Laudon, 1943.

Camerata with dicyclic base. *M.Ord.-U.Ord.; M.Sil.-L.Carb.*

Order Monobathrida Moore & Laudon, 1943.

Camerata with monocyclic base. *?L.Ord.; M.Ord.-U.Perm.*

Subclass Inadunata Wachsmuth & Springer, 1885.

Crinoidea with aboral cup composed of close-sutured plates; fixed brachials and interbrachials lacking (exception in a few primitive forms); anal plates commonly present in aboral cup; mouth subtegminal; arms pinnulate or nonpinnulate. *L.Ord.-U.Perm.; M.Trias.*

Order Disparida Moore & Laudon, 1943.

Monocyclic inadunates with weak to very prominent bilateral symmetry developed in planes other than through *A* ray and *CD* interray; radials commonly compound; arms typically nonpinnulate. *L.Ord.-U.Perm.*

Order Hybocrinida Jaekel, 1918.

Monocyclic inadunates, with undivided radials; radianal present; arms 5, uniserial, atomous, nonpinnulate, that may be lacking in *B* and *E* rays, or are recumbent or reduced to ambulacral grooves on cup plates. *L.Ord.-U.Ord.*

Order Coronata Jaekel, 1918.

Monocyclic inadunates, with highly pentamerous theca; tegmen composed of 5 peristomials, 5 interradially located large plates (?orals), and 10 elongated plates covering ambulacrum; no anal plates; radials and tegminal interradially located plates prolonged adorally into high coronal processes; arms attached to adoral end of each radial plate, with fixed small primaxil. *M.Ord.-U.Ord.; M.Sil.-U.Sil.*

Order Cladida Moore & Laudon, 1943.

---

[1] See footnote 2 on p. *T362*.

[2] PARSLEY & MINTZ (1975) set aside this class as a new subphylum, Paracrinozoa.

[3] Although, according to SPRINKLE (1973), these forms may possibly be related to the original stocks of crinoids, in my view the structure of the only known single genus and species of this subclass does not correspond to the definition of a crinoid.—G. UBAGHS.

Dicyclic inadunates, having mostly 3, 2, or 1 anal plate in cup, rarely none; anal sac generally prominent; arms branched or unbranched, nonpinnulate or pinnulate. *L. Ord.-U.Perm.; M.Trias.*

Subclass Flexibilia Zittel, 1895.

Crinoidea with cup plates mostly not rigidly united; aboral cup ordinarily including fixed brachials, interbrachials, and anal plates; tegmen flexible, with exposed mouth and food grooves; arms uniserial, nonpinnulate. *M.Ord.; L.Sil.-U.Perm.*

Order Taxocrinida Springer, 1913.

Flexibilia with elongate crown and relatively weak calyx; anal *X* not closely united to adjacent plates, and followed by series of anals bordered by many-plated tegument. *M.Ord.; L.Sil.-U.Penn.*

Order Sagenocrinida Springer, 1913.

Flexibilia with crown generally subglobular; calycal plates rather firmly united; anal *X* joined by close suture to adjacent plates; no series of anals bordered by many-plated tegument. *L.Sil.-U.Perm.*

Subclass Articulata Zittel, 1879.

Crinoidea with dicyclic or generally cryptodicyclic cup; basals generally small, reduced or even missing in some cases; no anal plates or compound radials in postlarval stage; tegmen flexible, with exposed mouth and ambulacral grooves; arms uniserial and pinnulate; articulations between radial and arm and between some or all brachials muscular; radials and brachials perforate. *L.Trias.-Holo.*

Order Millericrinida Sieverts-Doreck, in Moore, Lalicker, & Fischer, 1952.

Cup large, with 5 basals and 5 radials; infrabasals generally missing; fixed brachials and interbrachials may be present; column without nodals and cirri; articular face of columnals entirely covered with crenulae; proximal part of column commonly modified, 5 sided, or circular and forming a conical transition to cup; proximale incorporated in cup ordinarily present. *M. Trias.; L.Jur.-L.Cret.; L.Paleoc.; Holo.*

Order Cyrtocrinida Sieverts-Doreck, in Moore, Lalicker, & Fischer, 1952.

Cup composed of stout radials, with or without aboral part interpreted as fused basals or as proximale, and articulated to short column or directly to attachment disc; no cirri; arms short, commonly protected by interradial projections from cup or by stout proximal brachials. *L.Jur.-Mio.; Holo.*

Order Bourgueticrinida Sieverts-Doreck, 1953.

Cup small, lacking infrabasals; basals and radials united by closed sutures or fused. Columnals circular or elliptical in cross section, with synarthrial articulations; variable number of proximal columnals united by synostosis or fused to a proximale incorporated in cup; no cirri. *U.Cret.-L.Paleoc., ?M.Paleoc.; Eoc.-Mio.; Holo.*

Order Isocrinida Sieverts-Doreck, in Moore, Lalicker, & Fischer, 1952.

Cup rather small, dicyclic, cryptodicyclic, or lacking infrabasals; radial facet wide; column generally long, pentalobate, pentagonal or circular in cross section; nodals and cirri invariably present; no proximale. *L.Trias.-Mio.; Holo.*

Order Comatulida A. H. Clark, 1908.

Larval column typically obliterated except for cirriferous uppermost columnal or fused uppermost columnals enlarging and forming centrodorsal incorporated in cup. *L.Jur.-L.Paleoc.; Eoc.-Holo.*

Order Uintacrinida Broili in von Zittel, 1921.

Cup stemless, very large, spheroidal, composed of thin plates including centrale; no cirri or attachment organ; infrabasals may be present; proximal brachials and interbrachials incorporated in cup; arms very long. *U.Cret.*

Order Roveacrinida Sieverts-Doreck, in Moore, Lalicker, & Fischer, 1952.

Cup stemless, very small, composed of radials, small discrete or fused basals, and in some specimens, a centrale; arms well developed or absent. *M.Trias.-U.Trias.; U.Jur.-U.Cret.*

Subphylum Asterozoa Haeckel in Zittel, 1895.

Free-living radiate echinoderms in which a radially divergent pattern of growth produces projecting rays and star-shaped body; mouth on underside. *L.Ord.-Holo.*

Class Stelleroidea Lamarck, 1816.

Characters of subphylum. *L.Ord.-Rec.*

Subclass Somasteroidea Spencer, 1951.

Asterozoans with shallow ambulacral channel formed by double series of ambulacrals, each of which typically gives rise to transverse series (metapinnules) of rodlike ossicles (virgalia); radial water vessel enclosed to varying extent between ambulacrals. *L.Ord.-Holo.*

Order Goniactinida Spencer, 1951.

Characters of subclass. *L.Ord.-Rec.*

Subclass Asteroidea de Blainville, 1830.

Asterozoans with relatively broad, hollow arms which contain large lobes of body cavity and enclosed organs; arms normally not separated from central disc; radial water vessel on outside of ambulacral skeleton. *L.Ord.-Holo.*

Order Platyasterida Spencer, 1951.

Ambulacrals, adambulacrals and inferomarginals in regular transverse series recalling metapinnules. *M.Ord.-Holo.*

Order Paxillosida Perrier, 1884.

Mouth frame of adambulacral type; mouth-angle prominent; marginal frame (when present) separated from mouth frame by interradial areas with small ossicles; dorsal side usually covered with paxillae. *L.Ord.-Holo.*

Order Valvatida Perrier, 1884.

Mouth frame of adambulacral type; mouth-angle plate relatively inconspicuous; infero- and superomarginals (if present) normally equal in number and without intermarginal channels. *L.Ord.-Holo.*

Order Spinulosida Perrier, 1884.

Mouth frame of adambulacral type; mouth-angle plate prominent, not keeled; marginal frame ordinarily wanting; aboral skeleton reticulate, imbricate or absent; dorsal and oral sides with spines, mostly in groups on prominences from plates; simple pedicellariae may be present. *M.Ord.-Holo.*

Order Forcipulatida Perrier, 1884.

Mouth frame of ambulacral type; marginal plates generally inconspicuous; typical pedicellariae, when present, straight or crossed; disc generally small; arms ordinarily elongated, with rounded sides. *L.Ord.-Holo.*

Subclass Ophiuroidea Gray, 1840.

Asterozoans with slender, simple or branched arms sharply separated from disc, and typically supported by internal row of ambulacral ossicles; generally without open ambulacral groove; respiration by means of gills typically located in interrays. *L.Ord.-Holo.*

Order Stenurida Spencer, 1951.

Basins for tube feet shared, usually subequally, by 2 ambulacrals; buccal slits commonly present. *L.Ord.-U.Dev.*

Order Oegophiurida Matsumoto, 1915.

No oral or radial shields, dorsal or ventral arm plates, genital plates or bursae; disc covered by skin or imbricating scales; narrow frame at disc margin may be present; gastric coeca entering arms. *L.Ord.-Holo.*

Order Phrynophiurida Matsumoto, 1915.

Disc and arms covered with skin; radial shields and genital plates articulating by simple facet or transverse ridge; peristomial plates large; oral frame entire, without well-developed lateral wings; dorsal and ventral arm plates absent or rudimentary; lateral arm plates small, occupying only lower side edge of arms; ambulacrals with hourglass-shaped articulations. *L.Dev.-Holo.*

Order Ophiurida Müller & Troschel, 1840.

Radial shields, genital plates, and buccal shields generally present; ambulacral groove closed by growth of lateral arm plates on ventral side toward midline of arms; ambulacrals opposite and fused in pairs; dorsal and ventral arm plates typically present; ambulacrals with zygospondylous articulations. *Sil.-Holo.*

Subphylum Echinozoa Haeckel in Zittel, 1895.

Echinoderms mostly radiate, with globoid, cylindroid, or discoid body that typically lacks arms, brachioles, or outspread rays. *L.Cam.-Holo.*

Class Helicoplacoidea Durham & Caster, 1963.

Free-living nonradiate echinoderms with helically organized fusiform to pyriform test; mouth at one end of body. *L.Cam.*

Subclass Helicoplacida Durham & Caster, 1963.

Test built of columns of plates. *L.Cam.*

Subclass Polyplacida Durham, 1967.

Test built of mosaic of small plates. *L.Cam.*

Class Camptostromatoidea Durham, 1966.

Apparently medusaeform, radially symmetrical echinoderms with plated appendages attached to periphery of test; mouth and anus at opposite poles; sutural pores abundant except on region of aboral pole. *L.Cam.*

Class Edrioasteroidea Billings, 1858.

Sedentary radiate (generally quinqueradiate) echinoderms, with domal, clavate, or globose polyplated theca that lacks stalk; ambulacra endothecal, formed by floor plates and cover plates; anus on adoral surface. *L.Cam.-U.Penn.*

Order Stromatocystitoidea Termier & Termier, 1969.

Domal or semiconvex theca; aboral surface plated (at least in some genera); adoral surface ordinarily with frame of submarginal or (?) marginal plates; skirt of small plates forming margin may be present; ambulacra limited to adoral surface, with biserial floor plates and sutural passageways; interambulacral plates small, numerous, stellate with many sutural pores. *L.Cam.-M.Cam., ?U.Dev.*

Order Isophorida Bell, 1976.

Domal or clavate theca; aboral surface nonplated; ambulacra ordinarily limited to adoral surface, with uniserial floor plates, without sutural passageways; oral frame formed by proximal floor plates; cover plates with intraambulacral and/or intrathecal extensions. *M.Ord.-U.Penn.*

Order Edrioasterida Bell, 1976.

Edrioasteroidea with semigloboid theca; plated adoral surface extending below ambitus onto aboral side of theca; ambulacra passing onto aboral surface; ambulacral floor plates biserial with sutural passageways; cover plates without intraambulacral or intrathecal extensions. *M.Ord.*

Class Edrioblastoidea Fay, 1962.

Calyx made of numerous plates arranged into circlets, provided with a 5-part stem; quinqueradiate symmetry well developed; most large calycal plates with sutural indentations apparently penetrating to calyx interior; ambulacra 5, long, composed of halves representing deltoid limbs; food groove flanked by rows of pores and covered by biserial set of cover plates. *M.Ord.*

Class Cyclocystoidea Miller & Gurley, 1895.

Disk-shaped body with submarginal frame of thick, ornamented plates, enclosing central areas covered by thinly plated membranes, and with skirt of small plates around margin; sutural pores present on one central area; submarginal plates provided with large cuplike pores and small inner canals leading to interior of body. *M.Cam.-M.Dev.*

Class Ophiocistioidea Sollas, 1899.

Free-living quinqueradiate echinoderms with plated aboral face and plated or nonplated adoral face; peristome central, with 5 jaws; ambulacra confined to adoral face, composed of 3 plate columns; interambulacra with single plate column; appendages (? gigantic podia) located on adoral face, covered by imbricated scales; periproct aboral, not opposite peristome. *L.Ord.-L.Carb.*

Class Echinoidea Leske, 1778.

Free-living quinqueradiate echinoderms with plated test, bearing movable appendages (spines, pedicellariae, spheridia) externally; mouth directed toward substrate, primarily with, secondarily without masticatory apparatus; 2 principal systems of plates, 1) an apical system invariably including 5 ocular plates and 5 or fewer genital plates; 2) a coronal system composed of 5 ambulacral and 5 interambulacral areas, each one composed of 1, 2, or more meridional columns; ambulacral plates perforated for passage of tube feet. *Ord.-Holo.*

Subclass Perischoechinoidea M'Coy, 1849.

Ambulacrum composed of 2 to many columns; interambulacrum with 1 to many columns; ambulacral plates not compound; anus within apical system (endocyclic); perignathic girdle absent or composed of apophyses only. *M.Ord.-Holo.*

Order Bothriocidaroida Zittel, 1879.

Test rigid, plates not imbricating; ambulacrum of 2 columns, interambulacrum with single column; no genital plates. *Ord.*

Order Echinocystitoida Jackson, 1912.

Test flexible, plates strongly imbricating; ambulacral plates bevel under interambulacra; ambulacrum of 2 or more columns; interambulacrum of more than 2 columns. *U.Ord.-Perm.*

Order Palaechinoida Haeckel, 1866.

Test rigid, plates slightly imbricating; ambulacral plates bevel over interambulacrum; interambulacrum of 1 or more than 2 columns. *Sil.-Perm.*

Order Cidaroida Claus, 1880.

Test rigid or flexible; ambulacra conspicuously narrower than interambulacra; primary interambulacral tubercles conspicuous; ambulacrum of 2 columns, interambulacrum of 2 or more columns. *?Sil., U.Dev.-Holo.*

Subclass Euechinoidea Bronn, 1860.

External gills present or lost; ambulacrum and interambulacrum of 2 columns each; anus within apical system (endocyclic) or outside apical system (exocyclic); complete perignathic girdle present or lost. *?Carb., U.Trias.-Holo.*

Superorder Diadematacea Duncan, 1889.

Primary tubercles perforate; lantern with unkeeled teeth; perignathic girdle complete; gill slits present. *?Carb., U.Trias.-Holo.*

Order Echinothurioida Claus, 1880.

Test flexible; anus endocyclic; spines with hollow axis; ambulacral plates simple on peristome, compound in test; *U.Jur.-Rec.*

Order Diadematoida Duncan, 1889.

Test rigid or flexible; anus endocyclic; spines with axis; peristomial membrane with 10 buccal plates; ambulacrals simple or compound. *?L.Carb., U.Trias.-Holo.*

Order Pedinoida Mortensen, 1939.

Test rigid; anus endocyclic; spines with solid axis; peristomial membrane with 10 buccal plates; ambulacrals simple to compound. *U.Trias.-Holo.*

Order Pygasteroida Durham & Melville, 1957.

Similar to Pedinoida, but with anus exocyclic and consistently simple ambulacral plates. *L.Jur.-U.Cret.*

Superorder Echinacea Claus, 1876.

Test rigid; anus endocyclic; spines with solid axis; teeth keeled; perignathic girdle complete; gill slits present. *U.Trias.-Rec.*

Order Salenioida Delage & Hérouard, 1903.

Lantern stirodont; ambulacral plates simple or compound in diadematoid manner; apical system with one or more suranal plates; each interambulacral plate with single, large, usually crenulate, primary tubercle. *?U.Trias., L.Jur.-Holo.*

Order Hemicidaroida Beurlen, 1937.

Lantern stirodont; ambulacral plate diadematoid, simple adapically in some; primary tubercles perforate, mostly crenulate. *U.Trias.-U.Cret.*

Order Phymosomatoida Mortensen, 1904.

Lantern stirodont; ambulacral plates simple throughout or more usually diadematoid; primary tubercles inperforate. *L.Jur.-Holo.*

Order Arbacioida Gregory, 1900.

Lantern stirodont; test commonly with prominences simulating tubercles; ambulacral plates simple or arbacioid; primary tubercles imperforate, noncrenulate; periproct with conspicuous anal valves. *M.Jur.-Holo.*

Order Temnopleurida Mortensen, 1942.

Lantern camarodont; test usually sculptured, if not, gill slits deep; ambulacral plates compound, diadematoid or echinoid. *L.Jur.-Holo.*

Order Echinoida Claus, 1876.

Lantern camarodont; test not sculptured; ambulacral plates echinoid; gill slits shallow. *?U.Cret., Paleoc.-Holo.*

Order Plesiocidaroida Duncan, 1889.

Lantern unknown; apical system very large; gill slits absent (or indistinct); tubercles small, noncrenulate. *U.Trias.*

Superorder Gnathostomata Zittel, 1879.

Test rigid; anus exocyclic; lantern and girdle usually present in adult, with keeled teeth; apical system and peristome approximately opposite; spines hollow; primary tubercles usually perforate and crenulate. *L.Jur.-Holo.*

Order Holectypoida Duncan, 1889.

Ambulacra petaloid or not; narrower than interambulacra; apical system with genital plates fused or distinct; teeth with lateral flanges; anus supramarginal to inframarginal. *L.Jur.-Holo.*

Order Clypeasteroida A. Agassiz, 1872.

Ambulacra petaloid, as wide or wider than interambulacra on adoral surface; apical system with genital plates fused; lantern without compass; teeth without lateral flanges; small accessory tube feet outside of petals. *U.Cret.-Holo.*

Superorder Atelostomata Zittel, 1879.

Test rigid; anus exocyclic; lantern, girdle and gill slits absent in adult; apical system and peristome rarely opposite; primary spines hollow. *L.Jur.-Holo.*

Order Cassiduloida Claus, 1880.

Ambulacra petaloid adapically; phyllodes and bourrelet usually present; fascioles absent. *L.Jur.-Holo.*

Order Holasteroida Durham & Melville, 1957.

Apical system typically elongate or disjunct; no genitals; paired petals not impressed; plastron lacking to meridosternous; no floscelle. *L.Jur.-Holo.*

Order Spatangoida Claus, 1876.

Apical system compact; plastron amphisternous; phyllodes present, but no bourrelets; fascioles generally present. *L.Cret.-Holo.*

Order Neolampadoida Philip, 1963.

Apical system mono- or tetrabasal; ambulacra nonpetaloid, with pores simple or lacking adapically; floscelle absent or weakly developed. *U.Eoc.-Holo.*

Superorder Uncertain.

Order Orthopsida Mortensen, 1942.

Lantern camarodont; test rigid; ambulacral plates simple or with a few triads; tubercles perforate, noncrenulate. *L.Jur.-U.Cret.*

Class Holothuroidea de Blainville, 1834.

Echinoderms mostly free-living, usually with body elongated in oral-aboral axis and secondary bilateral symmetry; mouth encircled by tentacles, pharynx surrounded by calcareous ring; ambulacral grooves lacking; skeleton ordinarily consisting of microscopic sclerites embedded in body wall; gonad single. *Ord.-Holo.*

Subclass Dendrochirotacea Brandt, 1835.

Tentacles without ampullae; pharyngeal retractor muscles present; podia and respiratory trees usually present; madreporite free in body cavity. *Ord.-Holo.*

Order Dendrochirotida Brandt, 1835.

Tentacles richly branched, 10 to 30 in number. *Ord.-Holo.*

Order Dactylochirotida Pawson & Fell, 1965.

Tentacles digitiform or digitate 8 to 30 in number; body enclosed by a test comprising imbricate plates. *Holo.*

Subclass Aspidochirotacea Brandt, 1835.

Tentacles shield-shaped, 10 to 30 in number; pharyngeal retractor muscles wanting; body with conspicuous bilateral symmetry. *L.Carb.-Holo.*

Order Aspidochirotida Brandt, 1835.

Respiratory trees present. *L.Carb.-Holo.*

Order Elasipodida Theel, 1882.

Respiratory trees lacking. *Dev.-Holo.*

Subclass Apodacea Brandt, 1835.

Tentacles simple, digitate or pinnate; podia reduced, or, more usually, lacking; pharyngeal retractor muscles lacking; sclerites comprising anchor and anchor plates. *Miss.-Holo.*

Order Apodida Brandt, 1835.

Body cylindrical; respiratory trees and anal papillae lacking; deposits commonly including wheels. *Miss.-Holo.*

Order Molpadiida Müller, 1850.

Body fusiform, commonly with tapering caudal portion; respiratory trees present; wheels lacking. *Miss.-Holo.*

Subclass Undeterminate.

Order Arthrochirotida Seilacher, 1961.

Tentacles with articulated axial skeleton; sclerites stout and imperforate. *L.Dev.*

OUTLINE OF CLASSIFICATION OF CRINOIDEA

The following outline of the Crinoidea summarizes taxonomic relationships, geologic occurrence, and numbers of recognized genera and subgenera in each suprageneric group from class to subfamily. A single number refers to genera; where two numbers are given; the second indicates subgenera additional to nominotypical ones.

Main Divisions of Crinoidea

Crinoidea *(class)* (1009;6). *M.Cam.; L.Ord.-Holo.*

A Echmatocrinea *(subclass)* (1). *M.Cam.*

1. Echmatocrinida *(order)* (1). *M.Cam.*

Echmatocrinidae (1). *M.Cam.*

B Camerata *(subclass)* (209;2). *L.Ord.-U.Perm.*

1. Diplobathrida *(order)* (52). *M.Ord.-U.Ord.; M. Sil.-L.Carb.*

Zygodiplobathrina *(suborder)* (2). *M.Ord.; L.Dev.*

Cleiocrinidae (1). *M.Ord.*
Spyridiocrinidae (1). *L.Dev.*
Eudiplobathrina (*suborder*) (50). *M.Ord.-U. Ord.; M.Sil.-L.Carb.*
Rhodocrinitacea (*superfamily*) (37). *M.Ord.-U.Ord.; M.Sil.-L.Carb.*
Reteocrinidae (3). *M.Ord.-U.Ord.*
Opsiocrinidae (1). *M.Dev.*
Archaeocrinidae (5). *M.Ord.-U.Ord.*
Anthracocrinidae (3). *M.Ord.*
Anthemocrinidae (2). *M.Sil.-U.Sil.*
Rhodocrinitidae (21). *M.Ord.-U.Ord.; M. Sil.; L.Dev.-L.Carb.*
Family Uncertain (2). *M.Sil.*
Dimerocrinitacea (*superfamily*) (12). *U.Ord.; M.Sil.-U.Dev.*
Dimerocrinitidae (8). *U.Ord.; M.Sil.-U.Dev.*
Lampterocrinidae (2). *M.Sil.*
Gazacrinidae (1). *M.Sil.*
Orthocrinidae (1). *L.Dev.-M.Dev.*
Nyctocrinacea (*superfamily*) (1). *M.Sil*
Nyctocrinidae (1). *M.Sil.*
2. Monobathrida (*order*) (145;2). *?L.Ord., M.Ord.-U.Perm.*
Compsocrinina (*suborder*) (91). *?L.Ord., U. Ord.-U.Perm.*
Xenocrinacea (*superfamily*) (5). *?L.Ord., U. Ord.; U.Sil.*
Xenocrinidae (1). *U.Ord.*
Tanaocrinidae (3). *?L.Ord., U.Ord.*
Abacocrinidae (1). *U.Sil.*
Periechocrinacea (*superfamily*) (39). *Sil., ?L. Perm.-U.Perm.*
Periechocrinidae (14). *Sil.-Miss.*
Paragaricocrinidae (4). *U.Carb.; U.Perm.*
Amphoracrinidae (3). *L.Carb.(Tournais.)-U.Carb.(Namur.).*
Actinocrinitidae (18). *L.Carb.(L.Miss.)-U. Carb., ?Perm.*
Actinocrinitinae (7). *L.Carb.(L.Miss.)-U. Carb., ?Perm.*
Eumorphocrininae (4). *L.Carb.*
Cactocrininae (5). *L.Carb.*
Physetocrininae (2). *L.Carb.*
Carpocrinacea (*superfamily*) (22). *M.Sil.-U. Sil.; M.Dev.-L.Carb.*
Carpocrinidae (7). *M.Sil.-U.Sil.; M.Dev.*
Batocrinidae (10). *L.Miss.-U.Miss.*
Coelocrinidae (4). *M.Dev.-L.Miss.(Osag.).*
Family Uncertain (1). *L.Miss.*
Hexacrinitacea (*superfamily*) (25). *U.Sil.-U. Perm.*
Hexacrinitidae (5). *U.Sil.-U.Dev., ?Penn.*
Parahexacrinidae (3). *L.Dev.*
Dichocrinidae (7). *L.Miss.-U.Perm.*
Acrocrinidae (10). *L.Miss.-U.Penn., Up.L. Carb.-M.Carb.*
Acrocrininae (3). *U.Miss.(Chester.) or up. L.Carb.(Visean)-L.Penn.(Morrow.).*
Planacrocrininae (1). *L.Penn.(Morrow.).*
Globacrocrininae (6). *L.Miss.-U.Penn.(Missour.).*
Glyptocrinina (*suborder*) (54;2). *M.Ord.-U. Perm.*
Glyptocrinacea (*superfamily*) (3). *M.Ord.-U. Sil.*
Glyptocrinidae (3). *M.Ord.-U.Sil.*
Melocrinitacea (*superfamily*) (10). *U.Ord.-U. Dev.*
Scyphocrinitidae (3). *?L.Sil., U.Sil.-L.Dev.*
Paramelocrinidae (1). *U.Sil.*
Melocrinitidae (5). *U.Ord.-U.Dev.*
Family Uncertain (1). *M.Sil.(Niagaran).*
Eucalyptocrinitacea (*superfamily*) (13). *M. Sil.-M.Dev.*
Clonocrinidae (2). *U.Sil.-L.Dev.*
Eucalyptocrinitidae (2). *M.Sil.-M.Dev.*
Dolatocrinidae (4). *L.Dev.-M.Dev.*
Polypeltidae (4). *U.Sil.-L.Dev.*
Family Uncertain (1). *M.Sil.*
Patelliocrinacea (*superfamily*) (8). *U.Ord.-M.Dev.*
Stelidiocrinidae (1). *U.Sil.*
Patelliocrinidae (7). *U.Ord.-M.Dev.*
Platycrinitacea (*superfamily*) (20;2). *?U.Ord., Sil.-Perm.*
Marsupiocrinidae (1;2). *M.Sil.-L.Dev.*
Hapalocrinidae (12). *?U.Ord., Sil.-M.Dev., ?U.Perm.*
Platycrinitidae (7). *?U.Sil., Dev.-Perm.*
Order, Suborder, Superfamily, and Family Uncertain (12). *M.Ord.-L.Dev.; Miss.; M.Penn.*
C. Inadunata (*subclass*) (482). *L.Ord.-U.Perm.; M. Trias.*
1. Disparida (*order*) (93). *L.Ord.-U.Perm.*
Homocrinacea (*superfamily*) (6). *M.Ord.-U. Ord.; M.Sil.*
Homocrinidae (6). *M.Ord.-U.Ord.; M.Sil.*
Calceocrinacea (*superfamily*) (15). *M.Ord.-L.Miss.; L.Perm.*
Calceocrinidae (15). *M.Ord.-L.Miss.; L.Perm.*
Pisocrinacea (*superfamily*) (6). *M.Sil.(Wenlock., Niagaran)-U.Dev.(Chemung.).*
Pisocrinidae (6). *M.Sil.-U.Dev.*
Allagecrinacea (*superfamily*) (23). *Ord.; ?Sil., Dev.-Perm.*
Allagecrinidae (9). *U.Dev.-U.Perm.*
Catillocrinidae (11). *L.Ord.; M.Dev.; L.Miss.-M.Penn.; U.Perm.*
Anamesocrinidae (1). *M.Dev.*
Haplocrinitidae (1). *?Sil., Dev., ?L.Carb.*
Tunguskocrinidae (1). *Ord.*
Heterocrinacea (*superfamily*) (7). *M.Ord. (Mohawk.)-U.Ord.(Cincinnat.).*
Heterocrinidae (7). *M.Ord.(Mohawk.)-U. Ord.(Cincinnat.).*
Myelodactylacea (*superfamily*) (11). *L.Ord.-L.Dev., ?U.Dev.*
Myelodactylidae (5). *L.Sil.-L.Dev.(Helderberg.), ?U.Dev.(Chemung.).*
Iocrinidae (2). *?L.Ord., M.Ord.-U.Ord.*
Eustenocrinidae (4). *L.Ord.-M.Ord.*
Anomalocrinacea (*superfamily*) (3). *M.Ord.-U.Ord.*

Anomalocrinidae (3). *M.Ord.-U.Ord.*

Belemnocrinacea *(superfamily)* (18). *M.Sil.-U.Carb.; U.Perm.*

Belemnocrinidae (2). *L.Miss.(Osag.).*

Holynocrinidae (1). *M.Dev.*

Perissocrinidae (3). *M.Dev.*

Pygmaeocrinidae (2). *U.Sil.-M.Dev.*

Synbathocrinidae (7). *M.Sil.-U.Carb.; U.Perm.*

Zophocrinidae (3). *U.Sil.-M.Dev.*

Perittocrinacea *(superfamily)* (2). *Up.L.Ord. or low.M.Ord.*

Perittocrinidae (2). *Up.L.Ord. or low.M.Ord.*

Superfamily Uncertain (2). *L.Ord., U.Perm.*

Paradoxocrinidae (1). *U.Perm.*

Family Uncertain (1). *L.Ord.*

2. Hybocrinida *(order)* (7). *L.Ord.-U.Ord.*

Hybocrinidae (3). *M.Ord.*

Baerocrinidae (1). *L.Ord.-M.Ord.*

Hybocystitidae (1). *M.Ord.*

Cornucrinidae (2). *M.Ord.-U.Ord.*

3. Coronata *(order)* (5). *M.Ord.-U.Ord.; M.Sil.-U.Sil.*

Stephanocrinidae (5). *M.Ord.-U.Ord.; M.Sil.-U.Sil.*

4. Cladida *(order)* (377). *L.Ord.(Tremadoc.)-U.Perm.; M.Trias.*

Cyathocrinina *(suborder)* (77). *M.Ord.-U.Perm.*

Cyathocrinitacea *(superfamily)* (18). *M.Ord.-U.Miss.; U.Perm.*

Cyathocrinitidae (4). *?M.Ord.; M.Sil.-L.Miss.; U.Perm.*

Barycrinidae (3). *L.Dev.(Ems.); L.Miss.(Osag.)-U.Miss.(Meramec.).*

Euspirocrinidae (7). *M.Ord.(Mohawk.)-L.Carb.(Tournais.).*

Lecythocrinidae (4). *M.Dev.(Givet.)-L.Miss.(Osag.).*

Gasterocomacea *(superfamily)* (21). *M.Ord.(Blackriver.)-U.Ord.; M.Sil.-M.Dev.(Givet.).*

Gasterocomidae (7). *L.Dev.-M.Dev.*

Sphaerocrinidae (3). *M.Ord.; M.Sil.; M.Dev.*

Porocrinidae (2). *M.Ord.(Mohawk.)-U.Ord.(Richmond.).*

Carabocrinidae (1). *M.Ord.-U.Ord.*

Crotalocrinitidae (7). *M.Sil.-M.Dev.(Givet.).*

Petalocrinidae (1). *M.Sil.(Niagaran, Wenlock.).*

Codiacrinacea *(superfamily)* (36). *M.Sil.; L.Dev.-U.Perm.*

Codiacrinidae (18). *M.Sil.; L.Dev.-M.Dev.; L.Miss.-U.Perm.*

Codiacrininae (13). *M.Sil.; L.Dev.-M.Dev.; L.Miss.-U.Perm.*

Bolbocrininae (2). *L.Perm.-U.Perm.*

Thetidicrininae (3). *L.Carb.(Tournais.); U.Perm.*

Sycocrinitidae (5). *L.Carb.(Tournais.); L.Perm.-U.Perm.*

Streblocrinidae (13). *M.Dev.-U.Perm.*

Streblocrininae (7). *M.Dev.-L.Penn.; Perm.*

Pentececrininae (6). *?U.Dev., L.Miss.-L.Penn.; L.Perm.-U.Perm.*

Superfamily and Family Uncertain (2). *M.Sil.; L.Dev.*

Dendrocrinina *(suborder)* (46). *L.Ord.-L.Miss.*

Dendrocrinacea *(superfamily)* (19). *M.Ord.-U.Dev.*

Dendrocrinidae (7). *M.Ord.-U.Dev.*

Botryocrinidae (12). *?M.Ord.; M.Sil.-U.Dev.*

Mastigocrinacea *(superfamily)* (19). *L.Ord.-M.Ord.; M.Sil.-L.Miss.*

Aethocrinidae (1). *L.Ord.(Tremadoc.-Arenig.).*

Thenarocrinidae (1). *M.Sil.*

Mastigocrinidae (17). *M.Ord.; M.Sil.-L.Miss.*

Merocrinacea *(superfamily)* (7). *?L.Ord., M.Ord.-U.Ord.; U.Sil.-L.Dev.*

Merocrinidae (1). *M.Ord.-U.Ord.*

Cupulocrinidae (1). *M.Ord.-U.Ord.*

Ontariocrinidae (1). *M.Ord.*

Ottawacrinidae (1). *?L.Ord., M.Ord., ?U.Ord.*

Metabolocrinidae (3). *M.Ord., U.Sil.-L.Dev.*

Superfamily and Family Uncertain (1). *?M.Dev.*

Poteriocrinina *(suborder)* (251). *L.Dev.-U.Perm.; M.Trias.*

Poteriocrinitacea *(superfamily)* (5). *L.Dev.-L.Penn., ?M.Penn.-?U.Perm.*

Poteriocrinitidae (5). *L.Dev.-L.Penn., ?M.Penn.-?U.Perm.*

Rhenocrinacea *(superfamily)* (10). *L.Dev.-U.Miss.; M.Penn.*

Rhenocrinidae (9). *L.Dev.-U.Miss.; M.Penn.(Atokan).*

Proctothylacocrinidae (1). *M.Dev.*

Scytalocrinacea *(superfamily)* (43). *M.Dev.-U.Perm.*

Scytalocrinidae (22). *M.Dev.-U.Perm.*

Blothrocrinidae (10). *L.Miss.-L.Perm.*

Cercidocrinidae (3). *L.Miss.*

Aphelecrinidae (3). *L.Miss.(Kinderhook.)-U.Miss.(Chester.).*

Corythocrinidae (1). *L.Miss.(Osag.).*

Spaniocrinidae (4). *L.Miss.; L.Penn.-U.Perm.*

Cupressocrinitacea *(superfamily)* (2). *L.Dev.-U.Dev.*

Cupressocrinitidae (2). *L.Dev.-U.Dev.*

Mollocrinacea *(superfamily)* (3). *L.Penn.-U.Perm.*

Mollocrinidae (3). *L.Penn.-U.Perm.*

Lophocrinacea *(superfamily)* (25). *L.Miss.-U.Perm.*

Lophocrinidae (1). *Up.L.Carb.*

Pelecocrinidae (6). *L.Miss.-U.Miss.; M.Penn.-U.Penn.; U.Perm.*

Indocrinidae (3). *L.Perm.-U.Perm.*

Laudonocrinidae (6). *U.Miss.-L.Perm.*

Stellarocrinidae (7). *U.Miss.(Chester.)-L.Perm.*

Pachylocrinidae (2). *L.Miss.-L.Perm.*
Agassizocrinacea *(superfamily)* (28). *L.Miss.-U.Perm.*
Bursacrinidae (3). *L.Miss.*
Ampelocrinidae (9). *L.Miss.-U.Perm.*
Sundacrinidae (6). *L.Perm.-U.Perm.*
Anobasicrinidae (3). *L.Penn.(Morrow.)-U.Perm.*
Agassizocrinidae (5). *U.Miss.-L.Perm.*
Cricocrinidae (1). *M.Penn.(Desmoines.).*
Trimerocrinidae (1). *L.Perm.-U.Perm.*
Decadocrinacea *(superfamily)* (9). *M.Dev.-U.Penn.*
Decadocrinidae (8). *M.Dev.-U.Penn.*
Clathrocrinidae (1). *U.Penn.(Missour.).*
Cromyocrinacea *(superfamily)* (26). *U.Miss.-Perm.*
Eupachycrinidae (2). *U.Miss.(Chester.).*
Phanocrinidae (6). *U.Miss.(Chester.); L.Carb.(Visean)-U.Carb.(Namur.).*
Cromyocrinidae (13). *L.Carb.(U.Miss.)-Perm.*
Ulocrinidae (4). *L.Carb.(Visean); M.Penn.-Perm.*
Cadocrinidae (1). *U.Perm.(Basleo beds).*
Hydreionocrinacea *(superfamily)* (3). *Up.L.Carb., U.Miss.*
Hydreionocrinidae (3). *Up.L.Carb., U.Miss.*
Erisocrinacea *(superfamily)* (29). *L.Carb.(Miss.)-U.Perm.; M.Trias.*
Erisocrinidae (3). *U.Miss.-L.Perm.*
Graphiocrinidae (5). *Low.L.Carb.(Miss.); M.Penn.-U.Penn.; U.Perm.*
Paradelocrinidae (5). *U.Carb.; Penn.(Morrow.); U.Perm.(Basleo beds).*
Arkacrinidae (1). *L.Penn.(Morrow.).*
Diphuicrinidae (2). *L.Penn.(Morrow.)-L.Perm.*
Protencrinidae (2). *U.Carb.(Moscov.); M.Penn.(Atokan)-L.Perm.(Artinsk.).*
Catacrinidae (8). *L.Penn.(Morrow.)-U.Perm.(Basleo beds).*
Stachyocrinidae (2). *U.Perm.(Basleo beds).*
Encrinidae (1). *M.Trias.*
Apographiocrinacea *(superfamily)* (2). *L.Penn.-U.Perm.*
Apographiocrinidae (2). *L.Penn.-(Morrow.)-U.Perm.*
Pirasocrinacea *(superfamily)* (26). *L.Miss.-U.Perm.*
Pirasocrinidae (25). *U.Miss.-U.Perm.*
Adinocrinidae (1). *L.Miss.*
Texacrinacea *(superfamily)* (19). *L.Miss.-U.Perm.*
Texacrinidae (2). *L.Penn.-L.Perm.*
Galateacrinidae (1). *M.Penn.(Desmoines.)-U.Penn.(Virgil.).*
Sellardsicrinidae (1). *M.Penn.(Desmoines.).*
Cymbiocrinidae (8). *U.Miss.-U.Penn.*
Staphylocrinidae (7). *L.Miss.-U.Miss.; M.Penn.-U.Perm.*
Zeacrinitacea *(superfamily)* (17). *L.Miss.-U.Perm.*
Zeacrinitidae (8). *L.Miss.-U.Perm.*
Exocrinidae (3). *L.Penn.(Morrow.)-L.Perm.(Wolfcamp.).*
Timorechinidae (5). *U.Perm.(Basleo beds).*
Scotiacrinidae (1). *Up.L.Carb.*
Calceolispongiacea *(superfamily)* (2). *L.Perm.-U.Perm.*
Calceolispongiidae (2). *L.Perm.(Sakmar.-Artinsk.)-U.Perm.(Word. equiv.).*
Superfamily and Family Uncertain (2). *L.Carb.(Visean).*
Order and Superfamily Uncertain (3). *Ord.; L.Miss.-M.Miss.*
Paractocrinidae (2). *Ord.*
Family Uncertain (1). *L.Miss.-M.Miss.*

D. Flexibilia *(subclass)* (61). *M.Ord.-U.Perm.*
1. Taxocrinida *(order)* (10). *M.Ord.-U.Penn.*
Taxocrinacea *(superfamily)* (10). *M.Ord.(Trenton.)-U.Penn.(Missour.).*
Taxocrinidae (6). *M.Ord.(Trenton.)-U.Miss.(Chester.).*
Synerocrinidae (4). *L.Miss.-U.Penn.; L.Carb.-U.Carb.*
2. Sagenocrinida *(order)* (50). *L.Sil.-U.Perm.(Basleo beds).*
Lecanocrinacea *(superfamily)* (22). *U.Sil.-U.Perm.*
Lecanocrinidae (4). *U.Sil.-M.Dev.*
Nipterocrinidae (4). *U.Sil.-L.Dev.; U.Dev.-L.Miss.*
Mespilocrinidae (5). *L.Miss.(low. L.Carb.)-U.Perm.*
Calycocrinidae (3). *M.Dev.; L.Penn.; U.Perm.*
Gaulocrinidae (1). *L.Miss.*
Prophyllocrinidae (3). *U.Perm.*
Palaeoholopodidae (2). *U.Perm.*
Icthyocrinacea *(superfamily)* (5). *L.Sil.-L.Miss. or low.L.Carb.*
Icthyocrinidae (5). *L.Sil.(Medinan)-L.Miss.(Osag.).*
Sagenocrinitacea *(superfamily)* (23). *U.Sil.-U.Perm.*
Homalocrinidae (3). *U.Sil.*
Sagenocrinitidae (3). *U.Sil.; Miss.; L.Perm.*
Dactylocrinidae (10). *U.Sil.-L.Miss.; L.Penn.-U.Perm.*
Euryocrinidae (7). *M.Dev.-U.Penn.*
Order Uncertain (1). *L.Dev.-M.Dev.*
Edriocrinidae (1). *L.Dev.-M.Dev.*

E. Articulata *(subclass)* (256;4). *L.Trias.-Holo.*
1. Millericrinida *(order)* (15). *M.Trias.; L.Jur.-L.Cret.; L.Paleoc.; Holo.*
Millericrinina *(suborder)* (9). *M.Trias.; L.Jur.-L.Cret.*
Dadocrinidae (1). *M.Trias.*
Millericrinidae (5). *L.Jur.-U.Jur.*
Apiocrinitidae (2). *L.Jur.-L.Cret.*
Cyclocrinidae (1). *L.Jur.-L.Cret.*

Hyocrinina *(suborder)* (6). *L.Tert.(Dan.); Holo.*
Hyocrinidae (6). *L.Tert.; Holo.*
Hyocrininae (5). Recent.
Calamocrininae (1). *L.Tert.(Dan.); Holo.*
2. Cyrtocrinida *(order)* (20;4). *L.Jur.-Mio; Holo.*
Cyrtocrinina *(suborder)* (12;4). *L.Jur.-U.Cret.*
Plicatocrinidae (2). *L.Jur.-U.Jur.*
Sclerocrinidae (2). *U.Jur.(Oxford.)-L.Cret. (Hauteriv.).*
Hemicrinidae (1;2). *U.Jur.(Tithon.)-L.Cret. (Alb.).*
Eugeniacrinitidae (5). *M.Jur.(Bathon.)-U. Cret.(Campan.).*
Phyllocrinidae (2;2). *M.Jur.(Bajoc.)-L.Cret. (Neocom.).*
Holopodina *(suborder)* (7). *L.Jur.; U.Jur.-Mio.; Holo.*
Eudesicrinidae (2). *L.Jur.*
Hemibrachiocrinidae (3). *L.Cret.(Valangin.-Barrem.).*
Holopodidae (2). *U.Jur.(Tithon.)-Mio.; Holo.*
Suborder and Family Uncertain (1). *U.Jur.-L. Cret.(Neocom.).*
3. Bourgueticrinida *(order)* (11). *U.Cret.(Turon.)-L.Paleoc.(Dan.); ?M.Paleoc.(Heers.); Eoc.-Mio.; Holo.*
Bourgueticrinidae (1). *U.Cret.(Turon.)-L. Paleoc.(Dan.), ?M.Paleoc.(Heers.); Eoc.*
Bathycrinidae (6). *U.Cret.(Maastricht.)-L. Paleoc.(Dan.); Eoc.-Mio.; Holo.*
Phrynocrinidae (2). Recent.
Porphyrocrinidae (2). Recent.
4. Isocrinida *(order)* (23). *L.Trias.-Mio.; Holo.*
Holocrinidae (2). *M.Trias.*
Isocrinidae (15). *Trias.-Mio.; Holo.*
Genera Dubia (3). *U.Jur., Tert.(Oligo.-Mio.)*, recent.
Pentacrinitidae (2). *L.Jur.-U.Jur.*
Proisocrinidae (1). Recent.
5. Comatulida *(order)* (169). *L.Jur.-L.Paleoc.; Eoc.-Holo.*
Paracomatulacea *(superfamily)* (5). *L.Jur. (Toarc.)-L.Paleoc.; Holo.*
Paracomatulidae (1). *L.Jur.(Toarc.).-L.Cret. (Hauteriv.).*
Atelecrinidae (4). *U.Cret.-L.Paleoc.; Holo.*
Solanocrinitacea *(superfamily)* (11). *L.Jur. (Pliensbach.)-U.Cret.(Coniac.).*
Solanocrinitidae (4). *L.Jur.(Pliensbach.)-U. Cret.(Coniac.).*
Decameridae (3). *L.Cret.*
Thiolliericrinidae (4). *M.Jur.(Bathon.)-L. Cret.(Hauteriv.).*
Comasteracea *(superfamily)* (20). *Tert.(Eoc., Mio.-Plio.); Holo.*
Comasteridae (20). *Tert.(Eoc., Mio.-Plio.); Holo.*
Comasterinae (4). *U.Tert.(Mio.-Plio.)*, recent.
Capillasterinae (12). *L.Tert.(Eoc.)*; recent.
Comactiniinae (4). Recent.
Mariametracea *(superfamily)* (34). *Tert.(Eoc.-Mio.)*; recent.
Mariametridae (7). Recent.
Zygometridae (2). Recent.
Eudiocrinidae (1). Recent.
Himerometridae (6). *Tert.(Eoc.-Mio.)*; recent.
Colobometridae (18). Recent.
Tropiometracea *(superfamily)* (38). *U.Jur.; U.Cret.-L.Paleoc.; Eoc.-Plio.*; recent.
Tropiometridae (1). Recent.
Pterocomidae (2). *U.Jur.; U.Cret.-L.Paleoc.*
Conometridae (5). *U.Cret.(Cenoman.)-L. Paleoc.; Eoc.-Mio.*
Calometridae (5). Recent.
Ptilometridae (1). Recent.
Asterometridae (2). *U.Tert.(Plio.)*; recent.
Thalassometridae (14). *U.Tert.(Mio.)*; recent.
Charitometridae (8). Recent.
Notocrinacea *(superfamily)* (7). *M.Jur.(Bathon.)-U.Cret.(Maastricht.); ?Eoc.*; recent.
Notocrinidae (6). *M.Jur.(Bathon.)-U.Cret. (Maastricht.), ?Eoc.*, recent.
Aporometridae (1). Recent.
Antedonacea *(superfamily)* (54). *Cret.-L. Paleoc.; Eoc.; Mio.; Pleist.-Holo.*
Antedonidae (52). *Cret.(Alb.-Maastricht.)-L. Paleoc.; Eoc.; Mio.; Pleist.-Holo.*
Antedoninae (11). *Eoc.; Mio.; Pleist.-Holo.*
Perometrinae (4). Recent.
Thysanometrinae (2). Recent.
Zenometrinae (15). *L.Tert.(Eoc.)*; recent.
Bathymetrinae (11). Recent.
Heliometrinae (8). *L.Cret.-L.Paleoc.; Mio.*; recent.
Isometrinae (1). Recent.
Pentametrocrinidae (2). Recent.
6. Uintacrinida *(order)* (2). *U.Cret.(Santon.).*
Uintacrinidae (1). *U.Cret.(Santon.).*
Marsupitidae (1). *U.Cret.(Santon.).*
7. Roveacrinida *(order)* (14). *M.Trias.-U.Trias.; U.Jur.-U.Cret.*
Roveacrinidae (11). *M.Trias.-U.Trias.; L. Cret.-U.Cret.*
Roveacrininae (7). *L.Cret.-U.Cret.*
Somphocrininae (4). *M.Trias.-U.Trias.*
Saccocomidae (3). *U.Jur.-Cret.*
Saccocominae (2). *U.Jur.-Cret.*
Pseudosaccocominae (1). *U.Jur.-L.Cret.*
Order and Family Uncertain (2). *L.Jur.(Sinemur.)-U.Jur.(Oxford.).*

STRATIGRAPHIC DISTRIBUTION

The stratigraphic distribution of subclasses, orders, suborders, superfamilies, families, and subfamilies of Crinoidea recognized in the *Treatise* is indicated graphically in Table 5 (compiled by JACK D. KEIM).

TABLE 5. *Stratigraphic Distribution of the Crinoidea.*

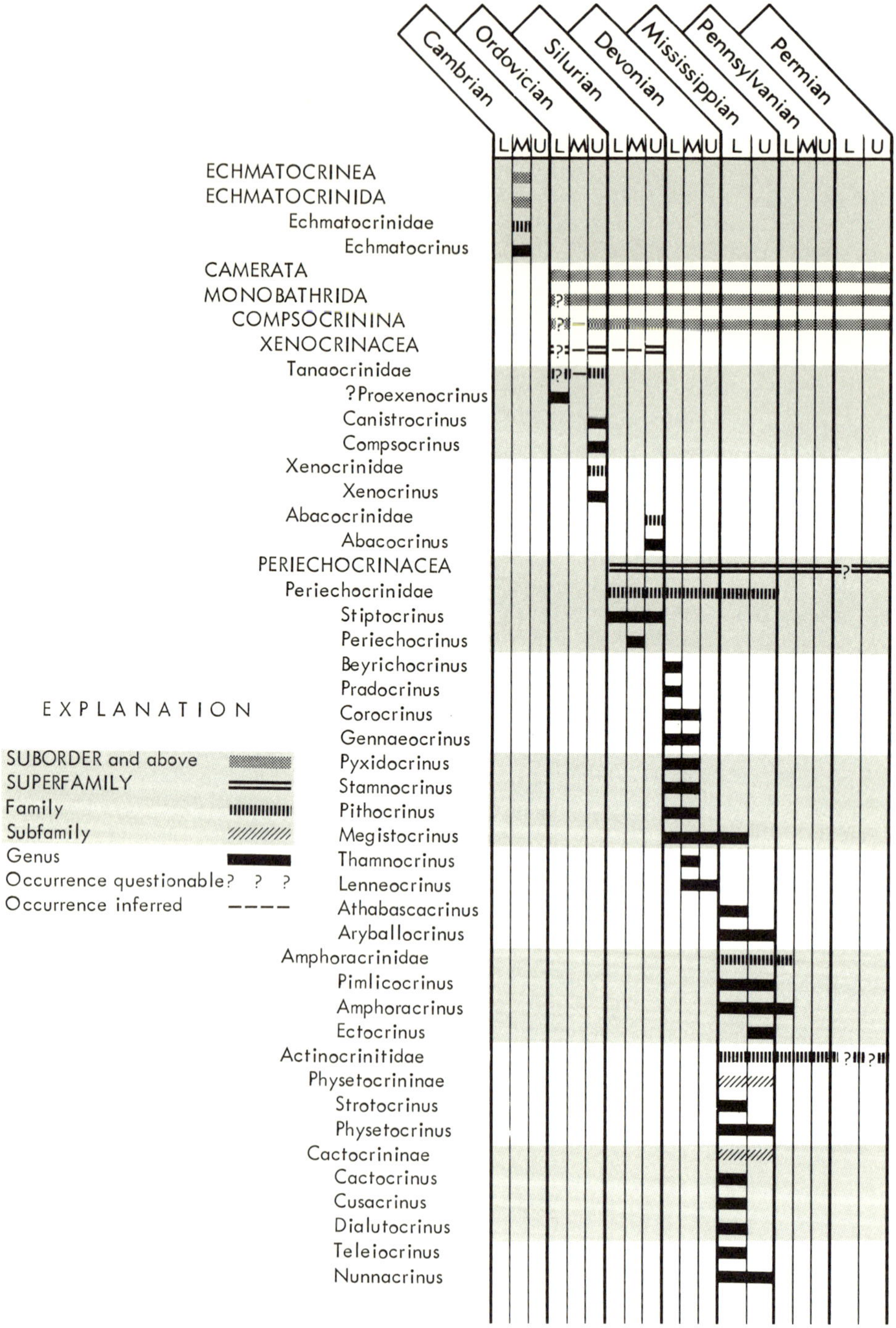

TABLE 5. *(Continued)*

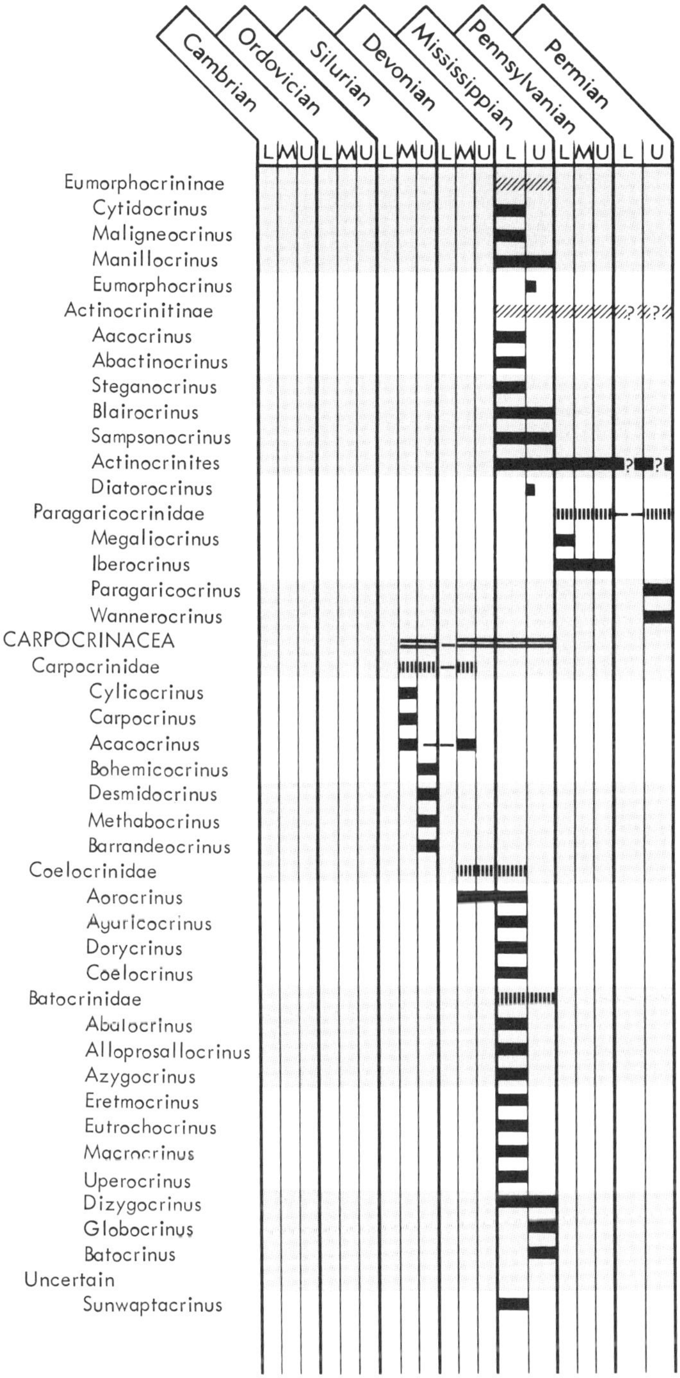

TABLE 5. *(Continued)*

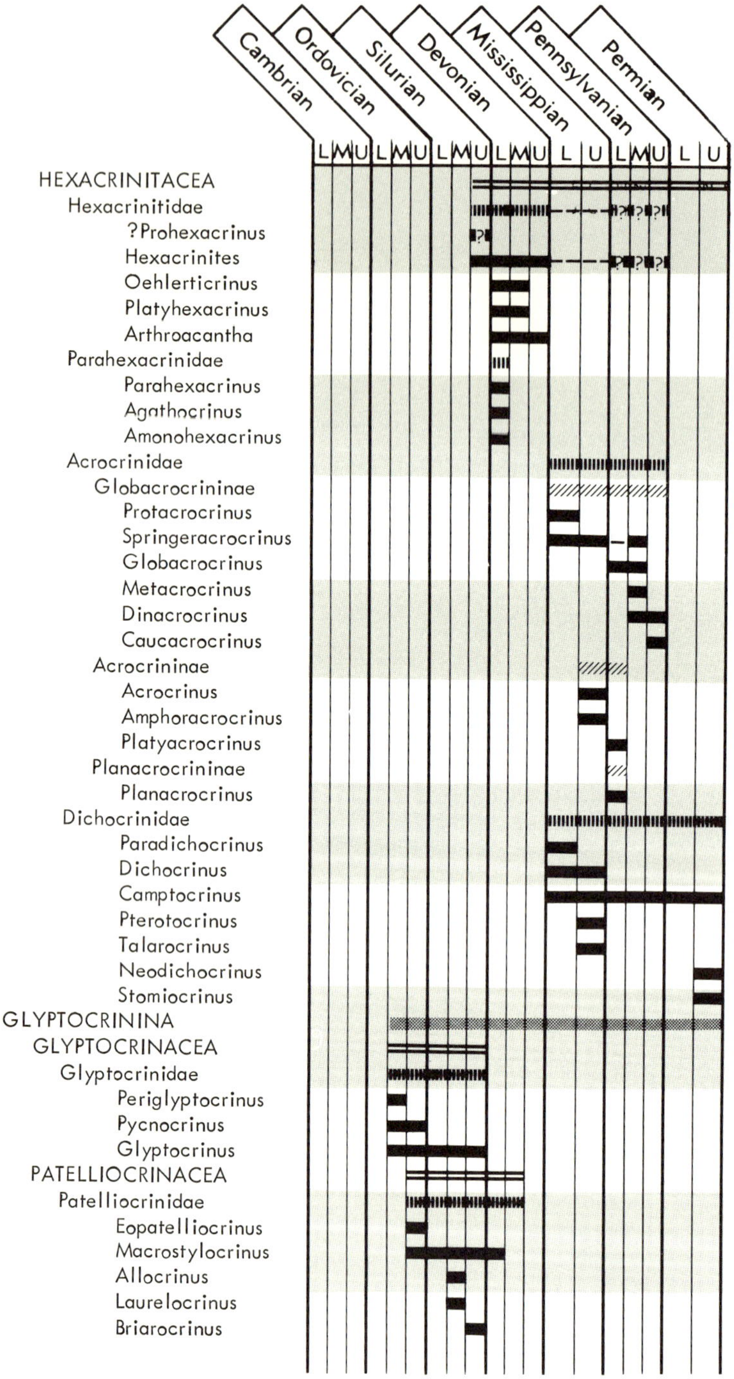

TABLE 5. *(Continued)*

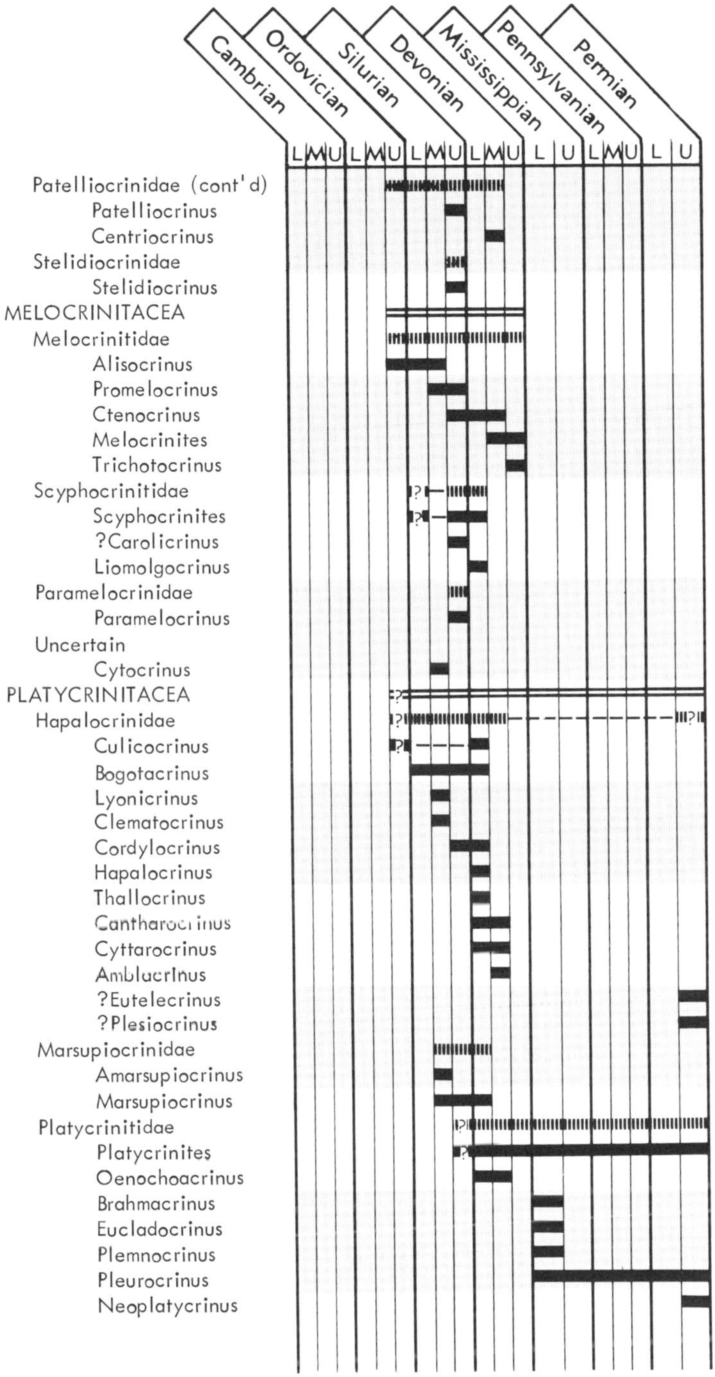

TABLE 5. *(Continued)*

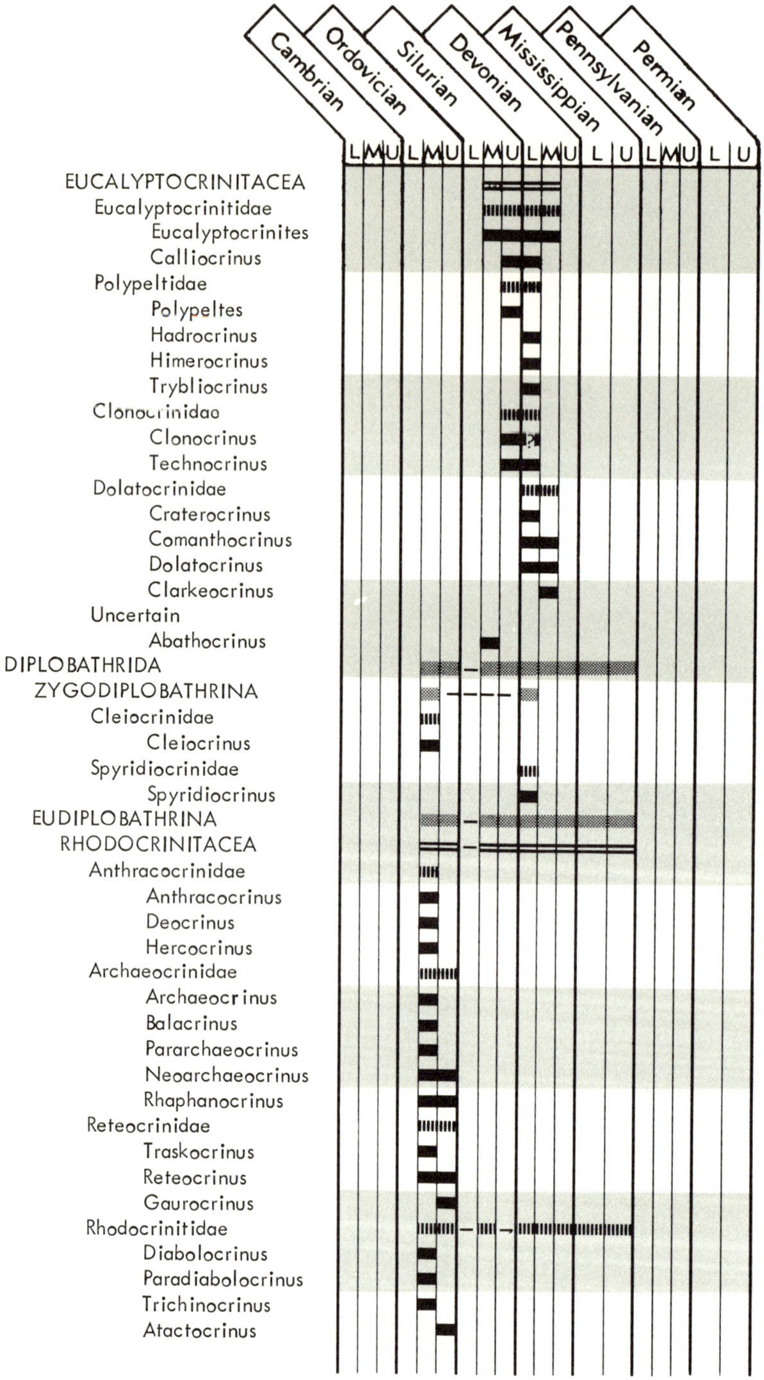

TABLE 5. *(Continued)*

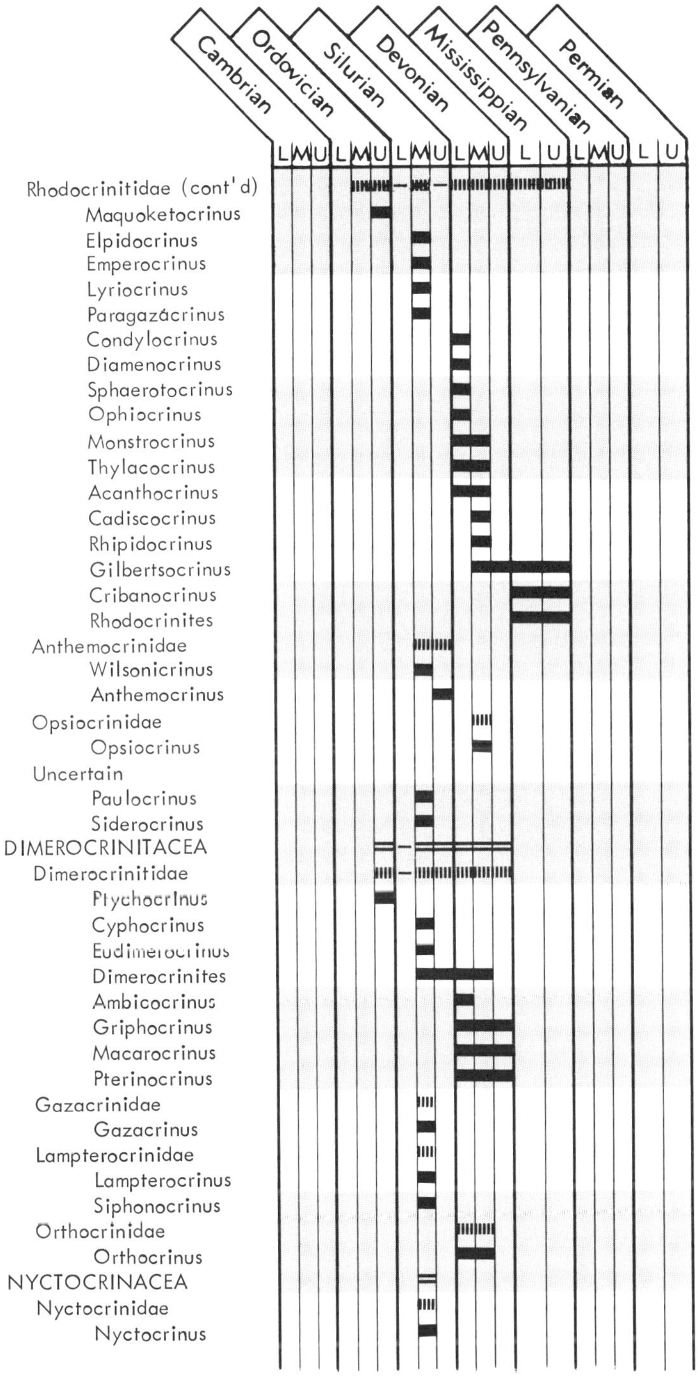

TABLE 5. *(Continued)*

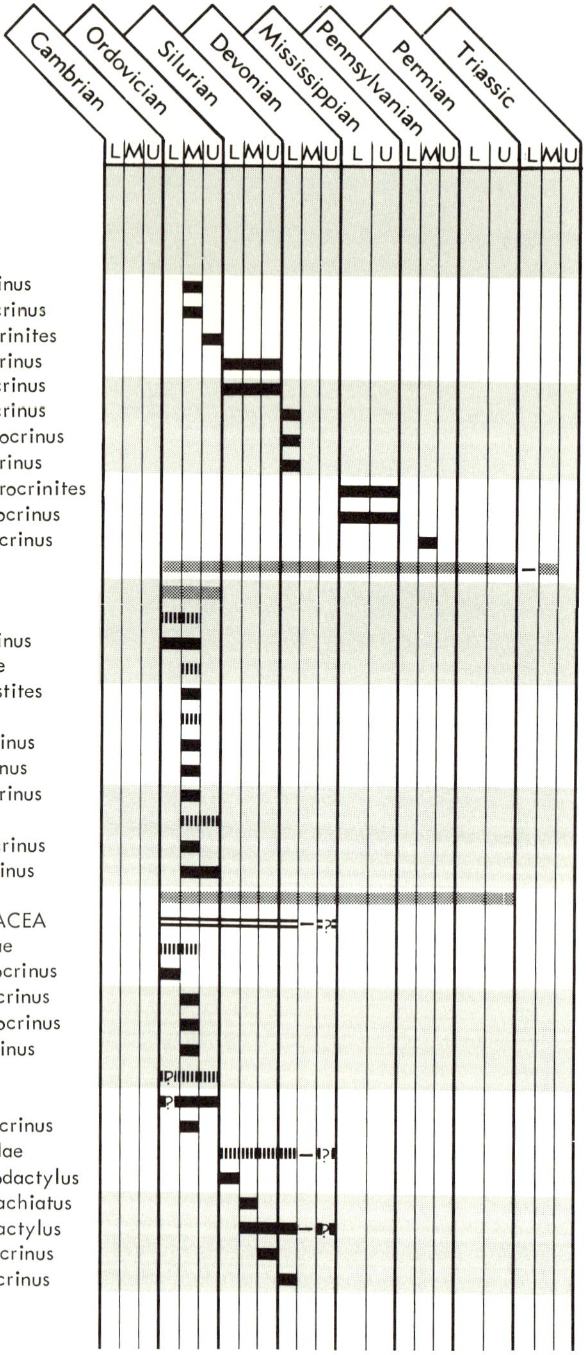

UNCERTAIN
UNCERTAIN
UNCERTAIN
Uncertain
Mitrocrinus
Pachyocrinus
Trochocrinites
Helicocrinus
Laubeocrinus
Coronocrinus
Craspedocrinus
Katarocrinus
Tetramerocrinites
Phillipsocrinus
Becharocrinus
INADUNATA
HYBOCRINIDA
Baerocrinidae
Baerocrinus
Hybocystitidae
Hybocystites
Hybocrinidae
Hoplocrinus
Hybocrinus
Revalocrinus
Cornucrinidae
Tripatocrinus
Cornucrinus
DISPARIDA
MYELODACTYLACEA
Eustenocrinidae
Ramseyocrinus
Eustenocrinus
Peniculocrinus
Ristnacrinus
Iocrinidae
Iocrinus
Caleidocrinus
Myelodactylidae
Eomyelodactylus
Crinobrachiatus
Myelodactylus
Herpetocrinus
Brachiocrinus

TABLE 5. *(Continued)*

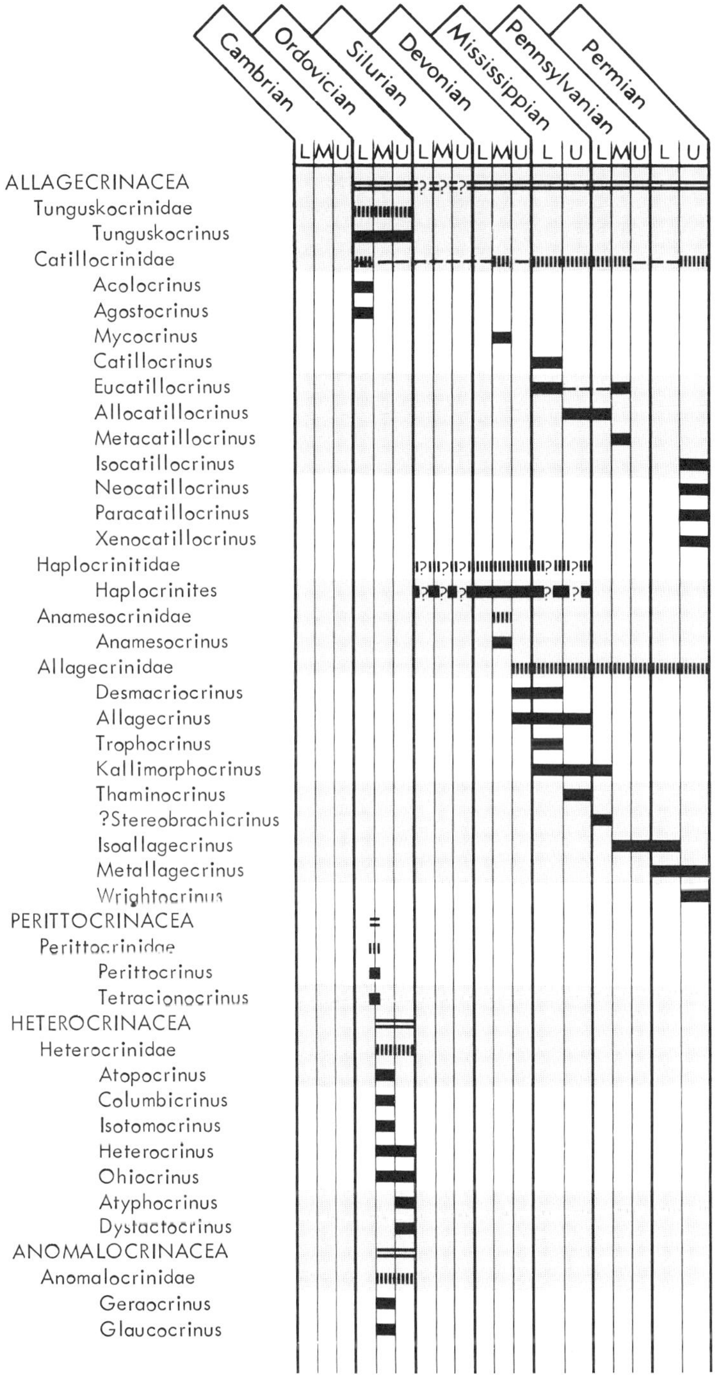

For *Atopocrinus* (Heterocrinidae), read *Othneiocrinus*.

TABLE 5. *(Continued)*

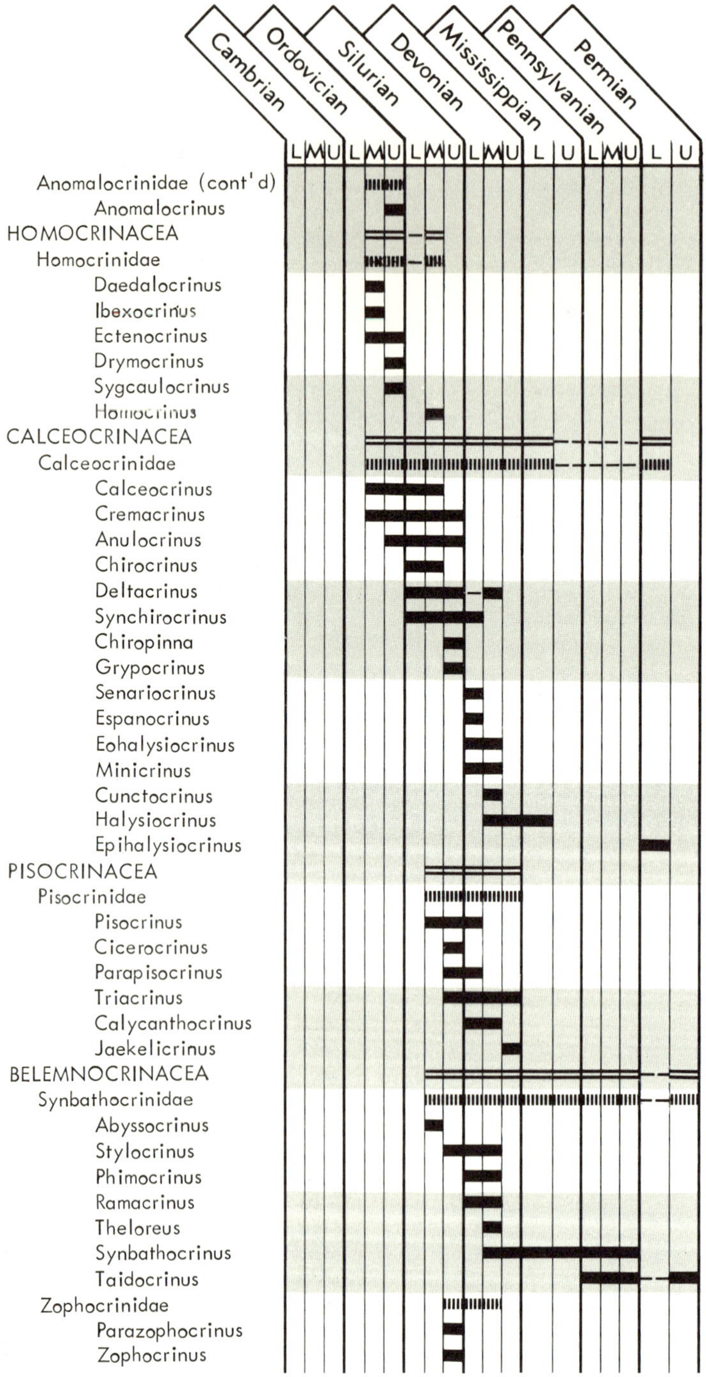

TABLE 5. *(Continued)*

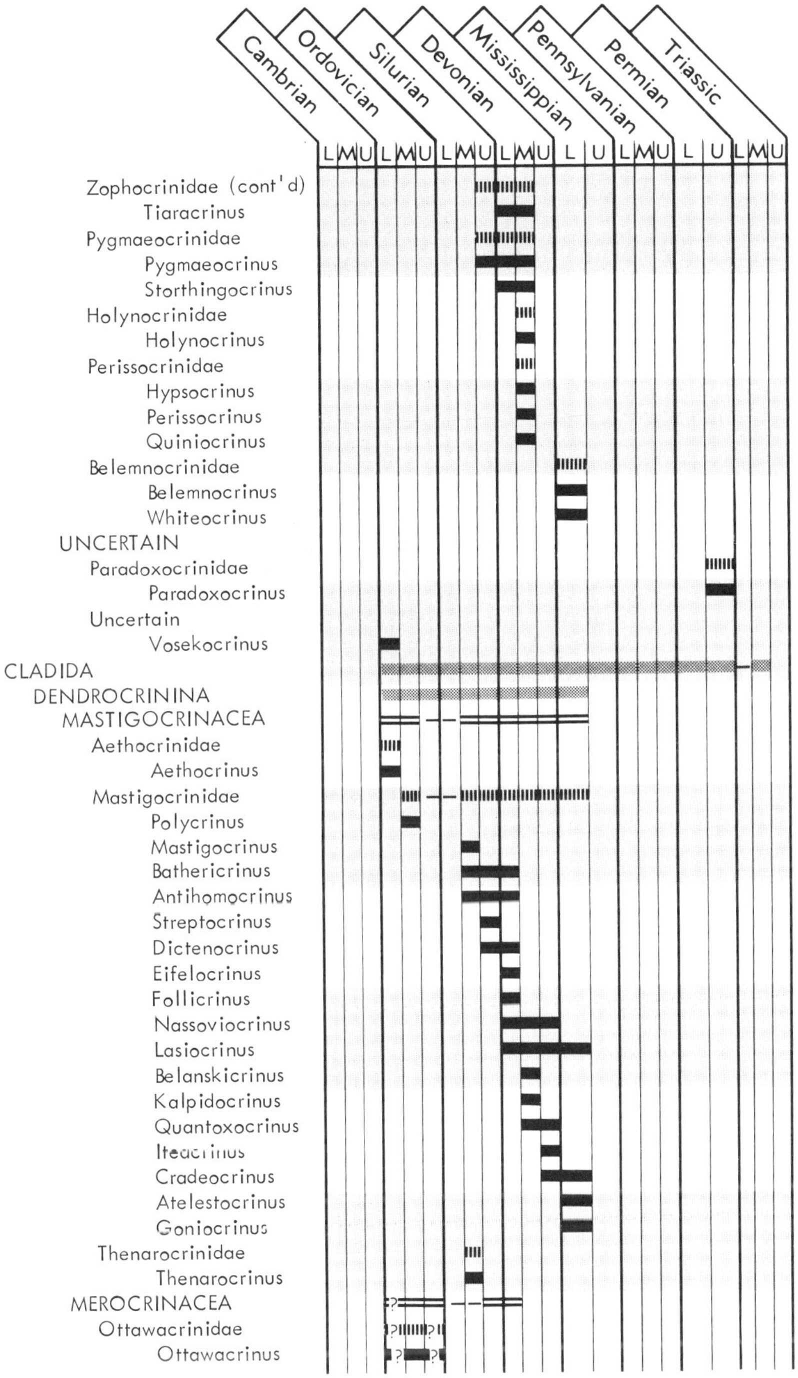

TABLE 5. *(Continued)*

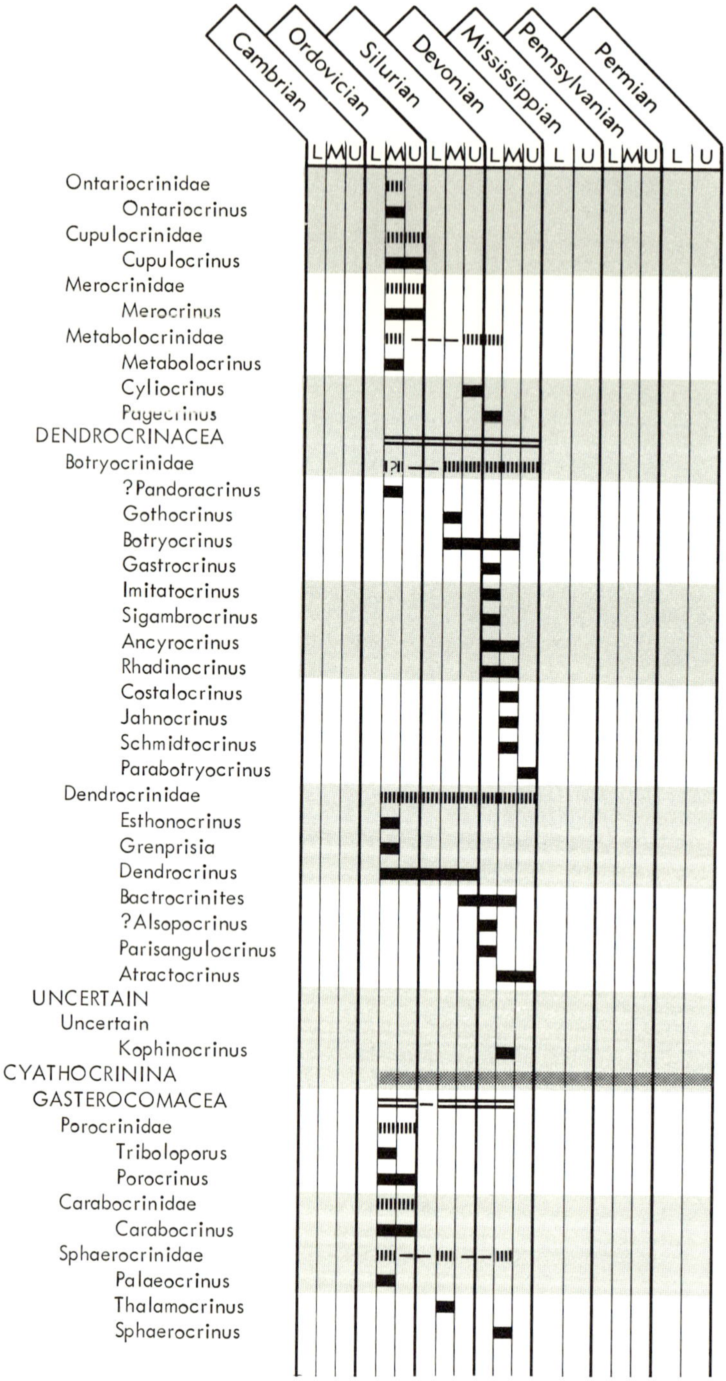

TABLE 5. *(Continued)*

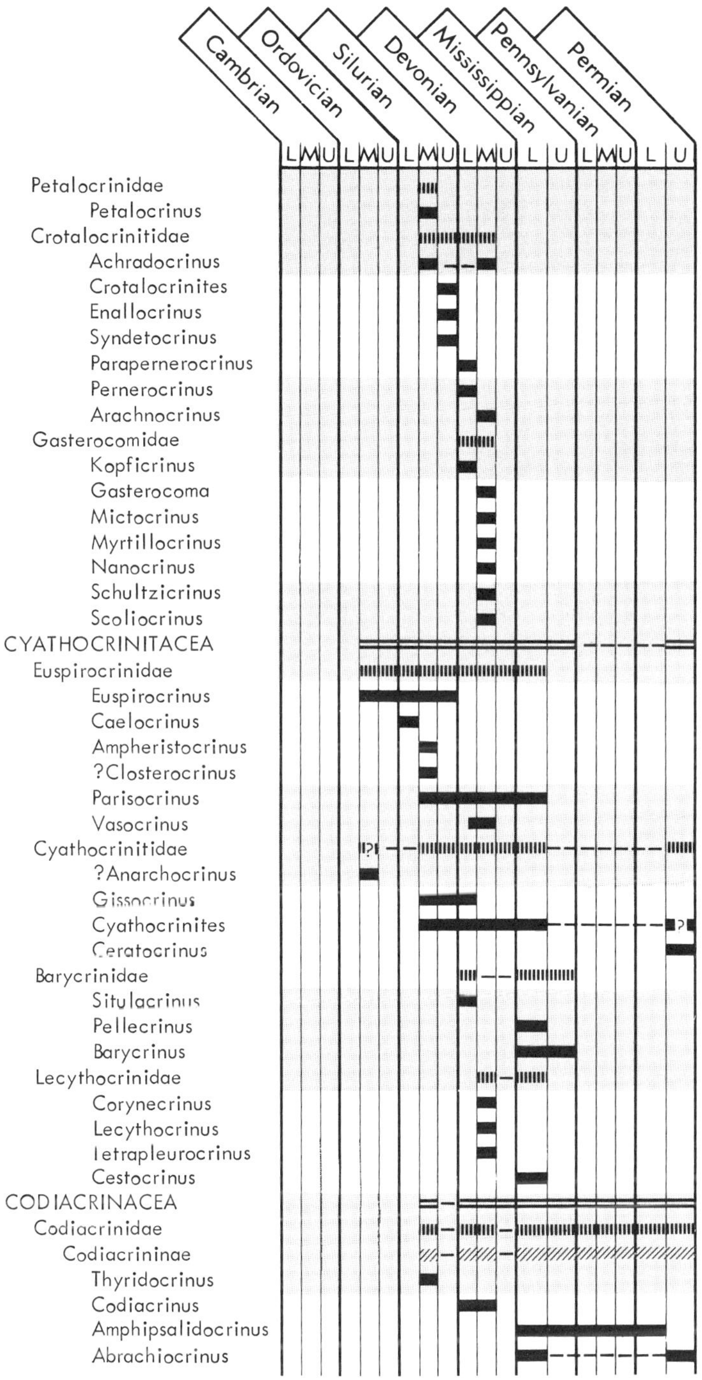

TABLE 5. *(Continued)*

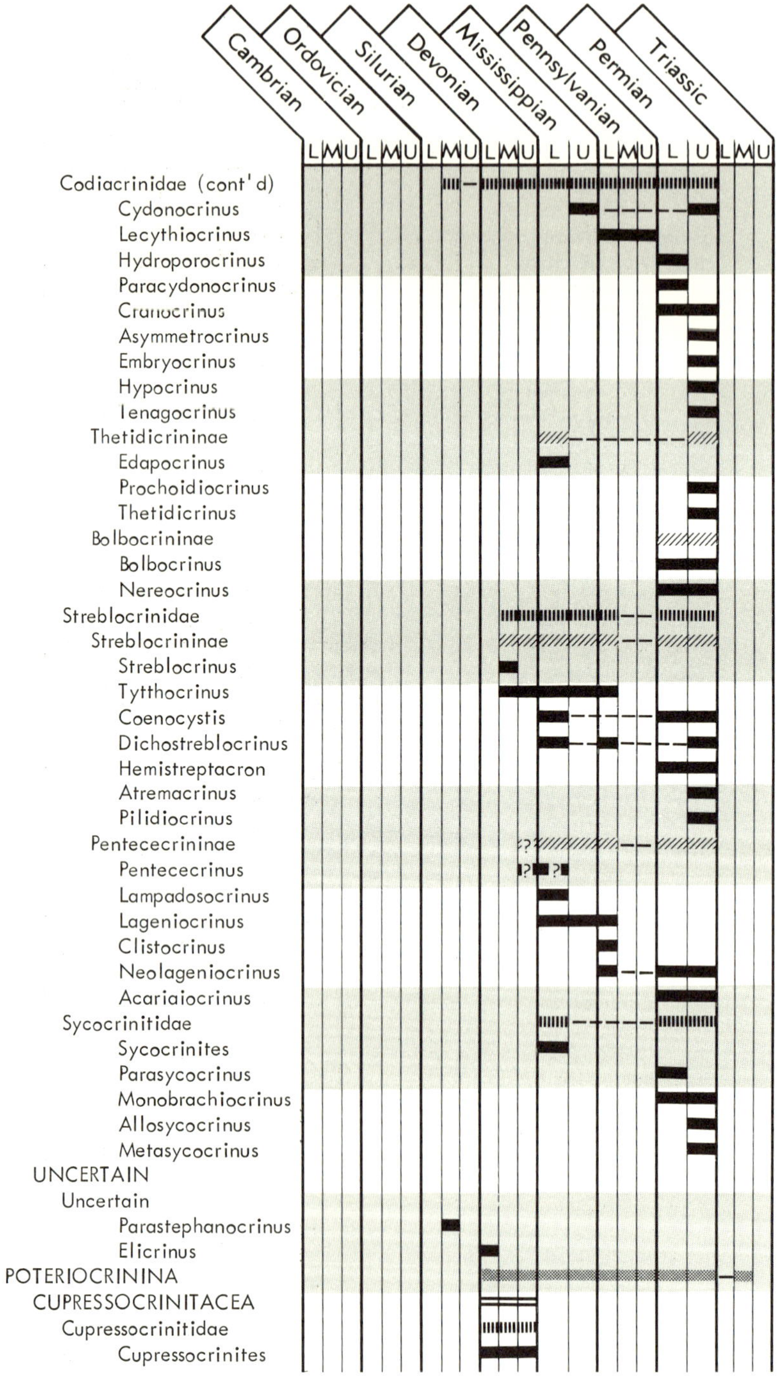

TABLE 5. *(Continued)*

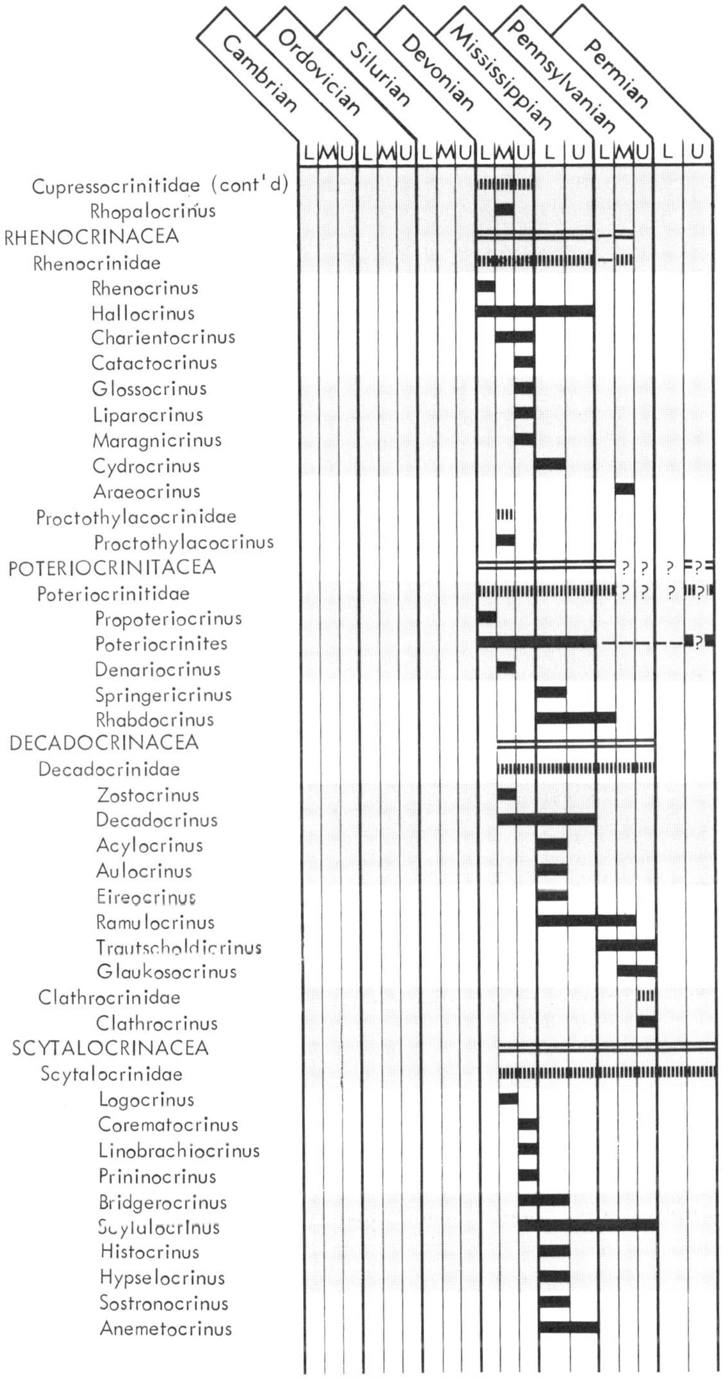

TABLE 5. *(Continued)*

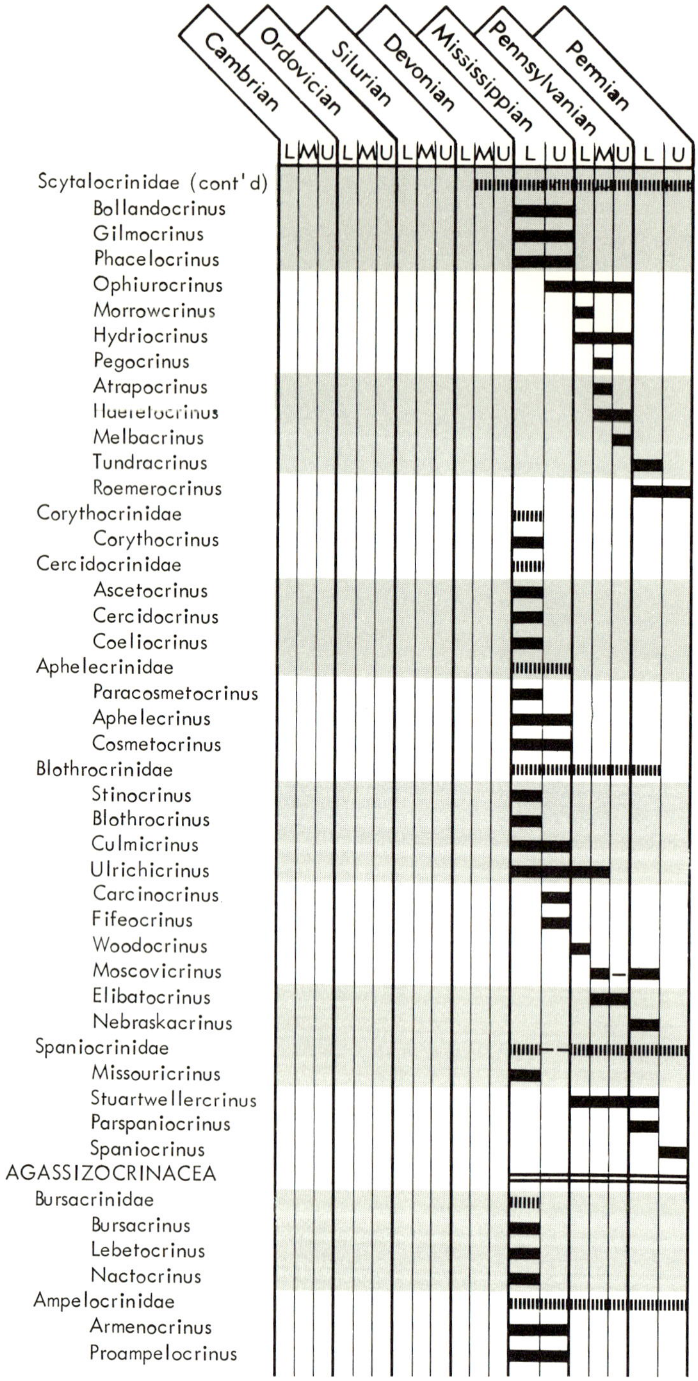

TABLE 5. *(Continued)*

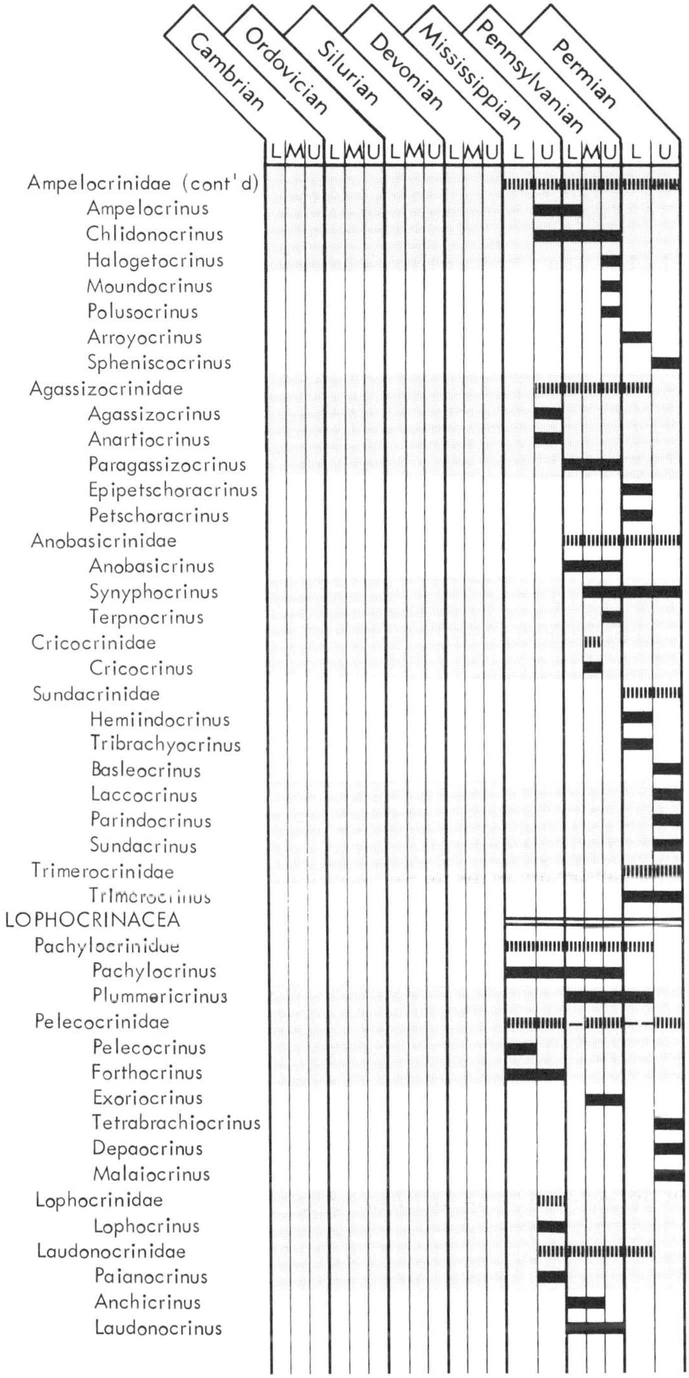

TABLE 5. *(Continued)*

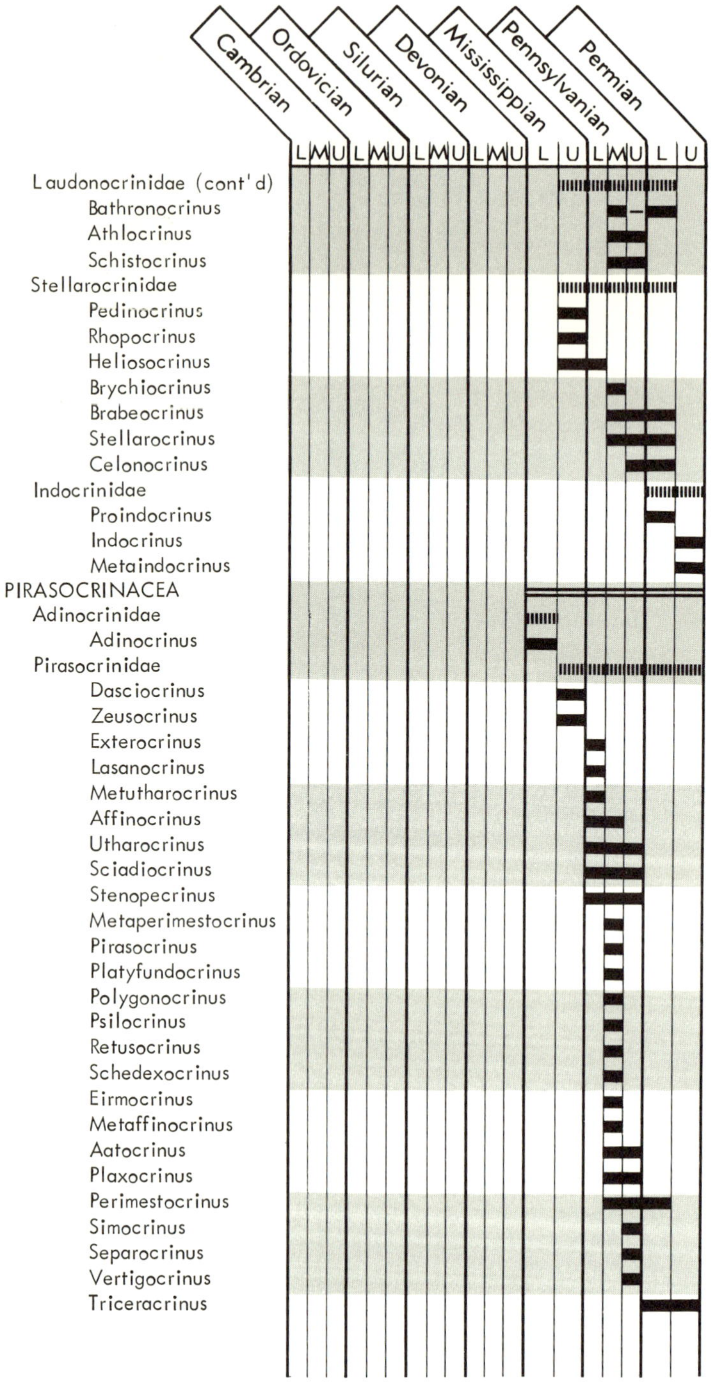

TABLE 5. *(Continued)*

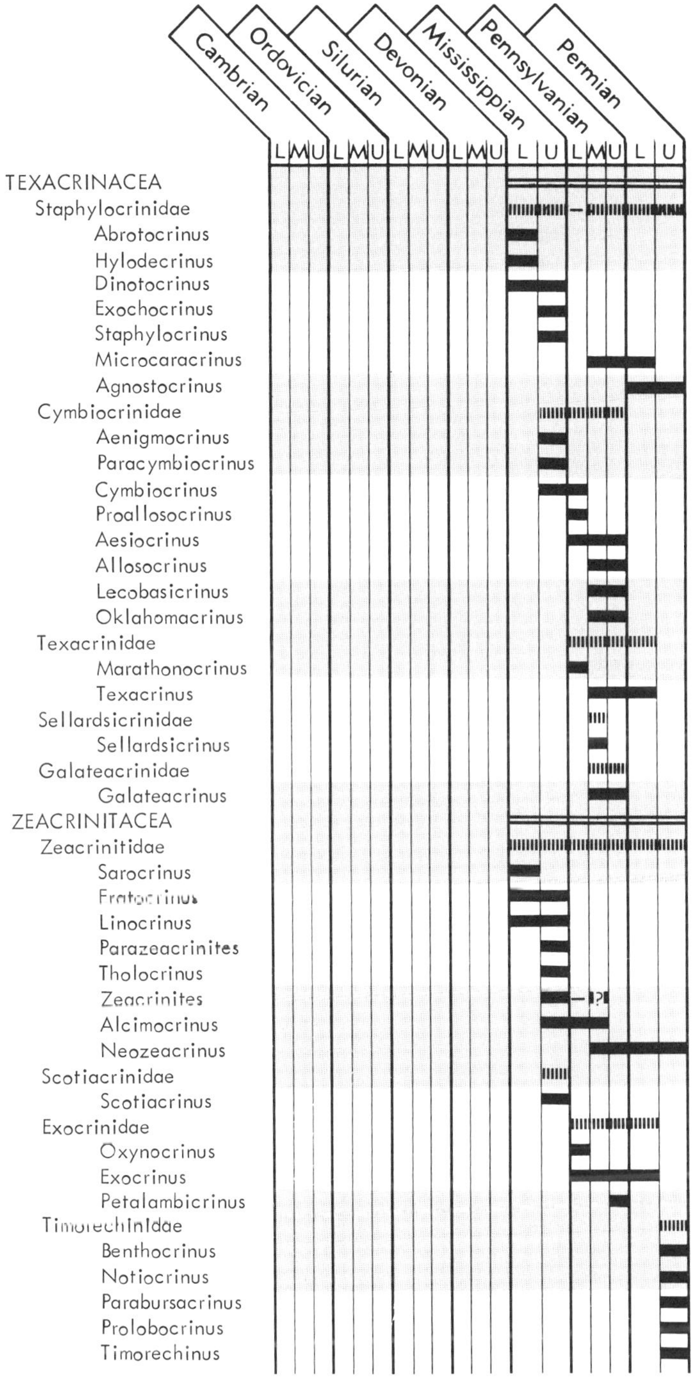

TABLE 5. *(Continued)*

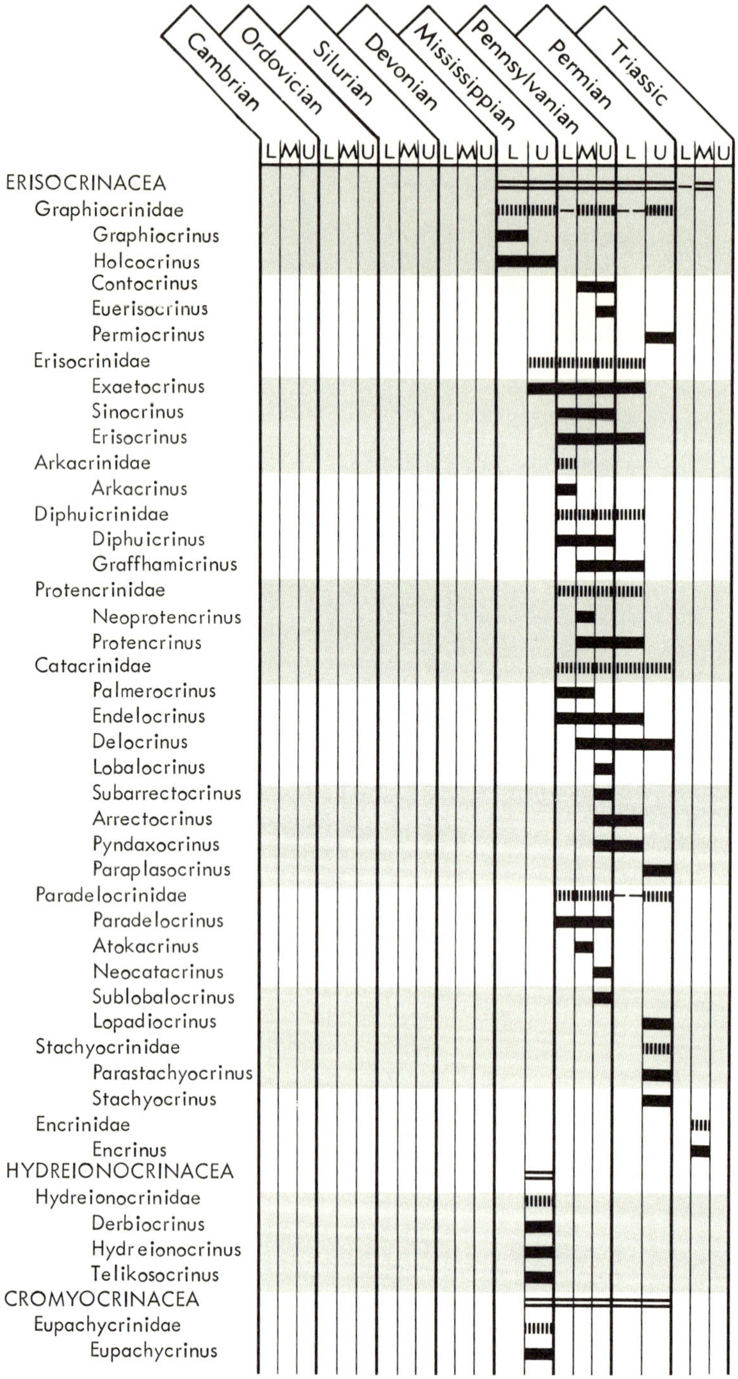

TABLE 5. *(Continued)*

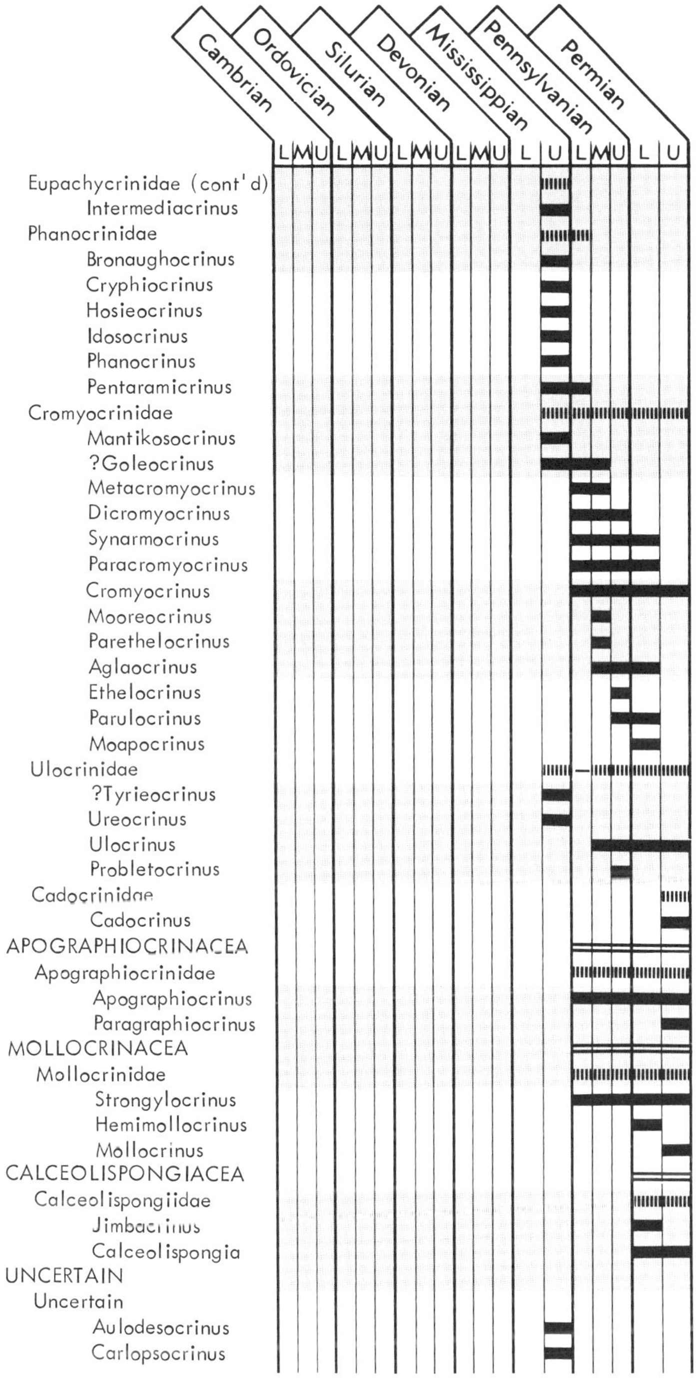

TABLE 5. *(Continued)*

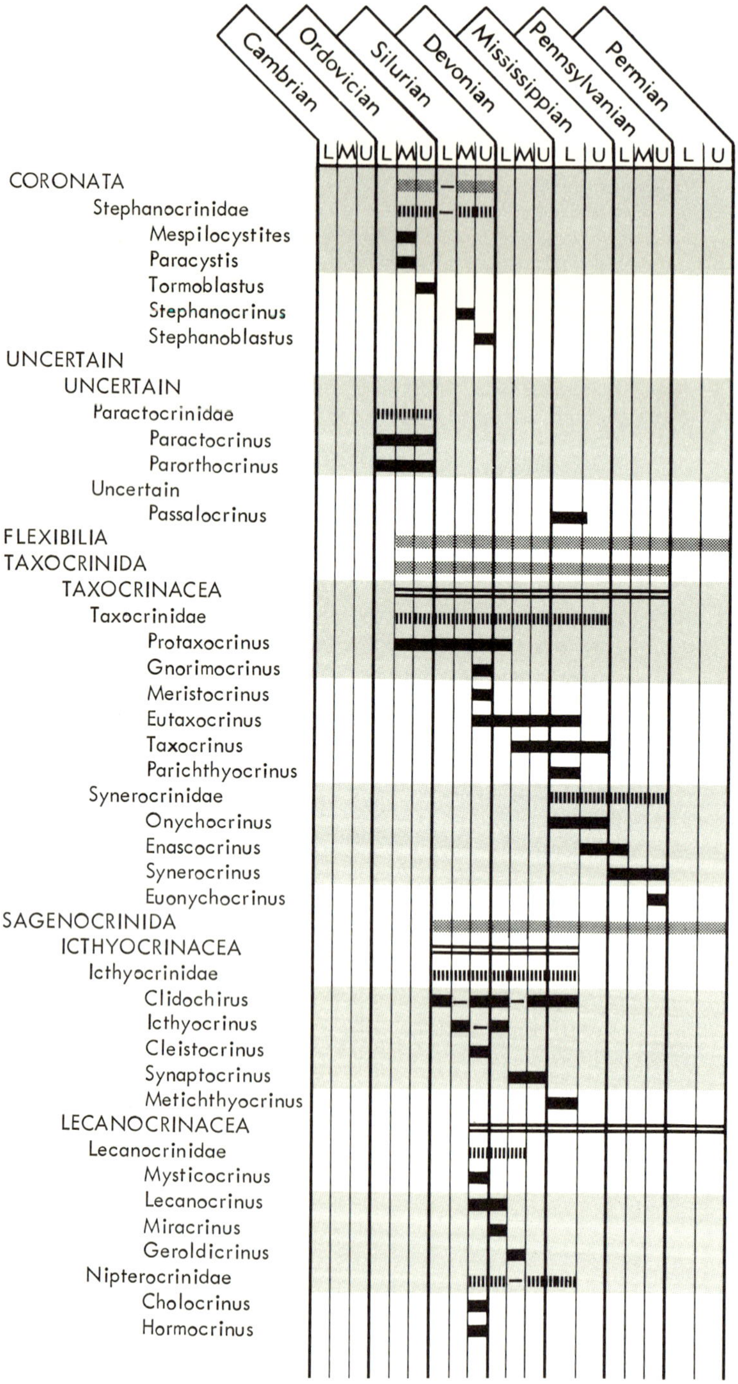

TABLE 5. *(Continued)*

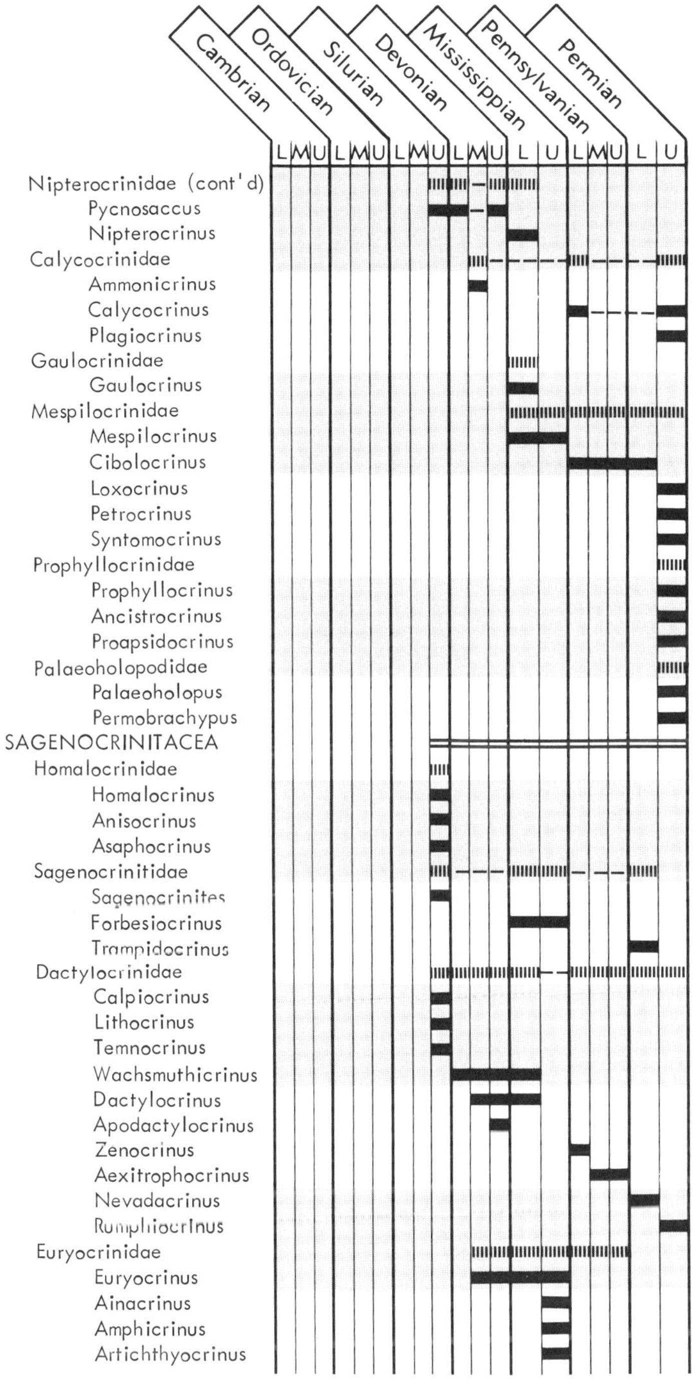

TABLE 5. *(Continued)*

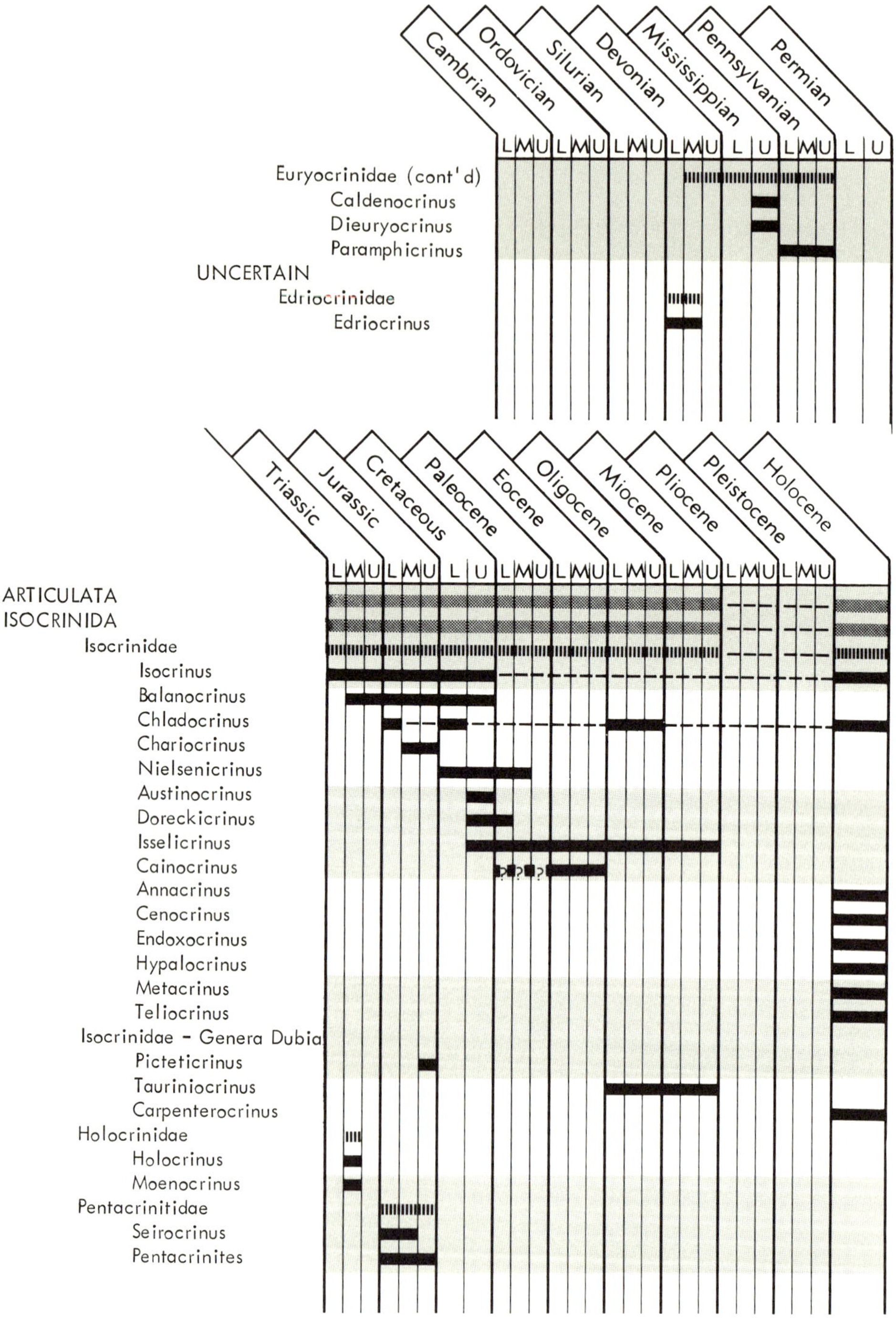

TABLE 5. *(Continued)*

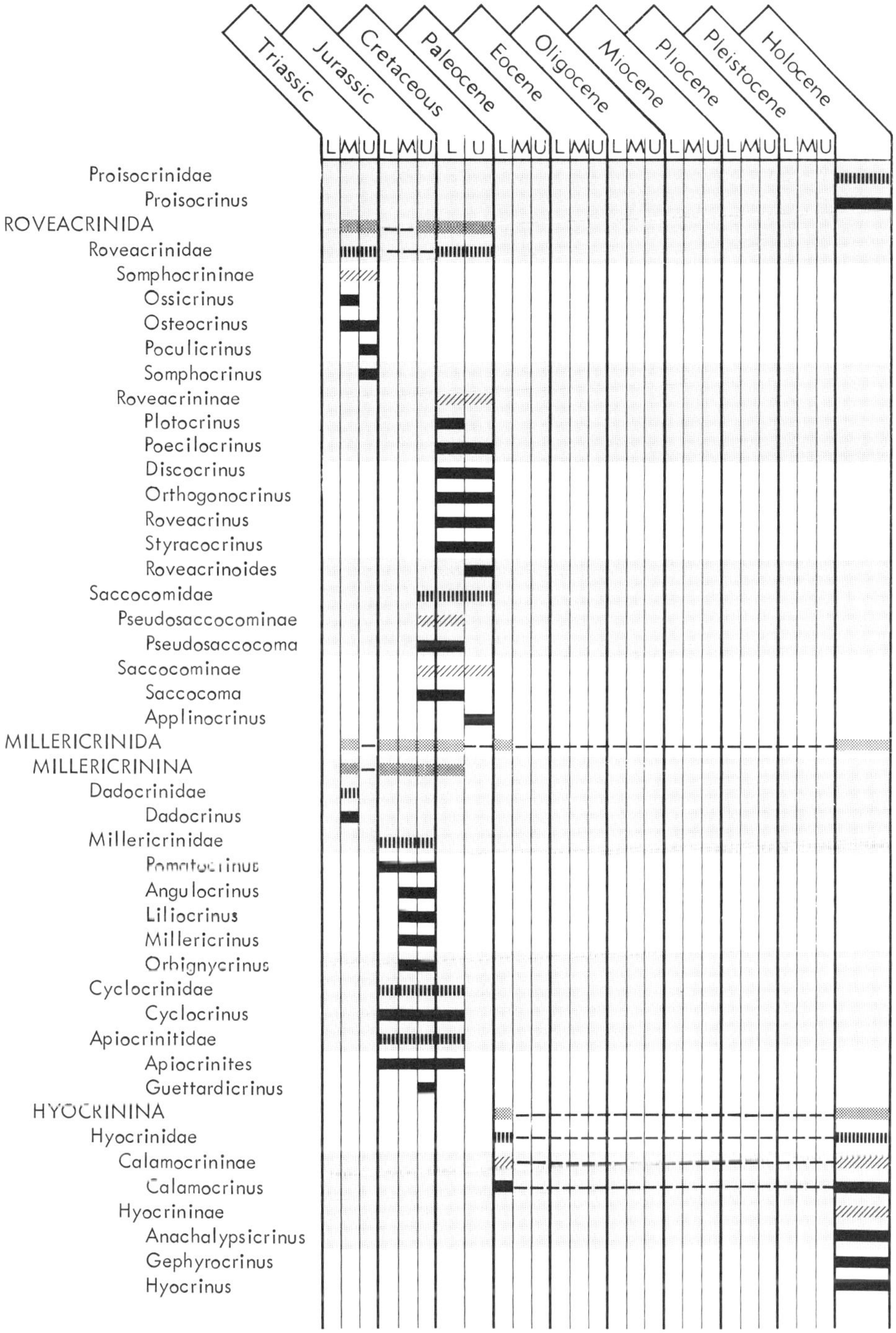

TABLE 5. *(Continued)*

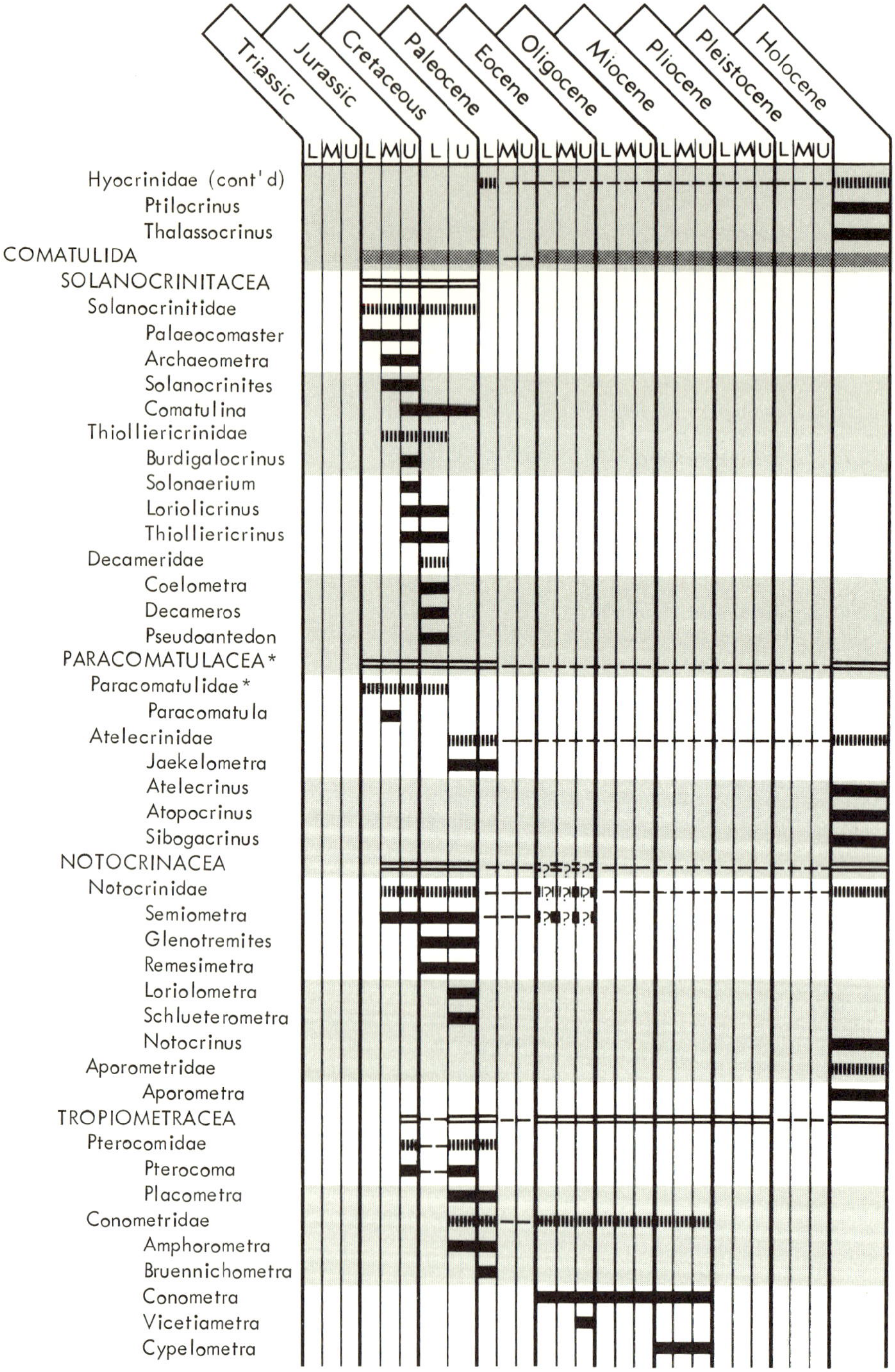

* Range extended, based upon two new unpublished genera (see p. *T*870).

TABLE 5. *(Continued)*

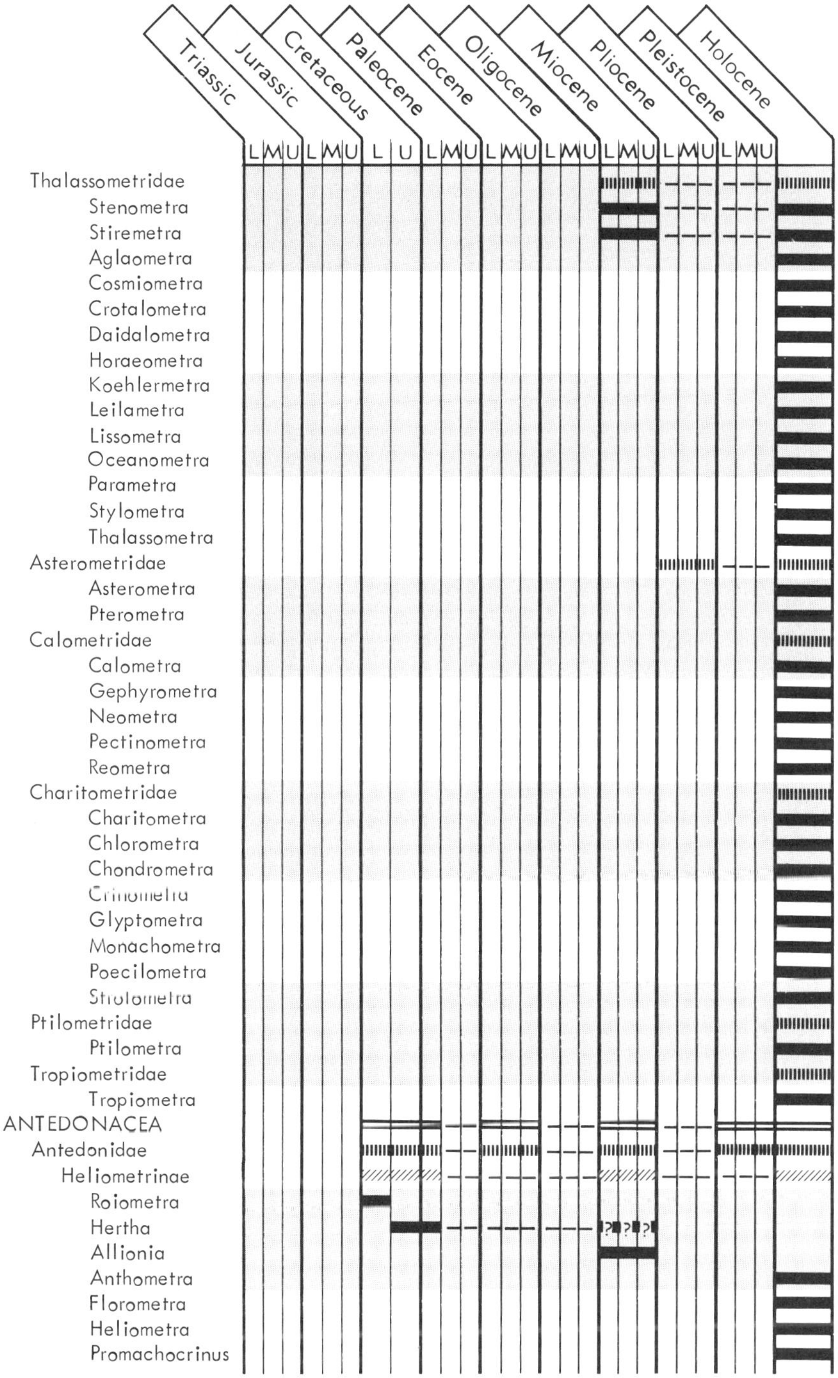

TABLE 5. *(Continued)*

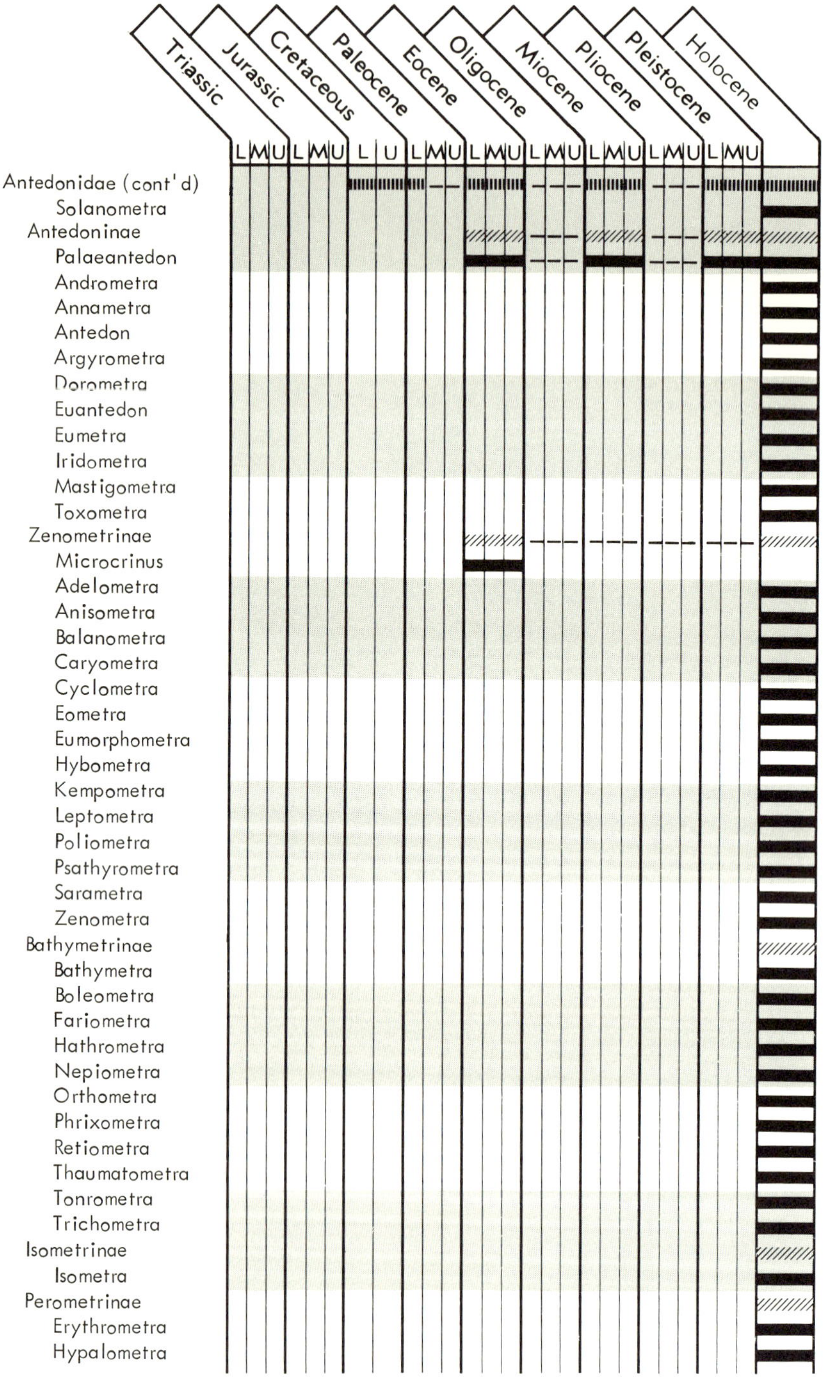

TABLE 5. *(Continued)*

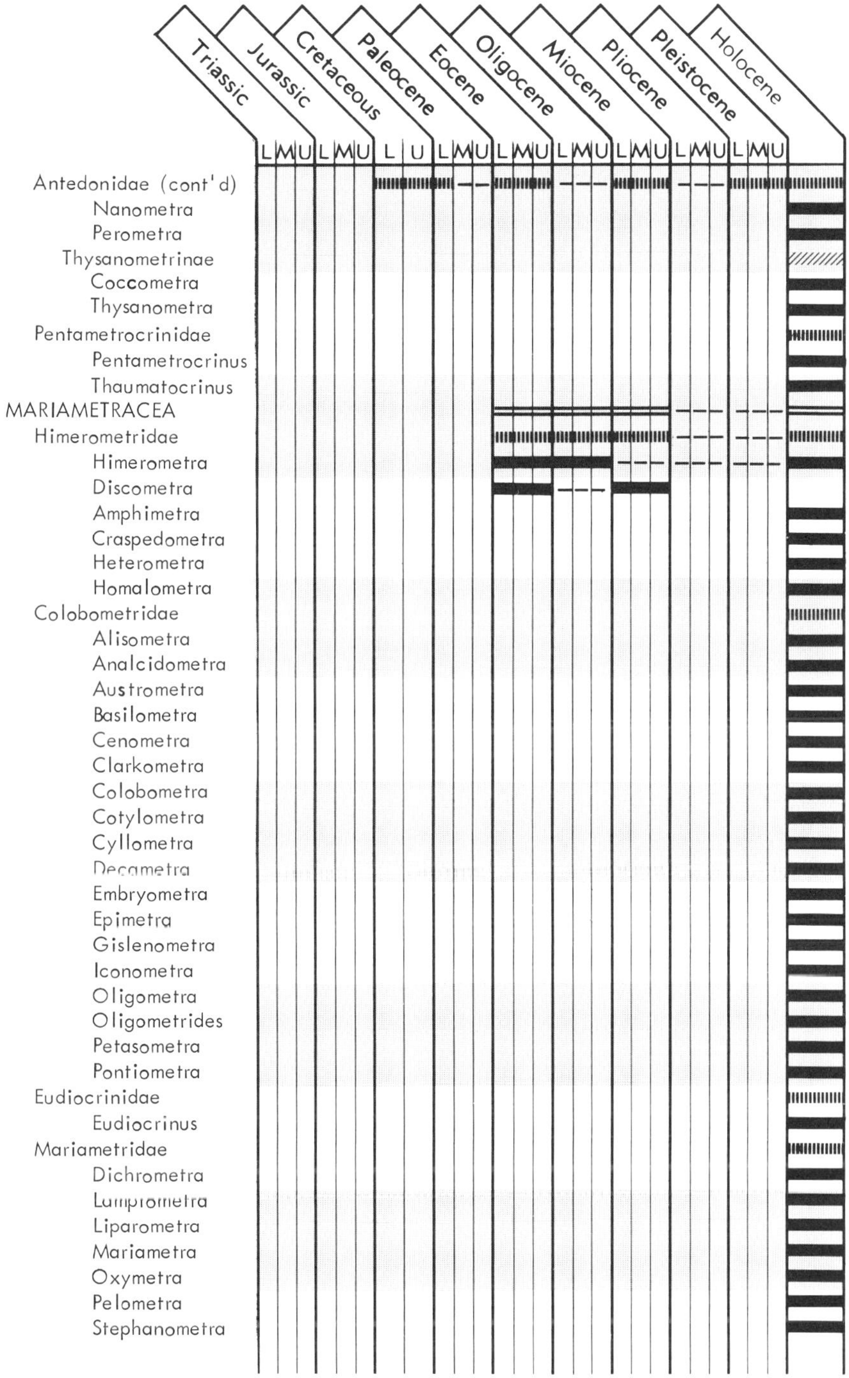

For *Discometra* (Himerometridae), read Tert.(Mio.).

TABLE 5. *(Continued)*

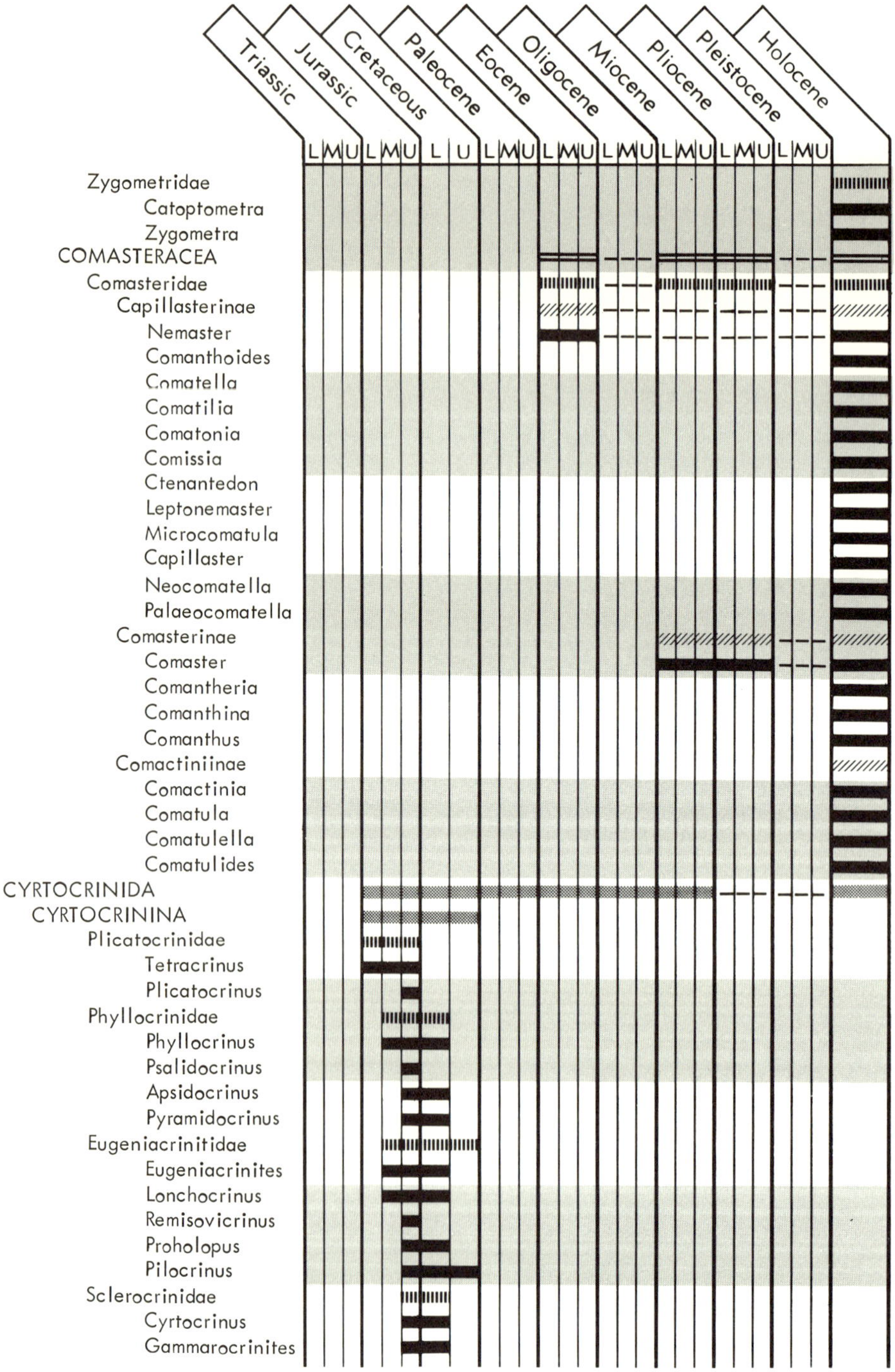

TABLE 5. *(Concluded)*

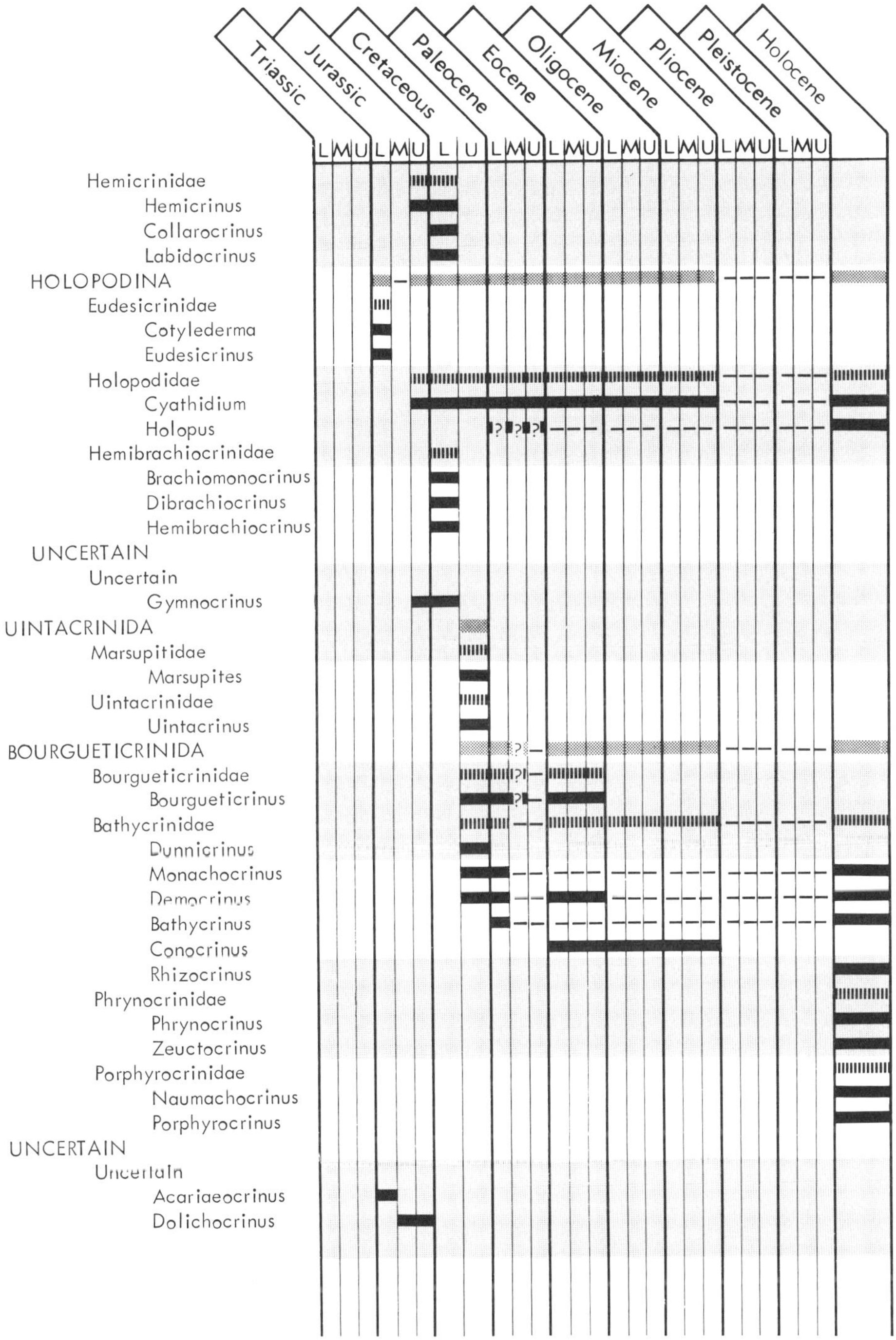